STABLE NON-GAUSSIAN RANDOM PROCESSES

STOCHASTIC MODELING

G. Samorodnitsky and M.S. Taqqu
Stable Non-Gaussian Processes: Stochastic Models with Infinite Variance

K. Sigman
Stationary Marked Point Processes: An Intuitive Approach

STABLE NON-GAUSSIAN RANDOM PROCESSES

Stochastic Models with Infinite Variance

STOCHASTIC MODELING

Gennady Samorodnitsky
Murad S. Taqqu

CHAPMAN & HALL
New York • London

Chapman & Hall
One Penn Plaza
New York, NY 10119

Published in Great Britain by
Chapman & Hall
2-6 Boundary Row
London SE1 8HN

Printed in the United States of America

Library of Congress Cataloging in Publication Data

Samorodnitsky, Gennady.

Stable non-Gaussian random processes : stochastic models with infinite variance / Gennady Samorodnitsky, Murad S. Taqqu.

p. cm.

Includes bibliographical references and indexes.

ISBN 0-412-05171-0

1. Gaussian processes. 2. Gaussian distribution.
I. Taqqu, Murad S. II. Title.

QA274.4.S26 1994

519.2—dc20 94-13685

CIP

British Library Cataloguing in Publication Data available

Please send your order for this or any Chapman & Hall book to **Chapman & Hall, 29 West 35th Street, New York, NY 10001, Attn: Customer Service Department.** You may also call our Order Department at 1-212-244-3336 or fax your purchase order to 1-800-248-4724.

For a complete listing of Chapman & Hall's titles, send your requests to **Chapman & Hall, Dept. BC, One Penn Plaza, New York, NY 10119.**

To Julia, Eric and Danny

and

to Rachelle, Yael and Jonathan

Contents

Preface

This book has been written to fill a gap that we, as teachers and researchers in the field in probability, have increasingly felt. Stable processes, which have attracted growing interest in recent years, are not the single subject of any monograph or comprehensive overview. In this book, we hope to make this important branch of probability widely accessible and provide both an introduction and a basic reference text.

The central limit theorem which offers the fundamental justification for approximate normality points to the importance of the stable distributions: they are the only limiting distributions of normalized sums of independent, identically distributed random variables, and perforce include the Gaussian as distinguished elements. Gaussian distributions and processes have long been well understood and their utility as both stochastic modeling constructs and analytical tools is well-accepted. However, they do not allow for large fluctuations and are thus often inadequate for modeling high variability. Non-Gaussian stable models, on the other hand, do not share such limitations. In general, the upper and lower tails of their marginal distributions decrease like a power function. The rate of decay depends on a number α, which takes a value between 0 and 2. The smaller α, the slower the decay and the heavier the tails. The distributions always have infinite variance and when $\alpha \leq 1$, they have an infinite mean as well.

In the last two or three decades, data with "heavy tails" have been collected in fields as diverse as economics, telecommunications, hydrology and physics of condensed matter, which suggests using non-Gaussian stable processes as possible models. Such models offer the additional merit of flexibility and variety when compared to Gaussian processes. The latter are completely specified by their mean and autocovariance functions, whereas non-Gaussian stable processes command a much richer parameterization. Gaussian distributions, moreover, are always symmetric around their mean; the non-Gaussian stable ones can have an arbitrary degree of skewness.

In this book, we emphasize the *probabilistic* approach over the analytic one. We talk of tails, moments and dependence structures and focus on multivariate properties and sample paths. The book will be useful to a wide spectrum of researchers in probability, applied probability and statistics as well as to graduate students. As background, we require only a first year graduate course in probability. Our goal has been to write a very readable text and to keep the wider context in clear perspective. Proofs are presented in detail. Each chapter begins with a brief summary and concludes with a wide range of exercises at varying levels of difficulty. To guide students on the more challenging problems, we have supplied detailed hints.

In Chapter 1 we introduce the one-dimensional stable random variables and provide a foundation for the subsequent chapters. We review classical material from the first part of this century, and then present more recent developments, including sections on simulation and on the series representation. That representation, which provides a link between the classical theory and the new, more modern approaches, is used extensively in the subsequent chapters.

Much of the rich structure of the stable non-Gaussian world becomes already apparent in Chapter 2 where multivariate stable distributions are presented and fundamental notions like *spectral measure, covariation*, and *codifference* are introduced. While the reader may notice that some of the Gaussian tools remain useful, this chapter is more a story of contrast than of similarity between Gaussian and non-Gaussian stable multivariate distributions. One important difference is that the components of a stable non-Gaussian vector cannot, in general, be expressed as a finite linear combination of independent random variables. Instead, one has to use stable stochastic integrals as a representation. These have enough built-in "independence" and are sufficiently versatile to represent non-Gaussian stable processes. Stable stochastic integrals are defined in several ways in Chapter 3, where the reader first encounters some of the most important classes of stable processes.

In Chapter 4 we study in greater detail the dependence structure of multivariate stable distributions: the effect of conditioning, order statistics, existence of joint moments, association and codifference are some of the topics that are discussed. Simple regression is analyzed in depth in Chapter 5. While non-Gaussian stable distributions have only moments of order less than α, they may have *conditional* moments of order greater than α. In particular, a regression of one symmetric α-stable random variable on another may be well defined even with $\alpha \leq 1$; it is then linear as in the Gaussian case. When the stable distributions involved are no longer symmetric, the situation is more complicated and the regression, in this case, is typically non-linear. We provide exact forms for the regression.

Complex-valued symmetric stable random variables, briefly introduced in Chapter 2, are discussed in greater detail in Chapter 6. In that chapter we develop the theory of complex-valued stable integrals and stochastic processes, together with their series representation, and introduce the harmonizable stable processes. Harmonizable or "spectral" representations are commonly used in Gaussian processes; in the non-Gaussian stable context, they characterize a special class of stochastic processes.

In Chapter 7 we discuss self-similar processes, also called "random fractals". These processes are invariant in distribution under judicious scaling of time and space. They are important in probability because of their connection to limit theorems and they are of great interest in modeling because they can display "long-range dependence" or "long memory." Some aspects of self-similarity appear in geophysics, hydrology, turbulence, economics, communications and in relation to "1/*f* noises." Self-similar processes are also used in physics, particularly in connection to the so-called "renormalization group theory" and "critical phenomena." We start by introducing the Gaussian ones, namely the fractional Brownian motions, and then concentrate on the much richer family of stable non-Gaussian self-similar processes. We also discuss autoregressive moving-average (ARMA) and fractional ARIMA models with stable innovations. The latter can display high variability together with long-range dependence.

In Chapter 8 we consider Chentsov random fields. We show how one can extend Chentsov's geometric construction of Lévy Brownian motion to construct a variety of self-similar stable random fields.

In the following four chapters we consider the sample path properties of stable processes. Chapter 9 is of an introductory nature. Its purpose is to help a reader with little prior experience to gain familiarity with the basic notions in the field, such as *versions*, *separability*, *zero-one laws*, etc. As a result, much of the discussion in Chapter 9 is not specific to stable processes but is applicable to any stochastic process. An advanced reader may wish to skip the first three sections of that chapter.

In Chapter 10 we discuss the most basic sample path properties of stable processes, namely sample boundedness and continuity. Some stable processes are continuous, others have discontinuities of the first type, and there are some that are unbounded in any finite interval. Self-similar stable processes provide a number of examples of these different types of sample paths. We also describe the upper tails of the distribution of suprema of stable processes and include a discussion on oscillations. In Chapter 11 we consider the measurability and integrability of the sample paths. One application is a formula for the inversion of the harmonizable process which is similar to the one for ordinary Fourier transforms.

The description of sample boundedness and continuity of stable processes given in Chapter 10 is complete only when $0 < \alpha < 1$. In Chapter 12 we offer a different approach for the case $1 \leq \alpha < 2$, one based on *metric entropy*. Although Chapter 12 does not contain a full solution to the problem, it provides new insight into boundedness and continuity in the stable case.

A promise made to the reader in Chapter 3 is kept in Chapter 13, where a detailed discussion of integral representation of α-stable processes is given. Finally, Chapter 14 contains historical notes, references and extensions to the material presented in the text.

There are many ways to read this book. Chapters 1, 2 and 3 are prerequisites for all the subsequent ones. Together with Chapter 4, they provide a good overview of the subject. The remaining chapters contain more specialized material and can be read independently. Chapters 9 through 12 which cover the sample paths' behavior have a more abstract flavor. Chapter 13 on integral representation requires only Chapter 3 as background. In a one-semester course one can cover the introductory chapters, 1 through 4, as well as Chapter 7 on self-similar processes.

We are grateful to the Weizmann Institute and the Guggenheim Foundation. The first offered a fellowship to Gennady Samorodnitsky, the second to Murad Taqqu. These fellowships allowed us to start the project. The National Science Foundation, the Air Force Office of Scientific Research and the Office of Naval Research provided us with grants to investigate stable processes. We thank also Albert Paulson for allowing us to reproduce some stable distribution tables from an unpublished technical report and the following publishers for their kind permission to reproduce some material from their publications: Birkhäuser, Blackwell Publishers, Academic Press, Elsevier Press, the American Statistical Association and the Institute for Mathematical Statistics.

Many people read parts of the manuscript and provided helpful suggestions. We would like to thank Laurence Baxter, Stamatis Cambanis, Daniel Chambers, Jan Kallsen, Mark Kon, Michel Ledoux, Makoto Maejima, Joop Mijnheer, Jolanta Misiewicz, Svetlozar (Zari) Rachev, Balram Rajput, Sid Resnick, Yumiko Sato, Shigeo Takenaka, Wim Vervaat and Tomasz Zak. We are also indebted to our students Renata Cioczek-Georges and Piotr Kokoszka, who were closely involved with many substantive aspects of this project, and Vadim Teverovsky who designed most of the figures. Alex Kasman, supported by the NSF grant USE-8953023, developed "calcgraphics," which allowed us to combine Mathematica and LaTeX. At the same time, Ognian Enchev showed us how to work around some of LaTeX's idiosyncrasies. We had the good fortune to have Tom Orowan, a superb typist and friend, type this manuscript at Boston University. And finally, we thank our wives, Julia and Rachelle, for their support and understanding.

Abbreviations

a.s.	almost surely
ARIMA	autoregressive integrated moving average process
ARMA	autoregressive moving average process
CHFSM	complex harmonizable fractional stable motion
FARIMA	fractional autoregressive integrated moving average process
FBM	fractional Brownian motion
FGN	fractional Gaussian noise
H-ss	self-similar with index H
H-sssi	self-similar with index H and with stationary increments
H-sssis	self-similar with index H and with stationary increments in the strong sense
LFSM	linear fractional stable motion
log-FSM	log-fractional stable motion
RHFSM	real harmonizable fractional stable motion
SαS	symmetric α-stable

Notation

$\stackrel{d}{=}$	equality in distribution
1_A or $1(A)$	indicator function of the set A
$a^{<p>}$	signed power, equal to $\lvert a\rvert^p \mathrm{sign}\, a$
$a_\lambda \sim b_\lambda$	$\lim_{\lambda\to\infty} a_\lambda/b_\lambda = 1$
$a_\lambda \approx b_\lambda$	$\lim_{\lambda\to\infty} a_\lambda/b_\lambda = k$ for some $k \neq 0$
$\delta(x_0)$ or $\delta_{\{x_0\}}$	measure that gives unit mass to the point x_0
plim	limit in probability
Leb	Lebesgue measure
$N(\mu, \nu^2)$	normal distribution with mean μ and variance ν^2
$\mathrm{sign}(x)$	1 if $x > 0$, 0 if $x = 0$, -1 if $x < 0$
$S_\alpha(\sigma, \beta, \mu)$	stable distribution with index α, scale parameter σ, skewness parameter β and shift parameter μ; in particular, $S_2(\sigma, 0, \mu) = N(\mu, 2\sigma^2)$
S_d	unit sphere in $\mathbb{R}^d$; it is a $(d-1)$-dimensional surface
$X \sim S_\alpha(\sigma, \beta, \mu)$	X has the distribution $S_\alpha(\sigma, \beta, \mu)$
$\lVert X\rVert_\alpha$	scale parameter of a $S\alpha S$ random variable X
x_+	x if $x \geq 0$, 0 if $x \leq 0$

STABLE NON-GAUSSIAN RANDOM PROCESSES

Chapter 1

Stable random variables on the real line

This chapter concerns univariate stable distributions. Because these distributions are described in several textbooks and monographs, we do not include the proof of some well known results. Instead, we refer the reader to Feller (1971) or Gnedenko and Kolmogorov (1954) where such proofs can be found. Our goal is to provide a clear summary of the classical theory. Modern developments, such as the series representation, are presented in detail.

We define univariate stable distributions in four equivalent ways. The first two definitions concern the "stability" property: the family of stable distributions is preserved under convolution. The third concerns the role of stable distributions in the context of the central limit theorem. Stable distributions approximate the distribution of normalized sums of i.i.d. random variables making them useful in modeling the contribution of many small random effects. The fourth definition specifies the characteristic function of a stable random variable. We will use the characteristic function extensively because few stable density functions are known in closed form.

A univariate stable distribution is characterized by four parameters. These are the index of stability α, the scale parameter σ, the skewness parameter β and the shift parameter μ. The stable distribution is Gaussian when $\alpha = 2$, and in this case, σ is proportional to the standard deviation, β can be taken to be zero and μ is the mean.

In Section 1.2 we present some basic properties of stable distributions. We clarify the role played by the four parameters. Stable distributions with $\alpha < 2$ share many properties with the Gaussian distribution, but they also differ from the Gaussian in significant ways. When $\alpha < 2$, for example, the tails of the distributions decay like a power function. This means that a stable random variable exhibits much more variability than a Gaussian one: it is much more likely to take values far away from the median. Mandelbrot (1982) referred to this as "Noah effect," an allusion to the biblical figure who lived through a very severe flood. The high variability of the stable distributions is one of the reasons they play an important role in modeling. Stable distributions have been used to model such diverse phenomena as gravitational fields of stars, temperature distributions in nuclear reactors, stresses in crystalline lattices, stock market prices and annual rainfall.

In Section 1.3, we discuss *symmetric* stable distributions, which are α-stable with β and μ zero. Their characteristic function takes a particularly simple form, very similar to that of a centered Gaussian distribution.

We establish in Sections 1.4 and 1.5 the series representation for a stable random variable. We show that any stable random variable can be expressed as an infinite sum which involves the arrival times of a Poisson process. This representation clarifies the role played by the index of stability α. Analogous representations will be established for stable random vectors and stable stochastic processes.

In Section 1.6, we display graphs of univariate stable density and cumulative distribution functions and provide a guide to numerical tables. We show, in Section 1.7, how to simulate stable random variables.

1.1 Equivalent definitions of a stable distribution

The theory of univariate stable distributions was essentially developed in the 1920s and 1930s by Paul Lévy and Aleksander Yakovlevich Khinchine. It is covered in detail in such classics as Gnedenko and Kolmogorov (1954) and Feller (1971), and it is the object of a more recent monograph of Zolotarev (1986). The main features of the theory are outlined in many graduate textbooks in probability.

We give four equivalent definitions of a stable distribution.

Definition 1.1.1 A random variable X is said to have a *stable* distribution if for any positive numbers A and B, there is a positive number C and a real number D such that

$$AX_1 + BX_2 \stackrel{d}{=} CX + D, \tag{1.1.1}$$

where X_1 and X_2 are independent copies of X, and where "$\stackrel{d}{=}$" denotes equality in distribution.

Note that a random variable X concentrated at one point is always stable. This degenerate case is of no special interest and, unless stated explicitly, we always assume that X is non-degenerate. A random variable X is called *strictly* stable if (1.1.1) holds with $D = 0$. A stable random variable X is called *symmetric* stable if its distribution is symmetric, that is, if X and $-X$ have the same distribution. A symmetric stable random variable is obviously strictly stable.

Warning. In the older literature, e.g. Lévy (1954) and Feller (1966), the terms "stable" and "strictly stable" are replaced, respectively, by "quasi-stable" and "stable." Feller (1971) uses, respectively, "stable (in the broad sense)" and "stable (in the strict sense)."

Theorem 1.1.2 *For any stable random variable X, there is a number $\alpha \in (0, 2]$ such that the number C in (1.1.1) satisfies*

$$C^{\alpha} = A^{\alpha} + B^{\alpha}. \tag{1.1.2}$$

See Feller (1971), Section VI.1, for a proof. The number α is called the *index of stability* or *characteristic exponent.* A stable random variable X with index α is called α*-stable.*

Example 1.1.3 If X is a Gaussian random variable with mean μ and variance ν^2 ($X \sim N(\mu, \nu^2)$), then X is stable with $\alpha = 2$ because

$$AX_1 + BX_2 \sim N((A + B)\mu,\ (A^2 + B^2)\nu^2),$$

i.e., (1.1.1) holds with $C = (A^2 + B^2)^{1/2}$ and $D = (A + B - C)\mu$.

We now turn to the second definition.

Definition 1.1.4 (*equivalent to Definition 1.1.1*). A random variable X is said to have a stable distribution if for any $n \geq 2$, there is a positive number C_n and a real number D_n such that

$$X_1 + X_2 + \cdots + X_n \stackrel{d}{=} C_n X + D_n, \tag{1.1.3}$$

where $X_1, X_2, \ldots, X_n$ are independent copies of X.

If X is stable according to the Definition 1.1.1, then, by induction, it is also stable according to Definition 1.1.4. It is easy to show that the reverse implication is also true (Feller 1971, Section VI.1). Therefore the two definitions are equivalent.

It turns out (Feller 1971, Theorem VI.1.1), that in (1.1.3) we have, necessarily,

$$C_n = n^{1/\alpha} \tag{1.1.4}$$

for some $0 < \alpha \leq 2$. This is of course the same α which appears in (1.1.2).

Remarks

1. If we want to use Definition 1.1.4, is it necessary to verify that (1.1.3) holds for every $n \geq 2$? Would it be sufficient to verify (1.1.3) for fewer ns ? It turns out that the requirement $X_1 + X_2 \stackrel{d}{=} C_2X + D_2$ alone is not sufficient, but a random variable satisfying both $X_1 + X_2 \stackrel{d}{=} C_2X + D_2$ and $X_1 + X_2 + X_3 \stackrel{d}{=} C_3X + D_3$ is necessarily stable (Zolotarev (1986), p. 14).

2. Consider a sequence $\mathbf{X} = \{X_i\}_{i=-\infty}^{\infty}$ of random variables. Fix a number $\delta > 0$. For each $n \geq 1$, define the transformation T_n that takes $\mathbf{X}$ into the new sequence of random variables $T_n\mathbf{X} = \{(T_nX)_i\}_{i=-\infty}^{\infty}$ given by

$$(T_nX)_i = \frac{1}{n^\delta} \sum_{j=in}^{(i+1)n-1} X_j.$$

To visualize the action of T_n, divide the real line in blocks of size n. In each block, sum the random variables whose index belongs to the block and renormalize that sum by a power function of the length of the block. The index n of T_n refers to the length of the block and δ is the exponent of the power function.

The transformations $T_n, n \geq 1$, are called *renormalization group transformations* with *critical exponent* δ. More general renormalization group transformations occur in physics in the context of critical phenomena, and in applied mathematics.

The family of transformations $\{T_n, n \geq 1\}$ forms a semi-group. Indeed, $T_{mn} = T_mT_n$, i.e., the transformation T_n followed by the transformation T_m is equivalent to the transformation T_{mn}. A sequence $\mathbf{X} = \{X_i\}_{i=-\infty}^{\infty}$ is said to be a *fixed point* of the renormalization group transformation if $T_n\mathbf{X} \stackrel{d}{=} \mathbf{X}$ for all $n \geq 1$, i.e., if the distribution of $\mathbf{X}$ is invariant under T_n for any $n \geq 1$.

The sequence $\mathbf{X} = \{X_i\}_{i=-\infty}^{\infty}$ of i.i.d. strictly α-stable random variables is a fixed point of the T_ns with $\delta = 1/\alpha$ because

$$(T_n\mathbf{X})_i = \frac{1}{n^{1/\alpha}}(X_{in} + X_{in+1} + \cdots + X_{i(n+1)-1}) \stackrel{d}{=} X_i,$$

by (1.1.3), and because the independence of the X_is implies the independence of the $(T_n\mathbf{X})_i$s.

We will see in Chapter 7 that there exist *dependent* stable sequences which are also invariant under renormalization group transformations. The corresponding critical exponent δ will then depend not only on the index of stability α but also on an additional parameter associated with the dependence.

The third definition states that stable distributions are the only distributions that can be obtained as limits of normalized sums of i.i.d. random variables.

Definition 1.1.5 (*equivalent to Definitions 1.1.1 and 1.1.4*). A random variable X is said to have a stable distribution if it has a *domain of attraction*, i.e., if there is a sequence of i.i.d. random variables $Y_1, Y_2, \dots$ and sequences of positive numbers $\{d_n\}$ and real numbers $\{a_n\}$, such that

$$\frac{Y_1 + Y_2 + \cdots + Y_n}{d_n} + a_n \stackrel{d}{\Rightarrow} X. \tag{1.1.5}$$

The notation $\stackrel{d}{\Rightarrow}$ denotes convergence in distribution.

It is clear that the previous definitions yield (1.1.5), e.g., by taking the Y_is to be independent and distributed like X. The converse is easy to show (Gnedenko & Kolmogorov 1954, p. 162). When X is Gaussian and the Y_is are i.i.d. with finite variance, then (1.1.5) is the statement of the ordinary central limit theorem. The Y_is are said to belong to the *normal* domain of attraction of X when $d_n = n^{1/\alpha}$. In general, $d_n = n^{1/\alpha}h(n)$ where $h(x)$, $x > 0$, is a slowly varying function at infinity, that is, $\lim_{x\to\infty} h(ux)/h(x) = 1$ for all $u > 0$ (Feller 1971, XVII.5). The function $h(x) = \ln x$, for example, is slowly varying at infinity.

The fourth definition specifies the characteristic function of a stable random variable.

Definition 1.1.6 (*equivalent to Definitions 1.1.1, 1.1.4 and 1.1.5*). A random variable X is said to have a stable distribution if there are parameters $0 < \alpha \le 2$, $\sigma \ge 0$, $-1 \le \beta \le 1$, and μ real such that its characteristic function has the following form:

$$E\exp i\theta X = \begin{cases} \exp\left\{-\sigma^\alpha|\theta|^\alpha\left(1 - i\beta(\operatorname{sign}\theta)\tan\frac{\pi\alpha}{2}\right) + i\mu\theta\right\} & \text{if } \alpha \ne 1, \\ \exp\left\{-\sigma|\theta|\left(1 + i\beta\frac{2}{\pi}(\operatorname{sign}\theta)\ln|\theta|\right) + i\mu\theta\right\} & \text{if } \alpha = 1. \end{cases} \tag{1.1.6}$$

The parameter α is the index of stability and

$$\operatorname{sign}\theta = \begin{cases} 1 & \text{if } \theta > 0, \\ 0 & \text{if } \theta = 0, \\ -1 & \text{if } \theta < 0. \end{cases}$$

The parameters σ, β and μ are unique (β is irrelevant when $\alpha = 2$).

It is easy to check that Definition 1.1.6 implies Definition 1.1.4. The proof of the converse can be found in many textbooks, e.g. Gnedenko and Kolmogorov (1954), Section 34. The idea of the proof is to use Definition 1.1.5 to obtain the *Lévy-Khinchine representation* of the characteristic function, namely

$$E\exp i\theta X = \exp\{iM\theta - \sigma^2\theta^2\}$$

if $\alpha = 2$, and

$$E\exp i\theta X = \exp\left\{ iM\theta + P\int_0^\infty \psi(\theta, x)\frac{dx}{x^{1+\alpha}} + Q\int_{-\infty}^0 \psi(\theta, x)\frac{dx}{|x|^{1+\alpha}} \right\} \tag{1.1.7}$$

if $\alpha < 2$. Here M is real, $\sigma \geq 0$, P and Q are non-negative numbers, and

$$\psi(\theta, x) = e^{i\theta x} - 1 - \frac{i\theta x}{1 + x^2}. \tag{1.1.8}$$

Then, in the case $\alpha < 2$, when X is non-degenerate $(P + Q > 0)$, one sets

$$\beta = \frac{P - Q}{P + Q}$$

and evaluates the integrals as in Feller (1966), p. 542.

There is a great deal of flexibility in the choice of the function ψ. Another choice, which involves the unit ball $[-1, 1]$ of $\mathbb{R}^1$, is

$$\psi(\theta, x) = e^{i\theta x} - 1 - i\theta x 1_{[-1,1]}(x).$$

(Changing ψ affects the values of the constants M, P, Q.)

The measure

$$L(dx) = \frac{P}{x^{1+\alpha}}1_{(0,\infty)}(x)dx + \frac{Q}{|x|^{1+\alpha}}1_{(-\infty,0)}(x)dx$$

in (1.1.7), is called the *Lévy measure*.[1] We will see in Section 3.12 that the representation (1.1.7) has an interpretation in terms of limits of compound Poisson random variables with $L(dx)$ playing the role of the intensity of "jumps" of size x.

[1] $L(dx)$ is sometimes called "spectral measure" in the literature. In order to avoid any possible confusion with the "spectral measure" of Chapter 2, we refer to $L(dx)$ as the "Lévy measure."

Remarks

1. When $\alpha = 2$, the characteristic function (1.1.6) becomes $E \exp i\theta X = \exp\{-\sigma^2\theta^2 + i\mu\theta\}$. This is the characteristic function of a Gaussian random variable with mean μ and variance $2\sigma^2$. Note that σ is not equal to the standard deviation. Note also that although the value of β is not specified because $\beta \tan \pi = 0$, one typically associates the Gaussian distribution with the choice $\beta = 0$.

2. When $\alpha = 1$, the imaginary term in (1.1.6) contains the factor $\ln |\theta|$. The presence of this logarithm is the source of many difficulties associated with the case $\alpha = 1$. In the sequel, the case $\alpha = 1$ will often have to be treated separately.

3. In (1.1.6), β appears with a negative sign when $\alpha \neq 1$ and with a positive sign with $\alpha = 1$. This minor point has been the source of great confusion in the literature (see Hall (1980) for a discussion).

4. The characteristic function (1.1.6) can be written as

$$E \exp i\theta X = \exp\{\sigma^\alpha(-|\theta|^\alpha + i\theta\omega(\theta, \alpha, \beta)) + i\mu\theta\},$$

where

$$\omega(\theta, \alpha, \beta) = \begin{cases} \beta|\theta|^{\alpha-1} \tan \frac{\pi\alpha}{2} & \text{if } \alpha \neq 1, \\ -\beta \frac{2}{\pi} \ln |\theta| & \text{if } \alpha = 1. \end{cases}$$

The function $\omega(\theta, \alpha, \beta)$ is not continuous at $\alpha = 1$ and $\beta \neq 0$. However, as Zolotarev (1986) remarks, setting

$$\mu_1 = \begin{cases} \mu + \beta\sigma^\alpha \tan \frac{\pi\alpha}{2} & \text{if } \alpha \neq 1, \\ \mu & \text{if } \alpha = 1, \end{cases} \tag{1.1.9}$$

yields the expression

$$E \exp i\theta X = \exp\{\sigma^\alpha(-|\theta|^\alpha + i\theta\omega_1(\theta, \alpha, \beta)) + i\mu_1\theta\}, \tag{1.1.10}$$

where

$$\omega_1(\theta, \alpha, \beta) = \begin{cases} \beta(|\theta|^{\alpha-1} - 1) \tan \frac{\pi\alpha}{2} & \text{if } \alpha \neq 1, \\ -\beta \frac{2}{\pi} \ln |\theta| & \text{if } \alpha = 1, \end{cases} \tag{1.1.11}$$

is a function which is jointly continuous in α and β. This last expression makes plausible the presence of the logarithm in the case $\alpha = 1$.

If one insists on having convergence in distribution as $\alpha \to \alpha_0$, $\sigma \to \sigma_0$, $\beta \to \beta_0$, $\mu \to \mu_0$, then one should change the parameterization and characterize an α-stable random variable as having a characteristic function of the form

$$E \exp i\theta X = \exp\Big\{\sigma^\alpha\Big(-|\theta|^\alpha + i\theta\beta(|\theta|^{\alpha-1} - 1)\tan\frac{\pi\alpha}{2}\Big) + i\mu_1\theta\Big\},$$

where $\alpha \in (0, 2]$, $\sigma \geq 0$, $|\beta| \leq 1$, $\mu_1 \in \mathbb{R}^1$, and where the characteristic function for $\alpha = 1$ is defined by letting $\alpha \to 1$. The drawback is that μ_1 does not have the nice properties of μ given in the next section. We shall, therefore, continue to use the form (1.1.6) of the characteristic function.

5. When $\alpha \neq 1$, the characteristic function (1.1.6) is sometimes written as

$$Ee^{i\theta X} = \exp\{-\sigma_2^\alpha |\theta|^\alpha e^{i\omega_2(\theta) + i\theta\mu}\}, \tag{1.1.12}$$

where

$$\omega_2(\theta) = \omega_2(\theta, \alpha, \beta) = -\beta_2\,(\text{sign }\theta)\frac{\pi}{2}K(\alpha),$$

and where β_2 is such that

$$\tan\left(\beta_2 \frac{\pi K(\alpha)}{2}\right) = \beta \tan\frac{\pi\alpha}{2},$$

and $K(\alpha)$ will be defined below. The new scale parameter σ_2 is related to σ as follows:

$$\sigma_2 = \sigma c_2,$$

where

$$c_2 = \left(\cos \beta_2 \frac{\pi}{2} K(\alpha)\right)^{1/\alpha},$$

or, in terms of β,

$$c_2 = \left(1 + \beta^2 \tan^2 \frac{\pi\alpha}{2}\right)^{-1/(2\alpha)}.$$

Zolotarev (1986), p. 12, chooses[2]

$$K(\alpha) = \alpha - 1 + \text{sign }(1-\alpha) = \begin{cases} \alpha & \text{if } \alpha < 1, \\ \alpha - 2 & \text{if } \alpha > 1. \end{cases}$$

[2]Some authors, e. g., DuMouchel (1971) and Chambers, Mallows, and Stuck (1976), choose

$$K(\alpha) = 1 - |1-\alpha| = \min(\alpha,\ 2-\alpha) = \begin{cases} \alpha & \text{if } \alpha < 1, \\ 2-\alpha & \text{if } \alpha > 1. \end{cases}$$

With this choice, β_2 has the same sign as β if $\alpha < 1$, but has the opposite sign if $\alpha > 1$.

With this choice,

$$\beta_2 = \begin{cases} \frac{2}{\pi\alpha}\text{Arctan}\left(\beta \tan \frac{\pi\alpha}{2}\right) & \text{if } 0 < \alpha < 1, \\ \frac{2}{\pi(2-\alpha)}\text{Arctan}\left(\beta \tan \frac{\pi(\alpha-2)}{2}\right) & \text{if } 1 < \alpha < 2. \end{cases}$$

For fixed α, the index β_2 is a monotone increasing function of β and $\beta_2 = -1,\ 0,\ 1$ if and only if $\beta = -1,\ 0,\ 1$, respectively.

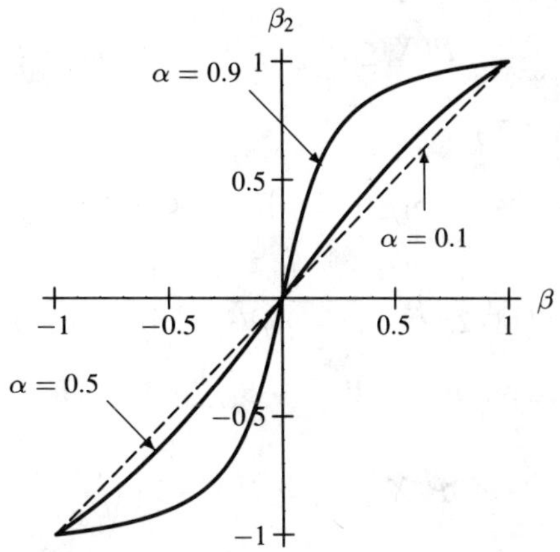

Figure 1.1: β_2 versus β. If $\alpha > 1$, replace the α in the figure by $2 - \alpha$.

Notation. Since (1.1.6) is characterized by four parameters

$$\alpha \in (0, 2],\ \sigma \geq 0,\ \beta \in [-1, 1],\ \mu \in \mathbb{R}^1,$$

we will denote stable distributions by $S_\alpha(\sigma, \beta, \mu)$ and write

$$X \sim S_\alpha(\sigma, \beta, \mu)$$

to indicate that X has the stable distribution $S_\alpha(\sigma, \beta, \mu)$. We also write

$$X \sim S\alpha S$$

when X is symmetric α-stable, i.e., as shown in Property 1.2.5 below, when $\beta = \mu = 0$.

The probability densities of α-stable random variables exist and are continuous but, with a few exceptions, they are not known in closed form (Zolotarev 1986). The exceptions are:

(a) *The Gaussian distribution* $S_2(\sigma, 0, \mu) = N(\mu, 2\sigma^2)$, whose density is

$$(2\sigma\sqrt{\pi})^{-1}\, e^{-(x-\mu)^2/4\sigma^2}.$$

(b) The *Cauchy distribution* $S_1(\sigma, 0, \mu)$, whose density is

$$\frac{\sigma}{\pi((x-\mu)^2+\sigma^2)}. \tag{1.1.13}$$

If $X \sim S_1(\sigma, 0, 0)$, then for $x > 0$,

$$P(X \le x) = \frac{1}{2} + \frac{1}{\pi}\,\text{Arctan}\left(\frac{x}{\sigma}\right). \tag{1.1.14}$$

(See Exercise 1.7.)

(c) The *Lévy distribution* $S_{1/2}(\sigma, 1, \mu)$, whose density

$$\left(\frac{\sigma}{2\pi}\right)^{1/2} \frac{1}{(x-\mu)^{3/2}} \exp\left\{-\frac{\sigma}{2(x-\mu)}\right\} \tag{1.1.15}$$

is concentrated on (μ, ∞). If $X \sim S_{1/2}(\sigma, 1, 0)$, then for $x > 0$,

$$P(X \le x) = 2\left(1 - \Phi\left(\sqrt{\frac{\sigma}{x}}\right)\right), \tag{1.1.16}$$

where Φ denotes the cumulative distribution function of the $N(0, 1)$ distribution. (See Exercise 1.9.)

(d) A *constant* μ which has the degenerate distribution $S_\alpha(0, 0, \mu)$ for any $0 < \alpha \le 2$. Our convention is to exclude degenerate distributions because they have unusual properties. For example, all moments of a degenerate distribution are finite, whereas a non-degenerate α-stable distribution with $0 < \alpha < 2$ has infinite second moments (Property 1.2.16).

1.2 Properties of stable random variables

A useful tool for studying α-stable distributions is the characteristic function (1.1.6). We shall use it to derive some basic properties of stable random variables and to obtain an interpretation of the parameters α, σ, β and μ.

Property 1.2.1 *Let X_1 and X_2 be independent random variables with $X_i \sim S_\alpha(\sigma_i, \beta_i, \mu_i)$, $i = 1, 2$. Then $X_1 + X_2 \sim S_\alpha(\sigma, \beta, \mu)$, with*

$$\sigma = (\sigma_1^\alpha + \sigma_2^\alpha)^{1/\alpha},\ \beta = \frac{\beta_1\sigma_1^\alpha + \beta_2\sigma_2^\alpha}{\sigma_1^\alpha + \sigma_2^\alpha},\ \mu = \mu_1 + \mu_2.$$

PROOF: We verify this for $\alpha \neq 1$. By independence,

$$\begin{aligned}\ln E \exp i\theta(X_1 + X_2) &= \ln(E \exp i\theta X_1) + \ln(E \exp i\theta X_2)\\ &= -(\sigma_1^\alpha + \sigma_2^\alpha)|\theta|^\alpha + i|\theta|^\alpha \operatorname{sign}(\theta)\left(\tan\frac{\pi\alpha}{2}\right)(\beta_1\sigma_1^\alpha + \beta_2\sigma_2^\alpha) + i\theta(\mu_1 + \mu_2)\\ &= -(\sigma_1^\alpha + \sigma_2^\alpha)|\theta|^\alpha\left[1 - i\frac{\beta_1\sigma_1^\alpha + \beta_2\sigma_2^\alpha}{\sigma_1^\alpha + \sigma_2^\alpha}\operatorname{sign}(\theta)\tan\frac{\pi\alpha}{2}\right] + i\theta(\mu_1 + \mu_2).\end{aligned}$$

The proof for $\alpha = 1$ is similar. ∎

The parameter μ is a *shift parameter* because of

Property 1.2.2 *Let $X \sim S_\alpha(\sigma, \beta, \mu)$ and let a be a real constant. Then $X + a \sim S_\alpha(\sigma, \beta, \mu + a)$.*

This follows trivially from the form of the characteristic function (1.1.6).

Property 1.2.3 *Let $X \sim S_\alpha(\sigma, \beta, \mu)$ and let a be a non-zero real constant. Then*

$$\begin{array}{lll} aX \sim S_\alpha(|a|\sigma,\ \operatorname{sign}(a)\beta,\ a\mu) & \text{if } \alpha \neq 1 \\ aX \sim S_1(|a|\sigma,\ \operatorname{sign}(a)\beta,\ a\mu - \frac{2}{\pi}a(\ln|a|)\sigma\beta) & \text{if } \alpha = 1. \end{array} \tag{1.2.1}$$

PROOF: By (1.1.6), we have for $\alpha \neq 1$,

$$\begin{aligned}&\ln\{E \exp i\theta(aX)\}\\ &\quad = -|\theta a|^\alpha\sigma^\alpha\left(1 - i\beta\operatorname{sign}(a\theta)\tan\frac{\pi\alpha}{2}\right) + i\mu(\theta a)\\ &\quad = -(\sigma|a|)^\alpha|\theta|^\alpha\left(1 - i\beta\operatorname{sign}(a)\operatorname{sign}(\theta)\tan\frac{\pi\alpha}{2}\right) + i(\mu a)\theta.\end{aligned}$$

The proof for $\alpha = 1$ is similar. ∎

The parameter σ is called the *scale parameter*. Observe that when $\alpha = 1$, multiplication by a constant affects the shift parameter in a non-linear way. Hence, the name "scale parameter" for σ is a misnomer when $\alpha = 1$ and $\beta \neq 0$. When $\mu = 0$, we have

Property 1.2.4 *For any $0 < \alpha < 2$,*

$$X \sim S_\alpha(\sigma, \beta, 0) \iff -X \sim S_\alpha(\sigma, -\beta, 0). \tag{1.2.2}$$

The following property identifies β as a *skewness parameter*.

Property 1.2.5 *$X \sim S_\alpha(\sigma, \beta, \mu)$ is symmetric if and only if $\beta = 0$ and $\mu = 0$. It is symmetric about μ if and only if $\beta = 0$.*

PROOF: For a random variable to be symmetric, it is necessary and sufficient that its characteristic function be real. By (1.1.6), this is the case if and only if $\beta = 0$ and $\mu = 0$. The second statement follows from Property 1.2.2. ∎

A symmetric stable random variable is strictly stable, but a strictly stable random variable is not necessarily symmetric. In fact,

Property 1.2.6 *Let $X \sim S_\alpha(\sigma, \beta, \mu)$ with $\alpha \neq 1$. Then X is strictly stable if and only if $\mu = 0$.*

PROOF: Let X_1, X_2 be independent copies of X and let A and B be arbitrary positive constants. By Properties 1.2.1 and 1.2.3,

$$AX_1 + BX_2 \sim S_\alpha\big(\sigma(A^\alpha + B^\alpha)^{1/\alpha}, \beta, \mu(A+B)\big).$$

We must set $C = (A^\alpha + B^\alpha)^{1/\alpha}$ in (1.1.1). By Properties 1.2.2 and 1.2.3,

$$CX + D \sim S_\alpha\big(\sigma(A^\alpha + B^\alpha)^{1/\alpha}, \beta, \mu(A^\alpha + B^\alpha)^{1/\alpha} + D\big),$$

and, therefore, we have $AX_1 + BX_2 \stackrel{d}{=} CX + D$ with $D = 0$ if and only if $\mu = 0$. The proof is now complete. ∎

Corollary 1.2.7 *Let $X \sim S_\alpha(\sigma, \beta, \mu)$ with $\alpha \neq 1$. Then $X - \mu$ is strictly stable.*

PROOF: Use Properties 1.2.2 and 1.2.6. ∎

Thus, any α-stable random variable with $\alpha \neq 1$ can be made strictly stable by shifting. This is not true when $\alpha = 1$, as the next property indicates.

Property 1.2.8 *$X \sim S_1(\sigma, \beta, \mu)$ is strictly stable if and only if $\beta = 0$.*

PROOF: Let X_1 and X_2 be independent copies of X and let $A > 0$, $B > 0$. Then, by Properties 1.2.3 and 1.2.1,

$$AX_1 + BX_2 \sim S_1\Big((A+B)\sigma,\ \beta,\ (A+B)\mu - \frac{2}{\pi}\sigma\beta(A \ln A + B \ln B)\Big),$$

whereas

$$(A+B)X \sim S_1\Big((A+B)\sigma,\ \beta,\ (A+B)\mu - \frac{2}{\pi}\sigma\beta(A+B)\ln(A+B)\Big).$$

Therefore $D = 0$ in (1.1.1) if and only if $AX_1 + BX_2 \stackrel{d}{=} (A+B)X$, i.e., if and only if

$$\beta(A \ln A + B \ln B) = \beta(A+B)\ln(A+B)$$

for any $A > 0, B > 0$. It is thus necessary and sufficient that $\beta = 0$. ∎

Corollary 1.2.9 *If $X_1, \ldots, X_n$ are i.i.d. $S_\alpha(\sigma, \beta, \mu)$, then*

$$X_1 + \cdots + X_n \stackrel{d}{=} n^{1/\alpha} X_1 + \mu(n - n^{1/\alpha})$$

if $\alpha \neq 1$, and

$$X_1 + \cdots + X_n \stackrel{d}{=} n X_1 + \frac{2}{\pi} \sigma \beta n \ln n$$

if $\alpha = 1$.

Corollary 1.2.10 *(i) No non-strictly* 1*-stable random variable can be made strictly stable by shifting.*

(ii) Every strictly 1*-stable random variable can be made symmetric by shifting.*

Of the four parameters α, σ, β and μ, the parameter μ is the least important because it affects only location. We shall often assume for simplicity that $\mu = 0$.

Let us now focus on the skewness parameter β. As noted in Property 1.2.4, $X \sim S_\alpha(\sigma, -\beta, 0) \iff -X \sim S_\alpha(\sigma, \beta, 0)$. The distribution $S_\alpha(\sigma, \beta, 0)$ is said to be *skewed to the right* if $\beta > 0$ and *to the left* if $\beta < 0$. It is said to be *totally skewed to the right* if $\beta = 1$ and *totally skewed to the left* if $\beta = -1$.

This terminology is somewhat misleading because it really refers, not to the support of the distribution, but to the parameters P and Q in (1.1.7) that weigh the Lévy measure. In the case $\alpha < 1$ and $\beta = 1$, however, the support of the distribution $S_\alpha(\sigma, 1, 0)$ is in fact the positive half-line. The following proposition establishes that the support is contained in the positive half-line by representing the random variable $X \sim S_\alpha(\sigma, 1, 0)$ as a limit of a sequence of positive compound Poisson random variables. The random variable $X \sim S_\alpha(\sigma, 1, 0)$ with $0 < \alpha < 1$ is called a *stable subordinator*.

Proposition 1.2.11 *Fix $0 < \alpha < 1$, $\delta > 0$ and let N_δ be a Poisson random variable with mean $EN_\delta = \delta^{-\alpha}$ and let $Y_{\delta,k}$, $k = 1, 2, \ldots,$ be i.i.d. positive random variables, independent of N_δ, with distribution*

$$P(Y_{\delta,k} > \lambda) = \begin{cases} \delta^\alpha \lambda^{-\alpha} & \text{if } \lambda > \delta, \\ 1 & \text{if } \lambda \leq \delta. \end{cases}$$

Then the compound Poisson random variable

$$X_\delta = \sum_{k=1}^{N_\delta} Y_{\delta,k}$$

converges in distribution as $\delta \to 0$ to the stable subordinator

$$X \sim S_\alpha(\sigma, 1, 0)$$

with

$$\sigma^\alpha = \Gamma(1-\alpha)\cos(\pi\alpha/2).$$

Moreover, the Laplace transform of X is given by

$$Ee^{-\gamma X} = e^{-a^\alpha \gamma^\alpha},\ \gamma \geq 0, \tag{1.2.3}$$

with $a > 0$ and

$$a^\alpha = \Gamma(1-\alpha) = \sigma^\alpha / \cos(\pi\alpha/2). \tag{1.2.4}$$

PROOF: For any complex number r,

$$Er^{N_\delta} = \sum_{k=0}^{\infty} \exp(-\delta^{-\alpha}) \frac{(\delta^{-\alpha} r)^k}{k!} = \exp\{\delta^{-\alpha}(r-1)\}.$$

Therefore

$$\begin{aligned} E\exp(i\theta X_\delta) &= E\,E[\exp(i\theta X_\delta)|N_\delta] \\ &= E[E\exp(i\theta Y_{\delta,1})]^{N_\delta} \\ &= \exp\{\delta^{-\alpha}(E\exp(i\theta Y_{\delta,1})-1)\} \\ &= \exp\Big\{\delta^{-\alpha}\int_\delta^\infty \delta^\alpha \alpha\lambda^{-(\alpha+1)}(e^{i\theta\lambda}-1)d\lambda\Big\}. \end{aligned}$$

Letting $\delta \to 0$, we obtain

$$\lim_{\delta\to 0} E\exp(i\theta X_\delta) = \exp\Big\{\int_0^\infty \alpha\lambda^{-(\alpha+1)}(e^{i\theta\lambda}-1)d\lambda\Big\}. \tag{1.2.5}$$

Evaluating this integral (see the Appendix of Feller (1966), Chapter XVII.4), we obtain

$$\begin{aligned} &\lim_{\delta\to 0} E\exp(i\theta X_\delta) \\ &\quad = \exp\Big\{-|\theta|^\alpha\Gamma(1-\alpha)\Big(\cos\frac{\pi\alpha}{2} - i\,\mathrm{sign}\,(\theta)\sin\frac{\pi\alpha}{2}\Big)\Big\} \\ &\quad = \exp\Big\{-|\theta|^\alpha\Gamma(1-\alpha)\cos\frac{\pi\alpha}{2}\Big(1 - i\,\mathrm{sign}\,(\theta)\tan\frac{\pi\alpha}{2}\Big)\Big\}, \end{aligned}$$

which proves that X_δ converges in distribution to

$$X \sim S_\alpha\bigg(\Big(\Gamma(1-\alpha)\cos\frac{\pi\alpha}{2}\Big)^{1/\alpha}, 1, 0\bigg).$$

Similarly, if $\gamma > 0$,

$$E\exp(-\gamma X_\delta) = \exp\Big\{\delta^{-\alpha}\int_\delta^\infty \delta^\alpha \alpha\lambda^{-(\alpha+1)}(e^{-\gamma\lambda}-1)d\lambda\Big\}.$$

Letting $\delta \to 0$,

$$\begin{aligned} E\exp(-\gamma X) &= \exp\Big\{\int_0^\infty \alpha\lambda^{-(\alpha+1)}(e^{-\gamma\lambda}-1)d\lambda\Big\} \\ &= \exp\Big\{\gamma^\alpha \int_0^\infty \alpha x^{-(\alpha+1)}(e^{-x}-1)dx\Big\} \\ &= \exp\Big\{-\gamma^\alpha \int_0^\infty x^{-\alpha}e^{-x}dx\Big\} \\ &= \exp\big\{-\gamma^\alpha\Gamma(1-\alpha)\big\}, \end{aligned}$$

which establishes (1.2.3). ■

Remarks

1. Relation (1.2.5) is the Lévy-Khinchine representation (1.1.7) with $Q = 0$. (The term $\int_0^\infty \frac{x}{1+x^2}\frac{dx}{x^{1+\alpha}}$ in (1.1.7) converges for $\alpha < 1$ and can thus be absorbed in the constant M.)

2. Proposition 1.2.11 shows only that the support of a $S_\alpha(\sigma, 1, 0)$ random variable, $0 < \alpha < 1$, is *contained* in $\mathbb{R}_+$. Exercise 1.14 shows that the support is in fact the whole of $\mathbb{R}_+$.

3. Proposition 1.2.11 shows that the Laplace transform $Ee^{-\gamma X}$, $\gamma \geq 0$ of $X \sim S_\alpha(\sigma, 1, 0)$, $0 < \alpha < 1$, exists and equals $e^{-a^\alpha\gamma^\alpha}$. Because of the rapid decay of the left tail in the distribution of X when $1 \leq \alpha < 2$ (see (1.2.11) and (1.2.12) below), the "Laplace transform" exists for all $0 < \alpha \leq 2$ (Gupta & Waymire 1990).

Proposition 1.2.12 *The "Laplace transform" $Ee^{-\gamma X}$, $\gamma \geq 0$, of the random variable $X \sim S_\alpha(\sigma, 1, 0)$, $0 < \alpha \leq 2$, $\sigma > 0$, equals*

$$Ee^{-\gamma X} = \exp\Big\{-\frac{\sigma^\alpha}{\cos\frac{\pi\alpha}{2}}\gamma^\alpha\Big\} \qquad \text{if } \alpha \neq 1,$$

and

$$Ee^{-\gamma X} = \exp\Big\{\sigma\frac{2}{\pi}\gamma\ln\gamma\Big\} \qquad \text{if } \alpha = 1.$$

PROOF: Use Proposition 1.2.11 if $0 < \alpha < 1$ and Exercise 1.15 if $1 \leq \alpha \leq 2$ to establish the result. ■

Remark. The constant $-\sigma^\alpha(\cos\frac{\pi\alpha}{2})^{-1}$ is negative if $0 < \alpha < 1$ and is positive if $1 < \alpha \leq 2$. It equals σ^2 when $\alpha = 2$.

Random variables that are totally skewed to the right can be regarded as basic building blocks because of the following:

Property 1.2.13 *Let X have distribution $S_\alpha(\sigma,\beta,0)$ with $\alpha < 2$. Then there exist two i.i.d. random variables Y_1 and Y_2 with common distribution $S_\alpha(\sigma,1,0)$ such that*

$$X \stackrel{d}{=} \left(\frac{1+\beta}{2}\right)^{1/\alpha} Y_1 - \left(\frac{1-\beta}{2}\right)^{1/\alpha} Y_2 \text{ if } \alpha \neq 1, \tag{1.2.6}$$

and

$$X \stackrel{d}{=} \left(\frac{1+\beta}{2}\right) Y_1 - \left(\frac{1-\beta}{2}\right) Y_2 + \sigma\left(\frac{1+\beta}{\pi}\ln\frac{1+\beta}{2} - \frac{1-\beta}{\pi}\ln\frac{1-\beta}{2}\right) \tag{1.2.7}$$

if $\alpha = 1$.

PROOF: This is a direct consequence of Properties 1.2.1, 1.2.2 and 1.2.3. (See Exercise 1.17.) ■

Since $-Y_2 \sim S_\alpha(\sigma,-1,0)$ is concentrated on the negative real line when $\alpha < 1$, we have, by (1.2.6),

Property 1.2.14 *For $\alpha < 1$ and fixed σ, the family of distributions $S_\alpha(\sigma,\beta,0)$ is stochastically ordered in $-1 \le \beta \le 1$; i.e., if $X_\beta \sim S_\alpha(\sigma,\beta,0)$ and $\beta_1 \le \beta_2$, then $P\{X_{\beta_1} \ge x\} \le P\{X_{\beta_2} \ge x\}$ for all x. Moreover, $S_\alpha(\sigma,\beta,0)$ has support on the whole real line for $-1 < \beta < 1$.*

The figures in Section 1.6 below show that when $\alpha \ge 1$, the family $\{S_\alpha(\sigma,\beta,0),\ -1 \le \beta \le 1\}$ is no longer stochastically ordered. Note also that the support of $S_\alpha(\sigma,\beta,0)$ is the whole real line even for $\beta = \pm 1$. The tails of the distribution, however, are affected by the skewness parameter β as Property 1.2.15 below indicates. Property 1.2.15 concerns the asymptotic behavior of the tail probabilities $P\{X > \lambda\}$ and $P\{X < -\lambda\}$ as $\lambda \to \infty$.

In the Gaussian case $\alpha = 2$,

$$P(X < -\lambda) = P(X > \lambda) \sim \frac{1}{2\sqrt{\pi}\sigma\lambda} e^{-\lambda^2/(4\sigma^2)}$$

as $\lambda \to \infty$ (Feller 1957). When $\alpha < 2$, however, the tail probabilities behave like $\lambda^{-\alpha}$. (If a_λ and b_λ are real numbers, we use the notation $a_\lambda \sim b_\lambda$ to denote $\lim_{\lambda\to\infty} a_\lambda/b_\lambda = 1$.)

Property 1.2.15 *Let $X \sim S_\alpha(\sigma,\beta,\mu)$ with $0 < \alpha < 2$. Then*

$$\begin{cases} \lim_{\lambda\to\infty} \lambda^\alpha P\{X > \lambda\} &= C_\alpha \frac{1+\beta}{2}\sigma^\alpha, \\ \lim_{\lambda\to\infty} \lambda^\alpha P\{X < -\lambda\} &= C_\alpha \frac{1-\beta}{2}\sigma^\alpha, \end{cases} \tag{1.2.8}$$

where

$$C_\alpha = \left(\int_0^\infty x^{-\alpha} \sin x dx \right)^{-1} = \begin{cases} \frac{1-\alpha}{\Gamma(2-\alpha)\cos(\pi\alpha/2)} & \text{if } \alpha \neq 1, \\ 2/\pi & \text{if } \alpha = 1. \end{cases} \tag{1.2.9}$$

One can use Theorem 1.4.5 below to obtain this result (Exercise 1.27). One can also apply a central limit theorem type argument, as in Feller (1971), Theorem XVII.5.1.

In the case $\alpha < 1$, Property 1.2.15 follows easily from a Tauberian theorem and (1.2.6). Indeed, let $X \sim S_\alpha(\sigma, 1, 0)$. By Proposition 1.2.12, $Ee^{-\gamma X} = \exp\{-a^\alpha \gamma^\alpha\}$ with $a^\alpha = \sigma^\alpha / \cos \frac{\pi\alpha}{2}$. Since

$$\begin{aligned} \int_0^\infty e^{-\gamma\lambda} P\{X > \lambda\} d\lambda &= \frac{1 - Ee^{-\gamma X}}{\gamma} \qquad \text{(integration by parts)} \\ &= \frac{1 - \exp\{-a^\alpha \gamma^\alpha\}}{\gamma} \\ &\sim a^\alpha \gamma^{\alpha - 1} \end{aligned}$$

as $\gamma \to 0$, we obtain, by the Tauberian theorem (Feller 1971, Theorem XIII.5.4),

$$\begin{aligned} P\{X > \lambda\} &\sim \frac{a^\alpha}{\Gamma(1-\alpha)} \lambda^{-\alpha} \\ &= \frac{\sigma^\alpha}{\Gamma(1-\alpha)\cos(\pi\alpha/2)} \lambda^{-\alpha} \\ &= \sigma^\alpha C_\alpha \lambda^{-\alpha} \end{aligned}$$

as $\lambda \to \infty$. This yields the result for $\alpha < 1$ and $\beta = 1$. We use (1.2.6) to obtain it for $\alpha < 1$ and all $-1 \leq \beta \leq 1$. We can then use Proposition 1.3.1 below to obtain it for all $\alpha \geq 1$ and $\beta = 0$ (Exercise 1.28).

As a special case, if $X \sim S_\alpha(\sigma, 0, 0)$, then as $\lambda \to \infty$,

$$P(X > \lambda) \sim \sigma^\alpha \frac{C_\alpha}{2} \lambda^{-\alpha}. \tag{1.2.10}$$

Suppose $X \sim S_\alpha(\sigma, -1, 0)$. Since $\beta = -1$, Property 1.2.15 gives $\lim_{\lambda \to \infty} \lambda^\alpha P(X > \lambda) = 0$, i.e., $P(X > \lambda)$ tends to 0 faster than $\lambda^{-\alpha}$ as $\lambda \to \infty$. What is the true rate? When $\alpha < 1$, X is totally skewed to the left and hence $P(X > \lambda) = 0$ for all $\lambda > 0$. It turns out, that when $\alpha > 1$, as $\lambda \to \infty$,

$$\begin{aligned} &P(X > \lambda) \\ &\sim \frac{1}{\sqrt{2\pi\alpha(\alpha-1)}} \left(\frac{\lambda}{\alpha\hat{\sigma}_\alpha} \right)^{-\alpha/(2(\alpha-1))} \exp\left(-(\alpha - 1) \left(\frac{\lambda}{\alpha\hat{\sigma}_\alpha} \right)^{\alpha/(\alpha-1)} \right), \end{aligned} \tag{1.2.11}$$

where

$$\widehat{\sigma}_\alpha = \sigma\Big(\cos\frac{\pi}{2}(2-\alpha)\Big)^{-1/\alpha}.$$

When $\alpha = 1$,

$$P(X > \lambda) \sim \frac{1}{\sqrt{2\pi}} \exp\left(-\frac{(\pi/2\sigma)\lambda - 1}{2} - e^{(\pi/2\sigma)\lambda - 1}\right) \qquad (1.2.12)$$

(see Zolotarev (1986), Theorem 2.5.3.). The same formulas apply to the left tail when $\beta = 1$ because $P(X < -\lambda) = P(-X > \lambda)$ and $X \sim S_\alpha(\sigma, 1, 0)$ implies $-X \sim S_\alpha(\sigma, -1, 0)$.

The tail behavior (1.2.8) is a widely used property of α-stable distributions. Since $E|X|^r = \int_0^\infty P\{|X|^r > \lambda\}d\lambda$, we have:

Property 1.2.16 *Let $X \sim S_\alpha(\sigma, \beta, \mu)$ with $0 < \alpha < 2$. Then*

$$E|X|^p < \infty \quad \text{for any} \quad 0 < p < \alpha\,,$$
$$E|X|^p = \infty \quad \text{for any} \quad p \geq \alpha\,.$$

The fact that α-stable random variables with $\alpha < 2$ have an infinite second moment means that many of the techniques valid for the Gaussian case do not apply. An added complication stems from the fact that even $E|X|^\alpha$ is infinite. When $\alpha \leq 1$, one also has $E|X| = \infty$, precluding the use of expectations.

Property 1.2.17 *Let $X \sim S_\alpha(\sigma, \beta, 0)$ with $0 < \alpha < 2$ and $\beta = 0$ in the case $\alpha = 1$. Then, for every $0 < p < \alpha$, there is a constant $c_{\alpha,\beta}(p)$ such that*

$$(E|X|^p)^{1/p} = c_{\alpha,\beta}(p)\sigma. \qquad (1.2.13)$$

The constant $c_{\alpha,\beta}(p)$ equals $(E|X_0|^p)^{1/p}$ where $X_0 \sim S_\alpha(1, \beta, 0)$.

PROOF: This follows immediately from the fact that $X \stackrel{d}{=} \sigma X_0$. ■

One can show (Hardin Jr. 1984) that

$$(c_{\alpha,\beta}(p))^p$$
$$= \frac{2^{p-1}\Gamma(1-\frac{p}{\alpha})}{p\int_0^\infty u^{-p-1}\sin^2 u\,du}\left(1+\beta^2\tan^2\frac{\alpha\pi}{2}\right)^{p/2\alpha}\cos\left(\frac{p}{\alpha}\arctan\Big(\beta\tan\frac{\alpha\pi}{2}\Big)\right).$$

Note that for fixed σ and p and $\alpha \neq 1$, $E|X|^p$ is an even function of β and increases in $|\beta|$. Note also that if $X \sim S\alpha S$, $0 < \alpha \leq 2$, then the moment $E|X|^p$ determines the scale parameter σ of X and, therefore, the whole distribution.

Property 1.2.18 *Let $X \sim S_\alpha(\sigma, \beta, \mu)$. Then*

$$\lim_{r\uparrow\alpha}(\alpha - r)E|X|^r = \alpha C_\alpha \sigma^\alpha, \tag{1.2.14}$$

$$\lim_{r\uparrow\alpha}(\alpha - r)EX^{<r>} = \alpha C_\alpha \beta \sigma^\alpha, \tag{1.2.15}$$

where C_α is given by (1.2.9) and $a^{<b>} := |a|^b$ sign (a).

PROOF: Clearly, for any $r < \alpha$, $EX_+^r = \int_0^\infty P(X > t^{1/r})dt$. By Property 1.2.15, for every $\epsilon > 0$, there is an $M = M(\epsilon) \in (1, \infty)$ such that, for every $u > M$,

$$u^\alpha P(X > u) \in \left(C_\alpha \frac{1+\beta}{2}\sigma^\alpha - \epsilon,\ C_\alpha \frac{1+\beta}{2}\sigma^\alpha + \epsilon\right).$$

Fixing an $\epsilon > 0$, we have

$$\begin{aligned}
&\limsup_{r\uparrow\alpha}(\alpha - r)EX_+^r \\
&\quad = \limsup_{r\uparrow\alpha}(\alpha - r)\left(\int_0^{M^\alpha} P(X > t^{1/r})dt + \int_{M^\alpha}^\infty P(X > t^{1/r})dt\right) \\
&\quad = \limsup_{r\uparrow\alpha}(\alpha - r)r\int_{M^{\alpha/r}}^\infty u^\alpha P(X > u)u^{r-\alpha-1}du,
\end{aligned}$$

with a similar relation for the limit inferior. Hence

$$\begin{aligned}
\alpha\left(C_\alpha \frac{1+\beta}{2}\sigma^\alpha - \epsilon\right) &\le \liminf_{r\uparrow\alpha}(\alpha - r)EX_+^r \\
&\le \limsup_{r\uparrow\alpha}(\alpha - r)EX_+^r \le \alpha\left(C_\alpha \frac{1+\beta}{2}\sigma^\alpha + \epsilon\right).
\end{aligned}$$

Since this is true for every $\epsilon > 0$, we conclude that

$$\lim_{r\uparrow\alpha}(\alpha - r)EX_+^r = \alpha C_\alpha \frac{1+\beta}{2}\sigma^\alpha. \tag{1.2.16}$$

Similarly,

$$\lim_{r\uparrow\alpha}(\alpha - r)EX_-^r = \alpha C_\alpha \frac{1-\beta}{2}C^\alpha. \tag{1.2.17}$$

Our claim, (1.2.14) and (1.2.15), follows now from (1.2.16) and (1.2.17) and the obvious relations $|X|^r = X_+^r + X_-^r$ and $X^{<r>} = X_+^r - X_-^r$. ∎

Property 1.2.19 *When $1 < \alpha \le 2$, the shift parameter μ equals the mean.*

PROOF: Let $X \sim S_\alpha(\sigma, \beta, \mu)$, $1 < \alpha \le 2$. The random variable X has finite mean (by Property 1.2.16 in the case $1 < \alpha < 2$, and because X is Gaussian when $\alpha = 2$.) Moreover, $X - \mu$ is strictly stable by Corollary 1.2.7. Let X_1 and X_2 be two independent copies of X. By Definition 1.1.1 of stability and (1.1.2), the relation

$$A(X_1 - \mu) + B(X_2 - \mu) \stackrel{d}{=} (A^\alpha + B^\alpha)^{1/\alpha}(X - \mu)$$

holds for any positive A and B. Taking expectations of both sides gives

$$A(EX - \mu) + B(EX - \mu) = (A^\alpha + B^\alpha)^{1/\alpha}(EX - \mu),$$

and thus $EX = \mu$. ■

For an alternative proof see Exercise 1.19.

1.3 Symmetric α-stable random variables

We have seen that X is $S\alpha S$ (symmetric α-stable) if and only if $X \sim S_\alpha(\sigma, 0, 0)$ (Property 1.2.5). Using (1.1.6), we observe that if X is $S\alpha S$, then its characteristic function takes the particularly simple form

$$E \exp i\theta X = e^{-\sigma^\alpha |\theta|^\alpha}. \tag{1.3.1}$$

Hence a $S\alpha S$ distribution is characterized only by the scale parameter σ. A random variable X is called *standard* $S\alpha S$ if $\sigma = 1$. Note that a standard $S\alpha S$ random variable with $\alpha = 2$ is $N(0, 2)$ because, when $\alpha = 2$, we have $\sigma^2 = \frac{1}{2}\text{Var}\, X$.

$S\alpha S$ random variables will play an important role in the sequel. The following result shows that one can always transform a $S\alpha' S$ random variable into a $S\alpha S$ random variable for any $0 < \alpha < \alpha'$.

Proposition 1.3.1 *Let $X \sim S_{\alpha'}(\sigma, 0, 0)$ with $0 < \alpha' \le 2$ and let $0 < \alpha < \alpha'$. Let A be an α/α'-stable random variable totally skewed to the right with Laplace transform*

$$E \exp(-\gamma A) = \exp\{-\gamma^{\alpha/\alpha'}\},\ \gamma > 0,$$

i.e., $A \sim S_{\alpha/\alpha'}\left((\cos \frac{\pi\alpha}{2\alpha'})^{\alpha'/\alpha}, 1, 0\right)$, and assume X and A to be independent. Then

$$Z = A^{1/\alpha'} X \sim S_\alpha(\sigma, 0, 0).$$

PROOF: For a real θ, we have, by (1.3.1),

$$\begin{aligned} E \exp i\theta Z &= E \exp\{i\theta A^{1/\alpha'} X\} \\ &= E(E \exp\{i(\theta A^{1/\alpha'})X\}|A) \end{aligned}$$

$$
\begin{aligned}
&= E\exp\{-\sigma^{\alpha'}|\theta|^{\alpha'}A\} \\
&= \exp\{-(\sigma^{\alpha'}|\theta|^{\alpha'})^{\alpha/\alpha'}\} \\
&= \exp\{-\sigma^{\alpha}|\theta|^{\alpha}\}. \quad \blacksquare
\end{aligned}
$$

In particular, this implies that if X is a zero mean Gaussian random variable and if A is a positive $\frac{\alpha}{2}$-stable random variable independent of X, then

$$Z = A^{1/2}X \sim S\alpha S.$$

This shows that every $S\alpha S$ random variable is conditionally Gaussian.[3]

A result similar to Proposition 1.3.1 holds for skewed $X \sim S_{\alpha'}(\sigma', \beta', 0)$ if $\alpha' \neq 1$, namely

$$Z = A^{1/\alpha'}X \sim \begin{cases} S_\alpha(\sigma, \beta, 0) & \text{if } \alpha \neq 1, \\ S_1(\sigma, 0, \mu) & \text{if } \alpha = 1. \end{cases}$$

The random variable Z is skewed if $\alpha \neq 1$ and it has non-zero shift if $\alpha = 1$. The parameters in its distribution are complicated functions of σ' and β' (Hardin Jr. 1984); however, in the special case $\alpha' < 1$ and $\beta' = 1$, the parameter β equals 1 (see Exercise 1.21).

1.4 Series representation

We wish now to show that an α-stable random variable X with $0 < \alpha < 2$ can be represented as a convergent sum of random variables involving arrival times of a Poisson process.

Let $N(t)$ represent the number of "customer arrivals" in the time interval $[0, t]$. The process $\{N(t), t \geq 0\}$ is a *Poisson process* with rate λ if the interarrival times $\tau_{i+1} - \tau_i$, $i \geq 1$, are independent and exponentially distributed with mean $1/\lambda$. In that case, $EN(t) = \lambda t$, i.e., the rate λ equals the mean number of arrivals per unit time.

Poisson processes have many special properties. Here are two of them. If each arrival is eliminated with probability $0 \leq p < 1$, independently of the others, then the resulting process is Poisson with rate $\lambda(1-p)$. If $\{N_1(t), t \geq 0\}$ and $\{N_2(t), t \geq 0\}$ are two independent Poisson processes, with rates λ_1 and λ_2, respectively, then the superposed process $\{N_1(t) + N_2(t), t \geq 0\}$ is Poisson with rate $\lambda_1 + \lambda_2$ (Ross 1985). These properties make intuitive sense. The second one turns out to be the main ingredient in the proof of the following result.

[3] If $X \sim N(0, \sigma^2)$, then the $S\alpha S$ random variable $Z = A^{1/2}X$ can be viewed informally as $N(0, \sigma^2 A)$, i.e., normal with the random variance $\sigma^2 A$.

Proposition 1.4.1 *Let $\{\tau_i\}$ denote the arrival times of a Poisson process with rate 1 and let $\{R_i\}$ be i.i.d. random variables, independent of the sequence $\{\tau_i\}$. If the series*

$$\sum_{i=1}^{\infty} \tau_i^{-1/\alpha} R_i \tag{1.4.1}$$

converges a.s., then it converges to a strictly α-stable random variable.

PROOF: Assume that $\sum_{i=1}^{\infty} \tau_i^{-1/\alpha} R_i$ converges a.s. to a random variable X and set $X' = \sum_{i=1}^{\infty} (\tau_i')^{-1/\alpha} R_i'$ and $X'' = \sum_{i=1}^{\infty} (\tau_i'')^{-1/\alpha} R_i''$ where $\{\tau_i'\}$, $\{\tau_i''\}$, $\{R_i'\}$, $\{R_i''\}$ are independent sequences and are copies of $\{\tau_i\}$ and $\{R_i\}$, respectively. For any $A > 0, B > 0, A^\alpha + B^\alpha = 1$, we have

$$\begin{aligned} AX' + BX'' &= \sum_{i=1}^{\infty} (A^{-\alpha}\tau_i')^{-1/\alpha} R_i' + \sum_{i=1}^{\infty} (B^{-\alpha}\tau_i'')^{-1/\alpha} R_i'' \\ &\stackrel{d}{=} \sum_{i=1}^{\infty} \tau_i^{-1/\alpha} R_i. \end{aligned} \tag{1.4.2}$$

To verify this last relation, note firstly that $\{A^{-\alpha}\tau_i'\}$ is the sequence of arrival times of a Poisson process with rate A^α since the interarrival times $A^{-\alpha}(\tau_{i+1} - \tau_i), i \geq 1$, are i.i.d. and satisfy

$$P\{A^{-\alpha}(\tau_{i+1} - \tau_i) > x\} = P\{\tau_{i+1} - \tau_i > A^\alpha x\} = e^{-A^\alpha x}$$

for $x > 0$, i.e., they are exponentially distributed with mean $A^{-\alpha}$. Similarly, $\{B^{-\alpha}\tau_i''\}$ is the sequence of arrival times of an (independent) Poisson process with rate B^α. Now let us superpose these two Poisson processes. To do this, we list on the same time axis the arrival times of both processes and view the combined arrival times as being generated by a single process. This superposed process is Poisson with rate $A^\alpha + B^\alpha = 1$. If $\tau_1, \tau_2, \ldots$ denote its arrival times, then each τ_i is either $A^{-\alpha}\,\tau_m'$ or $B^{-\alpha}\,\tau_n''$ for some $m \geq 1$ or $n \geq 1$. Set $R_i = R_m'$ if $\tau_i = A^{-\alpha}\tau_m'$ and $R_i = R_n''$ if $\tau_i = B^{-\alpha}\tau_n''$. The resulting R_is are obviously i.i.d., and thus (1.4.2) holds. The random variable X is strictly α-stable because $AX' + BX'' \stackrel{d}{=} X$. ■

The preceding proposition suggests that a stable random variable X can be represented as infinite series of the type (1.4.1). It is necessary, however, to add some assumptions on α and on the distribution of R_i in order to ensure that the series converges a.s.. We will require $0 < \alpha < 2$ and that the distribution of R_i have finite absolute α^{th} moment and be symmetric, so that the resulting X is $S\alpha S$. In fact, we will set $R_i = \epsilon_i W_i$ where $W_i = |R_i|$ and where $\epsilon_i = \pm 1$ is an independent random variable denoting the sign of R_i. Also, for convenience, we write Γ_i instead of τ_i. This is because the ith arrival in a Poisson process is the sum of i independent exponentials and hence has a gamma distribution. These comments together with Proposition 1.4.1 motivate the following theorem. We begin by stating our assumptions.

Let $\{\epsilon_1, \epsilon_2, \dots\}$, $\{W_1, W_2, \dots\}$, $\{\Gamma_1, \Gamma_2, \dots\}$ be three independent sequences of random variables. Here $\epsilon_1, \epsilon_2, \dots$ is an i.i.d. sequence of Rademacher variables, i.e.,

$$\epsilon_i = \begin{cases} 1 & \text{with probability } 1/2, \\ -1 & \text{with probability } 1/2\,. \end{cases}$$

$W_1, W_2, \dots$ is an i.i.d. sequence of random variables (not necessarily positive) with finite absolute α^{th} moment. Finally, let $\Gamma_1, \Gamma_2, \dots$ be a sequence of arrival times of a Poisson process with unit arrival rate, i.e., $\Gamma_i = \sum_{j=1}^{i} e_j$, where the e_js are i.i.d. exponential random variables with $Ee_j = 1$. The random variables $\Gamma_1, \Gamma_2, \dots$ are, therefore, *dependent* and they are *not* identically distributed. The random variable Γ_i has a gamma distribution with parameter i and mean $E\Gamma_i = i$.

Theorem 1.4.2 *Suppose* $0 < \alpha < 2$. *Then the sum* $\sum_{i=1}^{\infty} \epsilon_i \Gamma_i^{-1/\alpha} W_i$ *converges almost surely to a random variable* X *whose distribution is* $S_\alpha((C_\alpha^{-1} E|W_1|^\alpha)^{1/\alpha}, 0, 0)$ *where* C_α *is the constant defined in (1.2.9).*

PROOF: We proceed in three steps.

Step 1. Let $U_1, U_2, \dots$ be an i.i.d. sequence of uniform random variables on (0,1), independent of the sequences $\{\epsilon_1, \epsilon_2, \dots\}$ and $\{W_1, W_2, \dots\}$. Set

$$Y_i = \epsilon_i U_i^{-1/\alpha} W_i\,,\ i = 1, 2, \dots.$$

Because of the ϵ_is, $Y_1, Y_2, \dots$ is a sequence of i.i.d. symmetric random variables. Let us compute their tail probability. Letting $F_{|W|}$ denote the distribution of $|W|$, we have for $\lambda > 0$,

$$\begin{aligned} P(|Y_i| > \lambda) &= P(U_i^{\ 1/\alpha}|W_i| > \lambda) \\ &= P(U_i < \lambda^{-\alpha}|W_i|^\alpha) \\ &= \int_0^\infty P(U_i < \lambda^{-\alpha} w^\alpha) F_{|W|}(dw) \\ &= \int_0^\lambda \lambda^{-\alpha} w^\alpha F_{|W|}(dw) + \int_\lambda^\infty F_{|W|}(dw) \\ &= \lambda^{-\alpha} \int_0^\lambda w^\alpha F_{|W|}(dw) + P\{|W_i| > \lambda\}. \end{aligned}$$

Therefore

$$\lim_{\lambda\to\infty} \lambda^\alpha P(|Y_i| > \lambda) = E|W_i|^\alpha + \lim_{\lambda\to\infty} \lambda^\alpha P\{|W_i| > \lambda\} = E|W_i|^\alpha,$$

so that Y_i is in the domain of (normal) attraction of a symmetric α-stable random variable, i.e.,

$$\frac{1}{n^{1/\alpha}} \sum_{i=1}^{n} Y_i \overset{d}{\Rightarrow} X, \tag{1.4.3}$$

where $X \sim S_\alpha(\sigma, 0, 0)$ with $\sigma = (C_\alpha^{-1} E|W_i|^\alpha)^{1/\alpha}$ (see, for example, Feller (1971), page 581).

Step 2. By writing $n^{-1/\alpha} \sum_{i=1}^{n} \epsilon_i U_i^{-1/\alpha} W_i$ in a different way, we shall show that it has the same limiting distribution as $\sum_{i=1}^{\infty} \epsilon_i \Gamma_i^{-1/\alpha} W_i$, where the Γ_is are the gamma random variables appearing in the statement of the theorem.

To see the connection between the U_is and the Γ_is consider the "arrival" times in a Poisson process with unit rate. Given Γ_{n+1}, the time of the $(n+1)^{th}$ arrival, the previous n arrivals are uniformly distributed in the interval $(0, \Gamma_{n+1})$. Since the Γ_is are ordered, $\Gamma_1 \le \Gamma_2 \le \cdots \le \Gamma_n$, the (conditional) distribution of $(\Gamma_1, \Gamma_2, \ldots, \Gamma_n)$ is that of the order statistic of n i.i.d. $U(0, \Gamma_{n+1})$ random variables. Thus, given Γ_{n+1}, the vector $(\Gamma_i/\Gamma_{n+1},\ i = 1, 2, \ldots, n)$ is distributed as G_n, the order statistic distribution of n i.i.d. $U(0,1)$ random variables.

We now proceed by taking advantage of two helpful features. Firstly, the conditional distribution of $(\Gamma_i/\Gamma_{n+1},\ i = 1, 2, \ldots, n)$ given Γ_{n+1} is equal to the unconditional distribution. (This is because the distribution G_n does not depend on Γ_{n+1}.) Secondly, any permutation of the U_is in $\sum_{i=1}^{n} \epsilon_i U_i^{-1/\alpha} W_i$ preserves the distribution since the $\epsilon_i W_i$s are i.i.d.. This means that we can replace the U_is in that sum by the ordered U_is without affecting the distribution of the sum. Consequently,

$$n^{-1/\alpha} \sum_{i=1}^{n} Y_i = n^{-1/\alpha} \sum_{i=1}^{n} \epsilon_i U_i^{-1/\alpha} W_i \overset{d}{=} n^{-1/\alpha} \sum_{i=1}^{n} \epsilon_i \left(\frac{\Gamma_i}{\Gamma_{n+1}} \right)^{-1/\alpha} W_i.$$

Using (1.4.3), we obtain

$$\left(\frac{\Gamma_{n+1}}{n} \right)^{1/\alpha} \sum_{i=1}^{n} \epsilon_i \Gamma_i^{-1/\alpha} W_i \overset{d}{\Rightarrow} X \sim S_\alpha(\sigma, 0, 0). \tag{1.4.4}$$

Step 3. It remains to show that the left-hand side of (1.4.4) converges a.s. to $\sum_{i=1}^{\infty} \epsilon_i \Gamma_i^{-1/\alpha} W_i$.

By the strong law of large numbers, the event $A = \{\lim_{n \to \infty} \frac{1}{n} \Gamma_n = 1\} \cap \{\Gamma_1 > 0\}$ has probability 1, since Γ_n is a sum of n i.i.d. exponential random variables with mean 1. To prove that $\sum_{i=1}^{\infty} \epsilon_i \Gamma_i^{-1/\alpha} W_i$ converges a.s., it is enough to show that it converges a.s. for each fixed sequence $\{\Gamma_i\}$ belonging to the event A. Fix such a sequence. Then the summands $\epsilon_i \Gamma_i^{-1/\alpha} W_i$ are independent and $C_1 i \le \Gamma_i \le C_2 i$ for some positive constants C_1 and C_2 and $i = 1, 2, \ldots$. We can apply the three series theorem (Feller 1971, Theorem IX.9.3). For each $\lambda > 0$, we have:

(i)

$$\begin{aligned}\sum_{i=1}^{\infty} P\{|\epsilon_i \Gamma_i^{-1/\alpha} W_i| > \lambda\} &= \sum_{i=1}^{\infty} P\{|W_i|^\alpha > \lambda^\alpha \Gamma_i\} \\ &\leq \sum_{i=1}^{\infty} P\{|W_i|^\alpha > \lambda^\alpha C_1 i\} \\ &< \infty\end{aligned}$$

since $E|W_i|^\alpha < \infty$,

(ii)

$$\sum_{i=1}^{\infty} E\epsilon_i \Gamma_i^{-1/\alpha} W_i 1\{|\epsilon_i \Gamma_i^{-1/\alpha} W_i| \leq \lambda\} = 0$$

since each summand equals 0,

(iii)

$$\begin{aligned}&\sum_{i=1}^{\infty} E\left(\epsilon_i \Gamma_i^{-1/\alpha} W_i 1\{|\epsilon_i \Gamma_i^{-1/\alpha} W_i| \leq \lambda\}\right)^2 \\ &\quad \leq C_1^{-2/\alpha} \sum_{i=1}^{\infty} i^{-2/\alpha} \int_0^\infty w^2 1\{w \leq \lambda C_2^{1/\alpha} i^{1/\alpha}\} F_{|W|}(dw) \\ &\quad \leq C \int_0^\infty x^{-2/\alpha} dx \int_0^{\lambda C_2^{1/\alpha} x^{1/\alpha}} w^2 F_{|W|}(dw) \\ &\quad = C \int_0^\infty w^2 F_{|W|}(dw) \int_{\lambda^{-\alpha} C_2^{-1} w^\alpha}^\infty x^{-2/\alpha} dx \\ &\quad = C' \int_0^\infty w^\alpha F_{|W|}(dw)\end{aligned}$$

which is finite (C and C' are positive constants). Therefore $\sum_{i=1}^n \epsilon_i \Gamma_i^{-1/\alpha} W_i$ converges a.s. to $\sum_{i=1}^\infty \epsilon_i \Gamma_i^{-1/\alpha} W_i$ as $n \to \infty$ and we conclude that

$$\sum_{i=1}^{\infty} \epsilon_i \Gamma_i^{-1/\alpha} W_i \stackrel{d}{=} X. \quad \blacksquare$$

Let X and Y be two random variables, possibly defined on different probability spaces. We say that X has the *representation* Y if $X \stackrel{d}{=} Y$.

Corollary 1.4.3 *(Series representation). Any $S\alpha S$, $0 < \alpha < 2$, random variable with distribution $S_\alpha(\sigma, 0, 0)$ has the series representation*

$$\sigma \left(\frac{C_\alpha}{E|W_1|^\alpha} \right)^{1/\alpha} \sum_{i=1}^{\infty} \epsilon_i \Gamma_i^{-1/\alpha} W_i. \tag{1.4.5}$$

Remarks

1. We did not use Proposition 1.4.1 in the proof of Theorem 1.4.2. This proposition, together with step 3 of the preceding proof establish that X is (symmetric) α-stable but do not provide the value of σ.

2. The distribution of $\sum_{i=1}^{\infty} \epsilon_i \Gamma_i^{-1/\alpha} W_i$ depends only on $E|W_1|^\alpha$. The distribution W_i is immaterial as long as $E|W_1|^\alpha < \infty$. For example, the W_is can be of $U(0, 1)$ or $N(0, 1)$.

3. The series representation can, in principle, be used to generate a $S\alpha S$ random variable. Although the Γ_is are not independent, they are easily generated as successive sums of exponential random variables. This is not, however, a good way to generate $S\alpha S$ random variables because the convergence is too slow. The method of Chambers et al. discussed in Section 1.7 is much better.

The series representation provides some insight into $S\alpha S$ random variables. The absolute values $|\epsilon_i \Gamma_i^{-1/\alpha} W_i|$ of the successive summands decrease stochastically as i increases because $\Gamma_1^{-1/\alpha} \geq \Gamma_2^{-1/\alpha} \geq \cdots$. The first summand $\epsilon_1 \Gamma_1^{-1/\alpha} W_1$ is stochastically the greatest in absolute value. Its probability tail (that is, the tail of its distribution function) has the same asymptotic rate of growth as an α-stable random variable. Indeed,

$$\begin{aligned} P[|\epsilon_1 \Gamma_1^{-1/\alpha} W_1| > \lambda] &= \int_0^\infty P[\Gamma_1 < w^\alpha \lambda^{-\alpha}] F_{|W_1|}(dw) \\ &= \int_0^\infty (1 - e^{-w^\alpha \lambda^{-\alpha}}) F_{|W_1|}(dw) \\ &\sim E|W_1|^\alpha \lambda^{-\alpha} \end{aligned}$$

as $\lambda \to \infty$. The other summands $\epsilon_i \Gamma_i^{-1/\alpha} W_i$, $i \geq 2$ provide the necessary corrections for the whole sum to have an α-stable distribution.

In fact, Corollary 1.4.3 states that $X \sim S_\alpha(\sigma, 0, 0)$ can be represented as $X \stackrel{d}{=} C_\alpha^{1/\alpha} \sum_{i=1}^{\infty} \epsilon_i \Gamma_i^{-1/\alpha} W_i$ with $E|W_i|^\alpha = \sigma^\alpha$. Both X and the first term of the series, $C_\alpha^{1/\alpha}\epsilon_1\Gamma_1^{1/\alpha}W_1$, have the same asymptotic behavior, because on one hand,

$$P(X > \lambda) \sim \frac{1}{2}C_\alpha\sigma^\alpha\lambda^{-\alpha} \text{ as } \lambda \to \infty \tag{1.4.6}$$

(Property 1.2.15), and on the other hand,

$$P(C_\alpha^{1/\alpha}\epsilon_1\Gamma_1^{-1/\alpha}W_1 > \lambda) \sim \frac{1}{2}C_\alpha\sigma^\alpha\lambda^{-\alpha} \text{ as } \lambda \to \infty. \tag{1.4.7}$$

The probability tail of the remainder of the series, $\sum_{i=2}^\infty \epsilon_i\Gamma_i^{-1/\alpha}W_i$, has a smaller order. In fact, if $E|W_1|^{\alpha+\epsilon} < \infty$ for some $\epsilon > 0$, then

$$E\Big|\sum_{i=2}^\infty \epsilon_i\Gamma_i^{-1/\alpha}W_i\Big|^\alpha < \infty \tag{1.4.8}$$

and, hence,

$$P\Big(\sum_{i=2}^\infty \epsilon_i\Gamma_i^{-1/\alpha}W_i > \lambda\Big) = o(\lambda^{-\alpha})$$

as $\lambda \to \infty$. But if we only assume $E|W_1|^\alpha < \infty$, then (1.4.8) may not hold (see Exercise 1.25). Nevertheless, one always has

Property 1.4.4

$$P\Big(\sum_{i=2}^\infty \epsilon_i\Gamma_i^{-1/\alpha}W_i > \lambda\Big) = o(\lambda^{-\alpha}) \textit{ as } \lambda \to \infty. \tag{1.4.9}$$

PROOF: For any $M > 0$, write

$$\sum_{i=2}^\infty \epsilon_i\Gamma_i^{-1/\alpha}W_i = \sum_{i=2}^\infty \epsilon_i\Gamma_i^{-1/\alpha}W_i 1(|W_i| \le M) + \sum_{i=2}^\infty \epsilon_i\Gamma_i^{-1/\alpha}W_i 1(|W_i| > M).$$

Since $W_i 1(|W_i| < M)$ has moments of all order,

$$P\Big(\sum_{i=2}^\infty \epsilon_i\Gamma_i^{-1/\alpha}W_i 1(|W_i| \le M) > \lambda\Big) = o(\lambda^{-\alpha})$$

follows from (1.4.8). Hence,

$$\begin{aligned}
&\overline{\lim}_{\lambda\to\infty}\lambda^\alpha P\Big(\sum_{i=2}^\infty \epsilon_i\Gamma_i^{-1/\alpha}W_i > \lambda\Big) \\
&\le \overline{\lim}_{\lambda\to\infty}\lambda^\alpha P\Big(\sum_{i=2}^\infty \epsilon_i\Gamma_i^{-1/\alpha}W_i 1(|W_i| > M) > \lambda\Big) \\
&\le 2\lim_{\lambda\to\infty}\lambda^\alpha P\Big(\sum_{i=1}^\infty \epsilon_i\Gamma_i^{-1/\alpha}W_i 1(|W_i| > M) > \lambda\Big)
\end{aligned}$$

using the relation

$$P(U+V>\lambda)\geq P(U>\lambda,\ V>0)=P(U>\lambda)P(V>0),$$

valid for any two independent random variables U and V. Here $V=\epsilon_1 V_1$ and $P(\epsilon_1 W_1\geq 0)\geq \frac{1}{2}$. We conclude by (1.4.6) that

$$\overline{\lim}_{\lambda\to\infty}\lambda^\alpha P\Big(\sum_{i=2}^{\infty}\epsilon_i\Gamma_i^{-1/\alpha}W_i>\lambda\Big)\leq E|W_1|^\alpha 1(|W_i|>M).$$

We establish the claim by letting $M\to\infty$. ■

The series representation given in (1.4.5) holds for a $S\alpha S$ random variable. We now provide a corresponding representation for a skewed α-stable random variable.

Consider again two independent sequences $\{W_1,W_2,\ldots\}$ and $\{\Gamma_1,\Gamma_2,\ldots\}$. Here the sequence $W_1,W_2,\ldots$ is an i.i.d. sequence of random variables with

$$E|W_1|^\alpha<\infty \qquad \text{if } 0<\alpha<2,\ \alpha\neq 1,$$

$$E|W_1\ln|W_1||<\infty \qquad \text{if } \alpha=1.$$

The sequence $\Gamma_1,\Gamma_2,\ldots$ is defined as above.

Theorem 1.4.5 *The series*

$$\sum_{i=1}^{\infty}\left(\Gamma_i^{-1/\alpha}W_i-k_i^{(\alpha)}\right) \tag{1.4.10}$$

with

$$k_i^{(\alpha)}=\begin{cases}0 & \text{if } 0<\alpha<1,\\ E\left(W_1\int_{|W_1|/i}^{|W_1|/(i-1)}x^{-2}\sin x\,dx\right) & \text{if } \alpha=1,\\ \frac{\alpha}{\alpha-1}\left(i^{\frac{\alpha-1}{\alpha}}-(i-1)^{\frac{\alpha-1}{\alpha}}\right)EW_1 & \text{if } \alpha>1,\end{cases}$$

$i=1,2,\ldots,$ *converges a.s. to a* $S_\alpha(\sigma,\beta,0)$ *random variable with*

$$\sigma^\alpha=\frac{E|W_1|^\alpha}{C_\alpha},$$

where C_α *is the constant defined in (1.2.9) and*

$$\beta=\frac{E|W_1|^\alpha\operatorname{sign}W_1}{E|W_1|^\alpha}.$$

Moreover, in the case $\alpha = 1$, *the series*

$$\sum_{i=1}^{\infty} \left(\Gamma_i^{-1} W_i - EW_1 \int_{1/i}^{1/(i-1)} \frac{\sin x}{x^2} dx \right) \tag{1.4.11}$$

converges a.s. to a $S_1(\sigma, \beta, \mu)$ *random variable, where* σ *and* β *are as above and*

$$\mu = -EW_1 \ln |W_1|.$$

The proof is essentially similar to that of Theorem 1.4.2 but is much more technical. It will be given in the following section.

Remarks

1. This theorem can be viewed as a representation theorem. A random variable $X \sim S_\alpha(\sigma, \beta, \mu), 0 < \alpha < 2$ can be represented as

$$X \stackrel{d}{=} \mu + \sum_{i=1}^{\infty} \left(\Gamma_i^{-1/\alpha} W_i - k_i^{(\alpha)} \right)$$

   where the sequence of i.i.d. random variables W_i satisfies

$$E|W_1|^\alpha = C_\alpha \sigma^\alpha$$

   and

$$E|W_1|^\alpha \text{ sign } W_1 = C_\alpha \beta \sigma^\alpha.$$

   There is, therefore, a wide latitude in the choice of the distribution of the W_is. This can be useful in applications.

2. To recover the $S\alpha S$ case, replace each W_i by $\epsilon_i W_i$, where $P(\epsilon_i = 1) = P(\epsilon_i = -1) = 1/2$ and the sequence $\{\epsilon_i\}$ is independent of the sequence $\{W_i\}$.

3. The theorem can be used to derive (1.2.8) because the tail of the first term of the series dominates the tail of the other terms (Exercise 1.27).

4. When $0 < \alpha < 1$, the sum $\sum_{i=1}^{\infty} \Gamma_i^{-1/\alpha} W_i$ converges a.s.. For example, let $W_i = 1$. Then $\sum_{i=1}^{\infty} \Gamma_i^{-1/\alpha}$ converges because $\sum_{i=1}^{\infty} i^{-1/\alpha} < \infty$. By Theorem 1.4.5, we have

$$\sum_{i=1}^{\infty} \Gamma_i^{-1/\alpha} \sim S_\alpha \left(C_\alpha^{-1/\alpha}, 1, 0 \right)$$

   when $0 < \alpha < 1$. Hence, a stable subordinator $X \sim S_\alpha(\sigma, 1, 0)$ has the representation $\sigma C_\alpha^{1/\alpha} \sum_{i=1}^{\infty} \Gamma_i^{-1/\alpha}$.

5. The sum $\sum_{i=1}^{\infty} \Gamma_i^{-1/\alpha} W_i$ generally diverges when $\alpha \geq 1$ and the W_i are not symmetric. The compensation $k_i^{(\alpha)}$ when $\alpha > 1$ is asymptotically equal, as $i \to \infty$, to $E\Gamma_i^{-1/\alpha} W_i$, since

$$E\Gamma_i^{-1/\alpha} W_i = E\Gamma_i^{-1/\alpha} EW_1 \sim i^{-1/\alpha} EW_1 \sim k_i^{(\alpha)}.$$

When $\alpha = 1$, an asymptotic compensation by the mean $E\Gamma_i^{-1} W_i$ is also adequate to ensure convergence of the sum (1.4.11) because, as $i \to \infty$,

$$E\Gamma_i^{-1} W_i \sim i^{-1} EW_1 \sim \int_{1/i}^{1/(i-1)} \frac{\sin x}{x^2} dx \, EW_i.$$

6. Suppose $\alpha = 1$. Although the series (1.4.10) converges to a stable random variable with zero shift parameter, it has the disadvantage of not being linear in the W_is. The series (1.4.11), on the other hand, is linear in the W_is and it will be used later, in the context of stable integrals.

1.5 Series representation of skewed α-stable random variables

Here, we prove Theorem 1.4.5.

For $0 < \alpha < 2$ and $i \geq 1$ let

$$C_i^{(\alpha)} = \begin{cases} 0 & \text{if } 0 < \alpha < 1, \\ \int_{1/i}^{1/(i-1)} x^{-2} \sin x \, dx & \text{if } \alpha = 1, \\ \frac{\alpha}{\alpha-1}\left(i^{\frac{\alpha-1}{\alpha}} - (i-1)^{\frac{\alpha-1}{\alpha}}\right) & \text{if } 1 < \alpha < 2. \end{cases}$$

We observe firstly that

$$P\Big(\sum_{i=1}^{\infty} (\Gamma_i^{-1/\alpha} - C_i^{(\alpha)}) \text{ converges}\Big) = 1. \tag{1.5.1}$$

Indeed, (1.5.1) is trivial when $0 < \alpha < 1$, and, in the case $1 \leq \alpha < 2$, it follows simply from

$$\sum_{i=1}^{n} (\Gamma_i^{-1/\alpha} - C_i^{(\alpha)}) = \sum_{i=1}^{n} (\Gamma_i^{-1/\alpha} - i^{-1/\alpha}) + \sum_{i=1}^{n} (i^{-1/\alpha} - C_i^{(\alpha)})$$

and the following lemma.

Lemma 1.5.1 *Let $\Gamma_1, \Gamma_2, \ldots$ be arrival times of a Poisson process with unit arrival rate. Then for any $\alpha > 0$,*

$$\limsup_{i\to\infty} \frac{|\Gamma_i^{-1/\alpha} - i^{-1/\alpha}|}{i^{-1/\alpha-1/2}\sqrt{2\ln\ln i}} = \alpha \text{ a.s.}$$

PROOF: Recall that $E\Gamma_i = i$ and Var $\Gamma_i = i$. By the strong law of large numbers, $\Gamma_i/i \to 1$ a.s. and

$$\lim_{i\to\infty} \frac{(\frac{\Gamma_i}{i})^{-1/\alpha}|(\frac{\Gamma_i}{i})^{1/\alpha} - 1|}{\alpha^{-1}\frac{|\Gamma_i - i|}{i}} = \lim_{i\to\infty} \frac{|(1+\frac{\Gamma_i - i}{i})^{1/\alpha} - 1|}{\alpha^{-1}\frac{|\Gamma_i - i|}{i}} = 1 \text{ a.s.}$$

But by the law of the iterated logarithm,

$$\limsup_{i\to\infty} \frac{|\Gamma_i - i|}{\sqrt{i2\ln\ln i}} = 1 \text{ a.s.}$$

The statement of the lemma follows. ∎

We will now prove that the series (1.4.10) converges a.s. when $\alpha \neq 1$, and that the series (1.4.11) converges a.s. when $\alpha = 1$.

(a) Case $0 < \alpha < 1$. It is enough to prove that $\sum_{i=1}^{\infty} |W_i| i^{-1/\alpha} < \infty$ a.s., and we do this by using the three series theorem. For any $\lambda > 0$,

$$\sum_{i=1}^{\infty} P(|W_i| i^{-1/\alpha} > \lambda) = \sum_{i=1}^{\infty} P(|W_1|^\alpha > i\lambda^\alpha) < \infty$$

because $E|W_1|^\alpha < \infty$. Further, for every $i \geq 1$,

$$E|W_i| i^{-1/\alpha} 1(|W_i| i^{-1/\alpha} \leq 1) \leq i^{-1/\alpha} \int_0^{i^{1/\alpha}} P(|W_i| > x)dx,$$

and

$$\begin{aligned} &\sum_{i=3}^{\infty} i^{-1/\alpha} \int_{3^{1/\alpha}}^{i^{1/\alpha}} P(|W_1| > x)dx \\ &= \frac{1}{\alpha} \sum_{i=3}^{\infty} i^{-1/\alpha} \int_3^i y^{(1-\alpha)/\alpha} P(|W_1|^\alpha > y)dy \\ &\leq C \int_3^\infty P(|W_1|^\alpha > y)dy < \infty \end{aligned}$$

because $E|W_1|^\alpha < \infty$. Therefore

$$\sum_{i=1}^{\infty} E(|W_i| i^{-1/\alpha} 1(|W_i| i^{-1/\alpha} \leq 1)) < \infty,$$

which concludes the proof for the case $0 < \alpha < 1$.

(b) Case $1 < \alpha < 2$. From (1.5.1), it is enough to prove that the series $\sum_{i=1}^{\infty} \Gamma_i^{-1/\alpha}(W_i - EW_1)$ converges a.s.. Let $Y_i = W_i - EW_1$, $i = 1, 2, \dots$, and observe that $Y_1, Y_2, \dots$ are i.i.d., with $E|Y_1|^\alpha < \infty$ and $EY_1 = 0$. Let

$$A = \left\{ \lim_{n\to\infty} \frac{1}{n}\Gamma_n = 1 \right\} \cap \{\Gamma_1 > 0\}. \tag{1.5.2}$$

It is enough to prove that the series $\sum_{i=1}^{\infty} \Gamma_i^{-1/\alpha} Y_i$ converges a.s. for any fixed $\Gamma_1, \Gamma_2, \dots$ in A because $P(A^c) = 0$. We use once again the three series theorem. Let

$$T_i = \Gamma_i^{-1/\alpha} Y_i 1(\Gamma_i^{-1/\alpha}|Y_i| \le 1),\ i = 1, 2, \dots.$$

Since the convergence of the sum $\sum_{i=1}^{\infty} P(\Gamma_i^{-1/\alpha}|Y_i| > \lambda)$ for any $\lambda > 0$ follows as before from $E|Y_1|^\alpha < \infty$, it is sufficient to show

$$\sum_{i=1}^{\infty} ET_i < \infty \tag{1.5.3}$$

and

$$\sum_{i=1}^{\infty} \text{Var } T_i < \infty. \tag{1.5.4}$$

Observe firstly, that for each $i = 1, 2, \dots,$

$$\begin{aligned} ET_i &= \Gamma_i^{-1/\alpha} \int_{-\Gamma_i^{1/\alpha}}^{\Gamma_i^{1/\alpha}} y F_{Y_1}(dy) \\ &= -\Gamma_i^{-1/\alpha} \int_{-\infty}^{-\Gamma_i^{1/\alpha}} y F_{Y_1}(dy) - \Gamma_i^{-1/\alpha} \int_{\Gamma_i^{1/\alpha}}^{\infty} y F_{Y_1}(dy), \end{aligned}$$

because $EY_1 = 0$. Therefore (1.5.3) follows from the convergence of the series $\sum_{i=1}^{\infty} \Gamma_i^{-1/\alpha} \int_{\Gamma_i^{1/\alpha}}^{\infty} y F_{Y_1}(dy)$ and $\sum_{i=1}^{\infty} \Gamma_i^{-1/\alpha} \int_{-\infty}^{-\Gamma_i^{1/\alpha}} y F_{Y_1}(dy)$. The first series converges because $\Gamma_1, \Gamma_2, \dots$ belongs to the set A and, for some constant $c > 0$,

$$\begin{aligned} \sum_{i=1}^{\infty} \Gamma_i^{-1/\alpha} \int_{\Gamma_i^{1/\alpha}}^{\infty} y F_{Y_1}(dy) &= \int_0^{\infty} y F_{Y_1}(dy) \sum_{i=1}^{\infty} \Gamma_i^{-1/\alpha} 1(\Gamma_i \le y^\alpha) \\ &\le c^{-1} \int_0^{\infty} y^\alpha F_{Y_1}(dy) < \infty. \end{aligned}$$

The second series converges for a similar reason. This proves (1.5.3). For (1.5.4), it is enough to show $\sum_{i=1}^{\infty} ET_i^2 < \infty$. For some $c < \infty$,

$$\sum_{i=1}^{\infty} ET_i^2 = \sum_{i=1}^{\infty} \Gamma_i^{-2\alpha} \int_{-\Gamma_i^{1/\alpha}}^{\Gamma_i^{1/\alpha}} y^2 F_{Y_1}(dy)$$

$$= \int_{-\infty}^{\infty} y^2 F_{Y_1}(dy) \sum_{i=1}^{\infty} \Gamma_i^{-2/\alpha} 1(\Gamma_i > y^\alpha)$$

$$\leq \int_{-\infty}^{\infty} y^2 F_{Y_1}(dy) c \min(1, y^{-2+\alpha}) < \infty,$$

because $E|Y_1|^\alpha < \infty$. This establishes (1.5.4).

(c) Case $\alpha = 1$. In the notation of the case $1 < \alpha < 2$, it is enough to show that the series $\sum_{i=1}^{\infty} \Gamma_i^{-1} Y_i$ converges a.s. for any $\Gamma_1, \Gamma_2, \ldots$ in the set A, and thus we must again verify (1.5.3) and (1.5.4). Since the preceding proof of (1.5.4) applies to the case $\alpha = 1$, only (1.5.3) needs to be proved. Proceeding as before, it is enough to show

$$\sum_{i=1}^{\infty} \Gamma_i^{-1} \int_{\Gamma_i}^{\infty} y F_{Y_1}(dy) < \infty.$$

Since $\Gamma_1, \Gamma_2, \ldots$ are in A, for some $c > 0$, one has

$$\sum_{i=1}^{\infty} \Gamma_i^{-1} \int_{\Gamma_i}^{\infty} y F_{Y_1}(dy) = \int_{0}^{\infty} y F_{Y_1}(dy) \sum_{i=1}^{\infty} \Gamma_i^{-1} 1(\Gamma_i \leq y)$$

$$\leq c^{-1} \int_{0}^{\infty} y F_{Y_1}(dy)(|\log y| + 1) < \infty,$$

because $E|Y_1 \log |Y_1|| < \infty$.

We have now shown that the series (1.4.10) converges a.s. when $\alpha \neq 1$, and the series (1.4.11) converges a.s. when $\alpha = 1$. Since

$$\sum_{i=1}^{\infty} \left(EW_1 \int_{1/i}^{1/(i-1)} \frac{\sin x}{x^2} dx - EW_1 \int_{|W_1|/i}^{|W_1|/(i-1)} \frac{\sin x}{x^2} dx \right)$$

$$= \lim_{n \to \infty} EW_1 \int_{1/n}^{|W_1|/n} \frac{\sin x}{x^2} dx = EW_1 \log |W_1|, \tag{1.5.5}$$

the series (1.4.10) converges a.s. in the case $\alpha = 1$ as well.

It remains to determine the distribution of the series (1.4.10) and (1.4.11). The argument is similar to the one used in the proof of Theorem 1.4.2. Let $U_1, U_2, \ldots$ be a sequence of i.i.d. random variables, uniform on $(0, 1)$, and independent of the sequence $W_1, W_2, \ldots$. Setting $Y_i = U_i^{-1/\alpha} W_i$, $i = 1, 2, \ldots$, we immediately obtain, as in the proof of Theorem 1.4.2,

$$\lim_{\lambda \to \infty} \lambda^\alpha P(Y_1 > \lambda) = E(W_1)_+^\alpha$$

and

$$\lim_{\lambda \to \infty} \lambda^\alpha P(Y_1 < -\lambda) = E(W_1)_-^\alpha.$$

Therefore Y_1 is in the domain of normal attraction of an α-stable law. In the case $0 < \alpha < 1$, we have

$$\frac{1}{n^{1/\alpha}} \sum_{i=1}^{n} Y_i \overset{d}{\Rightarrow} X,$$

where $X \sim S_\alpha(\sigma, \beta, 0)$, with σ and β as in Theorem 1.4.2. In the case $1 < \alpha < 2$, we have

$$\frac{1}{n^{1/\alpha}} \sum_{i=1}^{n} (Y_i - EY_i) \overset{d}{\Rightarrow} X,$$

where $X \sim S_\alpha(\sigma, \beta, 0)$, with σ and β as above. Finally, in the case $\alpha = 1$,

$$\frac{1}{n} \sum_{i=1}^{n} Y_i - nE \sin \frac{Y_1}{n} \overset{d}{\Rightarrow} X,$$

where $X \sim S_1(\sigma, \beta, 0)$, with σ and β as above (Feller 1971, Theorem XVII.5.3).

Suppose first that $0 < \alpha < 1$. As in the proof of Theorem 1.4.2,

$$n^{-1/\alpha} \sum_{i=1}^{n} Y_i \overset{d}{=} \left(\frac{\Gamma_{n+1}}{n} \right)^{1/\alpha} \sum_{i=1}^{n} \Gamma_i^{-1/\alpha} W_i \to \sum_{i=1}^{\infty} \Gamma_i^{-1/\alpha} W_i \text{ a.s.}$$

as $n \to \infty$, which shows that the series (1.4.10) has the claimed distribution in the case $0 < \alpha < 1$.

Similarly, if $1 < \alpha < 2$,

$$\begin{aligned}
n^{-1/\alpha} \sum_{i=1}^{n} (Y_i - EY_1) &= n^{-1/\alpha} \sum_{i=1}^{n} Y_i - \frac{\alpha}{\alpha - 1} n^{1-1/\alpha} EW_1 \\
&\overset{d}{=} \left(\frac{\Gamma_{n+1}}{n} \right)^{1/\alpha} \sum_{i=1}^{n} \Gamma_i^{-1/\alpha} W_i - \frac{\alpha}{\alpha - 1} n^{1-1/\alpha} EW_1 \\
&= \left(\frac{\Gamma_{n+1}}{n} \right)^{1/\alpha} \sum_{i=1}^{n} \left(\Gamma_i^{-1/\alpha} W_i - EW_1 \frac{\alpha}{\alpha - 1} \left(i^{(\alpha-1)/\alpha} - (i-1)^{(\alpha-1)/\alpha} \right) \right) \\
&\quad + \frac{\alpha}{\alpha - 1} n^{(\alpha-1)/\alpha} \left((n^{-1}\Gamma_{n+1})^{1/\alpha} - 1 \right) EW_1 \\
&\to \sum_{i=1}^{\infty} (\Gamma_i^{-1/\alpha} W_i - k_i^{(\alpha)}) \text{ a.s.}
\end{aligned}$$

as $n \to \infty$, which proves that the series (1.4.10) has the claimed distribution in the case $1 < \alpha < 2$ as well.

Finally, we consider the case $\alpha = 1$:

$$\begin{aligned}
\frac{1}{n}\sum_{i=1}^{n} Y_i - nE\sin\frac{Y_1}{n} &\stackrel{d}{=} \frac{\Gamma_{n+1}}{n}\sum_{i=1}^{n}\Gamma_i^{-1}W_i - E\int_{1/n}^{\infty}\frac{\sin W_1 x}{x^2}dx \\
&= \frac{\Gamma_{n+1}}{n}\sum_{i=1}^{n}\Gamma_i^{-1}W_i - EW_1\int_{|W_1|/n}^{\infty}\frac{\sin y}{y^2}dy \\
&= \frac{\Gamma_{n+1}}{n}\sum_{i=1}^{n}\left(\Gamma_i^{-1}W_i - EW_1\int_{|W_1|/i}^{|W_1|/(i-1)}\frac{\sin y}{y^2}dy\right) \\
&\quad + \left(\frac{\Gamma_{n+1}}{n} - 1\right)EW_1\int_{|W_1|/n}^{\infty}\frac{\sin y}{y^2}dy \\
&\to \sum_{i=1}^{\infty}\left(\Gamma_i^{-1}W_i - EW_1\int_{|W_1|/i}^{|W_1|/(i-1)}\frac{\sin y}{y^2}dy\right) \quad \text{a.s.}
\end{aligned}$$

as $n \to \infty$, proving that the series (1.4.10) has also the claimed distribution in the case $\alpha = 1$.

This, together with (1.5.5), completes the proof of Theorem 1.4.5.

1.6 Graphs and tables of α-stable densities and c.d.f.'s

Let X be a $S_\alpha(1, \beta, 0)$ random variable and let

$$F(x;\alpha,\beta) = P(X \le x), \quad -\infty < x < \infty,$$

and

$$p(x;\alpha,\beta) = \frac{d}{dx}F(x;\alpha,\beta)$$

denote, respectively, the cumulative distribution function and the probability density function of X. The figures on the following pages display the cumulative distribution function $F(x;\alpha,\beta)$ and the density function $p(x;\alpha,\beta)$.

Tables

Tables for the upper-tail probabilities $1 - F(x;\alpha,\beta)$ and the fractiles $x_f = F^{-1}(f;\alpha,\beta)$ are given in DuMouchel (1971) for [4]

$$\begin{aligned}
\alpha &= 0.1\ (0.1)\ 1.9, 1.95, \\
\beta &= -1\ (0.25)\ 1.
\end{aligned}$$

[4] $a(b)c$ denotes a sequence from a to c with increments equal to b.

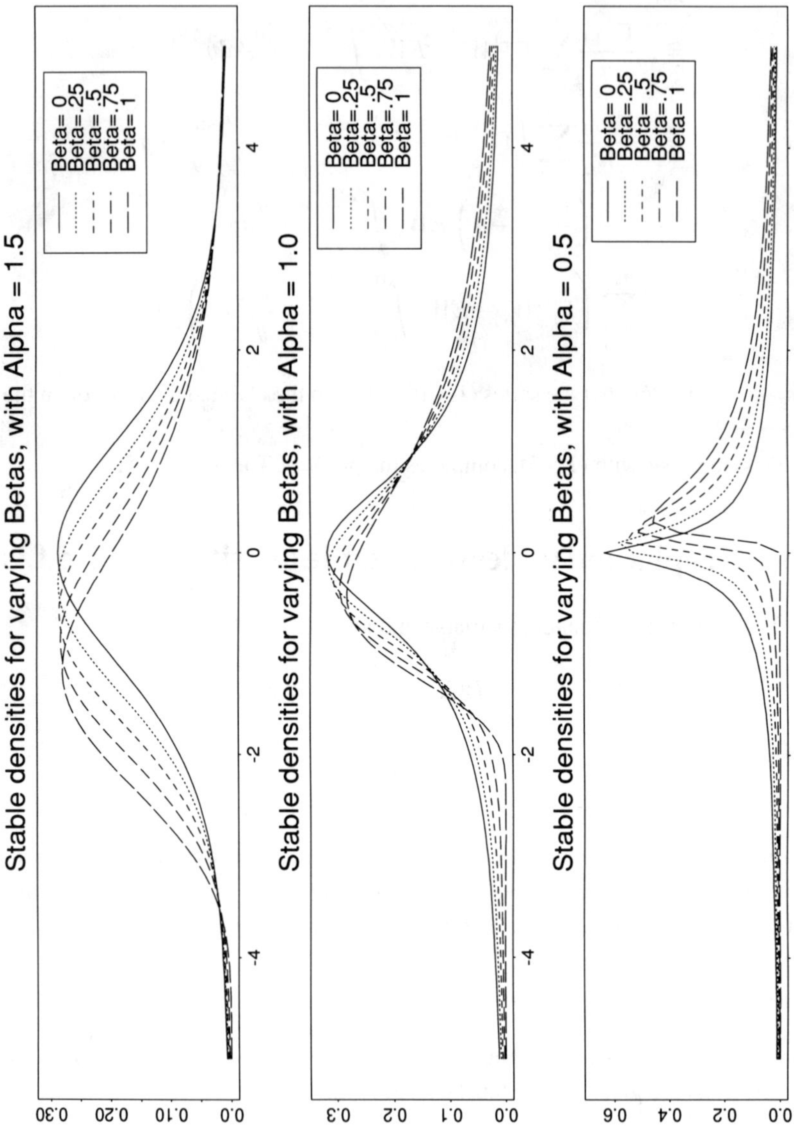
Stable densities for varying Betas, with Alpha = 1.5
Beta= 0
Beta=.25
Beta= .5
Beta=.75
Beta= 1
Stable densities for varying Betas, with Alpha = 1.0
Beta= 0
Beta=.25
Beta= .5
Beta=.75
Beta= 1
Stable densities for varying Betas, with Alpha = 0.5
Beta= 0
Beta=.25
Beta= .5
Beta=.75
Beta= 1

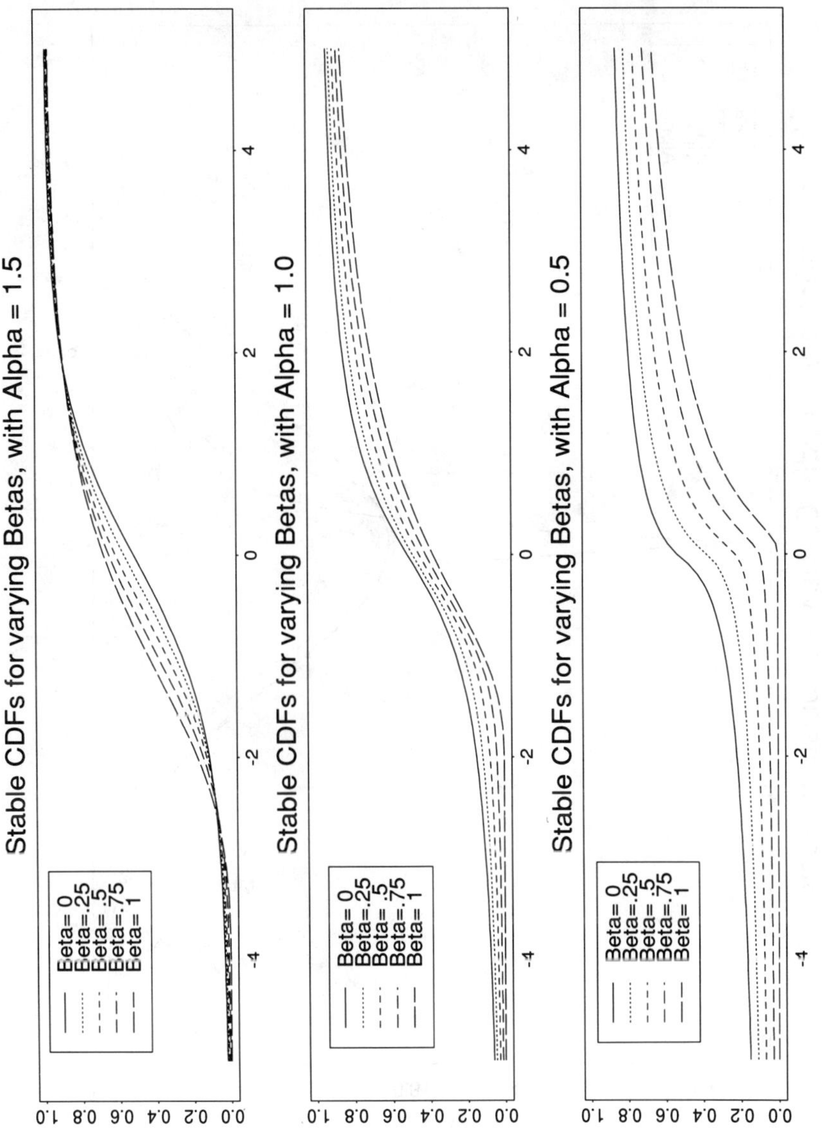
Stable CDFs for varying Betas, with Alpha = 1.5
Beta= 0
Beta=.25
Beta= .5
Beta=.75
Beta= 1
Stable CDFs for varying Betas, with Alpha = 1.0
Beta= 0
Beta=.25
Beta= .5
Beta=.75
Beta= 1
Stable CDFs for varying Betas, with Alpha = 0.5
Beta= 0
Beta=.25
Beta= .5
Beta=.75
Beta= 1

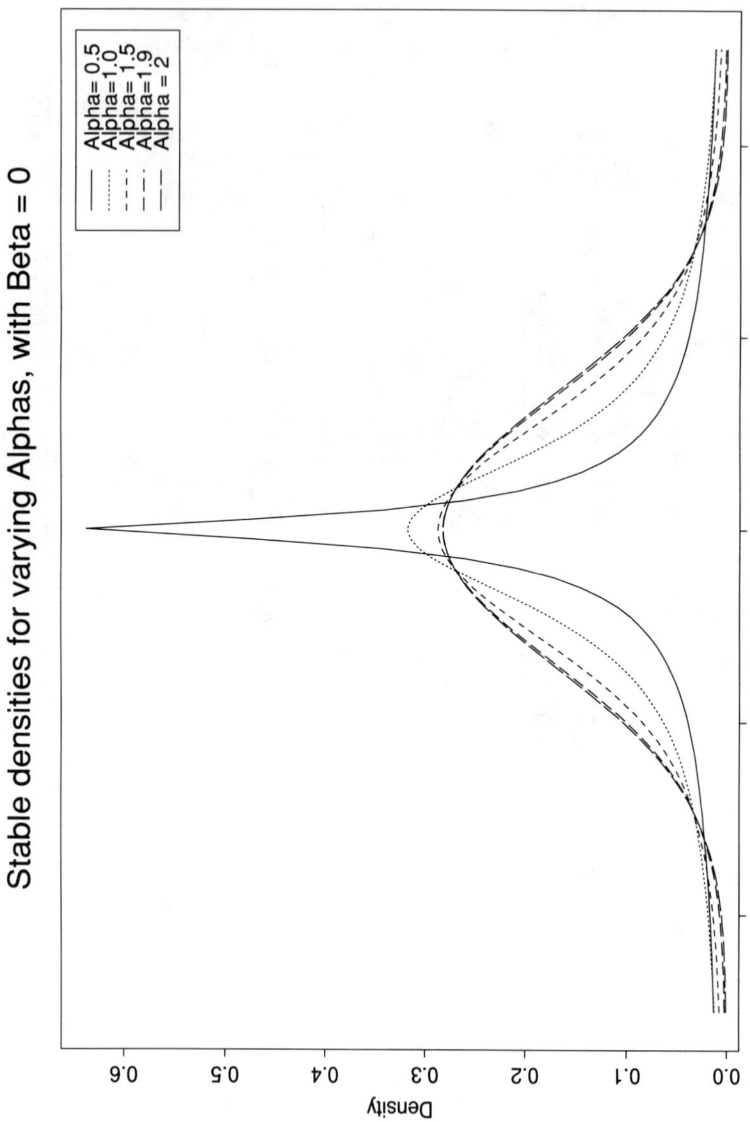
Stable densities for varying Alphas, with Beta = 0
Alpha= 0.5
Alpha=1.0
Alpha= 1.5
Alpha=1.9
Alpha = 2
Density
0.0
0.1
0.2
0.3
0.4
0.5
0.6
-4
-2
0
2
4

Tables for $p(x;\alpha,\beta)$ can be found in Holt and Crow (1973) for

$$\begin{aligned} \alpha &= 0.25\ (0.25)\ 1.75, \\ \beta &= -1\ (0.25)\ 1. \end{aligned}$$

Fractile tables can also be found in Brothers, DuMouchel and Paulson(1983) for[5]:

$$\begin{aligned} \alpha &= 0.1\ (0.1)\ 1.9, \\ \beta &= 0\ (0.1)\ 1, \\ f &= 0.001, 0.002, 0.005\ (0.005) \\ & \quad 0.030 (0.010) 0.100\ (0.025) \\ & \quad 0.900 (0.010) 0.970\ (0.005) \\ & \quad 0.995, 0.998, 0.999. \end{aligned}$$

That above-cited article includes also the following extreme upper fractiles:

$$\begin{aligned} \alpha &= 0.5\ (0.1)\ 1.9, \\ \beta &= -1\ (0.25)\ 1, \\ f &= 0.9990\ (0.0001)\ 0.9999. \end{aligned}$$

We have reproduced the fractiles in Appendix A for the symmetric case $\beta = 0$, with the kind permission of Albert Paulson.

The evaluation of these fractiles is based, for the most part, on Zolotarev's integral representation for the α-stable cumulative distribution function, which takes the following form in the $S\alpha S$, $\alpha \neq 1$, case:

Proposition 1.6.1 *For $0 < \alpha \leq 2,\ \alpha \neq 1,\ 0 < \gamma < \pi/2$, define*

$$U_\alpha(\gamma) = \left(\frac{\sin \alpha\gamma}{\cos \gamma} \right)^{\alpha/(1-\alpha)} \frac{\cos((1-\alpha)\gamma)}{\cos \gamma},$$

and let $X \sim S_\alpha(1,0,0)$. Then, for $x \geq 0$,

$$\frac{1}{\pi} \int_0^{\pi/2} \exp\{-x^{\alpha/(\alpha-1)} U_\alpha(\gamma)\} d\gamma = \begin{cases} P(0 \leq X \leq x) & \text{if } 0 < \alpha < 1, \\ P(X \geq x) & \text{if } 1 < \alpha \leq 2. \end{cases} \tag{1.6.1}$$

[5]The definition of β in Holt and Crow (1973) and in Brothers, DuMouchel and Paulson (1983) differs from ours. The β indicated here corresponds to our definition.

α	C_α	$\ln(C_\alpha/2)$	Intercept	Slope
0.5	0.7978840	−0.9189392	−0.9197074	−0.5000001
0.6	0.7669871	−0.9584325	−0.9604577	−0.5998732
0.7	0.7362977	−0.9992680	−0.9851851	−0.7016126
0.8	0.7048938	−1.0428553	−1.0320764	−0.8014147
0.9	0.6719293	−1.0907493	−1.0822671	−0.9012536
1.0	0.6366203	−1.1447290	−1.1372527	−1.0012093
1.1	0.5981977	−1.2069812	−1.1996404	−1.1012519
1.2	0.5559184	−1.2802809	−1.2708688	−1.2016571
1.3	0.5090762	−1.3683048	−1.3556695	−1.3023241
1.4	0.4569742	−1.4762756	−1.4576651	−1.4036502
1.5	0.3989431	−1.6120837	−1.5863361	−1.5053023
1.6	0.3343501	−1.7887137	−1.7489705	−1.6088360
1.7	0.2626145	−2.0302154	−1.9661427	−1.7155793
1.8	0.1832279	−2.3901720	−2.2768235	−1.8305965
1.9	0.0957816	−3.0388319	−2.7849723	−1.9798950

Table 1.1: C_α is defined in (1.2.9). Intercept and slope are for the least squares regression of $\ln(1-f)$ versus $\ln x_f$ where f and x_f are the numbers given in the table for extreme fractiles in Appendix A.

PROOF: This follows from Zolotarev (1986), Remark 1, page 78. ∎

Zolotarev (1986), in equation (2.2.30) on page 79, gives the following expression for the α-stable cumulative distribution function at $x = 0$:

$$F(0;\alpha,\beta) = \begin{cases} \frac{1}{2}(1-\beta_2) & \text{if } 0<\alpha<1, \\ \frac{1}{2}(1+\beta_2\frac{2-\alpha}{2}) & \text{if } 1<\alpha<2, \end{cases}$$

where β_2 is the skewness parameter appearing in the representation (1.1.12) of the α-stable characteristic function. Recall that β_2 is a monotone increasing function of β. Hence, as β increases, $F(0;,\alpha,\beta)$, viewed as a function of β, decreases for $0<\alpha<1$ but increases for $1<\alpha<2$. This is reflected in the figures of the α-stable cumulative distribution function.

Suppose $X \sim S_\alpha(1,0,0)$. One quick way to estimate α is to evaluate the slope in the log-log plot of $P\{X \ge x\}$ versus x. This technique is based on the asymptotic relation $P\{X \ge x\} \sim (C_\alpha/2)x^{-\alpha}$ (see (1.2.10)). Let us see how good this technique is, in the best of circumstances, namely when the distribution of X is known, the probabilities $P\{X \ge x\}$ are taken from a table and the xs are very large. We shall use the entries of the extreme upper fractiles in Appendix A to plot $\ln(1-f)$ versus $\ln x_f$. The results are given in Table 1.1, where

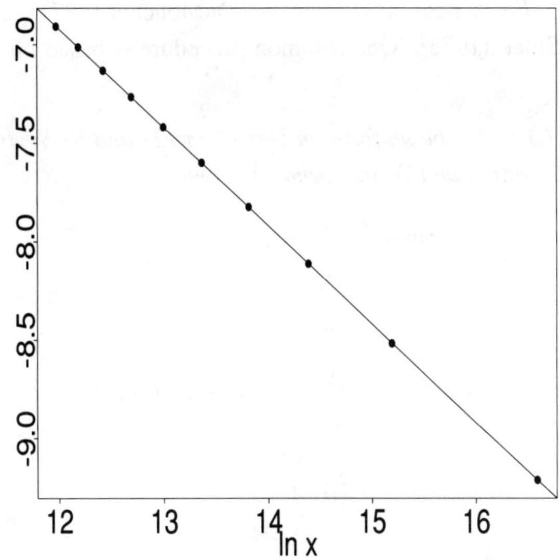

Figure 1.2: Display in the case $\alpha = 0.5$ of the least squares regression of $\ln(1-f)$ versus $\ln x_f$,where f and x_f are the numbers given in the table for extreme fractiles in Appendix A.

"intercept" estimates $\ln(C_\alpha/2)$ and "slope" estimates $-\alpha$. Even though we are using extremely large values of x, the technique does not give a good estimate at $\alpha = 1.9$. It improves as α decreases and is excellent at $\alpha = 0.5$, at these large values of x (see Figure 1.2). Note though, that the smaller the value of α, the larger the xs involved.

1.7 Simulation

To generate the *Cauchy distribution* $S_1(\sigma, 0, \mu)$ in (1.1.13), one can generate

$$\sigma \tan \gamma + \mu, \tag{1.7.1}$$

where γ is uniform on $(-\pi/2, \pi/2)$. To generate the *Lévy distribution* $S_{1/2}(\sigma, 1, \mu)$ in (1.1.15) one can generate

$$\sigma Z^{-2} + \mu, \tag{1.7.2}$$

where Z is $N(0,1)$. (Exercises 1.7 and 1.8.) It is easy also to generate the distributions $S_\alpha(1,0,0)$ and $S_\alpha(1,1,0)$ for $\alpha = 2^{-k}$, $k \geq 1$ (see Brown and Tukey (1946)).

In general, however, the generation of stable random variables is quite an involved affair. For a good discussion see DuMouchel (1971) and Chambers, Mallows, and Stuck (1976). One common procedure is based on the following result:

Proposition 1.7.1 *Let γ be uniform on $(-\pi/2,\ \pi/2)$ and let W be exponential with mean 1. Assume γ and W independent. Then*

$$X = \frac{\sin \alpha\gamma}{(\cos\gamma)^{1/\alpha}} \left(\frac{\cos((1-\alpha)\gamma)}{W}\right)^{(1-\alpha)/\alpha} \tag{1.7.3}$$

is $S_\alpha(1,0,0)$.

PROOF: When γ takes a value in $(0,\pi/2)$, the right-hand side of (1.7.3) can be expressed as

$$\left(\frac{a(\gamma)}{W}\right)^{(1-\alpha)/\alpha},$$

where

$$a(\gamma) = \left(\frac{\sin\alpha\gamma}{\cos\gamma}\right)^{\alpha/(1-\alpha)} \cos((1-\alpha)\gamma).$$

(a) *Case* $0 < \alpha < 1$.

$$\begin{aligned} P(0 \le X \le x) &= P(0 \le X \le x,\ \gamma > 0) \\ &= P(0 \le (a(\gamma)/W)^{(1-\alpha)/\alpha} \le x,\ \gamma > 0) \\ &= P(W \ge x^{-\alpha/(1-\alpha)} a(\gamma),\ \gamma > 0) \\ &= E \exp\{-x^{-\alpha/(1-\alpha) a(\gamma)}\} 1_{\{\gamma > 0\}} \\ &= \frac{1}{\pi}\int_0^{\pi/2} \exp\{-x^{-\alpha/(1-\alpha)} a(\gamma)\} d\gamma. \end{aligned}$$

Now use (1.6.1) to conclude that $X \sim S_\alpha(1,0,0)$.

(b) *Case* $1 < \alpha \le 2$. Start with $P(X \ge x) = P(X \ge x,\ \gamma > 0)$ and proceed as above, making use of the inequality $1 - \alpha < 0$.

(c) *Case* $\alpha = 1$. Relation (1.7.3) reduces to $X = \tan\gamma$ whose distribution is Cauchy (Exercise 1.7). ■

There is a corresponding result for skewed α-stable random variables (see Chambers, Mallows and Stuck(1976)). In the Gaussian case $\alpha = 2$, (1.7.3) reduces to $W^{1/2}\sin 2\gamma/\cos\gamma = 2W^{1/2}\sin\gamma$, which is the Box-Muller method for generating a $N(0,2)$ random variable (Box & Muller 1958).

In order to generate $S_\alpha(\sigma,\beta,\mu)$ random variables, it is only necessary, in principle, to generate $S_\alpha(1,1,0)$ random variables because, firstly, if $X \sim S_\alpha(1,\beta,0)$

can be obtained by generating two independent $S_\alpha(1,1,0)$ random variables (Property 1.2.13). Secondly, by Properties 1.2.2 and 1.2.3,

$$\sigma X + \mu \sim S_\alpha(\sigma, \beta, \mu) \qquad \text{if } \alpha \neq 1,$$

and

$$\sigma X + \frac{2}{\pi}\beta\sigma \ln \sigma + \mu \sim S_\alpha(\sigma, \beta, \mu) \quad \text{if } \alpha = 1.$$

Chambers, Mallows and Stuck (1976) describe a method for generating α-stable random variables for any $0 < \alpha < 2$ and $-1 \leq \beta \leq 1$ based on formulas of the type (1.7.3). To obtain a numerically accurate representation near $\alpha = 1$, they use the form (1.1.10) of the characteristic function. They simulate a random variable Y with characteristic function

$$E \exp i\theta Y = \exp\{-|\theta|^\alpha + i\theta\omega_1(\theta, \alpha, \beta)\}$$

where ω_1 is defined in (1.1.11). Observe that

$$\begin{cases} Y \sim S_\alpha(1, \beta, -\beta \tan \frac{\pi\alpha}{2}) & \text{if } \alpha \neq 1, \\ Y \sim S_\alpha(1, \beta, 0) & \text{if } \alpha = 1. \end{cases} \tag{1.7.4}$$

Thus, $X = Y + \overline{\mu} \sim S_\alpha(1, \beta, 0)$, where

$$\overline{\mu} = \begin{cases} \beta \tan \frac{\pi\alpha}{2} & \text{if } \alpha \neq 1, \\ 0 & \text{if } \alpha = 1. \end{cases}$$

Observe also that the shift $-\overline{\mu}$ in Y can be very large when $\beta \neq 0$ and α is close to 1. For example, whereas the support of X is $[0, \infty)$ if $\beta = 1$ and $\alpha < 1$, the support of Y is $[-\overline{\mu}, \infty)$ and may start far to the left of 0 if α is close to 1. The advantage of Y over X is that the distributions of Y vary continuously as α and β vary.

The Fortran program *rstab* (for "random stable") which generates Y is listed in Chambers, Mallows and Stuck (1976). S-PLUS[6] is a powerful statistical package which contains an updated version of *rstab*.

Ten thousand realizations of the random variable Y defined in (1.7.4) were simulated using the *rstab* program in S-PLUS, in the cases $\alpha = 1$, $\beta = 1$ and $\alpha = 0.99$, $\beta = 1$. The following figures show the scatter plots, the empirical cumulative distribution function and the corresponding histograms (empirical density functions). Although most realizations of Y are small, a number of them

[6]S-PLUS is marketed by StatSci, a division of MathSoft, Inc. It is based on S which is marketed by AT&T.

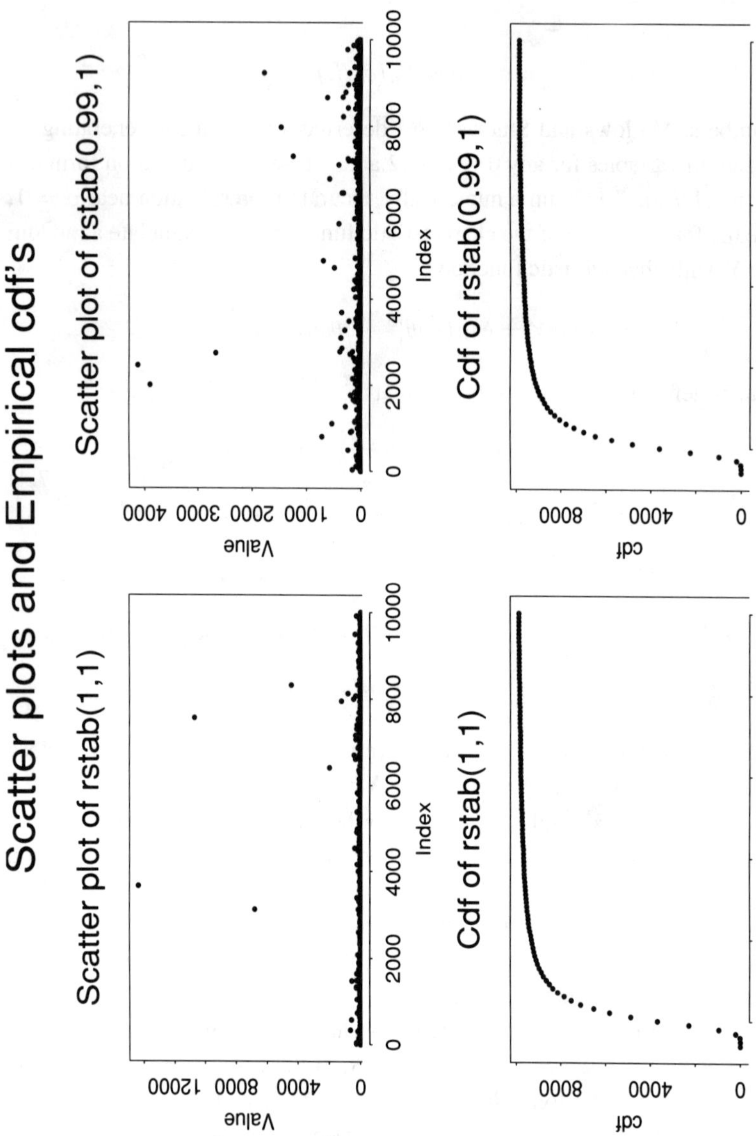
Scatter plots and Empirical cdf's
Scatter plot of rstab(1,1)
Scatter plot of rstab(0.99,1)
Cdf of rstab(1,1)
Cdf of rstab(0.99,1)
Value
Index
cdf
x
0
2000
4000
6000
8000
10000
1000
3000
12000
10
20
30
40
50

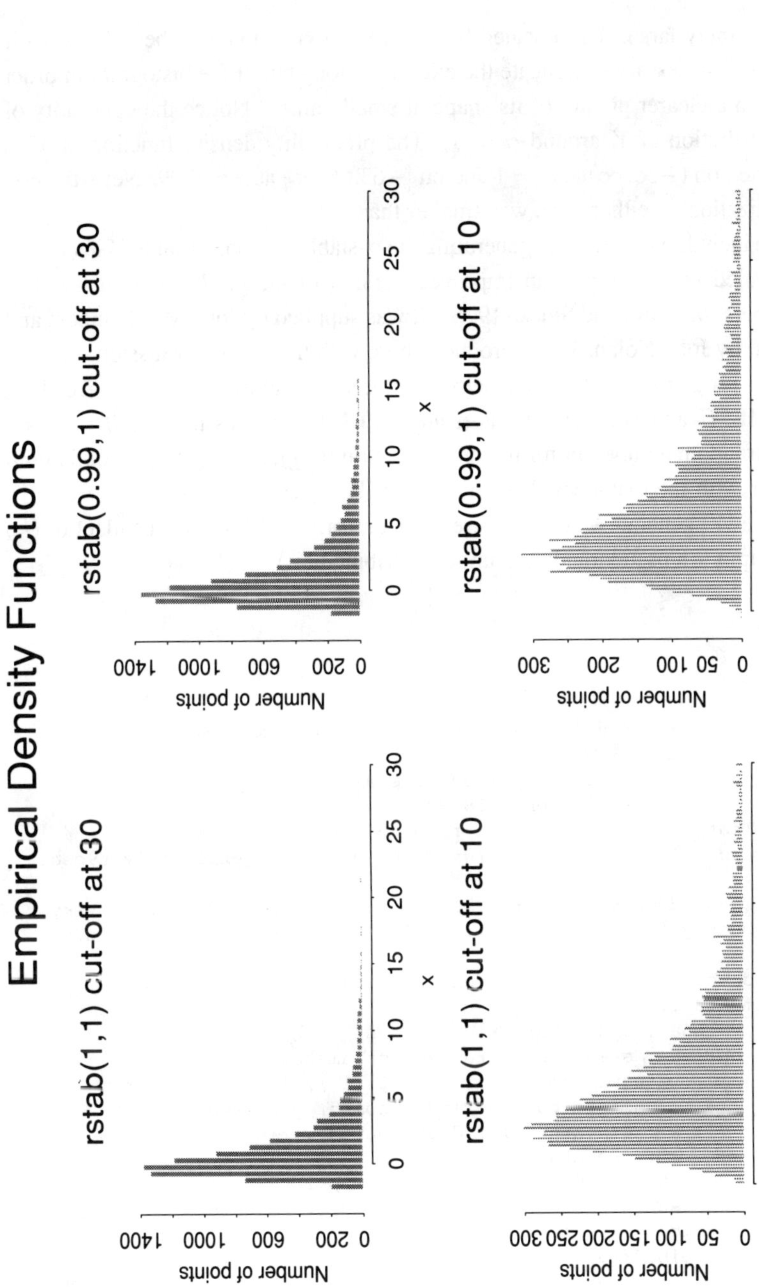
Empirical Density Functions
rstab(1,1) cut-off at 30
rstab(0.99,1) cut-off at 30
rstab(1,1) cut-off at 10
rstab(0.99,1) cut-off at 10
Number of points
x

are extremely large. Untruncated histograms provide no detail because of their scale. It is necessary to truncate the extremely long tail of the histogram in order to obtain a clearer picture of its shape at small values. Notice the continuity of the distribution of Y around $\alpha = 1$. The probability density function of Y is supported on $(-\infty, \infty)$ at $\alpha = 1$ and on $[-63.66, \infty)$ at $\alpha = 0.99$. Nevertheless, no realization, in either case, was smaller than -3.

We include a program for generating the α-stable random variable Y in (1.7.4). It is called *rstable* and is an improved version of the *rstab* program listed in Chambers, Mallows and Stuck (1976). It was supplied by John M. Chambers and adapted by John Nolan. It is reproduced here with their kind permission.

The program *rstable* listed below requires the following inputs: ALPHA, BPRIME, U and W. ALPHA is α and BPRIME is β (as in (1.1.10)), U is a pseudorandom number uniformly distributed on $(0, 1)$, and W is a pseudorandom number with the exponential distribution $P[W \leq x] = 1 - e^{-x}$, $x \geq 0$. (To generate W, generate $-\ln U'$, where U' is uniform on $(0, 1)$, independent of U.) The output is a pseudorandom number distributed as Y in (1.7.4).

```
C  rstable  --  random stable standardized form
C
C  Adapted from a file supplied by Chambers, et al.
C        by John Nolan on 5/18/92.
C     That file was converted from RATFOR to FORTRAN
C     and all variables and functions have been explicitly typed.
C
C     Changed to double precision on 3/15/93 by Vadim Teverovsky,
C     mostly for ease in dealing with other routines and S-PLUS.
C
      double precision function rstable(alpha,bprime,u,w, st)
C arguments ..
C       alpha : characteristic exponent
C      bprime : skewness in revised parameterization
C           u : uniform variate on (0.,1.), for example from a
C             uniform pseudo-random number generator
C           w : exponentially distributed variate
C           st:  return value
      double precision alpha,bprime,u,w,phiby2,thr1,a,eps,piby2,b,
     *  bb,tau,a2,a2p,b2,b2p,alogz,tan2d,z,d2d,d,st
C      double precision da,db
      data piby2/1.57079633/
      data thr1/0.99/

      eps = 1.0-alpha
C compute some tangents
      phiby2 = piby2*(u-0.5)
      a = phiby2*tan2d(phiby2)
```

```
      bb = tan2d(eps*phiby2)
      b = eps*phiby2*bb
      if (eps.gt.(-0.99))
     *   tau = bprime/(tan2d(eps*piby2)*piby2)
      if (eps.le.(-0.99))
     *   tau = bprime*piby2*eps*(1.-eps)*tan2d((1.-eps)*piby2)
C compute some necessary subexpressions
C if phi near pi by 2, use double precision.
C       if (a.gt.thr1) then
C double precision
C           da = dble(a)**2
C           db = dble(b)**2
C           a2 = 1.d0-da
C           a2p = 1.d0+da
C           b2 = 1.d0-db
C           b2p = 1.d0+db
C       else
C single precision
          a2 = a**2
          a2p = 1.+a2
          a2 = 1.-a2
          b2 = b**2
          b2p = 1.+b2
          b2 = 1.-b2
C       endif
C compute coefficient
      z = a2p*(b2+2.*phiby2*bb*tau)/(w*a2*b2p)
C compute the exponential-type expression
      alogz = dlog(z)
      d = d2d(eps*alogz/(1.-eps))*(alogz/(1.-eps))
C compute stable
      rstable = (1.+eps*d)*2.*((a-b)*(1.+a*b)-phiby2*tau*bb*
     *          (b*a2-2.*a))/(a2*b2p)+tau*d
      st = rstable
      return
      end



C d2d         evaluate (exp(x)-1)/x
      double precision function d2d(z)
      double precision z
      double precision p1,p2,q1,q2,q3,pv,zz
      data p1,p2,q1,q2,q3/.84006 68525 36483 239 d3,
     *  .20001 11415 89964 569 d2,
     *  .16801 33705 07926 648 d4,
     *  .1801 33704 07390 023 d3,1.d0/
C the approximation 1801 from hart et al (1968, p. 213)
      if (abs(z).gt.0.1) then
              d2d = (exp(z)-1.0)/z
      else
              zz = z*z
              pv = p1+zz*p2
              d2d = 2.d0*pv/(q1+zz*(q2+zz*q3)-z*pv)
      endif
```

```
      return
      end

C mytand        tangent function
      double precision function mytand(xarg)
      logical neg,inv
      double precision p0,p1,p2,q0,q1,q2,piby4,piby2,pi,x,xarg,xx
      data  p0,p1,p2,q0,q1,q2/.129221035e+3,-.887662377e+1,
     * .528644456e-1,.164529332e+3,-.45 1320561e+2,1.0/
C the approximation 4283 from hart et al(1968, p. 251)
      data piby4/.785398163/,piby2/1.57079633/
      data pi/3.14159265/

      neg = .false.
      inv = .false.
      x = xarg
      neg = (x.lt.0.0)
      x = abs(x)
C perform range reduction if necessary
      if (x.gt.piby4) then
              x = dmod(x,pi)
              if (x.gt.piby2) then
                      neg = .not. neg
                      x = pi-x
              endif
              if (x.gt.piby4) then
                      inv = .true.
                      x = piby2-x
              endif
              endif
      x = x/piby4
C convert to range of rational
      xx = x*x
      mytand = x*(p0+xx*(p1+xx*p2))/(q0+xx*(q1+xx*q2))
      if (neg) mytand = -mytand
      if (inv) mytand = 1./mytand
      return
      end

C  tan2d        compute tan(x)/x
C     function defined only for abs(xarg).le.pi by 4
C     for other arguments returns tan(x)/x, computed directly
      double precision function tan2d(xarg)
      double precision mytand,xarg,p0,p1,p2,q0,q1,q2,piby4,x,xx
      data  p0,p1,p2,q0,q1,q2/.129221035e+3,-.887662377e+1,
     * .528644456e-1,.164529332e+3,-.45 1320561e+2,1.0/
C the approximation 4283 from hart et al(1968, p. 251)
      data piby4/.785398163/
      x = abs(xarg)
      if (x.gt.piby4) then
              tan2d = mytand(xarg)/xarg
      else
              x = x/piby4
C convert to range of rational
              xx = x*x
```

```
            tan2d = (p0+xx*(p1+xx*p2))/(piby4*(q0+xx*(q1+xx*q2)))
      endif
      return
      end
```

Remark. Because S-PLUS allows calls to external C or Fortran functions, it can use the *rstable* program given here. It is important that the external program uses double precision (*rstable* does that).

1.8 Exercises

Exercise 1.1 Show that Definition 1.1.6 implies Definitions 1.1.1 and 1.1.4.

Exercise 1.2 Show that Definition 1.1.4 implies Definition 1.1.1.
Hint: Suppose X is non-degenerate. Then $C_n \to \infty$, $C_{n+1}/C_n \to 1$ as $n \to \infty$. For arbitrary constants $0 < a_2 < a_1$, there is a sequence of integers $\{m_n\}$ such that as $n \to \infty$, $C_{m_n}/C_n \to a_2/a_1$. Express $\sum_{i=1}^{n+m_n} X_i$ normalized in terms of a sum of $\sum_{i=1}^{n} X_i$ normalized and $\sum_{i=n+1}^{n+m_n} X_i$ normalized, and let $n \to \infty$.

Exercise 1.3 Prove that $w_1(\theta, \alpha, \beta)$ in (1.1.11) is jointly continuous in α and β but is not continuous in θ, the characteristic function argument. Show that it is not jointly continuous in θ and α even outside a neighborhood of $\theta = 0$. (Let $\theta \to \theta_0 \neq 0$ and $\alpha \to 1$.)

Exercise 1.4 Prove the statements about the representation (1.1.12)

Exercise 1.5 Prove that α-stable random variables have infinitely differentiable densities.
Hint: Keep integrating the characteristic function.

Exercise 1.6 Find the asymptotic behavior of $P(X > \lambda)$ as $\lambda \to \infty$ for $X \sim N(0, \nu^2)$.

Exercise 1.7 Let γ be uniformly distributed on the interval $(-\frac{\pi}{2}, \frac{\pi}{2})$. Show that

$$X = \sigma \tan \gamma + \mu \sim S_1(\sigma, 1, \mu),$$

where $S_1(\sigma, 1, \mu)$ is the Cauchy distribution whose density function is given in Relation (1.1.13). Use this to check the cumulative distribution function (1.1.14) of the Cauchy.

Exercise 1.8 Let $Z \sim N(0,1)$. Show that

$$X = \sigma Z^{-2} + \mu \sim S_{1/2}(\sigma, 1, \mu),$$

where $S_{1/2}(\sigma, 1, \mu)$ is the Lévy distribution whose density function is given in Relation (1.1.15).

Exercise 1.9 Let $\{X(t),\ t \geq 0\}$ be a standard Brownian motion with a.s. continuous paths. Let T_a denote the first time it reaches the level $a > 0$. Show that T_a has the Lévy distribution $S_{1/2}(a^2, 1, 0)$ given in (1.1.15). Use this to check the cumulative distribution function (1.1.16) of the Lévy distribution.
Hint: By the reflection principle, $P[T_a < t] = 2P[X(t) > a]$. Now use Exercise 1.8.

Exercise 1.10 Prove Property 1.2.1 for $\alpha = 1$.

Exercise 1.11 Prove Property 1.2.3 for $\alpha = 1$.

Exercise 1.12 If $X_1, \ldots, X_n$ are i.i.d. $S_\alpha(\sigma, \beta, \mu)$ and $S_n = X_1 + \cdots + X_n$, then the limiting distribution of

$$\begin{array}{ll} n^{-1/\alpha} S_n & \text{if } 0 < \alpha < 1, \\ n^{-1}(S_n - 2\pi^{-1}\sigma\beta n \ln n) - \mu & \text{if } \alpha = 1, \\ n^{-1/\alpha}(S_n - n\mu) & \text{if } 1 < \alpha \leq 2, \end{array}$$

as $n \to \infty$, is $S_\alpha(\sigma, \beta, 0)$.

Exercise 1.13 The following theorem (Mijnheer (1975)) characterizes the domain of attraction of stable distributions. Verify it for $X_i \sim S_\alpha(1, \beta, 0),\ i = 1, \ldots, n$. (A slowly varying function L at infinity is a non-negative function satisfying $\lim_{t\to\infty} L(xt)/L(x) = 1$ for all $t > 0$.)

- **Theorem 1.8.1** *Let $X_1, \ldots, X_n$ be i.i.d. with cumulative distribution function F. There exist $a_n > 0, b_n \in \mathbb{R},\ n = 1, 2, \ldots,$ such that the distribution of $a_n^{-1}[(X_1 + \cdots + X_n) - b_n]$ converges as $n \to \infty$ to $S_\alpha(1, \beta, 0)$ if and only if both*

  (i) $x^\alpha[1 - F(x) + F(-x)] = L(x)$ is slowly varying at infinity.

  (ii) $\frac{F(-x)}{1+F(-x)+F(-x)} \to \frac{1-\beta}{2}$ as $x \to \infty$.

  The a_n must satisfy

$$\lim_{n\to\infty} \frac{nL(a_n)}{a_n^\alpha} = \begin{cases} \Gamma(1-\alpha)\cos(\pi\alpha/2) & \textit{if } 0 < \alpha < 1, \\ 2/\pi & \textit{if } \alpha = 1, \\ \frac{\Gamma(2-\alpha)}{\alpha-1}|\cos\frac{\pi\alpha}{2}| & \textit{if } 1 < \alpha < 2. \end{cases}$$

The b_n may be chosen as follows:

$$b_n = \begin{cases} 0 & \text{for } 0 < \alpha < 1, \\ na_n \int_{-\infty}^{\infty} \sin(x/a_n)dF(x) & \text{for } \alpha = 1, \\ n \int_{-\infty}^{\infty} xdF(x) & \text{for } 1 < \alpha < 2. \end{cases}$$

In all cases, $a_n = n^{1/\alpha}L_0(n)$ where L_0 is slowly varying at infinity.

Exercise 1.14 This exercise complements Proposition 1.2.11, which shows that the support of a totally skewed to the right α-stable random variable, $0 < \alpha < 1$ (i.e., a stable subordinator) is contained in $\mathbb{R}_+$. Here, the goal is to prove that this support actually coincides with $\mathbb{R}_+$.

The support of a random variable X is defined as the smallest closed set A such that $P(X \in A) = 1$.

Let $X \sim S_\alpha(\sigma, 1, 0)$ with $0 < \alpha < 1$. Let f be the continuous density function of X. Prove that $f(x) > 0$ for every $x > 0$ by going through the following steps:

(i) Let $\sum = \{x \in \mathbb{R}: f(x) > 0\}$. Show that for any $A, B > 0$ such that $A^\alpha + B^\alpha = 1$, one has

$$A\Sigma + B\Sigma = \Sigma,$$

where $A\Sigma + B\Sigma = \{Ax + By: x \in \Sigma,\ y \in \Sigma\}$.

(ii) Conclude that either $\Sigma = \mathbb{R}$ or $\Sigma = \mathbb{R}_+$.

(iii) Use Proposition 1.2.11 to conclude that $\Sigma = \mathbb{R}_+$.

Exercise 1.15 Prove Proposition 1.2.12 for $0 < \alpha \le 2$.
Hints:

1. Using the Schwarz reflection principle (Ahlfors (1979), p. 172), prove the following lemma:

   Let f be a complex-valued function defined on $D \cap \mathbb{R} \neq \emptyset$, where D is a region and $\mathbb{R}$ is the real line. If f is continuable to $D^+ = \{z \in D : \text{Im } z > 0\}$, i.e. if there exists a function $\tilde{f}$ analytic in D^+, continuous in $D^+ \cup (D \cap \mathbb{R})$, which satisfies $f(t) = \tilde{f}(t)$ for $t \in D \cap \mathbb{R}$, then this continuation is unique.

2. Show, using (1.2.8), and (1.2.11), (1.2.12) for $\beta = 1$, that $Ee^{-\xi X}$ exists and is continuous for Re $\xi \ge 0$, and is analytic for Re $\xi > 0$.

3. For $0 < \alpha < 2, \alpha \neq 1, \beta = 1$ and $\theta > 0$ the characteristic function (1.1.6) can be written as $\phi(\theta) = E\exp(i\theta X) = \exp(-\sigma^\alpha \theta^\alpha e^{-i\frac{\pi}{2}\alpha}/\cos\frac{\pi\alpha}{2})$ and can be extended to an analytic single-valued function $\phi(z) = \exp(-\sigma^\alpha(-iz)^\alpha/\cos\frac{\pi\alpha}{2})$, where z belongs to the whole complex plane with the exception of some half-line (say arg $z = -\frac{3}{4}\pi$). Verify that using formula (1.1.6) for $\theta < 0$ you obtain the same extension.

4. By the above lemma, the functions $\phi(z)$ and $Ee^{-(-iz)X}$ coincide for $\text{Im } z \geq 0$.

5. Substitute $-iz = \gamma$ in the formula for ϕ to establish Proposition 1.2.12 for $\alpha \neq 1$.

6. For $\alpha = 1$ start with the other part of (1.1.6).

7. The proof is even simpler for $\alpha = 2$.

Alternative method: Establish Step 2. By Theorem 11.2.1 of Lukacs (1970), p. 311, the Lévy-Khinchine representation (1.1.7) of the characteristic function of X holds for $\text{Im } z \geq 0$. Show that the passage from the Lévy-Khinchine formulas to the form (1.1.6) is also valid for $z = it,\ t > 0$. When $\alpha = 1$, you need also to prove

$$\int_0^\infty \frac{\cos x - e^{-x}}{x} dx = 0.$$

Exercise 1.16 If X is α-stable then the random variable $Y = e^X$ is called *log-stable*. Y is *log-normal* if $\alpha = 2$. Suppose $0 < \alpha < 2$ and show the following. If $\beta \neq -1$, then $P(Y > y)$ is asymptotically proportional to $(\ln y)^{\alpha - 1}$ as $y \to \infty$, and consequently, Y has no finite moments. On the other hand, if $\beta = -1$, all moments of Y are finite.

Log-stable distributions with $\beta = -1$ are used in modeling multifractals, partly because of the finiteness of their moments. See for example Evertsz and Mandelbrot (1991), (1992) and Schertzer and Lovejoy (1993).

Exercise 1.17 Prove Property 1.2.13.

Exercise 1.18 Let $X \sim S_1(\sigma, \beta, 0)$ with $0 < \sigma < K$. Show that there is a finite constant γ_K which depends only on K and a positive number a, independent of σ, β and K, such that $P(X > \gamma_K) \geq a$.
Hint: Use Property 1.2.13 and set $a = P(Y_1 \geq 0) \cdot P(Y_2 \leq 0) > 0$ where Y_1 and Y_2 are i.i.d. $S_1(1, 1, 0)$ random variables.

Exercise 1.19 Using the weak law of large numbers to prove that when $1 < \alpha \leq 2$, the shift parameter μ equals the mean.

Exercise 1.20 Formulas for the density functions of the Cauchy distribution $S_1(\sigma, 0, \mu)$ and for the Lévy distribution $S_{1/2}(\sigma, 1, \mu)$ are given at the end of the first section. Use Property 1.2.15 to verify that the normalization constants in these formulas are correct.

Exercise 1.21 Let $X \sim S_{\alpha'}(\sigma', 1, 0)$, and $A \sim S_{\alpha/\alpha'}(\sigma_A, 1, 0), 0 < \alpha < \alpha' < 1$ be independent. Show that $Z = A^{1/\alpha'} X$ has a $S_\alpha(\sigma, 1, 0)$ distribution for some $\sigma > 0$.

Exercise 1.22 Let Γ_i have a gamma distribution with parameters i and $E\Gamma_i$, as in Theorem 1.4.2.

(a) Find the asymptotic behavior as $\lambda \to \infty$ of $P\{\Gamma_i^{-1/\alpha} > \lambda\}$ for $0 < \alpha \leq 2$.

(b) Let $\epsilon_i \Gamma_i^{-1/\alpha} W_i$ be defined as in Theorem 1.4.2. Find the asymptotic behavior of $P\{|\epsilon_i \Gamma_i^{-1/\alpha} W_i| > \lambda\}$ as $\lambda \to \infty$.

(c) What are the positive numbers p for which $E\Gamma_i^{-p} < \infty$?

(d) If $p > 1$, show that there is a constant K such that

$$E\Gamma_i^{-p} = \Gamma(i-p)/(i-1)! \leq K i^{-p}.$$

Exercise 1.23 Let $Y_i = \epsilon_i U_i^{-1/\alpha} W_i$ be defined as in Theorem 1.4.2 and suppose that $\alpha = 2$. Find the normalization d_n such that $d_n^{-1} \sum_{i=1}^n Y_i$ converges in distribution to a non-degenerate random variable Z. What is the distribution of Z?

Exercise 1.24 Let $X_1, X_2, \ldots$ be a sequence of i.i.d. random variables and let $p > 0$. Show that

(i) $E|X_1|^p < \infty$ is and only if $\lim_{n\to\infty} n^{-1/p} X_n = 0$ a.s.,

(ii) $E|X_1|^p = \infty$ if and only if $\overline{\lim}_{n\to\infty} n^{-1/p}|X_n| = \infty$ a.s.

More generally, show that for $p > 0$ and $r \in \mathbb{R}$,

(iii) $E|X_1|^p(\log|X_1|)^r < \infty$ if and only if $\lim_{n\to\infty} n^{-1/p}(\log n)^{r/p} X_n = 0$ a.s.,

(iv) $E|X_1|^p(\log|X_1|)^r = \infty$ if and only if $\overline{\lim}_{n\to\infty} n^{-1/p}(\log n)^{r/p}|X_n| = \infty$ a.s.

Exercise 1.25 How large can the remainder of the series $\sum_{j=2}^\infty \epsilon_j \Gamma_j^{-1/\alpha} W_j$ be?

(i) Suppose that $E|W_j|^{\alpha+\epsilon} < \infty$ for some $\epsilon > 0$. Prove that $E|\sum_{j=2}^\infty \epsilon_j \Gamma_j^{-1/\alpha} W_j|^\alpha < \infty$.

(ii) Show by counterexample that $E|W_j|^\alpha < \infty$ does not imply that $E|\sum_{j=2}^\infty \epsilon_j \Gamma_j^{-1/\alpha} W_j|^\alpha < \infty$.

Hint: Take W_js such that $E|W_j|^\alpha < \infty$ but $E|W_j|^\alpha \log|W_j| = \infty$ and use Exercise 1.24.

Exercise 1.26 Let $\{X_i\}$ be independent random variables $X_i \sim S_\alpha(\sigma_i, \beta_i, \mu_i)$, $0 < \alpha \le 2$. Prove that the series $\sum_{i=1}^\infty X_i$ converges a.s. if and only if $\sum_{i=1}^\infty \mu_i$ converges and $\sum_{i=1}^\infty \sigma_i^\alpha < \infty$.

Using the three series theorem, prove that the series converges absolutely a.s. if and only if $\sum_{i=1}^\infty |\mu_i| < \infty$ and

$$\begin{cases} \sum_{i=1}^\infty \sigma_i^\alpha < \infty & \text{if } 0 < \alpha < 1, \\ \sum_{i=1}^\infty \sigma_i |\ln(\sigma_i \wedge 0.5)| < \infty & \text{if } \alpha = 1, \\ \sum_{i=1}^\infty \sigma_i < \infty & \text{if } \alpha > 1. \end{cases}$$

Exercise 1.27 Prove (1.2.8) using Theorem 1.4.5.
Hint: Use Exercise 1.22.

Exercise 1.28 Suppose $X \sim S_\alpha(\sigma, 0, 0)$. Use Proposition 1.3.1 to show that if (1.2.8) holds for $0 < \alpha < 1$ then it holds for $1 < \alpha < 2$.

Exercise 1.29 Let X_α denote a stable subordinator, i.e., $X_\alpha \sim S_\alpha(\sigma, 1, 0)$ with $\sigma > 0$ and $\alpha < 1$. Prove that as $\alpha \to 0$,

$$(X_\alpha)^\alpha \xrightarrow{d} \frac{1}{E},$$

where E is an exponential random variable with mean 1.
Hint: Use the series representation for X_α and the fact (prove it) that a sequence $\mathbf{a} = \{a_i\}_{i=1}^\infty$ in l^{p_0}, $p_0 \ge 1$ satisfies $\lim_{p\to\infty} \|\mathbf{a}\|_p = \|\mathbf{a}\|_\infty$.

Chapter 2

Multivariate stable distributions

The multivariate stable distribution, which is the distribution of a stable random vector, is defined in Section 2.1 by simply extending to $\mathbb{R}^d$ the definition of a stable random variable. Gaussian vectors, however, can also be defined in the following way: a random vector is Gaussian if and only if any linear combination of its components is a Gaussian random variable. As in the Gaussian case, any linear combination of the components of a stable random vector is a stable random variable, but the converse is not always true. The converse holds when the linear combinations are either strictly stable or when $\alpha \geq 1$. In Section 2.2, we demonstrate with a counterexample that the converse does not generally hold when $\alpha < 1$.

As in the univariate case, multivariate stable cumulative distribution functions or density functions are not usually known in closed form and therefore one works instead with the characteristic functions. The characteristic function of an α-stable random vector is described in Section 2.3. It involves a finite measure Γ on the unit sphere of $\mathbb{R}^d$ and a shift vector $\boldsymbol{\mu}^0$ which plays a role similar to the shift parameter in the univariate case. The measure Γ is called the *spectral measure*. It replaces both the scale and skewness parameter that enter in the description of the univariate stable distribution.

Conditions for strict stability of $\mathbf{X}$ are discussed in Section 2.4. When $\alpha \neq 1$, $\mathbf{X}$ is strictly stable if the shift vector $\boldsymbol{\mu}^0 = 0$. Conditions for strict stability when $\alpha = 1$ are more complicated because they involve the spectral measure Γ. Nevertheless, strict stability turns out to be a componentwise property: $\mathbf{X}$ is strictly stable if and only if all its components are strictly stable.

Section 2.4 also contains conditions for $\mathbf{X}$ to be symmetric, i.e., for the vectors

$\mathbf{X}$ and $-\mathbf{X}$ to be identically distributed. In contrast to strict stability, symmetry is not a componentwise property: $\mathbf{X}$ is symmetric if and only if the shift vector $\boldsymbol{\mu}^0$ is zero and the spectral measure Γ is symmetric (i.e., Γ gives equal weight to any two antipodal sets).

We noted in Chapter 1 that every symmetric α-stable random variable has the same distribution as that of a mean zero Gaussian random variable multiplied by $A^{1/2}$, where A is a totally skewed stable random variable. Only some symmetric α-stable random vectors have a similar property. Those that do are called *sub-Gaussian* and they are introduced in Section 2.5. Because they are obtained by multiplying a mean zero Gaussian vector by $A^{1/2}$, they inherit the dependence structure of the underlying Gaussian vector. The multivariate Cauchy is an example of a sub-Gaussian random vector.

When the spectral measure Γ is uniform (and $\boldsymbol{\mu}^0 = 0$) the symmetric α-stable random vector has a very special type of sub-Gaussian structure: it is conditionally i.i.d. Gaussian. Such a structure is used in Section 2.6, in the context of isotropic complex-valued random variables. Although we are interested in real-valued random objects, we will later encounter an important class of real-valued stochastic processes (the so called "harmonizable processes") which are typically defined in terms of complex random variables. A complex random variable can be regarded as a two-dimensional random vector whose components are the real and imaginary parts.

When $\alpha = 2$, the random vector is Gaussian and its dependence structure is completely specified by its autocovariance function. There is no such simple description when $\alpha < 2$. Covariances do not exist, but when $\alpha > 1$, one can introduce the notion of *covariation*. We do this in Section 2.7. The covariation shares some of the properties of the covariance. Unfortunately, it is neither symmetric nor additive in its second component. However, it becomes additive in its second component when the random variables that are being summed are independent. This feature will be useful later, in the context of stable stochastic integrals.

In Section 2.8, we introduce *covariation norms*. Like the covariance, the covariation gives rise to a norm. The covariation norm of an α-stable random variable turns out to be identical to the scale parameter of that random variable. It metrizes convergence in probability, i.e., the covariation norm of a sequence of α-stable random variables tends to zero if and only if that sequence converges to zero in probability.

In Section 2.9, we introduce *James orthogonality*. Two zero mean Gaussian random variables have zero covariance if and only if they are independent. If two symmetric α-stable random variables with $1 < \alpha < 2$ are independent, then they have zero covariation, but the converse is not always true. It is possible to relate "zero covariation" to "James orthogonality." James orthogonality is well defined in normed vector spaces that possess no scalar product. We define it in our context by using covariation norms. Unfortunately, it is not a symmetric relation; that is, if X is James orthogonal to Y, then Y is not necessarily James orthogonal to X. Although not as versatile as the usual notion of orthogonality, James orthogonality is often valuable. It is used, for instance, to characterize the family of sub-Gaussian random vectors.

The *codifference*, another measure of bivariate dependence, is introduced in Section 2.10. Like the covariation, the codifference reduces to the covariance when $\alpha = 2$. But whereas the covariation may not be defined for $\alpha \leq 1$, the codifference is defined for all $0 < \alpha \leq 2$. It is, moreover, symmetric in its arguments and the codifference of a $S\alpha S$ random vector is non-negative definite.

We will encounter more properties of multivariate stable distributions in later chapters.

2.1 Stable random vectors

The definition of stability in $\mathbb{R}^d$ is analogous to that in $\mathbb{R}^1$.

Definition 2.1.1 A random vector $\mathbf{X} = (X_1, X_2, \ldots, X_d)$ is said to be a *stable random vector in* $\mathbb{R}^d$ if for any positive numbers A and B there is a positive number C and a vector $\mathbf{D} \in \mathbb{R}^d$ such that

$$A\mathbf{X}^{(1)} + B\mathbf{X}^{(2)} \stackrel{d}{=} C\mathbf{X} + \mathbf{D}, \tag{2.1.1}$$

where $\mathbf{X}^{(1)}$ and $\mathbf{X}^{(2)}$ are independent copies of $\mathbf{X}$.

Instead of saying "$\mathbf{X} = (X_1, \ldots, X_d)$ is a *stable random vector in* $\mathbb{R}^d$," we will often say "$X_1, X_2, \ldots, X_d$ are *jointly stable*", or "$\mathbf{X}$ has a stable distribution in $\mathbb{R}^d$," or "the distribution of $\mathbf{X}$ is multivariate stable."

The vector $\mathbf{X}$ is called *strictly stable* if (2.1.1) holds with $\mathbf{D} = \mathbf{0}$ for any $A > 0$ and $B > 0$. The vector $\mathbf{X}$ is called *symmetric stable* if it is stable and satisfies in addition the relation

$$P\{\mathbf{X} \in A\} = P\{-\mathbf{X} \in A\}$$

for any Borel set A of $\mathbb{R}^d$. As in $\mathbb{R}^1$, a symmetric stable vector is strictly stable.

The preceding definitions impose conditions on the joint distribution of $(X_1, X_2, \ldots, X_d)$. What do they imply about the individual components $X_1, X_2, \ldots, X_d$? If $\mathbf{X}$ is a stable random vector, are its components stable random variables? Are linear combinations of its components stable random variables? The following theorem provides an answer.

Theorem 2.1.2 *Let $\mathbf{X} = (X_1, \ldots, X_d)$ be a stable (respectively, strictly stable, symmetric stable) vector in $\mathbb{R}^d$. Then there is a constant $\alpha \in (0, 2]$ such that, in (2.1.1), $C = (A^\alpha + B^\alpha)^{1/\alpha}$. Moreover, any linear combination of the components of $\mathbf{X}$ of the type $Y = \sum_{k=1}^d b_k X_k$ is an α-stable (respectively, strictly α-stable, symmetric α-stable) random variable.*

PROOF: Let $\mathbf{X}^{(1)}$ and $\mathbf{X}^{(2)}$ be independent copies of $\mathbf{X}$, let $\mathbf{b} = (b_1, b_2, \ldots, b_d)$ be a vector in $\mathbb{R}^d$, and define $Y_1 = \sum_{k=1}^d b_k X_k^{(1)}$ and $Y_2 = \sum_{k=1}^d b_k X_k^{(2)}$. The random variables Y_1 and Y_2 are independent copies of the random variable Y.

Now fix $A > 0$ and $B > 0$. Since $\mathbf{X}$ is stable, by (2.1.1), there are a $C > 0$ and $\mathbf{D} = (D_1, \ldots, D_d)$ such that $A\mathbf{X}^{(1)} + B\mathbf{X}^{(2)} \stackrel{d}{=} C\mathbf{X} + \mathbf{D}$. When two vectors are equal in distribution, linear combinations of their respective components are also equal in distribution, as can be seen by comparing the characteristic functions. Hence,

$$\begin{aligned} AY_1 + BY_2 &= \sum_{k=1}^d b_k (AX_k^{(1)} + BX_k^{(2)}) \\ &\stackrel{d}{=} \sum_{k=1}^d b_k (CX_k + D_k) \\ &= C\sum_{k=1}^d b_k X_k + \sum_{k=1}^d b_k D_k \\ &= CY + (\mathbf{b}, \mathbf{D}), \end{aligned}$$

where $(\mathbf{b}, \mathbf{D})$ denotes the scalar product of the vectors $\mathbf{b}$ and $\mathbf{D}$. This shows that Y is a stable random variable. By Theorem 1.1.2, there is a constant $\alpha = \alpha(\mathbf{b}) \in (0, 2]$ such that $C = (A^\alpha + B^\alpha)^{1/\alpha}$. In fact, α is independent of $\mathbf{b}$, because if α' is the index resulting from a vector $\mathbf{b}' \neq \mathbf{b}$, then we would have $(A^\alpha + B^\alpha)^{1/\alpha} = (A^{\alpha'} + B^{\alpha'})^{1/\alpha'}$ for all $A > 0$ and $B > 0$, which is impossible unless $\alpha = \alpha'$. ∎

As in the one-dimensional case, we obtain (see Exercise 2.2):

Corollary 2.1.3 *A random vector $\mathbf{X}$ is stable if and only if for any $n \geq 2$, there is an $\alpha \in (0, 2]$ and a vector $\mathbf{D}_n$ such that*

$$\mathbf{X}^{(1)} + \mathbf{X}^{(2)} + \cdots + \mathbf{X}^{(n)} \stackrel{d}{=} n^{1/\alpha}\mathbf{X} + \mathbf{D}_n, \tag{2.1.2}$$

where $\mathbf{X}^{(1)}, \mathbf{X}^{(2)}, \ldots, \mathbf{X}^{(n)}$ are independent copies of $\mathbf{X}$.

These results motivate the following definition:

Definition 2.1.4 A random vector $\mathbf{X}$ in $\mathbb{R}^d$ is called α-*stable* if (2.1.1) holds with $C = (A^\alpha + B^\alpha)^{1/\alpha}$, or, equivalently, if (2.1.2) holds. The index α is the *index of stability* or the *characteristic exponent* of the vector $\mathbf{X}$.

The second part of Theorem 2.1.2 states that if $\mathbf{X}$ is an α-stable vector, then all linear combinations $(\mathbf{b}, \mathbf{X}) = \sum_{i=1}^d b_i X_i$ are α-stable random variables.

Is the converse true? If all of linear combinations of the components of a random vector $\mathbf{X}$ are α-stable, is the vector $\mathbf{X}$ necessarily stable? The question is a natural one in view of the fact that in the Gaussian case $\alpha = 2$, it is well known that the answer is *yes*. Unfortunately, when $\alpha < 2$, the answer is, in general, *no* as we will show in the next section. It turns out, however, that the answer is *yes* if all the linear combinations are *strictly* stable, or if $\alpha \geq 1$.

Theorem 2.1.5 *Let $\mathbf{X}$ be a random vector in $\mathbb{R}^d$.*

(a) If all linear combinations $Y = \sum_{k=1}^d b_k X_k$ have strictly *stable distributions, then $\mathbf{X}$ is a strictly stable random vector in $\mathbb{R}^d$.*

(b) If all linear combinations are symmetric *stable, then $\mathbf{X}$ is a symmetric stable random vector in $\mathbb{R}^d$.*

(c) If all linear combinations are stable *with index of stability greater or equal to one, then $\mathbf{X}$ is a stable vector in $\mathbb{R}^d$.*

PROOF: The index of stability of a linear combination $Y_\mathbf{b} = \sum_{k=1}^d b_k X_k = (\mathbf{b}, \mathbf{X})$ may a priori depend on $\mathbf{b}$. We first show that if all linear combinations $Y_\mathbf{b}$ are stable, then those that are non-degenerate have the same index of stability.

Suppose that this is not so, that there are non-zero vectors $\mathbf{b} \in \mathbb{R}^d$ and $\mathbf{c} \in \mathbb{R}^d$, such that $Y_\mathbf{b} = (\mathbf{b}, \mathbf{X})$ and $Y_\mathbf{c} = (\mathbf{c}, \mathbf{X})$ are non-degenerate random variables, and such that $Y_\mathbf{b}$ has index of stability α_1 and $Y_\mathbf{c}$ has index of stability α_2 with $0 < \alpha_1 < \alpha_2 \leq 2$. Let ρ_1, ρ_2 be any non-zero real numbers and set $\mathbf{a} = \rho_1 \mathbf{b} + \rho_2 \mathbf{c}$. Consider the random variable $Y_\mathbf{a}$. By our assumptions, $Y_\mathbf{a}$ must be stable with some index of stability, say α_3. Then for any $\lambda > 0$,

$$
\begin{aligned}
P(|Y_\mathbf{a}| > \lambda) &= P(|(\mathbf{a}, \mathbf{X})| > \lambda) \\
&= P(|\rho_1(\mathbf{b}, \mathbf{X}) + \rho_2(\mathbf{c}, \mathbf{X})| > \lambda) \\
&= P(|\rho_1 Y_\mathbf{b} + \rho_2 Y_\mathbf{c}| > \lambda) \\
&\geq P(|\rho_1 Y_\mathbf{b}| > 2\lambda) - P(|\rho_2 Y_\mathbf{c}| > \lambda) \\
&\sim \text{const. } \lambda^{-\alpha_1} - \text{const. } \lambda^{-\alpha_2} \\
&\sim \text{const. } \lambda^{-\alpha_1}
\end{aligned}
$$

as $\lambda \to \infty$. (If $\alpha_2 = 2$, replace $\lambda^{-\alpha_2}$ by $\lambda^{-1} \exp\{-\lambda^2/4\sigma^2\}$.) This implies $\alpha_3 \leq \alpha_1 = \min(\alpha_1, \alpha_2)$. The index of stability of $Y_\mathbf{a}$ is, therefore, no larger than the smaller of α_1 and α_2. Now, since $Y_{\mathbf{b}}$ and $Y_{\mathbf{c}}$ are non-degenerate, there is a sequence of non-zero real numbers $\{A_n\}_{n=1}^{\infty}$, converging to 0 as $n \to \infty$, such that for all n, $Z_n = A_n Y_{\mathbf{b}} + Y_{\mathbf{c}}$ is not equal to zero in distribution. Let $Z_n \sim S_{\alpha(n)}(\sigma_n, \beta_n, \mu_n)$, $n = 1, 2, \dots$. By the above fact,

$$\alpha(n) \le \alpha_1 < \alpha_2,\ n = 1, 2, \dots.$$

Clearly, $Z_n \to Y_{\mathbf{c}}$ in distribution as $n \to \infty$. Convergence of the real part of the characteristic function of Z_n to that of $Y_{\mathbf{c}}$ implies

$$\lim_{n\to\infty} \sigma_n^{\alpha(n)} |\theta|^{\alpha(n)} = \sigma_2^{\alpha_2} |\theta|^{\alpha_2},\ \theta \in \mathbb{R},$$

where σ_2 is the scale parameter of $Y_{\mathbf{c}}$. Hence $\lim_{n\to\infty} \sigma_n^{\alpha(n)} |\theta|^{\alpha(n)-\alpha_2} = \sigma_2^{\alpha_2}$, $\theta \neq 0$, together with $\alpha(n) - \alpha_2 \le \alpha_1 - \alpha_2 < 0$, implies $\sigma_2 = 0$, contradicting the fact that $Y_{\mathbf{c}}$ is non-degenerate. All non-degenerate linear combinations of the form $(\mathbf{b}, \mathbf{X})$ must therefore have the same index of stability.

To establish (a), assume now that all linear combinations $(\mathbf{b}, \mathbf{X})$ are strictly stable random variables with index of stability α.

Fix any $A > 0$, $B > 0$, let $\mathbf{Z} = A\mathbf{X}^{(1)} + B\mathbf{X}^{(2)}$, and $\mathbf{W} = (A^\alpha + B^\alpha)^{1/\alpha}\mathbf{X}$, where $\mathbf{X}^{(1)}$ and $\mathbf{X}^{(2)}$ are independent copies of $\mathbf{X}$. We want to show that $\mathbf{Z} \stackrel{\mathbf{d}}{=} \mathbf{W}$. For any vector $\mathbf{b} \in \mathbb{R}^d$, we have

$$(\mathbf{b}, \mathbf{Z}) = A(\mathbf{b}, \mathbf{X}^{(1)}) + B(\mathbf{b}, \mathbf{X}^{(2)}) \stackrel{d}{=} (A^\alpha + B^\alpha)^{1/\alpha}(\mathbf{b}, \mathbf{X}) = (\mathbf{b}, \mathbf{W}).$$

Therefore, for any $\mathbf{b} \in \mathbb{R}^d$,

$$E \exp i(\mathbf{b}, \mathbf{Z}) = E \exp i(\mathbf{b}, \mathbf{W}),$$

showing that the vectors $\mathbf{Z}$ and $\mathbf{W}$ have identical characteristic functions. Thus, $\mathbf{Z} \stackrel{d}{=} \mathbf{W}$, i.e., $A\mathbf{X}^{(1)} + B\mathbf{X}^{(2)} \stackrel{d}{=} (A^\alpha + B^\alpha)^{1/\alpha}\mathbf{X}$. This proves that $\mathbf{X}$ is a strictly stable vector in $\mathbb{R}^d$.

(b) Assume that all linear combinations $(\mathbf{b}, \mathbf{X})$ are symmetric stable random variables. Since a symmetric stable random variable is strictly stable, we conclude by part (a) that $\mathbf{X}$ is a strictly stable vector. To prove part (b), one has only to show that $\mathbf{X}$ is a symmetric vector.

Let $\mathbf{b}$ be a fixed arbitrary vector in $\mathbb{R}^d$. Since $(\mathbf{b}, \mathbf{X})$ is a symmetric random variable, we have

$$E \exp i(\mathbf{b}, \mathbf{X}) = E \exp\{-i(\mathbf{b}, \mathbf{X})\} = E \exp i(\mathbf{b}, -\mathbf{X}),$$

i.e., the characteristic function of $\mathbf{X}$ equals that of $-\mathbf{X}$. Hence $\mathbf{X} \stackrel{d}{=} -\mathbf{X}$, establishing part (b).

(c) Assume that all linear combinations are stable random variables with index of stability $\alpha \geq 1$.

Suppose firstly $\alpha = 1$. By assumption, for every $\mathbf{b} \in \mathbb{R}^d$,

$$Y_{\mathbf{b}} = (\mathbf{b}, \mathbf{X}) \sim S_1(\sigma(\mathbf{b}),\ \beta(\mathbf{b}), \mu(\mathbf{b})) \tag{2.1.3}$$

for some $\sigma(\mathbf{b}) \geq 0$, $\beta(\mathbf{b}) \in [-1, 1]$ and $\mu(\mathbf{b})$ in $\mathbb{R}$. If σ_k, β_k and μ_k denote, respectively, $\sigma(\mathbf{b})$, $\beta(\mathbf{b})$, $\mu(\mathbf{b})$, corresponding to the vector $\mathbf{b} = (0, \ldots, 0, 1, 0, \ldots, 0)$ with 1 in the kth position, then

$$X_k \sim S_1(\sigma_k, \beta_k, \mu_k),\ k = 1, \ldots, d.$$

Let $\mathbf{X}^{(1)}, \mathbf{X}^{(2)}, \ldots$ be i.i.d. copies of $\mathbf{X}$ and define, for $n \geq 1$,

$$\mathbf{S}^{(n)} = \Big(\frac{1}{n}\sum_{j=1}^{n} \mathbf{X}^{(j)}\Big) - \frac{2}{\pi}(\ln n)(\boldsymbol{\sigma} \times \boldsymbol{\beta}), \tag{2.1.4}$$

where $\boldsymbol{\sigma} \times \boldsymbol{\beta} = (\sigma_1\beta_1, \sigma_2\beta_2, \ldots, \sigma_d\beta_d)$. Using Properties 1.2.1, 1.2.3 and 1.2.2, $X_k \sim S_1(\sigma_k, \beta_k, \mu_k)$ implies

$$\begin{aligned}
\sum_{j=1}^{n} X_k^{(j)} &\sim S_1(n\sigma_k, \beta_k, n\mu_k), \\
\frac{1}{n}\sum_{j=1}^{n} X_k^{(j)} &\sim S_1\Big(\sigma_k, \beta_k, \mu_k + \frac{2}{\pi}(\ln n)\sigma_k\beta_k\Big) \\
S_k^{(n)} &\stackrel{d}{=} X_k \sim S_1(\sigma_k, \beta_k, \mu_k)
\end{aligned}$$

for each $k = 1, \ldots, n$. Since the distribution of $S_k^{(n)}$ does not depend on n, the sequence $\{S_k^{(n)},\ n \geq 1\}$ is tight; i.e., for each $\epsilon > 0$, there exists a bounded set $[a, b]$ such that $P(a \leq S_k^{(n)} \leq b) > 1 - \epsilon$ holds for all n. The sequence of vectors $\{\mathbf{S}^{(n)},\ n \geq 1\}$ is therefore tight (Exercise 2.3), and as such, it has a subsequence $\{\mathbf{S}^{(n_i)},\ i \geq 1\}$ converging weakly to a probability measure on $\mathbb{R}^d$ (Billingsley 1986, Theorem 29.3). In particular, $(\mathbf{b}, \mathbf{S}^{(n_i)})$ converges weakly for every vector $\mathbf{b} = (b_1, \ldots, b_n)$ in $\mathbb{R}^d$. But (2.1.3) and (2.1.4) imply

$$(\mathbf{b}, \mathbf{S}^{(n)}) \sim S_1\Big(\sigma(\mathbf{b}), \beta(\mathbf{b}), \mu(\mathbf{b}) + \frac{2}{\pi}(\ln n)\Big(\sigma(\mathbf{b})\beta(\mathbf{b}) - \sum_{k=1}^{d} b_k\sigma_k\beta_k\Big)\Big). \tag{2.1.5}$$

In order for $(\mathbf{b}, \mathbf{S}^{(n_i)})$ to converge weakly, the coefficient of $\ln n$ must be zero, i.e.,

$$\sigma(\mathbf{b})\beta(\mathbf{b}) = \sum_{k=1}^{d} b_k\sigma_k\beta_k \ \text{ for every } \mathbf{b} \in \mathbb{R}^d. \tag{2.1.6}$$

Now (2.1.3), (2.1.4) and (2.1.6) imply $(\mathbf{b}, \mathbf{S}^{(n)}) \stackrel{d}{=} (\mathbf{b}, \mathbf{X})$ for every $\mathbf{b} \in \mathbb{R}^d$ and every $n \geq 1$. Therefore $\mathbf{X} \stackrel{d}{=} \mathbf{S}^{(n)}$ for every $n \geq 1$. Thus, $\mathbf{X}$ satisfies (2.1.2) with $\alpha = 1$ and so $\mathbf{X}$ is a 1-stable random vector.

A similar proof holds for the case $1 < \alpha \leq 2$. Setting

$$\mathbf{S}^{(n)} = \left(\frac{1}{n^{1/\alpha}} \sum_{j=1}^{n} \mathbf{X}^{(j)}\right) - n^{1-1/\alpha}\boldsymbol{\mu} + \boldsymbol{\mu},$$

we have

$$(\mathbf{b}, \mathbf{S}^{(n)}) \sim S_\alpha\left(\sigma(\mathbf{b}), \beta(\mathbf{b}),\, n^{1-1/\alpha}\left(\mu(\mathbf{b}) - \sum_{k=1}^{d} b_k\mu_k\right) + \sum_{k=1}^{d} b_k\mu_k\right).$$

This time, the result follows from the fact that $n^{1-1/\alpha} \to \infty$ as $n \to \infty$. ■

Remarks

1. The proof of part (c) fails when $0 < \alpha < 1$ because $n^{1-1/\alpha} \to 0$ as $n \to \infty$.

2. Here is an alternate proof of part (c) when $1 < \alpha < 2$. The idea is to make use of part (a). Observe that $E|X_k| < \infty$ for $k = 1, 2, \ldots, d$. Indeed, X_k is a particular linear combination of the components of $\mathbf{X}$, and since, by assumption, all linear combinations are stable with index greater than one, they all have a finite mean (Property 1.2.16). Let $Y_\mathbf{b} = \sum_{k=1}^{d} b_k X_k$, let $\widetilde{\mathbf{X}} = (\widetilde{X}_1, \widetilde{X}_2, \ldots, \widetilde{X}_d)$ where $\widetilde{X}_k = X_k - EX_k$, $k = 1, \ldots, d$ are zero mean random variables, and let $\widetilde{Y}_\mathbf{b} = \sum_{k=1}^{d} b_k \widetilde{X}_k$. By the linearity of the expectation,

$$\widetilde{Y}_\mathbf{b} = \sum_{k=1}^{d} b_k X_k - E\sum_{k=1}^{d} b_k X_k, \tag{2.1.7}$$

i.e., $\widetilde{Y}_\mathbf{b} = Y_\mathbf{b} - EY_\mathbf{b}$. $Y_\mathbf{b}$ is stable by assumption and therefore, $\widetilde{Y}_\mathbf{b}$ is strictly stable by Property 1.2.19 and Corollary 1.2.7. This implies by part (a) that $\widetilde{\mathbf{X}}$ is strictly stable and hence $\mathbf{X} = \widetilde{\mathbf{X}} + (EX_1, EX_2, \ldots, EX_d)$ is stable.

3. The key step in the preceding proof of part (c) was the use in (2.1.7) of the linearity of the expectation in order to reduce the general case to the strictly stable case. When $\alpha = 1$, compensation by shifting is not enough to make the components strictly stable (see Corollary 1.2.10). When $\alpha < 1$, one can still make the components strictly stable by shifting (Corollary 1.2.7), but the shift for each component is not equal to the expectation (the expectation is infinite when $\alpha < 1$). As a result, linearity is lost: the shift for a given linear combination is not equal to the linear combination of the shifts. This means that a computation similar to (2.1.7) is not possible when $\alpha < 1$.

When $\alpha < 1$, one can construct a counterexample which shows that the stability of all the linear combinations $(\mathbf{b}, \mathbf{X})$ does not ensure the stability of the vector $\mathbf{X}$.

2.2 A counterexample for $0 < \alpha < 1$

We shall exhibit a counterexample due to David J. Marcus which demonstrates that there exists a non-stable vector $\mathbf{X} = (X_1, X_2)$ in $\mathbb{R}^2$ such that all linear combinations of its components are α-stable random variables with $\alpha < 1$.

Fix $0 < \alpha \leq 1$ and $\rho > 0$ and let

$$\Psi(\theta_1, \theta_2) = \exp\{-r^\alpha + i\rho r \cos(3\phi)\} \tag{2.2.1}$$

where

$$\theta_1 = r\cos\phi \text{ and } \theta_2 = r\sin\phi.$$

For sufficiently small $\rho > 0$, $\Psi(\theta_1, \theta_2)$ is the characteristic function of a random vector $\mathbf{X} = (X_1, X_2)$ in $\mathbb{R}^2$ (Marcus 1983).

We now proceed in two steps.

Step 1. We first show that for any vector $\mathbf{b} = (b_1, b_2)$ in $\mathbb{R}^2$, the linear combination $(\mathbf{b}, \mathbf{X}) = b_1 X_1 + b_2 X_2$ is a α-stable.

The characteristic function of the random variable $(\mathbf{b}, \mathbf{X})$ is

$$E\exp\{i\theta(\mathbf{b}, \mathbf{X})\} = E\exp\{i(\theta\mathbf{b}, \mathbf{X})\} = \Psi(\theta b_1, \theta b_2)$$

where θ is real valued. Setting $b_1 = r_\mathbf{b}\cos\phi_\mathbf{b}$, $b_2 = r_\mathbf{b}\sin\phi_\mathbf{b}$, we can write

$$\begin{aligned} \theta b_1 &= |\theta| r_\mathbf{b} \cos(\phi_\mathbf{b} + \pi 1(\theta < 0)), \\ \theta b_2 &= |\theta| r_\mathbf{b} \sin(\phi_\mathbf{b} + \pi 1(\theta < 0)), \end{aligned}$$

so that by (2.2.1),

$$\begin{aligned} E\exp\{i\theta(\mathbf{b}, \mathbf{X})\} &= \exp\{-(|\theta| r_b)^\alpha + i\rho|\theta| r_\mathbf{b} \cos[3(\phi_\mathbf{b} + \pi 1(\theta < 0))]\} \\ &= \exp\{-|\theta|^\alpha r_\mathbf{b}^\alpha + i\rho r_\mathbf{b}|\theta| \operatorname{sign}\theta \cos(3\phi_\mathbf{b})\} \\ &= \exp\{-r_\mathbf{b}^\alpha|\theta|^\alpha + i\theta(\rho r_\mathbf{b}\cos 3\phi_\mathbf{b})\}. \end{aligned}$$

By Definition 1.1.6, this is the characteristic function of $S_\alpha(r_\mathbf{b}, 0, \rho r_\mathbf{b}\cos 3\phi_\mathbf{b})$, the α-stable distribution with scale parameter $r_\mathbf{b}$, skewness parameter equal to 0 and shift parameter $\rho r_\mathbf{b}\cos 3\phi_\mathbf{b}$.

Step 2. We now show that the vector $\mathbf{X}$ with characteristic function $\Psi(\theta_1, \theta_2)$ is *not* a stable random vector in $\mathbb{R}^2$ when $0 < \alpha < 1$.

Suppose, to the contrary, that that $\mathbf{X}$ is a stable random vector in $\mathbb{R}^2$. It is then stable with index α because, by Step 1, any linear combination of its components is α-stable. Applying (2.1.1) with $A = B = 1$ and $C = (A^\alpha + B^\alpha)^{1/\alpha} = 2^{1/\alpha}$, we obtain

$$\mathbf{X}^{(1)} + \mathbf{X}^{(2)} \stackrel{d}{=} 2^{1/\alpha}\mathbf{X} + \mathbf{D} \tag{2.2.2}$$

for some vector $\mathbf{D} = (D_1, D_2)$ in $\mathbb{R}^2$. (The vectors $\mathbf{X}^{(1)}$ and $\mathbf{X}^{(2)}$ are independent copies of $\mathbf{X}$.) Relation (2.2.2) implies that the characteristic function of $\mathbf{X}_1 + \mathbf{X}_2$ must equal the characteristic function of $2^{1/\alpha}\mathbf{X} + D$.

Let $\boldsymbol{\theta} = (\theta_1, \theta_2)$ and $\theta_1 = r\cos\phi$ and $\theta_2 = r\sin\phi$. The characteristic function of $\mathbf{X}^{(1)} + \mathbf{X}^{(2)}$ is

$$\begin{aligned} E\exp i(\boldsymbol{\theta}, \mathbf{X}^{(1)} + \mathbf{X}^{(2)}) &= E\exp i(\boldsymbol{\theta}, \mathbf{X}^{(1)}) E\exp i(\boldsymbol{\theta}, \mathbf{X}^{(2)}) \\ &= [E\exp i(\boldsymbol{\theta}, \mathbf{X})]^2 \\ &= [\Psi(\theta_1, \theta_2)]^2 \\ &= \exp\{-2r^\alpha + i2\rho r\cos(3\phi)\}, \end{aligned}$$

whereas the characteristic function of $2^{1/\alpha}\mathbf{X} + \mathbf{D}$ is

$$\begin{aligned} &E\exp i(\boldsymbol{\theta}, 2^{1/\alpha}\mathbf{X} + \mathbf{D}) \\ &\quad = \exp\{i(\boldsymbol{\theta}, \mathbf{D})\} E\exp i(2^{1/\alpha}\boldsymbol{\theta}, \mathbf{X}) \\ &\quad = \exp\{i(\boldsymbol{\theta}, \mathbf{D})\}\Psi(2^{1/\alpha}\theta_1, 2^{1/\alpha}\theta_2) \\ &\quad = \exp\{-2r^\alpha + i[\rho(2^{1/\alpha}r)\cos(3\phi) + \theta_1 D_1 + \theta_2 D_2]\}. \end{aligned}$$

Equating the two characteristic functions yields

$$2\rho r\cos(3\phi) = 2^{1/\alpha}\rho r\cos(3\phi) + \theta_1 D_1 + \theta_2 D_2,$$

for all real θ_1 and θ_2, i.e.,

$$(2 - 2^{1/\alpha})\rho\cos 3\phi = D_1\cos\phi + D_2\sin\phi$$

for all $\phi \in [0, 2\pi)$. Setting $\phi = \pi/2$ yields $D_2 = 0$. Therefore,

$$(2 - 2^{1/\alpha})\rho\cos 3\phi = D_1\cos\phi, \tag{2.2.3}$$

for all $\phi \in [0, 2\pi)$. Since this is clearly impossible, we conclude that the vector $\mathbf{X} = (X_1, X_2)$ does not have a stable distribution.

Remarks

1. The function $\Psi(\theta_1, \theta_2)$ given in (2.2.1) is a characteristic function even when $\alpha = 1$ (provided that the parameter $\rho > 0$ is small enough) but, by Theorem 2.1.5, $\mathbf{X}$ is a 1-stable vector in this case.

2. The family of infinitely divisible distributions includes the stable distributions. When $\alpha < 1$, the vector $\mathbf{X}$ with characteristic function $\Psi(\theta_1, \theta_2)$ is not even infinitely divisible. This follows from the following theorem of Giné and Hahn (1983).

   Theorem 2.2.1 *Let $\mathbf{X}$ be a random vector in $\mathbb{R}^d$ such that all linear combinations of its components are stable. If $\mathbf{X}$ is also infinitely divisible, then $\mathbf{X}$ is stable.*

2.3 Characteristic function of an α-stable random vector

Let $\mathbf{X} = (X_1, X_2, \ldots, X_d)$ be an α-stable random vector in $\mathbb{R}^d$ and let

$$\Phi_\alpha(\boldsymbol{\theta}) = \Phi_\alpha(\theta_1, \theta_2, \ldots, \theta_d) = E\exp\{i(\boldsymbol{\theta}, \mathbf{X})\} = E\exp\Big\{i\sum_{k=1}^{d} \theta_k X_k\Big\}$$

denote its characteristic function. ($\Phi_\alpha(\theta_1, \ldots, \theta_d)$ is also called the *joint characteristic function* of the random variables $X_1, X_2, \ldots, X_d$.)

The expression for $\Phi_\alpha(\boldsymbol{\theta})$ given in the following theorem involves an integration over $S_d = \{\mathbf{s}\colon \|\mathbf{s}\| = 1\}$, the unit sphere in $\mathbb{R}^d$. Observe that S_d is a $(d-1)$-dimensional surface. For example, S_1 is the two point set $\{-1, 1\}$ and S_2 is the unit circle.

Theorem 2.3.1 *Let $0 < \alpha < 2$. Then $\mathbf{X} = (X_1, X_2, \ldots, X_d)$ is an α stable random vector in $\mathbb{R}^d$ if and only if there exists a finite measure Γ on the unit sphere S_d of $\mathbb{R}^d$ and a vector $\boldsymbol{\mu}^0$ in $\mathbb{R}^d$ such that:*
(a) If $\alpha \neq 1$,

$$\Phi_\alpha(\boldsymbol{\theta}) = \exp\Big\{-\int_{S_d} |(\boldsymbol{\theta}, \mathbf{s})|^\alpha \Big(1 - i \operatorname{sign}((\boldsymbol{\theta}, \mathbf{s})) \tan\frac{\pi\alpha}{2}\Big)\Gamma(d\mathbf{s}) + i(\boldsymbol{\theta}, \boldsymbol{\mu}^0)\Big\}. \tag{2.3.1}$$

(b) If $\alpha = 1$,

$$\Phi_\alpha(\boldsymbol{\theta}) = \exp\Big\{-\int_{S_d} |(\boldsymbol{\theta}, \mathbf{s})| \Big(1 + i\frac{2}{\pi} \operatorname{sign}((\boldsymbol{\theta}, \mathbf{s})) \ln|(\boldsymbol{\theta}, \mathbf{s})|\Big)\Gamma(d\mathbf{s}) + i(\boldsymbol{\theta}, \boldsymbol{\mu}^0)\Big\}. \tag{2.3.2}$$

The pair $(\Gamma, \boldsymbol{\mu}^0)$ is unique.

Definition 2.3.2 The vector $\mathbf{X}$ in Theorem 2.3.1 is said to have spectral representation $(\Gamma, \boldsymbol{\mu}^0)$. The measure Γ is called the *spectral measure* of the α-stable random vector $\mathbf{X}$.

Warning. As we will see in Example 2.3.4, the components of the vector $\boldsymbol{\mu}^0$, in the case $\alpha = 1$, are *not* equal to the shift parameters of the components $X_1, \ldots, X_d$ of $\mathbf{X}$.

The proof of Theorem 2.3.1 is sketched in the following remarks; for details, see for example Kuelbs (1973).

Remarks

1. It follows easily from Definition 2.1.1 that a random vector $\mathbf{X}$ is α-stable if it has characteristic function $\Phi_\alpha(\boldsymbol{\theta})$.

2. Theorem 2.3.1 continues to hold if $\mathbb{R}^d$ is replaced by a separable Hilbert space H with scalar product $(\cdot, \cdot)$. For example, H can be ℓ^2, the space of real sequences $\mathbf{X} = \{x_1, x_2, \ldots\}$ with $\|\mathbf{x}\|^2 = \sum_{k=1}^\infty x_k^2 < \infty$ and scalar product $(\mathbf{x}, \mathbf{y}) = \sum_{k=1}^\infty x_k y_k$. Theorem 2.3.1 for $H = \ell^2$ can be used to show that stable processes always have an integral representation.

3. To prove Theorem 2.3.1 in H (or in $H = \mathbb{R}^d$), one shows that the characteristic function of an infinitely divisible law in H with no Gaussian component has the form

$$\exp\left\{\int_H \left[e^{i(\boldsymbol{\theta},\mathbf{x})} - 1 - i\frac{(\boldsymbol{\theta}, \mathbf{x})}{1 + \|\mathbf{x}\|^2}\right] L(d\mathbf{x}) + i(\boldsymbol{\theta}, \boldsymbol{\mu}^0)\right\}$$

   (see, e.g., Varadhan (1962) or Laha and Rohatgi (1979), p. 499.) Here, L is the (unique) Lévy measure corresponding to that infinitely divisible law, i.e., L is a σ-finite measure, finite outside each neighborhood of zero, that satisfies $L(\{0\}) = 0$ and $\int_{\|\mathbf{x}\|\le 1} \|\mathbf{x}\|^2 L(d\mathbf{x}) < \infty$. Express $\mathbf{x}$ in terms of its radial and angular components by setting $\mathbf{x} = r\mathbf{s}$, where $r = \|\mathbf{x}\|$ and where $\mathbf{s} = \mathbf{x}/\|\mathbf{x}\|$ belongs to the unit sphere S of H. Then, using α-stability, prove that there is a measure σ on S such that $L(d\mathbf{x}) = \frac{1}{r^{1+\alpha}} dr\, \sigma(d\mathbf{s})$. The characteristic function becomes

$$\exp\left\{\int_S \int_0^\infty \left[e^{ir(\mathbf{s},\mathbf{x})} - 1 - i\frac{r(\mathbf{s}, \mathbf{x})}{1 + r^2}\right] \frac{dr}{r^{1+\alpha}} \sigma(d\mathbf{s}) + i(\boldsymbol{\theta},\ \boldsymbol{\mu}^0)\right\}.$$

   Observe that the inner integral is the same as in the case $H = \mathbb{R}^1$ (see (1.1.7)). After evaluating the inner integral, one obtains the final result, formulated in (2.3.1) and (2.3.2) in the case $H = \mathbb{R}^d$.

Example 2.3.3 Suppose $d = 1$. Then S_1 consists of the two points $\{-1\}$ and $\{1\}$, and the spectral measure Γ is concentrated on them. Write $\Gamma(1) = \Gamma(\{1\})$ and $\Gamma(-1) = \Gamma(\{-1\})$. If $\alpha \neq 1$, $\Phi_\alpha(\theta)$ becomes

$$\begin{aligned} E\exp\{i\theta\mathbf{X}\} = \exp\Big\{ & -|\theta 1|^\alpha \Big(1 - i \text{ sign } (\theta 1) \tan \frac{\pi\alpha}{2}\Big)\Gamma(1) \\ & -|\theta(-1)|^\alpha \Big(1 - i \text{ sign } (\theta(-1)) \tan \frac{\pi\alpha}{2}\Big)\Gamma(-1) + i\mu^0\theta \Big\} \\ = \exp\Big\{ & -|\theta|^\alpha \Big[(\Gamma(1)+\Gamma(-1)) - i \text{ sign } (\theta)(\Gamma(1)-\Gamma(-1)) \tan \frac{\pi\alpha}{2}\Big] + i\mu^0\theta\Big\}. \end{aligned}$$

It is the characteristic function of an α-stable random variable $X \sim S_\alpha(\sigma, \beta, \mu)$ with parameters

$$\sigma = (\Gamma(1) + \Gamma(-1))^{1/\alpha}, \quad \beta = \frac{\Gamma(1) - \Gamma(-1)}{\Gamma(1) + \Gamma(-1)}, \quad \mu = \mu^0.$$

The skewness parameter β is zero if the spectral measure Γ is symmetric. An analogous result holds for $\alpha = 1$.

Example 2.3.4 Let $\mathbf{X}$ be an α-stable random vector with characteristic function given in Theorem 2.3.1. We know by Theorem 2.1.2 that any linear combination $Y_\mathbf{b} = \sum_{k=1}^d b_k X_k$ has an α-stable distribution $S_\alpha(\sigma_\mathbf{b}, \beta_\mathbf{b}, \mu_\mathbf{b})$. To determine the parameters $\sigma_\mathbf{b}, \beta_\mathbf{b}$ and $\mu_\mathbf{b}$ characterizing that distribution, let γ be any real number and set $\boldsymbol{\theta} = \gamma\mathbf{b}$ in (2.3.1) and (2.3.2). The resulting function of γ is the characteristic function of the random variable $Y_\mathbf{b}$, and one obtains

$$\begin{aligned} & E\exp\{i\gamma Y_\mathbf{b}\} = E\exp\{i(\gamma\mathbf{b}, \mathbf{X})\} \\ & = \begin{cases} \exp\{-\int_{S_d} |(\gamma\mathbf{b}, \mathbf{s})|^\alpha (1 - i\,[\text{sign }(\gamma\mathbf{b}, \mathbf{s})](\tan \frac{\pi\alpha}{2})\Gamma(d\mathbf{s}) + i(\gamma\mathbf{b}, \boldsymbol{\mu}^0))\}, \\ \\ \exp\{-\int_{S_d} |\gamma\mathbf{b}, \mathbf{s}|(1 + i\frac{2}{\pi} \text{ sign }(\gamma\mathbf{b}, \mathbf{s})(\ln|(\gamma\mathbf{b}, s)|)\Gamma(d\mathbf{s}) + i(\gamma\mathbf{b}, \boldsymbol{\mu}^0))\}, \end{cases} \\ & = \begin{cases} \exp\Big\{-|\gamma|^\alpha \Big[\int_{S_d} |(\mathbf{b}, \mathbf{s})|^\alpha \Gamma(d\mathbf{s}) \\ \quad - i\,(\text{ sign }\gamma)(\tan\frac{\pi\alpha}{2}) \int_{S_d} |(\mathbf{b}, \mathbf{s})|^\alpha \text{ sign }(\mathbf{b}, \mathbf{s})\Gamma(d\mathbf{s})\Big] + i\gamma(\mathbf{b}, \boldsymbol{\mu}^0)\Big\}, \\ \\ \exp\Big\{-|\gamma| \Big[\int_{S_d} |(\mathbf{b}, \mathbf{s})|\Gamma(d\mathbf{s}) \\ \quad + i\,\frac{2}{\pi}(\text{ sign }\gamma) \int_{S_d} (\mathbf{b}, \mathbf{s}) \ln|\gamma(\mathbf{b}, \mathbf{s})|\Gamma(d\mathbf{s})\Big] + i\gamma(\mathbf{b}, \boldsymbol{\mu}^0)\Big\}, \end{cases} \end{aligned}$$

for $\alpha \neq 1$ and $\alpha = 1$ respectively, so that

$$\sigma_\mathbf{b} = \left(\int_{S_d} |(\mathbf{b}, \mathbf{s})|^\alpha \Gamma(d\mathbf{s})\right)^{1/\alpha}, \tag{2.3.3}$$

$$\beta_\mathbf{b} = \frac{\int_{S_d} |(\mathbf{b}, \mathbf{s})|^\alpha \text{ sign }(\mathbf{b}, \mathbf{s})\Gamma(d\mathbf{s})}{\int_{S_d} |(\mathbf{b}, \mathbf{s})|^\alpha \Gamma(d\mathbf{s})}, \tag{2.3.4}$$

$$\mu_{\mathbf{b}} = \begin{cases} (\mathbf{b}, \boldsymbol{\mu}^0) & \text{if } \alpha \neq 1, \\ (\mathbf{b}, \boldsymbol{\mu}^0) - \frac{2}{\pi} \int_{S_d} (\mathbf{b}, \mathbf{s}) \ln |(\mathbf{b}, \mathbf{s})| \Gamma(d\mathbf{s}) & \text{if } \alpha = 1. \end{cases} \tag{2.3.5}$$

By choosing suitable **b**s one can easily obtain the (marginal) distributions of the components X_k, $k = 1, 2, \ldots, d$, of the vector $\mathbf{X}$. It is easy to see, for example, that the shift parameter of X_k is μ_k^0 when $\alpha \neq 1$ but is $\mu_k^0 - \frac{2}{\pi} \int_{S_d} s_k \ln |s_k| \Gamma(d\mathbf{s})$ when $\alpha = 1$.

Example 2.3.5 An α-stable random vector $\mathbf{X} = (X_1, X_2, \ldots, X_d)$ has independent components if and only if its spectral measure Γ is discrete and concentrated on the intersection of the axes with the sphere S_d. It is easy to verify this fact by using the uniqueness of the spectral measure Γ. Suppose, for example, that $d = 2$ and consider independent random variables $X_i \sim S_\alpha(\sigma_i, \beta_i, \mu_i)$, $i = 1, 2$. Choose Γ concentrated on the points $(1, 0), (-1, 0), (0, 1)$ and $(0, -1)$ by setting

$$\Gamma = a_1 \delta((1, 0)) + a_2 \delta((-1, 0)) + a_3 \delta((0, 1)) + a_4 \delta((0, -1)),$$

where a_1, a_2, a_3, a_4 are non-negative numbers and where $\delta(\mathbf{s}_0)$ assigns unit mass to the point $\mathbf{s}_0$. Because of the uniqueness of the spectral measure, it is enough to verify that this choice of Γ is adequate. For convenience, assume $\alpha \neq 1$. The characteristic function (2.3.1) with the preceding choice of Γ is

$$\begin{aligned} &\Phi_\alpha(\theta_1, \theta_2) \\ &\quad = \exp\Big\{ - \sum_{(s_1, s_2)} |\theta_1 s_1 + \theta_2 s_2|^\alpha \Big(1 - i \operatorname{sign}(\theta_1 s_1 + \theta_2 s_2) \tan \frac{\pi\alpha}{2}\Big) \Gamma(s_1, s_2) \\ &\quad\quad + i(\theta_1 \mu_1^0 + \theta_2 \mu_2^0) \Big\}, \end{aligned}$$

where $\sum_{(s_1, s_2)}$ denotes summation over the four points $(1, 0)$, $(-1, 0)$, $(0, 1)$, $(0, -1)$. Equating this characteristic function to

$$\begin{aligned} &E \exp\{i(\theta_1 X_1 + \theta_2 X_2)\} \\ &\quad = \exp\Big\{ -\sigma_1^\alpha |\theta_1|^\alpha \Big(1 - i\beta_1 \operatorname{sign}(\theta_1) \tan \frac{\pi\alpha}{2}\Big) + i\theta_1 \mu_1 \\ &\quad\quad - \sigma_2^\alpha |\theta_2|^\alpha \Big(1 - i\beta_2 \operatorname{sign}(\theta_2) \tan \frac{\pi\alpha}{2}\Big) + i\theta_2 \mu_2 \Big\} \end{aligned}$$

yields the system of equations

$$\begin{aligned} a_1 + a_2 &= \sigma_1^\alpha, & \beta_1 \sigma_1^\alpha &= a_1 - a_2, & \mu_1 &= \mu_1^0, \\ a_3 + a_4 &= \sigma_2^\alpha, & \beta_2 \sigma_2^\alpha &= a_3 - a_4, & \mu_2 &= \mu_2^0, \end{aligned}$$

whose solution

$$a_1 = \sigma_1^\alpha \frac{1 + \beta_1}{2}, \quad a_2 = \sigma_1^\alpha \frac{1 - \beta_1}{2}, \quad a_3 = \sigma_2^\alpha \frac{1 + \beta_2}{2}, \quad a_4 = \sigma_2^\alpha \frac{1 - \beta_2}{2}$$

provides the values of the weights in the expression of the spectral measure Γ. The same result holds in the case $\alpha = 1$.

Example 2.3.6 Let $Y_k \sim S_\alpha(\sigma_k, \beta_k, \mu_k)$, $k = 1, \dots, m$, be independent random variables and let $A = \{a_{jk}\}$, $j = 1, \dots, d$, $k = 1, \dots, m$, be a real matrix. Consider the random vector $\mathbf{X} = (X_1, X_2, \dots, X_d)$ whose components are linear combinations

$$X_j = \sum_{k=1}^{m} a_{jk} Y_k, \ j = 1, \dots, d,$$

of the Y_ks. The vector $\mathbf{X}$ is α-stable with characteristic function

$$\begin{aligned} E\exp\Big\{i\sum_{j=1}^{d}\theta_j X_j\Big\} &= E\exp\Big\{i\sum_{k=1}^{m}\Big(\sum_{j=1}^{d}\theta_j a_{jk}\Big)Y_k\Big\} \\ &= \prod_{k=1}^{m} E\exp\Big\{i\sum_{j=1}^{d}\theta_j a_{jk} Y_k\Big\}. \end{aligned} \tag{2.3.6}$$

If $\alpha \neq 1$, this is equal to

$$\exp\Big\{-\sum_{k=1}^{m}\sigma_k^\alpha\Big|\sum_{j=1}^{d}\theta_j a_{jk}\Big|^\alpha\Big(1 - i\beta_k \operatorname{sign}\Big(\sum_{j=1}^{d}\theta_j a_{jk}\Big)\tan\frac{\pi\alpha}{2}\Big)$$

$$+ i\sum_{k=1}^{m}\mu_k\sum_{j=1}^{d}\theta_j a_{jk}\Big\}$$

$$= \exp\Big\{-\int_{S_d}\Big|\sum_{j=1}^{d}\theta_j s_j\Big|^\alpha\Big(1 - i\operatorname{sign}\Big(\sum_{j=1}^{d}\theta_j s_j\Big)\tan\frac{\pi\alpha}{2}\Big)\Gamma(ds) + i\sum_{j=1}^{d}\theta_j\mu_j^0\Big\},$$

where

$$\Gamma = \sum_{k=1}^{m}\Bigg[\frac{1+\beta_k}{2}\sigma_k^\alpha\Big(\sum_{j=1}^{d}a_{jk}^2\Big)^{\alpha/2}\delta\Big(\Big(\frac{a_{1k}}{(\sum_{j=1}^{d}a_{jk}^2)^{1/2}}, \dots, \frac{a_{dk}}{(\sum_{j=1}^{d}a_{jk}^2)^{1/2}}\Big)\Big)$$

$$+ \frac{1-\beta_k}{2}\sigma_k^\alpha\Big(\sum_{j=1}^{d}a_{jk}^2\Big)^{\alpha/2}\delta\Big(\Big(\frac{-a_{1k}}{(\sum_{j=1}^{d}a_{jk}^2)^{1/2}}, \dots, \frac{-a_{dk}}{(\sum_{j=1}^{d}a_{jk}^2)^{1/2}}\Big)\Big)\Bigg]$$

and

$$\mu_j^0 = \sum_{k=1}^{m} a_{jk}\mu_k.$$

If $\alpha = 1$, Relation (2.3.6) becomes

$$\exp\Big\{-\sum_{k=1}^{m}\sigma_k\Big|\sum_{j=1}^{d}\theta_j a_{jk}\Big|\Big(1+i\Big(\frac{2}{\pi}\Big)\beta_k \text{ sign }\Big(\sum_{j=1}^{d}\theta_j a_{jk}\Big)\ln\Big|\sum_{j=1}^{d}\theta_j a_{jk}\Big|\Big)$$

$$+ i\sum_{k=1}^{m}\mu_k\sum_{j=1}^{d}\theta_j a_{jk}\Big\}$$

$$= \exp\Big\{-\int_{S_d}\Big|\sum_{j=1}^{d}\theta_j s_j\Big|(1+i(\frac{2}{\pi})\text{ sign }(\sum_{j=1}^{d}\theta_j s_j)\ln\Big|\sum_{j=1}^{d}\theta_j s_j\Big|\Gamma(ds))+i\sum_{j=1}^{d}\theta_j\mu_j^0\Big\},$$

with Γ as above and

$$\mu_j^0 = \sum_{k=1}^{m} a_{jk}\Big(\mu_k - \frac{1}{\pi}\sigma_k\beta_k \ln\sum_{n=1}^{d} a_{nk}^2\Big).$$

Note that the spectral measure Γ of the α-stable random vector $(X_1, \dots, X_d)$ is *discrete* and concentrated on m symmetric pairs of points $(\mathbf{s}^{(k)}, -\mathbf{s}^{(k)})$, $k = 1, \dots, m$, of S_d.

Conversely, an α-stable random vector $(X_1, \dots, X_d)$ with a discrete spectral measure Γ concentrated on points $(\mathbf{s}^{(k)}, -\mathbf{s}^{(k)})$ as above can always be represented as

$$(X_1, \dots, X_d) \stackrel{d}{=} \Big(\sum_{k=1}^{m} a_{1k}Y_k, \dots, \sum_{k=1}^{m} a_{dk}Y_k\Big),$$

where $Y_k \sim S_\alpha(\sigma_k, \beta_k, \mu_k)$, $k = 1, \dots, m$, are independent random variables, and $A = \{a_{jk}\}$ is a matrix of real numbers. We leave the details to Exercise 2.9.

Example 2.3.6 and Exercise 2.9 yield:

Proposition 2.3.7 *The spectral measure Γ_d of an α-stable vector $(X_1, \dots, X_d)$ is concentrated on a finite number of points on the unit circle S_d if and only if $(X_1, \dots, X_d)$ can be expressed as a linear transformation of independent α-stable random variables.*

Note that the representation of the characteristic function of an α-stable random vector in (2.3.1) and (2.3.2) involves integration over the unit sphere $S_d = \{\mathbf{x}\colon \|\mathbf{x}\| = 1\}$ of $\mathbb{R}^d$ where $\|\cdot\|$ denotes the *Euclidean norm*. This is the usual way of representing the joint characteristic function of an α-stable vector. However, there are many other norms in $\mathbb{R}^d$, and it is possible to use the unit sphere relative to any one of them to represent the joint characteristic function of an α-stable random vector.

Specifically, let S_d be the unit sphere relative to the Euclidean norm and let Γ be the corresponding spectral measure. Let $\|\cdot\|$ now be an *arbitrary* norm in $\mathbb{R}^d$ and let $S_d^{\|\cdot\|}$ denote the corresponding unit sphere. We want to determine $\Gamma_{\|\cdot\|}$, the finite Borel measure on $S_d^{\|\cdot\|}$ that corresponds to spectral measure Γ on S_d.

Proposition 2.3.8 *Let $\widetilde{\Gamma}_{\|\cdot\|}$ be a finite Borel measure on S_d equivalent to Γ with $\widetilde{\Gamma}_{\|\cdot\|}(d\mathbf{s}) = \|\mathbf{s}\|^\alpha\Gamma(d\mathbf{s})$ and let $T_{\|\cdot\|}\colon S_d \to S_d^{\|\cdot\|}$ be given by $T\mathbf{s} = \mathbf{s}/\|\mathbf{s}\|$. Define*

$$\Gamma_{\|\cdot\|} = \widetilde{\Gamma}_{\|\cdot\|} \circ T_{\|\cdot\|}^{-1}$$

and

$$\boldsymbol{\mu}^0_{\|\cdot\|} = \begin{cases} \boldsymbol{\mu}^{(0)} & \text{if } \alpha \neq 1, \\ \boldsymbol{\mu}^{(0)} + \widetilde{\boldsymbol{\mu}}_{\|\cdot\|} & \text{if } \alpha = 1, \end{cases} \tag{2.3.7}$$

where

$$(\widetilde{\mu}_{\|\cdot\|})_j = -\frac{2}{\pi}\int_{S_d} s_j \ln\|\mathbf{s}\|\Gamma(d\mathbf{s}),\ j = 1,\dots,d.$$

Then the joint characteristic function $\Phi_\alpha(\boldsymbol{\theta})$ of the α-stable random vector $\mathbf{X}$ in $\mathbb{R}^d$ is also given by (2.3.1) and (2.3.2) with $(S_d, \Gamma, \boldsymbol{\mu}^{(0)})$ replaced by $(S_d^{\|\cdot\|}, \Gamma_{\|\cdot\|}, \boldsymbol{\mu}^0_{\|\cdot\|})$ and $(\boldsymbol{\theta}, \mathbf{s}) = \sum_{j=1}^d \theta_j s_j$.

PROOF: A change of variables in (2.3.1) and (2.3.2) shows that they hold with $(S_d, \Gamma, \boldsymbol{\mu}^0)$ replaced by $(S_d^{\|\cdot\|}, \Gamma_{\|\cdot\|}, \boldsymbol{\mu}^0_{\|\cdot\|})$. The uniqueness of $(\Gamma, \boldsymbol{\mu}^0)$ implies the uniqueness of $(\Gamma_{\|\cdot\|}, \boldsymbol{\mu}^0_{\|\cdot\|})$. ∎

We will usually work with the Euclidean norm in $\mathbb{R}^d$ but we shall encounter other norms as well. The L^∞ norm $\|\mathbf{x}\| = \max_{i=1,\dots,d}|x_i|$, for example, is a particularly natural one when studying the maxima of stable vectors. The corresponding unit sphere is the boundary of the cube $[-1, 1]^d$.

Suppose that the spectral measure and the shift vector of an α-stable distribution on $\mathbb{R}^d$ are given. The following proposition shows how to derive formally the spectral measure and the shift vector of any marginal distribution.

Proposition 2.3.9 *If $(X_1, X_2, \dots, X_d)$ is an α-stable random vector with spectral measure Γ_d and shift $\boldsymbol{\mu}^0_d = (\mu^0_1, \mu^0_2, \dots, \mu^0_d)$, then $(X_1, X_2, \dots, X_n)$, $n \le d$, is an α-stable random vector with spectral measure Γ_n and shift $\boldsymbol{\mu}^0_n$. The spectral measure Γ_n can be expressed as*

$$\Gamma_n = h(\widehat{\Gamma}_d) \equiv \widehat{\Gamma}_d \circ h^{-1},$$

where

$$h : S'_d \equiv \{(s_1, \dots, s_n,\ s_{n+1}, \dots, s_d) \in S_d : s_1^2 + \dots + s_n^2 > 0\} \to S_n,$$

$$h(s_1,\dots,s_n,\,s_{n+1},\dots,s_d) = \left(\frac{s_1}{(\sum_{j=1}^n s_j^2)^{1/2}},\dots,\frac{s_n}{(\sum_{j=1}^n s_j^2)^{1/2}}\right)$$

and

$$\widehat{\Gamma}_d(d\mathbf{s}) = \Big(\sum_{j=1}^n s_j^2\Big)^{\alpha/2}\Gamma_d(d\mathbf{s}), \quad \mathbf{s}\in S_d'.$$

The jth component, $j=1,\dots,n$, of the shift $\boldsymbol{\mu}_n^0$ equals

$$\begin{cases} \mu_j^0 & \text{if } \alpha\neq 1,\\ \mu_j^0 - \frac{1}{\pi}\int_{S_d'} s_j \ln(\sum_{k=1}^n s_k^2)\Gamma_d(d\mathbf{s}) & \text{if } \alpha = 1.\end{cases}$$

PROOF: We prove the result in the more complicated case $\alpha = 1$. The proof in the case $\alpha\neq 1$ is similar. The joint characteristic function of $X_1, X_2, \dots, X_n$ satisfies

$$-\ln E\exp\Big\{i\sum_{j=1}^n \theta_j X_j\Big\} = -\ln E\exp\Big\{i\Big(\sum_{j=1}^n \theta_j X_j + \sum_{j=n+1}^d 0\cdot X_j\Big)\Big\}$$

$$= \int_{S_d'} |\sum_{j=1}^n \theta_j s_j|\Big(1+i\frac{2}{\pi}\,\text{sign}\,\Big(\sum_{j=1}^n\theta_j s_j\Big)\ln\Big|\sum_{j=1}^n\theta_j s_j\Big|\Big)\Gamma_d(d\mathbf{s}) - i\sum_{j=1}^n\theta_j\mu_j^0$$

$$= \int_{S_d'}\Big|\sum_{j=1}^n\theta_j\frac{s_j}{(\sum_{k=1}^n s_k^2)^{1/2}}\Big|\left[1+i\frac{2}{\pi}\,\text{sign}\,\Big(\sum_{j=1}^n\theta_j\frac{s_j}{(\sum_{k=1}^n s_k^2)^{1/2}}\Big)\right.$$

$$\left.\cdot\Big(\ln\Big|\sum_{j=1}^n\theta_j\frac{s_j}{(\sum_{k=1}^n s_k^2)^{1/2}}\Big| + \ln\Big(\sum_{k=1}^n s_k^2\Big)^{1/2}\Big)\right]\Big(\sum_{k=1}^n s_k^2\Big)^{1/2}\Gamma_d(d\mathbf{s}) - i\sum_{j=1}^n\theta_j\mu_j^0$$

$$= \int_{S_n} |\sum_{j=1}^n\theta_j t_j|\Big(1+i\frac{2}{\pi}\,\text{sign}\,\Big(\sum_{j=1}^n\theta_j t_j\Big)\Big)\ln|\sum_{j=1}^n\theta_j t_j|)\Gamma_n(d\mathbf{t})$$

$$-i\sum_{j=1}^n\theta_j\Big(\mu_j^0 - \frac{1}{\pi}\int_{S_d'} s_j\ln\Big(\sum_{k=1}^n s_k^2\Big)\Gamma_d(d\mathbf{s})\Big). \quad\blacksquare$$

2.4 Strictly stable and symmetric stable random vectors

The expressions (2.3.1), (2.3.2) for the characteristic function $\Phi_\alpha(\boldsymbol{\theta})$ of an α-stable random vector involve the spectral measure Γ and the shift vector $\boldsymbol{\mu}^0$. Let us find conditions that Γ and $\boldsymbol{\mu}^0$ must satisfy for $\Phi_\alpha(\boldsymbol{\theta})$ to be the characteristic function of a *strictly* (respectively, *symmetric*) α-stable random vector.

When is the characteristic function $\Phi_\alpha(\boldsymbol{\theta})$ that of a *strictly* α-stable random vector $\mathbf{X}$ in $\mathbb{R}^d$? We know by Theorem 2.1.5 that it is necessary and sufficient that $Y_{\mathbf{b}} = \sum_{k=1}^d b_k X_k$ be a strictly α-stable random variable for any $\mathbf{b} \in \mathbb{R}^d$. When $\alpha \neq 1$, this amounts to requiring $0 = \mu_{\mathbf{b}} = (\mathbf{b}, \boldsymbol{\mu}^0)$ for any $\mathbf{b}$, that is, $\boldsymbol{\mu}^0 = 0$. When $\alpha = 1$, the condition for strict stability is $\beta_{\mathbf{b}} = 0$ for any $\mathbf{b}$, i.e.,

$$0 = \int_{S_d} |(\mathbf{b}, \mathbf{s})| \operatorname{sign}((\mathbf{b}, \mathbf{s}))\Gamma(d\mathbf{s}) = \int_{S_d} (\mathbf{b}, \mathbf{s})\Gamma(d\mathbf{s})$$

for any $\mathbf{b}$ by (2.3.4), i.e., $\int_{S_d} s_k \Gamma(d\mathbf{s}) = 0$ for $k = 1, 2, \ldots, d$. There are no conditions on $\boldsymbol{\mu}^0$. Therefore,

Theorem 2.4.1 $\mathbf{X}$ *is a strictly α-stable vector in $\mathbb{R}^d$ with $0 < \alpha \leq 2$ if and only if*

(a) $\alpha \neq 1$:

$$\boldsymbol{\mu}^0 = 0,$$

(b) $\alpha = 1$:

$$\int_{S_d} s_k \Gamma(d\mathbf{s}) = 0 \quad \text{for } k = 1, 2, \ldots, d. \tag{2.4.1}$$

As a result, we obtain

Corollary 2.4.2 *Suppose that $\mathbf{X} = (X_1, X_2, \ldots, X_d)$ is an α-stable vector in $\mathbb{R}^d$ with $0 < \alpha \leq 2$. Then $\mathbf{X}$ is strictly stable if and only if all its components X_k, $k = 1, \ldots, d$, are strictly stable random variables.*

When is the characteristic function $\Phi_\alpha(\boldsymbol{\theta})$ that of a *symmetric* α-stable random vector $\mathbf{X}$ in $\mathbb{R}^d$? A necessary and sufficient condition is that $\boldsymbol{\mu}^0 = 0$ and Γ be a symmetric measure on S_d (i.e., $\Gamma(A) = \Gamma(-A)$ for any Borel set A of S_d). In fact,

Theorem 2.4.3 $\mathbf{X}$ *is a symmetric α-stable vector in $\mathbb{R}^d$ with $0 < \alpha < 2$ if and only if there exists a unique symmetric finite measure Γ on the unit sphere S_d such that*

$$E \exp\{i(\boldsymbol{\theta}, \mathbf{X})\} = \exp\Big\{-\int_{S_d} |(\boldsymbol{\theta}, \mathbf{s})|^\alpha \Gamma(d\mathbf{s})\Big\}. \tag{2.4.2}$$

Γ is the spectral measure of the symmetric α-stable random vector $\mathbf{X}$.

PROOF: A necessary and sufficient condition for $\mathbf{X}$ to be symmetric is that its characteristic function be real, i.e., (2.4.2) holds. Now define $\widetilde{\Gamma}$ by $\widetilde{\Gamma}(A) = \frac{1}{2}(\Gamma(A) + \Gamma(-A))$ for all Borel sets A of S_d and observe that $\widetilde{\Gamma}$ is a symmetric measure. Since (2.4.2) also holds if we replace Γ by $\widetilde{\Gamma}$, we must have $\Gamma = \widetilde{\Gamma}$ by unicity (Theorem 2.3.1) and hence Γ is symmetric. ■

A symmetric α-stable distribution in $\mathbb{R}^d$ is denoted $S\alpha S$. We also say that the vector $\mathbf{X} = (X_1, X_2, \ldots, X_d)$ is $S\alpha S$ in $\mathbb{R}^d$ or that the random variables $X_1, X_2, \ldots, X_d$ are *jointly* $S\alpha S$. Figure 2.1 illustrates a $S\alpha S$ distribution with $\alpha = 1.5$ and spectral measure Γ with mass 0.2 at 0° and 180° and mass 0.8 at 90° and 270°. It displays the probability density function and the integral of that density from the point $(-2, -2)$.

Example 2.4.4 If $d = 1$, one has $S_1 = \{-1, 1\}$, $\Gamma(\{1\}) = \Gamma(\{-1\})$ and hence an α-stable $S\alpha S$ random variable X has distribution $S_\alpha(\sigma, 0, 0)$ with

$$\sigma = \left(\int_{S_1} |s|^\alpha \Gamma(ds)\right)^{1/\alpha} = (2\Gamma(\{1\}))^{1/\alpha}.$$

Remarks

1. Can a strictly 1-stable random vector $\mathbf{X}$ be made symmetric by shifting? To answer this question, we must consider two cases. Suppose, firstly, that $\mathbf{X}$ has a symmetric spectral measure Γ. Then shifting $\mathbf{X}$ by $\boldsymbol{\mu}$ transforms $\mathbf{X}$ into a symmetric random vector $\mathbf{X} - \boldsymbol{\mu}$. Suppose, now, that the spectral measure Γ of $\mathbf{X}$ is not symmetric. This can never happen when $d = 1$ because the condition for strict stability $0 = \int_{S_1} s\,\Gamma(ds) = \Gamma(\{1\}) - \Gamma(\{-1\})$ requires Γ to be symmetric. When $d > 1$, the conditions for strict stability are $\int_{S_d} s_k \Gamma(ds) = 0$ for $k = 1, 2, \ldots, d$. They can be satisfied by a non-symmetric spectral measure Γ, for example, by

$$m = \delta((1,0)) + \delta((0,1)) + \sqrt{2}\delta\left(\left(-\frac{1}{\sqrt{2}}, -\frac{1}{\sqrt{2}}\right)\right)$$

if $d = 2$. If $\mathbf{X}$ is a strictly 1-stable random vector with a non-symmetric Γ, then $\mathbf{X} - \boldsymbol{\mu}$ will not be symmetric, because shifting $\mathbf{X}$ has no effect on Γ: the vector $\mathbf{X} - \boldsymbol{\mu}$ and $\mathbf{X}$ have the same non-symmetric spectral measure Γ. Hence

Corollary 2.4.5 *Not every strictly* 1*-stable random vector in* $\mathbb{R}^d$ *with* $d > 1$ *can be made symmetric by shifting.*

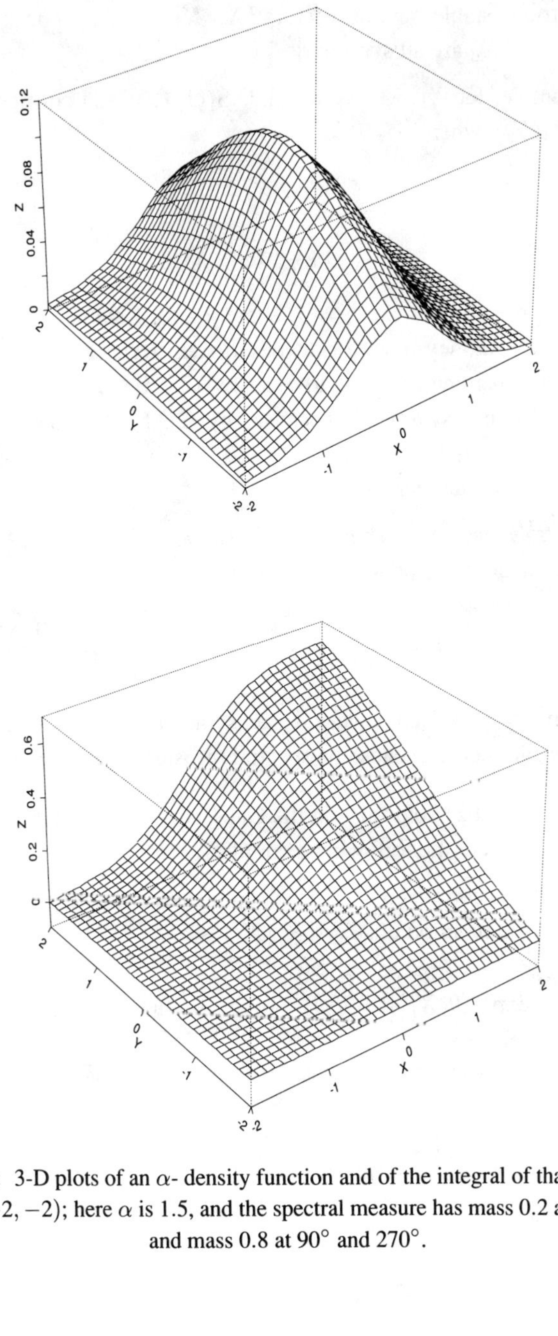

Figure 2.1: 3-D plots of an α- density function and of the integral of that density from the point $(-2, -2)$; here α is 1.5, and the spectral measure has mass 0.2 at 0° and 180° and mass 0.8 at 90° and 270°.

2. Unlike Corollary 2.4.2, the symmetry of an α-stable random vector cannot be regarded as a componentwise property. There are non-symmetric α-stable random vectors $(X_1, X_2, \ldots, X_d)$ whose components $X_1, X_2, \ldots, X_d$ are all symmetric.

   For example, let X_1, X_2, X_3 be i.i.d. $S_1(1, 1, 0)$ and consider the vector $\mathbf{Y} = (Y_1, Y_2)$, where

$$\begin{cases} Y_1 = X_1 - X_2, \\ Y_2 = X_2 - X_3. \end{cases}$$

   The random variables $Y_j,\ j = 1, 2$, are jointly 1-stable and they are symmetric because they have a skewness parameter equal to zero by (2.3.4) and a shift parameter equal to zero by (2.3.5). However, the random vector $\mathbf{Y}$ is not symmetric because, by (2.3.5), the linear combination $\theta_1 Y_1 + \theta_2 Y_2 = \theta_1 X_1 + (\theta_2 - \theta_1) X_2 - \theta_2 X_3$ has a non-zero shift parameter if θ_1 and θ_2 are non-zero and unequal.

   To illustrate the point when $\alpha \neq 1$, suppose that $\mathbf{X} = (X_1, X_2)$ is an α-stable random vector with spectral measure

$$\Gamma = \delta((1,0)) + \delta((0,1)) + 2^{\alpha/2}\delta\Big(\Big(-\frac{1}{\sqrt{2}}, -\frac{1}{\sqrt{2}}\Big)\Big),$$

   and shift $\boldsymbol{\mu}^0 = \mathbf{0}$. Then X_1 and X_2 are symmetric by (2.3.4) and (2.3.5) but $\mathbf{X}$ is not symmetric because its spectral measure is not symmetric.

3. Expression (2.4.2) holds also in the Gaussian case $\alpha = 2$, but then Γ is not unique. Consider the following example:

   Let X_1, X_2 be correlated $N(0, 1)$ random variables with $\rho = EX_1X_2$. The characteristic function of $\mathbf{X} = (X_1, X_2)$ is

$$E\exp\{i(\theta_1 X_1 + \theta_2 X_2)\} = \exp\Big\{-\frac{1}{2}(\theta_1^2 + 2\rho\theta_1\theta_2 + \theta_2^2)\Big\}.$$

   There are many symmetric measures Γ on S_2 such that

$$\frac{1}{2}(\theta_1^2 + 2\rho\theta_1\theta_2 + \theta_2^2) = \int_{S_2} (s_1\theta_1 + s_2\theta_2)^2 \Gamma(ds).$$

   In particular, any Γ such that

$$\Gamma = \sum_k \frac{1}{4} a_k \{\delta((b_k, (1 - b_k^2)^{1/2})) + \delta((-b_k, -(1 - b_k^2)^{1/2}))\},$$

with $a_k > 0, 0 \le b_k \le 1$ and

$$\sum_k a_k b_k^2 = 1, \ \sum_k a_k b_k (1-b_k^2)^{1/2} = \rho, \ \sum_k a_k (1-b_k^2) = 1,$$

satisfy this identity, for instance,

$$\Gamma = \frac{1}{4}\Big\{(1-\rho)\Big[\delta((1,0)) + \delta((-1,0)) + \delta((0,1)) + \delta((0,-1))\Big] + 2\rho\Big[\delta\Big(\Big(\frac{1}{\sqrt{2}}, \frac{1}{\sqrt{2}}\Big)\Big) + \delta\Big(\Big(-\frac{1}{\sqrt{2}}, -\frac{1}{\sqrt{2}}\Big)\Big)\Big]\Big\},$$

or[1]

$$\Gamma = \frac{1}{4}\Big\{(1-\rho^2)^{1/2}\Big[\delta((0,1)) + \delta((0,-1))\Big] + (1+\rho^2)\Big[\delta\Big(\Big(\frac{1}{(1+\rho^2)^{1/2}}, \frac{\rho}{(1+\rho^2)^{1/2}}\Big)\Big) + \delta\Big(\Big(\frac{1}{(1+\rho^2)^{1/2}}, \frac{-\rho}{(1+\rho^2)^{1/2}}\Big)\Big)\Big]\Big\}.$$

The spectral measure is not a useful concept in the Gaussian case $\alpha = 2$, because it is not unique.[2]

4. Let $X_1, X_2, \ldots, X_d$ be jointly $S\alpha S$ with spectral measure Γ_d. Then $X_1, X_2, \ldots, X_n$, $n \le d$, are jointly $S\alpha S$ with spectral measure Γ_n defined in Proposition 2.3.9.

5. Unlike the Gaussian case $\alpha = 2$, the distribution of a $S\alpha S$ random vector on $\mathbb{R}^d$, $d \ge 3$, is not determined by its two-dimensional marginals. (See Exercise 2.13.)

2.5 Sub-Gaussian random vectors

We have seen in Section 1.3 that the random variable $X = A^{1/2}G$ is $S\alpha S$, $\alpha < 2$, if $G \sim N(0, \sigma^2)$ and A is an $\alpha/2$-stable random variable totally skewed to the right and independent of G. This result extends to random vectors $\mathbf{X}$ as follows. Choose

$$A \sim S_{\alpha/2}\Big(\Big(\cos\frac{\pi\alpha}{4}\Big)^{2/\alpha}, 1, 0\Big) \tag{2.5.1}$$

[1] This second Γ can be readily obtained from the characteristic function of (X_1, X_2) by representing X_2 as $\rho X_1 + (1-\rho^2)^{1/2} Y$ where Y is a $N(0,1)$ random variable independent of X_1.

[2] The spectral measure Γ should not be confused with a measure F, also sometimes called spectral measure, which appears in the context of a stationary Gaussian process $\{X_t\}$. The measure F is defined by $EX_0X_t = \int_{-\infty}^{\infty} \exp\{i\lambda t\} dF(\lambda)$.

with $\alpha < 2$, so that its Laplace transform is

$$Ee^{-\gamma A} = e^{-\gamma^{\alpha/2}}, \ \gamma > 0, \tag{2.5.2}$$

by Proposition 1.2.12. Let

$$\mathbf{G} = (G_1, G_2, \ldots, G_d)$$

be a zero mean Gaussian vector in $\mathbb{R}^d$ independent of A. Then the random vector

$$\mathbf{X} = (A^{1/2}G_1, A^{1/2}G_2, \ldots, A^{1/2}G_d) \tag{2.5.3}$$

has a $S\alpha S$ distribution in $\mathbb{R}^d$ because, for any real numbers $b_1, b_2, \ldots, b_d$, the linear combination $\sum_{k=1}^d b_k A^{1/2} G_k = A^{1/2} \sum_{k=1}^d b_k G_k$ is a $S\alpha S$ random variable (Proposition 1.3.1) and hence, by Theorem 2.1.5, $\mathbf{X}$ is $S\alpha S$.

Definition 2.5.1 Any vector $\mathbf{X}$ distributed as in (2.5.3) is called a *sub-Gaussian $S\alpha S$ random vector in $\mathbb{R}^d$* with underlying Gaussian vector $\mathbf{G}$.[3] It is also said to be *subordinated* to $\mathbf{G}$.

The characteristic function of a sub-Gaussian random vector has the following special structure:

Proposition 2.5.2 *The sub-Gaussian symmetric α-stable random vector $\mathbf{X}$ defined in (2.5.3) has characteristic function*

$$E\exp\Big\{i\sum_{k=1}^d \theta_k X_k\Big\} = \exp\Big\{-\Big|\frac{1}{2}\sum_{i=1}^d\sum_{j=1}^d \theta_i\theta_j R_{ij}\Big|^{\alpha/2}\Big\}, \tag{2.5.4}$$

where $R_{ij} = EG_iG_j$, $i,j = 1, \ldots, d$, are the covariances of the underlying Gaussian random vector $(G_1, G_2, \ldots, G_d)$.

PROOF: By conditioning on A we obtain

$$\begin{aligned} E\exp\Big\{i\sum_{k=1}^d \theta_k X_k\Big\} &= EE\Big[\exp\Big\{iA^{1/2}\sum_{k=1}^d \theta_k G_k\Big\}\Big|A\Big] \\ &= E\exp\Big\{-A\frac{1}{2}\sum_{i=1}^d\sum_{j=1}^d R_{ij}\theta_i\theta_j\Big\} \\ &= \exp\Big\{-\Big|\frac{1}{2}\sum_{i=1}^d\sum_{j=1}^d R_{ij}\theta_i\theta_j\Big|^{\alpha/2}\Big\}, \end{aligned}$$

since $E\exp\{-\gamma A\} = \exp\{-\gamma^{\alpha/2}\}$, $\gamma > 0$. ∎

[3] Some authors use the term *sub-Gaussian* to refer to a class of stochastic processes with exponential moments. We do not use the term *sub-Gaussian* in that context.

Example 2.5.3 The characteristic function of a *multivariate Cauchy distribution* in $\mathbb{R}^d$ is

$$\phi(\boldsymbol{\theta}) = \exp\{-(\boldsymbol{\theta}^T \Sigma \boldsymbol{\theta})^{1/2} + i(\boldsymbol{\theta}, \boldsymbol{\mu}^0)\},$$

where T denotes a transpose and Σ is a $d \times d$ positive-definite symmetric matrix. The multivariate Cauchy distribution is thus a *shifted* $S1S$ *sub-Gaussian distribution.* Its density function is

$$f(\mathbf{x}) = \frac{c|\Sigma|^{-1/2}}{[1 + (\mathbf{x} - \boldsymbol{\mu}^0)\Sigma^{-1}(\mathbf{x} - \boldsymbol{\mu}^0)]^{(d+1)/2}}, \ \mathbf{x} \in \mathbb{R}^d,$$

where $|\Sigma|$ denotes the determinant of Σ and

$$c = \frac{1}{\pi^{(d+1)/2}} \Gamma\left(\frac{d+1}{2}\right).$$

(See Press (1972).)

Corollary 2.5.4 *Let* $\mathbf{X}$ *be sub-Gaussian* $S\alpha S$ *with underlying Gaussian vector* $\mathbf{G}$. *Then there is a one-to-one correspondence between the probability distribution of* $\mathbf{G}$ *and that of* $\mathbf{X}$.

PROOF: Let $\mathbf{G}'$ be a mean zero Gaussian vector on $\mathbb{R}^d$ with covariances R'_{ij}, $i, j = 1, \ldots, d$. We must show that if $\mathbf{X}' = A^{1/2}\mathbf{G}'$ has the same distribution as $\mathbf{X} = A^{1/2}\mathbf{G}$, then $\mathbf{G}'$ has the same distribution as $\mathbf{G}$. By Proposition 2.5.2, $\mathbf{X}' \stackrel{d}{=} \mathbf{X}$ implies

$$\sum_{i=1}^{d} \sum_{j=1}^{d} \theta_i \theta_j R'_{ij} = \sum_{i=1}^{d} \sum_{j=1}^{d} \theta_i \theta_j R_{ij}$$

for any real $\theta_1, \theta_2, \ldots, \theta_d$. This identity between two polynomials in the θ_is implies equality of the coefficients, i.e.,

$$R'_{ij} = R_{ij}, \ i, j = 1, \ldots, d.$$

Hence $\mathbf{G}' \stackrel{d}{=} \mathbf{G}$. ■

We noted in Theorem 2.4.3 that a $S\alpha S$ random vector has a symmetric spectral measure Γ. This merely means that Γ assigns equal weights to antipodal sets on the sphere S_d. We now want to characterize the $S\alpha S$ random vectors in $\mathbb{R}^d$ that have a *uniform* spectral measure Γ, i.e., for which there is a constant $C \geq 0$ such that $\Gamma(B) = C|B|$ for any Borel set B of S_d.

Proposition 2.5.5 *Let* $\mathbf{X}$ *be a* $S\alpha S$, $\alpha < 2$, *random vector in* $\mathbb{R}^d$. *Then the following three statements are equivalent:*

(a) $\mathbf{X}$ *is sub-Gaussian with an underlying Gaussian vector having i.i.d.* $N(0,\sigma^2)$ *components.*

(b) *The characteristic function of* $\mathbf{X}$ *is of the form*

$$E\exp\Big\{i\sum_{k=1}^{d}\theta_k X_k\Big\} = \exp\{-2^{-\alpha/2}\sigma^\alpha|\boldsymbol{\theta}|^\alpha\},$$

i.e. it depends only on the magnitude of the vector $\boldsymbol{\theta} = (\theta_1, \ldots, \theta_d)$.

(c) *The spectral measure of* $\mathbf{X}$ *is uniform.*

PROOF: (a) is equivalent to (b) because if $\mathbf{X} = (X_1, X_2, \ldots, X_d)$ is a sub-Gaussian $S\alpha S$ random vector whose underlying vector has components $G_1, G_2, \ldots, G_d$ i.i.d. $N(0,\sigma^2)$, then by Proposition 2.5.2, $\mathbf{X}$ has characteristic function

$$E\exp\Big\{i\sum_{k=1}^{d}\theta_k X_k\Big\} = \exp\Big\{-\Big(\frac{\sigma^2}{2}\sum_{k=1}^{d}\theta_k^2\Big)^{\alpha/2}\Big\}. \qquad (2.5.5)$$

To see that (c) $\Rightarrow$ (b), suppose that $\mathbf{X}$ is $S\alpha S$ with uniform spectral measure Γ on S_d. Then $\mathbf{X}$ has characteristic function

$$E\exp\Big\{i\sum_{k=1}^{d}\theta_k X_k\Big\} = \exp\{-f(\theta_1, \theta_2, \ldots, \theta_d)\} \qquad (2.5.6)$$

with

$$f(\theta_1, \theta_2, \ldots, \theta_d) = \int_{S_d} \Big|\sum_{k=1}^{d}\theta_k s_k\Big|^\alpha \Gamma(d\mathbf{s}).$$

Since Γ is uniform, f depends only on the Euclidean norm $|\boldsymbol{\theta}|$ of the vector $\boldsymbol{\theta} = (\theta_1, \theta_2, \ldots, \theta_d) \in \mathbb{R}^d$. Moreover,

$$f(c\theta_1, c\theta_2, \ldots, c\theta_d) = c^\alpha f(\theta_1, \theta_2, \ldots, \theta_d)$$

for any $c > 0$. This means that

$$f(\theta_1, \theta_2, \ldots, \theta_d) = C\Big(\sum_{k=1}^{d}\theta_k^2\Big)^{\alpha/2}$$

for some $C \geq 0$. Choosing σ such that $C = 2^{-\alpha/2}\sigma^\alpha$, we may equate the characteristic functions (2.5.5) and (2.5.6). This proves (c)$\Rightarrow$(b). The converse follows from the uniqueness of the spectral measure Γ. ∎

Example 2.5.6 Suppose that $\mathbf{X} = (X_1, X_2)$ has characteristic function

$$\phi(\theta_1, \theta_2) = e^{-\sigma\sqrt{\theta_1^2+\theta_2^2}}, \quad -\infty < \theta_1, \theta_2 < \infty.$$

Then $\mathbf{X}$ is a sub-Gaussian $S1S$ random vector with a uniform spectral measure. The density function of $\mathbf{X}$ is

$$f(x_1, x_2) = \frac{\sigma}{2\pi(x_1^2 + x_2^2 + \sigma^2)^{3/2}}, \quad -\infty < x_1, x_2 < \infty.$$

This is the *isotropic Cauchy density* in $\mathbb{R}^2$.

The components $G_1, G_2, \ldots, G_d$ of the underlying Gausssian vector $\mathbf{G}$ are in general not i.i.d. but, being Gaussian, they can always be expressed as a linear combination of i.i.d. $N(0, 1)$ random variables. Therefore,

Proposition 2.5.7 *Let* $\mathbf{Z}$ *be a* $S\alpha S$ *sub-Gaussian random vector in* R^d *with underlying Gaussian vector having i.i.d.* $N(0, 1)$ *components. Then for any* $S\alpha S$ *sub-Gaussian random vector* $\mathbf{X}$ *in* $\mathbb{R}^d$*, there is a lower-triangular* $d \times d$ *matrix* Λ *such that*

$$\mathbf{X} \stackrel{d}{=} \Lambda \mathbf{Z}.$$

The matrix Λ *is of full rank if the components of* $\mathbf{X}$ *are linearly independent.*

The proof is straightforward and is left as an exercise (Exercise 2.16).

What is the general form of the spectral measure Γ of a sub-Gaussian $S\alpha S$ random vector $\mathbf{X}$? We noted in Proposition 2.5.5 that it is uniform if the underlying Gaussian vector $\mathbf{G}$ has i.i.d. components $G_1, G_2, \ldots, G_d$. This fact will enable us to show that the spectral measure Γ of a general sub-Gaussian vector is a transform of the uniform measure on S_d.

Let Γ_0 be the uniform measure on S_d satisfying

$$\int_{S_d} |\sum_{j=1}^{d} \theta_j s_j|^\alpha \Gamma_0(ds) = \left(\frac{1}{2}\sum_{j=1}^{d} \theta_j^2\right)^{\alpha/2}, \tag{2.5.7}$$

and let $\mathbf{X} = A^{1/2}\mathbf{G}$ be a sub-Gaussian $S\alpha S$ random vector in $\mathbb{R}^d$ with spectral measure Γ. Let $\Sigma = ((R_{ij},\ i, j = 1, \ldots, d))$ be the covariance matrix of the underlying Gaussian vector $\mathbf{G}$. Since Σ is non-negative definite, there is a (lower-triangular) $d \times d$ matrix Λ such that

$$\Sigma = \Lambda\,\Lambda^T, \tag{2.5.8}$$

where T denotes a transpose. Our goal is to find a transformation

$$h : S_d \to S_d$$

such that

$$\Gamma = h(\Gamma_0) = \Gamma_0 \circ h^{-1}.$$

We do this in three steps:

(1) The measure Γ_0 can be regarded as a measure on $\mathbb{R}^d$ concentrated on the unit sphere S_d. In this first step, we let

$$h_1 : \mathbb{R}^d \to \mathbb{R}^d$$

be the transformation induced by the matrix Λ and

$$m_1 = h_1(\Gamma_0) \equiv \Gamma_0 \circ h_1^{-1}$$

be the induced measure on $\mathbb{R}^d$.

(2) In the second step we transform m_1 into

$$m_2(d\mathbf{s}) = \Big(\sum_{k=1}^{d} s_k^2\Big)^{\alpha/2} m_1(d\mathbf{s}),$$

which is also a measure on $\mathbb{R}^d$. Write $m_2 = h_2(m_1)$.

(3) In the third step, we map $\mathbb{R}^d \backslash \{0\}$ into the unit sphere S_d by applying the transformation

$$h_3 : \mathbb{R}^d \backslash \{0\} \to S_d,$$

where

$$h_3(s_1, s_2, \ldots, s_d) = \Big(\frac{s_1}{(\sum_{k=1}^d s_k^2)^{1/2}}, \ldots, \frac{s_d}{(\sum_{k=1}^d s_k^2)^{1/2}} \Big),$$

and we define

$$m_3 = h_3(m_2).$$

We then have $m_3 = h(\Gamma_0)$ where

$$h = h_3 \circ h_2 \circ h_1. \tag{2.5.9}$$

We claim that m_3 is, in fact, Γ, the spectral measure of the sub-Gaussian random vector $\mathbf{X}$.

Proposition 2.5.8 *The spectral measure Γ of a sub-Gaussian $S\alpha S$ random vector in $\mathbb{R}^d$ has the form*

$$\Gamma = h(\Gamma_0),$$

where Γ_0 is the uniform measure on S_d defined in (2.5.7) and where $h : S_d \to S_d$ is defined in (2.5.9).

PROOF: By Theorem 2.4.3, we have

$$E\exp\{i(\boldsymbol{\theta},\mathbf{X})\} = \exp\Big\{-\int_{S_d} |\sum_{j=1}^{d}\theta_j s_j|^\alpha \Gamma(d\mathbf{s})\Big\}.$$

Since the spectral measure is unique, it is enough to show that we can replace Γ by $m_3 = h(\Gamma_0)$. Now,

$$\begin{aligned}
\int_{S_d} |\sum_{j=1}^{d}\theta_j s_j|^\alpha m_3(d\mathbf{s}) &= \int_{\mathbb{R}^d\setminus\{0\}} |\sum_{j=1}^{d}\theta_j \frac{s_j}{(\sum_{k=1}^{d} s_k^2)^{1/2}}|^\alpha m_2(d\mathbf{s}) \\
&= \int_{\mathbb{R}^d\setminus\{0\}} |\sum_{j=1}^{d}\theta_j s_j|^\alpha m_1(d\mathbf{s}) = \int_{\mathbb{R}^d} |\sum_{j=1}^{d}\theta_j s_j|^\alpha m_1(d\mathbf{s}) \\
&= \int_{\mathbb{R}^d} |(\boldsymbol{\theta},\Lambda\mathbf{s})|^\alpha \Gamma_0(d\mathbf{s}) = \int_{S_d} |(\boldsymbol{\theta},\Lambda\mathbf{s})|^\alpha \Gamma_0(d\mathbf{s}) \\
&= \int_{S_d} |(\Lambda^T\boldsymbol{\theta},\mathbf{s})|^\alpha \Gamma_0(d\mathbf{s}) = \Big(\frac{1}{2}\sum_{j=1}^{d}(\Lambda^T\boldsymbol{\theta})_j^2\Big)^{\alpha/2} \qquad \text{(using (2.5.7))} \\
&= \Big(\frac{1}{2}(\Lambda^T\boldsymbol{\theta},\Lambda^T\boldsymbol{\theta})\Big)^{\alpha/2} = \Big(\frac{1}{2}(\boldsymbol{\theta},\Lambda\Lambda^T\boldsymbol{\theta})\Big)^{\alpha/2} \\
&= \Big(\frac{1}{2}(\boldsymbol{\theta},\Sigma\boldsymbol{\theta})\Big)^{\alpha/2} = \Big(\frac{1}{2}\sum_{i=1}^{d}\sum_{j=1}^{d} R_{ij}\theta_i\theta_j\Big)^{\alpha/2}
\end{aligned}$$

which, by (2.5.4), is the exponent of the characteristic function of $\mathbf{X}$. This proves that $\Gamma = m_3 = h(\Gamma_0)$. ∎

Corollary 2.5.9 *Let $\mathbf{X} = (X_1, X_2)$ be a sub-Gaussian $S\alpha S$ random vector in $\mathbb{R}^2$ with underlying Gaussian vector $\mathbf{G} = (G_1, G_2)$. Then X_1 and X_2 are dependent unless $G_1 = 0$ a.s. or $G_2 = 0$ a.s.*

PROOF: As before, we let $\Gamma = h_3(m_2)$, $m_2 = h_2(m_1)$ and $m_1 = h_1(\Gamma_0)$. Suppose, to the contrary, that $G_1 \neq 0$ a.s., $G_2 \neq 0$ a.s. and that X_1 and X_2 are independent. Then the spectral measure $\Gamma \not\equiv 0$ of $\mathbf{X}$ is concentrated on the points $(0,1), (0,-1), (1,0), (-1,0)$ of the unit circle S_2 (see Example 2.3.5). Therefore m_2 is concentrated on the axes of $\mathbb{R}^2$, and so is m_1. Since $m_1 = h_1(\Gamma_0)$, where Γ_0 is the uniform measure on S_2, we can find two linearly independent vectors $\mathbf{s}^{(1)}$ and $\mathbf{s}^{(2)}$ on S_2 such that $h_1(\mathbf{s}^{(1)})$ and $h_1(\mathbf{s}^{(2)})$ are both concentrated on either the "x axis" or the "y axis" of $\mathbb{R}^2$. Suppose without loss of generality that they are concentrated on the "x axis." Since h_1 mapes two linearly independent vectors into that subspace, it maps the whole of $\mathbb{R}^2$ into that subspace. This means that the matrix Λ in (2.5.8) has the form

$$\Lambda = \begin{pmatrix} 0 & 0 \\ a & b \end{pmatrix},$$

so

$$\Sigma = \Lambda\Lambda^T = \begin{pmatrix} 0 & 0 \\ a & b \end{pmatrix} \begin{pmatrix} 0 & a \\ 0 & b \end{pmatrix} = \begin{pmatrix} 0 & 0 \\ 0 & a^2 + b^2 \end{pmatrix},$$

and hence Var $G_1 = 0$. This implies $G_1 = 0$ a.s., contradicting the hypothesis. The proof is now complete. ■

Not all symmetric α-stable random vectors are sub-Gaussian. For example, by Corollary 2.5.9, a vector with independent non-zero components cannot be sub-Gaussian. We will see in Section 6.7 that the components of a sub-Gaussian $S\alpha S$ random vector are, in fact, strongly dependent.

2.6 Complex $S\alpha S$ random variables

Although we are mainly interested in real-valued random elements, we will later encounter an important class of real-valued stable processes (the so-called "harmonizable processes"), which are typically defined in terms of complex $S\alpha S$ random variables. Let X_1 and X_2 be real random variables defined on the same probability space. It is the joint distribution of X_1 and X_2 that characterizes the complex random variable $X = X_1 + iX_2$.

Definition 2.6.1 A complex random variable $X = X_1 + iX_2$ is called *symmetric α-stable* ($S\alpha S$) if the random vector (X_1, X_2) is $S\alpha S$ in $\mathbb{R}^2$.

Definition 2.6.2 A complex $S\alpha S$ random variable $X = X_1 + iX_2$ is *rotationally invariant* (or *isotropic*) if

$$e^{i\phi}X \stackrel{d}{=} X$$

for any $\phi \in [0, 2\pi)$.

Warning. Some authors use the term complex $S\alpha S$ random variables to mean random variables that are both complex $S\alpha S$ and isotropic.

Since (complex) isotropic $S\alpha S$ random variables play an important role in the context of so-called "harmonizable processes," we shall examine in greater detail the structure of the joint distribution of X_1 and X_2 in the isotropic case. Isotropy is a strong condition. For any $\phi \in [0, 2\pi)$, one must have

$$(\cos\phi + i\sin\phi)(X_1 + iX_2) \stackrel{d}{=} (X_1 + iX_2),$$

i.e.,

$$(X_1 \cos\phi - X_2 \sin\phi,\ X_2 \cos\phi + X_1 \sin\phi) \stackrel{d}{=} (X_1, X_2)$$

or

$$R_\phi^T \mathbf{X} \stackrel{d}{=} \mathbf{X},$$

where

$$R_\phi = \begin{pmatrix} \cos\phi & \sin\phi \\ -\sin\phi & \cos\phi \end{pmatrix}$$

is the matrix corresponding to the rotation by an angle ϕ, T denotes transpose, and $\mathbf{X} = (X_1, X_2)$.

If $\alpha = 2$, the isotropy condition $e^{i\phi}X \stackrel{d}{=} X$ is satisfied by $X = X_1 + iX_2$ with X_1, X_2 i.i.d. $N(0, \sigma^2)$, since

$$E(X_1 \cos\phi - X_2 \sin\phi)(X_2 \cos\phi + X_1 \sin\phi) = 0 = EX_1X_2,$$

$$E(X_1 \cos\phi - X_2 \sin\phi)^2 = E(X_2 \cos\phi + X_1 \sin\phi)^2 = \sigma^2,$$

and, in fact, one can easily verify that if X_1, X_2 are not i.i.d., then $X = X_1 + iX_2$ is not isotropic (see Exercise 2.18.) We will see below that when $\alpha < 2$, the real and imaginary parts of an isotropic $S\alpha S$ random variable are dependent.

Let us suppose $\alpha < 2$. The following theorem shows that isotropy implies a very special spectral measure for $\mathbf{X} = (X_1, X_2)$.

Theorem 2.6.3 *Let $\alpha < 2$. A complex $S\alpha S$ random variable $X = X_1 + iX_2$ is isotropic if and only if (X_1, X_2) has a uniform spectral measure.*

PROOF: Let $\boldsymbol{\theta} = (\theta_1, \theta_2)$. The characteristic function of $\mathbf{X} = (X_1, X_2)$,

$$E \exp\{i(\boldsymbol{\theta}, \mathbf{X})\} = \exp\Big\{-\int_{S_2} |(\boldsymbol{\theta}, \mathbf{s})|^\alpha \Gamma(d\mathbf{s})\Big\}$$

must be equal to the characteristic function of $R_\phi^T \mathbf{X}$, i.e., to

$$\begin{aligned} E \exp\{i(\boldsymbol{\theta}, R_\phi^T \mathbf{X})\} &= E \exp\Big\{i(R_\phi \boldsymbol{\theta}, \mathbf{X})\Big\} \\ &= \exp\Big\{-\int_{S_2} |(R_\phi \boldsymbol{\theta}, \mathbf{s})|^\alpha \Gamma(d\mathbf{s})\Big\} \\ &= \exp\Big\{-\int_{S_2} |(\boldsymbol{\theta}, R_\phi^T \mathbf{s})|^\alpha \Gamma(d\mathbf{s})\Big\} \\ &= \exp\Big\{-\int_{S_2} |(\boldsymbol{\theta}, \mathbf{t})|^\alpha \Gamma_\phi(d\mathbf{t})\Big\} \end{aligned}$$

where we set $\mathbf{t} = R_\phi^T \mathbf{s}$ and

$$\Gamma_\phi(B) = \Gamma((R_\phi^T)^{-1}(B)) = \Gamma(R_\phi(B))$$

for any Borel set B of S_2. Because of the uniqueness of the spectral measure, this occurs if and only if

$$\Gamma = \Gamma_\phi\,,\ \forall \phi \in [0, 2\pi),$$

that is, if and only if Γ is a uniform measure on S_2. ■

In view of Proposition 2.5.5, we have

Corollary 2.6.4 *Let $\alpha < 2$. A complex $S\alpha S$ random variable $X = X_1 + iX_2$ is isotropic if and only if there are two i.i.d. zero mean normal random variables G_1 and G_2 and a random variable $A \sim S_{\alpha/2}((\cos \pi\alpha/4)^{2/\alpha}, 1, 0)$ independent of (G_1, G_2) such that (X_1, X_2) is sub-Gaussian with underlying vector (G_1, G_2).*

Thus $(X_1, X_2) \stackrel{d}{=} (A^{1/2}G_1, A^{1/2}G_2)$. In other words, every complex isotropic $S\alpha S$ random variable with $\alpha < 2$ is of the form

$$X = A^{1/2}(G_1 + iG_2).$$

Corollary 2.5.9 implies that the real and imaginary parts of X are always dependent (unless G_1 and G_2 are degenerate).

Corollary 2.6.5 *Let $\alpha < 2$ and suppose that the complex $S\alpha S$ random variable $X = X_1 + iX_2$ is isotropic. Then*

$$Ee^{i(\theta_1 X_1 + \theta_2 X_2)} = e^{-c_0 \Gamma(S_2)|\boldsymbol{\theta}|^\alpha},$$

where Γ is the spectral measure of (X_1, X_2) and

$$c_0 = \frac{1}{2\pi}\int_0^{2\pi} |\cos\phi|^\alpha d\phi.$$

PROOF: The result follows from Corollary 2.6.4 and Proposition 2.5.5, or directly by noting that the characteristic function of $\mathbf{X} = (X_1, X_2)$ is

$$E\exp\{i(\boldsymbol{\theta}, \mathbf{X})\} = \exp\Big\{-\int_{S_2} |(\boldsymbol{\theta}, \mathbf{s})|^\alpha \Gamma(d\mathbf{s})\Big\},$$

where Γ is a uniform measure on S_2. Letting κ and ϕ denote, respectively, the arguments of the vectors $\boldsymbol{\theta}$ and $\mathbf{s}$, we have

$$\begin{aligned} E\exp\{i(\boldsymbol{\theta}, \mathbf{X})\} &= \exp\Big\{-|\boldsymbol{\theta}|^\alpha \int_0^{2\pi} |\cos(\kappa - \phi)|^\alpha \frac{\Gamma(S_2)d\phi}{2\pi}\Big\} \\ &= \exp\{-c_0\Gamma(S_2)|\boldsymbol{\theta}|^\alpha\}. \end{aligned}$$ ■

This shows that the characteristic function of $\mathbf{X} = (X_1, X_2)$ depends only on the modulus of $\boldsymbol{\theta}$.

When $\alpha = 2$, the necessity part of Theorem 2.6.3 fails because Γ is no longer unique. In fact, we noted at the beginning of this section that when $\alpha = 2$, the real and imaginary parts of X are independent. This is a major difference between the cases $\alpha = 2$ and $\alpha < 2$.

2.7 Covariation

The covariance function is an extremely powerful tool in the study of Gaussian random elements, but it is not defined when $\alpha < 2$. The *covariation* is designed to replace the covariance when $1 < \alpha < 2$. Unfortunately, it is not as powerful a tool, because it lacks some of the desirable properties of the covariance. It is, however, a useful quantity and it appears naturally in many settings, for example, in the context of linear regression.

In this section, we define the covariation of two jointly $S\alpha S$ random variables with $1 < \alpha \le 2$, and compare its properties with those of the covariance. We start with the definition of "signed power." Let a and p be real numbers. The *signed power* $a^{<p>}$ equals

$$a^{<p>} = |a|^p \operatorname{sign} a = \begin{cases} a^p & \text{if } a \ge 0, \\ -|a|^p & \text{if } a < 0. \end{cases} \tag{2.7.1}$$

For example, $(-2)^{<1/2>} = -\sqrt{2}$ and $(-2)^{<2>} = -4$. In general, $a^{<1>} = a$, $a^{<2>} \neq a^2$ and $(ab)^{<p>} = a^{<p>}b^{<p>}$.

Definition 2.7.1 Let X_1 and X_2 be jointly $S\alpha S$ with $\alpha > 1$ and let Γ be the spectral measure of the random vector (X_1, X_2). The *covariation* of X_1 on[4] X_2 is the real number

$$[X_1, X_2]_\alpha = \int_{S_2} s_1 s_2^{<\alpha-1>}\Gamma(d\mathbf{s}). \tag{2.7.2}$$

Example 2.7.2 If $\alpha = 2$, the characteristic function of (X_1, X_2) is

$$\begin{aligned} E\exp\{i(\theta_1 X_1 + \theta_2 X_2)\} &= \exp\Big\{-\int_{S_2} (\theta_1 s_1 + \theta_2 s_2)^2 \Gamma(d\mathbf{s})\Big\} \\ &= \exp\Big\{-\Big(\theta_1^2 \int_{S_2} s_1^2\Gamma(d\mathbf{s}) + 2\theta_1\theta_2\,[X_1, X_2]_2 + \theta_2^2 \int_{S_2} s_2^2\Gamma(d\mathbf{s})\Big)\Big\}. \end{aligned}$$

[4]As we shall see, the covariation is not symmetric in its arguments. Nevertheless, $[X_1, X_2]_\alpha$ is often called "the covariation of X_1 and X_2."

On the other hand,

$$E\exp\{i(\theta_1X_1+\theta_2X_2)\}$$
$$=\exp\Big\{-\frac{1}{2}(\theta_1^2 \operatorname{Var}X_1+2\theta_1\theta_2\operatorname{Cov}(X_1,X_2)+\theta_2^2\operatorname{Var}X_2)\Big\}.$$

Setting first $\theta_1=0$ and then $\theta_2=0$ yields

$$\int_{S_2} s_j^2\Gamma(ds)=\frac{1}{2}\operatorname{Var}X_j,\ j=1,2,$$

and hence

$$[X_1,X_2]_2=\frac{1}{2}\operatorname{Cov}(X_1,X_2).$$

The definition of the covariation $[X_1,X_2]_\alpha$ involves the spectral measure Γ which appears in the spectral representation of the characteristic function of the vector (X_1,X_2). We will see in the next chapter that the covariation $[X_1,X_2]_\alpha$ can also be expressed in terms of the so-called "integral representation" of (X_1,X_2).

It is possible, however, to give a definition of the covariation which is not related to representations and which can be used when the representations are difficult to obtain explicitly. To state this second definition, proceed as follows.

Let (X_1,X_2) be jointly $S\alpha S$, $1<\alpha\le 2$, and consider the $S\alpha S$ random variable

$$Y=\theta_1X_1+\theta_2X_2.$$

where θ_1 and θ_2 are real numbers. Let $\sigma(\theta_1,\theta_2)$ be the scale parameter of the random variable Y.

Definition 2.7.3 (*equivalent to Definition 2.7.1*). The covariation $[X_1,X_2]_\alpha$ of (X_1,X_2) is

$$[X_1,X_2]_\alpha=\frac{1}{\alpha}\frac{\partial\sigma^\alpha(\theta_1,\theta_2)}{\partial\theta_1}\bigg|_{\theta_1=0,\theta_2=1}. \qquad (2.7.3)$$

To prove that this definition is equivalent to Definition 2.7.1, let Γ denote the spectral measure of (X_1,X_2). Then by (2.3.3),

$$\sigma^\alpha(\theta_1,\theta_2)=\int_{S_2}|\theta_1s_1+\theta_2s_2|^\alpha\Gamma(ds).$$

It is easy to check that the finiteness of Γ and the fact that $\alpha>1$ imply that

$$\frac{\partial\sigma^\alpha(\theta_1,\theta_2)}{\partial\theta_1}=\alpha\int_{S_2}s_1(\theta_1s_1+\theta_2s_2)^{<\alpha-1>}\Gamma(ds).$$

Substituting $\theta_1=0$ and $\theta_2=1$ yields (2.7.3). This shows that Definitions 2.7.1 and 2.7.3 are equivalent.

Example 2.7.4 Let $(G_1, G_2, \ldots, G_n)$ be mean zero jointly Gaussian random variables with covariance $R_{ij} = EG_iG_j,\ i,j = 1,\ldots,n$. Fix $1 < \alpha < 2$ and let $A \sim S_{\alpha/2}\left((\cos\frac{\pi\alpha}{4})^{2/\alpha},\ 1, 0\right)$ be independent of $(G_1, G_2, \ldots, G_n)$.

The random vector $\mathbf{X} = (X_1, X_2, \ldots, X_n)$ with

$$X_k = A^{1/2} G_k\ ,\quad k = 1, 2, \ldots, n,$$

is sub-Gaussian (see Section 2.5). Let us compute the covariations $[X_i, X_j]_\alpha,\ i,j = 1,\ldots,n$, of the components of $\mathbf{X}$. Since $\mathbf{X}$ has characteristic function

$$E\exp\Big\{i\sum_{k=1}^n \theta_k X_k\Big\} = \exp\Big\{-2^{-\alpha/2}|\sum_{i=1}^n\sum_{j=1}^n \theta_i\theta_j R_{ij}|^{\alpha/2}\Big\}$$

(Proposition 2.5.2), we see that the scale parameter $\sigma(\theta_i, \theta_j)$ of $Y = \theta_i X_i + \theta_j X_j$, $i,j = 1,\ldots,n$, satisfies

$$\sigma^\alpha(\theta_i,\theta_j) = 2^{-\alpha/2}\left(\theta_i^2 R_{ii} + 2\theta_i\theta_j R_{ij} + \theta_j^2 R_{jj}\right)^{\alpha/2}.$$

Using (2.7.3), we see that

$$[X_i, X_j]_\alpha = \frac{1}{\alpha}\frac{\partial\sigma^\alpha(\theta_i,\theta_j)}{\partial\theta_i}\Big|_{\theta_i=0,\theta_j=1} = 2^{-\alpha/2} R_{ij} R_{jj}^{(\alpha-2)/2}. \tag{2.7.4}$$

(Observe that here $[X_i, X_j]_\alpha = [X_j, X_i]_\alpha$ if $R_{ii} = R_{jj}$.)

As an application of (2.7.3), we have

Lemma 2.7.5 *Let $\mathbf{X} = (X_1, X_2, \ldots, X_n)$ be a $S\alpha S$ random vector with $\alpha > 1$ and spectral measure $\Gamma_{\mathbf{X}}$, and let $Y = \sum_{k=1}^n a_k X_k$ and $Z = \sum_{k=1}^n b_k X_k$. Then*

$$[Y, Z]_\alpha = \int_{S_n}\left(\sum_{k=1}^n a_k s_k\right)\left(\sum_{k=1}^n b_k s_k\right)^{<\alpha-1>}\Gamma_{\mathbf{X}}(ds).$$

PROOF: The scale parameter $\sigma(\theta_1, \theta_2)$ of the linear combination $\theta_1 Y + \theta_2 Z$ is

$$\Big(\int_{S_n}|\sum_{k=1}^n(\theta_1 a_k + \theta_2 b_k)s_k|^\alpha \Gamma_{\mathbf{X}}(ds)\Big)^{1/\alpha}$$

because

$$\begin{aligned}\exp\{-\sigma^\alpha(\theta_1,\theta_2)\} &= E\exp\{i(\theta_1 Y + \theta_2 Z)\}\\ &= E\exp\Big\{i\sum_{k=1}^n(\theta_1 a_k + \theta_2 b_k)X_k\Big\}\\ &= \exp\Big\{-\int_{S_n}|\sum_{k=1}^n(\theta_1 a_k + \theta_2 b_k)s_k|^\alpha\Gamma_{\mathbf{X}}(ds)\Big\}.\end{aligned}$$

Applying (2.7.3) yields the desired result. ∎

The following corollary expresses the covariation of two components of a random vector in terms of the spectral measure of the vector.

Corollary 2.7.6 *Under the conditions of Lemma 2.7.5,*

$$[X_1, X_2]_\alpha = \int_{S_n} s_1 s_2^{<\alpha-1>} \Gamma_{\mathbf{X}}(ds)$$

and

$$[X_1, X_1]_\alpha = \int_{S_n} |s_1|^\alpha \Gamma_{\mathbf{X}}(ds) = \sigma_{X_1}^\alpha$$

where σ_{X_1} is the scale parameter of the $S\alpha S$ random variable X_1.

We now list a number of properties of the covariation. We suppose throughout that $1 < \alpha \leq 2$. The first two properties follow directly from Lemma 2.7.5.

Property 2.7.7 *(Additivity in the first argument). Let (X_1, X_2, Y) be jointly $S\alpha S$. Then*

$$[X_1 + X_2, Y]_\alpha = [X_1, Y]_\alpha + [X_2, Y]_\alpha.$$

Property 2.7.8 *(Scaling). Let (X, Y) be jointly $S\alpha S$ and let a and b be real numbers. Then*

$$[aX, bY]_\alpha = ab^{<\alpha-1>}[X, Y]_\alpha.$$

The preceding properties point to important differences between covariation and covariance. For example,

Corollary 2.7.9 *Although the covariation is linear in its first argument, it is in general not linear in its second argument.*

The linearity in the first argument,

$$[a_1 X_1 + a_2 X_2, Y]_\alpha = a_1[X_1, Y]_\alpha + a_2[X_2, Y]_\alpha,$$

follows from the preceding properties. Observe that the covariation may not even be additive in its second argument, i.e.,

$$[X, Y_1 + Y_2]_\alpha \neq [X, Y_1]_\alpha + [X, Y_2]_\alpha,$$

since one has

$$[X, X + X]_\alpha = 2^{\alpha-1}[X, X]_\alpha \neq 2[X, X]_\alpha = [X, X]_\alpha + [X, X]_\alpha$$

for $\alpha < 2$ and $[X, X]_\alpha \neq 0$. (By Corollary 2.7.6, this last requirement is equivalent to $X \neq 0$ a.s.)

Corollary 2.7.10 *The covariation is, in general, not symmetric in its arguments.*

To verify that, in general,

$$[X,Y]_\alpha \neq [Y,X]_\alpha,$$

let $\alpha < 2, [X,X] \neq 0$ and let a and b be two different non-zero real numbers. Then

$$[aX,bX]_\alpha = ab^{<\alpha-1>}[X,X]_\alpha \neq a^{<\alpha-1>}b[X,X]_\alpha = [bX,aX]_\alpha.$$

The next property states that like the covariance, the covariation is zero when its arguments are independent.

Property 2.7.11 *If X and Y are jointly $S\alpha S$ and independent, then*

$$[X,Y]_\alpha = 0.$$

PROOF: We saw in Example 2.3.5 that the independence of X_1 and X_2 implies that the spectral measure Γ of (X_1, X_2) must be concentrated on the points $(1,0), (-1,0), (0,1)$ and $(0,-1)$ of S_2. We then have

$$[X,Y]_\alpha = \int_{S_2} s_1 s_2^{<\alpha-1>} \Gamma(ds) = 0$$

since the support of Γ is such that either s_1 or s_2 is zero. ∎

Example 2.7.12 Unlike the Gaussian case $\alpha = 2$, it is possible in the case $1 < \alpha < 2$ to have $[X,Y]_\alpha = 0$ with *dependent* X and Y. For example, let $X = A^{1/2}G_1$, $Y = A^{1/2}G_2$ be jointly sub-Gaussian, with G_1, G_2 i.i.d. (non-degenerate) standard normal random variables. It follows from (2.7.4) that $[X,Y]_\alpha = [Y,X]_\alpha = 0$. But X and Y cannot be independent by Corollary 2.5.9.

We observed that, in general,

$$[X, Y_1 + Y_2]_\alpha \neq [X, Y_1]_\alpha + [X, Y_2]_\alpha.$$

We shall now prove that equality holds when Y_1 and Y_2 are independent. This property turns out to be very useful in the context of stochastic integrals. We start with the following lemmas.

Lemma 2.7.13 *Let u and v be real numbers. If $0 < p \leq 1$, then*

$$|u+v|^p \leq |u|^p + |v|^p \tag{2.7.5}$$

with equality, if $p < 1$, if and only if either $u = 0$ or $v = 0$. If $1 \leq p < \infty$, then

$$|u+v|^p \leq 2^{p-1}\left(|u|^p + |v|^p\right). \tag{2.7.6}$$

PROOF: Since $(|u+v|)^p \leq (|u|+|v|)^p$, we may suppose $u \geq 0$ and $v \geq 0$. If $0 < p < 1$, then, for fixed $v > 0$ and for any $u \geq 0$, we have $g_v(u) := u^p + v^p - (u+v)^p \geq 0$ with equality only at $u = 0$, because $g_v(0) = 0$ and $g_v'(u) > 0$ for $u > 0$. If $1 \leq p < \infty$, then, by Jensen's inequality

$$(u+v)^p = 2^p \left(\frac{u+v}{2}\right)^p \leq 2^p \frac{1}{2}(u^p + v^p). \quad \blacksquare$$

Lemma 2.7.14 *Let $(E, \mathcal{E}, m)$ be an arbitrary σ-finite measure space, and let $f_i : E \to \mathbb{R}$, $i = 1, 2$, be two functions in $L^\alpha(E, \mathcal{E}, m)$.*

(1) If $0 < \alpha < 1$, then either of the relations

$$\int_E |f_1(x) + f_2(x)|^\alpha m(dx) = \int_E |f_1(x)|^\alpha m(dx) + \int_E |f_2(x)|^\alpha m(dx)$$

or

$$\int_E |f_1(x) - f_2(x)|^\alpha m(dx) = \int_E |f_1(x)|^\alpha m(dx) + \int_E |f_2(x)|^\alpha m(dx)$$

implies

$$f_1(x) f_2(x) = 0 \ \ m\text{-a.e.}$$

(2) If $1 \leq \alpha < 2$, then

$$\begin{aligned}\int_E |f_1(x) + f_2(x)|^\alpha m(dx) &= \int_E |f_1(x) - f_2(x)|^\alpha m(dx) \\ &= \int_E |f_1(x)|^\alpha m(dx) + \int_E |f_2(x)|^\alpha m(dx)\end{aligned}$$

implies

$$f_1(x) f_2(x) = 0 \ \ m\text{-a.e.}$$

PROOF: (1) Suppose $0 < \alpha < 1$. By Lemma 2.7.13,

$$\int_E |f_1(x) + f_2(x)|^\alpha m(dx) \leq \int_E |f_1(x)|^\alpha m(dx) + \int_E |f_2(x)|^\alpha m(dx), \quad (2.7.7)$$

with equality if and only if $f_1(x) f_2(x) = 0$ m-a.e. Replacing f_2 by $-f_2$ in (2.7.7) completes the proof of the first part of the lemma.
(2) Suppose, now, $1 \leq \alpha < 2$. Adding the equations, we obtain

$$\int_E |f_1(x) + f_2(x)|^\alpha m(dx) + \int_E |f_1(x) - f_2(x)|^\alpha m(dx)$$

$$= 2\left(\int_E |f_1(x)|^\alpha m(dx) + \int_E |f_2(x)|^\alpha m(dx)\right). \quad (2.7.8)$$

We prove that (2.7.8) implies the conclusion of the proposition. Since $0 < \alpha < 2$, the function $r(u) = u^{\alpha/2}$, $u \geq 0$ is concave. Therefore, for fixed $x \in E$,

$$\begin{aligned}
&|f_1(x) + f_2(x)|^\alpha + |f_1(x) - f_2(x)|^\alpha \\
&\quad = ([f_1(x) + f_2(x)]^2)^{\alpha/2} + ([f_1(x) - f_2(x)]^2)^{\alpha/2} \\
&\quad = r([f_1(x) + f_2(x)]^2) + r([f_1(x) - f_2(x)]^2) \\
&\quad = 2\frac{r([f_1(x) + f_2(x)]^2) + r([f_1(x) - f_2(x)]^2)}{2} \\
&\quad \leq 2r\left(\frac{[f_1(x) + f_2(x)]^2 + [f_1(x) - f_2(x)]^2}{2}\right) \qquad \text{(concavity)} \quad (2.7.9) \\
&\quad = 2r(f_1^2(x) + f_2^2(x)) \\
&\quad = 2\left(f_1^2(x) + f_2^2(x)\right)^{\alpha/2} \\
&\quad \leq 2\left((f_1^2(x))^{\alpha/2} + (f_2^2(x))^{\alpha/2}\right) \qquad \text{(exponent } \alpha/2 < 1) \\
&\quad = 2\left(|f_1(x)|^\alpha + |f_2(x)|^\alpha\right)
\end{aligned}$$

with equality in the preceding relations equivalent either to $f_1(x) = 0$ or to $f_2(x) = 0$. (To verify this, notice that since r is strictly concave, the first inequality in (2.7.9) becomes an equality if and only if $(f_1 + f_2)^2 = (f_1 - f_2)^2$, that is, $f_1 f_2 = 0$. For the second inequality, use Lemma 2.7.13.)

Now (2.7.9) implies that the left-hand side of (2.7.8) is always less than or equal to the right-hand side of (2.7.8) and, if they are equal, then necessarily (2.7.9) holds with equality for m-almost any $x \in E$. Therefore, for m-almost any $x \in E$, we have $f_1(x)f_2(x) = 0$. ∎

We are now in position to prove

Property 2.7.15 *Let (X, Y_1, Y_2) be jointly $S\alpha S$, $\alpha > 1$, with Y_1 and Y_2 independent. Then*

$$[X, Y_1 + Y_2]_\alpha = [X, Y_1]_\alpha + [X, Y_2]_\alpha.$$

PROOF: Let Γ be the spectral measure of (X, Y_1, Y_2). The characteristic function of (Y_1, Y_2) is

$$E \exp\{i(\theta_1 Y_1 + \theta_2 Y_2)\} = \exp\Big\{-\int_{S_3} |0 \cdot s_1 + \theta_1 s_2 + \theta_2 s_3|^\alpha \Gamma(ds)\Big\},$$

but, by independence of Y_1 and Y_2, it is also equal to

$$\begin{aligned}
&E \exp\{i\theta_1 Y_1\} E \exp\{i\theta_2 Y_2\} \\
&= \exp\Big\{-\int_{S_3} |0 \cdot s_1 + \theta_1 s_2 + 0 \cdot s_3|^\alpha \Gamma(ds)\Big\} \exp\Big\{-\int_{S_3} |0 \cdot s_1 + 0 \cdot s_2 + \theta_2 s_3|^\alpha \Gamma(ds)\Big\}.
\end{aligned}$$

Therefore

$$\int_{S_3} |\theta_1 s_2 + \theta_2 s_3|^\alpha \Gamma(ds) = |\theta_1|^\alpha \int_{S_3} |s_2|^\alpha \Gamma(ds) + |\theta_2|^\alpha \int_{S_3} |s_3|^\alpha \Gamma(ds)$$

holds for all θ_1 and θ_2. By Lemma 2.7.14, this implies

$$s_2 s_3 = 0 \quad \Gamma \text{ a.e.}$$

and hence,

$$\int_{S_3} s_1(s_2+s_3)^{<\alpha-1>}\Gamma(ds) = \int_{S_3} s_1 s_2^{<\alpha-1>}\Gamma(ds) + \int_{S_3} s_1 s_3^{<\alpha-1>}\Gamma(ds).$$

Using Lemma 2.7.5, we obtain

$$[X, Y_1 + Y_2]_\alpha = [X, Y_1]_\alpha + [X, Y_2]_\alpha. \quad \blacksquare$$

The covariation $[X, Y]_\alpha$ is related to the joint moment $EXY^{<p-1>}$.

Lemma 2.7.16 *Let* (X, Y) *be* $S\alpha S$ *with* $\alpha > 1$. *Then for all* $1 < p < \alpha$,

$$\frac{EXY^{<p-1>}}{E|Y|^p} = \frac{[X,Y]_\alpha}{\|Y\|_\alpha^\alpha},$$

where $\|Y\|_\alpha$ *denotes the scale parameter of* Y.

PROOF: Let Y_0 be a $S\alpha S$ random variable with scale parameter $\|Y_0\|_\alpha = 1$, and let Γ be the spectral measure of the vector (X, Y). Since the random variable $\lambda X + Y$ has scale parameter $\|\lambda X + Y\|_\alpha = (\int_{S_2} |\lambda s_1 + s_2|^\alpha d\Gamma)^{1/\alpha}$, we have, by Property 1.2.17,

$$E|\lambda X + Y|^p = \|\lambda X + Y\|_\alpha^p E|Y_0|^p. \tag{2.7.10}$$

Now

$$\frac{d}{d\lambda} E|\lambda X + Y|^p = pE(\lambda X + Y)^{<p-1>}X$$

and

$$\frac{d}{d\lambda}\|\lambda X + Y\|_\alpha^p = \frac{d}{d\lambda}\left(\int_{S_2} |\lambda s_1 + s_2|^\alpha d\Gamma\right)^{p/\alpha}$$

$$= \frac{p}{\alpha}\left(\int_{S_2} |\lambda s_1 + s_2|^\alpha d\Gamma\right)^{(p/\alpha)-1} \alpha \int_{S_2} (\lambda s_1 + s_2)^{<\alpha-1>} s_1 d\Gamma.$$

Use (2.7.10) and set $\lambda = 0$ to obtain

$$\begin{aligned} EXY^{<p-1>} &= \left(\int_{S_2} |s_2|^\alpha d\Gamma\right)^{(p/\alpha)-1} \left(\int_{S_2} s_1 s_2^{<\alpha-1>} d\Gamma\right) E|Y_0|^p \\ &= \|Y\|_\alpha^{p-\alpha} [X,Y]_\alpha^\alpha E|Y_0|^p \\ &= \|Y\|_\alpha^{-\alpha} [X,Y]_\alpha^\alpha E|Y|^p, \end{aligned}$$

which proves the result. ∎

2.8 Covariation norm

We know that any finite linear combination of $S\alpha S$ random variables is $S\alpha S$. Let $\mathcal{S}_\alpha$ be a linear space of jointly $S\alpha S$ random variables. When $\alpha > 1$, the covariation induces a norm on $\mathcal{S}_\alpha$.

Definition 2.8.1 The *covariation norm* of $X \in \mathcal{S}_\alpha,\ \alpha > 1$, is

$$\|X\|_\alpha = ([X,X]_\alpha)^{1/\alpha}.$$

Corollary 2.7.6 implies that the covariation norm is equal to the scale parameter:

Property 2.8.2 *If $X \sim S_\alpha(\sigma,0,0)$ with $\alpha > 1$, then*

$$\|X\|_\alpha = \sigma.$$

Proposition 2.8.3 *$\|\cdot\|_\alpha$ is a norm on $\mathcal{S}_\alpha$. Convergence in $\|\cdot\|_\alpha$ is equivalent to convergence in probability and convergence in L^p for any $p < \alpha$.*

PROOF: (a) Let $X \sim S_\alpha(\sigma_X,0,0)$. To see that $\|\cdot\|_\alpha$ is a norm, note that

1. $\|X\|_\alpha = 0$ if and only if $X = 0$ a.s. (by Property 2.8.2).
2. $aX \sim S_\alpha(|a|\sigma_X,0,0)$, so that $\|aX\|_\alpha = |a|\sigma_X = |a|\ \|X\|_\alpha$.
3. If X_1 and X_2 are jointly $S\alpha S$ with spectral measure Γ, then

$$\begin{aligned} \|X_1 + X_2\|_\alpha &= \sigma_{X_1+X_2} \\ &= \left(\int_{S_2} |s_1 + s_2|^\alpha \Gamma(ds)\right)^{1/\alpha} \\ &\le \left(\int_{S_2} |s_1|^\alpha \Gamma(ds)\right)^{1/\alpha} + \left(\int_{S_2} |s_2|^\alpha \Gamma(ds)\right)^{1/\alpha} \\ &= \sigma_{X_1} + \sigma_{X_2} \\ &= \|X_1\|_\alpha + \|X_2\|_\alpha. \end{aligned}$$

(b) To prove that convergence in the covariance norm is equivalent to convergence in probability, let $X_1, X_2, \ldots \in \mathcal{S}_\alpha$, $Y \in \mathcal{S}_\alpha$. By Property 2.8.2, $\lim_{n\to\infty} \|X_n - Y\|_\alpha = 0$ is equivalent to $\lim_{n\to\infty} \sigma_{X_n - Y} = 0$, where σ_{X_n-Y} is the scale parameter of $X_n - Y$. Fix $\epsilon > 0$. Then, if $Z \sim S_\alpha(1,0,0)$ (that is, Z is standard symmetric α-stable), we have, by scaling,

$$P(|X_n - Y| > \epsilon) = P(\sigma_{X_n-Y}|Z| > \epsilon) = P\Big(|Z| > \frac{\epsilon}{\sigma_{X_n-Y}}\Big) \to 0$$

as $n \to \infty$ if and only if

$$\lim_{n\to\infty} \sigma_{X_n-Y} = 0.$$

(c) To show that this is equivalent to convergence in L^p for $p < \alpha$, note that

$$E|X_n - Y|^p = E(\sigma_{X_n-Y}|Z|)^p = \sigma^p_{X_n-Y} E|Z|^p \to 0$$

as $n \to \infty$ if and only if $\lim_{n\to\infty} \sigma_{X_n-Y} = 0$. ∎

Moreover, we have

Property 2.8.4 *Let $(X,Y) \sim S\alpha S$ with $1 < \alpha \le 2$. Then*

$$|[X,Y]_\alpha| \le \|X\|_\alpha \|Y\|_\alpha^{\alpha-1}.$$

PROOF: If Γ is the spectral measure of (X,Y) then, by the Hölder inequality

$$\begin{aligned} |[X,Y]_\alpha| &= \Big|\int_{S_2} s_1 s_2^{<\alpha-1>} \Gamma(ds)\Big| \\ &\le \Big(\int_{S_2} |s_1|^\alpha \Gamma(ds)\Big)^{1/\alpha} \Big(\int_{S_2} |s_2|^{(\alpha-1)(1-1/\alpha)^{-1}} \Gamma(ds)\Big)^{1-1/\alpha} \\ &= \sigma_X \, \sigma_Y^{\alpha(1-1/\alpha)} \\ &= \|X\|_\alpha \, \|Y\|_\alpha^{\alpha-1}. \end{aligned}$$ ∎

Remark. Whereas the covariation norm is not defined when $\alpha \le 1$, the scale parameter σ exists for all possible values of α. The proof of Proposition 2.8.3 allows us to conclude that for any $0 < \alpha \le 2$, $\sigma_{X_n-Y} \to 0$ is equivalent to convergence of X_n to Y in probability and in L^p for $p < \alpha$.

This result, in fact, is a natural one. For given α, the distribution of $X \sim S_\alpha(\sigma_X, 0, 0)$ depends only on the scale parameter σ_X. Since $\sigma_{aX} = |a|\sigma_X$, one has

$$E|X|^p = \sigma_X^p E|X_0|^p, \tag{2.8.1}$$

where $X_0 \sim S_\alpha(1,0,0)$, so that convergence of σ_X to zero is equivalent to convergence in probability or in L^p.

It is sometimes convenient to use the **notation**

$$\|X\|_\alpha = \sigma_X, \quad 0 < \alpha \le 2,$$

to denote the scale parameter of a $S\alpha S$ random variable X.

2.9 James orthogonality

Property 2.7.11 states that if two jointly $S\alpha S$ random variables X and Y are independent, then their covariation $[X,Y]_\alpha$ is zero. The converse holds in the Gaussian case $\alpha = 2$ since $[X,Y]_2 = \frac{1}{2}\text{Cov}(X,Y) = 0$ implies that X and Y are independent, but in the non-Gaussian stable case $\alpha < 2$, $[X,Y]_\alpha = 0$ does not imply independence (see Example 2.7.12). What, then, is the meaning of $[X,Y]_\alpha = 0$ when $1 < \alpha < 2$?

Suppose, firstly, that $\alpha = 2$. Jointly Gaussian mean zero random variables X and Y can be viewed as vectors in the space $L_0^2(\Omega, \mathcal{F}, P)$ of all random variables with zero mean and finite second moment. The space $L_0^2(\Omega, \mathcal{F}, P)$ is a Hilbert space with scalar product

$$(X,Y) := \text{Cov}(X,Y) = EXY.$$

The Gaussian random variables X and Y are independent if and only if they are orthogonal in $L_0^2(\Omega, \mathcal{F}, P)$.

Now let $1 < \alpha < 2$ and consider $\mathcal{S}_\alpha$, a linear space of $S\alpha S$ random variables. We noted in the preceding section that $\mathcal{S}_\alpha$ is a normed space, endowed with the covariation norm $\|\cdot\|_\alpha$. One has $\|X\|_\alpha^\alpha = [X,X]_\alpha$, but the covariation $[X,Y]_\alpha$ does not generally define a scalar product in $\mathcal{S}_\alpha$. The space $\mathcal{S}_\alpha$ is, therefore, not a pre-Hilbert space (i.e., a linear (vector) space endowed with a scalar product).

The classical notion of orthogonality is valid only in (pre-)Hilbert spaces. Several alternative definitions of "orthogonality" can be introduced in normed vector spaces. One of them is James orthogonality (James (1947)).

Definition 2.9.1 Let E be a normed vector space. A vector $x \in E$ is said to be *James orthogonal* to a vector $y \in E$ ($x \perp_J y$) if for any real λ

$$\|x + \lambda y\| \ge \|x\|.$$

In general, $x \perp_J y$ does not imply $y \perp_J x$. The relation $\|x + \lambda y\| \ge \|x\|$ for all real λ means that the vectors $x + \lambda y$, $\lambda \in \mathbb{R}$, belong to a supporting hyperplane of the sphere $\{s : \|s\| = \|x\|\}$ at the point x.

To see that James orthogonality is a natural extension of the usual notion of orthogonality, suppose that the space E is pre-Hilbert and hence has a scalar product. Let $x, y \in E$ and write $x \perp y$ if x is orthogonal to y in the usual sense. Then

$$\begin{aligned} x \perp y \quad &\Rightarrow \quad \|x + \lambda y\|^2 = \|x\|^2 + \|\lambda y\|^2 \geq \|x\|^2 \quad \text{for any } \lambda \\ &\Rightarrow \quad x \perp_J y. \end{aligned}$$

Observe, moreover, that if x is not orthogonal to y, then $\|x + \lambda y\|$ can be greater or smaller than $\|x\|$, depending on the value of λ. Hence,

$$x \perp y \Leftrightarrow x \perp_J y$$

when E is (pre-)Hilbert.

If E is a normed vector space which is not pre-Hilbert, $x \perp y$ is not defined, but $x \perp_J y$ is well defined because the definition of James orthogonality involves relations between norms and not between scalar products.

Now, let E be $\mathcal{S}_\alpha$, $1 < \alpha < 2$, endowed with covariation norm $\|\cdot\|_\alpha$.

Proposition 2.9.2 *Let X and Y be jointly $S\alpha S$ random variables with $\alpha > 1$. Then*

$$[X, Y]_\alpha = 0$$

if and only if Y is James orthogonal to X (symbolically $Y \perp_J X$), i.e.,

$$\|\lambda X + Y\|_\alpha \geq \|Y\|_\alpha$$

for every real λ.

PROOF: Suppose $[X, Y]_\alpha = 0$ and let λ be arbitrary. Then, by Property 2.8.4,

$$|[\lambda X + Y, Y]_\alpha| \leq \|\lambda X + Y\|_\alpha \|Y\|_\alpha^{\alpha-1}.$$

On the other hand, using Property 2.7.7, we have

$$[\lambda X + Y, Y]_\alpha = \lambda [X, Y]_\alpha + [Y, Y]_\alpha = \|Y\|_\alpha^\alpha,$$

since $[X, Y]_\alpha = 0$. ($\|\cdot\|_\alpha^\alpha$ denotes the αth power of the covariation norm.) Therefore,

$$\|Y\|_\alpha^\alpha \leq \|\lambda X + Y\|_\alpha \|Y\|_\alpha^{\alpha-1},$$

i.e.,

$$\|\lambda X + Y\|_\alpha \geq \|Y\|_\alpha.$$

To prove the converse, let Y be James orthogonal to X and let Γ be the spectral measure of the $S\alpha S$ vector (X,Y). Suppose that $[X,Y]_\alpha < 0$. Then

$$\int_{S_2} s_1 s_2^{<\alpha-1>} \Gamma(ds) < 0. \tag{2.9.1}$$

For real λ, define

$$g(\lambda) = \|\lambda X + Y\|_\alpha^\alpha = \int_{S_2} |\lambda s_1 + s_2|^\alpha \Gamma(ds).$$

We have, clearly,

$$g'(\lambda) = \alpha \int_{S_2} s_1 (\lambda s_1 + s_2)^{<\alpha-1>} \Gamma(ds).$$

The fact that $\alpha > 1$ implies that $g'(\lambda)$ is a continuous function of λ. Since (2.9.1) can be restated in the form $g'(0) < 0$, we conclude that $g'(\lambda) < 0$ in some interval $(0, a)$. Therefore, for any $0 < \lambda < a$, we have

$$\|\lambda X + Y\|_\alpha^\alpha = g(\lambda) < g(0) = \|Y\|_\alpha^\alpha,$$

contradicting the James orthogonality of Y to X.

The case $[X,Y]_\alpha > 0$ is similar. ∎

The following proposition characterizes the $S\alpha S$ random variables for which James orthogonality is a symmetric property.

Proposition 2.9.3 *Let $1 < \alpha < 2$ and let $\mathcal{S}_\alpha$ be a linear space of jointly $S\alpha S$ random variables with $\dim \mathcal{S}_\alpha \geq 3$. Then the following four statements are equivalent:*

(i) $\mathcal{S}_\alpha$ has the property:

$$X, Y \in \mathcal{S}_\alpha\,,\ [X,Y]_\alpha = 0 \ \Rightarrow\ [Y,X]_\alpha = 0.$$

(ii) $\mathcal{S}_\alpha$ has the property:

$$X, Y, Z \in \mathcal{S}_\alpha,\ [X,Y]_\alpha = 0,\ [X,Z]_\alpha = 0 \ \Rightarrow\ [X, Y+Z]_\alpha = 0.$$

(iii) There is an inner product $(.,.)$ on $\mathcal{S}_\alpha$ such that

$$\|X\|_\alpha = (X,X)^{1/2}, \quad X \in \mathcal{S}_\alpha. \tag{2.9.2}$$

(iv) $\mathcal{S}_\alpha$ consists of jointly sub-Gaussian $S\alpha S$ random variables.

PROOF: Using Proposition 2.9.2, we can rephrase statements (i) and (ii) as follows:

(i) James orthogonality in $\mathcal{S}_\alpha$ is symmetric.

(ii) James orthogonality in $\mathcal{S}_\alpha$ is additive on the right.

Since $\dim \mathcal{S}_\alpha \geq 3$, we conclude using Theorems 1 and 2 of James (1947), that (i), (ii) and (iii) are equivalent. We shall now prove that (iii) implies (iv).

Set $R_{X,Y} = (X,Y),\ X,Y \in \mathcal{S}_\alpha$, and let $\{G_X,\ X \in \mathcal{S}_\alpha\}$ be a zero mean Gaussian process having $R_{X,Y}$ as a covariance function.

Let A be a positive $\alpha/2$-random variable, independent of the process $\{G_X,\ X \in \mathcal{S}_\alpha\}$, having Laplace transform

$$E\exp(-\gamma A) = \exp\{-\gamma^{\alpha/2}\}.\quad \gamma \geq 0.$$

Define

$$Z_X = \sqrt{2}\, A^{1/2}\, G_X,\quad X \in \mathcal{S}_\alpha.$$

We want to show that

$$\{Z_X,\ X \in \mathcal{S}_\alpha\} \stackrel{d}{=} \{X,\ X \in \mathcal{S}_\alpha\}. \tag{2.9.3}$$

Let $X_1, X_2, \ldots, X_n \in \mathcal{S}_\alpha$, and let $b_1, b_2, \ldots, b_n$ be real numbers. Note that

$$\mathrm{Var}\Big(G_{\sum_{j=1}^n b_j X_j} - \sum_{j=1}^n b_j G_{X_j}\Big)$$

$$= \Big(\sum_{j=1}^n b_j X_j,\ \sum_{j=1}^n b_j X_j\Big) - 2\sum_{k=1}^n\Big(\sum_{j=1}^n b_j X_j,\ b_k X_k\Big) + \sum_{j=1}^n\sum_{k=1}^n (b_j X_j, b_k X_k)$$

$$= 2\sum_{j=1}^n\sum_{k=1}^n b_j b_k (X_j, X_k) - 2\sum_{j=1}^n\sum_{k=1}^n b_j b_k (X_j, X_k) = 0.$$

Therefore, $G_{\sum_{j=1}^n b_j X_j} = \sum_{j=1}^n b_j G_{X_j}$ a.s., so that

$$Z_{\sum_{j=1}^n b_j X_j} = \sum_{j=1}^n b_j Z_{X_j} \text{ a.s.} \tag{2.9.4}$$

for any $X_1, \ldots, X_n \in \mathcal{S}_\alpha$, and any real $b_1, \ldots, b_n$.

To prove (2.9.3), it is enough to show that the distribution of any linear combination of the random variables on the left-hand side of (2.9.3) coincides with the distribution of the corresponding linear combination of the random variables on the right-hand side of (2.9.3). Because of (2.9.4) we need only to show that

$Z_X \stackrel{d}{=} X$ for any $X \in \mathcal{S}_\alpha$. Since both X and Z_X are $S\alpha S$ random variables, we have only to show that

$$\|X\|_\alpha = \|Z_X\|_\alpha\,,\; X \in \mathcal{S}_\alpha.$$

On one hand,

$$\|X\|_\alpha = (X, X)^{1/2} = R_{X,X}^{1/2}$$

by (2.9.2). To compute $\|Z_X\|_\alpha$, recall that if G is zero mean normal, then $\|A^{1/2}G\|_\alpha^\alpha = 2^{-\alpha/2}(\text{Var } G)^{\alpha/2}$ (see Example 2.7.4). Hence,

$$\|Z_X\|_\alpha = \|\sqrt{2}A^{1/2}G_X\|_\alpha = (\text{Var } G_X)^{1/2} = R_{XX}^{1/2},$$

proving that $\|X\|_\alpha = \|Z_X\|_\alpha$. Therefore, $\mathcal{S}_\alpha$ consists of jointly sub-Gaussian $S\alpha S$ random variables.

Finally, if (iv) holds, then (i) follows directly from (2.7.4). This completes the proof. ∎

Remark. When $\dim \mathcal{S}_\alpha = 2$, then Proposition 2.9.3 fails. More precisely, (ii) is in general not equivalent to either (i), (iii) or (iv). For example, let

$$\mathcal{S}_\alpha = \{aU + bV : a, b \text{ real}\},$$

where U and V are i.i.d. $S\alpha S$ with $0 < \alpha < 2$. Let $X = a_1U + b_1V$ and $Y = a_2U + b_2V$ with non-zero coefficients a_1, b_1, a_2, b_2. Then

$$\begin{aligned} [X,Y]_\alpha &= [a_1U + b_1V, a_2U + b_2V]_\alpha \\ &= a_1a_2^{<\alpha-1>} + b_1b_2^{<\alpha-1>} \end{aligned}$$

implies

$$[X,Y]_\alpha = 0 \Leftrightarrow \frac{b_2}{a_2} = -\left(\frac{a_1}{b_1}\right)^{<\frac{1}{\alpha-1}>}. \tag{2.9.5}$$

Since $\alpha \neq 2$ in general,

$$\begin{aligned} [X,Y]_\alpha = 0 \Leftrightarrow \frac{b_2}{a_2} = -\left(\frac{a_1}{b_1}\right)^{<\frac{1}{\alpha-1}>} &\not\Leftrightarrow \frac{b_1}{a_1} = -\left(\frac{a_2}{b_2}\right)^{<\frac{1}{\alpha-1}>} \\ &\Leftrightarrow [Y,X]_\alpha = 0 \end{aligned}$$

and hence (i) fails. Therefore, (iii) and (iv) fail as well because they both imply (i). On the other hand, by (2.9.5), for a fixed $X \in \mathcal{S}_\alpha$ and hence for a fixed a_1 and b_1, the collection of all random variables W belonging to $\mathcal{S}_\alpha$ such that $[X, W]_\alpha = 0$ is a (one-dimensional) linear subspace $\mathcal{L}$ of $\mathcal{S}_\alpha$. Therefore,

$$[X,Y]_\alpha = 0 \text{ and } [X,Z]_\alpha = 0 \Rightarrow Y \in \mathcal{L},\; Z \in \mathcal{L}$$

$$\Rightarrow Y + Z \in \mathcal{L} \Rightarrow [X, Y + Z]_\alpha = 0$$

and hence (ii) holds.

Note that one has (iii) $\Leftrightarrow$ (iv) even when $\dim \mathcal{S}_\alpha = 2$, because the argument establishing this equivalence in Proposition 2.9.3 does not use dimensionality.

If $\dim \mathcal{S}_\alpha = 2$, then either (iii) or (iv) imply (i), but we do not know whether (i) is equivalent to (iii) or to (iv).

The following proposition characterizes the spaces $\mathcal{S}_\alpha$ of $S\alpha S$ random variables in which the normalized covariation is symmetric.

Proposition 2.9.4 *Let $1 < \alpha < 2$ and let $\mathcal{S}_\alpha$ be a linear space of $S\alpha S$ random variables. Then the following two statements are equivalent:*

(i) $\mathcal{S}_\alpha$ has the property

$$X, Y \in \mathcal{S}_\alpha,\ \|X\|_\alpha = \|Y\|_\alpha \ \Rightarrow\ [X, Y]_\alpha = [Y, X]_\alpha.$$

(ii) $\mathcal{S}_\alpha$ consists of jointly sub-Gaussian random $S\alpha S$ variables.

PROOF: Let

$$(X, Y) = \|Y\|_\alpha^{2-\alpha}[X, Y]_\alpha.$$

It is easy to check that this is an inner product on $\mathcal{S}_\alpha$ if (i) holds. We can then apply the same argument as in the proof of Proposition 2.9.3. Finally, by using Example 2.7.4, we conclude that (ii) implies (i). ■

The next result due to Cambanis, Hardin and Weron (1988) is an immediate consequence of Proposition 2.9.3 if $\dim \mathcal{S}_\alpha \geq 3$. It holds also, however, if $\dim \mathcal{S}_\alpha = 2$.

Proposition 2.9.5 *Let $1 < \alpha \leq 2$, and let $\mathcal{S}_\alpha$ be a linear space of $S\alpha S$ random variables with $\dim \mathcal{S}_\alpha \geq 2$. Then the following two statements are equivalent:*

(i) $\mathcal{S}_\alpha$ has the property

$$X, Y \in \mathcal{S}_\alpha,\ [X, Y]_\alpha = 0 \Rightarrow X, Y \text{ are independent}.$$

(ii) $\alpha = 2$, i.e., $\mathcal{S}_\alpha$ consists of zero mean Gaussian random variables.

PROOF: Suppose that (i) holds and choose an arbitrary non-zero $X \in \mathcal{S}_\alpha$. Since $\dim \mathcal{S}_\alpha > 1$ we may find a $Z \in \mathcal{S}_\alpha$ such that $Z \neq \lambda X$ for any real λ. Let $\beta = [Z, X]_\alpha / \|X\|_\alpha^\alpha$. Then $[Z - \beta X, X]_\alpha = 0$ so that, by assumption, $Z - \beta X$ is independent of X. Since both $Z - \beta X$ and X are $S\alpha S$ and non-zero, they can differ only by the value of their scale parameter. There is, therefore, a non-zero constant b such that $Y = b(Z - \beta X)$ has the same distribution as X, so that X and Y are i.i.d. random variables. Note that by Properties 2.7.7 and 2.7.15,

$$\begin{aligned}&[X+Y, X-Y]_\alpha \\ &\quad = [X,X]_\alpha + [Y,X]_\alpha - [X,Y]_\alpha - [Y,Y]_\alpha \\ &\quad = \|X\|_\alpha^\alpha - \|Y\|_\alpha^\alpha = 0.\end{aligned}$$

Therefore $X+Y$ and $X-Y$ are also independent. But $X \sim S_\alpha(\sigma, 0, 0)$ for some $\sigma > 0$. Therefore, by the independence of $X+Y$ and $X-Y$ and of X and Y,

$$\begin{aligned}\exp\{-2^\alpha \sigma^\alpha\} &= E\exp i(2X) \\ &= E\exp i[(X+Y)+(X-Y)] \\ &= E\exp i(X+Y)\, E\exp i(X-Y) \\ &= (E\exp iX)^4 \\ &= \exp\{-4\sigma^\alpha\},\end{aligned}$$

which implies $2^\alpha = 4$ and $\alpha = 2$. ∎

2.10 Codifference

The covariation was defined in Section 2.7. Here, we introduce a different measure of bivariate dependence called the codifference which, like the covariation, reduces to the covariance when $\alpha = 2$. Whereas the covariation may not be defined for $\alpha \le 1$, the codifference is defined for all $0 < \alpha \le 2$.

Definition 2.10.1 The *codifference* of two jointly $S\alpha S$, $0 < \alpha \le 2$, random variables X and Y equals

$$\tau_{X,Y} = \|X\|_\alpha^\alpha + \|Y\|_\alpha^\alpha - \|X-Y\|_\alpha^\alpha, \tag{2.10.1}$$

where $\|X\|_\alpha$, $\|Y\|_\alpha$ and $\|X-Y\|_\alpha$ denote, respectively, the scale parameters of X, Y and $X-Y$.

We shall be mainly interested in investigating the asymptotic behavior of $\tau(t) = \tau_{X(t),X(0)}$ as $t \to \infty$, where $\{X(t),\ t \in \mathbb{R}\}$ is a stationary $S\alpha S$ stochastic process. In this section, however, we consider some elementary properties of $\tau_{X,Y}$, where (X,Y) is a given $S\alpha S$ random vector. Observe firstly that, in contrast to the covariation, we have:

Property 2.10.2

$$\tau_{X,Y} = \tau_{Y,X}.$$

Like the covariation, the codifference reduces to the covariance when $\alpha = 2$:

Property 2.10.3 *If* $\alpha = 2$, *then* $\tau_{X,Y} = \operatorname{Cov}(X,Y)$.

PROOF: When $\alpha = 2$, $\|X\|_2^2 = \frac{1}{2}\text{Var}\, X$ and hence $\tau_{X,Y} = \frac{1}{2}\{\text{Var}\, X + \text{Var}\, Y - \text{Var}\,(X - Y)\} = \text{Cov}\,(X, Y)$. ■

The codifference, like the covariation, vanishes when the random variables are independent:

Property 2.10.4 *If X and Y are independent, then $\tau_{X,Y} = 0$. Conversely, if $\tau_{X,Y} = 0$ and $0 < \alpha < 1$, then X and Y are independent.*

PROOF: If X and Y are independent, then $s_1 s_2 = 0$, $\Gamma_{X,Y}$ a.e. (Example 2.3.5), where $\Gamma_{X,Y}$ is the spectral measure of (X, Y). Hence,

$$\begin{aligned} &\|X - Y\|_\alpha^\alpha \\ &= \int_{S_2} |s_1 - s_2|^\alpha \Gamma_{X,Y}(ds) \\ &= \int_{S_2} |s_1|^\alpha \Gamma_{X,Y}(ds) + \int_{S_2} |s_2|^\alpha \Gamma_{X,Y}(ds) \qquad (s_1 s_2 = 0,\ \Gamma_{X,Y} \text{ a.e.}) \\ &= \|X\|_\alpha^\alpha + \|Y\|_\alpha^\alpha, \end{aligned}$$

i.e., $\tau_{X,Y} = 0$.

Further, if $0 < \alpha < 1$, then $|s_1 - s_2|^\alpha \le |s_1|^\alpha + |s_2|^\alpha$, with equality only when $s_1 s_2 = 0$. The preceding computation shows that if $\tau_{X,Y} = 0$, then $s_1 s_2 = 0$ $\Gamma_{X,Y}$-a.e., and so X and Y are independent. ■

When $1 \le \alpha < 2$, $\tau_{X,Y} = 0$ does not imply that X and Y are independent (Exercise 2.27).

Property 2.10.5

$$\tau_{X,Y} \le \|X\|_\alpha^\alpha + \|Y\|_\alpha^\alpha$$

and

$$\tau_{X,Y} \ge \begin{cases} 0 & \text{if } 0 < \alpha \le 1, \\ (1 - 2^{\alpha-1})(\|X\|_\alpha^\alpha + \|Y\|_\alpha^\alpha) & \text{if } 1 \le \alpha \le 2. \end{cases}$$

PROOF: The upper bound follows from $\|X - Y\|_\alpha \ge 0$.

The lower bound on $\tau_{X,Y}$ depends on α. If $0 < \alpha \le 1$, then $\|\cdot\|_\alpha^\alpha$ satisfies $\|X - Y\|_\alpha^\alpha \le \|X\|_\alpha^\alpha + \|Y\|_\alpha^\alpha$ implying $\tau_{X,Y} \ge 0$. If $1 \le \alpha \le 2$, $\|\cdot\|_\alpha$ is a norm, so $\|X - Y\|_\alpha^\alpha \le (\|X\|_\alpha + \|Y\|_\alpha)^\alpha \le 2^{\alpha-1}(\|X\|_\alpha^\alpha + \|Y\|_\alpha^\alpha)$ by (2.7.6), implying $\tau_{X,Y} \ge (1 - 2^{\alpha-1})(\|X\|_\alpha^\alpha + \|Y\|_\alpha^\alpha)$. ■

The upper bound is achieved when $X = Y$. The lower bound is achieved for X and Y independent if $0 < \alpha \le 1$, and for $X = -Y$ if $1 \le \alpha \le 2$. If X and Y

are both standardized ($\|X\|_\alpha = \|Y\|_\alpha = 1$), then

$$0 \le \tau_{X,Y} \le 2 \quad \text{if} \quad 0 < \alpha \le 1,$$
$$2(1-2^{\alpha-1}) \le \tau_{X,Y} \le 2 \quad \text{if} \quad 1 \le \alpha \le 2.$$

When $\alpha = 2$, this is exactly equivalent to $-1 \le$ Corr $(X,Y) \le 1$, where Corr denotes the correlation coefficient.

The next property shows the significance of the codifference.

Property 2.10.6 *Let (X,Y) and (X',Y') be two $S\alpha S$, $0 < \alpha \le 2$, random vectors with $\|X\|_\alpha = \|Y\|_\alpha = \|X'\|_\alpha = \|Y'\|_\alpha$. Then*

$$\tau_{X,Y} \le \tau_{X',Y'}$$

implies that for every $c > 0$

$$P\{|X-Y| > c\} \ge P\{|X'-Y'| > c\}.$$

Conversely, if the last relation holds for some $c > 0$, then $\tau_{X,Y} \le \tau_{X',Y'}$.

PROOF: Suppose that $\|X-Y\|_\alpha \ne 0$ and $\|X'-Y'\|_\alpha \ne 0$ (the property holds trivially otherwise). The inequality $\tau_{X,Y} \le \tau_{X',Y'}$ is equivalent to $\|X-Y\|_\alpha \ge \|X'-Y'\|_\alpha$. This in turn is equivalent to

$$\begin{aligned} P\{|X-Y|>c\} &= P\{\|X-Y\|_\alpha^{-1}|X-Y| > \|X-Y\|_\alpha^{-1}c\} \\ &= P\{\|X'-Y'\|_\alpha^{-1}|X'-Y'| > \|X-Y\|_\alpha^{-1}c\} \\ &= P\{|X'-Y'| > \|X'-Y'\|_\alpha\|X-Y\|_\alpha^{-1}c\} \\ &\ge P\{|X'-Y'|>c\}. \quad \blacksquare \end{aligned}$$

The inequality $P\{|X-Y| > c\} \ge P\{|X'-Y'| > c\}$ has the following interpretation: the random variables X' and Y' are less likely to differ than X and Y and so are "more dependent." Thus, the larger τ, the "greater" the dependence.

The codifference shares an additional property with the covariance.

Property 2.10.7 *Let $(X_1,\ldots,X_d)$ be a $S\alpha S$ random vector. Then the matrix $\{\tau_{X_i,X_j},\ i,j=1,\ldots,d\}$ is non-negative definite.*

PROOF: We must show that for any integer d and any $u_1,\ldots,u_d \in \mathbb{R}$, $\sum_{i=1}^d\sum_{j=1}^d \tau_{X_i,X_j}\ u_iu_j \ge 0$. Let Γ be the spectral measure of the $S\alpha S$ random vector $(X_1,\ldots,X_d)$. Then

$$\sum_{i=1}^d\sum_{j=1}^d \tau_{X_i,X_j}u_iu_j = \sum_{i=1}^d\sum_{j=1}^d(\|X_i\|_\alpha^\alpha + \|X_j\|_\alpha^\alpha - \|X_i-X_j\|_\alpha^\alpha)u_iu_j$$

$$= \int_{S_d} \sum_{i=1}^{d} \sum_{j=1}^{d} (|s_i|^\alpha + |s_j|^\alpha - |s_i - s_j|^\alpha) u_i u_j \Gamma(d\mathbf{s}) \geq 0$$

because of the following lemma:

Lemma 2.10.8 *For any $0 < \alpha \leq 2$ and $n \geq 1$, the function*

$$\{A(\mathbf{s}_1, \mathbf{s}_2) = |\mathbf{s}_1|^\alpha + |\mathbf{s}_2|^\alpha - |\mathbf{s}_1 - \mathbf{s}_2|^\alpha\}, \quad \mathbf{s}_1, \mathbf{s}_2 \in \mathbb{R}^n,$$

is non-negative definite.

PROOF: We must show that for any $\mathbf{s}_1, \ldots, \mathbf{s}_m \in \mathbb{R}^n$ and real $u_1, \ldots, u_m$, $\sum_{i=1}^m \sum_{j=1}^m A(\mathbf{s}_i, \mathbf{s}_j) u_i u_j \geq 0$. The proof is in three steps.

Step 1. View each u_i as a mass at the point $\mathbf{s}_i$, $i = 1, \ldots, m$. Now add a mass $u_0 = -\sum_{i=1}^m u_i$ at the origin $\mathbf{s}_0 = 0$. Then $\sum_{i=0}^m u_i = 0$ and

$$\sum_{i=1}^m \sum_{j=1}^m \{|\mathbf{s}_i|^\alpha + |\mathbf{s}_j|^\alpha - |\mathbf{s}_i - \mathbf{s}_j|^\alpha\} u_i u_j = -\sum_{i=0}^m \sum_{j=0}^m |\mathbf{s}_i - \mathbf{s}_j|^\alpha u_i u_j$$

since, for example,

$$\begin{aligned} \sum_{i=1}^m \sum_{j=1}^m |\mathbf{s}_i|^\alpha u_i u_j &= \sum_{i=1}^m |\mathbf{s}_i|^\alpha u_i \sum_{j=1}^m u_j \\ &= -\sum_{i=1}^m |\mathbf{s}_i|^\alpha u_i u_0 \\ &= -\sum_{i=0}^m |\mathbf{s}_i - \mathbf{s}_0|^\alpha u_i u_0. \end{aligned}$$

Step 2. For any constant $c > 0$,

$$\begin{aligned} \sum_{i=0}^m \sum_{j=0}^m e^{-c|\mathbf{s}_i - \mathbf{s}_j|^\alpha} u_i u_j &= \sum_{i=0}^m \sum_{j=0}^m (e^{-c|\mathbf{s}_i - \mathbf{s}_j|^\alpha} - 1) u_i u_j \\ &= -c \sum_{i=0}^m \sum_{j=0}^m |\mathbf{s}_i - \mathbf{s}_j|^\alpha u_i u_j + o(c), \end{aligned}$$

as c tends to 0. It is then sufficient to prove that

$$\sum_{i=0}^m \sum_{j=0}^m e^{-c|\mathbf{s}_i - \mathbf{s}_j|^\alpha} u_i u_j \geq 0.$$

Step 3. The function $\phi(\boldsymbol{\theta}) = \exp\{-c|\boldsymbol{\theta}|^\alpha\}$, $0 < \alpha \leq 2$, $\boldsymbol{\theta} = (\theta_1, \ldots, \theta_n)$ is a characteristic function (of a sub-Gaussian $S\alpha S$ random vector (Proposition 2.5.5)). It is, therefore, a non-negative definite function of the variable $\boldsymbol{\theta}$, i.e., $\sum_{i=0}^m \sum_{j=0}^m \phi(\boldsymbol{\theta}_i - \boldsymbol{\theta}_j) u_i u_j \geq 0$. ∎

2.11 Exercises

Exercise 2.1 Let $Y = \rho_1 X_1 + \rho_2 X_2$ where ρ_1 and ρ_2 are non-zero real numbers and $X_i \sim S_{\alpha_i}(\sigma_i, \beta_i, \mu_i)$, $i = 1, 2$ with $\alpha_1 < \alpha_2$. Find C and α_3, such that

$$P(|Y| > \lambda) \sim C\lambda^{-\alpha_3}$$

as $\lambda \to \infty$.

Exercise 2.2 Prove Corollary 2.1.3.
Hint: Compare with Definition 1.1.4 and Relation (1.1.3).

Exercise 2.3 Let $\{\mathbf{Y}^{(n)} = (Y_1^{(n)}, \ldots, Y_d^{(n)}),\ n \geq 1\}$ be a sequence of random vectors in $\mathbb{R}^d$. Prove that $\{\mathbf{Y}^{(n)},\ n \geq 1\}$ is tight if the sequence of components $\{Y_k^{(n)},\ n \geq 1\}$ is tight for each $k = 1, \ldots, d$.
Definition. The sequence $\{\mathbf{Y}^{(n)},\ n \geq 1\}$ is *tight* if for every $\epsilon > 0$ there is a bounded rectangle A such that $P\{\mathbf{Y}^{(n)} \in A\} > 1 - \epsilon$ for all n.

Exercise 2.4 Using the results of Section 2.2, show that the function $\Psi(\theta_1, \theta_2)$ in 2.2.1 cannot be a characteristic function for $1 < \alpha \leq 2$.

Exercise 2.5 Let $X_i \sim S_\alpha(\sigma_i, \beta_i, 0)$, $i = 1, 2, \ldots d$, be independent random variables. Characterize the spectral measure of the α-stable random vector $(X_1, X_2, \ldots, X_d)$.

Exercise 2.6 Prove Proposition 2.3.9 in the case $\alpha \neq 1$.

Exercise 2.7 Let X_1, X_2, X_3 be i.i.d. $S_1(1, 1, 0)$ and let $Y_1 = X_1 - X_2$ and $Y_2 = X_2 - X_3$. Let Γ be the random measure of $\mathbf{Y} = (Y_1, Y_2)$. What are the points on which Γ is concentrated? Do these points have equal weight? What is the shift parameter of the random variable $\theta_1 Y_1 + \theta_2 Y_2$?

Exercise 2.8 Let X_1 and X_2 be jointly $S\alpha S$ with spectral measure Γ. Find the spectral measure of the vector $(aX_1 + bX_2, cX_1 + dX_2)$.

Exercise 2.9 Let $\mathbf{s}^{(k)} = (s_1^{(k)}, \ldots, s_d^{(k)}) \in S_d$, $k = 1, \ldots, m$, and suppose that the α-stable random vector $\mathbf{X} = (X_1, \ldots, X_d)$ has a shift vector $\boldsymbol{\mu}^0$ and a spectral measure Γ concentrated on the points $\mathbf{s}^{(k)}$, $k = 1, \ldots, m$, and $-\mathbf{s}^{(k)}$, $k = 1, \ldots, m$. Example 2.3.6 states that for suitably chosen triples $(\sigma_k, \beta_k, \mu_k)$, $k = 1, \ldots, m$, and a real matrix $A = (a_{jk})$, $j = 1, \ldots, d$, $k = 1, \ldots, m$, we have

$$(X_1, \ldots, X_d) \stackrel{d}{=} \Big(\sum_{k=1}^m a_{1k} Y_k, \ldots, \sum_{k=1}^m a_{dk} Y_k \Big),$$

where $Y_k \sim S_\alpha(\sigma_k, \beta_k, \mu_k)$, $k = 1, \dots, m$ are independent random variables. Verify that the choice

$$\sigma_k = 1,\ \beta_k = \frac{W_k^+ - W_k^-}{W_k^+ + W_k^-},\ k = 1, \dots, m,$$

and

$$a_{jk} = (W_k^+ + W_k^-)^{1/\alpha} s_j^{(k)},\ j = 1, \dots, d,\ k = 1, \dots, m,$$

will work, where

$$W_k^+ = \Gamma(\{\mathbf{s}^{(k)}\}),\ W_k^- = \Gamma(\{-\mathbf{s}^{(k)}\}),\ k = 1, \dots, m.$$

How should the μ_k, $k = 1, \dots, m$, be chosen?

Exercise 2.10 Find two symmetric (around 0) random variables X and Y, whose sum $X + Y$ is symmetric, but not around 0.
Hint: Let $Z_j,\ j = 1, 2, 3$ be independent with an $S_1(1, 1, 0)$ distribution and set $X = Z_1 - Z_2$ and $Y = Z_1 - Z_3$ (Chen & Shepp 1983).

Exercise 2.11 Find another example of a strictly 1-stable random vector which is not the translate of a symmetric 1-stable random vector.
Hint: Build the measure Γ in (2.3.1) concentrated on a few points of S_d. Reduce (2.4.1) to a system of linear equations.

Exercise 2.12 Let $\mathbf{X}$ be an α-stable random vector in $\mathbb{R}^d$ with characteristic function

$$E \exp\{i(\boldsymbol{\theta}, \mathbf{X})\} = \exp\Big\{-(\boldsymbol{\theta}^T \Sigma \boldsymbol{\theta})^{\alpha/2} + i \sum_{j=1}^{d} \theta_j \mu_j^0\Big\},$$

where T denotes transpose, Σ is a positive-definite symmetric $d \times d$ matrix and where $\boldsymbol{\mu}^0 = (\mu_1^0, \mu_2^0, \dots, \mu_d^0) \neq 0$. Is its spectral measure Γ symmetric? Is $\mathbf{X}$ strict? Is $\mathbf{X}$ symmetric? (Distinguish between the cases $\alpha \neq 1$ and $\alpha = 1$.) Is $\mathbf{X} - \boldsymbol{\mu}^0$ strict? Is it symmetric?

Exercise 2.13 Consider the following two symmetric measures on S_3:

$$\begin{aligned}
\Gamma_1 &= 3^{\alpha/2}\Big[\delta\Big(\frac{\sqrt{3}}{3}, \frac{\sqrt{3}}{3}, \frac{\sqrt{3}}{3}\Big) + \delta\Big(-\frac{\sqrt{3}}{3} - \frac{\sqrt{3}}{3}, -\frac{\sqrt{3}}{3}\Big)\Big] \\
&+ \big[\delta(1,0,0) + \delta(-1,0,0)\big] \\
&+ \big[\delta(0,1,0) + \delta(0,-1,0)\big] \\
&+ \big[\delta(0,0,1) + \delta(0,0,-1)\big]
\end{aligned}$$

and

$$\begin{aligned}\Gamma_2 &= 2^{\alpha/2}\left[\delta\left(\frac{\sqrt{2}}{2},\frac{\sqrt{2}}{2},0\right)+\delta\left(-\frac{\sqrt{2}}{2},-\frac{\sqrt{2}}{2},0\right)\right]\\ &+ 2^{\alpha/2}\left[\delta\left(\frac{\sqrt{2}}{2},0,\frac{\sqrt{2}}{2}\right)+\delta\left(-\frac{\sqrt{2}}{2},0,-\frac{\sqrt{2}}{2}\right)\right]\\ &+ 2^{\alpha/2}\left[\delta\left(0,\frac{\sqrt{2}}{2},\frac{\sqrt{2}}{2}\right)+\delta\left(0,-\frac{\sqrt{2}}{2},-\frac{\sqrt{2}}{2}\right)\right].\end{aligned}$$

Show that the two $S\alpha S$ distributions having, respectively, Γ_1 and Γ_2 as their spectral measures have identical two-dimensional marginals.

Exercise 2.14 The characteristic function $\phi(\theta_1,\theta_2)$ and density function $f(x_1,x_2)$ of an isotropic Cauchy random vector $\mathbf{X}=(X_1,X_2)$ are given in Example 2.5.6. Derive the density f from the characteristic function ϕ by using three different methods:

(a) Use the inversion formula for characteristic functions.

(b) Use the fact that $(X_1,X_2) = A^{1/2}(G_1,G_2)$ where G_1 and G_2 are i.i.d. Gaussian random variables and where $A^{1/2}$ is independent of (G_1,G_2) and has a Lévy distribution $S_{1/2}(\sigma,1,0)$ whose density is given in Chapter 1.

(c) Apply the formula for the density function of a multivariate Cauchy distribution given in Example 2.5.3.

Exercise 2.15 Let $\phi(\boldsymbol{\theta})$ and $f(\mathbf{x})$ be, respectively, the characteristic function and the density function of the multivariate Cauchy distribution, as given in Example 2.5.3. Derive $f(\mathbf{x})$ starting from $\phi(\boldsymbol{\theta})$.

Exercise 2.16 Prove Proposition 2.5.7.

Exercise 2.17 Let $X = X_1 + iX_2$ be a complex $S\alpha S$ random variable which is isotropic, and let $|X|$ be its modulus. Find the asymptotic behavior of $P\{|X|>\lambda\}$ as $\lambda\to\infty$.

Exercise 2.18 Show that the complex random variable $X = X_1 + iX_2$ is not isotropic if X_1 and X_2 are dependent mean zero Gaussian random variables.

Exercise 2.19 Suppose (X_1,X_2) is a $S\alpha S$ vector in $\mathbb{R}^2$ with uniform spectral measure Γ such that $\Gamma(S_2) = 1$. Use polar coordinates to evaluate the joint characteristic function of (X_1,X_2).

Exercise 2.20 Prove Properties 2.7.7 and 2.7.8 by using Definition 2.7.1.

Exercise 2.21 Let $\mathcal{S}_\alpha$ be a linear space of $S\alpha S$ random variables with $0 < \alpha \leq 1$. For $X \in \mathcal{S}_\alpha$, define

$$\sigma(X) = \text{scale parameter of } X.$$

Show that $\sigma(X)^\alpha$ is a norm on $\mathcal{S}_\alpha$ if $\alpha = 1$, and a quasi-norm (i.e., the triangular inequality holds) if $0 < \alpha < 1$.

Exercise 2.22 One can define the covariation for $\alpha = 1$ as

$$[X, Y]_1 = \int_{S_2} s_1 \text{ sign } (s_2) \Gamma(ds),$$

where Γ is the spectral measure of the jointly $S1S$ stable random variables X and Y. Using this definition, verify that all the results in Section 2.7 also hold for $\alpha = 1$, but that the alternative definition 2.7.3 is not valid.

Exercise 2.23 Define the covariation norm for $\alpha = 1$ using the definition in the previous exercise. Verify that all the results in Section 2.8 also hold for $\alpha = 1$.

Exercise 2.24 Using the previous exercise show that the "only if" part of Proposition 2.9.2 holds for $\alpha = 1$. Does the "if" part hold as well?

Exercise 2.25 Prove Proposition 2.9.5 when $\dim \mathcal{S}_\alpha \geq 3$ by using Proposition 2.9.3.

Exercise 2.26 Show that $\tau_{X,Y} = 0$ if X and Y are independent $S\alpha S$, without using the spectral measure.

Exercise 2.27 For a given $1 \leq \alpha < 2$ construct an example of a $S\alpha S$ random vector (X, Y) such that $\tau_{X,Y} = 0$ but X and Y are not independent.
Hint: Choose $\Gamma_{X,Y}$ so that $s_1 s_2 = 0$ $\Gamma_{X,Y}$-a.e.

Exercise 2.28 Let $A(x, y) = |x|^\alpha + |y|^\alpha - |x - y|^\alpha$, $0 < \alpha \leq 2$, $x, y \in \mathbb{R}$. Here is another way to show that A is non-negative definite.

Step 1. For some $c > 0$,

$$|\theta|^\alpha = c \int_{-\infty}^{\infty} [1 - \cos(\theta s)] |s|^{-(1+\alpha)} ds.$$

Step 2. The function

$$B(a, b) = \cos(a - b) - \cos a - \cos b + 1, \ a, b \in \mathbb{R},$$

is non-negative definite.
Hint: $2B(a, b) = g(a)\overline{g(b)} + \overline{g(a)}g(b)$ where $g(a) = 1 - e^{ia}$.

Step 3. Show that $A(x, y) = c \int_{-\infty}^{\infty} B(xs, ys)|s|^{-(1+\alpha)} ds$ and conclude that A is non-negative definite.

Chapter 3

Stable random processes and stochastic integrals

An α-stable stochastic process is a random element whose finite-dimensional distributions are α-stable. This definition is given in Section 3.1, and it is used to introduce the notion of α-stable stochastic integrals, i.e., integrals of non-random functions with respect to an α-stable random measure. It is convenient to view these integrals as α-stable stochastic processes parameterized by their integrands. In this way, we can define them by specifying their finite-dimensional distributions (Section 3.2). In order to show they are also bona fide integrals, we proceed in three steps: we introduce α-stable random measures (Section 3.3), construct integrals of simple functions with respect to such measures and show that a stable integral is a limit (in probability) of integrals of simple functions (Section 3.4.)

In Section 3.5, we develop some basic properties of stable integrals. We show, for example, in marked contrast to the Gaussian case $\alpha = 2$, that two α-stable stochastic integrals with $\alpha < 2$ are independent if and only if their integrands have disjoint support. We also state the representation theorem, whose proof can be found in Chapter 13. The representation theorem states that an α-stable random vector can be represented as an α-stable stochastic integral. This is why stochastic integrals are a convenient tool for studying α-stable random variables, vectors and, as we will see, processes as well.

Several important examples of stochastic integrals are presented in Section 3.6. These include the $S\alpha S$ Lévy motion, which is the counterpart of Brownian motion for $\alpha < 2$; the ($S\alpha S$) Ornstein-Uhlenbeck process, which is the α-stable extension of the Gaussian process with the same name; the reverse Ornstein-Uhlenbeck process, which is different from the previous one when $\alpha < 2$; the well-balanced linear fractional stable motion and the log-fractional stable motion, two self-similar (scaling) processes that will be discussed more thoroughly in Chapter 7. Sub-Gaussian processes and substable process are additional examples. They are introduced in Sections 3.7 and 3.8, respectively.

We mentioned earlier two definitions of stochastic integrals, one as a stochastic process, the second as a limit of integrals of simple functions. We consider also a third definition, in terms of a series representation. The series representation of an α-stable random measure is developed in Section 3.9, and it is used in Section 3.10 to obtain the series representation of the stochastic integral. This representation sheds light into the structure of α-stable stochastic processes. We use it, for example, in Section 3.11 to show that a broad class of $S\alpha S$ processes, those that satisfy the so-called *Condition S*, are in fact conditionally centered Gaussian processes. These processes always admit an integral representation.

There exists a fourth representation of stable stochastic integrals. Since a stable measure can be expressed in terms of a Poisson measure, it is possible to define stable stochastic integrals directly as integrals with respect to Poisson measures. This was Paul Lévy's original approach, and we present it in Section 3.12.

3.1 Stable stochastic processes

We start with the definition of a stable stochastic process $\{X(t), t \in T\}$, where T is an arbitrary set, e.g., $\mathbb{R}$, $\mathbb{R}^n$, or a set of functions. The finite-dimensional distributions of $\{X(t), t \in T\}$ are the distributions of the vectors

$$(X(t_1), X(t_2), \ldots, X(t_d)),\ t_1, t_2, \ldots, t_d \in T,\ d \geq 1.$$

Definition 3.1.1 A stochastic process $\{X(t), t \in T\}$ is *stable* if all its finite-dimensional distributions are stable. It is *strictly stable* or *symmetric stable* if all its finite-dimensional distributions are, respectively, strictly stable or symmetric stable.

If the finite-dimensional distributions are stable then, by consistency they must all have the same index of stability α. We use the term α-*stable* when we wish to specify the index of stability.

The following theorem is an immediate consequence of Theorem 2.1.2 and Theorem 2.1.5.

Theorem 3.1.2 *Let $\{X(t), t \in T\}$ be a stochastic process.*

(a) $\{X(t), t \in T\}$ is strictly stable if and only if all linear combinations

$$\sum_{k=1}^{d} b_k X(t_k),\ d \geq 1,\ t_1, t_2, \ldots, t_d \in T,\ b_1, b_2, \ldots, b_d \text{ real} \tag{3.1.1}$$

are strictly stable.

(b) $\{X(t), t \in T\}$ *is symmetric stable if and only if all linear combinations (3.1.1) are symmetric stable.*

(c) If $\alpha \geq 1$*, then* $\{X(t), t \in T\}$ *is* α*-stable if and only if all linear combinations (3.1.1) are* α*-stable.*

Example 3.1.3 *α-stable Lévy motion.* A stochastic process $\{X(t), t \geq 0\}$ is called (standard) α-stable Lévy motion if

(1) $X(0) = 0$ a.s.

(2) X has independent increments.[1]

(3) $X(t) - X(s) \sim S_\alpha((t-s)^{1/\alpha}, \beta, 0)$ for any $0 \leq s < t < \infty$ and for some $0 < \alpha \leq 2,\ -1 \leq \beta \leq 1$.[2]

Observe that the process X has stationary increments. It is Brownian motion when $\alpha = 2$. The α-stable Lévy motions are $S\alpha S$ when $\beta = 0$ and they are $1/\alpha$-self-similar (unless $\alpha = 1, \beta \neq 0$), i.e., for all $c > 0$, $\{X(ct), t \geq 0\}$ and $\{c^{1/\alpha}X(t),\ t \geq 0\}$ have the same finite-dimensional distributions.

The role that α-stable Lévy motion plays among stable processes is similar to the role that Brownian motion plays among Gaussian processes.[3] We will encounter many more examples of stable processes in the sequel.

3.2 Definition of stable integrals as a stochastic process

The "stable integral" of a non-random function f will be denoted $I(f)$. In this section, we define the family of stable integrals $\{I(f), f \in F\}$ as a stochastic process indexed by a set F of functions. The definition will ensure that the integral $I(f)$ is a stable random variable for each integrand f, and that $I(\cdot)$ is a linear functional, i.e., $I(a_1f_1 + a_2f_2) = a_1I(f_1) + a_2I(f_2)$ a.s. for any $f_1, f_2 \in F$ and $a_1, a_2 \in \mathbb{R}$.

In order to define the stochastic process $\{I(f), f \in F\}$, it is enough to specify its finite-dimensional distributions, show that they are consistent and then appeal to Kolmogorov's existence theorem (Billingsley (1986), Theorem 36.2) to conclude that $\{I(f), f \in F\}$ is a well-defined stochastic process.

[1] This means that the random variables $X(t_2) - X(t_1), X(t_3) - X(t_2), \cdots, X(t_n) - X(t_{n-1})$ are independent for any $t_1 \leq t_2 \leq \ldots \leq t_n$.

[2] Some authors also impose a condition on the path structure. However, in this chapter, all processes are defined only in terms of their finite-dimensional distributions.

[3] Some authors whose research focuses on α-stable Lévy motion call it "the stable process." Such a terminology is confusing and should be avoided. When $\alpha = 2$, one distinguishes between "Brownian motion" and a "Gaussian process." Similarly, when $\alpha < 2$, it is necessary to distinguish between "α-stable Lévy motion" and an "α-stable process."

We start with the specification of the finite-dimensional distributions. Let $(E, \mathcal{E}, m)$ be a measure space and let

$$\beta : E \to [-1, 1]$$

be a measurable function. Choose

$$F = \begin{cases} L^\alpha(E, \mathcal{E}, m) & \text{if } \alpha \neq 1, \\ \mathcal{F}(m, \beta) & \text{if } \alpha = 1, \end{cases} \tag{3.2.1}$$

where

$$L^\alpha(E, \mathcal{E}, m) = \{f : f \text{ is measurable}, \int_E |f(x)|^\alpha m(dx) < \infty\},$$

$$\mathcal{F}(m, \beta) = \{f : f \in L^1(E, \mathcal{E}, m) \text{ and } \int_E |f(x)\beta(x) \ln |f(x)|\, |m(dx) < \infty\}.$$

It is easy to check that F is a linear space. (See Exercise 3.3.) We may and will assume without loss of generality that m is σ-finite, because $f \in F$ implies that the support of f is contained in a region of E where m is σ-finite (see Exercise 3.1).

SPECIFICATION OF THE FINITE-DIMENSIONAL DISTRIBUTIONS:

Given $f_1, \ldots, f_d \in F$, we define a probability measure $P_{f_1,\ldots,f_d}$ in $\mathbb{R}^d$ by its characteristic function, as follows:

(i) If $\alpha \neq 1$:

$$\phi_{f_1,\ldots,f_d}(\theta_1, \ldots, \theta_d) =$$

$$\exp\Big\{-\int_E \Big|\sum_{j=1}^d \theta_j f_j(x)\Big|^\alpha \Big(1 - i\beta(x) \text{ sign } \Big(\sum_{j=1}^d \theta_j f_j(x)\Big) \tan \frac{\pi\alpha}{2}\Big) m(dx)\Big\}; \tag{3.2.2}$$

(ii) if $\alpha = 1$:

$$\phi_{f_1,\ldots,f_d}(\theta_1, \ldots, \theta_d) =$$

$$\exp\Big\{-\int_E \Big|\sum_{j=1}^d \theta_j f_j(x)\Big| \Big(1+i\frac{2}{\pi}\beta(x)\text{sign}\Big(\sum_{j=1}^d \theta_j f_j(x)\Big) \ln \Big|\sum_{j=1}^d \theta_j f_j(x)\Big|\Big) m(dx)\Big\}.$$

We must prove that $\phi_{f_1,\ldots,f_d}$ is, indeed, the characteristic function of a probability measure in $\mathbb{R}^d$; we will show that, in fact, it is the characteristic function of an α-stable distribution. This is accomplished by making a change of variables that will transform the integral over E to an integral over the unit sphere S_d in $\mathbb{R}^d$.

Note that here m is not necessarily the spectral measure of an α-stable law because E is not necessarily the unit sphere S_d. In fact, the advantage of the representation (3.2.2) is that the same measure m is used for all values of d and all functions $f_1, \ldots, f_d$ in F.

For notational convenience, set $\boldsymbol{\theta} = (\theta_1, \ldots, \theta_d)$, $\mathbf{f} = (f_1, \ldots, f_d)$ and let $u_\alpha(\boldsymbol{\theta}, \mathbf{f}(x), \beta(x))$ denote

$$\Big|\sum_{j=1}^d \theta_j f_j(x)\Big|^\alpha \Big(1 - i\beta(x) \text{ sign } \Big(\sum_{j=1}^d \theta_j f_j(x)\Big) \tan \frac{\pi\alpha}{2}\Big)$$

if $\alpha \neq 1$, and

$$\Big|\sum_{j=1}^d \theta_j f_j(x)\Big| \Big(1 + i\frac{2}{\pi}\beta(x) \text{ sign } \Big(\sum_{j=1}^d \theta_j f_j(x)\Big) \ln \Big|\sum_{j=1}^d \theta_j f_j(x)\Big|\Big)$$

if $\alpha = 1$. Set also

$$E_+ = \Big\{x \in E : \sum_{j=1}^d f_j(x)^2 > 0\Big\}.$$

Suppose, firstly, that $\alpha \neq 1$. Then

$$\begin{aligned}
&\phi_{f_1,\ldots,f_d}(\theta_1, \ldots, \theta_d) \\
&\quad = \exp\Big\{-\int_{E_+} u_\alpha(\boldsymbol{\theta}, \mathbf{f}(x), \beta(x)) m(dx)\Big\} \\
&\quad = \exp\Big\{-\int_{E_+} u_\alpha\Big(\boldsymbol{\theta}, \frac{\mathbf{f}(x)}{(\sum_{k=1}^d f_k(x)^2)^{1/2}}, \beta(x)\Big) \Big(\sum_{k=1}^d f_k(x)^2\Big)^{\alpha/2} m(dx)\Big\} \\
&\quad = \exp\Big\{-\int_{E_+} u_\alpha(\boldsymbol{\theta}, \mathbf{g}(x), \beta(x)) m_1(dx)\Big\} \qquad (3.2.3)
\end{aligned}$$

where

$$g_j(x) = \frac{f_j(x)}{(\sum_{k=1}^d f_k(x)^2)^{1/2}}, \; j = 1, \ldots, d,$$

and

$$m_1(dx) = \Big(\sum_{k=1}^d f_k(x)^2\Big)^{\alpha/2} m(dx).$$

Note that m_1 is a finite measure on $(E_+, \mathcal{E})$ since $f_k \in L^\alpha(E, \mathcal{E}, m)$ for $k = 1, \ldots, d$. Moreover, $\sum_{j=1}^d g_j(x)^2 = 1$, for all $x \in E_+$. Writing

$$\begin{aligned}\phi_{f_1,\ldots,f_d}(\theta_1,\ldots,\theta_d) &= \exp\Big\{-\int_{E_+} u_\alpha(\boldsymbol{\theta},\mathbf{g}(x),1)\frac{1+\beta(x)}{2}m_1(dx)\\ &\quad -\int_{E_+} u_\alpha(\boldsymbol{\theta},-\mathbf{g}(x),1)\frac{1-\beta(x)}{2}m_1(dx)\Big\},\end{aligned}$$

we can make the change of variables $s_j = g_j(x)$ in one integral and $s_j = -g_j(x)$ in the other, to conclude that $\phi_{f_1,\ldots,f_d}(\theta_1,\ldots,\theta_d)$ equals

$$\exp\Big\{-\int_{S_d} \Big|\sum_{j=1}^d \theta_j s_j\Big|^\alpha \Big(1 - i\ \mathrm{sign}\ \Big(\sum_{j=1}^d \theta_j s_j\Big)\tan\frac{\pi\alpha}{2}\Big)\Gamma(ds)\Big\},$$

where the finite measure Γ on S_d is given by

$$\Gamma(A) = \int_{g^{-1}(A)} \frac{1+\beta(x)}{2}m_1(dx) + \int_{g^{-1}(-A)} \frac{1-\beta(x)}{2}m_1(dx), \qquad (3.2.4)$$

where A is a Borel set in S_d and

$$g^{-1}(A) = \{x \in E_+ : (g_1(x),\ldots,g_d(x)) \in A\}.$$

This shows that (3.2.2) is the characteristic function of an α-stable law on $\mathbb{R}^d$ when $\alpha \neq 1$.

In the case $\alpha = 1$, we have to pay special attention to the presence of the logarithm. Since

$$\ln\Big|\sum_{j=1}^d \theta_j f_j(x)\Big| = \ln\Big|\sum_{j=1}^d \theta_j \frac{f_j}{(\sum_{k=1}^d f_k(x)^2)^{1/2}}\Big| + \ln\Big(\sum_{k=1}^d f_k(x)^2\Big)^{1/2},$$

Relation (3.2.3) becomes

$$\phi_{f_1,\ldots,f_d}(\theta_1,\ldots,\theta_d) = \exp\Big\{-\int_{E_+} u_\alpha(\boldsymbol{\theta},\mathbf{g}(x),\beta(x))m_1(dx) + i\sum_{j=1}^d \theta_j\mu_j^0\Big\},$$

where $\mathbf{g} = (g_1,\ldots,g_d)$ and m_1 are defined as above and where

$$\mu_j^0 = -\frac{1}{\pi}\int_{E_+} f_j(x)\beta(x)\ln\left(\sum_{k=1}^d f_k(x)^2\right)m(dx),\ j = 1,\ldots,d. \qquad (3.2.5)$$

Proceeding as in the case $\alpha \neq 1$, we conclude that $\phi_{f_1,\ldots,f_d}(\theta_1,\ldots,\theta_d)$ equals

$$\exp\Big\{-\int_{S_d} \Big|\sum_{j=1}^d \theta_j s_j\Big|\Big(1+i\frac{2}{\pi}\ \mathrm{sign}\ \Big(\sum_{j=1}^d \theta_j s_j\Big)\ln\Big|\sum_{j=1}^d \theta_j s_j\Big|\Big)\Gamma(ds)+i\sum_{j=1}^d \theta_j\mu_j^0\Big\}.$$

This shows that (3.2.2) is also the characteristic function of an α-stable law on $\mathbb{R}^d$ in the case $\alpha = 1$.

To verify the consistency of the probability measures $P_{f_1,\ldots,f_d}$, note that for any permutation $(\pi(1),\ldots,\pi(d))$ of $(1,\ldots,d)$, we have

$$\phi_{f_{\pi(1)},\ldots,f_{\pi(d)}}(\theta_{\pi(1)},\ldots,\theta_{\pi(d)}) = \phi_{f_1,\ldots,f_d}(\theta_1,\ldots,\theta_d)$$

and for any $n \le d$,

$$\phi_{f_1,\ldots,f_n}(\theta_1,\ldots,\theta_n) = \phi_{f_1,\ldots,f_n,\ldots,f_d}(\theta_1,\ldots,\theta_n,0,\ldots,0).$$

This proves consistency.

By Kolmogorov's existence theorem, there is a stochastic process which we denote $\{I(f), f \in F\}$ whose finite-dimensional distributions are given by (3.2.2).

$I(f)$ is called the *α-stable integral of f*. The measure m is called the *control measure* and the function β is called the *skewness intensity*.

We now establish some elementary properties of the integral. We will say that a function f is integrable if it belongs to the relevant index set F in (3.2.1).

Property 3.2.1 *For any integrable $f_1, f_2, \ldots, f_d$, the integrals $I(f_1), I(f_2), \ldots, I(f_d)$ are jointly α-stable with joint characteristic function given by (3.2.2); they are jointly $S\alpha S$ if the skewness intensity is zero.*

Property 3.2.2 *Let f be integrable. Then $I(f) \sim S_\alpha(\sigma_f, \beta_f, \mu_f)$ where*

$$\begin{aligned}
\sigma_f &= \left(\int_E |f(x)|^\alpha m(dx)\right)^{1/\alpha}, \\
\beta_f &= \frac{\int_E f(x)^{<\alpha>}\beta(x)m(dx)}{\int_E |f(x)|^\alpha m(dx)}, \\
\mu_f &= \begin{cases} 0 & \textit{if } \alpha \neq 1, \\ -\frac{2}{\pi}\int_E f(x)\beta(x)\ln|f(x)|m(dx) & \textit{if } \alpha = 1. \end{cases}
\end{aligned}$$

Property 3.2.1 follows from the definition of the process $\{I(f),\ f \in F\}$. Property 3.2.2 is straightforward (substitute $\theta_2 = \cdots = \theta_d = 0$ in (3.2.2)). Note that when $\alpha \neq 1$, $I(f)$ has zero shift parameter and is, therefore, strictly stable, but it is usually not strictly stable when $\alpha = 1$. $I(f)$ is symmetric α-stable when $\beta(\cdot) = 0$.

The next property states that the integral $I(\cdot)$ is a linear functional.

Property 3.2.3 *If f_1 and f_2 are integrable, then*

$$I(a_1 f_1 + a_2 f_2) = a_1 I(f_1) + a_2 I(f_2) \ \textit{a.s.}$$

for any real numbers a_1 and a_2.

PROOF: We must show that

$$I(a_1 f_1 + a_2 f_2) - a_1 I(f_1) - a_2 I(f_2) = 0 \text{ a.s.}$$

For any real θ we have

$$\begin{aligned} &E \exp\left\{i\theta[I(a_1 f_1 + a_2 f_2) - a_1 I(f_1) - a_2 I(f_2)]\right\} \\ &= E \exp\left\{i[\theta I(a_1 f_1 + a_2 f_2) - (a_1\theta) I(f_1) - (a_2\theta) I(f_2)]\right\} \\ &= 1 \end{aligned}$$

by (3.2.2), because

$$\theta(a_1 f_1 + a_2 f_2) - (a_1\theta) f_1 - (a_2\theta) f_2 = 0. \quad \blacksquare$$

3.3 α-stable random measures

It is most convenient to view a random measure as a stochastic process $M(\cdot)$ indexed by sets A. The term "random measure" captures the two main characteristics of M: the fact that $(M(A_1), \ldots, M(A_d))$ is a random vector and the fact that M is additive, i.e., $M(\cup_i A_i) = \sum_i M(A_i)$ a.s. if the sets A_i are disjoint. This last relation does not imply that M is an ordinary signed measure for almost every realization ω, since, for example, $M(\cdot, \omega)$ may not have bounded variation.[4] As we will see in the next section, M can nevertheless serve as integrator in an alternative definition $I(f) = \int_E f(x) M(dx)$ of the stable integral.

We now turn to the definition of an α-stable random measure M. Denote by $(\Omega, \mathcal{F}, P)$ the underlying probability space and by $L^0(\Omega)$ the set of all real random variables defined on it. Let $(E, \mathcal{E}, m)$ be a measure space,

$$\beta : E \to [-1, 1]$$

be a measurable function, and let

$$\mathcal{E}_0 = \{A \in \mathcal{E} : m(A) < \infty\}$$

be the subset of $\mathcal{E}$ that contains sets of finite m-measure.

Definition 3.3.1 An independently scattered σ-additive set function

$$M : \mathcal{E}_0 \to L^0(\Omega)$$

[4]This is why some authors prefer to call M "random noise" instead of "random measure."

such that for each $A \in \mathcal{E}_0$,

$$M(A) \sim S_\alpha\left((m(A))^{1/\alpha}, \frac{\int_A \beta(x)m(dx)}{m(A)}, 0\right)$$

is called an *α-stable random measure on $(E, \mathcal{E})$ with control measure m and skewness intensity β.*

Independently scattered means that if $A_1, A_2, \ldots, A_k$ belong to $\mathcal{E}_0$ and are disjoint, then the random variables $M(A_1), M(A_2), \ldots, M(A_k)$ are independent. *σ-additive* means that if $A_1, A_2, \ldots$ belong to $\mathcal{E}_0$, are disjoint and $\cup_{j=1}^\infty A_j \in \mathcal{E}_0$, then

$$M\Big(\bigcup_{j=1}^\infty A_j\Big) = \sum_{j=1}^\infty M(A_j) \text{ a.s.}$$

Note that we require the shift parameter μ to be 0 for each $M(A)$. This means that the random measure M is characterized by a (non-random) measure m and a function β.

In order to show that the α-stable random measure M exists, we define

$$\{M(A), A \in \mathcal{E}_0\}$$

as a stochastic process. Using the existence of a stochastic process $\{I(f),\ f \in F\}$ established in Section 3.2 and setting $M(A) = I(1_A)$ for $A \in \mathcal{E}_0$, yields the existence of a stochastic process $\{M(A),\ A \in \mathcal{E}_0\}$ with the following finite-dimensional distributions: for $A_1, A_2, \ldots, A_d \in \mathcal{E}_0$ and $\theta_1, \theta_2, \ldots, \theta_d$ real numbers,

$$\sum_{j=1}^d \theta_j M(A_j) \sim S_\alpha(\sigma, \beta, \mu), \tag{3.3.1}$$

where

$$\sigma^\alpha = \int_E \Big|\sum_{j=1}^d \theta_j 1_{A_j}(x)\Big|^\alpha m(dx),$$

$$\beta = \frac{\int_E \beta(x)\left(\sum_{j=1}^d \theta_j 1_{A_j}(x)\right)^{<\alpha>} m(dx)}{\int_E |\sum_{j=1}^d \theta_j 1_{A_j}(x)|^\alpha m(dx)},$$

$$\mu = \begin{cases} 0 & \text{if } \alpha \neq 1, \\ -\frac{2}{\pi}\int_E \sum_{j=1}^d \theta_j 1_{A_j}(x)\beta(x) \ln|\sum_{j=1}^d \theta_j 1_{A_j}(x)| m(dx) & \text{if } \alpha = 1. \end{cases}$$

Is such an M independently scattered? If the As in (3.3.1) are disjoint, then for $\alpha \neq 1$, by (3.3.1),

$$E\exp\Big\{i\sum_{j=1}^d \theta_j M(A_j)\Big\}$$

$$= \exp\Big\{-\int_E |\sum_{j=1}^d \theta_j 1_{A_j}(x)|^\alpha \Big(1 - i\beta(x)\ \text{sign}\ \Big(\sum_{j=1}^d \theta_j 1_{A_j}(x)\Big) \tan\frac{\pi\alpha}{2}\Big) m(dx)\Big\}$$

$$= \exp\Big\{-\sum_{j=1}^d |\theta_j|^\alpha m(A_j)\Big(1 - i\frac{\int_{A_j}\beta(x)m(dx)}{m(A_j)}\ \text{sign}\ (\theta_j)\tan\frac{\pi\alpha}{2}\Big)\Big\}$$

$$= \prod_{j=1}^d E\exp\{i\theta_j M(A_j)\}.$$

This shows that M is independently scattered. The argument for $\alpha = 1$ is similar. Finite additivity follows from the linearity of the integral (Property 3.2.3) and the equality $\sum_{j=1}^d 1_{A_j} = 1_{\cup_{j=1}^d A_j}$ for disjoint A_js. But is M σ-additive? To prove that it is, let $A_1, A_2, \ldots \in \mathcal{E}_0$, $B = \cup_{j=1}^\infty A_j \in \mathcal{E}_0$. We must establish that

$$M(B) = \sum_{j=1}^\infty M(A_j) \ \text{a.s.},$$

i.e.,

$$\lim_{n\to\infty}\sum_{j=1}^n M(A_j) = M(B) \ \text{a.s.} \tag{3.3.2}$$

or

$$\lim_{n\to\infty}\sum_{j=1}^n M(A_j) = M(B) \text{ in probability}, \tag{3.3.3}$$

because (3.3.2) is equivalent to (3.3.3), since the series $\sum_{j=1}^n M(A_j)$ has independent summands. By finite additivity, we have $\sum_{j=1}^n M(A_j) = M(\cup_{j=1}^n A_j)$ a.s. and, hence, almost surely,

$$M(B) - \sum_{j=1}^n M(A_j) = M(B) - M\Big(\bigcup_{j=1}^n A_j\Big) = M\Big(\bigcup_{j=n+1}^\infty A_j\Big),$$

by again using finite additivity. But $M(\cup_{j=n+1}^\infty A_j) \sim S_\alpha(\sigma_n, \beta_n, 0)$, where

$$\begin{aligned}\sigma_n^\alpha &= m\Big(\bigcup_{j=n+1}^\infty A_j\Big)\\ &= \sum_{j=n+1}^\infty m(A_j) \quad (\sigma\text{-additivity of the measure } m)\\ &< \infty \qquad\qquad (\text{because } \cup_{j=n+1}^\infty A_j \subset B \in \mathcal{E}_0).\end{aligned}$$

Since $\lim_{n\to\infty} \sigma_n = 0$, we have $M(\cup_{j=n+1}^{\infty} A_j) \stackrel{prob}{\to} 0$ as $n \to \infty$ and, hence,

$$M(B) - \sum_{j=1}^{n} M(A_j) \stackrel{prob}{\to} 0$$

as $n \to \infty$, proving that M is σ-additive.

Definition 3.3.2 M is called a $S\alpha S$ random measure if the skewness intensity β is zero.

Example 3.3.3 Let M be an α-stable random measure on $([0,\infty),\mathcal{B})$ with Lebesgue control measure and constant skewness intensity $\beta(x) = \beta,\ 0 \le x < \infty$. Let

$$X(t) = M([0,t]),\ 0 \le t < \infty.$$

$M([0,t])$ is well defined because $m([0,t]) < \infty$ for any $t > 0$. It is easy to check that the stochastic process $\{X(t),\ 0 \le t < \infty\}$ is α-stable Lévy motion.

3.4 Constructive definition of stable integrals

The α-stable integrals $I(f)$ were defined in Section 3.2 as a stochastic process indexed by the integrands f. In this section, we show that $I(f)$ can also be constructed as a bona fide integral, which we denote $\int_E f(x)M(dx)$, where M is some α-stable random measure. To define $\int_E f(x)M(dx)$, we proceed as follows: we approximate f by simple functions $f^{(n)}$, i.e., functions that take only a finite number of different values, we define $\int_E f^{(n)}(x)M(dx)$ and then take a limit in probability as $n \to \infty$. The limit is denoted $\int_E f(x)M(dx)$. We finally show that this definition coincides with the definition of the integral given in Section 3.2.

Let M be an α-stable random measure on $(E,\mathcal{E})$ with a control measure m and skewness intensity β, and let $\mathcal{E}_0 = \{A \in \mathcal{E} : m(A) < \infty\}$. Recall that we want to define

$$I(f) = \int_E f(x)M(dx)$$

for all measurable functions $f : E \to \mathbb{R}^1$ satisfying the condition

$$\int_E |f(x)|^\alpha m(dx) < \infty \tag{3.4.1}$$

and also the condition

$$\int_E \Big| f(x)\beta(x)\ln|f(x)| \Big| m(dx) < \infty \tag{3.4.2}$$

if $\alpha = 1$. We denote the collection of functions satisfying (3.4.1) and (3.4.2) by F, as in (3.2.1). Observe that F is a linear space (Exercise 3.3).

As usual, for a simple function of the form $f(x) = \sum_{j=1}^{n} c_j 1_{A_j}(x)$ where A_j are disjoint sets belonging to $\mathcal{E}_0$, we define

$$I(f) = \int_E f(x)M(dx) = \sum_{j=1}^{n} c_j M(A_j). \tag{3.4.3}$$

Since the random measure M is independently scattered, the α-stable random variables $M(A_1), \ldots, M(A_n)$ are independent. Applying Properties 1.2.1 and 1.2.3, we conclude that $I(f) \sim S_\alpha(\sigma_f, \beta_f, \mu_f)$ with

$$\sigma_f = \left(\int_E |f(x)|^\alpha m(dx) \right)^{1/\alpha}, \tag{3.4.4}$$

$$\beta_f = \frac{\int_E f(x)^{<\alpha>} \beta(x) m(dx)}{\int_E |f(x)|^\alpha m(dx)}, \tag{3.4.5}$$

$$\mu_f = \begin{cases} 0 & \text{if } \alpha \neq 1, \\ -\frac{2}{\pi} \int_E f(x)\beta(x) \ln |f(x)| m(dx) & \text{if } \alpha = 1. \end{cases} \tag{3.4.6}$$

The integral $I(f)$ is obviously linear in the simple functions f.

Consider, now, a general $f \in F$. We choose any sequence of simple functions $\{f^{(n)}\}_{n=1}^{\infty}$ possessing the following properties:

$$f^{(n)}(x) \to f(x) \text{ for almost every } x \in E, \tag{3.4.7}$$

$$|f^{(n)}(x)| \leq \theta(x), \text{ for every } n, x, \text{ and some } \theta \in F. \tag{3.4.8}$$

Such a sequence always exists. We can take, for example,

$$f^{(n)}(x) = \begin{cases} \frac{i}{n} & \text{if } \frac{i}{n} \leq f(x) < \frac{i+1}{n},\ i = 0, 1, \ldots, n^2 - 1, \\ -\frac{i}{n} & \text{if } -\frac{(i+1)}{n} < f(x) \leq -\frac{i}{n},\ i = 0, 1, \ldots, n^2 - 1, \\ 0 & \text{if } |f(x)| \geq n, \end{cases}$$

in which case $\theta = |f|$.

The sequence of the integrals $I(f^{(n)})$, $n = 1, 2, \ldots$, is well defined by (3.4.3), and we claim that it converges in probability. To show this, we must prove that $I(f^{(n)}) - I(f^{(m)})$ converges to zero in the metric of convergence in probability as $n, m \to \infty$. But

$$I(f^{(n)}) - I(f^{(m)}) = I(f^{(n)} - f^{(m)}) \sim S_\alpha(\sigma_{n,m}, \beta_{n,m}, \mu_{n,m}),$$

where

$$\sigma_{n,m} = \left(\int_E |f^{(n)}(x) - f^{(m)}(x)|^\alpha m(dx) \right)^{1/\alpha},$$

$-1 \le \beta_{n,m} \le 1$ and $\mu_{n,m}$ equals 0 if $\alpha \ne 1$ and

$$-\frac{2}{\pi} \int_E (f^{(n)}(x) - f^{(m)}(x)) \ln |f^{(n)}(x) - f^{(m)}(x)| \beta(x) m(dx)$$

if $\alpha = 1$. Therefore, convergence in probability of $\{I(f_n), n = 1, 2, \ldots\}$ will hold if we prove that $\sigma_{n,m} \to 0$ and $\mu_{n,m} \to 0$ as $n, m \to \infty$. Obviously, (3.4.8) implies that $|f^{(n)}(x) - f^{(m)}(x)| \le 2\theta(x)$, so that by the dominated convergence theorem, $\sigma_{n,m} \to 0$ as $n, m \to \infty$.

To show that $\mu_{n,m} \to 0$ in the case $\alpha = 1$, define

$$\psi(y) = \begin{cases} 0 & \text{if } y = 0, \\ y|\ln y| & \text{if } 0 < y \le e^{-1}, \\ e^{-1} & \text{if } e^{-1} < y \le \tau, \\ y \ln y & \text{if } y > \tau, \end{cases}$$

where τ is the solution of the equation $t \ln t = e^{-1}$. Clearly, ψ is a non-decreasing continuous function and, moreover,

$$\psi(y) \ge y|\ln y| \tag{3.4.9}$$

for any $y > 0$. Note that for any $\theta \in F$,

$$\begin{aligned}
&\int_E \psi(|\theta(x)|)|\beta(x)|m(dx) \\
&= \int_{\{|\theta(x)| \in (e^{-1}, \tau]\}} \psi(|\theta(x)|)|\beta(x)|m(dx) \\
&\quad + \int_{\{|\theta(x)| \notin (e^{-1}, \tau]\}} \psi(|\theta(x)|)|\beta(x)|m(dx) \\
&= e^{-1} \int_{\{|\theta(x)| \in (e^{-1}, \tau]\}} |\beta(x)|m(dx) \\
&\quad + \int_{\{|\theta(x)| \notin (e^{-1}, \tau]\}} |\theta(x)(\ln |\theta(x)|)\beta(x)|m(dx) \\
&\le e^{-1} m\{x : |\theta(x)| \in (e^{-1}, \tau]\} + \int_E |\theta(x)(\ln |\theta(x)|)\beta(x)|m(dx),
\end{aligned}$$

which is finite by (3.4.1) and (3.4.2). Now, the dominated convergence theorem implies that

$$\int_E \psi(|f^{(n)}(x) - f^{(m)}(x)|)|\beta(x)|m(dx) \to 0 \text{ as } n, m \to \infty,$$

and by (3.4.9) this implies that $\mu_{n,m} \to 0$ as $n, m \to \infty$. Therefore, the sequence $\{I(f^{(n)}), n = 1, 2, \ldots\}$ converges in probability, and we *define*

$$I(f) = \text{plim}_{n\to\infty} I(f^{(n)}),$$

where plim denotes limit in probability.

This definition does not depend on the choice of the approximating sequence $f^{(n)}$. Indeed, if both $f^{(n)}$ and $g^{(n)}$ converge to f and satisfy (3.4.7) and (3.4.8), then setting

$$h^{(n)} = \begin{cases} f^{(m)} & \text{if } n = 2m, \\ g^{(m)} & \text{if } n = 2m - 1 \end{cases}$$

yields $I(h^{(n)}) \overset{prob}{\to} I(f)$, which shows that the limits of $I(f^{(m)})$ and $I(g^{(m)})$ coincide.

Since convergence in probability implies convergence in distribution, we have

Proposition 3.4.1 *Let σ_f, β_f, μ_f be defined, respectively, as in (3.4.4), (3.4.5) and (3.4.6). Then*

$$I(f) \sim S_\alpha(\sigma_f, \beta_f, \mu_f),$$

i.e.,

(i) if $\alpha \neq 1$:

$E\exp\{i\theta I(f)\}$

$$= \exp\Big\{-\int_E |\theta f(x)|^\alpha \Big(1 - i\beta(x) \text{ sign } (\theta f(x)) \tan \frac{\pi\alpha}{2}\Big) m(dx)\Big\}; \quad (3.4.10)$$

(ii) if $\alpha = 1$:

$E\exp\{i\theta I(f)\}$

$$= \exp\Big\{-\int_E |\theta f(x)| \Big(1 + i\frac{2}{\pi}\beta(x) \text{ sign } (\theta f(x)) \ln |\theta f(x)|\Big) m(dx)\Big\}.$$

We now show that $I(f)$ is linear in f. Let $f \in F$, $g \in F$ and let $\{f^{(n)}\}_{n=1}^\infty$, $\{g^{(n)}\}_{n=1}^\infty$ be two sequences of simple functions satisfying (3.4.7) and (3.4.8) for f and g, respectively. Set $h = af + bg$ and $h^{(n)} = af^{(n)} + bg^{(n)}$, where a and b are two real numbers. Clearly, $\{h^{(n)}\}_{n=1}^{\infty}$ is a sequence of simple functions such that $h^{(n)}(x) \to h(x)$ as $n \to \infty$ for almost every x, and

$$\begin{aligned} |h^{(n)}(x)| &\leq |a||f^{(n)}(x)| + |b||g^{(n)}(x)| \\ &\leq |a|\theta_1(x) + |b|\theta_2(x) \end{aligned}$$

for some θ_1 and θ_2 in F. Therefore,

$$\begin{aligned} I(h) &= \operatorname{plim}_{n\to\infty} I(h^{(n)}) \\ &= \operatorname{plim}_{n\to\infty} (aI(f^{(n)}) + bI(g^{(n)})) \\ &= \operatorname{plim}_{n\to\infty} aI(f^{(n)}) + \operatorname{plim}_{n\to\infty} bI(g^{(n)}) \\ &= aI(f) + bI(g), \end{aligned}$$

proving linearity.

The linearity of the integral and Proposition 3.4.1 imply

Proposition 3.4.2 *For any $f_1, \ldots, f_d$ in F, the characteristic function of the random vector $(I(f_1), \ldots, I(f_d))$ is given by (3.2.2).*

This shows that the random vector $(I(f_1), \ldots, I(f_d))$ is, in fact, α-stable and, moreover, that the constructive definition of stable integrals is equivalent to that given in Section 3.2. Therefore,

Proposition 3.4.3 *For any $f_1, \ldots, f_d$ in F, the random vector $(I(f_1), \ldots, I(f_d))$ is α-stable with spectral measure Γ given by (3.2.4) and shift vector $\boldsymbol{\mu}^0$ equal to zero if $\alpha \neq 1$ and given by (3.2.5) if $\alpha = 1$.*

We adopt the terminology used in Section 3.2 and call $I(f) = \int_E f(x)M(dx)$ the integral of f with respect to an α-stable random measure M with control measure m and skewness intensity β.

Remark. If $\alpha = 1$ and the control measure m is finite, then the space $\mathcal{I}(m, \beta)$ of measurable functions satisfying (3.4.1) and (3.4.2) actually becomes a Banach space, as it is the intersection of the Banach space $L^1(m)$ and the Banach (Orlicz) space $L \log L(m_\beta)$ of functions satisfying

$$\int_E |f(x)| \ln_+ |f(x)| m_\beta(dx) < \infty,$$

where

$$\ln_+ a = \begin{cases} \ln a & \text{if } a \geq 1, \\ 0 & \text{if } 0 < a < 1, \end{cases}$$

and where

$$m_\beta(dx) = |\beta(x)|m(dx).$$

We equip $\mathcal{I}(m, \beta)$ with the maximum norm

$$\|f\|_{\mathcal{I}(m,\beta)} = \max\left(\|f\|_{L^1(m)}, \|f\|_{L \log L(m_\beta)}\right).$$

3.5 Properties of stable integrals

The following proposition relates the convergence of a sequence of α-stable integrals to the convergence of the sequence of integrands.

Proposition 3.5.1 *Let* $X_j = \int_E f_j(x)M(dx)$, $j = 1, 2, \ldots,$ *and* $X = \int_E f(x)M(dx)$, *where* M *is an* α*-stable random measure with control measure* m *and skewness intensity* β. *Then*

$$\operatorname{plim}_{j\to\infty} X_j = X$$

if and only if

$$\lim_{j\to\infty} \int_E |f_j(x) - f(x)|^\alpha m(dx) = 0 \tag{3.5.1}$$

and in the case $\alpha = 1$, *if also*

$$\lim_{j\to\infty} \int_E (f_j(x) - f(x)) \ln|f_j(x) - f(x)|\beta(x)m(dx) = 0. \tag{3.5.2}$$

PROOF: The argument is similar to the one used in the constructive definition of the integral. The convergence $\operatorname{plim}_{j\to\infty} X_j = X$ is equivalent to $\operatorname{plim}_{j\to\infty}(X_j - X) = 0$ and hence to convergence in distribution to zero of the sequence $\{X_j - X,\ j = 1, 2, \ldots\}$. But the linearity of the integral and Proposition 3.4.1 imply $X_j - X \sim S_\alpha(\sigma_j, \beta_j, \mu_j)$, where

$$\sigma_j = \left(\int_E |f_j(x) - f(x)|^\alpha m(dx)\right)^{1/\alpha}, \quad j = 1, 2, \ldots,$$

and

$$\mu_j = \begin{cases} 0, & \text{if } \alpha \neq 1 \\ -\frac{2}{\pi}\int_E (f_j(x) - f(x)) \ln|f_j(x) - f(x)|\beta(x)m(dx), & \text{if } \alpha = 1. \end{cases}$$

Therefore the convergence $\operatorname{plim}_{j\to\infty} X_j = X$ is equivalent to the convergence of the sequences $\{\sigma_j, j = 1, 2, \ldots\}$ and $\{\mu_j, j = 1, 2, \ldots\}$ to zero, establishing the proposition. ∎

Consider two α-stable integrals $X_1 = \int_E f_1(x)M(dx)$ and $X_2 = \int_E f_2(x)M(dx)$. The next proposition expresses the covariation of X_1 and X_2 in terms of f_1 and f_2.

Proposition 3.5.2 *Let $X_j = \int_E f_j(x)M(dx)$, $j = 1, 2$, where M is a $S\alpha S$ random measure with $1 < \alpha \le 2$ and control measure m. Then*

$$[X_1, X_2]_\alpha = \int_E f_1(x) f_2(x)^{<\alpha-1>} m(dx). \tag{3.5.3}$$

PROOF: Use the expression for the spectral measure Γ of (X_1, X_2) given in Proposition 3.4.3. Applying Definition 2.7.1 of the covariation, we obtain

$$[X_1, X_2]_\alpha = \int_{S_2} s_1 s_2^{<\alpha-1>} \Gamma(ds)$$

$$= \int_{E_+} \frac{f_1(x)}{(f_1^2(x) + f_2^2(x))^{1/2}} \left(\frac{f_2(x)}{(f_1^2(x) + f_2^2(x))^{1/2}} \right)^{<\alpha-1>} (f_1^2(x) + f_2^2(x))^{\alpha/2} m(dx)$$

$$= \int_{E_+} f_1(x) f_2(x)^{<\alpha-1>} m(dx) = \int_E f_1(x) f_2(x)^{<\alpha-1>} m(dx). \quad \blacksquare$$

Remarks

1. Relation (3.5.3) can also be obtained as follows. Suppose that f_1 and f_2 are simple functions, i.e., $f_1(x) = \sum_{i=1}^n u_i 1_{A_i}(x)$ and $f_2(x) = \sum_{i=1}^n v_i 1_{A_i}(x)$. By Property 2.7.7 (linearity in the first argument),

$$[I(f_1), I(f_2)]_\alpha = \sum_{i=1}^n u_i \Big[M(A_i), \sum_{j=1}^n v_j M(A_j)\Big]_\alpha.$$

Since the $M(A_j)$s are independent, we also have additivity in the second argument by Property 2.7.15. Hence,

$$[I(f_1), I(f_2)]_\alpha = \sum_{i=1}^n \sum_{j=1}^n u_i v_j^{<\alpha-1>} [M(A_i), M(A_j)]_\alpha.$$

Independence yields $[M(A_i), M(A_j)]_\alpha = 0$ if $i \neq j$, and $[M(A_i), M(A_j)]_\alpha = \|M(A_i)\|_\alpha^\alpha = m(A_i)$ if $i = j$. Therefore

$$\begin{aligned}[I(f_1), I(f_2)]_\alpha &= \sum_{i=1}^n u_i v_i^{<\alpha-1>} m(A_i) \\ &= \int_E f_1(x) f_2^{<\alpha-1>}(x) m(dx).\end{aligned}$$

To derive this relation for all f_1 and f_2 in $L^\alpha(m)$, approximate f_1 and f_2 by simple functions as in Section 3.4, use the Hölder inequality to deduce that

$$\int_E |f_1(x)||f_2(x)|^{\alpha-1} m(dx)$$
$$\leq \left(\int_E |f_1(x)|^\alpha m(dx)\right)^{1/\alpha} \left(\int_E |f_2(x)|^\alpha m(dx)\right)^{1-1/\alpha} < \infty,$$

and then apply the dominated convergence theorem.

For still another proof, see Exercise 3.4.

2. In the Gaussian case $\alpha = 2$, we have $[X_1, X_2]_2 = \frac{1}{2}\operatorname{Cov}(X_1, X_2)$, so that, when $\alpha = 2$, (3.5.3) becomes

$$\operatorname{Cov}(I(f_1), I(f_2)) = 2\int_E f_1(x) f_2(x) m(dx). \tag{3.5.4}$$

The independence of $I(f_1) = \int_E f_1(x) M(dx)$ and $I(f_2) = \int_E f_2(x) M(dx)$ is then equivalent to

$$\int_E f_1(x) f_2(x) m(dx) = 0, \tag{3.5.5}$$

a relation that can hold even when

$$m\{\text{support } (f_1) \cap \text{ support } (f_2)\} > 0. \tag{3.5.6}$$

(See Exercise 3.5.)

Independence of $I(f_1)$ and $I(f_2)$ in the case $\alpha < 2$ imposes a much stronger restriction on the functions f_1 and f_2: they must have m-almost disjoint supports. Indeed,

Theorem 3.5.3 *Let $X_1 = \int_E f_1(x) M(dx)$ and $X_2 = \int_E f_2(x) M(dx)$ be two α-stable integrals with respect to an α-stable random measure M with $0 < \alpha < 2$ and control measure m. Then X_1 and X_2 are independent if and only if*

$$f_1(x) f_2(x) \equiv 0 \;\; m\text{–a.e. on } E.$$

PROOF: For convenience, suppose $\alpha \neq 1$. X_1 and X_2 are independent if and only if for any real θ_1 and θ_2,

$$E\exp\{i(\theta_1 X_1 + \theta_2 X_2)\} = E\exp\{i\theta_1 X_1\} E\exp\{i\theta_2 X_2\}.$$

But $E\exp\{i(\theta_1X_1+\theta_2X_2)\}$ equals

$$\exp\Big\{-\int_E|\sum_{k=1}^{2}\theta_kf_k(x)|^\alpha\Big[1-i\beta(x)\text{ sign }\Big(\sum_{k=1}^{2}\theta_kf_k(x)\Big)\tan\frac{\pi\alpha}{2}\Big]m(dx)\Big\}, \tag{3.5.7}$$

whereas $E\exp\{i\theta_1X_1\}E\exp\{i\theta_2X_2\}$ equals

$$\begin{aligned}\exp\Big\{&-\int_E|\theta_1|^\alpha|f_1(x)|^\alpha\Big[1-i\beta(x)\text{ sign }(\theta_1f_1(x))\tan\frac{\pi\alpha}{2}\Big]m(dx)\\ &-\int_E|\theta_2|^\alpha|f_2(x)|^\alpha\Big[1-i\beta(x)\text{ sign }(\theta_2f_2(x))\tan\frac{\pi\alpha}{2}\Big]m(dx)\Big\}.\end{aligned} \tag{3.5.8}$$

Equating the moduli gives

$$\begin{aligned}&\int_E|\theta_1f_1(x)+\theta_2f_2(x)|^\alpha m(dx)\\ &\qquad=|\theta_1|^\alpha\int_E|f_1(x)|^\alpha m(dx)+|\theta_2|^\alpha\int_E|f_2(x)|^\alpha m(dx),\end{aligned}$$

which holds for all real θ_1 and θ_2 only if

$$f_1(x)f_2(x)=0\ m\text{-a.e.} \tag{3.5.9}$$

by Lemma 2.7.14. This proves that (3.5.9) is a necessary condition for the independence of X_1 and X_2. It is also sufficient because if (3.5.9) holds, then the imaginary parts of (3.5.7) and (3.5.8) are also equal. ■

The preceding result is very useful and will often be used in the sequel.

Corollary 3.5.4 *Let $X_j=\int_E f_jM(dx)$, $j=1,\dots,d$, be jointly α-stable. They are independent if and only if they are pairwise independent, i.e., if and only if $f_{k_1}(x)f_{k_2}(x)=0$ m-a.e. on E for any subset $\{k_1,k_2\}$ of $\{1,2,\dots,d\}$.*

PROOF: Independence clearly implies pairwise independence. To prove the converse, use the following fact: $X_1,\dots,X_d$ are independent if and only if any linear combination of $\{X_j,\ j\in J\}$ is independent of any linear combination of $\{X_j,\ j\in J^c\}$, where J is an arbitrary subset of $\{1,\dots,d\}$. Then apply Theorem 3.5.3 to conclude the proof.

More directly, pairwise independence implies by Theorem 3.5.3 that there are disjoint Borel sets $A_1,\dots,A_d$ such that

$$|\{x:f_j(x)\neq0\}\backslash A_j|=0$$

for each $j = 1, \ldots, d$. The joint characteristic function of $X_1, \ldots, X_d$ factorizes. For example, in the $S\alpha S$ case,

$$
\begin{aligned}
E\exp\Big\{i\sum_{j=1}^{d}\theta_j X_j\Big\} &= E\exp\Big\{i\sum_{j=1}^{d}\theta_j \int_E f_j(x)M(dx)\Big\} \\
&= \exp\Big\{-\int_E |\sum_{j=1}^{d}\theta_j f_j(x)|^\alpha m(dx)\Big\} \\
&= \exp\Big\{-\sum_{j=1}^{d}\int_{A_j} |\theta_j f_j(x)|^\alpha m(dx)\Big\} \\
&= \exp\Big\{-\sum_{j=1}^{d}\int_{E} |\theta_j f_j(x)|^\alpha m(dx)\Big\} \\
&= \prod_{j=1}^{d} E\exp\Big\{i\theta_j \int_E f_j(x)M(dx)\Big\} \\
&= \prod_{j=1}^{d} E\exp\{i\theta_j X_j\}.
\end{aligned}
$$

This proves that $X_1, \ldots, X_d$ are independent. ■

The following proposition shows the effect of a change of variable.

Proposition 3.5.5 *(Change of variable). Let M_m and M_ν be either two α-stable random measures with $0 < \alpha \le 2$, $\alpha \neq 1$ or two $S1S$ random measures, with identical skewness intensities and with control measures m and ν, respectively, satisfying*

$$\frac{m(dx)}{\nu(dx)} = (r(x))^\alpha, \quad x \in E,$$

where $r(x) \ge 0$. Then

$$\int_E f(x)M_m(dx) \stackrel{d}{=} \int_E f(x)r(x)M_\nu(dx) \tag{3.5.10}$$

for any $f \in L^\alpha(E, m)$.

PROOF: Let $\beta(x)$, $x \in E$ denote the skewness intensity of the random measures M_m and M_ν. The result follows from the fact that the left-hand side and right-hand side of (3.5.10) are both strictly α-stable with the same scale parameter

$$\Big(\int_E |f(x)|^\alpha m(dx)\Big)^{1/\alpha} = \Big(\int_E |f(x)r(x)|^\alpha \nu(dx)\Big)^{1/\alpha},$$

and same skewness parameter

$$\frac{\int_E(\text{ sign } f(x))|f(x)|^\alpha \beta(x)m(dx)}{\int_E |f(x)|^\alpha m(dx)}$$

$$= \frac{\int_E(\text{ sign } f(x)r(x))|f(x)r(x)|^\alpha \beta(x)\nu(dx)}{\int_E |f(x)r(x)|^\alpha \nu(dx)}. \quad \blacksquare$$

Remark. The conclusion of the proposition does not generally hold if $\alpha = 1$ and M_m is not symmetric stable because then the shift parameter of the left-hand side of (3.5.10),

$$-\frac{2}{\pi}\int_E f(x)\beta(x)\ln|f(x)|m(dx),$$

does not necessarily equal the shift parameter

$$-\frac{2}{\pi}\int_E f(x)r(x)\left(\ln|f(x)r(x)|\right)\beta(x)\nu(dx)$$

of the right-hand side of (3.5.10).

We say that a process $\{X(t), t \in T\}$ has the representation $\{Y(t), t \in T\}$ if $X \stackrel{d}{=} Y$, i.e., if the processes X and Y have the same finite-dimensional distributions. We now turn to the *representation theorem.*

Recall that the distribution of an α-stable vector is characterized by its spectral measure Γ on S_d and its shift vector $\boldsymbol{\mu}^0$ (Theorem 2.3.1). For a given α-stable vector $(X_1, \ldots, X_d)$ we want to find a measurable space $(E, \mathcal{E})$, an α-stable measure M defined on it, a sequence of measurable functions $f_1, \ldots, f_d$, and a constant vector $\boldsymbol{\eta}$ in $\mathbb{R}^d$, such that

$$(X_1, \ldots, X_d) \stackrel{d}{=} \left(\int_E f_1(x)M(dx), \ldots, \int_E f_d(x)M(dx)\right) + \boldsymbol{\eta}. \quad (3.5.11)$$

The next theorem shows that this is always possible. Moreover, the space, the random measure and the functions f_j can be chosen in a special way.

Theorem 3.5.6 *(Representation theorem in $\mathbb{R}^d$).*

Let $\mathbf{X} = (X_1, X_2, \ldots, X_d)$ be an α-stable random vector in $\mathbb{R}^d$.

(i) Let $\|\cdot\|$ be an arbitrary *norm on $\mathbb{R}^d$, $S_d^{\|\cdot\|} = \{\mathbf{s}\colon \|\mathbf{s}\| = 1\}$ the corresponding unit sphere and let $\Gamma_{\|\cdot\|}, \boldsymbol{\mu}^0_{\|\cdot\|}$ be the corresponding spectral measure and shift parameter as defined in Proposition 2.3.8. Then Relation (3.5.11) always holds with:*

$(E, \mathcal{E}) = (S_d^{\|\cdot\|}$, *Borel σ-algebra on* $S_d^{\|\cdot\|})$,

M has control measure $m = \Gamma_{\|\cdot\|}$ and skewness intensity $\beta(\cdot) \equiv 1$,

$f_j : S_d^{\|\cdot\|} \to \mathbb{R}$ is defined by $f_j((s_1, \ldots, s_d)) = s_j$, $j = 1, \ldots, d$,

$\boldsymbol{\eta} = \boldsymbol{\mu}^0_{\|\cdot\|}$,

so that

$$\mathbf{X} \stackrel{d}{=} \left(\int_{S_d^{\|\cdot\|}} s_1 M(d\mathbf{s}), \int_{S_d^{\|\cdot\|}} s_2 M(d\mathbf{s}), \ldots, \int_{S_d^{\|\cdot\|}} s_d M(d\mathbf{s}) \right) + \boldsymbol{\mu}^0_{\|\cdot\|}. \quad (3.5.12)$$

(ii) If $\mathbf{X}$ is $S\alpha S$, there are bounded measurable functions $f_1, f_2, \ldots, f_d$ such that

$$\mathbf{X} \stackrel{d}{=} \left(\int_0^1 f_1(x) M(dx), \int_0^1 f_2(x) M(dx), \cdots, \int_0^1 f_d(x) M(dx) \right),$$

where M is an $S\alpha S$ random measure with control measure $m(dx) = dx$.

(iii) If $\mathbf{X}$ is strictly α-stable with $\alpha \neq 1$, then there are bounded measurable functions $f_1, f_2, \ldots, f_d$ such that

$$\mathbf{X} \stackrel{d}{=} \left(\int_0^1 f_1(x) M(dx), \int_0^1 f_2(x) M(dx), \ldots, \int_0^1 f_d(x) M(dx) \right),$$

where M is an α-stable random measure with control measure $m(dx) = dx$ and skewness intensity $\beta(x) \equiv 1$.

(iv) If $\alpha = 1$, then there are bounded measurable functions $f_1, f_2, \ldots, f_d$ such that

$$\mathbf{X} \stackrel{d}{=} \left(\int_0^1 f_1(x) M(dx), \int_0^1 f_2(x) M(dx), \ldots, \int_0^1 f_d(x) M(dx) \right) + \boldsymbol{\eta},$$

where M is a 1-stable random measure with control measure $m(dx) = dx$ and skewness intensity $\beta(x) \equiv 1$, and where $\boldsymbol{\eta} = (\eta_1, \ldots, \eta_d)$ with

$$\eta_j = \frac{1}{\pi} \int_0^1 f_j(x) \ln\Big(\sum_{k=1}^d f_k(x)^2 \Big) dx + \mu_j^0, \ j = 1, 2, \ldots, d.$$

PROOF: *(Part (i))*. In view of Proposition 2.3.8 it is sufficient to prove (3.5.12) when $\|\cdot\|$ is the Euclidean norm on $\mathbb{R}^d$. Then $S_d^{\|\cdot\|}$, $\Gamma_{\|\cdot\|}$ and $\boldsymbol{\mu}^0_{\|\cdot\|}$ are the familiar S_d, Γ and $\boldsymbol{\mu}^0$. Use Proposition 3.4.3 to show that the joint characteristic function of the left-hand side and right-hand side of (3.5.12) coincide. (Observe in particular, that when $\alpha = 1$, the shift vector of the right-hand side equals

$$-\frac{1}{\pi}\int_{S_d} \mathbf{s} \ln\Big(\sum_{k=1}^{d} s_k^2\Big)\Gamma(d\mathbf{s}) + \boldsymbol{\mu}^0 = \boldsymbol{\mu}^0$$

since $\sum_{k=1}^d s_k^2 = 1$.) ∎

Parts (ii)-(iv) follow from a more general representation theorem, which will be formulated and proved in Chapter 13. The idea behind their proof is to start with the representation (3.5.12) on $S_d^{\|\cdot\|}$ and then make a change of variables in such a way that $S_d^{\|\cdot\|}$ is mapped to $(0, 1)$ and the control measure of M becomes Lebesgue.

Remarks

1. If $\|\cdot\|$ is the Euclidean norm, then $S_d^{\|\cdot\|}$, $\Gamma_{\|\cdot\|}$ and $\boldsymbol{\mu}^0_{\|\cdot\|}$ are the usual S_d, Γ and $\boldsymbol{\mu}^0$. Part (i) shows that one can always write down an integral representation by using the pair $(\Gamma, \boldsymbol{\mu}^0)$, which specifies the joint characteristic function.

2. A nice feature of the integral representation in parts (ii)-(iv) is that one uses the same space E for all $d \geq 1$. The measure space, moreover, is particularly simple since it equals $([0, 1], \mathcal{B}, dx)$.

3. Parts (ii)-(iv) effectively cover all cases of α-stable random vectors. In the case where $X_1, X_2, \dots, X_d$ are not strictly α-stable and $\alpha \neq 1$, the representation (3.5.11) can be derived from Part (iii) of the theorem because, by Property 1.2.6 and Corollary 2.4.2, there are constants $\eta_1, \eta_2, \dots, \eta_d$ such that $X_1 - \eta_1, X_2 - \eta_2, \dots, X_d - \eta_d$ are jointly strictly α-stable.

4. Since the skewness parameter β_f of $\int_E f(x)M(dx)$ satisfies

$$|\beta_f| = \frac{|\int_E \beta(x) f(x)^{<\alpha>} m(dx)|}{\int_E |f(x)|^\alpha m(dx)} \leq \sup_{x\in E} |\beta(x)|,$$

choosing $\beta(x) = C, -1 \leq C \leq 1$ in parts (i), (iii) and (iv) of the theorem, with C different from either 1 or -1 would unduly restrict the possible values for β_f. With the choice $\beta(x) \equiv 1$, one has

$$\beta_f = \frac{\int_E (\text{ sign } f(x))|f(x)|^\alpha m(dx)}{\int_E |f(x)|^\alpha m(dx)}.$$

5. The proof of the theorem in the Gaussian case $\alpha = 2$ is straightforward. If $X_1, X_2, \ldots, X_d$ are jointly Gaussian with mean zero, then for some constants $a_{jk},\ j = 1, \ldots, d,\ k = 1, \ldots, j$,

$$
\begin{aligned}
X_1 &= a_{11}Y_1, \\
X_2 &= a_{21}Y_1 + a_{22}Y_2, \\
X_3 &= a_{31}Y_1 + a_{32}Y_2 + a_{33}Y_3, \\
&\vdots \\
X_d &= a_{d1}Y_1 + a_{d2}Y_2 + a_{d3}Y_3 + \cdots + a_{dd}Y_d,
\end{aligned}
$$

where $Y_1, Y_2, \ldots, Y_d$ are i.i.d. $N(0,1)$. In this case the X_js are expressed as a sum of i.i.d. $N(0,1)$ random variables.

To express each $X_j,\ j = 1, 2, \ldots, d$, as an integral $\int_0^1 f_j(x)dM(x)$ where M is a Gaussian random measure with control measure $m(dx) = dx$, choose

$$f_j(x) = \sqrt{\frac{d}{2}} \sum_{s=1}^{j} a_{js}\, 1\left(\frac{s-1}{d} \le x < \frac{s}{d}\right),$$

so that

$$
\begin{aligned}
\text{Cov}\Big(\int_0^1 f_j(x)M(dx), \int_0^1 f_k(x)M(dx)\Big) &= 2\int_0^1 f_j(x)f_k(x)dx \\
&= \sum_{s=1}^{j\wedge k} a_{js}a_{ks} = \text{Cov}\,(X_j, X_k).
\end{aligned}
$$

6. The integral representation holds not only for α-stable random vectors in $\mathbb{R}^d$, but also for most α-stable stochastic process $\{X(t), t \in T\}$, as we will see in a Chapter 13. In that case, one may have to replace $([0,1], \mathcal{B}, dx)$ by some general measure space $(E, \mathcal{E}, m)$.

7. Property 3.2.1 suggests that an α-stable random vector of the form

$$X_k = \int_E f_k(x)M(dx),\ k = 1, \ldots, d,$$

where M is a skewed α-stable measure, is, in general, not symmetric. However, one can sometimes have X symmetric even with a non-symmetric M. For example, part (iii) of Theorem 3.5.6 implies that any strictly α-stable random vector with $\alpha \neq 1$ (in particular, any $S\alpha S$ random vector with $\alpha \neq 1$) has an integral representation as above with a non-symmetric M. A simple example (which includes $\alpha = 1$) is

$$X_k = \int_0^1 \left(1_{[k-1,k]}(x) - 1_{[-k,-k+1]}(x)\right) M(dx),\ k = 1, \ldots, d,$$

where M is an α-stable random measure with Lebesgue control measure and skewness intensity $\beta \equiv 1$. (To check that X is symmetric, verify that all linear combinations of the components of X are $S\alpha S$ random variables.)

Pairwise independence of the components of a Gaussian random vector implies independence of all the components. The same property holds for α-stable random vectors.

Corollary 3.5.7 *Let $X_1, X_2, \ldots, X_d$ be jointly α-stable. If they are pairwise independent, then they are independent.*

PROOF: By Theorem 3.5.6, we can assume without loss of generality that

$$X_k = \int_E f_k(x) M(dx),\ k = 1, \ldots, d,$$

where M is an α-stable random measure with control measure m and skewness intensity $\beta \equiv 1$. The result follows now from Corollary 3.5.4. ∎

3.6 Examples

Example 3.6.1 *The $S\alpha S$ Lévy motion.*

Let

$$X(t) = \int_0^\infty 1\{x \le t\} M(dx) = \int_0^t M(dx),\ t \ge 0,$$

where M is $S\alpha S$ on $[0, \infty)$ with control measure $m(dx) = dx$. Then

$$X(0) = 0 \text{ a.s.},$$

$$X(t) - X(s) = \int_s^t M(dx) = M([s,t]) \sim S_\alpha(|t-s|^{1/\alpha}, 0, 0).$$

If $0 \le t_1 < t_2 < \ldots < t_n$, then

$$\left(X(t_2) - X(t_1), X(t_3) - X(t_2), \ldots, X(t_n) - X(t_{n-1})\right)$$

$$= \left(\int_{t_1}^{t_2} M(dx), \int_{t_2}^{t_3} M(dx), \ldots, \int_{t_{n-1}}^{t_n} M(dx)\right).$$

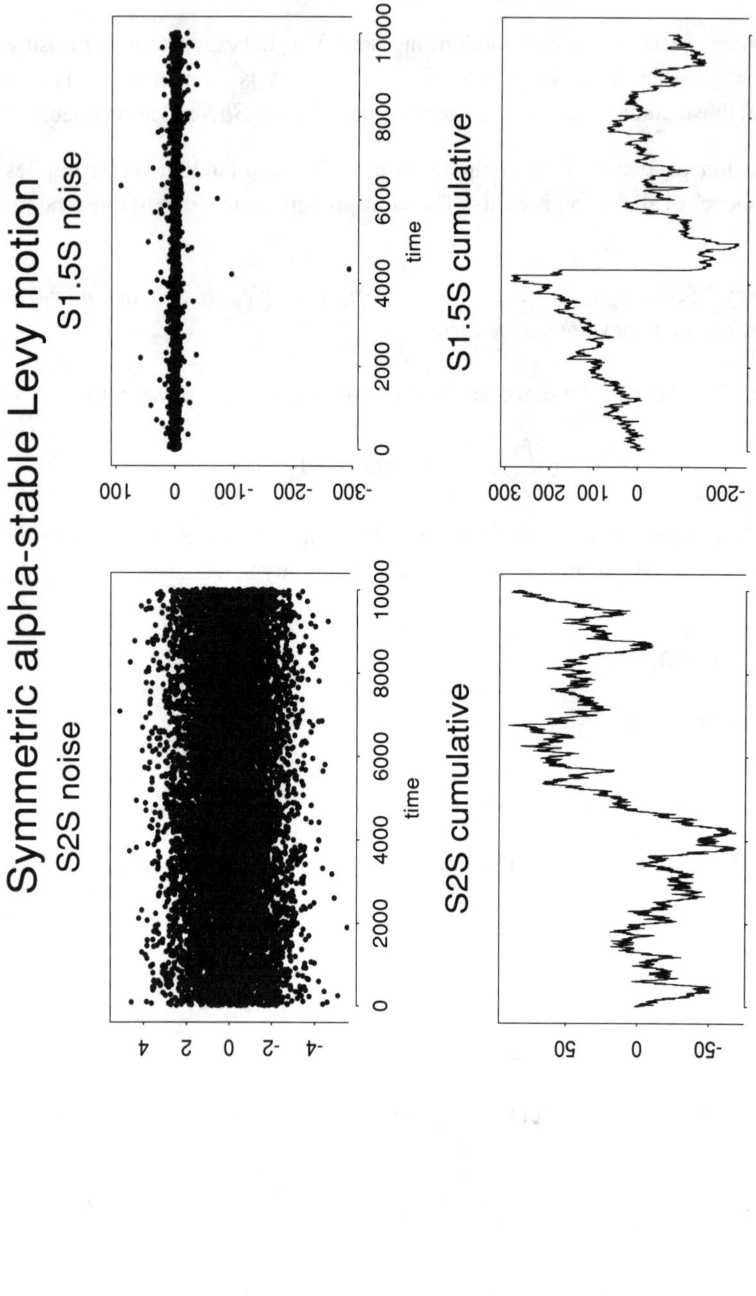
Symmetric alpha-stable Levy motion
S2S noise
S1.5S noise
S2S cumulative
S1.5S cumulative
time
time
time
time

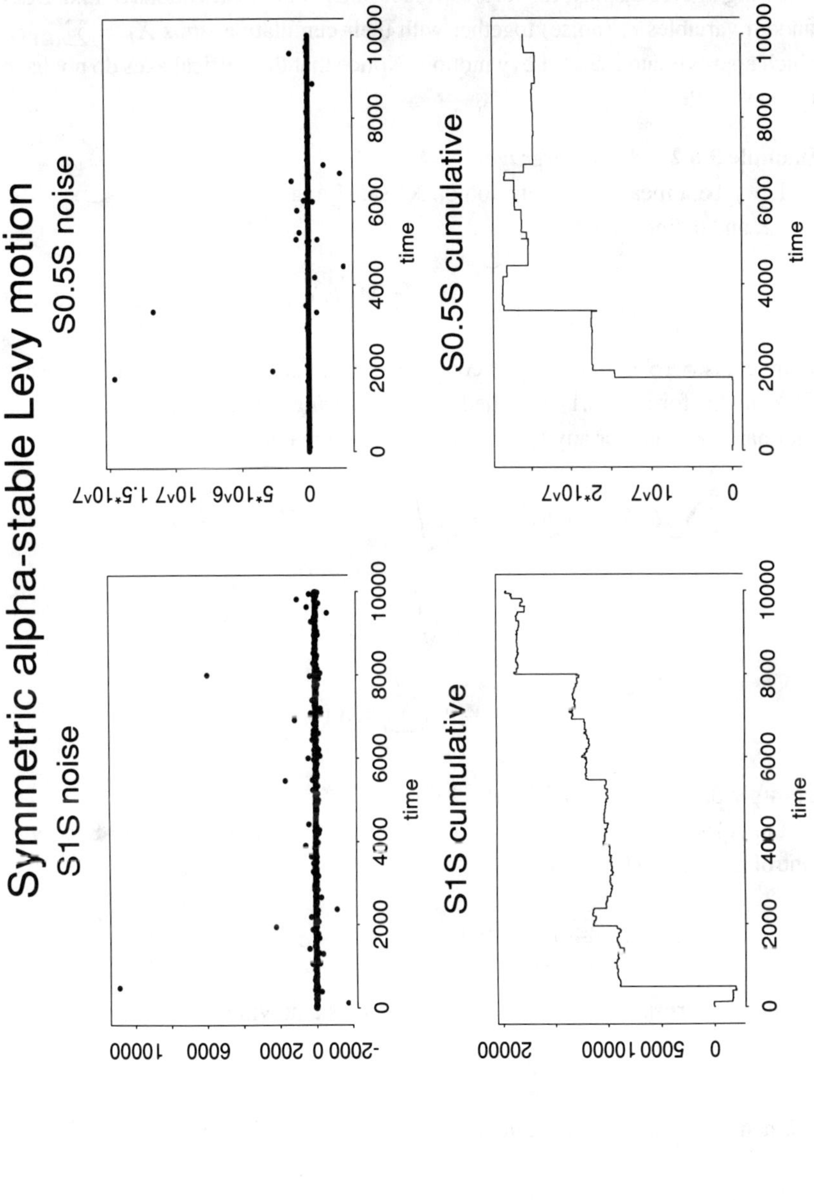
Symmetric alpha-stable Levy motion
S1S noise
S0.5S noise
S1S cumulative
S0.5S cumulative
time

The components of this random vector are independent because the integrands have disjoint supports. Hence, $\{X(t), t \geq 0\}$ is a process that starts at 0, has stationary independent increments and $S\alpha S$ finite-dimensional distributions. Therefore, $\{X(t), t \geq 0\}$ is the $S\alpha S$ Lévy motion (see Example 3.1.3). The following figures display, for various values of α, 10000 realizations of i.i.d. $S\alpha S$ random variables ϵ_i (noise) together with their cumulative sums $X_t = \sum_{i=1}^t \epsilon_i$, which approximate a $S\alpha S$ Lévy motion. Notice that the vertical axes do not have the same scale.

Example 3.6.2 *Moving averages*

Let f be a measurable function on $\mathbb{R}^1$ satisfying $\int_{-\infty}^{\infty} |f(x)|^\alpha dx < \infty$, $0 < \alpha \leq 2$, and define

$$X(t) = \int_{-\infty}^{\infty} f(t-x) M(dx), \; t \in \mathbb{R}^1, \tag{3.6.1}$$

where M is $S\alpha S$ with Lebesgue control measure (i.e. $m(dx) = dx$). A process $X(\cdot)$ of the form (3.6.1) is called a $S\alpha S$ *moving average process.* $X(\cdot)$ is stationary, because for any $t_1, \ldots, t_d$, $h \in \mathbb{R}$ and real $\theta_1, \ldots, \theta_d$,

$$\begin{aligned}\|\sum_{j=1}^d \theta_j X(t_j + h)\|_\alpha^\alpha &= \int_{-\infty}^{\infty} |\sum_{j=1}^d \theta_j f(t_j + h - x)|^\alpha dx \\ &= \int_{-\infty}^{\infty} |\sum_{j=1}^d \theta_j f(t_j - y)|^\alpha dy \\ &= \|\sum_{j=1}^d \theta_j X(t_j)\|_\alpha^\alpha.\end{aligned}$$

Example 3.6.3 *Ornstein-Uhlenbeck process*

Let $\lambda > 0$ and M be a $S\alpha S$ random measure, $0 < \alpha \leq 2$ with Lebesgue control measure. The process

$$X(t) = \int_{-\infty}^{t} e^{-\lambda(t-x)} M(dx), \; -\infty < t < \infty,$$

is called an *Ornstein-Uhlenbeck process.* $X(t)$ is a moving average process with

$$f(x) = e^{-\lambda x} 1_{[0,\infty)}(x), \; -\infty < x < \infty,$$

and, hence, by Example 3.6.2 it is well defined and stationary. Note that for any fixed $s < t$,

$$X(t) - e^{-\lambda(t-s)} X(s) = \int_{s}^{t} e^{-\lambda(t-x)} M(dx) \tag{3.6.2}$$

is a $S\alpha S$ random variable independent of any linear combination $\sum_{k=1}^{d} b_k X(u_k)$, $u_k \le s$, $k = 1, \ldots, n$. Therefore, $X(t) - e^{-\lambda(t-s)}X(s)$ is independent of $\sigma(X(u), u \le s)$, the σ-algebra generated by $\{X(u),\ u \le s\}$. This implies that the Ornstein-Uhlenbeck process is, in fact, a Markov process.

Recall that in the Gaussian case $\alpha = 2$, the Ornstein-Uhlenbeck process is the only continuous in probability stationary Markov Gaussian process (Breiman (1968)). In the case $0 < \alpha < 2$ there is at least one more stationary Markov $S\alpha S$ process. It is given in the following example.

Example 3.6.4 *Reverse Ornstein-Uhlenbeck process*

Let $\lambda > 0$ and M be a $S\alpha S$ random measure, $0 < \alpha \le 2$ with Lebesgue control measure. The process

$$X(t) = \int_t^{\infty} e^{-\lambda(x-t)} M(dx), \ -\infty < t < \infty,$$

is well defined and is called *reverse* (or *fully anticipating*) *Ornstein-Uhlenbeck process*. Note that for any $s > t$

$$X(t) - e^{-\lambda(s-t)}X(s) = \int_t^s e^{-\lambda(x-t)} M(dx). \tag{3.6.3}$$

Using the same argument as in the previous example, we conclude that the reverse Ornstein-Uhlenbeck process is also Markov.

We noted earlier that in the Gaussian case $\alpha = 2$, all stationary Markov processes are Ornstein-Uhlenbeck. This means that when $\alpha = 2$, the Ornstein-Uhlenbeck process and the reverse Ornstein-Uhlenbeck process are the same process. (One can also verify this fact by computing the autocovariance function.) It turns out that these two processes are different when $0 < \alpha < 2$. To see this, let X_1 and X_2 be, respectively, the Ornstein-Uhlenbeck and the reverse Ornstein-Uhlenbeck processes. Fixing $s < t$ and using (3.6.2), we have

$$\begin{aligned}
&E\exp\{i(\theta_1 X_1(s) + \theta_2 X_1(t))\} \\
&= E\exp\Big\{i\Big[\big(\theta_1 + \theta_2 e^{-\lambda(t-s)}\big)X_1(s) + \theta_2\big(X_1(t) - e^{-\lambda(t-s)}X_1(s)\big)\Big]\Big\} \\
&= E\exp\Big\{i\theta_2(X_1(t) - e^{-\lambda(t-s)}X_1(s))\Big\} \cdot E\exp\Big\{i(\theta_1 + \theta_2 e^{-\lambda(t-s)})X_1(s)\Big\} \\
&= \exp\Big\{-|\theta_2|^\alpha \int_s^t (e^{-\lambda(t-x)})^\alpha dx - |\theta_1 + \theta_2 e^{-\lambda(t-s)}|^\alpha \int_{-\infty}^s (e^{-\lambda(s-x)})^\alpha dx\Big\} \\
&= \exp\Big\{-\Big[\frac{1}{\alpha\lambda}(1 - e^{-\alpha\lambda(t-s)})|\theta_2|^\alpha + \frac{1}{\alpha\lambda}|\theta_1 + \theta_2 e^{-\lambda(t-s)}|^\alpha\Big]\Big\} \\
&= \exp\Big\{-\frac{1}{\alpha\lambda}\Big[A|\theta_2|^\alpha + B^{-\alpha}|B\theta_1 + Be^{-\lambda(t-s)}\theta_2|^\alpha\Big]\Big\},
\end{aligned}$$

where $A = 1 - e^{-\alpha\lambda(t-s)}$ and $B = |1 + e^{-2\lambda(t-s)}|^{-1/2}$. Hence, the spectral measure Γ_1 of the $S\alpha S$ random vector $(X_1(s), X_1(t))$ is given by

$$\Gamma_1 = \frac{1}{2\alpha\lambda}\Big[A(\delta_{\{0,1\}} + \delta_{\{0,-1\}}) + B^{-\alpha}(\delta_{\{B,Be^{-\lambda(t-s)}\}} + \delta_{\{-B,-Be^{-\lambda(t-s)}\}})\Big].$$

Similarly, the spectral measure Γ_2 of the $S\alpha S$ random vector $(X_2(s), X_2(t))$ is given by

$$\Gamma_2 = \frac{1}{2\alpha\lambda}\Big[A(\delta_{\{1,0\}} + \delta_{\{-1,0\}}) + B^{-\alpha}(\delta_{\{Be^{-\lambda(t-s)},B\}} + \delta_{\{-Be^{-\lambda(t-s)},-B\}})\Big].$$

Since $\Gamma_1 \neq \Gamma_2$, the uniqueness of the spectral measure implies that $(X_1(s), X_1(t)) \neq (X_2(s), X_2(t))$. Therefore the Ornstein-Uhlenbeck and the reverse Ornstein-Uhlenbeck processes are different for $0 < \alpha < 2$.

Example 3.6.5 *Well-balanced linear fractional stable motion*

Let M be $S\alpha S, 0 < \alpha \leq 2$, with Lebesgue control measure and consider

$$X(t) = \int_{-\infty}^{\infty} \left(|t-x|^{H-1/\alpha} - |x|^{H-1/\alpha}\right) M(dx), \ -\infty < t < \infty, \tag{3.6.4}$$

where $0 < H < 1$, $H \neq 1/\alpha$. The process X is called the *well-balanced (symmetric) linear fractional stable motion.* (We will encounter other linear fractional stable motions in Chapter 7.)

The process $X(t)$ has two important features. It is self-similar with parameter H, i.e., for any $c > 0$, $t_1, \ldots, t_d \in \mathbb{R}$, $(X(ct_1), \ldots, X(ct_d)) \stackrel{d}{=} (c^H X(t_1), \ldots, c^H X(t_d))$, and it has stationary increments, i.e., for any $\tau \in \mathbb{R}$,

$$\{X(t) - X(0),\ -\infty < t < \infty\} \stackrel{d}{=} \{X(\tau + t) - X(\tau),\ -\infty < t < \infty\}.$$

(See Exercise 3.9.)

In the Gaussian case $\alpha = 2$, the process (3.6.4) is known as *fractional Brownian motion.*

To verify that the process $X(t)$ is well defined, we must prove that

$$\int_{-\infty}^{\infty} \left||t-x|^{H-1/\alpha} - |x|^{H-1/\alpha}\right|^{\alpha} dx < \infty.$$

This integral may diverge at (i) $x = \pm\infty$ or at (ii) $x = 0$ or t. As $x \to \pm\infty$, the integrand behaves like $(|x|^{H-1/\alpha-1})^{\alpha} = |x|^{H\alpha-1-\alpha}$, which is integrable since $H < 1$. As $x \to 0$ or t, the integrand behaves like $(|x|^{H-1/\alpha})^{\alpha} = |x|^{H\alpha-1}$, which is integrable since $H > 0$. Hence the process (3.6.4) is well defined.

The process $X(t)$ will be discussed more thoroughly later on, in the context of self-similar processes.

Example 3.6.6 *Log-fractional stable motion*

Again suppose M to be $S\alpha S$ with Lebesgue control measure but assume that $1 < \alpha < 2$. The process

$$X(t) = \int_{-\infty}^{\infty} (\ln|t-x| - \ln|x|)M(dx), \ -\infty < t < \infty, \tag{3.6.5}$$

is called the *(symmetric) log-fractional stable motion.* It is self-similar with $H = 1/\alpha$ and has stationary increments (see Exercise 3.10). The value $\alpha = 2$ is also admissible, but then (3.6.5) and (3.6.4) become identical processes (see Exercise 3.11).

Recall that α-stable motion (Example 3.6.1) is also self-similar with $H = 1/\alpha$ and has stationary increments (see Example 3.1.3). Log-fractional stable motion, however, is a different process. To see this, note that the increments of α-stable motion are independent but, by Theorem 3.5.3, the increments of log-fractional stable motion are dependent since

$$X(t+1) - X(t) = \int_{-\infty}^{\infty} (\ln|t+1-x| - \ln|t-x|)M(dx)$$

and the functions $\ln|t+1+\cdot| - \ln|t-\cdot|$ and $\ln|t-\cdot| - \ln|\cdot|$ do not have disjoint supports.

The log-fractional stable motion will be discussed more thoroughly in the context of self-similar processes.

Example 3.6.7 *Real stationary $S\alpha S$ harmonizable process*

The finite-dimensional characteristic function of a real $S\alpha S$ harmonizable process $\{H(t),\ t \in T\}$ is

$$\begin{aligned} &E\exp\Big\{i(\theta_1 H(t_1) + \cdots + \theta_d H(t_d))\Big\} \\ &= \exp\Big\{-c_0 \int_{-\infty}^{\infty} \Big|\Big(\sum_{j=1}^d \theta_j \cos t_j x\Big)^2 + \Big(\sum_{j=1}^d \theta_j \sin t_j x\Big)^2\Big|^{\alpha/2} m(dx)\Big\} \\ &= \exp\Big\{-c_0 \int_{-\infty}^{\infty} \Big|\sum_{j=1}^d \sum_{k=1}^d \theta_j\theta_k \cos((t_j - t_k)x)\Big|^{\alpha/2} m(dx)\Big\}, \end{aligned} \tag{3.6.6}$$

where m is a finite measure on $(\mathbb{R}, \mathcal{B})$ and $c_0 = \frac{1}{2\pi}\int_0^{2\pi}(\cos\phi)^\alpha d\phi$ is the constant which appears in Corollary 2.6.5.[5] The simplest way to represent the process $\{H(t),\ t \in T\}$ is through the representation

$$H(t) = \text{Re} \int_{-\infty}^{\infty} e^{itx} M(dx),\ t \in T,$$

[5] The constant c_0 can be incorporated in m.

where M is some complex-valued measure. This representation is defined in Chapter 6.

3.7 Sub-Gaussian processes

Let $\{G(t), t \in T\}$ be a Gaussian process and let $A \sim S_{\alpha/2}((\cos\frac{\pi\alpha}{4})^{2/\alpha}, 1, 0)$ where $\alpha < 2$, so that $Ee^{-\gamma A} = e^{-\gamma^{\alpha/2}}$, $\gamma \geq 0$ (see (2.5.2)). Assume that the random variable A is independent of $\{G(t), t \in T\}$. The $S\alpha S$ process $\{X(t) = A^{1/2}G(t),\ t \in T\}$ is a *sub-Gaussian process* with an underlying Gaussian process $\{G(t), t \in T\}$. Its finite dimensional projections, $(X(t_1), \ldots, X(t_d))$, $d \geq 1$, are the sub-Gaussian $S\alpha S$ random vectors introduced in Section 2.5. What kind of integral representation does the process $\{X(t) = A^{1/2}G(t),\ t \in T\}$ have?

Let $(\Omega, \mathcal{F}, P)$ denote the probability space on which $\{G(t),\ t \in T\}$ is defined. We write $\{G(t), t \in T\} = \{G(t, \omega),\ t \in T,\ \omega \in \Omega\}$ in order to make the dependence on ω explicit. However, to avoid possible confusion, we use in the sequel the letter x instead of the letter ω. Now, let M be a $S\alpha S$ random measure on $(\Omega, \mathcal{F})$ with control measure P. The following proposition gives the integral representation of $X(t)$.

Proposition 3.7.1 *The sub-Gaussian process $\{X(t) = A^{1/2}G(t),\ t \in T\}$ has the representation*

$$\{X(t), t \in T\} \stackrel{d}{=} \left\{\frac{1}{d_\alpha\sqrt{2}} \int_\Omega G(t, x) M(dx),\ t \in T\right\}, \tag{3.7.1}$$

where $d_\alpha = (E|Z|^\alpha)^{1/\alpha}$ with $Z \sim N(0, 1)$.

Remarks

1. The right-hand side of (3.7.1) is somewhat unusual because $E = \Omega$ and $m = P$. Observe that although $G(t, x)$ is not random, M is random, and hence $X(t)$ is random.

2. The integral is well defined because $\int_\Omega |G(t, x)|^\alpha P(dx) = E|G(t)|^\alpha < \infty$.

3. Any $N(0, \sigma^2)$ random variable G satisfies

$$E|G|^\alpha = E\left|\frac{G}{\sigma}\right|^\alpha (\sigma^2)^{\alpha/2} = d_\alpha^\alpha (E|G|^2)^{\alpha/2}.$$

4. One has (see Exercise 3.12),

$$(d_\alpha)^\alpha = E|N(0, 1)|^\alpha = 2^{\alpha/2}\pi^{-1/2}\Gamma\left(\frac{\alpha + 1}{2}\right). \tag{3.7.2}$$

PROOF: For any $d \geq 1$, $t_1, \ldots, t_d \in T$ and $\theta_1, \ldots, \theta_d$ real,

$$
\begin{aligned}
E\exp\Big\{i\sum_{j=1}^{d}\theta_j \frac{d_\alpha^{-1}}{\sqrt{2}}\int_\Omega G(t_j,x)M(dx)\Big\}
&= \exp\Big\{-\int_\Omega d_\alpha^{-\alpha}\Big|\frac{1}{\sqrt{2}}\sum_{j=1}^{d}\theta_j G(t_j,x)\Big|^\alpha P(dx)\Big\} \\
&= \exp\Big\{-d_\alpha^{-\alpha}E\Big|\frac{1}{\sqrt{2}}\sum_{j=1}^{d}\theta_j G(t_j)\Big|^\alpha\Big\} \\
&= \exp\Big\{-\Big(E\Big|\frac{1}{\sqrt{2}}\sum_{j=1}^{d}\theta_j G(t_j)\Big|^2\Big)^{\alpha/2}\Big\} \\
&= \exp\Big\{-\Big|\frac{1}{2}\sum_{j=1}^{d}\sum_{k=1}^{d}\theta_j\theta_k EG(t_j)G(t_k)\Big|^{\alpha/2}\Big\} \\
&= E\exp\Big\{i\sum_{j=1}^{d}\theta_j X_j\Big\}, \qquad (3.7.3)
\end{aligned}
$$

by Proposition 2.5.2. ■

3.8 Sub-stable processes

The Gaussian process $G(t)$ is stable with index of stability $\alpha' = 2$. What happens if, in the preceding section, we replace $G(t)$ by a process $Y(t)$ that is stable with $\alpha' \leq 2$ and modify A accordingly?

Let $\{Y(t), t \in T\}$ be a $S\alpha' S$ process with $\alpha' < 2$. Fix $\alpha < \alpha'$ and let

$$
A \sim S_{\alpha/\alpha'}\left(\left(\cos\left(\frac{\pi\alpha}{2\alpha'}\right)\right)^{\alpha'/\alpha}, 1, 0\right)
$$

be independent of the process $\{Y(t), t \in T\}$. Relations (1.2.3) and (1.2.4) ensure that A has Laplace transform $Ee^{-\gamma A} = \exp\{-\gamma^{\alpha/\alpha'}\}$, $\gamma \geq 0$.

Proposition 3.8.1 *The process*

$$
X(t) = A^{1/\alpha'} Y(t),\ t \in T, \qquad (3.8.1)
$$

is $S\alpha S$.

PROOF: For any $d \geq 1$, $t_1, \ldots, t_d \in T$, $b_1, \ldots, b_d$ real, θ real,

$$\begin{aligned} E\exp\Big\{i\theta\sum_{j=1}^{d} b_j X(t_j)\Big\} &= EE\Big[\exp\{i\theta A^{1/\alpha'}\sum_{j=1}^{d} b_j Y(t_j)\}\Big| A\Big] \\ &= E\exp\Big\{-A\|\theta\sum_{j=1}^{d} b_j Y(t_j)\|_{\alpha'}^{\alpha'}\Big\} \\ &= \exp\Big\{-|\theta|^{\alpha}\|\sum_{j=1}^{d} b_j Y(t_j)\|_{\alpha'}^{\alpha}\Big\}. \end{aligned} \tag{3.8.2}$$

This is the characteristic function of a $S\alpha S$ random variable. (We use the notation $\|U\|_\alpha, \alpha \leq 2$ to denote the scale parameter of an α-stable random variable U.)

Since every finite linear combination of components of $\{X(t), t \in T\}$ is a $S\alpha S$ random variable, the process $\{X(t), t \in T\}$ is $S\alpha S$ by Theorem 2.1.5 and Definition 3.1.1. ∎

The $S\alpha S$ process $X(t)$ is called a *sub-stable process* or *sub-α-stable process* with underlying stable process $\{Y(t), t \in T\}$. It is sub-Gaussian when $\alpha' = 2$. To find its integral representation, we proceed as in the sub-Gaussian case. Let $(\Omega, \mathcal{F}, P)$ denote the probability space on which $\{Y(t), t \in T\}$ is defined, let M be a $S\alpha S$ random measure on $(\Omega, \mathcal{F})$ with control measure P and let $d_{p,\alpha'} = (E|Z|^p)^{1/p}$ where $Z \sim S_{\alpha'}(1, 0, 0)$.

Proposition 3.8.2 *The sub-stable process $X(t) = A^{1/\alpha'}Y(t), t \in T$ has the representation*

$$\{X(t), t \in T\} \stackrel{d}{=} \Big\{d_{\alpha,\alpha'}^{-1}\int_\Omega Y(t,x)M(dx),\ t \in T\Big\}. \tag{3.8.3}$$

PROOF: Proceeding as in the proof of Proposition 3.7.1 , we have

$$\begin{aligned} E\exp\Big\{id_{\alpha,\alpha'}^{-1}\int_\Omega\sum_{j=1}^{d}\theta_j Y(t_j,x)M(dx)\Big\} &= \exp\Big\{-d_{\alpha,\alpha'}^{-\alpha}E|\sum_{j=1}^{d}\theta_j Y(t_j)|^{\alpha}\Big\} \\ &= \exp\Big\{-\|\sum_{j=1}^{d}\theta_j Y(t_j)\|_{\alpha'}^{\alpha}\Big\} \end{aligned}$$

by (2.8.1), which is equal, by (3.8.2), to $E\exp\{i\sum_{j=1}^{d}\theta_j X(t_j)\}$. ∎

Remark. The process $\{X(t) = \int_\Omega W(t,x)M(dx),\ t \in T\}$ is well defined for any stochastic process $\{W(t), t \in T\}$ defined on $(\Omega, \mathcal{F}, P)$ with $E|W(t)|^\alpha < \infty$, and for any α-stable random measure M on $(\Omega, \mathcal{F})$ with control measure P. The random measure M does not even have to be symmetric. The resulting process $\{X(t),\ t \in T\}$ will then be α-stable, but it is not in general in the form $A^p W(t)$ where A is a random variable and p is a real constant.

3.9 Series representation for α-stable random measures

The series representation described in Section 1.4 for one-dimensional α-stable random variables can be extended to α-stable random measures, and it indicates that α-stable random measures have essentially a "discrete" structure.

Let M be an α-stable random measure on $(E, \mathcal{E})$ with finite control measure m and skewness intensity $\beta(\cdot)$. Denote by $\widehat{m}$ the normalized measure

$$\widehat{m} = \frac{m}{m(E)}.$$

Let $\{\Gamma_1, \Gamma_2, \ldots\}$ be a sequence of arrival times of a Poisson process with unit arrival rate, and let $\{(V_1, \gamma_1), (V_2, \gamma_2), \ldots\}$ be a sequence of i.i.d. random vectors, independent of $\{\Gamma_1, \Gamma_2, \ldots\}$, such that V_i has distribution $\widehat{m}$ on E and

$$P\Big(\gamma_i = 1 \Big| V_i\Big) = 1 - P\Big(\gamma_i = -1 \Big| V_i\Big) = \frac{1 + \beta(V_i)}{2}.$$

We have the following:

Theorem 3.9.1

$$\{M(A), A \in \mathcal{E}\} \stackrel{d}{=}$$

$$\Big\{(C_\alpha m(E))^{1/\alpha} \sum_{i=1}^{\infty} \Big[\gamma_i \Gamma_i^{-1/\alpha} 1(V_i \in A) - b_i^{(\alpha)} \int_A \beta(x) \widehat{m}(dx)\Big] + \eta_A,\ A \in \mathcal{E}\Big\} \tag{3.9.1}$$

where C_α is given in (1.2.9),

$$b_i^{(\alpha)} = \begin{cases} 0 & \textit{if } 0 < \alpha < 1, \\ \int_{1/i}^{1/(i-1)} x^{-2} \sin x \, dx & \textit{if } \alpha = 1, \\ \frac{\alpha}{\alpha - 1}\big(i^{\frac{\alpha-1}{\alpha}} - (i-1)^{\frac{\alpha-1}{\alpha}}\big) & \textit{if } 1 < \alpha < 2, \end{cases}$$

and

$$\eta_A = \begin{cases} 0 & \textit{if } \alpha \neq 1, \\ \frac{2}{\pi} \ln\big(\frac{2}{\pi} m(E)\big) \int_A \beta(x) m(dx) & \textit{if } \alpha = 1. \end{cases}$$

PROOF: Fix $A \in \mathcal{E}$ and let $W_i^{(A)} = \gamma_i 1(V_i \in A)$, $i = 1, 2, \ldots$. Clearly, $W_i^{(A)}, i = 1, 2 \ldots$, is a sequence of i.i.d. random variables taking values $-1, 0, 1$ and satisfying

$$E|W_1^{(A)}|^\alpha = P(V_i \in A) = \widehat{m}(A),$$

$$\begin{aligned} EW_1^{(A)} &= \int_A E(\gamma_1 | V_1 = x)\widehat{m}(dx) \\ &= \int_A \big[P(\gamma_1 = 1|V_1 = x) - P(\gamma_1 = -1|V_1 = x)\big]\widehat{m}(dx) \\ &= \int_A \beta(x)\widehat{m}(dx) \end{aligned}$$

and

$$EW_1^{(A)} \ln|W_1^{(A)}| = E\gamma_1 1(V_1 \in A) \ln 1(V_1 \in A) = 0.$$

We apply now Theorem 1.4.5 to conclude that the series

$$\sum_{i=1}^{\infty} \Big[\gamma_i \Gamma_i^{-1/\alpha} 1(V_i \in A) - b_i^{(\alpha)} \int_A \beta(x)\widehat{m}(dx)\Big]$$

converges a.s. to a $S_\alpha(\sigma_A, \beta_A, 0)$ random variable with

$$\sigma_A^\alpha = \frac{\widehat{m}(A)}{C_\alpha},$$

and

$$\beta_A = \frac{E(W_1^{(A)})^{<\alpha>}}{E|W_1^{(A)}|^\alpha} = \frac{EW_1^{(A)}}{E|W_1^{(A)}|^\alpha} = \frac{\int_A \beta(x)\widehat{m}(dx)}{\widehat{m}(A)} = \frac{\int_A \beta(x)m(dx)}{m(A)}.$$

Hence, by Property 1.2.3, the random variable

$$\widetilde{M}(A) = (C_\alpha m(E))^{1/\alpha} \sum_{i=1}^{\infty} \Big[\gamma_i \Gamma_i^{-1/\alpha} 1(V_i \in A) - b_i^{(\alpha)} \int_A \beta(x)\widehat{m}(dx)\Big] + \eta_A \tag{3.9.2}$$

has a $S_\alpha(\widehat{\sigma}_A, \widehat{\beta}_A, 0)$ distribution with

$$\widehat{\sigma}_A^\alpha = (C_\alpha m(E))\sigma_A^\alpha = m(A)$$

and

$$\widehat{\beta}_A = \beta_A = \frac{\int_A \beta(x)m(dx)}{m(A)}.$$

To see that it has zero shift when $\alpha = 1$, note that $C_1 = \frac{2}{\pi}$ and that η_A exactly compensates for the shift induced by the multiplicative constant $C_1 m(E)$. Therefore $M(A) \stackrel{d}{=} \widetilde{M}(A)$ for each $A \in \mathcal{E}$.

We now prove that

$$(M(A_1), \dots, M(A_d)) \stackrel{d}{=} (\widetilde{M}(A_1), \dots, \widetilde{M}(A_d)) \tag{3.9.3}$$

for disjoint sets $A_1, \dots, A_d$. Let $\theta_1, \dots, \theta_d$ be real numbers. By (3.9.2),

$$\sum_{j=1}^{d} \theta_j \widetilde{M}(A_j)$$

$$= (C_\alpha m(E))^{1/\alpha} \sum_{i=1}^{\infty} \left[\gamma_i \Gamma_i^{-1/\alpha} \sum_{j=1}^{d} \theta_j 1(V_i \in A_j) - b_i^{(\alpha)} \sum_{j=1}^{d} \theta_j \int_{A_j} \beta(x) \widehat{m}(dx) \right]$$

$$+ \sum_{j=1}^{d} \theta_j \eta_{A_j}.$$

Let $W_i = \gamma_i \sum_{j=1}^{d} \theta_j 1(V_i \in A_j) = \sum_{j=1}^{d} \theta_j W_i^{(A_j)}$, $i = 1, 2, \dots$. Then the sequence $\{W_i, i = 1, 2, \dots\}$ is a sequence of i.i.d. random variables satisfying

$$\begin{aligned} E|W_1|^\alpha &= \sum_{j=1}^{d} |\theta_j|^\alpha P(V_1 \in A_j) \\ &= \sum_{j=1}^{d} |\theta_j|^\alpha \widehat{m}(A_j), \\ E(W_1)^{<\alpha>} &= \sum_{j=1}^{d} \int_{A_j} E\big((\gamma_1 \theta_j)^{<\alpha>} | V_1 = x\big) \widehat{m}(dx) \\ &= \sum_{j=1}^{d} \theta_j^{<\alpha>} \int_{A_j} \beta(x) \widehat{m}(dx), \\ EW_1 &= \sum_{j=1}^{d} \theta_j \int_{A_j} \beta(x) \widehat{m}(dx), \end{aligned}$$

and

$$EW_1 \ln |W_1| = E \sum_{j=1}^{d} \theta_j W_1^{(A_j)} \ln \left| \sum_{j=1}^{d} \theta_j W_1^{(A_j)} \right|$$

$$= E \sum_{j=1}^{d} \theta_j W_1^{(A_j)} \ln |\theta_j W_1^{(A_j)}| = \sum_{j=1}^{d} \theta_j \ln |\theta_j| \int_{A_j} \beta(x) \widehat{m}(dx).$$

Applying Theorem 1.4.5 again, we conclude that the series

$$\sum_{i=1}^{\infty} \left[\gamma_i \Gamma_i^{-1/\alpha} \sum_{j=1}^{d} \theta_j 1(V_i \in A_j) - b_i^{(\alpha)} \sum_{j=1}^{d} \theta_j \int_{A_j} \beta(x) \widehat{m}(dx) \right]$$

has a $S_\alpha(\sigma_1, \beta_1, \mu_1)$ distribution, where

$$\begin{aligned}
\sigma_1^\alpha &= \frac{\sum_{j=1}^d |\theta_j|^\alpha \widehat{m}(A_j)}{C_\alpha}, \\
\beta_1 &= \frac{\sum_{j=1}^d \theta_j^{<\alpha>} \int_{A_j} \beta(x)\widehat{m}(dx)}{\sum_{j=1}^d |\theta_j|^\alpha \widehat{m}(A_j)} \\
&= \frac{\sum_{j=1}^d \theta_j^{<\alpha>} \int_{A_j} \beta(x)m(dx)}{\sum_{j=1}^d |\theta_j|^\alpha m(A_j)},
\end{aligned}$$

and

$$\mu_1 = \begin{cases} 0 & \text{if } \alpha \neq 1, \\ -\sum_{j=1}^d \theta_j \ln|\theta_j| \int_{A_j} \beta(x)\widehat{m}(dx) & \text{if } \alpha = 1. \end{cases}$$

Hence, by Property 1.2.3, $\sum_{j=1}^d \theta_j \widetilde{M}(A_j) \sim S_\alpha(\sigma_2, \beta_2, \mu_2)$, where

$$\sigma_2^\alpha = (C_\alpha m(E))\sigma_1^\alpha = \sum_{j=1}^d |\theta_j|^\alpha m(A_j),$$

$$\beta_2 = \beta_1 = \frac{\sum_{j=1}^d \theta_j^{<\alpha>} \int_{A_j} \beta(x)m(dx)}{\sum_{j=1}^d |\theta_j|^\alpha m(A_j)}$$

and

$$\mu_2 = \begin{cases} 0 & \text{if } \alpha \neq 1, \\ -\frac{2}{\pi}\sum_{j=1}^d \theta_j \ln|\theta_j| \int_{A_j} \beta(x)m(dx) & \text{if } \alpha = 1, \end{cases}$$

since $C_1 = \frac{2}{\pi}$. Recalling that M is independently scattered and that each $M(A)$ is α-stable with scale parameter $M(A)^{1/\alpha}$, skewness $m(A)^{-1}\int_A \beta(x)m(dx)$ and zero shift, we obtain by Property 1.2.1 that

$$\sum_{j=1}^d \theta_j M(A_j) \stackrel{d}{=} \sum_{j=1}^d \theta_j \widetilde{M}(A_j).$$

Since this is true for any real $\theta_1, \dots, \theta_2$, we conclude that Relation (3.9.3) holds for disjoint sets $A_1, \dots, A_d$. Because of the additivity of M and of $\widetilde{M}$ (for $\widetilde{M}$, see (3.9.2)), Relation (3.9.3) also holds for non-disjoint sets $A_1, \dots, A_d$, concluding the proof of the theorem. ■

3.10 A third definition of stable stochastic integrals using the series representation

Let M be an α-stable random measure with $0 < \alpha < 2$. The series representation for M given in Theorem 3.9.1 shows that the measure M is the sum of two components: a random component consisting of signed random point masses $\gamma_i \Gamma_i^{-1/\alpha}$ placed at random points V_i, and a non-random component proportional to the signed measure $m_s(dx) = \beta(x)m(dx)$. The non-random component is absent when $0 < \alpha < 1$ and is equal to zero when M is $S\alpha S$. Therefore, if f is an integrable function, its integral $I(f) = \int_E f(x)M(dx)$ should have a representation as a series whose terms are the sum of two components: a random term equal to $\gamma_i \Gamma_i^{-1/\alpha} f(V_i)$ and a non-random integral proportional to $\int_E f(x)m_s(dx)$. The following theorem states that this is indeed the case. The proof is similar to that of Theorem 3.9.1 and is left to the reader (see Exercise 3.14).

Theorem 3.10.1 *Let $(E, \mathcal{E}, m)$ be a finite measure space and let M be an α-stable random measure with $0 < \alpha < 2$, a finite control measure m and skewness intensity β. Let $\{\Gamma_1, \Gamma_2, \ldots\}$ be a sequence of arrival times of a Poisson process with unit arrival rate, and let $\{(V_1, \gamma_1), (V_2, \gamma_2), \ldots\}$ be a sequence of i.i.d. random vectors such that V_i has distribution $\widehat{m} = m/m(E)$ on E and*

$$P(\gamma_i = 1|V_i) = 1 - P(\gamma_i = -1|V_i) = \frac{1 + \beta(V_i)}{2}.$$

Assume also that the sequence $\{(V_1, \gamma_1), (V_2, \gamma_2), \ldots\}$ is independent of the sequence $\{\Gamma_1, \Gamma_2, \ldots\}$.

Then, for any function $f \in F$ (defined by (3.4.1) and (3.4.2)), we have

$$I(f) \stackrel{d}{=} S(f),$$

where $S(f)$ is an a.s. convergent random series defined as follows:
(i) If $0 < \alpha < 1$,

$$S(f) = (C_\alpha m(E))^{1/\alpha} \sum_{i=1}^{\infty} \gamma_i \Gamma_i^{-1/\alpha} f(V_i),$$

where C_α is given in (1.2.9).
(ii) If $\alpha = 1$,

$$S(f) = \frac{2}{\pi} m(E) \sum_{i=1}^{\infty} \left[\gamma_i \Gamma_i^{-1} f(V_i) - b_i^{(1)} \int_E f(x)\beta(x)\widehat{m}(dx) \right] + \eta_f,$$

where $b_i^{(1)}$ *is given in (3.9.1) and*

$$\eta_f = \frac{2}{\pi} \ln\left(\frac{2}{\pi} m(E)\right) \int_E f(x)\beta(x)m(dx). \tag{3.10.1}$$

(iii) If $1 < \alpha < 2$,

$$S(f) = (C_\alpha m(E))^{1/\alpha} \sum_{i=1}^{\infty} \Big[\gamma_i \Gamma_i^{-1/\alpha} f(V_i) - b_i^{(\alpha)} \int_E f(x)\beta(x)\widehat{m}(dx)\Big],$$

where $b_i^{(\alpha)}$ *is given in (3.9.1).*

We refer to $S(f)$ as the *series representation of the stable stochastic integral* $I(f)$.

It is clear that $S(f)$ is linear in $f \in F$. Therefore, for any $f_1, \ldots, f_d \in F$ and any real numbers $\theta_1, \ldots, \theta_d$,

$$\sum_{k=1}^{d} \theta_k S(f_k) = S\left(\sum_{k=1}^{d} \theta_k f_k\right) \stackrel{d}{=} I\left(\sum_{k=1}^{d} \theta_k f_k\right) = \sum_{k=1}^{d} \theta_k I(f_k).$$

We then immediately obtain the following:

Corollary 3.10.2 *If* $f_1, f_2, \ldots, f_d \in F$, *then*

$$(I(f_1), I(f_2), \ldots, I(f_d)) \stackrel{d}{=} (S(f_1), S(f_2), \ldots, S(f_d)).$$

The strength of this result allows us to regard $S(f)$ as an additional definition of stable stochastic integral.

Specifically, consider a finite measure space $(E, \mathcal{E}, m)$, a measurable function $\beta : E \to [-1, 1]$ and a measurable function $f : E \to \mathbb{R}$ satisfying

$$\int_E |f(x)|^\alpha dx < \infty,$$

and also, if $\alpha = 1$,

$$\int_E |f(x)(\ln |f(x)|)\beta(x)|m(dx) < \infty.$$

We then define $S(f)$ as in Theorem 3.10.1 and call it the *α-stable integral of* f. The measure m is called the *control measure* and the function β is called the *skewness intensity* of the integral. Corollary 3.10.2 implies that this definition of stable stochastic integral is equivalent to the definitions given in Sections 3.2 and 3.4.

Example 3.10.3 Here, we use the series representation of stable stochastic integrals to gain insight into the structure of Lévy α-stable motion.

Let $\{X(t), 0 \le t \le 1\}$ be standard Lévy α-stable motion on $[0, 1]$, i.e., with $X(1) \sim S_\alpha(1, \beta, 0)$. Example 3.1.3 shows that

$$\{X(t), 0 \le t \le 1\} \stackrel{d}{=} \left\{\int_0^1 1_{[0,t]}(x) M(dx),\ 0 \le t \le 1\right\},$$

where M is an α-stable random measure on $([0, 1], \mathcal{B})$ with Lebesgue control measure and skewness intensity $\beta(x) \equiv \beta$. Therefore Corollary 3.10.2 implies that, for $0 < \alpha < 1$,

$$\{X(t), 0 \le t \le 1\} \stackrel{d}{=} \left\{C_\alpha^{1/\alpha} \sum_{i=1}^{\infty} \gamma_i \Gamma_i^{-1/\alpha} 1(U_i \le t),\ 0 \le t \le 1\right\};$$

for $\alpha = 1$,

$$\{X(t), 0 \le t \le 1\} \stackrel{d}{=} \left\{\frac{2}{\pi} \sum_{i=1}^{\infty} (\gamma_i \Gamma_i^{-1} 1(U_i \le t) - \beta t b_i^{(1)}) + \beta t \frac{2}{\pi} \ln \frac{2}{\pi},\ 0 \le t \le 1\right\};$$

and for $1 < \alpha < 2$,

$$\{X(t), 0 \le t \le 1\} \stackrel{d}{=} \left\{C_\alpha^{1/\alpha} \sum_{i=1}^{\infty} (\gamma_i \Gamma_i^{-1/\alpha} 1(U_i \le t) - \beta t b_i^{(\alpha)}),\ 0 \le t \le 1\right\},$$

where $\{\gamma_1, \gamma_2, \dots\}$ is a sequence of i.i.d. random variables satisfying

$$P(\gamma_i = 1) = 1 - P(\gamma_i = -1) = \frac{1 + \beta}{2},$$

$\{\Gamma_1, \Gamma_2, \dots\}$ is a sequence of arrival times of a Poisson process with unit arrival rate and $\{U_1, U_2, \dots\}$ is a sequence of i.i.d. random variables uniformly distributed on $[0, 1]$. These three sequences are independent. Finally, the constants $b_i^{(\alpha)}$ are given by (3.9.1).

This means that we can regard α-stable Lévy motion with $0 < \alpha < 1$ and $S\alpha S$ Lévy motion with any $0 < \alpha < 2$ as pure jump processes. The instants U_is of the jumps are distributed uniformly over $[0, 1]$, the direction of a jump (γ_i) is upward with a probability $\frac{1+\beta}{2}$ and downward with a probability $\frac{1-\beta}{2}$, and the height of the jumps viewed in decreasing order is distributed as the $-1/\alpha$ power of arrival times of Poisson process with unit arrival rate. In particular, $\Gamma_1^{-1/\alpha}$ is the height of the highest jump, $\Gamma_2^{-1/\alpha}$ is the height of the second highest jump and so on. The terms $1(U_i \le t),\ i = 1, 2, \dots,$ indicate which jump sizes have occurred by time t.

Non-symmetric α-stable Lévy motions with $1 \le \alpha \le 2$ can be regarded as a pure jump process with a linear deterministic trend.

In the symmetric case, we have

Corollary 3.10.4 *If M, in Theorem 3.10.1, is $S\alpha S$, $0 < \alpha < 2$, then for any $f \in F$,*

$$I(f) \stackrel{d}{=} (C_\alpha m(E))^{1/\alpha} \sum_{i=1}^{\infty} \epsilon_i \Gamma_i^{-1/\alpha} f(V_i), \tag{3.10.2}$$

where $\{\epsilon_i, i \geq 1\}$ is an i.i.d. sequence satisfying $P(\epsilon_i = 1) = P(\epsilon_i = -1) = 1/2$, independent of $\{\Gamma_i, i \geq 1\}$ and $\{V_i, i \geq 1\}$.

3.11 Condition S

An important modification to the series representation for $S\alpha S$ processes given in Corollary 3.10.4 is obtained by replacing the Rademacher sequence $\{\epsilon_1, \epsilon_2, \ldots\}$ by the sequence $\{d_\alpha^{-1}G_1, d_\alpha^{-1}G_2, \ldots\}$ where $G_1, G_2, \ldots$ are i.i.d. standard normal random variables and $E|d_\alpha^{-1}G_1|^\alpha = 1$. This is, of course, permissible by Theorem 1.4.2. Now, $d_\alpha^\alpha = 2^{\alpha/2}\pi^{-1/2}\Gamma(\frac{\alpha+1}{2})$ by (3.7.2). If C_α is as in (1.2.9) and

$$C'_\alpha = d_\alpha^{-\alpha} C_\alpha = 2^{-\alpha/2}\pi^{1/2}\left(\Gamma\left(\frac{\alpha+1}{2}\right)\right)^{-1} C_\alpha, \tag{3.11.1}$$

then

$$I(f) \stackrel{d}{=} (C'_\alpha m(E))^{1/\alpha} \sum_{i=1}^{\infty} G_i \Gamma_i^{-1/\alpha} f(V_i). \tag{3.11.2}$$

Consider, now, a $S\alpha S$ stochastic process represented as

$$\{X(t) = \int_E f_t(x) M(dx),\ t \in T\}.$$

Arguing as in Corollary 3.10.2, we obtain

$$\{X(t),\ t \in T\} \stackrel{d}{=} \left\{(C'_\alpha m(E))^{1/\alpha} \sum_{i=1}^{\infty} G_i \Gamma_i^{-1/\alpha} f_t(V_i),\ t \in T\right\}. \tag{3.11.3}$$

Now view the series on the right-hand side of (3.11.3) as defined on the product of two probability spaces: $(\Omega_1, \mathcal{F}_1, P_1)$ on which the Gaussian sequence $\{G_1, G_2, \ldots\}$ is defined, and $(\Omega_2, \mathcal{F}_2, P_2)$ on which the sequences $\{\Gamma_1, \Gamma_2, \ldots\}$ and $\{V_1, V_2, \ldots, \}$ are defined. Therefore, if we "fix" the values of Γ_j, $j = 1, 2, \ldots$, and V_j, $j = 1, 2, \ldots$, the series on the right-hand side of (3.11.3) becomes a series of the form $\{\sum_{i=1}^{\infty} A_i(t) G_i,\ t \in T\}$, which represents a zero mean Gaussian process on the probability space $(\Omega_1, \mathcal{F}_1, P_1)$. We obtain, therefore, the following:

Proposition 3.11.1 *Let M be a $S\alpha S$ random measure on $(E, \mathcal{E})$ with a* finite *control measure m. Let $f_t \in L^\alpha(m)$, $t \in T$. Then the $S\alpha S$ stochastic process $X(t) = \int_E f_t(x)M(dx)$, $t \in T$ is conditionally centered Gaussian, i.e., it is a probability mixture of zero mean Gaussian processes.*

This statement allows us to relate $S\alpha S$ processes to Gaussian processes whose structure and properties have been extensively studied.

In Section 3.7, we introduced the sub-Gaussian $S\alpha S$ processes. These are processes of the form $\{X(t) = A^{1/2}G(t),\ t \in T\}$ where $G(t)$ is a centered Gaussian process. It is, therefore, obvious that sub-Gaussian $S\alpha S$ processes are conditionally centered Gaussian. Proposition 3.11.1 implies that not only sub-Gaussian $S\alpha S$ processes but also many other $S\alpha S$ processes are conditionally centered Gaussian; in fact, all $S\alpha S$ processes that can be represented in the form

$$\{X(t),\ t \in T\} \stackrel{d}{=} \left\{\int_E f_t(x)M(dx),\ t \in T\right\},\quad m \text{ finite}, \tag{3.11.4}$$

where M is a $S\alpha S$ random measure on $(E, \mathcal{E})$ with a *finite* control measure m.

But how general is the class of $S\alpha S$ processes $\{X(t),\ t \in T\}$ that admit the representation (3.11.4)? This class turns out to be very broad. We will spend the rest of this section discussing $S\alpha S$ processes that admit representation (3.11.4).

Definition 3.11.2 Let $\{X(t),\ t \in T\}$ be a stochastic process. We say that it satisfies *Condition S* if there is a countable subset $T_0 \subseteq T$ such that for every $t \in T$, $X(t)$ is a limit in probability as $n \to \infty$ of sums of the type $\sum_{j=1}^n a_{j,n}X(t_{j,n})$, for some $a_{j,n} \in \mathbb{R}$, $t_{j,n} \in T_0$, $j = 1, \ldots, n$, and some $n - 1, 2, \ldots$.

Remarks

1. The letter "S" stands for separability.

2. Condition S is equivalent to separability in probability, i.e., to the following:

   "There is a countable subset $T_0 \subseteq T$ such that for every $t \in T$, $X(t) = \operatorname{plim}_{k\to\infty} X(t_k)$, where $t_k \in T_0$, $k = 1, 2, \ldots$. " (see Exercise 3.21.)

3. We will see in Chapter 13 that all $S\alpha S$ processes satisfying Condition S do admit the representation (3.11.4).

We are now ready to present an undoubtedly partial list of $S\alpha S$ processes that admit the representation (3.11.4). For convenience, we present this list in the following:

Proposition 3.11.3 *A $S\alpha S$ process $\{X(t),\ t \in T\}$ admits the integral representation (3.11.4) if it satisfies at least one of the following conditions:*

(i) (T, d) is a separable metric space, and there is a countable subset $T_1 \subseteq T$ such that $\{X(t),\ t \in T \backslash T_1\}$ is continuous in probability.

(ii) $\{X(t),\ t \in T\} \stackrel{d}{=} \{\int_E g_t(x) M'(dx),\ t \in T\}$, where M' is a $S\alpha S$ random measure on $(E', \mathcal{E}')$ with a σ-finite control measure.

PROOF: We first show that if $\{X(t),\ t \in T\}$ satisfies Condition (i), then it satisfies Condition S. Since (T, d) is separable, we can choose a countable set $T_2 \subseteq T$ such that for any $t \in T$, there is a sequence $\{t_k\}_{k=1}^{\infty}$ in T_2 such that $d(t_k, t) \to 0$ as $k \to \infty$. Since $\{X(t),\ t \in T \backslash T_1\}$ is continuous in probability, for every $t \in T - T_1$, $X(t)$ is a limit in probability of $\{X(s_k)\}$, $s_k \in T_2$, $k = 1, 2, \ldots$. Hence, for every $t \in T$, $X(t)$ is a limit in probability of $\{X(s_k)\}$ where s_k, $k = 1, 2, \ldots$, belongs to the countable set $T_0 = T_1 \cup T_2$. The process $\{X(t),\ t \in T\}$, therefore, satisfies Condition S. As noted in the preceding remark, this implies that the process admits the representation (3.11.4).

Suppose, now, that $\{X(t),\ t \in T\}$ satisfies (ii). Let m' be the control measure of M' and let $A_1, A_2, \ldots$ be a partition of E' into $\mathcal{E}'$-measurable sets such that $0 < m'(A_i) < \infty$, $i = 1, 2, \ldots$. We now apply Proposition 3.5.5 to conclude that

$$\{X(t),\ t \in T\} \stackrel{d}{=} \left\{ \int_{E'} g_t(x) r(x) M(dx),\ t \in T \right\},$$

where, $r(x) = (2^i m'(A_i))^{1/\alpha}$ if $x \in A_i$, and M is a $S\alpha S$ random measure on $(E', \mathcal{E}')$ whose control measure $m(dx) = 2^{-i}(m'(A_i))^{-1} m'(dx)$, $x \in A_i$, is a probability measure. This completes the proof. ∎

The class of $S\alpha S$ processes described in Proposition 3.11.3 is broad enough to cover virtually every $S\alpha S$ process that we will encounter in this book.

Counterexample. Let $X(t)$, $t \in [0,1]$, be i.i.d. (non-degenerate) $S\alpha S$ random variables. The process $\{X(t),\ t \in [0,1]\}$ is well defined since its finite-dimensional distributions exist and form a consistent family. It clearly does not satisfy Condition S. One can also check directly that it does not have representation (3.11.4) with finite (or σ-finite) control measure since, if it did, the supports of f_t, $t \in [0,1]$, would all be disjoint and also have positive measure, which is not possible. Nevertheless, this process $\{X(t),\ t \in [0,1]\}$ is conditionally centered Gaussian since it can be represented as $X(t) \stackrel{d}{=} c\sqrt{A(t)}G(t)$, where $A(t)$, $t \in [0,1]$, are i.i.d. and $G(t)$, $t \in [0,1]$, are i.i.d. standard normal random variables independent of the $A(t)$s.

3.12 A fourth definition of stable stochastic integrals using a Poisson representation

Let M be an α-stable random measure on $(E, \mathcal{E})$ with control measure m (not necessarily finite) and skewness intensity β, and let f be any element of the space F defined in (3.2.1). Our goal is to represent the α-stable stochastic integral $I(f) = \int_E f(x)M(dx)$ as an integral with respect to a Poisson random measure N on $E \times \mathbb{R}_0$, where $\mathbb{R}_0 = \mathbb{R}\backslash\{0\}$.

Note that $f \in F$ implies that the control measure m is σ-finite on the support of f because $\int_E |f(x)|^\alpha m(dx) < \infty$ implies $m(E_i) < \infty$ where $E_i = \{x : 2^{-i} < |f(x)| < 2^{-i+1}\}$, $i = -\infty, \dots, +\infty$. We will therefore assume without loss of generality, that the control measure m is σ-finite. The σ-finiteness of m is an important requirement in the following construction of the integral.

The heuristic idea is to let

$$N(dx, du) = \text{number of randomly placed points in } dxdu\ ,\ x \in E,\ u \in \mathbb{R}_0, \tag{3.12.1}$$

have a Poisson distribution with a mean $n(dx, du) = EN(dx, du)$ adequately chosen, and to express M, in the case $\alpha < 1$, as $M(dx) = \int_{\mathbb{R}_0} uN(dx, du)$, i.e., as the sum of the heights of all the points located within a strip dx in E. The situation is more delicate when $1 \leq \alpha < 2$ because N has "too many" small jumps and the integral $\int_{\mathbb{R}_0} uN(dx, du)$ diverges. It is then necessary to "compensate" adequately for the jumps.

We start with a formal definition of a Poisson random measure. Let $(S, \mathcal{S}, n)$ be a measure space and let $\mathcal{S}_0 = \{A \in \mathcal{S} : n(A) < \infty\}$.

Definition 3.12.1 A *Poisson random measure* N on $(S, \mathcal{S}, n)$ is an independently scattered σ-additive set function $N : \mathcal{S}_0 \to L^0(\Omega)$ such that, for each set A in $\mathcal{S}_0$, the random variable $N(A)$ has a Poisson distribution with mean $n(A)$, i.e.,

$$P(N(A) = k) = e^{-n(A)} \frac{(n(A))^k}{k!},\ k = 0, 1, 2, \dots,$$

n is called the *control* (or *mean*) *measure* of N.

Note that this definition is analogous to the definition of an α-stable random measure given in Definition 3.3.1. The only difference is that the distribution is Poisson instead of α-stable. The fact that the Poisson distribution is integer-valued makes Poisson random measures somewhat more intuitive than their stable counterparts. Think of a Poisson random measure N as a point process on S: for each $A \in \mathcal{S}_0$, $N(A)$ can be regarded as the (random) number of points belonging to A (see (3.12.1)), which is why N is also called a counting measure. Then every realization of N is a (non-random) discrete measure on S and the integrals with respect to N can be defined for each realization. As all the measures involved are discrete, the integrals, in fact, are sums. Thus, in a sense, the idea of Poisson representation of stable stochastic integrals has much in common with the series representation of these integrals developed in Section 3.10.

Set $S = E \times \mathbb{R}_0$ and choose

$$n(dx, du) = EN(dx, du) = \begin{cases} (1+\beta(x))m(dx)\frac{du}{|u|^{\alpha+1}} & \text{if } u > 0, \\ (1-\beta(x))m(dx)\frac{du}{|u|^{\alpha+1}} & \text{if } u < 0, \end{cases} \tag{3.12.2}$$

where m and β are, respectively, the control measure and skewness intensity of the α-stable random measure M.

Let E_i, $i = 1, 2, \ldots,$ be a partition of E into $\mathcal{E}$-measurable sets of finite measure m, let $\delta > 0$ and let $(-\delta, \delta)^c$ denote the complement of the interval $(-\delta, \delta)$.

Theorem 3.12.2 *(Poisson Representation). The α-stable stochastic integral $I(f) = \int_E f(x)M(dx)$ has the representation:*
(a) $0 < \alpha < 1$:

$$I(f) \stackrel{d}{=} \left(2\alpha^{-1}(\Gamma(1-\alpha))\cos\frac{\pi\alpha}{2}\right)^{-1/\alpha} \int_E \int_{\mathbb{R}_0} f(x)uN(dx, du). \tag{3.12.3}$$

(b) $1 < \alpha < 2$:

$$I(f) \stackrel{d}{=} \left(2\frac{\Gamma(2-\alpha)}{\alpha(\alpha-1)}\left(-\cos\frac{\pi\alpha}{2}\right)\right)^{-1/\alpha}$$

$$\times \lim_{\delta\to 0}\sum_{i=1}^{\infty}\left(\int_{E_i}\int_{(-\delta,\delta)^c} f(x)uN(dx, du) - E\int_{E_i}\int_{(-\delta,\delta)^c} f(x)uN(dx, du)\right). \tag{3.12.4}$$

(c) $\alpha = 1$:

$$I(f) \stackrel{d}{=} \frac{1}{\pi}\lim_{\delta\to 0}\sum_{i=1}^{\infty}\left(\int_{E_i}\int_{(-\delta,\delta)^c} f(x)uN(dx, du)\right.$$

$$\left. -2\left(\ln_+\frac{1}{\delta}\right)\int_{E_i}\int_{(-\delta,\delta)^c} f(x)\beta(x)m(dx)\right) - \frac{2b}{\pi}\int_E f(x)\beta(x)m(dx) \tag{3.12.5}$$

where

$$\ln_+\frac{1}{\delta} = \begin{cases} 0 & \text{if } \delta \geq 1, \\ \ln\frac{1}{\delta} & \text{if } \delta \leq 1, \end{cases}$$

and

$$b = \ln \pi + \int_0^\infty \frac{\sin u - u1(u \le 1)}{u^2} du.$$

The expressions on the right-hand sides of (3.12.3), (3.12.4) and (3.12.5) converge a.s.

Remarks

1. As mentioned above, the stochastic integrals with respect to a Poisson random measure N are defined as sums: for a measurable set $A \subset E \times \mathbb{R}_0$ and a measurable function $g : A \to \mathbb{R}$, we have $\int_A g(x,u)N(dx,du) = \sum_{i=1}^\infty 1_A((X_i,U_i))g(X_i,U_i)$ whenever the series converges absolutely, where $\{(X_i,U_i),\ i = 1,2,\ldots\}$ is an enumeration of the points of the random measure N.

2. If $m(E) < \infty$, then set $E_1 = E$ and $E_2 = E_3 = \cdots = \emptyset$.

3. When $1 < \alpha < 2$, the integrals on the right-hand side of (3.12.4) can be written $\int_{E_i} \int_{(-\delta,\delta)^c} f(x)u\widetilde{N}(dx,du)$, where $\widetilde{N} = N - EN = N - n$.

4. When $\alpha = 1$,

$$2\Big(\ln_+ \frac{1}{\delta}\Big) \int_{E_i} f(x)\beta(x)m(dx) = \int_{E_i} \int_{(-\delta,\delta)^c} f(x)u1(u \le 1)\, n(dx,du).$$

   To verify this, use (3.12.2), $\ln_+ \frac{1}{\delta} = \int_\delta^\infty u\, 1(u \le 1)u^{-2}du$ and write

$$2\int_{E_i} f(x)\beta(x)m(dx) = \int_{E_i} [f(x)(1+\beta(x))+(-f(x))(1-\beta(x))]m(dx).$$

   We cannot replace $2(\ln_+ \frac{1}{\delta}) \int_{E_i} f(x)\beta(x)m(dx)$ on the right-hand side of (3.12.5) by $\int_{E_i} \int_{(-\delta,\delta)^c} f(x)u\, n(dx,du)$ because

$$\int_\delta^\infty u\, n(dx,du) = (1+\beta(x))m(dx) \int_\delta^\infty u \frac{du}{u^2} = \infty.$$

5. To understand heuristically why the Poisson integrals in the theorem are well defined, recall that the Poisson random measure $N(dx,du)$ has control measure $n(dx,du) = EN(dx,du) < 2m(dx)\frac{du}{|u|^{\alpha+1}}$. Fix a strip dx and consider the dependence on u. The mean density of points $|u|^{-\alpha-1}$ tends to infinity as $|u| \to 0$. For $a > 0$, the mean number of points in $(-a,a)^c$, $\int_{|u|>a} n(dx,du) \le 4m(dx) \int_a^\infty u^{-\alpha-1}du$ is finite. There are therefore only finitely many points in the strip $(dx, |u| > a)$ and the integral on N

with respect to $|u| > a$ is, in fact, a finite sum. Now consider the strip $(dx, |u| < a)$.

When $\alpha < 1$, the expected sum of absolute heights of the points in $(-a, a)$, $\int_{-a}^{a} |u| n(dx, du) \le 4m(dx) \int_0^a u^{-\alpha} du$, is finite. This is why we can use a Poisson integral with respect to N in $(-a, a)$ when $\alpha < 1$.[6]

When $1 \le \alpha < 2$, $\int_{-a}^{a} |u| n(dx, du)$ is infinite, but $\int_{-a}^{a} u^2 n(dx, du) \le 4m(dx) \int_0^a u^{1-\alpha} du$ is finite. This is why we can use a Poisson integral with respect to $\widetilde{N} = N - EN = N - n$ in $(-a, a)$ when $1 \le \alpha < 2$.[7]

Observe that when $1 < \alpha < 2$, the Poisson representation in (3.12.4) involves $\widetilde{N}$, even on $(-a, a)^c$. This is valid because

$$u1\{|u| > a\} d\widetilde{N} = u1\{|u| > a\} dN - u1\{|u| > a\} dn$$

and $\int_{-(a,a)^c} |u| n(dx, du) \le 4m(dx) \int_a^\infty u^{-\alpha} du$ is finite when $\alpha > 1$.

The proof of the theorem uses the following construction of a Poisson random measure on $E \times (\delta, \infty)$, $\delta > 0$.

Lemma 3.12.3 *Let $(E, \mathcal{E}, \mu)$ be a finite measure space, $0 < \alpha < 2$ and $\delta > 0$. Introduce*

1. *$Z_1, Z_2, \ldots,$ E-valued independent random variables with common law $\widehat{\mu} = \mu/\mu(E)$.*

2. *$Y_1, Y_2, \ldots,$ real-valued independent positive random variables with distribution*

$$P\{Y_i > \lambda\} = \begin{cases} \delta^\alpha \lambda^{-\alpha} & \text{if } \lambda > \delta, \\ 1 & \text{otherwise.} \end{cases} \tag{3.12.6}$$

[6]If N is a Poisson random measure on $(S, \mathcal{S}, n)$, then the Poisson integral $\int_S g(s) N(ds)$ is defined for all measurable functions g satisfying $\int_S |g(s)| n(ds) < \infty$. See Exercise 3.22. By the same exercise, the characteristic function of $\int_S g(s) N(ds)$ is

$$E \exp\left\{ i\theta \int_S g(s) N(ds) \right\} = \exp\left\{ \int_S (e^{i\theta g(s)} - 1) n(ds) \right\}$$

and $|e^{i\theta g(s)} - 1| \le |\theta g(s)|$.

[7]The Poisson integral $\int_S g(s) \widetilde{N}(ds)$ is defined for all g satisfying $\int_S g^2(s) n(ds) < \infty$ because the characteristic function of $\int_S g(s) \widetilde{N}(ds)$ is

$$E \exp\left\{ i\theta \int_S g(s) \widetilde{N}(ds) \right\} = \exp\left\{ \int_{-\infty}^{\infty} [e^{i\theta g(s)} - 1 - i\theta g(s)] n(ds) \right\}$$

and $|e^{i\theta g(s)} - 1 - i\theta g(s)| \le \frac{1}{2} \theta^2 g^2(s)$. See Exercise 3.23.

3. D, a Poisson random variable with mean $ED = \alpha^{-1}\delta^{-\alpha}\mu(E)$.

Assume $\{Z_i\}, \{Y_i\}, D$ *independent. Then the points* $(Z_j, Y_j),\ j = 1, \ldots, D$, *are a realization of a Poisson random measure* N *on* $E \times \mathbb{R}_0$ *with control measure*

$$EN(dz, dy) = \mu(dz)\frac{dy}{y^{1+\alpha}}1(y > \delta).$$

PROOF: Let $A_1, \ldots, A_n$ be disjoint measurable sets in $E \times \mathbb{R}_0$ and let

$$N(A_j) = \sum_{k=1}^{D} 1_{A_j}(Z_k, Y_k)$$

denote the number of points $(Z_k, Y_k),\ k = 1, \ldots, D$, in the set A_j. Then N is a counting measure and $N(A_1), \ldots, N(A_n)$ are independent Poisson random variables because they correspond to a splitting of the Poisson number of arrivals D. To compute the control measure of N, note that

$$\begin{aligned} EN(A) &= (ED)P\{(Z_1, Y_1) \in A\} \\ &= \alpha^{-1}\delta^{-\alpha}\mu(E)\int_A 1(y > \delta)\alpha\delta^{\alpha}y^{-(1+\alpha)}\widehat{\mu}(dz)dy \\ &= \int_A 1(y > \delta)y^{-(1+\alpha)}\mu(dz)dy. \quad \blacksquare \end{aligned}$$

The proof of Theorem 3.12.2 proceeds by splitting E into subspaces E_i with $m(E_i) < \infty$, defining for each i, two independent Poisson random measures, one on $E_i \times (\delta, \infty)$, the other on $E_i \times (-\infty, -\delta)$, and showing that the right-hand sides in Theorem 3.12.2 converge a.s. as $\delta \to 0$ to a random variable which has the same distribution as $I(f)$.

PROOF OF THEOREM 3.12.2:

(a) $0 < \alpha < 1$:

Set $J(f) = \int_E \int_{\mathbb{R}_0} f(x)uN(dx, du)$. If $(X_k, U_k),\ k =, 0, 1, \ldots$, is an enumeration of the points of the Poisson random measure N, then

$$J(f) = \sum_{k \geq 0} U_k f(X_k).$$

Our first task is to show that

$$H(f) = \sum_{k \geq 0} |U_k||f(X_k)| < \infty \ a.s.$$

It is obviously sufficient to consider the case $\beta(x) \equiv 1$. Then $n(dx, du) = 0$ for $u < 0$ by (3.12.2) and hence $U_k \geq 0$. We can write

$$H(f) = \sum_{k:(X_k,U_k)\in A_1} U_k|f(X_k)| + \sum_{m=1}^{\infty} \sum_{k:(X_k,U_k)\in B_m} U_k|f(X_k)|, \quad (3.12.7)$$

where $A_\lambda = \{(x,u) \in E \times \mathbb{R}^+ : u|f(x)| > \lambda\}$, $\lambda > 0$, and $B_j = A_{(j+1)^{-1}} \backslash A_{j^{-1}}$, $j = 1, 2, \ldots$. Observe that

$$\begin{aligned} n(A_\lambda) &= \int_{A_\lambda}\int 2m(dx)\frac{du}{u^{\alpha+1}} \\ &= 2\int_E m(dx)\int_{\lambda|f(x)|^{-1}}^{\infty} u^{-(1+\alpha)}du \\ &= 2\alpha^{-1}\lambda^{-\alpha}\int_E |f(x)|^\alpha m(dx) \\ &= C\lambda^{-\alpha}, \end{aligned} \quad (3.12.8)$$

and $n(B_j) = C((j+1)^\alpha - j^\alpha)$, where $C = 2\alpha^{-1}\int_E |f(x)|^\alpha m(dx) < \infty$. Since (3.12.8) implies that for almost any realization of N, only finitely many points are in A_1, the first term on the right-hand side of (3.12.7) is finite with probability 1. Moreover, if N_j equals the (random) number of points in B_j, then $EN_j = n(B_j)$ and

$$\begin{aligned} E\sum_{j=1}^{\infty} \sum_{k:(X_k,U_k)\in B_j} U_k|f(X_k)| &\leq E\sum_{j=1}^{\infty}\frac{1}{j}N_j = \sum_{j=1}^{\infty}\frac{1}{j}EN_j \\ &\leq C\sum_{j=1}^{\infty}\frac{(j+1)^\alpha - j^\alpha}{j} \leq C\alpha\sum_{j=1}^{\infty} j^{\alpha-2} < \infty \end{aligned}$$

since $0 < \alpha < 1$. The double sum in (3.12.7) is therefore finite with probability 1. This proves $H(f) < \infty$ a.s. and, thus, that $J(f)$ is well defined in the sense of a.s. convergence.

Our next step is to prove (3.12.3). Let

$$J_i(f) = \sum_{k:X_k\in E_i} U_k f(X_k), \quad i = 1, 2, \ldots ,$$

where $E_1, E_2, \ldots$ is a partition of E into $\mathcal{E}$-measurable sets of finite measure m. Clearly, $J_1(f), J_2(f), \ldots$ are independent random variables and

$$J(f) = \sum_{i=1}^{\infty} J_i(f) \text{ a.s.}, \quad (3.12.9)$$

since $\sum_{i=1}^{\infty} |J_i(f)| \le H(f) < \infty$. Now choose $\delta > 0$ and define for each $i = 1, 2, \ldots$

$$\begin{cases} J^+_{\delta,i}(f) = \sum_{k: X_k \in E_i, U_k > \delta} \ U_k f(X_k), \\ J^-_{\delta,i}(f) = \sum_{k: X_k \in E_i, U_k < -\delta} U_k f(X_k). \end{cases} \tag{3.12.10}$$

Then $J^+_{\delta,i}(f)$ and $J^-_{\delta,i}(f)$ are independent random variables and

$$J_i(f) = \lim_{\delta \to 0} (J^+_{\delta,i}(f) + J^-_{\delta,i}(f)) \text{ a.s.} \tag{3.12.11}$$

Now $J^+_{\delta,i}(f)$ is actually a sum over the points of a Poisson random measure on $E \times \mathbb{R}_0$ with finite control measure $m_\beta(dx) 1(x \in E_i) u^{-(1+\alpha)} 1(u > \delta)$, where

$$m_\beta(dx) = (1 + \beta(x)) m(dx).$$

Therefore, by Lemma 3.12.3, we can write

$$J^+_{\delta,i}(f) = \sum_{k=1}^{N^+_{\delta,i}} Y_k f(Z_k), \tag{3.12.12}$$

where $\{Y_1, Y_2, \ldots\}$ and $\{Z_1, Z_2, \ldots\}$ are two independent sequences of i.i.d. random variables: Y_k has the law (3.12.6), Z_k has the law $m_\beta(dx) 1_{E_i}(x) / m_\beta(E_i)$, and $N^+_{\delta,i}$ is a Poisson random variable, independent of the sequences $\{Y_k\}$ and $\{Z_k\}$, with mean $\alpha^{-1} \delta^{-\alpha} m_\beta(E_i)$. Arguing as in the proof of Proposition 1.2.11, we obtain

$$E \exp\{i\theta J^+_{\delta,i}(f)\} = \exp\{\alpha^{-1} \delta^{-\alpha} m_\beta(E_i) (E e^{i\theta Y_1 f(Z_1)} - 1)\}. \tag{3.12.13}$$

If ϕ_f denotes the characteristic function of $f(Z_1)$,

$$\begin{aligned} E \exp\{i\theta Y_1 f(Z_1)\} &= \int_\delta^\infty \phi_f(\theta\lambda) \alpha \delta^\alpha \lambda^{-(\alpha+1)} d\lambda \\ &= \int_\delta^\infty \alpha \delta^\alpha \lambda^{-(\alpha+1)} d\lambda \int_{E_i} e^{i\theta f(x)\lambda} (m_\beta(E_i))^{-1} m_\beta(dx) \\ &= \alpha \delta^\alpha (m_\beta(E_i))^{-1} \int_{E_i} m_\beta(dx) \int_\delta^\infty e^{i\theta f(x)\lambda} \lambda^{-(\alpha+1)} d\lambda. \end{aligned}$$

Substituting this relation in (3.12.13), we obtain

$$E \exp\{i\theta J^+_{\delta,i}(f)\} = \exp\Big\{ \int_{E_i} m_\beta(dx) \int_\delta^\infty (e^{i\theta f(x)\lambda} - 1) \lambda^{-(\alpha+1)} d\lambda \Big\}. \tag{3.12.14}$$

In the same way,

$$E\exp\{i\theta J^-_{\delta,i}(f)\} = \exp\Big\{\int_{E_i} m_{-\beta}(dx)\int_\delta^\infty (e^{-i\theta f(x)\lambda}-1)\lambda^{-(\alpha+1)}d\lambda\Big\}, \tag{3.12.15}$$

where $m_{-\beta}(dx) = (1-\beta(x))m(dx)$. By the independence of $J^+_{\delta,i}(f)$ and $J^-_{\delta,i}(f)$, (3.12.11) and (3.12.9), we obtain

$$\begin{aligned} &E\exp\{i\theta J(f)\} \\ &= \exp\Big\{\int_E m_\beta(dx)\int_0^\infty (e^{i\theta f(x)\lambda}-1)\lambda^{-(1+\alpha)}d\lambda \\ &\quad + \int_E m_{-\beta}(dx)\int_0^\infty (e^{-i\theta f(x)\lambda}-1)\lambda^{-(1+\alpha)}d\lambda\Big\}. \end{aligned} \tag{3.12.16}$$

The integrals over λ on the right-hand side of (3.12.16) are well known:

$$\begin{aligned} &\int_0^\infty (e^{\pm i\theta f(x)\lambda}-1)\lambda^{-(1+\alpha)}d\lambda \\ &= -\alpha^{-1}\Gamma(1-\alpha)\Big(\cos\frac{\pi\alpha}{2}\Big)|\theta f(x)|^\alpha\Big(1\mp i\,\text{sign}\,(\theta f(x))\tan\frac{\pi\alpha}{2}\Big) \end{aligned} \tag{3.12.17}$$

(see Feller (1966), p.542, and also Exercise 3.24). Substituting (3.12.17) in (3.12.16) and using $m_{\pm\beta}(dx) = (1\pm\beta(x))m(dx)$, we conclude $E\exp\{i\theta J(f)\}$ equals

$$\exp\Big\{-2\alpha^{-1}\Gamma(1-\alpha)\Big(\cos\frac{\pi\alpha}{2}\Big)|\theta|^\alpha\int_E|f(x)|^\alpha\Big(1-i\,\text{sign}\,(\theta f(x))\tan\frac{\pi\alpha}{2}\Big)m(dx)\Big\}.$$

Hence, $J(f)\sim S_\alpha(\sigma_f,\beta_f,0)$ with

$$\sigma_f^\alpha = 2\alpha^{-1}\Gamma(1-\alpha)\Big(\cos\frac{\pi\alpha}{2}\Big)\int_E|f(x)|^\alpha m(dx),$$

and β_f defined as in Property 3.2.2. Therefore, by Property 3.2.2,

$$J(f)\stackrel{d}{=}\Big(2\alpha^{-1}\Gamma(1-\alpha)\cos\frac{\pi\alpha}{2}\Big)^{-1}I(f).$$

(b) $1<\alpha<2$:

Let $J^+_{\delta,i}(f)$ and $J^-_{\delta,i}(f)$ be defined as in (3.12.10). We must prove

$$I(f)\stackrel{d}{=}\Big(2\frac{\Gamma(2-\alpha)}{\alpha(\alpha-1)}\Big(-\cos\frac{\pi\alpha}{2}\Big)\Big)^{-1/\alpha}\widehat{J}(f), \tag{3.12.18}$$

where

$$\widehat{J}(f) = \lim_{\delta \to 0} \sum_{i=1}^{\infty} \Big(J_{\delta,i}^{+}(f) - EJ_{\delta,i}^{+}(f) + J_{\delta,i}^{-}(f) - EJ_{\delta,i}^{-}(f) \Big). \tag{3.12.19}$$

Our first task is to show that the right-hand side of (3.12.19) converges a.s. Observe that the random variables $J_{\delta,i}^{+}(f),\ i = 1,2,\ldots,\ J_{\delta,i}^{-}(f),\ i = 1,2,\ldots,$ are independent. Relation (3.12.12) expresses $J_{\delta,i}^{+}(f)$ in terms of $\{Y_k\}, \{Z_k\}$ and $N_{\delta,i}^{+}$. Since $\alpha > 1$, $EY_1 = \int_{\delta}^{\infty} \lambda \alpha \delta^{\alpha} \lambda^{-\alpha-1} d\lambda = \alpha\delta/(\alpha - 1) < \infty$ and

$$\begin{aligned} EJ_{\delta,i}^{+}(f) &= EN_{\delta,i}^{+} EY_1 Ef(Z_1) \\ &= (\alpha^{-1}\delta^{-\alpha} m_{\beta}(E_i))(\alpha(\alpha-1)^{-1}\delta) m_{\beta}(E_i)^{-1} \int_{E_i} f(x) m_{\beta}(dx) \\ &= (\alpha - 1)^{-1} \delta^{1-\alpha} \int_{E_i} f(x) m_{\beta}(dx). \end{aligned}$$

A similar expression holds for $EJ_{\delta,i}^{-}(f)$ with $\delta^{1-\alpha}$ replaced by $-\delta^{1-\alpha}$ and m_{β} by $m_{-\beta}$. Define

$$\widehat{J}_{\delta,i}^{\pm}(f) = J_{\delta,i}^{\pm}(f) - EJ_{\delta,i}^{\pm}(f)$$

and

$$\widehat{J}_{\delta}(f) = \sum_{i=1}^{\infty} (\widehat{J}_{\delta,i}^{+}(f) + \widehat{J}_{\delta,i}^{-}(f)).$$

We claim that this series converges a.s. Since it is a sum of independent random variables, we need only to prove convergence in distribution. By (3.12.14) and (3.12.15),

$$\begin{aligned} E \exp\{i\theta \widehat{J}_{\delta,i}^{\pm}(f)\} &= \exp\Big\{ \int_{E_i} m_{\pm\beta}(dx) \int_{\delta}^{\infty} (e^{\pm i\theta f(x)\lambda} - 1) \lambda^{-(1+\alpha)} d\lambda \\ &\qquad \mp i\theta \frac{\delta^{1-\alpha}}{\alpha - 1} \int_{E_i} f(x) m_{\pm\beta}(dx) \Big\} \\ &= \exp\Big\{ \int_{E_i} m_{\pm\beta}(dx) \int_{\delta}^{\infty} [e^{\pm i\theta f(x)\lambda} - 1 \mp i\theta f(x)\lambda] \lambda^{-(1+\alpha)} d\lambda \Big\}. \end{aligned}$$

The exponent in $E\exp\{i\theta(\widehat{J}_{\delta,i}(f) + \widehat{J}_{\delta,i}^{-}(f))\}$ equals

$$\begin{aligned} A_i &= \int_{E_i} m_{\beta}(dx) \int_{\delta}^{\infty} [e^{i\theta f(x)\lambda} - 1 - i\theta f(x)\lambda] \lambda^{-(1+\alpha)} d\lambda \\ &\quad + \int_{E_i} m_{-\beta}(dx) \int_{\delta}^{\infty} [e^{-i\theta f(x)\lambda} - 1 + i\theta f(x)\lambda] \lambda^{-(1+\alpha)} d\lambda, \end{aligned}$$

since the $\widehat{J}_{\delta,i}^{\pm}(f)$ are independent random variables.

With the changes of variables in the inner integrals, $u = \lambda|f(x)|$ and $u = -\lambda|f(x)|$, we obtain

$$\begin{aligned} |A_i| &\le 2\int_{E_i} |f(x)|^\alpha m(dx) \int_{-\infty}^{\infty} |e^{i\theta u} - 1 - i\theta u||u|^{-(1+\alpha)}du \\ &\le C\int_{E_i} |f(x)|^\alpha m(dx), \end{aligned}$$

where C is a finite constant. Since $E = \bigcup_{i=1}^{\infty} E_i$, this implies $\sum_{i=1}^{\infty}|A_i| < \infty$ and, therefore, $\widehat{J}_\delta(f)$ is well defined in the sense of a.s. convergence,

$$\widehat{J}(f) = \lim_{\delta\to 0} \widehat{J}_\delta(f)$$

exists in distribution, and

$$E\exp\{i\theta\widehat{J}(f)\} = \exp\Big\{\int_E m_\beta(dx)\int_0^\infty [e^{i\theta f(x)\lambda} - 1 - i\theta f(x)\lambda]\lambda^{-(1+\alpha)}d\lambda$$

$$+ \int_E m_{-\beta}(dx)\int_0^\infty [e^{-i\theta f(x)\lambda} - 1 + i\theta f(x)\lambda]\lambda^{-(1+\alpha)}d\lambda\Big\}. \quad (3.12.20)$$

To see that $\widehat{J}(f)$ is defined a.s., note that the summands in

$$\widehat{J}(f) = \sum_{j=1}^{\infty}(\widehat{J}_{2^{-j}}(f) - \widehat{J}_{2^{-j+1}}(f)) + \widehat{J}_1(f) \quad (3.12.21)$$

are independent random variables and hence the convergence in distribution of the sum implies a.s. convergence.

It remains to prove (3.12.18). To do this, we need to compute the integrals in (3.12.20). An integration by parts of (3.12.17) (see Exercise 3.24) gives

$$\int_0^\infty (e^{\pm i\theta f(x)\lambda} - 1 \mp i\theta f(x)\lambda)\lambda^{-(1+\alpha)}d\lambda$$

$$= -|\theta f(x)|^\alpha \frac{\Gamma(2-\alpha)}{\alpha(\alpha-1)}\Big(-\cos\frac{\pi\alpha}{2}\Big)\Big(1 \mp i \text{ sign }(\theta f(x)) \tan\frac{\pi\alpha}{2}\Big).$$

Substituting this relation in (3.12.20) and using $m_{\pm\beta}(dx) = (1 \pm \beta(x))m(dx)$ yields

$$E\exp\{i\theta\widehat{J}(f)\} =$$

$$\exp\Big\{2\frac{\Gamma(2-\alpha)}{\alpha(\alpha-1)}\Big(\cos\frac{\pi\alpha}{2}\Big)|\theta|^\alpha \int_E |f(x)|^\alpha\Big(1 - i \text{ sign }(\theta f(x)) \tan\frac{\pi\alpha}{2}\Big)m(dx)\Big\}.$$

Hence, $\widehat{J}(f) \sim S_\alpha(\sigma_f, \beta_f, 0)$ with $\sigma_f^\alpha = 2\frac{\Gamma(2-\alpha)}{\alpha(\alpha-1)}(-\cos\frac{\pi\alpha}{2})\int_E |f(x)|^\alpha m(dx)$ and β_f is defined as in Property 3.2.2. Therefore Property 3.2.2 implies Relation (3.12.18).

(c) $\alpha = 1$:

Because the main steps are the same as in the case $1 < \alpha < 2$, we consider only differences in computation. Once more, let $J^+_{\delta,i}(f)$ and $J^-_{\delta,i}(f)$ be defined as in (3.12.10). We must prove

$$I(f) \stackrel{d}{=} \frac{1}{\pi}\widehat{J}(f) - \frac{2b}{\pi}\int_E f(x)\beta(x)m(dx), \tag{3.12.22}$$

where this time

$$\widehat{J}(f) = \lim_{\delta\to 0}\sum_{i=1}^{\infty}\Big(J^+_{\delta,i}(f) + J^-_{\delta,i}(f) - 2\Big(\ln_+\frac{1}{\delta}\Big)\int_{E_i} f(x)\beta(x)m(dx)\Big). \tag{3.12.23}$$

Define

$$\widehat{J}^{\pm}_{\delta,i}(f) = J^{\pm}_{\delta,i}(f) \mp (\ln_+ \delta^{-1})\int_{E_i} f(x)(1 \pm \beta(x))m(dx)$$

and $\widehat{J}_\delta(f) = \sum_{i=1}^{\infty}(\widehat{J}^+_{\delta,i}(f) + \widehat{J}^-_{\delta,i}(f))$. We firstly prove that this sum converges in distribution and, hence a.s. We have

$$\begin{aligned} E\exp\{i\theta\widehat{J}^{\pm}_{\delta,i}(f)\} &= \exp\Big\{\int_{E_i}(1 \pm \beta(x))m(dx)\int_\delta^\infty (e^{\pm i\theta f(x)\lambda} - 1)\lambda^{-2}d\lambda \\ &\qquad \mp i\theta(\ln_+ \delta^{-1})\int_{E_i} f(x)(1 \pm \beta(x))m(dx)\Big\} \\ &= \exp\Big\{\int_{E_i}(1 \pm \beta(x))m(dx)\int_\delta^\infty [e^{\pm i\theta f(x)} - 1 \mp i\theta f(x)\lambda 1(\lambda \le 1)]\lambda^{-2}d\lambda\Big\} \end{aligned}$$

since $\ln_+ \delta^{-1} = \int_\delta^\infty \lambda^{-1}1(\lambda \le 1)d\lambda$. The exponent in $E\exp\{i\theta(\widehat{J}^+_{\delta,i}(f) + \widehat{J}^-_{\delta,i}(f))\}$ equals

$$\begin{aligned} &\int_{E_i} m(dx)\int_\delta^\infty (e^{i\theta f(x)} + e^{-i\theta f(x)} - 2)\lambda^{-2}d\lambda \\ &\quad + \int_{E_i}\beta(x)m(dx)\int_\delta^\infty [e^{i\theta f(x)\lambda} - e^{-i\theta f(x)\lambda} - 2i\theta f(x)\lambda 1(\lambda \le 1)]\lambda^{-2}d\lambda. \end{aligned}$$

This can be expressed as $A_i + B_i$, where

$$A_i = 2\int_{E_i} m(dx)\int_\delta^\infty [\cos(\theta f(x)\lambda) - 1]\lambda^{-2}d\lambda,$$

$$B_i = 2i\int_{E_i}\beta(x)m(dx)\int_\delta^\infty [\sin(\theta f(x)\lambda) - \theta f(x)\lambda 1(\lambda \le 1)]\lambda^{-2}d\lambda.$$

Now with the change of variables $u = |\theta f(x)|\lambda$ in the inner integral,

$$|A_i| \le 2|\theta| \int_{E_i} |f(x)| m(dx) \int_0^\infty |1 - \cos u| u^{-2} du,$$

$$|B_i| \le 2|\theta| \int_{E_i} |f(x)\beta(x)| m(dx) L(f(x)),$$

where, for $\delta \le 1$,

$$\begin{aligned} L(f(x)) &= \int_0^\infty \big| \sin u - u 1(u \le 1) \big| u^{-2} du \\ &\quad + \int_0^\infty u \big| 1(u \le 1) - 1(u \le |\theta f(x)|) \big| u^{-2} du \\ &= C + \Big| \int_{|\theta f(x)|}^1 u^{-1} du \Big| \le C + \big| \ln |f(x)| \big|, \end{aligned}$$

for some $C < \infty$, so that

$$|B_i| \le 2C|\theta| \int_{E_i} |f(x)| m(dx) + 2|\theta| \int_{E_i} |f(x)\beta(x) \ln |f(x)|| m(dx).$$

$E = \bigcup_{i=1}^\infty E$ and (3.2.1) imply $\sum_{i=1}^\infty (|A_i| + |B_i|) < \infty$. Therefore $\widehat{J}_\delta(f)$ is well defined in the sense of a.s. convergence, $\widehat{J}(f) = \lim_{\delta \to 0} \widehat{J}_\delta(f)$ exists in distribution and

$$\begin{aligned} &E \exp\{i\theta \widehat{J}(f)\} \\ &= \exp\Big\{ -2 \int_E m(dx) \int_0^\infty [1 - \cos(\theta f(x)\lambda)] \lambda^{-2} d\lambda \\ &\quad + 2i \int_E \beta(x) m(dx) \int_0^\infty [\sin(\theta f(x)\lambda) - \theta f(x)\lambda 1(\lambda \le 1)] \lambda^{-2} d\lambda \Big\}. \end{aligned} \tag{3.12.24}$$

Since (3.12.21) holds also when $\alpha = 1$, we conclude that $\widehat{J}(f)$ is defined a.s.. It remains to evaluate the integrals in (3.12.24). We have

$$\int_0^\infty [1 - \cos(\theta f(x)\lambda)] \lambda^{-2} d\lambda = |\theta f(x)| \int_0^\infty (1 - \cos u) u^{-2} du = \frac{\pi}{2} |\theta f(x)|$$

and

$$\begin{aligned} &\int_0^\infty [\sin(\theta f(x)\lambda) - \theta f(x)\lambda 1(\lambda \le 1)] \lambda^{-2} d\lambda \\ &\quad = \theta f(x) \int_0^\infty [\sin u - u 1(u \le |\theta f(x)|)] u^{-2} du \end{aligned}$$

$$= \theta f(x) \int_0^\infty [\sin u - u1(u \le 1)]u^{-2}du$$
$$+ \int_0^\infty u^{-1}[1(u \le 1) - 1(u \le |\theta f(x)|)]du$$
$$= \theta f(x)\Big(a + \int_{|\theta f(x)|}^1 u^{-1}du\Big) = \theta f(x)(a - \ln|\theta f(x)|),$$

where

$$a = \int_0^\infty [\sin u - u1(u \le 1)]u^{-2}du.$$

Substituting the preceding relations in (3.12.24) yields

$$E\exp\Big\{i\theta\frac{1}{\pi}\widehat{J}(f)\Big\} = E\exp\Big\{i\theta\widehat{J}\Big(\frac{f}{\pi}\Big)\Big\}$$
$$= \exp\Big\{-|\theta|\int_E |f(x)|m(dx) + \frac{2}{\pi}i\theta\Big[a\int_E f(x)\beta(x)m(dx)$$
$$- \ln|\theta|\int_E f(x)\beta(x)m(dx) - \int_E f(x)\beta(x)\ln\Big|\frac{1}{\pi}f(x)\Big|m(dx)\Big]\Big\}$$
$$= \exp\Big\{-\sigma_f|\theta|(1 + i\beta_f\frac{2}{\pi}\ln|\theta|) + i\mu_f\theta\Big\}$$

where $\sigma_f = \int_E |f(x)|m(dx)$. β_f is defined as in Property 3.2.2 and

$$\mu_f = \frac{2}{\pi}(a + \ln\pi)\int_E f(x)\beta(x)m(dx) - \frac{2}{\pi}\int_E f(x)\beta(x)\ln|f(x)|m(dx).$$

Hence, $\frac{1}{\pi}\widehat{J}(f) \sim S_1(\sigma_f, \beta_f, \mu_f)$. Property 3.2.2 implies Relation (3.12.22). ∎

3.13 Exercises

Exercise 3.1 Let F be defined as in (3.2.1). Prove that $f \in F$ implies that the support of f is contained in a region of E where m is σ-finite.
Hint: Integrate $|f|$ over $2^{-n} < f < 2^{-(n+1)}$.

Exercise 3.2 Show that the stochastic process $\{M(A), A \in \mathcal{E}_0\}$ with finite-dimensional distributions given by (3.3.1) satisfies

$$M\Big(\bigcup_{j=1}^d A_j\Big) = \sum_{j=1}^d M(A_j) \text{ a.s.}$$

for any disjoint $A_1, \dots, A_d$ in $\mathcal{E}_0$. (This is finite additivity for a random measure.)

Exercise 3.3 Show that the set of function F defined in (3.2.1) is a linear space.

Exercise 3.4 Prove Proposition 3.5.2 using Definition 2.7.3.

Exercise 3.5 Find square integrable functions f_1 and f_2 such that both (3.5.5) and (3.5.6) hold. (This illustrates that Theorem 3.5.3 is false in the Gaussian case $\alpha = 2$.)

Exercise 3.6 Let M be a $S\alpha S$ random measure on $(E, \mathcal{E})$ with control measure m. Show that there is an isometry between the space of all $S\alpha S$ random variables of the form

$$X = \int_E f(x)M(dx)$$

equipped with the covariation norm and $L^\alpha(E, \mathcal{E}, m)$.

Exercise 3.7 Let X_1, X_2 be jointly α-stable random variables in $\mathbb{R}^2$ with $\alpha \neq 1$ and let f_1 and f_2 be the functions that appear in their integral representation. What conditions must f_1 and f_2 satisfy if X_1 is $S\alpha S$, X_2 is $S\alpha S$ but (X_1, X_2) is not $S\alpha S$?

Exercise 3.8 Let $\{X_2(t), -\infty < t < \infty\}$ be the reverse Ornstein-Uhlenbeck process. Verify that the spectral measure Γ_2 of the vector $(X_2(s), X_2(t))$, $s < t$, has the form given in Example 3.6.4.

Exercise 3.9 Prove that the linear fractional stable motion in Example 3.6.5 is self-similar and has stationary increments. To prove self-similarity, use the change of variables result given in Proposition 3.5.5.

Exercise 3.10 Show that the log-fractional stable motion given in Example 3.6.6 is well defined, is self-similar with $H = 1/\alpha$ and has stationary increments.

Exercise 3.11 Prove that when $\alpha = 2$, the log-fractional stable motion (3.6.5) has the same autocovariance function as the well-balanced fractional stable motion (3.6.4). The process with $\alpha = 2$ is called *fractional Brownian motion.*

Exercise 3.12 Show that $(d_\alpha)^\alpha = E|Z|^\alpha = 2^{\alpha/2}\pi^{-1/2}\Gamma(\frac{\alpha+1}{2})$ where $Z \sim N(0,1)$.

Exercise 3.13 Say that a $S\alpha S$ random vector $\mathbf{X}$ is *sub-sub-stable* if, for some $0 < \alpha < \alpha' < \alpha'' < 2$, it is sub-stable, i.e., $\mathbf{X} \stackrel{d}{=} A^{1/\alpha'}\mathbf{Y}$, and if $\mathbf{Y}$ is also sub-stable, i.e., $\mathbf{Y} \stackrel{d}{=} A^{1/\alpha''}\mathbf{Z}$. Here $A \sim S_{\alpha'/\alpha}(\sigma_A, 1, 0)$ is independent of $\mathbf{Y} \sim S\alpha' S$ and $B \sim S_{\alpha''/\alpha'}(\sigma_B, 1, 0)$ is independent of $\mathbf{Z} \sim S\alpha'' S$.

(1) Show that a sub-sub-stable random vector is sub-stable.

(2) Express σ_C as a function of σ_A and σ_B.

Hint for (1): Show that there is a random variable $C \sim S_{\alpha''/\alpha}(\sigma_C, 1, 0)$ independent of $\mathbf{Z}$ such that $\mathbf{X} \stackrel{d}{=} C^{1/\alpha''}\mathbf{Z}$. Take $C^{1/\alpha''} = A^{1/\alpha'}B^{1/\alpha''}$ and verify that C has the required distribution.

Exercise 3.14 Prove Theorem 3.10.1.

Exercise 3.15 Show that a $S\alpha S$ random vector $\mathbf{X}$ in $\mathbb{R}^d$, $0 < \alpha < 2$, can be represented as

$$X \stackrel{d}{=} \sum_{i=1}^{\infty} \epsilon_i \Gamma_i^{-1/\alpha} \mathbf{W}_i,$$

where $\{\epsilon_i\}, \{\Gamma_i\}, \{\mathbf{W}_i\}$ are independent sequences, the ϵ_is and Γ_is are defined as in Theorem 3.10.1 and the $\mathbf{W}_i = (W_i^{(1)}, \dots, W_i^{(d)})$s are i.i.d. vectors in $\mathbb{R}^d$ satisfying $E\|\mathbf{W}_1\|^\alpha < \infty$ or, equivalently, $E|W_1^{(j)}|^\alpha < \infty, j = 1, \dots, d.$

Exercise 3.16 Let $\{\Gamma_1, \Gamma_2, \dots\}$ be a sequence of arrival times of a Poisson process with unit rate and let $\{G_i(t), t \in T\}$, $i = 1, 2, \dots$, be i.i.d. Gaussian processes with mean zero and independent of $\{\Gamma_1, \Gamma_2, \dots\}$. Prove that the series

$$X(t) = 2^{-\alpha/2} C_\alpha^{1/\alpha} d_\alpha^{-\alpha} \sum_{i=1}^{\infty} G_i(t) \Gamma_i^{-1/\alpha}$$

converges a.s. for each $t \in T$ and $\{X(t),\ t \in T\}$ is a sub-Gaussian process. (The constant C_α is defined in (1.2.9) and $d_\alpha = (E|N(0,1)|^\alpha)^{1/\alpha}$.)

Exercise 3.17 A subset C of a metric space (T, d) is said to be separable with respect to d-convergence if there is a countable subset $C_0 \subseteq C$ such that for every $u \in C$, there is a sequence $\{u_k\}_{k=1}^\infty$ in C_0 such that $\lim_{k\to\infty} d(u, u_k) = 0$.

Prove that a subset of a separable set is again separable.

Exercise 3.18 Show that every non-random function $\{\phi(t),\ t \in T\}$, regarded as a (degenerate) stochastic process, satisfies Condition S.
Hint: The subset $\{y = \phi(t),\ t \in T\}$ of $\mathbb{R}$ is separable.

Exercise 3.19 Show that a stochastic process $\{X(t),\ t \in T\}$ satisfies Condition S if and only if the linear span L_X of the process is separable with respect to convergence in probability. The linear span L_X is the linear space of the type $\sum_{j=1}^n a_j X(t_j),\ a_j \in \mathbb{R},\ t_j \in T,\ j = 1, 2, \dots, n,\ n = 1, 2, \dots$.

Exercise 3.20 Let $\{X(t),\ t \in T\}$ and $\{Y(t),\ t \in T\}$ be two stochastic processes defined on the same probability space and satisfying Condition S. Then $\{X(t) + Y(t),\ t \in T\}$ also satisfies Condition S.

Exercise 3.21 Use Exercises 3.19 and 3.17 to show that Condition S is equivalent to the following, apparently stronger condition, that $\{X(t),\ t \in T\}$ is separable in probability, i.e., that there is a countable set $T_0 \subset T$ such that for every $t \in T$, there is a sequence $\{t_k\}_{k=1}^\infty$ in T_0 such that $X(t) = \text{plim}_{k\to\infty} X(t_k)$.

Exercise 3.22 Let N be a Poisson random measure on $(S, \mathcal{S}, n)$, and let g be a measurable function: $S \to \mathbb{R}$ satisfying $\int_S |g(s)| n(ds) < \infty$. Then the Poisson stochastic integral $\int_S g(s) N(ds)$ converges a.s., and its characteristic function is given by

$$E \exp\{i\theta \int_S g(s) N(ds)\} = \exp\{\int_S (e^{i\theta g(s)} - 1) n(ds)\}.$$

Hint: The control measure n must be σ-finite on the support of g. Partition this support into sets A_i satisfying $n(A_i) < \infty$, so that $\sum_{k \geq 0} |g(X_k)| = \sum_{i=1}^{\infty} \sum_{k \geq 0} 1_{A_i}(X_k) |g(X_k)|$, with $\{X_k\}$ being a realization of N. For each $i = 1, 2, \ldots,$ devise a construction of N restricted to A_i in the spirit of Lemma 3.12.3, and use an argument similar to the one used to prove that $H(f) < \infty$ a.s. in (3.12.7).

Exercise 3.23 Let N be as in Exercise 3.22, and let g this time satisfy $\int_S |g(s)|^2 n(ds) < \infty$. Then

$$\int_S g(s) \widetilde{N}(ds) := \sum_{i=1}^{\infty} \Big(\int_{E_i} g(s) N(ds) - E \int_{E_i} g(s) N(ds) \Big)$$

is well defined in the sense of a.s. convergence. Here, $\{E_i\}$ is an arbitrary partition of S into sets of finite measure n. Show that $\int_S g(s) \widetilde{N}(ds)$ does not depend on the chosen partition.
Hint: Argue as in Exercise 3.22 and use the fact that a series of independent random variables converges a.s. if and only if it converges in distribution.

Exercise 3.24 (1) Prove that for $0 < \alpha < 1$,

$$\begin{aligned} &\int_0^{\infty} (e^{ia\lambda} - 1) \lambda^{-(1+\alpha)} d\lambda \\ &= \; -|a|^{\alpha} \alpha^{-1} \Gamma(1-\alpha) \cos \frac{\pi\alpha}{2} \Big(1 - i \,(\text{sign } a) \tan \frac{\pi\alpha}{2} \Big). \qquad (3.13.1) \end{aligned}$$

Hint: Suppose first $a > 0$ and view the integral as the limit as ν tends to 0 of $\int_0^{\infty} (e^{ia\lambda - \nu\lambda} - 1) \lambda^{-(1+\alpha)} d\lambda$. After integration by parts, obtain $(ia - \nu)\alpha^{-1} \int_0^{\infty} e^{ia\lambda} e^{-\nu\lambda} \lambda^{-\alpha} d\lambda$, an integral proportional to the characteristic function of a gamma distribution. Since

$$\int_0^{\infty} e^{iax} \frac{1}{\Gamma(t)} x^{t-1} e^{-\nu x} dx = \left(\frac{\nu}{1 - ia\nu} \right)^t,$$

the right-hand side of (3.13.1) equals $-(-ia)^{\alpha} \alpha^{-1} \Gamma(1-\alpha)$. If $a < 0$, take the complex conjugate. *Question:* Why introduce ν?

(2) Prove that for $1 < \alpha < 2$,

$$\int_0^\infty (e^{ia\lambda} - 1 - ia\lambda)\lambda^{-(1+\alpha)}d\lambda = |a|^\alpha \frac{\Gamma(2-\alpha)}{\alpha(\alpha-1)} \cos\frac{\pi\alpha}{2}\left(1 - i(\text{sign}\, a)\tan\frac{\pi\alpha}{2}\right). \tag{3.13.2}$$

Hint: Suppose first $a > 0$, integrate by parts to obtain $a^\alpha i\alpha^{-1}\int_0^\infty (e^{iu} - 1)u^{-\alpha}du$ and use (3.13.1) to obtain

$$-a^\alpha i\alpha^{-1}\left[(\alpha-1)^{-1}\Gamma(1-(\alpha-1))\cos\frac{\pi(\alpha-1)}{2}\left(1 - i\tan\frac{\pi(\alpha-1)}{2}\right)\right]$$

$$= a^\alpha \frac{\Gamma(2-\alpha)}{\alpha(\alpha-1)}\cos\frac{\pi\alpha}{2}\left(1 - i\tan\frac{\pi\alpha}{2}\right).$$

If $a < 0$, take the complex conjugate.

Chapter 4

Dependence structures of multivariate stable distributions

In this chapter we describe a number of properties of multivariate stable distributions. We start with conditional expectations. Let (X_1, X_2) be symmetric α-stable with $1 < \alpha < 2$. We show in Section 4.1 that, as in the Gaussian case, $E(X_2|X_1)$ is linear in X_1. In fact, $E(X_2|X_1) = cX_1$, and the coefficient c equals the normalized covariation of X_2 on X_1. (See Chapter 5 for extensions to $\alpha \leq 1$ and to the case where (X_1, X_2) is skewed α-stable).

The analogy to the Gaussian case breaks down if one conditions on more than one variable. Suppose that $(X_1, \dots, X_n)$ is symmetric α-stable, $1 < \alpha < 2$. Then $E(X_n|X_1, \dots, X_{n-1})$ is usually not linear. It is linear in some special cases, for example when $X_1, \dots, X_{n-1}$ are independent. If $(X_1, \dots, X_n)$ is sub-Gaussian, then, as in the Gaussian case, any random variable in the linear span of $X_1, \dots, X_n$ is linear.

If $(X_1, \dots, X_n)$ is Gaussian, then the conditional laws are also Gaussian. Unfortunately, very little is known about the conditional laws $\mathcal{L}\{X_n|X_1, \dots, X_{n-1}\}$ when $X_1, \dots, X_n$ are jointly α-stable. The conditional law is, in general, not α-stable and it is, in general, not even symmetric around the conditional mean. We provide in Section 4.2 a necessary and sufficient condition for the symmetry of $\mathcal{L}\{X_2|X_1\}$ around the conditional mean cX_1. In the special case where the conditional law $\mathcal{L}(X_2|X_1)$ is α-stable, symmetry around the conditional mean cX_1 occurs if and only if X_1 and X_2 are linearly dependent, i.e., if $X_2 = cX_1 + Y$, where Y is $S\alpha S$ and independent of X_1 (Section 4.3).

In Section 4.4 we analyze the asymptotic behavior of order statistics. Suppose that $X_1, \dots, X_n$ are jointly α-stable, $0 < \alpha < 2$. We show that the tail $P\{X^{(k)} > \lambda\}$, $k = 1, \dots, n$, of the distribution of the order statistics $X^{(k)}$ is always asymptotically proportional to $\lambda^{-\alpha}$ as $\lambda \to \infty$, i.e., it has the same type of asymptotic behavior as the tail of any of the X_js, $j = 1, \dots, n$. The coefficients of proportionality, however, depend on the joint distribution.

In Section 4.5 we investigate the existence of joint moments of α-stable random variables $X_1, \dots, X_n$, $0 < \alpha < 2$. For what (non-negative) exponents $p_1, \dots, p_n$ is the joint moment $E|X_1|^{p_1} \dots |X_n|^{p_n}$ finite? If the X_js are independent, then, necessarily, $p_j < \alpha$ for all j. If the X_js are all equal, then $p_1 + \dots + p_n < \alpha$. We give a necessary and sufficient condition on $p_1, \dots, p_n$ for $E|X_1|^{p_1} \dots |X_n|^{p_n}$ to be finite when $p_1, \dots, p_n$ are arbitrary non-negative numbers.

Association is a dependence structure that has been widely used in the finite variance case. Jointly normal random variables, for example, are associated if and only if their correlations are all non-negative. The corresponding result in the α-stable case, $0 < \alpha < 2$, is given in Section 4.6.

The codifference between two stable random variables was defined in Chapter 2. We listed there some of its basic properties and noted in particular that the codifference is identical to the covariance when $\alpha = 2$. In Section 4.7, we consider a stationary $S\alpha S$ process $X = \{X(t),\ t \in \mathbb{R}\}$ with index $0 < \alpha \le 2$ and study the codifference between $X(0)$ and $X(t)$. We investigate the asymptotic behavior of this codifference as the lag t tends to infinity. The codifference always tends to zero when the process X is a moving average. For example, it tends to zero exponentially fast if the process is $S\alpha S$ Ornstein-Uhlenbeck.

The mean number of times that a stationary sub-Gaussian process crosses a given level u is computed in Section 4.8. It decreases like $u^{-\alpha}$ as $u \to \infty$.

4.1 Linear regression

The regression of a random variable X_2 on a random variable X_1 is *linear* if there exists a constant c such that $E(X_2|X_1) = cX_1$ a.s. The following lemma provides a criterion for linearity in terms of the joint characteristic function $\phi(\theta_1, \theta_2) = E\exp\{i(\theta_1 X_1 + \theta_2 X_2)\}$ of X_1 and X_2.

Lemma 4.1.1 *Suppose that $E|X_1| < \infty$ and $E|X_2| < \infty$. Then $E(X_2|X_1) = cX_1$ a.s. is equivalent to*

$$\left.\frac{\partial\phi(\theta_1,\theta_2)}{\partial\theta_2}\right|_{\theta_2=0} = c\left.\frac{\partial\phi(\theta_1,\theta_2)}{\partial\theta_1}\right|_{\theta_2=0}. \tag{4.1.1}$$

PROOF: Setting $\phi(\theta_1, \theta_2) = E\exp\{i(\theta_1 X_1 + \theta_2 X_2)\}$ in (4.1.1) yields

$$EX_2e^{i\theta_1X_1} = EcX_1e^{i\theta_1X_1}$$

for all real θ_1, i.e.,

$$\int_{-\infty}^{\infty}\int_{-\infty}^{\infty}(cx_1 - x_2)e^{i\theta x_1}dF(x_1, x_2) = 0 \tag{4.1.2}$$

for all real θ, where F is the distribution function of the vector (X_1, X_2). The left-hand side of (4.1.2), which we denote $\widehat{\mu}(\theta)$, can be regarded as the Fourier transform

$$\widehat{\mu}(\theta) = \int_{-\infty}^{\infty} e^{i\theta x}\mu(dx), \ -\infty < \theta < \infty,$$

of the signed measure μ on $\mathbb{R}^1$ given by

$$\mu(B) = \int_B\int_{-\infty}^{\infty}(cx_1 - x_2)dF(x_1, x_2). \tag{4.1.3}$$

Relation (4.1.2) implies $\mu = 0$. To verify this, we can appeal to the uniqueness of characteristic functions. Indeed, let us write $\mu = \mu_1 - \mu_2$ where μ_1 and μ_2 are two non-negative finite measures whose Fourier transforms are $\widehat{\mu}_1$ and $\widehat{\mu}_2$, respectively. Relation (4.1.2) implies $\widehat{\mu}_1 - \widehat{\mu}_2 = \widehat{\mu} = 0$, i.e. $\widehat{\mu}_1 = \widehat{\mu}_2$. Now, either $\mu_1 = \mu_2 = 0$ or neither μ_1 nor μ_2 is a zero measure. In the latter case, dividing μ_1 and μ_2 by the constant $\mu_1(\mathbb{R}^1) = \widehat{\mu}_1(0) = \widehat{\mu}_2(0) = \mu_2(\mathbb{R}^1)$ converts them to probability measures and we have $\mu_1 = \mu_2$ by the uniqueness theorem for characteristic functions. Hence $\mu = 0$.

Since for any $A \in \sigma(X_1)$,

$$\int_A E(cX_1 - X_2|X_1)dP = \int_A(cX_1 - X_2)dP,$$

the condition $\mu = 0$ is equivalent to $E(cX_1 - X_2|X_1) = 0$ a.s., i.e., $E(X_2|X_1) = cX_1$ a.s. ■

As is well known, the regression is linear when X_1 and X_2 are jointly Gaussian mean zero random variables, and in that case $c = EX_2X_1/EX_1^2$. Now suppose that X_1 and X_2 are jointly $S\alpha S$ with $\alpha > 1$. (The assumption $\alpha > 1$ ensures that expectations exist.) The following theorem shows that the regression is still linear.

Theorem 4.1.2 *Let X_1 and X_2 be two jointly $S\alpha S$ random variables with $1 < \alpha \le 2$. Then*

$$E(X_2|X_1) = \frac{[X_2, X_1]_\alpha}{\|X_1\|_\alpha^\alpha}X_1 \ \ a.s. \tag{4.1.4}$$

PROOF: It is sufficient to verify that (4.1.1) holds with $c = [X_2, X_1]_\alpha / \|X_1\|^\alpha$. The joint characteristic function of (X_1, X_2) is $\phi(\theta_1, \theta_2) = \exp\{-\sigma^\alpha(\theta_1, \theta_2)\}$, where

$$\sigma^\alpha(\theta_1, \theta_2) = \int_{S_2} |\theta_1 s_1 + \theta_2 s_2|^\alpha \Gamma(ds)$$

and where Γ is the spectral measure. By (2.7.2),

$$\begin{aligned} \left.\frac{\partial \sigma^\alpha(\theta_1, \theta_2)}{\partial \theta_1}\right|_{\theta_2=0} &= \alpha \int_{S_2} s_1 (\theta_1 s_1)^{<\alpha-1>} \Gamma(ds) \\ &= \alpha \theta_1^{<\alpha-1>} \|X_1\|_\alpha^\alpha \end{aligned}$$

and

$$\begin{aligned} \left.\frac{\partial \sigma^\alpha(\theta_1, \theta_2)}{\partial \theta_2}\right|_{\theta_2=0} &= \alpha \int_{S_2} s_2 (\theta_1 s_1)^{<\alpha-1>} \Gamma(ds) \\ &= \alpha \theta_1^{<\alpha-1>} [X_2, X_1]_\alpha. \end{aligned}$$

Setting

$$c = \frac{[X_2, X_1]_\alpha}{\|X_1\|_\alpha^\alpha},$$

we obtain

$$\left.\frac{\partial \sigma^\alpha(\theta_1, \theta_2)}{\partial \theta_2}\right|_{\theta_2=0} = c \left.\frac{\partial \sigma^\alpha(\theta_1, \theta_2)}{\partial \theta_1}\right|_{\theta_2=0}$$

and, hence, (4.1.1). ∎

What about multiple regression? Whereas multiple regression is always linear in the Gaussian case, this is not so in the $S\alpha S$, $\alpha < 2$ case. Necessary and sufficient conditions for linearity are known, but they are complicated. (See Miller (1978).) The necessary conditions for linearity given in the following corollary can be used to construct counterexamples.

Corollary 4.1.3 *Let $X_1, X_2, \ldots, X_n$ be jointly $S\alpha S$ with $1 < \alpha < 2$. If*

$$E(X_n | X_1, \cdots, X_{n-1}) = a_1 X_1 + \cdots + a_{n-1} X_{n-1} \ \ a.s., \tag{4.1.5}$$

then the coefficients $a_1, \ldots, a_{n-1}$ satisfy the system of linear equations

$$\sum_{i=1}^{n-1} a_i [X_i, X_j]_\alpha = [X_n, X_j]_\alpha\,, \ \ j = 1, \ldots, n-1. \tag{4.1.6}$$

Moreover, if $n = 3$ and if X_1 and X_2 are linearly independent, then the system (4.1.6) uniquely determines the coefficients.

PROOF: Set $c_{ij} = [X_i, X_j]_\alpha$ and suppose that (4.1.5) holds. Then, for any $j = 1, \ldots, n-1$,

$$
\begin{aligned}
\frac{c_{nj}}{c_{jj}} X_j &= E(X_n | X_j) \\
&= E[E(X_n | X_1 + \cdots + X_j + \cdots + X_{n-1}) | X_j] \\
&= E(a_1 X_1 + \cdots + a_j X_j + \cdots + a_{n-1} X_{n-1} | X_j) \\
&= a_j X_j + \sum_{\substack{i=1 \\ i \neq j}}^{n-1} a_i E(X_i | X_j) \\
&= a_j X_j + \sum_{\substack{i=1 \\ i \neq j}}^{n-1} a_i \frac{c_{ij}}{c_{jj}} X_j \\
&= \frac{1}{c_{jj}} \sum_{i=1}^{n-1} a_i c_{ij} X_j,
\end{aligned}
$$

establishing (4.1.6).

When $n = 3$, the system has a unique solution unless $c_{11}c_{22} = c_{12}c_{21}$, i.e., unless

$$
\int_{S_2} |s_1|^\alpha \Gamma(ds) \int_{S_2} |s_2|^\alpha \Gamma(ds) = \int_{S_2} s_1 s_2^{<\alpha-1>} \Gamma(ds) \int_{S_2} s_2 s_1^{<\alpha-1>} \Gamma(ds). \tag{4.1.7}
$$

Applying the Hölder inequality to the right-hand side of (4.1.7) yields the left-hand side. This means that (4.1.7) is equivalent to $s_1 = \lambda s_2$ Γ-a.e. for some real λ, i.e., to $X_1 = \lambda X_2$ a.s. ∎

Remark. When $n \geq 3$, linear independence of $X_1, X_2, \ldots, X_{n-1}$ is obviously a necessary condition for (4.1.6) to have a unique solution. The converse is true when $n = 3$ (Corollary 4.1.3), but is not true when $n > 3$ (Exercise 4.3).

Example 4.1.4 *A $S\alpha S$ law with non-linear multiple regression*

Let X_1, X_2, X_3 be three i.i.d. $S\alpha S$ random variables with $1 < \alpha < 2$. If

$$
Y_1 = X_1 + X_2, \quad Y_2 = X_1 + X_3, \quad Y_3 = X_1,
$$

then $E(Y_3|Y_1, Y_2)$ is *not* linear in Y_1 and Y_2. To show this, we will use Corollary 4.1.3 and the fact that the covariation is not linear in its second argument.

Suppose, to the contrary, that $E(Y_3|Y_1, Y_2) = a_1 Y_1 + a_2 Y_2$ a.s.. Then, for any $\theta > 0$,

$$
\frac{[Y_3, Y_1 + \theta Y_2]_\alpha}{\|Y_1 + \theta Y_2\|_\alpha^\alpha}(Y_1 + \theta Y_2)
$$

$$
\begin{aligned}
&= E(Y_3|Y_1+\theta Y_2)\\
&= E[E(Y_3|Y_1,Y_2)|Y_1+\theta Y_2]\\
&= a_1E(Y_1|Y_1+\theta Y_2)+a_2E(Y_2|Y_1+\theta Y_2)\\
&= (a_1[Y_1,Y_1+\theta Y_2]_\alpha+a_2[Y_2,Y_1+\theta Y_2]_\alpha)\frac{Y_1+\theta Y_2}{\|Y_1+\theta Y_2\|_\alpha^\alpha},
\end{aligned}
$$

so that

$$[Y_3,Y_1+\theta Y_2]_\alpha = a_1[Y_1,Y_1+\theta Y_2]_\alpha + a_2[Y_2,Y_1+\theta Y_2]_\alpha.$$

Setting $c_{ij} = [Y_i, Y_j]_\alpha$ and replacing the coefficients a_1 and a_2 with the solution to the system of equations in Corollary 4.1.3, yields

$$
\begin{aligned}
&[Y_3,Y_1+\theta Y_2]_\alpha\\
&\quad=\frac{(c_{31}c_{22}-c_{21}c_{32})[Y_1,Y_1+\theta Y_2]_\alpha+(c_{11}c_{32}-c_{31}c_{12})[Y_2,Y_1+\theta Y_2]_\alpha}{c_{11}c_{22}-c_{21}c_{12}}.
\end{aligned}
$$

All these covariations can be simplified; for example,

$$
\begin{aligned}
&[Y_1,Y_1+\theta Y_2]_\alpha\\
&\quad=[X_1+X_2,X_1+X_2+\theta(X_1+X_3)]_\alpha\\
&\quad=[X_1,(1+\theta)X_1+X_2+\theta X_3]_\alpha+[X_2,(1+\theta)X_1+X_2+\theta X_3]_\alpha\\
&\quad=(1+\theta)^{\alpha-1}\sigma^\alpha+\sigma^\alpha,
\end{aligned}
$$

where $\sigma^\alpha = [X_i, X_i]_\alpha = \|X_i\|_\alpha^\alpha$, $i = 1, 2, 3$. After simplification, one obtains

$$(1+\theta)^{\alpha-1} = \frac{1}{3}(1+\theta^{\alpha-1}+2(1+\theta)^{\alpha-1})$$

or $(1+\theta)^{\alpha-1} = 1+\theta^{\alpha-1}$, which holds for any $\theta > 0$ if and only if $\alpha = 2$. Hence $E(Y_3|Y_1, Y_2)$ is not linear in Y_1 and Y_2 when $\alpha < 2$.

The following corollary shows that $E(X_n|X_1, \ldots, X_{n-1})$ is linear if the random variables $X_1, \ldots, X_{n-1}$ are independent.

Corollary 4.1.5 *Let $X_1, \ldots, X_{n-1}, X_n$ be jointly $S\alpha S$ with $n > 2$ and $1 < \alpha < 2$, and suppose that $X_1, \ldots, X_{n-1}$ are independent. Then*

$$E(X_n|X_1,\ldots,X_{n-1}) = \sum_{k=1}^{n-1} c_kX_k$$

where

$$c_k = \frac{[X_n,X_k]_\alpha}{\|X_k\|_\alpha^\alpha}.$$

PROOF: We can assume by the representation theorem 3.5.6 that

$$X_k = \int_E f_k(x) M(dx), \quad k = 1, \ldots, n,$$

where M is a $S\alpha S$ random measure with control measure m and $f_k \in L^\alpha(E, m)$. Moreover, since $X_1, \ldots, X_{n-1}$ are independent, we can assume that the functions $f_1, \ldots, f_{n-1}$ have disjoint supports (see Theorem 3.5.3). Therefore there are disjoint sets $A_1, \ldots, A_{n-1}$ such that $f_k(x) = 0$ for any $x \notin A_k$, $k = 1, \ldots, n-1$.

On one hand, for any $k = 1, \ldots, n-1$,

$$\begin{aligned}
&E\left(\int_{A_k} f_n(x) M(dx) \,\middle|\, X_1, \ldots, X_{n-1}\right) \\
&= E\left(\int_{A_k} f_n(x) M(dx) \,\middle|\, \int_{A_j} f_j(x) M(dx),\ j = 1, \ldots, n-1\right) \\
&= E\left(\int_{A_k} f_n(x) M(dx) \,\middle|\, \int_{A_k} f_k(x) M(dx)\right) \quad \text{(independence)} \\
&= c_k \int_{A_k} f_k(x) M(dx) \quad \text{(Theorem 4.1.2)} \\
&= c_k X_k,
\end{aligned}$$

where, by Proposition 3.5.2,

$$\begin{aligned}
c_k &= \frac{1}{\|X_k\|_\alpha^\alpha} \int_{A_k} f_n(x) f_k(x)^{<\alpha-1>} m(dx) \\
&= \frac{1}{\|X_k\|_\alpha^\alpha} \int_E f_n(x) f_k(x)^{<\alpha-1>} m(dx) \\
&= \frac{[X_n, X_k]_\alpha}{\|X_k\|_\alpha^\alpha}.
\end{aligned}$$

On the other hand, $\int_{(\cup_{k=1}^{n-1} A_k)^c} f_n(x) M(dx)$ is independent of $X_1, \ldots, X_{n-1}$ and therefore

$$\begin{aligned}
&E\left(\int_{(\bigcup_{k=1}^{n-1} A_k)^c} f_n(x) M(dx) \,\middle|\, X_1, \ldots, X_{n-1}\right) \\
&= E \int_{(\bigcup_{k=1}^{n-1} A_k)^c} f_n(x) M(dx) \\
&= 0.
\end{aligned}$$

Hence

$E(X_n | X_1, \ldots, X_{n-1})$

$$= E\left(\sum_{k=1}^{n-1}\int_{A_k} f_n(x)M(dx) + \int_{(\bigcup_{k=1}^{n-1} A_k)^c} f_n(x)M(dx)\,\middle|\, X_1,\ldots,X_{n-1}\right)$$

$$= \sum_{k=1}^{n-1} c_k X_k \quad \blacksquare$$

We now want to characterize those $S\alpha S$ vectors having linear spans where *all* multiple regressions are linear. Let $\mathbf{X} = (X_1, X_2, \ldots, X_n)$ be a $S\alpha S$ random vector with $1 < \alpha \leq 2$ and let sp($\mathbf{X}$) denote the linear span of $\mathbf{X}$, i.e. , the vector space of all finite linear combinations $\sum_{k=1}^{n} \lambda_k X_k$, $-\infty < \lambda_1, \ldots, \lambda_n < \infty$.

Definition 4.1.6 The vector $\mathbf{X}$ has the *multiple regression property* if for all random variables $Y_1, Y_2, \ldots, Y_n$ in sp($\mathbf{X}$),

$$E(Y_n | Y_1, \ldots, Y_{n-1}) \in \mathrm{sp}\{(Y_1, \ldots, Y_{n-1})\}.$$

It is well known that every Gaussian vector has the multiple regression property. What about $S\alpha S$ random vectors with $1 < \alpha < 2$? Theorem 4.1.2 shows that the property holds when $\dim(\mathrm{sp}(\mathbf{X})) = 2$. But, as we shall see, only sub-Gaussian vectors have the property when $\dim(\mathrm{sp}(\mathbf{X})) \geq 3$.

Proposition 4.1.7 *Let* $\mathbf{X}$ *be a* $S\alpha S$ *random vector with* $1 < \alpha < 2$ *and suppose that* $\dim(\mathrm{sp}(\mathbf{X})) \geq 3$. *Then the following three statements are equivalent:*

(i) $\mathbf{X}$ *has the multiple regression property,*

(ii) $\mathbf{X}$ *is sub-Gaussian,*

(iii) James orthogonality in sp($\mathbf{X}$) *is additive on the left, i.e.,*

$$Y_1, Y_2, Y_3 \in \mathrm{sp}(\mathbf{X}),\ [Y_3, Y_1]_\alpha = 0,\ [Y_3, Y_2]_\alpha = 0 \ \Rightarrow\ [Y_3, Y_1 + Y_2]_\alpha = 0.$$

PROOF: The equivalence (ii)⇔(iii) follows from Proposition 2.9.3. To prove (i) ⇒ (iii), suppose (i) and let Y_1, Y_2,Y_3 be elements of sp($\mathbf{X}$), satisfying

$$[Y_3, Y_1]_\alpha = [Y_3, Y_2]_\alpha = 0.$$

Obviously, $[Y_3, Y_1 + Y_2]_\alpha = 0$ if Y_1 and Y_2 are a multiple of each other. If Y_1 and Y_2 are linearly independent,

$$\begin{aligned}
(Y_1 + Y_2)\frac{[Y_3, Y_1 + Y_2]_\alpha}{\|Y_1 + Y_2\|_\alpha^\alpha} &= E(Y_3 | Y_1 + Y_2) \\
&= E\big(E(Y_3|Y_1, Y_2)\big| Y_1 + Y_2\big) \\
&= E(a_1 Y_1 + a_2 Y_2 | Y_1 + Y_2) \\
&= 0,
\end{aligned}$$

since, by Corollary 4.1.3, $a_1 = a_2 = 0$. Hence we obtain again $[Y_3, Y_1+Y_2]_\alpha = 0$.

We now prove that (ii)⇒(i). Let $Y_1, \dots, Y_n$ be in sp($\mathbf{X}$). Since $\mathbf{X}$ is sub-Gaussian, so is $(Y_1, \dots, Y_n)$, i.e., $Y_j = A^{1/2}G_j$, $j = 1, \dots, n$, where $A \sim S_{\alpha/2}(\sigma, 1, 0)$ and $(G_1, \dots, G_n)$ is a mean zero Gaussian vector. Then

$$\begin{aligned}
&E(Y_n |\, Y_1, \dots, Y_{n-1}) \\
&\quad = E(A^{1/2}G_n |\, A^{1/2}G_1, \dots, A^{1/2}G_{n-1}) \\
&\quad = E\Big(E(A^{1/2}G_n | A, G_1, \dots, G_{n-1}) \;\Big|\; A^{1/2}G_1, \dots, A^{1/2}G_{n-1}\Big) \\
&\quad = E\Big(A^{1/2}E(G_n | G_1, \dots, G_{n-1}) \;\Big|\; A^{1/2}G_1, \dots, A^{1/2}G_{n-1}\Big) \\
&\quad = E\Big(\sum_{i=1}^{n-1} \frac{\text{Cov}\,(G_i, G_n)}{\text{Var}\, G_i} A^{1/2}G_i \;\Big|\; A^{1/2}G_1, \dots, A^{1/2}G_{n-1}\Big) \\
&\quad = \sum_{i=1}^{n-1} \frac{\text{Cov}\,(G_i, G_n)}{\text{Var}\, G_i} A^{1/2}G_i \\
&\quad = \sum_{i=1}^{n-1} \frac{\text{Cov}\,(G_i, G_n)}{\text{Var}\, G_i} Y_i,
\end{aligned}$$

establishing (i). ∎

4.2 Conditional laws that are symmetric around the conditional mean

We saw in Theorem 4.1.2 that for pairs of jointly $S\alpha S$ random variables, $\alpha > 1$, the conditional expectation is linear and behaves as in the Gaussian case. However, many properties of conditional distributions for pairs of Gaussian random variables do not extend to the stable case, e.g., symmetry. When (X_1, X_2) are jointly Gaussian with zero mean, the conditional law $X_2|X_1$ is symmetric around the conditional mean because $E(X_2|X_1) = cX_1$ and $X_2|X_1 \sim N(cX_1, \cdot)$. We will see that when (X_1, X_2) are jointly $S\alpha S$, $1 < \alpha < 2$, the conditional law $X_2|X_1$ is not always symmetric around the conditional mean.

Suppose (X_1, X_2) jointly $S\alpha S$, $1 < \alpha < 2$. We know from Theorem 4.1.2 that there is a constant c such that $E(X_2|X_1) = cX_1$. If the conditional distribution of $X_2|X_1$ is symmetric, it must necessarily be symmetric around its conditional mean cX_1, i.e., for every Borel set B_2,

$$P(X_2 - cX_1 \in B_2|X_1) = P(cX_1 - X_2 \in B_2|X_1) \text{ a.s.} \tag{4.2.1}$$

Integrating this relation over the event $\{X_1 \in B_1\}$ yields

$$P(X_1 \in B_1, X_2 - cX_1 \in B_2) = P(X_1 \in B_1, cX_1 - X_2 \in B_2) \tag{4.2.2}$$

for all Borel sets B_1 and B_2. Of course, (4.2.2) also implies (4.2.1). Therefore $X_2|X_1$ is conditionally symmetric (around its conditional mean cX_1) if and only if

$$(X_1, X_2 - cX_1) \stackrel{d}{=} (X_1, cX_1 - X_2)$$

and hence (Exercise 4.8) if and only if

$$(X_1, X_2) \stackrel{d}{=} (X_1, 2cX_1 - X_2). \tag{4.2.3}$$

This last expression states that the distribution of the (column) vector (X_1, X_2) is invariant under the transformation

$$\begin{pmatrix} 1 & 0 \\ 2c & -1 \end{pmatrix}.$$

It also shows that (4.2.1) holds if and only if the spectral measures of the $S\alpha S$ random vectors (X_1, X_2) and $(X_1, 2cX_1 - X_2)$ coincide. This last criterion would enable us, for example, to verify that a $S\alpha S$ random vector (X_1, X_2) with spectral measure

$$\begin{aligned}\Gamma = &\frac{1}{2}\left[\delta\left(\left(\frac{\sqrt{2}}{2}, \frac{\sqrt{2}}{2}\right)\right) + \delta\left(\left(-\frac{\sqrt{2}}{2}, -\frac{\sqrt{2}}{2}\right)\right)\right] \\ &+\frac{1}{2}\left[\delta\left(\left(-\frac{\sqrt{2}}{2}, \frac{\sqrt{2}}{2}\right)\right) + \delta\left(\left(\frac{\sqrt{2}}{2}, -\frac{\sqrt{2}}{2}\right)\right)\right]\end{aligned} \tag{4.2.4}$$

satisfies (4.2.3), whereas a $S\alpha S$ random vector (X_1, X_2) with the spectral measure

$$\begin{aligned}\Gamma = &\left[\delta\left(\left(\frac{\sqrt{2}}{2}, \frac{\sqrt{2}}{2}\right)\right) + \delta\left(\left(-\frac{\sqrt{2}}{2}, -\frac{\sqrt{2}}{2}\right)\right)\right] \\ &+\frac{1}{2}\left[\delta\left(\left(-\frac{\sqrt{2}}{2}, \frac{\sqrt{2}}{2}\right)\right) + \delta\left(\left(\frac{\sqrt{2}}{2}, -\frac{\sqrt{2}}{2}\right)\right)\right]\end{aligned} \tag{4.2.5}$$

does not (see Exercise 4.9). The criterion, unfortunately, is not particularly easy to visualize because it involves a linear transformation of Γ on the circle S_2. Theorem 4.2.2 below gives a more natural criterion, involving the translation of a measure on the real line.

We start with an alternative representation for the joint characteristic function of (X_1, X_2).

Lemma 4.2.1 *Let X_1 and X_2 be jointly $S\alpha S$ random variables, $0 < \alpha < 2$. Then there is a unique constant b and measure σ on $\mathbb{R}^1$, satisfying $\int_{-\infty}^{\infty} |t|^\alpha \sigma(dt) < \infty$, such that*

$$E\exp\{i(\theta_1 X_1 + \theta_2 X_2)\} = \exp\left\{-\int_{-\infty}^{\infty} |\theta_1 + \theta_2 t|^\alpha \sigma(dt) - b|\theta_2|^\alpha\right\}. \tag{4.2.6}$$

PROOF: By Theorem 2.4.3,

$$
\begin{aligned}
&E\exp\{i(\theta_1 X_1+\theta_2 X_2)\}\\
&= \exp\Big\{-\int_{S_2}|\theta_1 s_1+\theta_2 s_2|^\alpha\Gamma(ds)\Big\}\\
&= \exp\Big\{-\int_{S_2'}|\theta_1 s_1+\theta_2 s_2|^\alpha\Gamma(ds)-|\theta_2|^\alpha\{\Gamma((0,-1))+\Gamma((0,1))\}\Big\}\\
&= \exp\Big\{-\int_{S_2'}\Big|\theta_1+\theta_2\frac{s_2}{s_1}\Big|^\alpha m_1(ds)-b|\theta_2|^\alpha\Big\},
\end{aligned}
$$

where $S_2' = S_2\backslash\{(0,-1),(0,1)\}$, $m_1(ds) = |s_1|^\alpha\Gamma(ds)$ and $b = \Gamma((0,-1)) + \Gamma((0,1)) = 2\Gamma((0,1))$. The transformation $h(s_1,s_2) = s_2/s_1$ from S_2' to $\mathbb{R}^1$ induces the measure $\sigma = h(m_1) \equiv m_1\circ h^{-1}$ on $\mathbb{R}^1$. Setting $t = s_2/s_1$ yields (4.2.6). Note that

$$
\int_{-\infty}^{\infty}|t|^\alpha\sigma(dt)\le\int_{S_2}|s_2|^\alpha\Gamma(ds)\le\int_{S_2}\Gamma(ds)<\infty.
$$

The representation (4.2.6) is uniquely characterized by the pair (σ, b) because the spectral measure Γ is unique and can be recovered from (σ, b). ∎

Our criterion will now be stated in terms of the representation (4.2.6), i.e. in terms of σ (and b).

Theorem 4.2.2 *A jointly $S\alpha S$ pair of random variables X_1, X_2, $1<\alpha<2$, has a symmetric conditional law $X_2|X_1$, i.e., it satisfies (4.2.3) for some c, if and only if its representation (4.2.6) involves a measure σ which is symmetric (around c).*

PROOF: Using the representation (4.2.6) we obtain

$$
\begin{aligned}
&E\exp\{i(\theta_1 X_1+\theta_2 X_2)\}\\
&\quad = E\exp i\{\theta_1 X_1+\theta_2(2cX_1-X_2)\}\\
&\quad = E\exp\{i(\theta_1+2c\theta_2)X_1-\theta_2 X_2\}\\
&\quad = \exp\Big\{-\int_{-\infty}^{\infty}|\theta_1+2c\theta_2-\theta_2 t|^\alpha\sigma(dt)-b|\theta_2|^\alpha\Big\}\\
&\quad = \exp\Big\{-\int_{-\infty}^{\infty}|\theta_1+\theta_2(2c-t)|^\alpha\sigma(dt)-b|\theta_2|^\alpha\Big\}\\
&\quad = \exp\Big\{-\int_{-\infty}^{\infty}|\theta_1+\theta_2 s|^\alpha\sigma_1(ds)-b|\theta_2|^\alpha\Big\}
\end{aligned}
$$

by setting $s = h(t) \equiv 2c-t$ and letting σ_1 be the induced measure $h(\sigma)\equiv\sigma\circ h^{-1}$. Hence (X_1, X_2) has also the representation (4.2.6) but with measure σ_1. Since the representation is unique, (4.2.3) is equivalent to $\sigma = \sigma_1$. Since the transformation $t \to h(t) = 2c - t$ reflects points of $\mathbb{R}^1$ around c, the relation $\sigma = \sigma_1$ means that σ is a symmetric measure around c. ∎

There exist measures σ that are not symmetric around any point, e.g., $\sigma(dx) = 1\{x > 0\}e^{-x}dx$. If (X_1, X_2) has the representation (4.2.6) with such a σ and some b, then the conditional distribution $X_2|X_1$ will not be symmetric. Hence,

Corollary 4.2.3 *There are $S\alpha S$ vectors (X_1, X_2), $1 < \alpha < 2$ such that the conditional laws $X_2|X_1$ are not symmetric (around the conditional mean).*

On the other hand, if σ is symmetric around some point c, then the distribution of $X_2|X_1$ is symmetric around cX_1, and, by Theorem 4.1.2, c will be equal to $[X_2, X_1]_\alpha / \|X_1\|_\alpha$.

The following corollary replaces the symmetry condition on σ by a condition on (X_1, X_2).

Corollary 4.2.4 *Let X_1, X_2, $1 < \alpha < 2$, be jointly $S\alpha S$ random variables. Then the conditional distribution $X_2|X_1$ is symmetric (around cX_1) if and only if*

$$(X_1, X_2) \stackrel{d}{=} (Y_1, Y_2) + (\widetilde{Y}_1, 2c\widetilde{Y}_1 - \widetilde{Y}_2), \tag{4.2.7}$$

where Y_1, Y_2 are jointly $S\alpha S$ random variables and $(\widetilde{Y}_1, \widetilde{Y}_2)$ is an independent copy of (Y_1, Y_2).

PROOF: To prove sufficiency, note that the vectors

$$\begin{aligned}(X_1, X_2) &= (Y_1, Y_2) + (\widetilde{Y}_1, 2c\widetilde{Y}_1 - \widetilde{Y}_2)\\ &= (Y_1 + \widetilde{Y}_1, 2c\widetilde{Y}_1 + Y_2 - \widetilde{Y}_2)\end{aligned}$$

and

$$\begin{aligned}(X_1, 2cX_1 - X_2) &= (Y_1 + \widetilde{Y}_1, 2c(Y_1 + \widetilde{Y}_1) - (2c\widetilde{Y}_1 + Y_2 - \widetilde{Y}_2))\\ &= (Y_1 + \widetilde{Y}_1, 2cY_1 + \widetilde{Y}_2 - Y_2)\end{aligned}$$

have the same distribution.

To prove necessity, let (Z_1, Z_2) and $(\widetilde{Z}_1, \widetilde{Z}_2)$ be independent copies of (X_1, X_2). By the definition of stability and (4.2.3),

$$\begin{aligned}(X_1, X_2) &\stackrel{d}{=} 2^{-1/\alpha}((Z_1, Z_2) + (\widetilde{Z}_1, \widetilde{Z}_2))\\ &\stackrel{d}{=} 2^{-1/\alpha}((Z_1, Z_2) + (\widetilde{Z}_1, 2c\widetilde{Z}_1 - \widetilde{Z}_2))\\ &= (Y_1, Y_2) + (\widetilde{Y}_1, 2c\widetilde{Y}_1 - \widetilde{Y}_2),\end{aligned}$$

where $(Y_1, Y_2) = 2^{-1/\alpha}(Z_1, Z_2)$ and $(\widetilde{Y}_1, \widetilde{Y}_2) = 2^{-1/\alpha}(\widetilde{Z}_1, \widetilde{Z}_2)$. ∎

4.3 Linear dependence

A random variable X_2 is *linearly dependent* on a random variable X_1 if there is a constant c and a random variable Y independent of X_1 such that

$$X_2 = cX_1 + Y. \tag{4.3.1}$$

Jointly Gaussian random variables are always linearly dependent. As we will see, linear dependence in the $S\alpha S$ case, $1 < \alpha < 2$, is related to the non-skewness *and* stability of the conditional laws. (We use the term *non-skewed* to characterize a law which is symmetric around some point on the real line.)

Let X_1 and X_2 be $S\alpha S$, $\alpha < 2$. If X_2 is linearly dependent on X_1 then all the conditional laws $X_2|X_1 = x_1$, $-\infty < x_1 < \infty$, are not only symmetric (around cx_1) but also α-stable. (In Section 4.2, we did not require that the conditional laws be stable.) The following lemma shows that in the case $1 < \alpha < 2$, non-skewness and stability of the conditional laws hold only under linear dependence. The case $\alpha \leq 1$ is still open.

Lemma 4.3.1 *Let X_1, X_2 be jointly $S\alpha S$, $1 < \alpha < 2$. If all the conditional laws of X_2 given X_1 are non-skewed α-stable, then*

$$X_2 \stackrel{d}{=} cX_1 + Y$$

for some real c and a $S\alpha S$ random variable Y, independent of X_1.

PROOF: Since all the conditional laws of X_2 given X_1 are α-stable and non-skewed, the conditional characteristic function of X_2 given X_1 is of the form

$$E\{\exp(i\theta_2 X_2)|X_1\} = \exp\{-|\theta_2|^\alpha M(X_1) + i\theta_2 N(X_1)\},$$

where $M \geq 0$ is a measurable function and, where, by Theorem 4.1.2,

$$N(X_1) = E(X_2|X_1) = cX_1, \text{ a.s..}$$

Let

$$Z = cX_1 + (M(X_1))^{1/\alpha} Z_0, \tag{4.3.2}$$

where $Z_0 \sim S_\alpha(1, 0, 0)$ is a random variable independent of X_1. Since all conditional laws of Z given X_1 coincide with the conditional laws of X_2 given X_1, we have

$$(Z, X_1) \stackrel{d}{=} (X_2, X_1). \tag{4.3.3}$$

This shows that Z and X_1 are jointly $S\alpha S$, and thus the linear combination $(M(X_1))^{1/\alpha} Z_0 = Z - cX_1$ is a $S\alpha S$ random variable. There is, therefore, some $\sigma \geq 0$ such that, for any real θ,

$$\begin{aligned} \exp\{-|\theta|^\alpha \sigma^\alpha\} &= E\exp\{i\theta M(X_1)^{1/\alpha} Z_0\} \\ &= EE\{\exp(i\theta M(X_1)^{1/\alpha} Z_0 | Z_0)\} \\ &= E\exp\{-|\theta|^\alpha M(X_1)\}. \end{aligned}$$

$M(X_1) = \sigma^\alpha$ a.s. is a solution of this equation and it is the only solution by the uniqueness of the Laplace transform. Using (4.3.2) and (4.3.3), we obtain

$$X_2 \stackrel{d}{=} cX_1 + \sigma Z_0,$$

proving the lemma. ∎

The following result shows that the $S\alpha S$ case, $1 < \alpha < 2$, differs greatly from the Gaussian case.

Proposition 4.3.2 *Let X_1, X_2 be jointly $S\alpha S$, $1 < \alpha < 2$. Then both the conditional law of $X_2|X_1$ and the conditional law of $X_1|X_2$ are non-skewed α-stable if and only if X_1 and X_2 are either independent or proportional to each other.*

PROOF: Sufficiency is obvious. Suppose that the conditional laws of $X_2|X_1$ and $X_1|X_2$ are non-skewed and α-stable. Then, by Lemma 4.3.1, there is a $S\alpha S$ random variable Y_1 independent of X_1 and a $S\alpha S$ random variable Y_2 independent of X_2 such that

$$X_2 \stackrel{d}{=} cX_1 + Y_1$$

and

$$X_1 \stackrel{d}{=} bX_2 + Y_2,$$

where c and b are constants.

Consider the first relation. Since $\theta_1 X_1 + \theta_2 X_2 = (\theta_1 + \theta_2 c)X_1 + \theta_2 Y_2$, we see, by writing down the joint characteristic function of (X_1, X_2), that the spectral measure of (X_1, X_2) is concentrated on the set

$$\left\{\pm\left(\frac{1}{\sqrt{1+c^2}}, \frac{c}{\sqrt{1+c^2}}\right)\right\} \cup \{\pm(0,1)\}$$

of the unit circle S_2 (draw a picture; refer also to Example 2.3.4). Similarly, the relation $X_1 = bX_2 + Y_2$, implies that the spectral measure of (X_1, X_2) is concentrated on

$$\left\{\pm\left(\frac{b}{\sqrt{1+b^2}}, \frac{1}{\sqrt{1+b^2}}\right)\right\} \cup \{\pm(1,0)\}.$$

Since the spectral measure is unique, these two sets must coincide. Clearly, this happens if and only if the spectral measure is concentrated either on the four poles of S^2 or on two diagonally opposite points (draw pictures!). The first case corresponds to independence (Example 2.3.5), the second to $X_2 \stackrel{d}{=} cX_1$. ∎

Remark. This proposition is clearly false in the Gaussian case because it would read: "Any pair of Gaussian random variables has correlation zero or one." The proof fails in the Gaussian case because the spectral measure is not unique when $\alpha = 2$ (see Section 2.4).

4.4 Probability tails of order statistics

We study in this section the probability tails of order statistics of jointly strictly α-stable random variables $X_1, X_2, \ldots, X_n$. Because of the representation theorem 3.5.6, we can always suppose that the random variables are given by their integral representation $X_k = \int_E f_k(x)M(dx)$, $k = 1, \ldots, n$, where M is an α-stable random measure. The order statistics of $X_1, X_2, \ldots, X_n$ are the random variables $X^{(1)} \geq X^{(2)} \geq \cdots \geq X^{(n)}$ obtained by arranging $X_1, X_2, \ldots, X_n$ in decreasing order. We consider firstly $X^{(n)}$, the "minimum" order statistic.

Theorem 4.4.1 *Let $X_k = \int_E f_k(x)M(dx)$, $k = 1, \ldots, n$, where M is an α-stable random measure with control measure m and skewness intensity β. Then*

$$\begin{aligned}\lim_{\lambda\to\infty} \lambda^\alpha P\left(\min_{k=1,\ldots,n} X_k > \lambda\right) &= \frac{1}{2}C_\alpha\left\{\int_E\left(\min_{k=1,\ldots,n}[f_k(x)]_+^\alpha\right)(1+\beta(x))m(dx)\right. \\ &\quad \left. + \int_E\left(\min_{k=1,\ldots,n}[f_k(x)]_-^\alpha\right)(1-\beta(x))m(dx)\right\}, \qquad (4.4.1)\end{aligned}$$

where C_α is the constant defined in (1.2.9).

Notation. Herc,

$$x_+ = \begin{cases} x & \text{if } x \geq 0 \\ 0 & \text{if } x < 0 \end{cases} \quad \text{and } x_- = \begin{cases} 0 & \text{if } x > 0 \\ -x & \text{if } x \leq 0 \end{cases} = (-x)_+ \geq 0$$

denote, respectively, the positive and negative part of x.

We know (see Property 1.2.15) that if $X_k \sim S_\alpha(\sigma_k, \beta_k, 0)$, then $\lim_{\lambda\to\infty} \lambda^\alpha P(X_k > \lambda) = (1/2)C_\alpha(1+\beta_k)\sigma_k^\alpha$. Theorem 4.4.1 shows that

$P(\min(X_1,\ldots,X_n)>\lambda)$ and $P(X_k>\lambda)$, $k=1,\ldots,n$, have the same asymptotic behavior up to a constant of proportionality.

Let us consider the constant of proportionality. If $X_k=\int_E f_k(x)M(dx)$, then, by Properties 3.2.2 and 1.2.15,

$$\lim_{\lambda\to\infty}\lambda^\alpha P\{X_k>\lambda\}$$
$$=\frac{1}{2}C_\alpha\left\{\int_E [f_k(x)]_+^\alpha(1+\beta(x))m(dx)+\int_E [f_k(x)]_-^\alpha(1-\beta(x))m(dx)\right\}.$$

Comparing this with (4.4.1), we see that $\lim_{\lambda\to\infty}\lambda^\alpha P\{\min_{k=1,\ldots,n}X_k>\lambda\}$ is obtained by replacing $[f_k(x)]_\pm$ by the pointwise minimum $\min_{k=1,\ldots,n}[f_k(x)]_\pm$.

Remarks

1. Relation (4.4.1) can be expressed in terms of the spectral measure Γ of the α-stable random vector $(X_1,\ldots,X_n)$ as follows:

$$\lim_{\lambda\to\infty}\lambda^\alpha P(\min_{i=1,\ldots,n}X_i>\lambda)=C_\alpha\int_{S_n}\min_{i=1,\ldots,n}[s_i]_+^\alpha\Gamma(ds). \quad (4.4.2)$$

Indeed, it is easy to verify that the right-hand side of (4.4.2) and of (4.4.1) coincide (see Exercise 4.11).

2. If M is $S\alpha S$, then (4.4.1) becomes

$$\lim_{\lambda\to\infty}\lambda^\alpha P\{\min_{k=1,\ldots,n}X_k>\lambda\} \quad (4.4.3)$$
$$=\frac{1}{2}C_\alpha\left\{\int_E\left(\min_{k=1,\ldots,n}[f_k(x)]_+^\alpha+\int_E\min_{k=1,\ldots,n}[f_k(x)]_-^\alpha\right)m(dx)\right\}.$$

3. The proof of Theorem 4.4.1 uses the series representation. The idea is to show that the tail behavior of the first term in the series representation dominates the tail behavior of all other terms. Using Lemma 4.4.2 below, we conclude that only the first term contributes to the limit. The behavior of the first term is $K_\alpha\lambda^{-\alpha}$ where K_α equals the right-hand side of (4.4.3)

Note that the series representation requires the control measure to be finite. Our proof addresses this requirement by replacing the original control measure m with a finite measure $\widetilde{m}$, obtained from m by a judicious "rescaling."

Lemma 4.4.2 *Suppose that X is a random variable with a regularly varying tail, i.e., there is a number $\theta>0$ such that for every number $a>1$,*

$$\lim_{x\to\infty}\frac{P(X>ax)}{P(X>x)}=a^{-\theta}.$$

Suppose also that the tail of X dominates the tail of a positive random variable Y in the sense that

$$\lim_{x\to\infty} \frac{P(Y > x)}{P(X > x)} = 0.$$

Then

$$\lim_{x\to\infty} \frac{P(X+Y > x)}{P(X > x)} = \lim_{x\to\infty} \frac{P(X-Y > x)}{P(X > x)} = 1.$$

PROOF: For $x > 0$ and any σ, $0 < \sigma < 1$, we have

$$\begin{aligned} &P(X+Y > x) \\ &\quad = P(X+Y > x,\ Y > \sigma x) + P(X+Y > x,\ Y \le \sigma x) \\ &\quad \le P(Y > \sigma x) + P(X > (1-\sigma)x) \end{aligned}$$

so that

$$\begin{aligned} \limsup_{x\to\infty} \frac{P(X+Y > x)}{P(X > x)} &\le \limsup_{x\to\infty} \frac{P(Y > \sigma x) + P(X > (1-\sigma)x)}{P(X > x)} \\ &\le \limsup_{x\to\infty} \frac{P(Y > \sigma x)}{P(X > \sigma x)} \frac{P(X > \sigma x)}{P(X > x)} + \limsup_{x\to\infty} \frac{P(X > (1-\sigma)x)}{P(X > x)} \\ &= (1-\sigma)^{-\theta}. \end{aligned}$$

Since σ is arbitrary and since, obviously,

$$\liminf_{x\to\infty} \frac{P(X+Y > x)}{P(X > x)} \ge 1,$$

the first part of the lemma follows. The second part can be proved in a similar way. ∎

PROOF OF THEOREM 4.4.1: Let $f^*(x) = \max_{k=1,\dots,n} |f_k(x)|$, $x \in E$. Clearly $f^* \in L^\alpha(m)$. Let $\widetilde{m}(dx) = f^*(x)^\alpha m(dx)$. Then $\widetilde{m}$ is a finite measure on $(E, \mathcal{E})$ and we denote by $\widetilde{M}$ an α-stable random measure on $(E, \mathcal{E})$ with control measure $\widetilde{m}$ and the same skewness intensity β as M. If $g_k(x) = f_k(x)/f^*(x)$, $k = 1, \dots, n$,[1] then

$$(X_1, X_2, \dots, X_n) \stackrel{d}{=} (Y_1, Y_2, \dots, Y_n), \tag{4.4.4}$$

where

$$Y_k = \int_E g_k(x) \widetilde{M}(dx) + \mu_k,\ \ k = 1, \dots, n,$$

[1]We assume without loss of generality that $f^*(x) \neq 0$ for all $x \in E$. Alternatively, one may set $g_k(x) = 0$ when $f^*(x) = 0$.

and

$$\mu_k = \begin{cases} 0 & \text{if } \alpha \neq 1, \\ \frac{2}{\pi} \int_E f_k(x)\beta(x) \ln |f^*(x)| m(dx) & \text{if } \alpha = 1. \end{cases}$$

Note that $|g_k(x)| \le 1$ for all $k = 1, \dots, n$ and $x \in E$. The Y_ks have series representation (see Theorem 3.10.1):

$$Z_k = (C_\alpha \widetilde{m}(E))^{1/\alpha} \sum_{i=1}^{\infty} \Big(\gamma_i \Gamma_i^{-1/\alpha} g_k(V_i) - b_i(\alpha, g_k) \Big), \ k = 1, \dots, n,$$

where the V_is are i.i.d. with distribution $m_0 = \widetilde{m}/\widetilde{m}(E)$ and the $b_i(\alpha, g_k)$ are constants. We will prove

$$\lim_{\lambda \to \infty} \lambda^\alpha P\{ \min_{k=1,\dots,n} Z_k > \lambda \} = \frac{1}{2} C_\alpha \Big(\int_E h_+(x)^\alpha (1 + \beta(x)) m(dx) + \int_E h_-(x)^\alpha (1 - \beta(x)) m(dx) \Big) \tag{4.4.5}$$

where

$$h_+(x) = \min_{k=1,\dots,n} [f_k(x)]_+, \quad h_-(x) = \min_{k=1,\dots,n} [f_k(x)]_-. \tag{4.4.6}$$

By (4.4.4), this will imply (4.4.1).

For $k = 1, \dots, n$, set $Z_k = U_k + W_k$, where

$$U_k = (C_\alpha \widetilde{m}(E))^{1/\alpha} [\gamma_1 \Gamma_1^{-1/\alpha} g_k(V_1) - b_1(\alpha, g_k)],$$

$$W_k = (C_\alpha \widetilde{m}(E))^{1/\alpha} \sum_{i=2}^{\infty} [\gamma_i \Gamma_i^{-1/\alpha} g_k(V_i) - b_i(\alpha, g_k)].$$

Since

$$\min_{k=1,\dots,n} U_k - \max_{k=1,\dots,n} |W_k| \le \min_{k=1,\dots,n} Z_k \le \min_{k=1,\dots,n} U_k + \max_{k=1,\dots,n} |W_k|,$$

Relation (4.4.5) will follow (see Lemma 4.4.2) if we prove

$$\lim_{\lambda \to \infty} \lambda^\alpha P\{ \min_{k=1,\dots,n} U_k > \lambda \} = \frac{1}{2} C_\alpha \Big(\int_E h_+(x)^\alpha (1 + \beta(x)) m(dx) + \int_E h_-(x)^\alpha (1 - \beta(x)) m(dx) \Big) \tag{4.4.7}$$

and

$$\lim_{\lambda \to \infty} \lambda^\alpha P\{ \max_{k=1,\dots,n} |W_k| > \lambda \} = 0. \tag{4.4.8}$$

We first establish (4.4.7). We may obviously assume $b_1(\alpha, g_k) \equiv 0$. For $\lambda > 0$,

$$
\begin{aligned}
&P\{\min_{k=1,\dots,n} U_k > \lambda\} \\
&= P\Big\{(C_\alpha \widetilde{m}(E))^{1/\alpha}\Gamma_1^{-1/\alpha}\frac{\min_{k=1,\dots,n}\gamma_1 f_k(V_1)}{f^*(V_1)} > \lambda\Big\} \\
&= \int_E P\left((C_\alpha \widetilde{m}(E))^{1/\alpha}\Gamma_1^{-1/\alpha}\frac{h_+(x)}{f^*(x)} > \lambda\right)\frac{1+\beta(x)}{2}m_0(dx) \\
&+ \int_E P\left((C_\alpha \widetilde{m}(E))^{1/\alpha}\Gamma_1^{-1/\alpha}\frac{h_-(x)}{f^*(x)} > \lambda\right)\frac{1-\beta(x)}{2}m_0(dx) \\
&= \int_E \left(1 - \exp\Big\{-C_\alpha \widetilde{m}(E)\lambda^{-\alpha}\left(\frac{h_+(x)}{f^*(x)}\right)^\alpha\Big\}\right)\frac{1+\beta(x)}{2}m_0(dx) \\
&+ \int_E \left(1 - \exp\Big\{-C_\alpha \widetilde{m}(E)\lambda^{-\alpha}\left(\frac{h_-(x)}{f^*(x)}\right)^\alpha\Big\}\right)\frac{1-\beta(x)}{2}m_0(dx).
\end{aligned}
\tag{4.4.9}
$$

Now (4.4.7) follows from (4.4.9) and the dominated convergence theorem.

We now prove (4.4.8). Since $P\{\max_{k=1,\dots,n}|W_k| > \lambda\} \le \sum_{k=1}^n P\{|W_k| > \lambda\}$, it is enough to show (see Exercise 4.14) that $E|W_k|^\alpha < \infty$ for all $k = 1,\dots,n$. Let $\widetilde{W}_k$ be an independent copy of W_k given by

$$\widetilde{W}_k = (C_\alpha \widetilde{m}(E))^{1/\alpha}\sum_{i=2}^\infty[\widetilde{\gamma}_i\widetilde{\Gamma}_i^{-1/\alpha}g_k(\widetilde{V}_i) - b_i(\alpha, g_k)],$$

where $\{(\widetilde{\gamma}_i, \widetilde{\Gamma}_i, \widetilde{V}_i),\ i = 1,2,\dots\}$ is an independent copy of $\{(\gamma_i, \Gamma_i, V_i),\ i = 1,2,\dots\}$. Let

$$R_k = (C_\alpha \widetilde{m}(E))^{-1/\alpha}(W_k - \widetilde{W}_k) = \sum_{i=1}^\infty(\gamma_i\Gamma_i^{-1/\alpha}g_k(V_i) - \widetilde{\gamma}_i\widetilde{\Gamma}_i^{-1/\alpha}g_k(\widetilde{V}_i)).$$

It is clearly enough to prove that $E|R_k|^\alpha < \infty$. Write $R_k = R_k^{(1)} + R_k^{(2)}$, where

$$
\begin{aligned}
R_k^{(1)} &= \sum_{i=2}^\infty(\gamma_i g_k(V_i) - \widetilde{\gamma}_i g_k(\widetilde{V}_i))\Gamma_i^{-1/\alpha}, \\
R_k^{(2)} &= \sum_{i=2}^\infty(\Gamma_i^{-1/\alpha} - \widetilde{\Gamma}_i^{-1/\alpha})\widetilde{\gamma}_i g_k(\widetilde{V}_i).
\end{aligned}
$$

Then we need to show that $E|R_k^{(j)}|^\alpha < \infty$ for $j = 1,2$. Consider first $j = 1$. Let $\{\epsilon_i,\ i \ge 1\}$ be a sequence of i.i.d. Rademacher random variables independent of all other random variables in $R_k^{(1)}$ and let $i_0 = [2/\alpha] + 1$. Then

$$\begin{aligned} E & \left(\sum_{i=i_0}^{\infty} (\gamma_i g_k(V_i) - \widetilde{\gamma}_i g_k(\widetilde{V}_i))\Gamma_i^{-1/\alpha} \right)^2 \\ &= E \left(\sum_{i=i_0}^{\infty} \epsilon_i (\gamma_i g_k(V_i) - \widetilde{\gamma}_i g_k(\widetilde{V}_i))\Gamma_i^{-1/\alpha} \right)^2 \\ &= E \sum_{i=i_0}^{\infty} (\gamma_i g_k(V_i) - \widetilde{\gamma}_i g_k(\widetilde{V}_i))^2 \Gamma_i^{-2/\alpha} \\ &\le 4 \sum_{i=i_0}^{\infty} E\Gamma_i^{-2/\alpha} \le 4C \sum_{i=i_0}^{\infty} i^{-2/\alpha} < \infty \end{aligned}$$

using Exercise 4.15 and $|g_k(x)| \le 1$. Moreover, for any $i = 2, \dots, i_0 - 1$,

$$E\left(|\gamma_i g_k(V_i) - \widetilde{\gamma}_i g_k(\widetilde{V}_i)|\Gamma_i^{-1/\alpha} \right)^\alpha \le 2^\alpha E\Gamma_i^{-1} < \infty.$$

Hence $E|R_k^{(1)}|^\alpha$ is bounded by

$$i_0^\alpha \left[\sum_{i=2}^{i_0-1} E|(\gamma_i g_k(V_i) - \widetilde{\gamma}_i g_k(\widetilde{V}_i))\Gamma_i^{-1/\alpha}|^\alpha + E|\sum_{i=i_0}^{\infty} (\gamma_i g_k(V_i) - \widetilde{\gamma}_i g_k(\widetilde{V}_i))\Gamma_i^{-1/\alpha}|^\alpha \right]$$

which is finite. It now remains to prove that $E|R_k^{(2)}|^\alpha < \infty$. We use Exercise 4.16 to conclude that there is a j_1 such that for any $j > j_1$, $E\exp\{j^{-1/2}|\Gamma_j - j|\} \le 4$. Choose, now, a number $\theta > 2$ such that $\frac{1}{\theta} + \frac{1}{\alpha} > 1$. By Markov's inequality, for any $\lambda > 0$,

$$\begin{aligned} P(|\Gamma_j - j| > \lambda j^{1-1/\theta}) &= P(\exp\{j^{-1/2}|\Gamma_j - j|\} > \exp\{\lambda j^{1/2-1/\theta}\}) \\ &\le 4\exp\{-\lambda j^{1/2-\theta}\}. \end{aligned}$$

Therefore, for some constant $b_1 = b_1(\theta)$ and any $\lambda \ge 1$,

$$\begin{aligned} \sum_{j=j_1+1}^{\infty} P(|\Gamma_j - j| > \lambda j^{1-1/\theta}) &\le 4 \sum_{j=j_1+1}^{\infty} \exp\{-\lambda j^{1/2-1/\theta}\} \\ &\le 4 \int_{j_1}^{\infty} \exp\{-\lambda x^{1/2-1/\theta}\} dx \\ &\le b_1 \lambda^{-4} \end{aligned} \tag{4.4.10}$$

since $\lambda^4 e^{-\lambda x^\epsilon}$, $\epsilon > 0$, is integrable. By the mean value theorem,

$$|\Gamma_j^{-1/\alpha} - \widetilde{\Gamma}_j^{-1/\alpha}| = \frac{|\Gamma_j^{1/\alpha} - \widetilde{\Gamma}_j^{1/\alpha}|}{\Gamma_j^{1/\alpha}\widetilde{\Gamma}_j^{1/\alpha}} \le \frac{1}{\alpha} \frac{|\Gamma_j - \widetilde{\Gamma}_j|}{\min(\Gamma_j, \widetilde{\Gamma}_j)^{1+1/\alpha}},$$

and hence

$$\begin{aligned} P(|\Gamma_j^{-1/\alpha} - \widetilde{\Gamma}_j^{-1/\alpha}| > \lambda j^{-(\frac{1}{\theta}+\frac{1}{\alpha})}) \\ \le P(|\Gamma_j - \widetilde{\Gamma}_j| > \lambda\alpha \min(\Gamma_j, \widetilde{\Gamma}_j)^{1+1/\alpha} j^{-(\frac{1}{\theta}+\frac{1}{\alpha})}) \\ \le P(\min(\Gamma_j, \widetilde{\Gamma}_j) \le j\lambda^{-\frac{\alpha}{2(\alpha+1)}}) \\ +P(|\Gamma_j - \widetilde{\Gamma}_j| > \alpha\lambda^{1/2} j^{1-1/\theta}). \end{aligned} \tag{4.4.11}$$

Set $a = \alpha/2(\alpha+1)$. For any $\lambda > 0$ large enough, by Stirling's formula,

$$\begin{aligned} P(\min(\Gamma_j, \widetilde{\Gamma}_j) \le j\lambda^{-a}) &\le 2P(\Gamma_j \le j\lambda^{-a}) \\ &= 2\int_0^{j\lambda^{-a}} \frac{x^{j-1}}{(j-1)!} e^{-x} dx \\ &\le 2\frac{(j\lambda^{-a})^{j-1}}{(j-1)!} j\lambda^{-a} \\ &\le b_2 j^{3/2} \left(e\lambda^{-a}\right)^j, \end{aligned}$$

where b_2 is a finite constant. Therefore

$$\begin{aligned} \sum_{j=j_1+1}^{\infty} P(\min(\Gamma_j, \widetilde{\Gamma}_j) \le j\lambda^{-a}) &\le b_2 \sum_{j=j_1+1}^{\infty} j^{3/2} \left(e\lambda^{-a}\right)^j \\ &\le b_3 \left(e\lambda^{-a}\right)^{j_1+1} \frac{1}{1 - e\lambda^{-a}} \\ &\le b_4 \lambda^{-(j_1+1)a} \\ &\le b_4 \lambda^{-2} \end{aligned}$$

by choosing λ and j_1 large enough. Here b_i, $i = 3, 4$, are finite constants. Further, by the triangle inequality and (4.4.10), for any λ large enough,

$$\begin{aligned} &\sum_{j=j_1+1}^{\infty} P(|\Gamma_j - \widetilde{\Gamma}_j| > \alpha\lambda^{1/2} j^{1-1/\theta}) \\ &\le 2 \sum_{j=j_1+1}^{\infty} P\left(|\Gamma_j - j| > \frac{\alpha}{2}\lambda^{1/2} j^{1-1/\theta}\right) \le b_5 \lambda^{-2} \end{aligned}$$

for some finite constant b_5. Therefore, for large enough j_1, we obtain

$$\begin{aligned} &P\left(\sup_{j>j_1} \frac{|\Gamma_j^{-1/\alpha} - \widetilde{\Gamma}_j^{-1/\alpha}|}{j^{-(\frac{1}{\theta}+\frac{1}{\alpha})}} > \lambda\right) \\ &\le \sum_{j=j_1+1}^{\infty} P(|\Gamma_j^{-1/\alpha} - \widetilde{\Gamma}_j^{-1/\alpha} > \lambda j^{-(\frac{1}{\theta}+\frac{1}{\alpha})}) \le O(\lambda^{-2}), \end{aligned} \tag{4.4.12}$$

and, for such j_1,

$$\begin{aligned} P\Big(\Big|\sum_{j=j_1+1}^{\infty}(\Gamma_j^{-1/\alpha}-\widetilde{\Gamma}_j^{-1/\alpha})\widetilde{\gamma}_i g_k(\widetilde{V}_i)\Big|>\lambda\Big) &\\ \le P\Big(\sum_{j=j_1+1}^{\infty}|\Gamma_j^{-1/\alpha}-\widetilde{\Gamma}_j^{-1/\alpha}|>\lambda\Big) &\\ \le P\Big(\sup_{j>j_1}\frac{|\Gamma_j^{-1/\alpha}-\widetilde{\Gamma}_j^{-1/\alpha}|}{j^{-(\frac{1}{\theta}+\frac{1}{\alpha})}}\sum_{j=j_1+1}^{\infty}j^{-(\frac{1}{\theta}+\frac{1}{\alpha})}>\lambda\Big)\le O(\lambda^{-2}) & \end{aligned}$$

because $\frac{1}{\theta}+\frac{1}{\alpha}>1$. Since, for every $j\ge 2$,

$$E|(\Gamma_j^{-1/\alpha}-\widetilde{\Gamma}_j^{-1/\alpha})\widetilde{\gamma}_i g_k(\widetilde{V}_i)|^\alpha \le E|\Gamma_j^{-1/\alpha}-\widetilde{\Gamma}_j^{-1/\alpha}|^\alpha \le 2^{1+\alpha}E\Gamma_j^{-1}<\infty,$$

we conclude that $E|R_k^{(2)}|^\alpha<\infty$. This proves the theorem. ∎

Corollary 4.4.3 *Under the conditions of Theorem 4.4.1,*

$$\lim_{\lambda\to\infty}\lambda^\alpha P\left(\min_{k=1,\dots,n}|X_k|>\lambda\right)=C_\alpha\int_E \min_{k=1,\dots,n}|f_k(x)|^\alpha m(dx).$$

PROOF: Let $\mathcal{S}=\{(s_1,\dots,s_n): s_k=\pm1,\ k=1,\dots,n\}$ be the set of all n-tuple possible "signs." Since

$$\begin{aligned} P(\min_{k=1,\dots,n}|X_k|>\lambda) &= P(|X_1|>\lambda,\dots,|X_n|>\lambda)\\ &= \sum_{(s_1,\dots,s_n)\in\mathcal{S}} P(s_1X_1>\lambda,\dots,s_nX_n>\lambda), \end{aligned}$$

$\lim_{\lambda\to\infty}\lambda^\alpha(\min_{k=1,\dots,n}|X_k|>\lambda)$ equals the right-hand side of (4.4.1) with $\min_{k=1,\dots,n}[f_k(x)]_\pm$ replaced by $\sum_{(s_1,\dots,s_n)\in\mathcal{S}}\min_{k=1,\dots,n}[s_kf_k(x)]_\pm$. Fix now $x\in E$ and observe that the minimum is zero if $f_k(x)=0$ for some k. Suppose $f_k(x)\neq0$ for all ks. Then $\min_k[s_kf_k(x)]_+\neq0$ only if $s_k=\text{sign }(f_k)$ for all ks, in which case $\min_k[s_kf_k(x)]_+=\min_k|f_k(x)|$. Similarly, $\min_k[s_kf_k(x)]_-\neq0$ only if $s_k=-\text{ sign }(f_k)$ for all ks in which case we obtain again $\min_k[s_kf_k(x)]_-=\min_k|f_k(x)|$. This proves the corollary. ∎

Theorem 4.4.1 gives the tail behavior of the minimum. In order to obtain the tail behavior of any other order statistic, we use an inclusion-exclusion type formula (see Lemma 4.4.4 below) which relates any order statistics to the minimum.

Lemma 4.4.4 *Let $a_1, a_2, \ldots, a_n$ be real numbers and let $a^{(1)} \geq a^{(2)} \geq \cdots \geq a^{(n)}$ be a non-increasing permutation of these numbers. Then*

$$a^{(k)} = \sum_{j=k}^{n} (-1)^{j-k} \binom{j-1}{k-1} \sum_{1 \leq i_1 < i_2 < \cdots < i_j \leq n} \min(a_{i_1}, a_{i_2}, \ldots, a_{i_j}).$$

PROOF: See Exercise 4.18.

Theorem 4.4.5 *Let $X_1, X_2, \ldots, X_n$ be jointly strictly α-stable random variables, given in the form $X_k = \int_E f_k(x) M(dx)$, $k = 1, \ldots, n$, where M is an α-stable random measure with control measure m and skewness intensity β. Let $X^{(1)} \geq \cdots \geq X^{(n)}$ be $X_1, X_2, \ldots, X_n$ rearranged in nonincreasing order. Then, for any $k = 1, 2, \ldots, n$*

$$\lim_{\lambda \to \infty} \lambda^\alpha P(X^{(k)} > \lambda)$$

$$= \frac{1}{2} C_\alpha \left\{ \int_E h_{k,+}(x)^\alpha (1 + \beta(x)) m(dx) + \int_E h_{k,-}(x)^\alpha (1 - \beta(x)) m(dx) \right\}, \tag{4.4.13}$$

where

$$\begin{aligned} h_{k,+}(x) &= k^{\text{th}} \textit{ largest among } [f_i(x)]_+, \ i = 1, \ldots, n, \\ h_{k,-}(x) &= k^{\text{th}} \textit{ largest among } [f_i(x)]_-, \ i = 1, \ldots, n. \end{aligned} \tag{4.4.14}$$

PROOF: Set $a_k = I(X_k > \lambda)$ in Lemma 4.4.4 to obtain

$$\lim_{\lambda \to \infty} \lambda^\alpha P(X^{(k)} > \lambda)$$

$$= \sum_{j=k}^{n} (-1)^{j-k} \binom{j-1}{k-1} \sum_{1 \leq i_1 < \ldots < i_j \leq n} \lim_{\lambda \to \infty} \lambda^\alpha P(X_{i_1} > \lambda, \ldots, X_{i_j} > \lambda)$$

$$= \sum_{j=k}^{n} (-1)^{j-k} \binom{j-1}{k-1} \sum_{1 \leq i_1 < \ldots < i_j \leq n} \frac{1}{2} C_\alpha$$

$$\times \left\{ \int_E \min_{d=1,\ldots,j} [f_{i_d}(x)]_+^\alpha (1 + \beta(x)) m(dx) \right.$$

$$\left. + \int_E \min_{d=1,\ldots,j} [f_{i_d}(x)]_-^\alpha (1 - \beta(x)) m(dx) \right\}$$

$$= \frac{1}{2} C_\alpha \left\{ \int_E h_{k,+}(x)^\alpha (1 + \beta(x)) m(dx) + \int_E h_{k,-}(x)^\alpha (1 - \beta(x)) m(dx) \right\}.$$

We used Theorem 4.4.1 in the second step, and in the last step, we used Lemma 4.4.4 once more. ∎

Remark. The tail of the distribution of any order statistics behaves like $\lambda^{-\alpha}$ as $\lambda \to \infty$.

Corollary 4.4.6 *Let $|X|^{(1)} \geq \cdots \geq |X|^{(n)}$ be a non-increasing permutation of $|X_1|, \ldots, |X_n|$. Under the conditions of Theorem 4.4.5,*

$$\lim_{\lambda \to \infty} \lambda^\alpha P(|X|^{(k)} > \lambda) = C_\alpha \int_E h_k(x)^\alpha m(dx),$$

where $h_k(x)$ is the kth largest among $|f_i(x)|$, $i = 1, \ldots, n$.

The proof is similar to that of Corollary 4.4.3 and is left as an exercise.

As an application of the preceding results, we will illustrate the difference in behavior of $\lim_{\lambda \to \infty} P(X_2 > \lambda \,|X_1 > \lambda)$ between the Gaussian and stable cases.

Example 4.4.7 The conditional probability $P(X_2 > \lambda \,|X_1 > \lambda)$, $\lambda > 0$, is a measure of dependence. If X_1 large implies that X_2 is likely to be large, then X_1 and X_2 are "positively" dependent. If it implies that X_2 is unlikely to be large, then X_1 and X_2 are "negatively" dependent.

Suppose, firstly, that (X_1, X_2) is a mean zero Gaussian vector. Then $X_2 = cX_1 + Y$ where $Y \sim N(0, \sigma^2)$ is independent of X_1, and (see Exercise 4.20)

$$\lim_{\lambda \to \infty} P(X_2 > \lambda \,|X_1 > \lambda) = \begin{cases} 1 & \text{if } c > 1 \text{ or if } c = 1, \sigma = 0, \\ 0 & \text{if } c < 1, \\ 1/2 & \text{if } c = 1, \sigma > 0. \end{cases} \tag{4.4.15}$$

In the stable case, on the other hand, $\lim_{\lambda \to \infty} P(X_2 > \lambda \,|X_1 > \lambda)$ can be any number in [0,1]. Let $X_k = \int_E f_k(x) M(dx)$, $k = 1, 2$, where M is a $S\alpha S$ random measure with control measure m. By Theorem 4.4.1,

$$\lim_{\lambda \to \infty} P(X_2 > \lambda \,|X_1 > \lambda) = \lim_{\lambda \to \infty} \frac{P(\min(X_1, X_2) > \lambda)}{P(X_1 > \lambda)}$$

$$= \frac{\int_E \min((f_1(x))_+, (f_2(x))_+)^\alpha m(dx) + \int_E \min((f_1(x))_-, (f_2(x))_-)^\alpha m(dx)}{\int_E (f_1(x))_+^\alpha m(dx) + \int_E (f_1(x))_-^\alpha m(dx)}$$

$$= \frac{\int_{E_1} |f_1(x)|^\alpha m(dx) + \int_{E_2} |f_2(x)|^\alpha m(dx)}{\int_E |f_1(x)|^\alpha m(dx)} \tag{4.4.16}$$

where

$$E_1 = \left\{ x \in E : f_2(x) \neq 0,\ 0 < \frac{f_1(x)}{f_2(x)} \leq 1 \right\},$$

$$E_2 = \left\{ x \in E : f_2(x) \neq 0,\ \frac{f_1(x)}{f_2(x)} > 1 \right\}.$$

(Note that $f_1(x)$ and $f_2(x)$ have the same signs on E_1 and E_2.) By choosing f_1 and f_2 suitably, $\lim_{\lambda \to \infty} P(X_2 > \lambda \,|X_1 > \lambda)$ may take any value in [0,1].

The statement of Corollary 4.4.6 with $k = 1$ can be rewritten in the form

$$\lim_{\lambda\to\infty} \lambda^\alpha P(\|\mathbf{X}\| > \lambda) = C_\alpha \Gamma_{\|\cdot\|}(S_d^{\|\cdot\|}), \tag{4.4.17}$$

where $\|\cdot\|$ is the L^∞-norm on $\mathbb{R}^d$ (use Theorem 3.5.6, Part (i)). It turns out that (4.4.17) remains true for any norm on $\mathbb{R}^d$, and, in fact, the statement can be strengthened.

Theorem 4.4.8 *Let $\|\cdot\|$ be a norm on $\mathbb{R}^d$, and let $\mathbf{X} = (X_1, \ldots, X_d)$ be an α-stable random vector whose characteristic function is given in Proposition 2.3.8. Then for every Borel set $A \subseteq S_d^{\|\cdot\|}$ with $\Gamma_{\|\cdot\|}(\partial A) = 0$,*

$$\lim_{\lambda\to\infty} \lambda^\alpha P\Big(\|\mathbf{X}\| > \lambda, \frac{\mathbf{X}}{\|\mathbf{X}\|} \in A\Big) = C_\alpha \Gamma_{\|\cdot\|}(A). \tag{4.4.18}$$

PROOF: The proof is similar to that of Theorem 4.4.1. We first use Theorem 3.5.6 and Theorem 3.10.1 to conclude that $(X_1, X_2, \ldots, X_d) \stackrel{d}{=} (Z_1, Z_2, \ldots, Z_d)$ where

$$Z_k = (C_\alpha \Gamma_{\|\cdot\|}(S_d^{\|\cdot\|}))^{1/\alpha} \sum_{i=1}^{\infty} [\Gamma_i^{-1/\alpha} S_i^{(k)} - b_{i,k}(\alpha)],\ k = 1, 2, \ldots, d.$$

Here $\mathbf{S}_i = (S_i^{(1)}, \ldots, S_i^{(d)}),\ i = 1, 2, \ldots,$ are i.i.d. $S_d^{\|\cdot\|}$-valued random vectors with common law $\Gamma_{\|\cdot\|}/\Gamma_{\|\cdot\|}(S_d^{\|\cdot\|})$ and $b_{i,k}(\alpha),\ i \geq 1,\ k = 1, \ldots, d,$ are constants. If

$$\begin{aligned} U_k &= (C_\alpha \Gamma_{\|\cdot\|}(S_d^{\|\cdot\|}))^{1/\alpha} [\Gamma_1^{-1/\alpha} S_1^{(k)} - b_{1,k}(\alpha)],\ k = 1, \ldots, d, \\ W_k &= (C_\alpha \Gamma_{\|\cdot\|}(S_d^{\|\cdot\|}))^{1/\alpha} \sum_{i=2}^{\infty} [\Gamma_i^{-1/\alpha} S_i^{(k)} - b_{i,k}(\alpha)],\ k = 1, \ldots, d \end{aligned}$$

and $\mathbf{U} = (U_1, \ldots, U_d)$, $\mathbf{W} = (W_1, \ldots, W_d)$, then $\mathbf{Z} = \mathbf{U} + \mathbf{W}$.

For any $0 < \epsilon < 1/2$,

$$\begin{aligned} &P\Big(\|\mathbf{Z}\| > \lambda, \frac{\mathbf{Z}}{\|\mathbf{Z}\|} \in A\Big) \\ &\leq P\Big(\|\mathbf{U}\| > (1-\epsilon)\lambda, \|\mathbf{W}\| \leq \epsilon\lambda, \frac{\mathbf{U}+\mathbf{W}}{\|\mathbf{U}+\mathbf{W}\|} \in A\Big) + P(\|\mathbf{W}\| > \epsilon\lambda). \end{aligned} \tag{4.4.19}$$

Using the equivalence of norms in $\mathbb{R}^d$ and (4.4.8) we conclude that for some finite positive constant c,

$$\begin{aligned} &\overline{\lim}_{\lambda\to\infty} \lambda^\alpha P(\|\mathbf{W}\| > \epsilon\lambda) \\ &\leq \overline{\lim}_{\lambda\to\infty} \lambda^\alpha P(c\|\mathbf{W}\|_\infty > \epsilon\lambda) \\ &= c^\alpha \epsilon^{-\alpha} \lim_{\lambda\to\infty} \lambda^\alpha P(\max_{k=1,\ldots,d} |W_k| > \lambda) = 0. \end{aligned} \tag{4.4.20}$$

Suppose, now, that the set A is closed, and for $\sigma > 0$, let A_σ denote the closed σ-neighborhood of A. Since

$$\left\| \frac{\mathbf{U}+\mathbf{W}}{\|\mathbf{U}+\mathbf{W}\|} - \frac{\mathbf{U}}{\|\mathbf{U}\|} \right\| \leq \frac{2\|\mathbf{W}\|}{\|\mathbf{U}+\mathbf{W}\|},$$

we conclude that

$$P\Big(\|\mathbf{U}\| > (1-\epsilon)\lambda, \|\mathbf{W}\| \leq \epsilon\lambda, \frac{\mathbf{U}+\mathbf{W}}{\|\mathbf{U}+\mathbf{W}\|} \in A, \frac{\mathbf{U}}{\|\mathbf{U}\|} \notin A_{\epsilon'}\Big) = 0,$$

where $\epsilon' = 2\epsilon/(1-2\epsilon)$. Therefore

$$P\Big(\|\mathbf{U}\| > (1-\epsilon)\lambda, \|\mathbf{W}\| \leq \epsilon\lambda, \frac{\mathbf{U}+\mathbf{W}}{\|\mathbf{U}+\mathbf{W}\|} \in A\Big) \leq P\Big(\|\mathbf{U}\| > (1-\epsilon)\lambda, \frac{\mathbf{U}}{\|\mathbf{U}\|} \in A_{\epsilon'}\Big). \tag{4.4.21}$$

Setting $\mathbf{U}^{(0)} = (U_1^{(0)}, \dots, U_d^{(0)})$ with

$$U_k^{(0)} = \Big(C_\alpha \Gamma_{\|\cdot\|}(S_d^{\|\cdot\|})\Big)^{1/\alpha} \Gamma_1^{-1/\alpha} S_1^{(k)}, \ k = 1, \dots, d,$$

and $b^* = \max_{k=1,\dots,d} |b_{1,k}(\alpha)|$, we can choose λ large enough so that

$$\|\mathbf{U} - \mathbf{U}^{(0)}\| \leq \lambda\epsilon \quad \text{and} \quad \left| \frac{\mathbf{U}}{\|\mathbf{U}\|} - \frac{\mathbf{U}^{(0)}}{\|\mathbf{U}^{(0)}\|} \right| \leq \epsilon'.$$

For such λ, if $\epsilon'' = 2\epsilon' = 4\epsilon/(1-2\epsilon)$, the right-hand side of (4.4.21) is bounded above by

$$\begin{aligned} &P\Big(\|\mathbf{U}^{(0)}\| > (1-2\epsilon)\lambda, \frac{\mathbf{U}^{(0)}}{\|\mathbf{U}^{(0)}\|} \in A_{\epsilon''}\Big) \\ &= P\Big(\Gamma_1^{-1/\alpha} > (1-2\epsilon)(C_\alpha \Gamma_{\|\cdot\|}(S_d^{\|\cdot\|}))^{-1/\alpha}\lambda, \mathbf{S}_1 \in A_{\epsilon''}\Big) \\ &= P\Big(\Gamma_1 < (1-2\epsilon)^{-\alpha} C_\alpha \Gamma_{\|\cdot\|}(S_d^{\|\cdot\|})\lambda^{-\alpha}\Big) \frac{\Gamma_{\|\cdot\|}(A_{\epsilon''})}{\Gamma_{\|\cdot\|}(S_d^{\|\cdot\|})}. \end{aligned}$$

Therefore, for any $0 < \epsilon < 1$,

$$\begin{aligned} &\overline{\lim}_{\lambda\to\infty} \lambda^\alpha P\Big(\|\mathbf{Z}\| > \lambda, \frac{\mathbf{Z}}{\|\mathbf{Z}\|} \in A\Big) \\ &\leq \overline{\lim}_{\lambda\to\infty} \lambda^\alpha P\Big(\Gamma_1 < (1-2\epsilon)^{-\alpha} C_\alpha \Gamma_{\|\cdot\|}(S_d^{\|\cdot\|})\lambda^{-\alpha}\Big) \frac{\Gamma_{\|\cdot\|}(A_{\epsilon''})}{\Gamma_{\|\cdot\|}(S_d^{\|\cdot\|})} \\ &= (1-2\epsilon)^{-\alpha} C_\alpha \Gamma_{\|\cdot\|}(A_{\epsilon''}). \end{aligned} \tag{4.4.22}$$

Letting $\epsilon \downarrow 0$ and noting that $A_{\epsilon''} \downarrow A$ as $\epsilon \downarrow 0$, we conclude from (4.4.22) that for any closed A,

$$\overline{\lim}_{\lambda\to\infty}\lambda^\alpha P\Big(\|\mathbf{Z}\| > \lambda, \frac{\mathbf{Z}}{\|\mathbf{Z}\|} \in A\Big) \le C_\alpha \Gamma_{\|\cdot\|}(A). \tag{4.4.23}$$

On the other hand, we have $P(\|\mathbf{U}\| - \|\mathbf{W}\| > \lambda) \le P(\|\mathbf{Z}\| > \lambda) \le P(\|\mathbf{U}\| + \|\mathbf{W}\| > \lambda)$, and hence

$$\lim_{\lambda\to\infty} \lambda^\alpha P(\|\mathbf{Z}\| > \lambda) = C_\alpha \Gamma_{\|\cdot\|}(S_d^{\|\cdot\|}) \tag{4.4.24}$$

by Lemma 4.4.2. Clearly, (4.4.23) and (4.4.24) imply that as $\lambda \to \infty$, the measure P_λ defined by $P_\lambda(A) = \lambda^\alpha P(\|\mathbf{Z}\| > \lambda, \mathbf{Z}/\|\mathbf{Z}\| \in A)$ converges vaguely to $C_\alpha \Gamma_{\|\cdot\|}$. Therefore (4.4.18) holds. ■

The following is an extension to Property 1.2.18.

Corollary 4.4.9 *Let $\|\cdot\|$ be a norm on $\mathbb{R}^d$, and let $\mathbf{X} = (X_1, \ldots, X_d)$ be an α-stable random vector. For $0 < r < \alpha$, let $\Gamma^{(r)}_{\|\cdot\|}$ be the following finite Borel measure on $S_d^{\|\cdot\|}$:*

$$\Gamma^{(r)}_{\|\cdot\|} = (\alpha - r)(\alpha C_\alpha)^{-1} P^{(r)}_{\mathbf{X}} \circ T^{-1},$$

where $P^{(r)}_{\mathbf{X}}(d\mathbf{x}) = \|\mathbf{x}\|^r P_{\mathbf{X}}(d\mathbf{x})$, $P_{\mathbf{X}}$ is the law of $\mathbf{X}$, and $T: \mathbb{R}^d \backslash \{0\} \to S_d^{\|\cdot\|}$ is given by $T\mathbf{x} = \frac{\mathbf{x}}{\|\mathbf{x}\|}$. Then as $r \uparrow \alpha$, $\Gamma^{(r)}_{\|\cdot\|}$ converges to $\Gamma_{\|\cdot\|}$ vaguely, i.e., for every continuous function h on $S_d^{\|\cdot\|}$, we have

$$\int_{S_d^{\|\cdot\|}} h(\mathbf{s}) \Gamma^{(r)}_{\|\cdot\|}(d\mathbf{s}) \to \int_{S_d^{\|\cdot\|}} h(\mathbf{s}) \Gamma_{\|\cdot\|}(d\mathbf{s}) \ \text{ as } r \uparrow \alpha.$$

PROOF: Since the measures $\Gamma^{(r)}_{\|\cdot\|}$, $0 < r < \alpha$, and $\Gamma_{\|\cdot\|}$ are concentrated on $S_d^{\|\cdot\|}$, we only need to show that for every Borel set $A \subseteq S_d^{\|\cdot\|}$ with $\Gamma_{\|\cdot\|}(\partial A) = 0$, we have

$$\lim_{r\uparrow\alpha} \Gamma^{(r)}_{\|\cdot\|}(A) = \Gamma_{\|\cdot\|}(A). \tag{4.4.25}$$

But for any $0 < r < \alpha$ and any Borel set $A \subseteq S_d^{\|\cdot\|}$,

$$\begin{aligned}
\Gamma^{(r)}_{\|\cdot\|}(A) &= (\alpha - r)(\alpha C_\alpha)^{-1} \int_{\{\mathbf{x}\in\mathbb{R}^d\backslash\{0\}, \mathbf{x}/\|\mathbf{x}\|\in A\}} \|\mathbf{x}\|^r P_{\mathbf{X}}(d\mathbf{x}) \\
&= (\alpha - r)(\alpha C_\alpha)^{-1} E\|\mathbf{X}\|^r 1(\mathbf{X}/\|\mathbf{X}\| \in A) \\
&= (\alpha - r)(\alpha C_\alpha)^{-1} \int_0^\infty P(\|\mathbf{X}\| > t^{1/r}, \mathbf{X}/\|\mathbf{X}\| \in A) dt.
\end{aligned}$$

The proof concludes by following the argument in the proof of Property 1.2.18 and then using Theorem 4.4.8. ■

4.5 Joint moments

Let $\mathbf{X} = (X_1, \ldots, X_n)$ be an α-stable random vector in $\mathbb{R}^n$, $0 < \alpha < 2$, and let $p_1, \ldots, p_n$ be non-negative numbers. We want to find a necessary and sufficient condition on $p_1, \ldots, p_n$ for the joint moment $E|X_1|^{p_1} \cdots |X_n|^{p_n}$ to be finite. We suppose throughout that $\mathbf{X}$ is non-degenerate, i.e., that its distribution is not concentrated on a point.

If the X_j are independent, then $E|X_1|^{p_1} \cdots |X_n|^{p_n} = E|X_1|^{p_1} \cdots E|X_n|^{p_n} < \infty$ if and only if $p_j < \alpha$, $j = 1, \ldots, n$. If the X_j are all equal (say $X_j = X_1$, $j = 2, \ldots, n$), then $E|X_1|^{p_1} \cdots |X_n|^{p_n} = E|X_1|^{p_1 + \cdots + p_n} < \infty$ if and only if $p_1 + \cdots + p_n < \alpha$. We use the integral representation $X_j = \int_E f_j(x) M(dx) + \eta_j$, $j = 1, \ldots, n$, in (3.5.11) (see Theorem 3.5.6 (i)) to obtain a necessary and sufficient condition for $E|X_1|^{p_1} \cdots |X_n|^{p_n} < \infty$ in the general case. Suppose firstly that the X_js are *n-fold dependent*, i.e.,

$$m\{x \in E : f_1(x) f_2(x) \cdots f_n(x) \neq 0\} > 0.$$

When $n = 2$, n-fold dependence is equivalent to dependence by Theorem 3.5.3. When $n > 2$, the random variables can be dependent without being necessarily n-fold dependent.

Example 4.5.1 Let U, V, W be three disjoint subsets of E with positive m measure, and let $X_j = \int_E f_j(x) M(dx)$, $j = 1, 2, 3$, where f_1, f_2, f_3 have support $U \cup V, V \cup W, W \cup U$, respectively. The random variables X_1, X_2, X_3 are dependent, in fact pairwise dependent, because $m\{x \in E : f_i(x) f_j(x) \neq 0\} > 0$, $i \neq j$, but they are not 3-fold dependent because $m\{x \in E : f_1(x) f_2(x) f_3(x) \neq 0\} = 0$.

The spectral measure Γ can also be used to express the dependence structure of $(X_1, \ldots, X_n)$. The X_js are independent if and only if Γ is concentrated on the $2n$ points $(0, \ldots, 0, \pm 1, 0, \ldots, 0)$ of the unit sphere S_n of $\mathbb{R}^n$. They are n-fold dependent if and only if

$$\Gamma\{\mathbf{s} \in S_n : s_1 \neq 0, s_2 \neq 0, \ldots, s_n \neq 0\} > 0.$$

Lemma 4.5.2 *Let $X_1, \ldots, X_n$ be jointly α-stable and n-fold dependent. Then $E|X_1|^{p_1} \cdots |X_n|^{p_n} < \infty$ if and only if $p_1 + \cdots + p_n < \alpha$.*

PROOF: *Necessity*. Suppose $E|X_1|^{p_1} \cdots |X_n|^{p_n} < \infty$ and let $A = \{x \in E : f_1(x) \cdots f_n(x) \neq 0\}$. For each $j = 1, \ldots, n$, define $Y_j = \int_A f_j(x) M(dx)$ and $Z_j = \int_{A^c} f_j(x) M(dx)$. The vectors $(Y_1, \ldots, Y_n)$ and $(Z_1, \ldots, Z_n)$ are independent because A and A^c are disjoint. Therefore the hypothesis

$$\infty > E|X_1|^{p_1} \cdots |X_n|^{p_n} = E|Y_1 + Z_1 + \eta_1|^{p_1} \cdots |Y_n + Z_n + \eta_n|^{p_n}$$

implies there are constants $z_1, \ldots, z_n$ such that

$$\begin{aligned}\infty &> E|Y_1 + z_1|^{p_1} \cdots |Y_n + z_n|^{p_n} \\ &= \int_0^\infty \cdots \int_0^\infty P(|Y_1 + z_1|^{p_1} > x_1, \ldots, |Y_n + z_n|^{p_n} > x_n) dx_1 \ldots dx_n \\ &\geq J(\lambda)\end{aligned}$$

for any $\lambda > 0$ where $J(\lambda)$ equals

$$\int_{(\frac{1}{2}\lambda)^{p_1}}^{\lambda^{p_1}} \cdots \int_{(\frac{1}{2}\lambda)^{p_n}}^{\lambda^{p_n}} P(|Y_1 + z_1|^{p_1} > x_1, \ldots, |Y_n + z_n|^{p_n} > x_n) dx_1 \ldots dx_n. \tag{4.5.1}$$

Since $J(\lambda)$ is the tail of a convergent integral, we have

$$\lim_{\lambda \to \infty} J(\lambda) = 0.$$

On the other hand, replacing each x_j in (4.5.1) by the upper bound λ^{p_j}, $j = 1, \ldots, n$, yields

$$\begin{aligned}J(\lambda) &\geq C\lambda^{p_1+\cdots+p_n} P(|Y_1 + z_1| > \lambda, \ldots, |Y_n + z_n| > \lambda) \\ &\geq C\lambda^{p_1+\cdots+p_n} P(|Y_1| > 2\lambda, \ldots, |Y_n| > 2\lambda)\end{aligned}$$

for some positive constant C and for all large enough λ. By Corollary 4.4.3, there is a positive constant $C_1 < C$, such that $J(\lambda) \geq C_1 2^{-\alpha} C_\alpha \lambda^{p_1+\cdots+p_n-\alpha} \int_A (\min_{k=1,\ldots,n} |f_k(x)|^\alpha) m(dx)$ for all large enough λ. The integral on the right-hand side is positive because n-fold dependence implies $m(A) > 0$. Since $J(\lambda) \to 0$, we must have $p_1 + \cdots + p_n < \alpha$.

Sufficiency. If $p_1 + \cdots + p_n < \alpha$, then, by the Hölder inequality,

$$E|X_1|^{p_1} \cdots |X_n|^{p_n} \leq \prod_{k=1}^n (E|X_k|^{p_1+\cdots+p_n})^{p_k/(p_1+\cdots+p_n)} < \infty.$$

This completes the proof. ∎

Remark. The assumption of n-fold dependence is not used to prove sufficiency.

Example 4.5.3 If X_1 and X_2 are jointly α-stable, $\alpha < 2$, and not independent, then they are 2-fold dependent. Consequently, $E|X_1|^{p_1}|X_2|^{p_2} < \infty$ if and only if $p_1 + p_2 < \alpha$.

We introduce now the notion of maximal subset in order to analyze the general case. Let $\{k_1, \ldots, k_m\}$ denote a subset of size m of $\{1, 2, \ldots, n\}$ and let

$$A_{k_1,\ldots,k_m} = \{x \in E : f_{k_1}(x) \ldots f_{k_m}(x) \neq 0\} \tag{4.5.2}$$

be the intersection of the supports of $f_{k_1}, \ldots, f_{k_m}$.

Definition 4.5.4 The subset $\{k_1, \ldots, k_m\}$ is *maximal* if $m\{A_{k_1,\ldots,k_m}\} > 0$ and $m\{A_{k_1,\ldots,k_m,i}\} = 0$ for any $i \notin \{k_1, \ldots, k_m\}$.

Example 4.5.5 If X_1, X_2, X_3 are as in Example 4.5.1, then $\{1,2\}, \{2,3\}$ and $\{3,1\}$ are the three maximal subsets of $\{1,2,3\}$.

Theorem 4.5.6 *Let* $\mathbf{X} = (X_1, \ldots, X_n)$ *be a non-degenerate* α*-stable random vector. Then* $E|X_1|^{p_1} \cdots |X_n|^{p_n} < \infty$ *if and only if* $p_{k_1} + \cdots + p_{k_m} < \alpha$ *for every maximal subset* $\{k_1, \ldots, k_m\}$ *of* $\{1, 2, \ldots, n\}$.

PROOF: We suppose without loss of generality that

$$(X_1, \ldots, X_n) = \left(\int_E f_1(x) M(dx), \ldots, \int_E f_n(x) M(dx) \right).$$

Necessity. Suppose $E|X_1|^{p_1} \cdots |X_n|^{p_n} < \infty$ and let $\{k_1, \ldots, k_m\}$ be any maximal subset of $\{1, 2, \ldots, n\}$. We want to show that $p_{k_1} + \cdots + p_{k_m} < \alpha$. Define

$$Y_j = \int_{A_{k_1,\ldots,k_m}} f_j(x) M(dx) \quad \text{and} \quad Z_j = \int_{A^c_{k_1,\ldots,k_m}} f_j(x) M(dx),$$

$j = 1, \ldots, n$, where $A_{k_1,\ldots,k_m}$ is given in (4.5.2). Since $\{k_1, \ldots, k_m\}$ is maximal, for every $j \notin \{k_1, \ldots, k_m\}$ we have $f_j(x) \equiv 0$ m-almost everywhere on $A_{k_1,\ldots,k_m}$ and, therefore, $Y_j = 0$ a.s. By hypothesis,

$$\begin{aligned} \infty &> E|X_1|^{p_1} \cdots |X_n|^{p_n} \\ &= E|Y_{k_1} + Z_{k_1}|^{p_{k_1}} \cdots |Y_{k_m} + Z_{k_m}|^{p_{k_m}} \prod_{j \notin \{k_1,\ldots,k_m\}} |Z_j|^{p_j}. \end{aligned}$$

The independence of $(Y_1, \ldots, Y_n)$ and $(Z_1, \ldots, Z_n)$ implies $E|Y_{k_1} + z_{k_1}|^{p_{k_1}} \cdots |Y_{k_m} + z_{k_m}|^{p_{k_m}} < \infty$. Since $m(A_{k_1,\ldots.k_m}) > 0$, the vector $(Y_{k_1}, \ldots, Y_{k_m})$ is m-fold dependent and therefore, by Lemma 4.5.2, we have $p_{k_1} + \cdots + p_{k_m} < \alpha$.

Sufficiency. For every subset $\{k_1, \ldots, k_m\}$ of $\{1, \ldots, n\}$ let

$$\begin{aligned} &B(\{k_1, \ldots, k_m\}) \\ &\quad = \{x \in E : f_{k_1}(x) \cdots f_{k_m}(x) \neq 0,\ f_i(x) = 0, \forall i \notin \{k_1, \ldots, k_m\}\}. \end{aligned}$$

Since the sets $\{B(\{k_1, \ldots, k_m\}) : \{k_1, \ldots, k_m\} \in 2^{\{1,\ldots,n\}},\ 0 \le m \le n\}$[2] partition $\bigcup_{k=1}^n \operatorname{supp}(f_k)$, we can write

[2] $\{k_1, \ldots, k_m\}$ is expressed here as an element of $2^{\{1,\ldots,n\}}$, the set of all subsets of $\{1, \ldots, n\}$.

$$X_j = \sum_{\{k_1,\ldots,k_m\}}{}' X_j(\{k_1,\ldots,k_m\}),\ j = 1.\ldots,n,$$

where

$$X_j(\{k_1,\ldots,k_m\}) = \int_{B(\{k_1,\ldots,k_m\})} f_j(x)M(dx)$$

and Σ' means that the summation is over all subsets $\{k_1,\ldots,k_m\}$ satisfying $m\{B(\{k_1,\ldots,k_m\})\} > 0$. (To simplify the notation, we shall write in the sequel $B(k_1,\ldots,k_m)$ and $X(k_1,\ldots,k_m)$.) Since, for some constant $C_{\alpha,n}$,

$$E|X_1|^{p_1}\cdots|X_n|^{p_n} \le C_{\alpha,n}E\prod_{r=1}^{n}\sum_{\{k_1^{(r)},\ldots,k_{m_r}^{(r)}\}}{}' |X_r(k_1^{(r)},\ldots,k_{m_r}^{(r)})|^{p_r},$$

it is enough to prove

$$E|X_1(k_1^{(1)},\ldots,k_{m_1}^{(1)})|^{p_1}\cdots|X_n(k_1^{(n)},\ldots,k_{m_n}^{(n)})|^{p_n} < \infty \tag{4.5.3}$$

for all subsets $\{k_1^{(r)},\ldots,k_{m_r}^{(r)}\}$ of $\{1,\ldots,n\}$ satisfying $m(B(k_1^{(r)},\ldots,k_{m_r}^{(r)})) > 0$, $r = 1,\ldots,n$. Some of the subsets $\{k_1^{(r)},\ldots,k_{m_r}^{(r)}\}$, $r = 1,\ldots,n$, may coincide. Let K denote the collection of all different such subsets. Observe that the cardinality of K is at most n. Now, for each $\{k_1,\ldots,k_m\} \in K$, let $I(k_1,\ldots,k_m)$ denote the set of all rs as above such that $\{k_1^{(r)},\ldots,k_{m_r}^{(r)}\} = \{k_1,\ldots,k_m\}$. Since $X_i(k_1^{(i)},\ldots,k_{m_i}^{(i)})$ and $X_j(k_1^{(j)},\ldots,k_{m_j}^{(j)})$ are independent, if $\{k_1^{(i)},\ldots,k_{m_i}^{(i)}\} \ne \{k_1^{(j)},\ldots,k_{m_j}^{(j)}\}$,

$$E|X_1(k_1^{(1)},\ldots,k_{m_1}^{(1)})|^{p_1}\cdots|X_n(k_1^{(n)},\ldots,k_{m_n}^{(n)})|^{p_n}$$
$$= \prod_{\{k_1,\ldots,k_m\}\in K} E \prod_{j\in I(k_1,\ldots,k_m)} |X_j(k_1,\ldots,k_m)|^{p_j}.$$

It is therefore enough to prove that

$$E\prod_{j\in I(k_1,\ldots,k_m)}|X_j(k_1,\ldots,k_m)|^{p_j} < \infty \tag{4.5.4}$$

for any $\{k_1,\ldots,k_m\}$ satisfying $m(B(k_1,\ldots,k_m)) > 0$. Clearly, if $j \notin \{k_1,\ldots,k_m\}$, then $X_j(k_1,\ldots,k_m) = 0$, so we may assume in (4.5.4) that $j \in \{k_1,\ldots,k_m\}$ for every $j \in I(k_1,\ldots,k_m)$. In that case, the random variables $\{X_j(k_1,\ldots,k_m),\ j \in I(k_1,\ldots,k_m)\}$ are m-fold dependent. Since $\{k_1,\ldots,k_m\}$ is a subset of some maximal subset $\{k_1,\ldots,k_m,k_{m+1},\ldots,k_{m'}\}$ of $\{1,\ldots,n\}$, we have

$$\sum_{j\in I(k_1,\ldots,k_m)} p_j \le p_{k_1}+\cdots+p_{k_m}+p_{k_{m+1}}+\cdots+p_{k_{m'}} < \alpha$$

by the sufficiency hypothesis. Relation (4.5.4) therefore follows from Lemma 4.5.2. ∎

Example 4.5.7 If X_1, X_2, X_3 are as in Example 4.5.1, the three maximal subsets are $\{1,2\}, \{2,3\}$ and $\{3,1\}$ and therefore $E|X_1|^{p_1}|X_2|^{p_2}|X_3|^{p_3} < \infty$ if and only if $p_1 + p_2 < \alpha,\ p_2 + p_3 < \alpha$ and $p_3 + p_1 < \alpha$.

Remark. No inequality in the system

$$\Big\{p_{k_1} + \cdots + p_{k_m} < \alpha, \{k_1, \ldots, k_m\} \text{ is a maximal subset of } \{1, 2, \ldots, n\},\ 1 \le m \le n\Big\}$$

is redundant in the sense that no inequality is implied by the others. To verify this, consider a particular inequality, say $p_{k_1} + \cdots + p_{k_m} < \alpha$. The choice of values $p_{k_1} = \cdots = p_{k_m} = \frac{\alpha}{m},\ p_j = 0$ for $j \neq k_1, \ldots, k_m$ violates this inequality, but it violates no other in the system because $\{k_1, \ldots, k_m\}$ is a maximal subset and hence no other inequality in the system can include more than $m - 1$ of the p_is, $i = k_1, k_2, \ldots, k_m$.

4.6 Association of stable random variables

Random variables $X_1, \ldots, X_n$ are said to be *associated* if, for any functions $f, g : \mathbb{R}^n \to \mathbb{R}$ which are non-decreasing in each argument, one has

$$\text{Cov}\,\big(f(X_1, \ldots, X_n), g(X_1, \ldots, X_n)\big) \geq 0,$$

whenever the covariance exists. The following theorem gives necessary and sufficient conditions for α-stable random variables with $0 < \alpha < 2$ to be associated.

Theorem 4.6.1 *Let $X_1, \ldots, X_n$ be jointly α-stable random variables, $0 < \alpha < 2$, with spectral measure Γ on the unit sphere S_n of $\mathbb{R}^n$. Then $X_1, \ldots, X_n$ are associated if and only if the spectral measure Γ satisfies the condition*

$$\Gamma(S_n^-) = 0, \tag{4.6.1}$$

where

$$S_n^- = \Big\{(s_1, \ldots, s_n) \in S_n : \text{ for some } i, j \in \{1, \ldots, n\},\ s_i > 0 \text{ and } s_j < 0\Big\}.$$

PROOF: *Necessity.* Suppose to the contrary that the $X_1, \dots, X_n$ are associated and $\Gamma(S_n^-) > 0$. There then exist $i, j \in \{1, \dots, n\}$ such that

$$\Gamma\{(s_1, \dots, s_n) \in S_n : s_i > 0,\ s_j < 0\} > 0. \tag{4.6.2}$$

Without loss of generality we assume that $i = 1,\ j = 2$. Proposition 2.3.9 together with (4.6.2) imply that X_1, X_2 are jointly α-stable in $\mathbb{R}^2$ and the spectral measure $\Gamma_{1,2}$ of (X_1, X_2) satisfies

$$\Gamma_{1,2}\{(s_1, s_2) \in S_2 : s_1 > 0,\ s_2 < 0\} > 0. \tag{4.6.3}$$

Moreover X_1, X_2 are associated, being a subset of a set of associated random variables $\{X_1, \dots, X_n\}$. For any $\lambda > 0$, the association of X_1 and X_2 implies (see Exercise 4.23)

$$P(X_1 > \lambda, X_2 > -\lambda) \geq P(X_1 > \lambda)P(X_2 > -\lambda)$$

and hence

$$P(X_2 \leq -\lambda \,|\, X_1 > \lambda) \leq P(X_2 \leq -\lambda). \tag{4.6.4}$$

Clearly, the right-hand side of (4.6.4) goes to zero as λ goes to infinity. Moreover, as X_1 and $-X_2$ are also jointly α-stable, with spectral measure $\overline{\Gamma}_{1,2}$ defined by

$$\overline{\Gamma}_{1,2}(A) = \Gamma_{1,2}\{(s_1, s_2) \in S_2 : (s_1, -s_2) \in A\},$$

we conclude from (4.4.2) and (4.6.3) that

$$\begin{aligned}
&\lim_{\lambda\to\infty} P(X_2 \leq -\lambda \,|\, X_1 > \lambda) \\
&\quad = \lim_{\lambda\to\infty} P(-X_2 \geq \lambda \,|\, X_1 > \lambda) \\
&\quad = \frac{\int_{S_2} (\min([s_1]_+, [s_2]_+))^\alpha \overline{\Gamma}_{1,2}(ds)}{\int_{S_2} ([s_1]_+)^\alpha \overline{\Gamma}_{1,2}(ds)} \\
&\quad = \frac{\int_{\{s_1>0, s_2<0\}} (\min(s_1, -s_2))^\alpha \Gamma_{1,2}(ds)}{\int_{S_2} ([s_1]_+)^\alpha \Gamma_{1,2}(ds)} > 0.
\end{aligned}$$

The contradiction thus obtained proves the necessity part of the theorem.

Sufficiency. Suppose that (4.6.1) holds. Then the spectral measure Γ is concentrated on $S_n^{\text{pos}} \cup S_n^{\text{neg}}$, where

$$\begin{aligned}
S_n^{\text{pos}} &= \{(s_1, \dots, s_n) \in S_n :\ s_i \geq 0,\ i = 1, \dots, n\}, \\
S_n^{\text{neg}} &= \{(s_1, \dots, s_n) \in S_n :\ s_i \leq 0,\ i = 1, \dots, n\}.
\end{aligned}$$

We may and will assume until the end of the proof that the shift vector $\boldsymbol{\mu}^0$ in the characteristic functions of $(X_1, \dots, X_n)$ and all other α-stable random vectors defined later is zero, since subtracting the shift does not affect the association property.

Let $\mathbf{X}^+ = (X_1^+, \ldots, X_n^+)$ and $\mathbf{X}^- = (X_1^-, \ldots, X_n^-)$ be two independent α-stable random vectors in $\mathbb{R}^n$ with spectral measures Γ^+ and Γ^-, respectively given by

$$\Gamma^+(A) = \Gamma(A \cap S_n^{\text{pos}}), \quad \Gamma^-(A) = \Gamma((-A) \cap S_n^{\text{neg}})$$

for any Borel set A on S_n. Clearly,

$$(X_1^+, \ldots, X_n^+) - (X_1^-, \ldots, X_n^-) \stackrel{d}{=} (X_1, \ldots, X_n),$$

so that it is enough to prove both that $X_1^+, \ldots, X_n^+$ are associated and $X_1^-, \ldots, X_n^-$ are associated. Since both Γ^+ and Γ^- are concentrated on S_n^{pos}, we assume, to simplify the notation, that the spectral measure Γ of the α-stable vector $(X_1, \ldots, X_n)$ is concentrated on S_n^{pos}, and prove that the random variables $X_1, \ldots, X_n$ are associated.

Let M be an α-stable random measure on S_n^{pos} with control measure Γ and skewness intensity 1. Then $(X_1, \ldots, X_n) \stackrel{d}{=} (Y_1, \ldots, Y_n)$, where

$$Y_i = \int_{S_n^{\text{pos}}} s_i M(d\mathbf{s}), \; i = 1, \ldots, n.$$

By the definition of the integral, for each $i = 1, \ldots, n$, there is a sequence of simple Borel functions from S_n^{pos} to $\mathbb{R}$, denoted $\{f_m^{(i)}, \; m = 1, 2, \ldots\}$, such that $Y_m^{(i)}$ tends to Y_i in probability as $m \to \infty$ where $Y_m^{(i)} = \int_{S_n^{\text{pos}}} f_m^{(i)}(\mathbf{s}) M(d\mathbf{s})$, $m = 1, 2, \ldots$. Clearly, we may always choose the functions $f_m^{(i)}$ such that $f_m^{(i)}(\mathbf{s}) \geq 0$ for every m, i, and $\mathbf{s}$. Suppose that

$$f_m^{(i)}(\mathbf{s}) = \sum_{j=1}^{K_{m,i}} f_j^{(m,i)} 1(\mathbf{s} \in A_j^{(m,i)}), \; i = 1, \ldots, n, \; m = 1, 2, \ldots,$$

for some $f_j^{(m,i)} \geq 0, \quad i = 1, \ldots, n, \quad m = 1, 2, \ldots, \quad j = 1, \ldots, K_{m,i}$, where $A_1^{(m,i)}, \ldots, A_{K_{m,i}}^{(m,i)}$ is a partition of S_n^{pos} into Borel sets. Since the measure M is independently scattered, for each fixed $m = 1, 2, \ldots,$ the collection

$$\mathcal{C}_m = \{M(A_{j_1}^{(m,1)} \cap A_{j_2}^{(m,2)} \cap \ldots \cap A_{j_n}^{(m,n)}), \\ \text{where } j_1 = 1, \ldots, K_{m,1}; \ldots; j_n = 1, \ldots, K_{m,n}\}$$

is a collection of independent, and therefore associated, random variables. Since

$$Y_m^{(i)} = \sum_{j=1}^{K_{m,i}} f_j^{(m,i)} M(A_j^{(m,i)}), \; i = 1, \ldots, n,$$

we conclude that each $Y_m^{(i)}$ is a non-decreasing function of random variables from $\mathcal{C}_m$. Therefore $Y_m^{(1)}, \dots, Y_m^{(n)}$ are associated for any $m = 1, 2, \dots$ (see Exercise 4.25). But, as $m \to \infty$, $(Y_m^{(1)}, \dots, Y_m^{(n)})$ tends to $(Y_1, \dots, Y_n)$ in probability, and association is preserved under convergence in distribution. Therefore, $Y_1, \dots, Y_n$ are associated, which completes the proof of the theorem. ■

As a by-product of the proof of Theorem 4.6.1 we obtain an equivalence of several different types of positive dependence for jointly α-stable random variables, $0 < \alpha < 2$. Note that in the Gaussian case such equivalence follows directly from Pitt (1982).

Following the terminology of Brindley and Thompson (1972), we call random variables $X_1, \dots, X_n$ *$\overline{F}$-positive orthant dependent* ($\overline{\text{F}}$POD) if

$$P(X_1 > x_1, \dots, X_n > x_n) \geq P(X_1 > x_1) \cdots P(X_n > x_n)$$

for any $x_1, \dots, x_n$, and we call them *F-positive orthant dependent* (FPOD) if

$$P(X_1 \leq x_1, \dots, X_n \leq x_n) \geq P(X_1 \leq x_1) \cdots P(X_n \leq x_n)$$

for any $x_1, \dots, x_n$. Clearly, association implies $\overline{\text{F}}$POD and FPOD and, in general, these implications cannot be reversed (see Exercise 4.26).

Corollary 4.6.2 *Let $X_1, \dots, X_n$ be jointly α-stable. Then all the notions of positive dependence above are equivalent, and each of them is equivalent to (4.6.1).*

PROOF: By Theorem 4.6.1, it is enough to prove that each of $\overline{\text{F}}$POD and FPOD implies (4.6.1). The fact that $\overline{\text{F}}$POD implies (4.6.1) follows directly from the proof of the necessity part of Theorem 4.6.1 (in that proof, we did not use the full force of the association property, but only its consequence, the $\overline{\text{F}}$POD property.) The same argument applied to the vector $(-X_1, \dots, -X_n)$ shows that FPOD implies (4.6.1) as well. ■

Finally, we give necessary and sufficient conditions for the negative association of jointly α-stable random variables. We cannot simply reverse the definition of association because no non-degenerate random vector $\mathbf{X}$ can satisfy $\text{Cov}(f(\mathbf{X}), g(\mathbf{X})) \leq 0$ for all non-decreasing function f and g (to see this, take $f = g$). Following Alam and Saxena (1982), we call random variables $X_1, \dots, X_n$ *negatively associated* if, for any $1 < k < n$, any $f : \mathbb{R}^k \to \mathbb{R}$, $g : \mathbb{R}^{n-k} \to \mathbb{R}$, non-decreasing in each argument,

$$\text{Cov}\,(f(\mathbf{Y}), g(\mathbf{Z})) \leq 0$$

whenever the covariance exists, where $\mathbf{Y}$ and $\mathbf{Z}$ are any k- and $(n-k)$-dimensional random vectors, representing a partition of the set $(X_1, \ldots, X_n)$ into two subsets of sizes k and $n-k$, respectively.

Theorem 4.6.3 *Let $X_1, \ldots, X_n$ be jointly α-stable random variables, $0 < \alpha < 2$, with spectral measure Γ on the unit sphere S_n of $\mathbb{R}^n$. Then $X_1, \ldots, X_n$ are negatively associated if and only if the spectral measure Γ satisfies the condition*

$$\Gamma(S_n^+) = 0, \tag{4.6.5}$$

where $S_n^+ = \{(s_1, \ldots, s_n) \in S_n : \text{ for some } i \neq j,\ s_i s_j > 0\}$.

PROOF: The proof of the necessity is identical to that in Theorem 4.6.1. To establish sufficiency, note that (4.6.5) implies that Γ-almost every vector $\mathbf{s} = (s_1, \ldots, s_n) \in S_n^+$ has at most two of its coordinates different from zero. Thus, if $E_{i,j}$ is the subset of S_n, where only s_i and s_j are different from zero, and if D_i is the subset of S_n where only s_i is different from zero, then

$$(X_1, \ldots, X_n) \stackrel{d}{=} \sum_{i=1}^{n-1} \sum_{j=i+1}^{n} (X_1(i,j), \ldots, X_n(i,j)) + (W_1, \ldots, W_n), \tag{4.6.6}$$

where, using the notation introduced in the proof of Theorem 4.6.1,

$$X_k(i,j) = \int_{E_{i,j}} s_k M(ds),\ k = 1, \ldots, n;\ i = 1, \ldots, n-1;\ j = i+1, \ldots, n,$$

and

$$W_k = \int_{D_k} s_k M(ds),\ k = 1, \ldots, n.$$

Note that the random vectors appearing in the right-hand side of (4.6.6) are all independent, so that we will be done once we prove that each one of these random vectors consists of negatively associated random variables. The latter is trivially true for the random vector $(W_1, \ldots, W_n)$, which consists of independent random variables. Now, for each fixed (i,j), $X_k(i,j) = 0$ a.s. if $k \notin \{i,j\}$, which reduces our task to showing that $X_i(i,j)$ and $X_j(i,j)$ are negatively associated. But (4.6.5) implies $s_i s_j \leq 0$ Γ-a.e. on $E_{i,j}$, so that, by Theorem 4.6.1, $X_i(i,j)$ and $-X_j(i,j)$ are associated. Thus (see Exercise 4.27), $X_i(i,j)$ and $X_j(i,j)$ are negatively associated. The proof of the theorem is now complete. ∎

4.7 The codifference for stationary $S\alpha S$ processes

Let $X = \{X(t) = \int_E f(t,x) M(dx),\ t \in \mathbb{R}\}$ be a (strictly) stationary $S\alpha S$ process, i.e., for any $d \geq 1,\ t_1, \ldots, t_d \in \mathbb{R}$, the finite-dimensional distribution of the vector $(X(t_1+h),\ldots,X(t_d+h))$ does not depend on the choice of $h\in\mathbb{R}$. When the process X is Gaussian, the autocovariance function $EX(t)X(0)$ totally describes the dependence. When X is $S\alpha S$ with $0<\alpha<2$, covariances do not exist. One could use the covariation but this quantity is not always defined for $\alpha\le 1$. Here, we shall use the *codifference* defined in Section 2.10. Since X is a stationary process, the codifference between $X(t)$ and $X(0)$ equals

$$\begin{aligned}\tau(t) &= \tau_{X(t),X(0)}\\ &= 2\|X(0)\|_\alpha^\alpha-\|X(t)-X(0)\|_\alpha^\alpha\\ &= 2\|f(0,\cdot)\|_\alpha^\alpha-\|f(t,\cdot)-f(0,\cdot)\|_\alpha^\alpha,\end{aligned}\tag{4.7.1}$$

where $\|f(t,\cdot)\|_\alpha^\alpha=\int_E|f(t,x)|^\alpha m(dx)$ and where m denotes the control measure of M.

When comparing codifferences (see Property 2.10.6), it is convenient to standardize the process (i.e., consider $X(t)/\|X(0)\|_\alpha$) or, equivalently, to consider the standardized codifference

$$\frac{\tau(t)}{\|X(0)\|_\alpha^\alpha}=2-\frac{\|X(t)-X(0)\|_\alpha^\alpha}{\|X(0)\|_\alpha^\alpha}.$$

Example 4.7.1 We shall compute the codifference $\tau(t)$ for the *Ornstein-Uhlenbeck* process

$$X(t)=\int_{-\infty}^t e^{-\lambda(t-x)}M(dx),\ -\infty<t<\infty,$$

defined in Example 3.6.3. Recall that $\lambda>0$ and M is a $S\alpha S$ random measure with Lebesgue control measure. We have $\|X(0)\|_\alpha^\alpha=\int_{-\infty}^0 e^{\alpha\lambda x}dx=\frac{1}{\alpha\lambda}$ and, for $t>0$,

$$\begin{aligned}\|X(t)-X(0)\|_\alpha^\alpha &= \int_{-\infty}^t e^{\alpha\lambda x}|e^{-\lambda t}-1(x<0)|^\alpha dx=\int_{-\infty}^0+\int_0^t\\ &= \frac{1}{\alpha\lambda}(1-e^{-\lambda t})^\alpha+\frac{1}{\alpha\lambda}(1-e^{-\alpha\lambda t}),\end{aligned}$$

so that

$$\tau(t)=\frac{1}{\alpha\lambda}[1-(1-e^{-\lambda t})^\alpha+e^{-\alpha\lambda t}],\ t>0.$$

The standardized codifference equals

$$\frac{\tau(t)}{\|X(0)\|_\alpha^\alpha}=1-(1-e^{-\lambda t})^\alpha+e^{-\alpha\lambda t},\ t>0.$$

In particular,

$$\tau(t)=\frac{1}{\lambda}e^{-\lambda t},\ \text{if}\ \alpha=2$$

(which is the autocovariance function of the Gaussian Ornstein-Uhlenbeck process),

$$\tau(t) = \frac{2}{\lambda} e^{-\lambda t}, \text{ if } \alpha = 1$$

(this is the codifference of the Cauchy Ornstein-Uhlenbeck process) and, as $t \to \infty$,

$$\tau(t) \sim \begin{cases} \frac{1}{\lambda} e^{-\lambda t} & \text{if } 1 < \alpha \leq 2, \\ \frac{2}{\lambda} e^{-\lambda t} & \text{if } \alpha = 1, \\ \frac{1}{\alpha\lambda} e^{-\alpha\lambda t} & \text{if } 0 < \alpha < 1. \end{cases}$$

The standardized codifference $\tau(t)/\|X(0)\|_\alpha^\alpha = \alpha\lambda\tau(t)$ equals $2e^{-\lambda t}$ for $\alpha = 2$ and $\alpha = 1$, and as $t \to \infty$,

$$\frac{\tau(t)}{\|X(0)\|_\alpha^\alpha} \sim \begin{cases} \alpha e^{-\lambda t} & \text{if } 1 < \alpha < 2, \\ 2e^{-\lambda t} & \text{if } \alpha = 1, \\ e^{-\alpha\lambda t} & \text{if } 0 < \alpha < 1. \end{cases}$$

As a function of α, the asymptotic expression for the standardized codifference decreases as α goes from 0 to 1, is discontinuous at $\alpha = 1$, and then increases as α goes from 1 to 2. But, for any fixed $0 < \alpha \leq 2$, and for all $t > 0$,

$$\tau_\lambda(t) < \tau_{\lambda'}(t) \text{ if } \lambda > \lambda' > 0,$$

where $\tau_\lambda(t)$ and $\tau_{\lambda'}(t)$ are the τs that correspond to the standardized $S\alpha S$ Ornstein-Uhlenbeck processes X_λ and $X_{\lambda'}$ with parameters λ and λ', respectively. Therefore, by Property 2.10.6, if $\lambda > \lambda' > 0$,

$$P\{|X_\lambda(t) - X_\lambda(0)| > c\} \geq P\{|X_{\lambda'}(t) - X_{\lambda'}(0)| > c\} \text{ for all } c > 0,$$

i.e., $(X_{\lambda'}(t),\ X_{\lambda'}(0))$ is "more dependent" than $(X_\lambda(t),\ X_\lambda(0))$ for all $t > 0$.

Samorodnitsky and Taqqu (1994) show that the following relation holds for standardized $S\alpha S$ Ornstein-Uhlenbeck processes $\{X_\lambda(t),\ t \in \mathbb{R}\}$ and $\{X_{\lambda'}(t),\ t \in \mathbb{R}\}$ with $0 < \alpha < 2$: if $\lambda > \lambda' > 0$, then, for all $t_1, \ldots, t_d \in \mathbb{R}$ and all real numbers $c_1, \ldots, c_d$,

$$P\{X_\lambda(t_1) > c_1, \ldots, X_\lambda(t_d) > c_d\} \leq P\{X_{\lambda'}(t_1) > c_1, \ldots, X_{\lambda'}(t_d) > c_d\}. \tag{4.7.2}$$

This last relation has the following interpretation: the probability that the components of X_λ are all large is not greater than the corresponding probability for $X_{\lambda'}$. Hence, the components of $X_{\lambda'}$ are "more dependent" than those of X_λ.

The inequality (4.7.2) holds also for $\alpha = 2$, but in that case, it follows from the Slepian conditions $EX_\lambda^2(t_i) = EX_{\lambda'}^2(t_i)$ for every $i = 1,\dots,d$ and $EX_\lambda(t_i)X_\lambda(t_j) \le EX_{\lambda'}(t_i)X_{\lambda'}(t_j)$ for every $i,j = 1,\dots,d$ (see, for example, Adler (1990) or Ledoux and Talagrand (1991)). The Slepian conditions hold for X_λ and $X_{\lambda'}$ because these processes are standardized and $EX_\lambda(t_i)X_\lambda(t_j) = 2\exp\{-\lambda|t_i - t_j|\}$.

We shall be mostly interested in the asymptotic behavior of $\tau(t)$ as the lag t tends to infinity. We saw in the previous example that the codifference $\tau(t)$ for the Ornstein-Uhlenbeck process tends to zero as $t \to \infty$. This, in fact, always happens if the process is a $S\alpha S$ stationary moving average, i.e., a process of the form

$$X(t) = \int_{-\infty}^{\infty} f(t-x)M(dx), \quad t \in \mathbb{R},$$

where M is a $S\alpha S$ random measure with Lebesgue control measure. To prove it, we need the following lemma.

Lemma 4.7.2 *Let $f, g \in L^\alpha(E, \mathcal{E}, m)$, $0 < \alpha \le 2$. Then*

$$\int_E \Big| |f(x)|^\alpha - |g(x)|^\alpha \Big| m(dx) \le C \left(\int_E |f(x) - g(x)|^\alpha m(dx) \right)^{1 \wedge (1/\alpha)} \tag{4.7.3}$$

with

$$C = \begin{cases} 1 & \text{if } 0 < \alpha \le 1, \\ 2^{1/\alpha}\alpha(\|f\|_\alpha^{\alpha-1} + \|g\|_\alpha^{\alpha-1}) & \text{if } 1 < \alpha \le 2. \end{cases}$$

PROOF: Suppose, firstly, $0 < \alpha \le 1$. Then $|x+y|^\alpha \le |x|^\alpha + |y|^\alpha$, and thus $|x|^\alpha = |x-y+y|^\alpha \le |x-y|^\alpha + |y|^\alpha$, yielding $||x|^\alpha - |y|^\alpha| \le |x-y|^\alpha$. Hence, $\int_E ||f|^\alpha - |g|^\alpha| dm \le \int_E |f-g|^\alpha dm$, and so (4.7.3) holds with $C = 1$.

Suppose, now, $1 < \alpha \le 2$. The mean value theorem applied to $|x|^\alpha$ gives $||x|^\alpha - |y|^\alpha| \le \alpha|x-y|||x|^{\alpha-1} + |y|^{\alpha-1}|$. Hence, using the Hölder inequality (with conjugate exponents α and $\alpha/(\alpha-1)$) and (2.7.6), we obtain

$$\begin{aligned} &\int_E ||f|^\alpha - |g|^\alpha| dm \\ &\le \alpha \int_E |f-g|(|f|^{\alpha-1} + |g|^{\alpha-1}) dm \\ &\le \alpha\Big(\int_E |f-g|^\alpha dm\Big)^{1/\alpha} \Big(\int_E (|f|^{\alpha-1} + |g|^{\alpha-1})^{\frac{\alpha}{\alpha-1}} dm\Big)^{\frac{\alpha-1}{\alpha}} \\ &\le \alpha\Big(\int_E |f-g|^\alpha dm\Big)^{1/\alpha} \Big[2^{1/(\alpha-1)}\Big(\int_E |f|^\alpha dm + \int_E |g|^\alpha dm\Big)\Big]^{\frac{\alpha-1}{\alpha}} \end{aligned}$$

$$\leq \left(\int_E |f-g|^\alpha dm\right)^{1/\alpha} 2^{1/\alpha}\alpha(\|f\|_\alpha^{\alpha-1}+\|g\|_\alpha^{\alpha-1}). \quad \blacksquare$$

Theorem 4.7.3 *For a $S\alpha S, 0<\alpha\leq 2$, stationary moving average process,*

$$\lim_{t\to\infty} \tau(t) = 0.$$

PROOF: The moving average X is well defined if and only if $\|X(0)\|_\alpha^\alpha = \int_{-\infty}^{\infty}|f(x)|^\alpha dx < \infty$. Suppose, firstly, that f has compact support. Since, for large enough t, the supports of $f(t+\cdot)$ and $f(\cdot)$ are disjoint,

$$\begin{aligned}\lim_{t\to\infty}\|X(t)-X(0)\|_\alpha^\alpha &= \lim_{t\to\infty}\int_{-\infty}^{\infty}|f(t+x)-f(x)|^\alpha dx\\ &= \lim_{t\to\infty}\int_{-\infty}^{\infty}|f(t+x)|^\alpha dx+\int_{-\infty}^{\infty}|f(x)|^\alpha dx\\ &= 2\|X(0)\|_\alpha^\alpha\end{aligned}$$

and hence $\lim_{t\to\infty}\tau(t)=0$.

If f does not have compact support, let $\epsilon>0$ be arbitrary and choose a bounded interval K_ϵ such that $\int_{-\infty}^{\infty}|f(x)|^\alpha 1_{K_\epsilon^c}(x)dx<\epsilon$, where $1_{K_\epsilon^c}$ is the indicator function of the complement of K_ϵ. Now choose t large enough so that the functions $f(\cdot)1_{K_\epsilon}(\cdot)$ and $f(t+\cdot)1_{K_\epsilon}(t+\cdot)$ have disjoint support. For such t, $\tau_\epsilon(t)=0$, where τ_ϵ is τ with f replaced by $f1_{K_\epsilon}$. Moreover, letting $f_t(\cdot)=f(t+\cdot)$ and $(f1_{K_\epsilon})_t(\cdot)=f(t+\cdot)1_{K_\epsilon}(t+\cdot)$, we see by Lemma 4.7.2, that

$$\begin{aligned}&|\tau(t)| = |\tau(t)-\tau_\epsilon(t)|\\ &\leq 2\int_{-\infty}^{\infty}\Big||f|^\alpha-|f1_{K_\epsilon}|^\alpha\Big|dx+\int_{-\infty}^{\infty}\Big||f_t-f|^\alpha-|(f1_{K_\epsilon})_t-f1_{K_\epsilon}|^\alpha\Big|dx\\ &\leq 2\int_{-\infty}^{\infty}|f1_{K_\epsilon^c}|^\alpha dx+\Big[C\int_{-\infty}^{\infty}\Big|f_t-(f1_{K_\epsilon})_t-(f-f1_{K_\epsilon})\Big|^\alpha dx\Big]^{1\wedge(1/\alpha)}\\ &\leq 2\int_{-\infty}^{\infty}|f1_{K_\epsilon^c}|^\alpha dx+\Big[2C\int_{-\infty}^{\infty}|(f1_{K_\epsilon^c})_t|^\alpha dx+2C\int_{-\infty}^{\infty}|f1_{K_\epsilon}|^\alpha dx\Big]^{1\wedge(1/\alpha)}\\ &\leq 2\epsilon+(4C\epsilon)^{1\wedge(1/\alpha)}. \quad \blacksquare\end{aligned}$$

In the literature one can also find two quantities, parametrized by $-\infty<\theta_1,\theta_2<\infty$, which are closely related to the codifference $\tau(t)$. The first is

$$U(\theta_1,\theta_2;t) = Ee^{i(\theta_1X(t)+\theta_2X(0))}-Ee^{i\theta_1X(t)}Ee^{i\theta_2X(0)}, \qquad (4.7.4)$$

i.e., the joint characteristic function of $(X(t), X(0))$ minus the product of the marginal characteristic functions. The second is

$$I(\theta_1, \theta_2; t) = -\ln Ee^{i(\theta_1 X(t)+\theta_2 X(0))} + \ln Ee^{i\theta_1 X(t)} + \ln Ee^{i\theta_2 X(0)}, \quad (4.7.5)$$

which is related to $U(\theta_1, \theta_2; t)$ by

$$U(\theta_1, \theta_2; t) = K(\theta_1, \theta_2)\left(e^{-I(\theta_1,\theta_2;t)} - 1\right),$$

where $K(\theta_1, \theta_2) = (E \exp i\theta_1 X(t))(E \exp i\theta_2 X(0))$ does not depend on t since X is stationary. If $I(\theta_1, \theta_2; t) \to 0$ as $t \to \infty$, then

$$U(\theta_1, \theta_2; t) \sim -K(\theta_1, \theta_2) I(\theta_1, \theta_2; t), \quad (4.7.6)$$

i.e., U and I are asymptotically proportional.

Both U and I can be used if the process is skewed α-stable. If the process is $S\alpha S$, then

$$\tau(t) = -I(1, -1; t). \quad (4.7.7)$$

Remarks

1. If $\lim_{t\to\infty} I(\theta_1, \theta_2; t) = 0$ for any θ_1 and θ_2, then the process $\{X(t),\ t \in \mathbb{R}\}$ is *mixing*, that is,

$$\lim_{t\to\infty} \left| P(A \cap B) - P(A)P(B) \right| = 0$$

for every $A \in \sigma\{X(s),\ s \le 0\}$ and $B \in \sigma\{X(s),\ s \ge t\}$ (cf. Maruyama (1970) and Gross (1993)).

2. Gross (1993) also shows that for stationary $S\alpha S$ stable processes satisfying Condition S, the relation "$\lim_{t\to\infty} I(\theta_1, \theta_2; t) = 0$ for all θ_1 and θ_2" (and thus that $\{X(t),\ t \in \mathbb{R}\}$ is mixing) follows from $\lim_{t\to\infty} \tau(t) = 0$ when $0 < \alpha \le 1$, and from $\lim_{t\to\infty} \tau(t) = 0$ and $\lim_{t\to\infty} I(1, 1; t) = 0$ when $1 < \alpha < 2$.

The advantage of I over τ stems from the presence of the parameters θ_1 and θ_2. Consider, for example, two standardized stationary $S\alpha S$ stochastic processes $X = \{X(t),\ t \in \mathbb{R}\}$ and $X' = \{X'(t),\ t \in \mathbb{R}\}$ and let τ and I relate to X and τ' and I' relate to X'. One way of showing that X and X' are different processes is to show that $\tau(t)$ is not asymptotic to $\tau'(t)$ as $t \to \infty$. If $\tau(t)$ is asymptotic to $\tau'(t)$, one can then turn to I and I': the processes X and X' are different if $I(\theta_1, \theta_2; t)$ is not asymptotic to $I'(\theta_1, \theta_2; t)$ as $t \to \infty$ for some θ_1 and θ_2. We now illustrate this technique.

Suppose that $\{X(t),\ t \in \mathbb{R}\}$ is a stationary sub-Gaussian process, i.e.,

$$X(t) = A^{1/2}G(t), \quad -\infty < t < \infty,$$

where $\{G(t),\ -\infty < t < \infty\}$ is a stationary Gaussian process with variance $\sigma_0^2 = \text{Var}\, G(0)$ and autocorrelation function $\rho(t) = EG(t)G(0)/\sigma_0^2$. The vector A is independent of the process G and is distributed as in Relation (2.5.1). (Sub-Gaussian processes are defined in Section 3.7. See also Section 2.5 where sub-Gaussian random vectors are introduced.) Since

$$Ee^{i(\theta_1 X(t)+\theta_2 X(0))} = \exp\left\{-\left|\frac{1}{2}\text{Var}(\theta_1 G(t) + \theta_2 G(0))\right|^{\alpha/2}\right\}$$

(see (2.5.4)), we have

$$\|\theta_j X(0)\|_\alpha^\alpha = -\left|\frac{1}{2}\text{Var}(\theta_j G(t))\right|^{\alpha/2} = \left(\frac{\sigma_0^2}{2}\right)^{\alpha/2} |\theta_j|^\alpha,\ j = 1, 2,$$

and

$$\|\theta_1 X(t) + \theta_2 X(0)\|_\alpha^\alpha = \left(\frac{\sigma_0^2}{2}\right)^{\alpha/2} (\theta_1^2 + \theta_2^2 + 2\theta_1\theta_2\rho(t))^{\alpha/2}.$$

Hence,

$$I(\theta_1, \theta_2; t) = \left(\frac{\sigma_0^2}{2}\right)^{\alpha/2} \left[(\theta_1^2 + \theta_2^2 + 2\theta_1\theta_2\rho(t))^{\alpha/2} - |\theta_1|^\alpha - |\theta_2|^\alpha\right]. \quad (4.7.8)$$

Setting $\tau(t) = -I(1, -1; t)$, we obtain

Proposition 4.7.4 *For the stationary sub-Gaussian process* $\{X(t) = A^{1/2}G(t),\ t \in \mathbb{R}\}$,

$$\tau(t) = \left(\frac{\sigma_0^2}{2}\right)^{\alpha/2} \left[2 - (2 - 2\rho(t))^{\alpha/2}\right],\ t \in \mathbb{R}. \quad (4.7.9)$$

If, in addition, $\lim_{t\to\infty} \rho(t) = 0$, *then*

$$\lim_{t\to\infty} \tau(t) = \left(\frac{\sigma_0^2}{2}\right)^{\alpha/2} (2 - 2^{\alpha/2}) > 0 \ \textit{for}\ 0 < \alpha < 2.$$

Let us now prove that stationary $S\alpha S$ moving averages are different from stationary $S\alpha S$ sub-Gaussian processes when $0 < \alpha < 2$. Let $X(\cdot) = A^{1/2}G(\cdot)$ denote a stationary $S\alpha S$ sub-Gaussian process and let $Y(\cdot)$ be a stationary $S\alpha S$ moving average. Let $\rho(t)$ be the corelation of G. Since $\lim_{t\to\infty} \tau_Y(t) = 0$ (Theorem 4.7.3), it is sufficient to show that $\tau_X(t)$ (given in (4.7.9)) does not converge to 0 as $t \to \infty$. Suppose, to the contrary, that $\lim_{t\to\infty} \tau_X(t) = 0$. Then $\lim_{t\to\infty} \rho(t) = \rho_\infty$ where $2-(2-\rho_\infty)^{\alpha/2} = 0$, that is, $\rho_\infty = 1-2^{(2/\alpha)-1}$. Since $\rho_\infty < 0$, the correlation function of G cannot be non-negative definite (Exercise 4.31), which is a contradiction. We have therefore established

Theorem 4.7.5 *Stationary $S\alpha S$ moving averages are different from stationary $S\alpha S$ sub-Gaussian processes for any $0 < \alpha < 2$.*

We now turn to the $S\alpha S$ *sub-stable* processes $\{X(t) = A^{1/\alpha'}Y(t),\ t \in \mathbb{R}\}$, defined in Section 3.8. Here, Y is a $S\alpha' S$ stationary process and $0 < \alpha < \alpha' < 2$. Let $\tau_X(t)$ and $\tau_Y(t)$ denote, respectively, the codifference for the processes X and Y. Since $\|\sum_{j=1}^d b_j X(t_j)\|_\alpha^\alpha = \|\sum_{j=1}^d b_j Y(t_j)\|_{\alpha'}^\alpha$ (Relation (3.8.2)), we have $\tau_X(t) = 2\|Y(0)\|_{\alpha'}^\alpha - \|Y(t) - Y(0)\|_{\alpha'}^\alpha$. But $\tau_Y(t) = 2\|Y(0)\|_{\alpha'}^{\alpha'} - \|Y(t) - Y(0)\|_{\alpha'}^{\alpha'}$ and, therefore,

Proposition 4.7.6 *For the stationary sub-stable process $\{X(t) = A^{1/\alpha}Y(t),\ t \in \mathbb{R}\}$,*

$$\tau_X(t) = 2\|Y(0)\|_{\alpha'}^\alpha - (2\|Y(0)\|_{\alpha'}^{\alpha'} - \tau_Y(t))^{\alpha/\alpha'}.$$

If $\tau_Y(t) \to 0$ as $t \to \infty$ (e.g. if Y is a moving average), then

$$\lim_{t\to\infty} \tau_X(t) = \|Y(0)\|_{\alpha'}^\alpha (2 - 2^{\alpha/\alpha'}) > 0.$$

If, moreover, X is standardized ($\|X(0)\|_\alpha = 1$), then $\|Y(0)\|_{\alpha'} = 1$ and $\lim_{t\to\infty} \tau_X(t) = 2 - 2^{\alpha/\alpha'}$ which is different for different α'. This shows, for example, that $S\alpha S$ sub-stable processes with different α' are different (if $\tau_Y(t) \to 0$).

4.8 The expected number of level crossings for stationary sub-Gaussian processes

Let $\{X(t),\ t \geq 0\}$ be a $S\alpha S$ sub-Gaussian process, i.e., $\{X(t) = A^{1/2}G(t),\ t \geq 0\}$ defined in Section 3.7. We want to evaluate the mean number of crossings of a given level $u \in \mathbb{R}$ by the process X. We shall assume that $\{G(t),\ t \geq 0\}$ is a mean zero *stationary* Gaussian process so that X is stationary as well. The autocovariance function $R(t) = EG(t+s)G(s)$ of G can be expressed as $R(t) = \int_{-\infty}^{\infty} e^{i\lambda t} dS(\lambda)$ where S is the "spectral distribution function" (see, for example, Brockwell and Davis (1991)). Observe that the variance $R(0) = \int_{\infty}^{\infty} dS(\lambda)$ equals the total spectral mass. If R is differentiable at zero, then $R'(0) = 0$ because R is even. If R is twice differentiable at zero, then $-R''(0) = \int_{-\infty}^{\infty} \lambda^2 S(d\lambda) > 0$. The quantity $-R''(0)$ is called the second spectral moment. We will assume in the sequel that $-R''(0)$ exists. [3]

Let $C_u(T)$ denote the number of crossings of the level u by the $S\alpha S$ stationary sub-Gaussian process $\{X(t) = A^{1/2}G(t),\ t \geq 0\}$ during the time interval $[0, T]$.

[3] We also assume that $\{G(t),\ t \geq 0\}$ has continuous sample paths. The existence of $R'(0)$ is sufficient to ensure that such a version exists (Leadbetter, Lindgren & Rootzén 1983, p. 152).

(For a formal definition of level crossings see, for example, Cramér and Leadbetter (1967) or Leadbetter, Lindgren and Rootzén (1983).) We want to evaluate $EC_u(T)$. (By symmetry, $EC_u(T)/2$ equals the mean number of upcrossings.)

Since X is stationary, it is immediate that $EC_u(T) = TEC_u(1)$, and so from now on we shall only study $C_u := C_u(1)$.

The mean number of crossings EC_u^G of the level u by the Gaussian process G is given by the Rice formula

$$EC_u^G = \frac{1}{\pi}\left(\frac{-R''(0)}{R(0)}\right)^{1/2} \exp\left(-\frac{u^2}{2R(0)}\right) \tag{4.8.1}$$

(Rice (1945)). This formula shows that EC_u^G depends only on the variance $R(0)$ and on the second spectral moment $-R''(0)$ of G. If we denote the distribution function of A by $F_{\alpha/2}$, then it is immediate from the relation $X(t) = A^{1/2}G(t)$ and Rice's formula that for a stationary sub-Gaussian process,

$$EC_u = \frac{1}{\pi}\left(\frac{-R''(0)}{R(0)}\right)^{1/2} \int_0^\infty \exp\left(-\frac{u^2}{2R(0)a}\right) dF_{\alpha/2}(a). \tag{4.8.2}$$

Clearly, EC_u is a monotone decreasing function of $|u|$. Since, however, $F_{\alpha/2}$ is not generally known, (4.8.2) will not usually enable us to obtain an explicit formula for EC_u. In one special case of interest, the Cauchy case $\alpha = 1$, one can obtain EC_u explicitly because $F_{1/2}$ has the density

$$f_{1/2}(a) = \frac{e^{-1/4a}}{2\sqrt{\pi}a^{3/2}}$$

(see (1.1.15)). Substituting this into (4.8.2) and performing the integration leads to

Proposition 4.8.1 *If X is a stationary $S1S$ sub-Gaussian process, then*

$$EC_u = \frac{\sqrt{-R''(0)}}{\pi\sqrt{R(0) + 2u^2}}.$$

PROOF: Exercise 4.32. ∎

Observe that EC_u is asymptotically proportional to u^{-1} in the Cauchy case $\alpha = 1$. We shall now derive an asymptotic formula for EC_u for general sub-Gaussian $S\alpha S$ processes.

Theorem 4.8.2 *For a general sub-Gaussian $S\alpha S$ process, we have*

$$\lim_{u\to\infty} u^\alpha EC_u = \frac{2^{\alpha/2}\Gamma(1+\alpha/2)(-R''(0))^{1/2}}{\pi\Gamma(1-\alpha/2)(R(0))^{(1-\alpha/2)}}. \tag{4.8.3}$$

PROOF: We start with (4.8.2). Since $A \sim S_{\alpha/2}((\cos\frac{\pi\alpha}{4})^{2/\alpha}, 1, 0)$, by Property 1.2.15,

$$\begin{aligned}\lim_{a\to\infty} a^{\alpha/2}(1 - F_{\alpha/2}(a)) &= \lim_{a\to\infty} a^{\alpha/2} P(A > a) \\ &= \frac{1-\alpha/2}{\Gamma(2-\alpha/2)\cos(\pi\alpha/4)} \cos\frac{\pi\alpha}{4} \\ &= (\Gamma(1-\alpha/2))^{-1}.\end{aligned}$$

It therefore follows that

$$\begin{aligned}&\lim_{v\to\infty} v^{\alpha/2} \int_0^\infty \exp(-v/a) F_{\alpha/2}(da) \\ &= \lim_{v\to\infty} v^{\alpha/2} E(e^{-v/A}) \\ &= \lim_{v\to\infty} v^{\alpha/2} \int_0^\infty P\{e^{-v/A} > x\} dx \\ &= \lim_{v\to\infty} \int_0^\infty v^{\alpha/2} P\{A > v/t\} e^{-t} dt \\ &= \int_0^\infty \lim_{v\to\infty} v^{\alpha/2} P\{A > v/t\} e^{-t} dt \\ &= \frac{1}{\Gamma(1-\alpha/2)} \int_0^\infty t^{\alpha/2} e^{-t} dt \\ &= \frac{\Gamma(1+\alpha/2)}{\Gamma(1-\alpha/2)},\end{aligned}$$

where the interchange of limit and integration is justified by dominated convergence since $v^{\alpha/2} P\{A > v/t\}$ is bounded by $ct^{\alpha/2}$ for some finite c.

Substituting the above into (4.8.2) proves the theorem. ∎

Whereas EC_u decreases exponentially fast in the Gaussian case, it is asymptotically proportional to $u^{-\alpha}$ in the sub-Gaussian case with index $0 < \alpha < 2$.

4.9 Exercises

Exercise 4.1 Let M be a $S\alpha S$ random measure on $(E, \mathcal{E})$, $1 < \alpha < 2$, with control measure m. Show that

$$E(M(A_2)|M(A_1)) = \frac{m(A_1 \cap A_2)}{m(A_1)} M(A_1) \text{ a.s.}$$

Exercise 4.2 Let X_1, X_2 be jointly $S\alpha S$ with $1 < \alpha < 2$ and let $Y_1 = aX_1 + bX_2$ and $Y_2 = cX_1 + dX_2$. Find the constant k such that $E(Y_2|Y_1) = kY_1$ a.s.

Exercise 4.3 Show that the jointly $S\alpha S$ random variables (X_1, X_2, X_3, X_4) defined below are such that X_1, X_2, X_3 are linearly independent but the system (4.1.6) does not have a unique solution.

Here $X_i = \int_0^1 f_i(x)M(dx)$, $i = 1, 2, 3, 4$, M has Lebesgue control measure, and

$$f_1(x) \text{ is } \; 1 \;\text{ if } x \in \left(0, \frac{1}{3}\right), \text{ is } -1 \text{ if } x \in \left(\frac{1}{3}, \frac{2}{3}\right), \text{ is } 0 \text{ if } x \in \left(\frac{2}{3}, 1\right);$$

$$f_2(x) \text{ is } \; 1 \;\text{ if } x \in \left(0, \frac{2}{3}\right), \text{ is } -1 \text{ if } x \in \left(\frac{2}{3}, 1\right);$$

$$f_3(x) \text{ is } \; 1 \;\text{ if } x \in \left(0, \frac{1}{3}\right), -2^{1-\alpha} \text{ if } x \in \left(\frac{1}{3}, \frac{2}{3}\right), -4^{1-\alpha} \text{ if } x \in \left(\frac{2}{3}, 1\right).$$

Exercise 4.4 Let $X_1, X_2, \ldots, X_n$ be random variables whose first moments exist and with joint characteristic function ϕ. Show that

$$E(X_n | X_1, \ldots, X_{n-1}) = a_1 X_1 + \cdots + a_n X_n \;\text{ a.s.}$$

if and only if for all $\theta_1, \ldots, \theta_{n-1}$,

$$\left.\frac{\partial}{\partial \theta_n} \phi(\theta_1, \ldots, \theta_{n-1}, \theta_n)\right|_{\theta_n = 0}$$
$$= a_1 \frac{\partial}{\partial \theta_1} \phi(\theta_1, \ldots, \theta_{n-1}, 0) + \cdots + a_{n-1} \frac{\partial}{\partial \theta_{n-1}} \phi(\theta_1, \ldots, \theta_{n-1}, 0).$$

Exercise 4.5 Use Exercise 4.4 to show that $E(X_n | X_1, \ldots, X_{n-1})$ is always linear when $(X_1, \ldots, X_n)$ is sub-Gaussian with $1 < \alpha < 2$.

Exercise 4.6 Use Exercise 4.4 to show that if $X_1, \ldots, X_n$ are jointly $S\alpha S$, $1 < \alpha < 2$, with spectral measure Γ, then

$$E(X_n | X_1, \ldots, X_{n-1}) = a_1 X_1 + \cdots + a_{n-1} X_{n-1} \;\text{ a.s.}$$

if and only if for all $\theta_1, \ldots, \theta_{n-1}$,

$$\int_{S_n} (s_n - a_1 s_1 - \cdots - a_{n-1} s_{n-1})(\theta_1 s_1 + \cdots + \theta_{n-1} s_{n-1})^{<\alpha-1>} \Gamma(d\mathbf{s}) = 0.$$

Hint: See Miller (1978).

Exercise 4.7 Suppose that (X_1, X_2, X_3) are jointly $S\alpha S$ with $1 < \alpha < 2$. Use Exercise 4.6 to show that when $E(X_3 | X_1, X_2) = a_1 X_1 + a_2 X_2$, the coefficients a_1 and a_2 satisfy the system (4.1.6). Solve that system when X_1 and X_2 are not multiples of each other.

Exercise 4.8 Let c be a real number and X_1 and X_2 be random variables. Show that the relation $(X_1, X_2 - cX) \stackrel{d}{=} (X_1, cX_1 - X_2)$ is equivalent to $(X_1, X_2) \stackrel{d}{=} (X_1, 2cX_1 - X_2)$.

Exercise 4.9 Let (X_1, X_2) be $S\alpha S$, $0 < \alpha < 2$, with spectral measure Γ. Show that the relation $(X_1, X_2) \stackrel{d}{=} (X_1, 2cX_1 - X_2)$ is satisfied if Γ is given by (4.2.4), but it is not satisfied if Γ is given by (4.2.5).

Exercise 4.10 Fill all the details in the proof of Proposition 4.3.2.

Exercise 4.11 Prove that the right-hand side of Relation (4.4.3) coincides with the right-hand side of Relation (4.4.1).
Hint: Make a change of variables from S_n to E and use (3.2.4).

Exercise 4.12 Prove the second part of Lemma 4.4.2.

Exercise 4.13 Let X be a positive random variable, independent of an exponential random variable Γ_1 with mean 1. Then

$$\lim_{\lambda \to 0} \lambda^{-1} P(X\Gamma_1 < \lambda) = E(X^{-1}).$$

Exercise 4.14 Let $X > 0$ and $-\infty < p < \infty$. Show that $EX^p < \infty$ implies $\lim_{\lambda \to \infty} \lambda^p P\{X > \lambda\} = 0$.

Exercise 4.15 A gamma random variable Γ_r with parameters $r > 0$ and $\gamma > 0$ has density function $f(x) = \frac{\gamma^r}{\Gamma(r)} x^{r-1} e^{-\gamma x}$ for $x > 0$ and $f(x) = 0$ for $x \le 0$. Show that for any $-\infty < p < \infty$:

(1) $E(\Gamma_r)^p = \infty$ if $p \le -r$

(2) If $p > -r$, then

$$E(\Gamma_r)^p = \gamma^{-p} \frac{\Gamma(r+p)}{\Gamma(r)} \le Cr^p,$$

where $C = C(\gamma, r, p)$ is a positive constant.

(3) If $0 < \alpha < 2$,

$$E(\Gamma_i)^{-[2/\alpha]} = \infty \text{ and } i \le \left[\frac{2}{\alpha}\right] \text{ and } E(\Gamma_i)^{-[2/\alpha]} \le Ci^{-2/\alpha} \text{ for } i > \left[\frac{2}{\alpha}\right] + 1.$$

Exercise 4.16 Let Γ_j be a gamma random variable with mean $E\Gamma_j = j$. Show that

$$\overline{\lim}_{j \to \infty} E \exp\{j^{-1/\alpha}(\Gamma_j - j)\} \le 2e^{1/2}.$$

Exercise 4.17 Prove that for $\lambda > 0$ and $0 < \epsilon < 1$,

$$\begin{aligned}
P\Big(\min_{k=1,\ldots,n} (X_k + Y_k) > \lambda\Big) \\
\leq \; P\Big(\{\min_{k=1,\ldots,n} X_i > \lambda(1-\epsilon)\} \cup \{\max_{k=1,\ldots,n} Y_k > \lambda(1-\epsilon)\} \\
\cup \{\max_{k=1,\ldots,n} X_k > \lambda\epsilon\} \cup \{\max_{k=1,\ldots,n} Y_k > \lambda\epsilon\}\Big).
\end{aligned}$$

Exercise 4.18 Prove Lemma 4.4.4 by induction.

Exercise 4.19 Prove Corollary 4.4.6.

Exercise 4.20 Prove Relation (4.4.15).

Exercise 4.21 Let $X_1, \ldots, X_n$ be non-negative random variables. Prove by induction that

$$E(X_1 \cdots X_n) = \int_0^\infty \cdots \int_0^\infty P(X_1 > t_1, \ldots, X_n > t_n) dt_1 \ldots dt_n.$$

Exercise 4.22 Let $0 < \alpha < 2$. The spectral measure can be used to express the dependence structure of α-stable random variables $X_1, X_2, \ldots, X_n$.

(1) Show that the X_js are independent if and only if Γ is concentrated on the $2n$ points $(0, \ldots, 0, \pm 1, 0, \ldots, 0)$.

(2) Show that the X_js are n-fold dependent if and only if

$$\Gamma\{\mathbf{s} \in S_n : s_1 s_2 \ldots s_n \neq 0\} > 0.$$

(3) Find a necessary and sufficient condition in terms of Γ for $E|X_1|^{p_1} \ldots |X_n|^{p_n} < \infty$.

Exercise 4.23 Show that if U_1 and U_2 are associated random variables, then

$$P(U_1 > \lambda_1, U_2 > \lambda_2) \geq P(U_1 \geq \lambda_1) P(U_2 \geq \lambda_2).$$

Exercise 4.24 If two random variables are associated and their covariance is zero, then they are independent.
Hint: Prove it first for positive random variables.

Exercise 4.25 If $T_1, T_2, \ldots, T_k$ are associated random variables and if $G_i = g_i(T_1, T_2, \ldots, T_k)$, $i = 1, \ldots, n$, where for each i, the function g_i is non-decreasing in each of its arguments, then the random variables $G_1, G_2, \ldots, G_n$ are associated.

Exercise 4.26 Show that neither $\overline{F}$-positive orthant dependence nor F-positive orthant dependence imply association.

Exercise 4.27 Show that if X and $-Y$ are associated, then X and Y are negatively associated.

Exercise 4.28 Show that for all $0 < \alpha < 2$,

$$I(1,-1;t) + I(1,1;t) \leq 0.$$

Hint: Proof of Lemma 2.7.14.

Exercise 4.29 Compute $\tau(t)$ for the reverse Ornstein-Uhlenbeck process (defined in Example 3.6.4).

Exercise 4.30 Show that $I(\theta_1, \theta_2; t)$ (defined in Relation (4.7.5)) tends to 0 as $t \to \infty$ for all θ_1 and θ_2 if the process is a stationary $S\alpha S$, $0 < \alpha \leq 2$, moving average. Show that this is also the case if the measure M in $X(t) = \int_{-\infty}^{\infty} f(t-x)M(dx)$ is an α-stable (not necessarily symmetric) random measure.

Exercise 4.31 Let $\{G(t),\ t \in \mathbb{R}\}$ be a stationary Gaussian process with auto-correlation function $\rho(t)$ converging to ρ_∞ as $t \to \infty$. Prove that ρ_∞ cannot be negative[4].
Hint: Suppose $\rho_\infty < 0$ and choose time points $t_1, \ldots, t_n$ separated by long enough intervals so that the correlation matrix of the vector $(G(t_1), \ldots, G(t_n))$ consists of 1 on the diagonal and all off-diagonal entries are less than $\rho_\infty + c < 0$. Choose n sufficiently large to make all the entries of the correlation matrix add up to a negative number.

Exercise 4.32 Prove Proposition 4.8.1.

[4]Suggested by Aaron Gross.

Chapter 5

Non-linear regression

We proved in Chapter 4 that the bivariate regression is linear when the random variables are jointly $S\alpha S$ and $1 < \alpha \leq 2$. In this chapter, we give a complete picture of bivariate regression in the general, possibly asymmetric, case. We show that when skewness is present, regression can be either linear or nonlinear. We also show that the regression can be defined even when $0 < \alpha \leq 1$. Section 5.5 summarizes the main results of this chapter.

In Section 5.1 we give sufficient conditions for the finiteness of the conditional moment $E(|X_2|^p|X_1 = x)$ when $\alpha \leq p < 2\alpha + 1$. When $p < \alpha$, the conditional moment always exists because $E|X_2|^p < \infty$. However, the conditional moment can also exist when $p \geq \alpha$. This depends on the spectral measure Γ of the α-stable random vector (X_1, X_2). Γ is a measure on the unit circle. The smaller the mass concentrated around the north and south poles of the unit circle, the higher the value of p. These results are relevant to the regression problem. Setting $p = 1$, we obtain a sufficient condition for the existence of the regression in the case $0 < \alpha \leq 1$. This condition will become our standard assumption.

In Section 5.2 we derive the form of the bivariate regression. The formulas are valid for $1 < \alpha < 2$, and also, when the standard assumption holds, for $0 < \alpha \leq 1$. We give necessary and sufficient conditions for linearity. Interestingly, the regressions are typically asymptotically linear. In Section 5.3 we apply the bivariate regression results to important classes of α-stable laws, namely those of moving averages, sub-Gaussian and harmonizable vectors. The sometimes exotic behavior of the regression functions in the non-linear case is illustrated graphically in Section 5.4.

The regression in the non-linear case involves two integrals which cannot be computed analytically. To make matters worse, these integrals have features which render them intractable by standard numerical integration procedures. In

Section 5.5, we give algorithms for computing the regression formulas. Since a straightforward application of the midpoint trapezoidal or Simpson's rule gives a very poor approximation, we developed variants with variable step size and cutoffs. Although we make no claim that the procedures described here represent the best possible way to compute the regressions, they are the most appropriate among those examined. To facilitate applications of the regression formulas we rewrite them in Section 5.5 using the integral representation. We also refer to a software package for computing bivariate regressions and probability density function of α-stable random variables. A description of the package and the listing of the source code, written in the C language, can be found in Hardin, Samorodnitsky and Taqqu (1991*b*).

5.1 Conditional moments of order greater than or equal to α

Let (X_1, X_2) be an α-stable random vector. Clearly, the conditional moment $E(|X_2|^p|X_1 = x)$ is finite for almost every x if $p < \alpha$ because $E|X|^p < \infty$. What about $p \geq \alpha$? In this section some partial answers are provided.

It is well known that the existence of moments is related to the behavior of the characteristic function around the origin. We shall therefore consider the behavior of the conditional characteristic function $\phi_{X_2|x}(r) = E[e^{irX_2}|X_1 = x]$ as $r \to 0$.

We consider, firstly, a d-dimensional α-stable random vector $\mathbf{X} = (X_1, X_2, \ldots, X_d)$ and study

$$\phi_{X_d|x_1,\ldots,x_{d-1}}(r) = E[e^{irX_d}|X_1 = x_1, \ldots, X_{d-1} = x_{d-1}].$$

Let $\phi_{\mathbf{X}}(t_1, \ldots, t_{d-1}, r) = E\exp\{i(\sum_{j=1}^{d-1} t_jX_j + rX_d)\}$ denote the characteristic function of the vector $\mathbf{X}$ and let Γ be its spectral measure. Recall that Γ is a measure on the unit sphere S_d of $\mathbb{R}^d$. We start with the following simple result.

Lemma 5.1.1 *If $d \geq 2$ and for each $(t_1, \ldots, t_{d-1}) \in S_{d-1}$,*

$$\Gamma\{\mathbf{s} \in S_d : t_1s_1 + \cdots + t_{d-1}s_{d-1} \neq 0\} > 0, \tag{5.1.1}$$

then for each real r,

$$\int_{\mathbb{R}^{d-1}} |\phi_{\mathbf{X}}(t_1, \ldots, t_{d-1}, r)| dt_1 \ldots dt_{d-1} < \infty. \tag{5.1.2}$$

Remark. Note that S_1 consists of the two point set $\{-1, 1\}$. Hence Relation (5.1.1), in the case $d = 2$, is equivalent to $\Gamma\{(s_1, s_2) \in S_2 : s_1 \neq 0\} > 0$, i.e., to

$$\Gamma\{S_2 \backslash \{(0,1), (0,-1)\}\} > 0, \tag{5.1.3}$$

that is, to $X_1 \not\equiv 0$ a.s.

PROOF OF LEMMA 5.1.1: Define $M : \mathbb{R}^{d-1} \to \mathbb{R}_+$ by

$$M(t_1, \ldots, t_{d-1}) = \int_{S_d} |t_1 s_1 + \cdots + t_{d-1} s_{d-1}|^\alpha \Gamma(ds).$$

Then

$$M_0 \equiv \inf_{(t_1,\ldots,t_{d-1}) \in S_{d-1}} M(t_1, \ldots, t_{d-1}) > 0.$$

If $0 < \alpha < 1$, then, by (2.3.1),

$$\begin{aligned}
&|\phi_{\mathbf{X}}(t_1, \ldots, t_{d-1}, r)| \\
&= \exp\Big\{ -\int_{S_d} |t_1 s_1 + \cdots + t_{d-1} s_{d-1} + r s_d|^\alpha \Gamma(ds) \Big\} \\
&\leq \exp\Big\{ -\int_{S_d} (|t_1 s_1 + \cdots + t_{d-1} s_{d-1}|^\alpha - |r s_d|^\alpha) \Gamma(ds) \Big\} \\
&\leq e^{C|r|^\alpha} \exp\{ -M_0 (t_1^2 + \cdots + t_{d-1}^2)^{\alpha/2} \},
\end{aligned}$$

with $C > 0$, proving (5.1.2). The case $1 \leq \alpha < 2$ is similar. ■

Fix $d \geq 2$ and let $F_{X_1,\ldots,X_{d-1},X_d}$, $F_{X_1,\ldots,X_{d-1}}$ and $F_{X_d|x_1,\ldots,x_{d-1}}$ denote, respectively, the distribution function of $(X_1, \ldots, X_{d-1}, X_d)$, $(X_1, \ldots, X_{d-1})$ and the conditional distribution function of X_d given $X_1 = x_1, \ldots, X_{d-1} = x_{d-1}$. Further, let

$$\phi_{X_d|x_1,\ldots,x_{d-1}} = \int_{-\infty}^{\infty} e^{irx_d} dF_{X_d|x_1,\ldots,x_{d-1}}(x_d)$$

denote the corresponding characteristic function. We have by Fubini's theorem

$$\begin{aligned}
&\phi_{\mathbf{X}}(t_1, \ldots, t_{d-1}, r) \\
&= \int_{\mathbb{R}^d} e^{i(t_1 x_1 + \cdots + t_{d-1} x_{d-1} + r x_d)} dF_{X_1,\ldots,X_{d-1},X_d}(x_1, \ldots, x_{d-1}, x_d) \\
&= \int_{\mathbb{R}^{d-1}} e^{i(t_1 x_1 + \cdots + t_{d-1} x_{d-1})} \phi_{X_d|x_1,\ldots,x_{d-1}}(r) dF_{X_1,\ldots,X_{d-1}}(x_1, \ldots, x_{d-1}).
\end{aligned}$$

Therefore, for fixed r, we may regard $\phi_{\mathbf{X}}(t_1, \ldots, t_{d-1}, r)$ as the Fourier transform of a complex-valued finite measure G_r on $\mathbb{R}^{d-1}$ given by

$$G_r(A) = \int_A \phi_{X_d|x_1,\ldots,x_{d-1}}(r) dF_{X_1,\ldots,X_{d-1}}(x_1, \ldots, x_{d-1}) \tag{5.1.4}$$

for any Borel set A in $\mathbb{R}^{d-1}$.

Suppose, now, that (5.1.1) holds. Then both G_r and $F_{X_1,\ldots,X_{d-1}}$ have bounded continuous densities. Indeed, by Lemma 5.1.1, $\phi_{\mathbf{X}}(t_1, \ldots, t_{d-1}, r)$ is absolutely integrable in $t_1, \ldots, t_{d-1}$, so that, by Fourier inversion theorem in $\mathbb{R}^{d-1}$, G_r is absolutely continuous with respect to Lebesgue measure in $\mathbb{R}^{d-1}$, with a bounded continuous density $g_r(x_1, \ldots, x_{d-1})$ equal to

$$\frac{1}{(2\pi)^{d-1}} \int_{\mathbb{R}^{d-1}} e^{-i(t_1x_1+\cdots+t_{d-1}x_{d-1})} \phi_{\mathbf{X}}(t_1, \ldots, t_{d-1}, r) dt_1 \ldots dt_{d-1}. \quad (5.1.5)$$

Setting $r = 0$ in (5.1.5) shows that $F_{X_1,\ldots,X_{d-1}}$ also has a bounded continuous density $f_{X_1,\ldots,X_{d-1}}$ with respect to Lebesgue measure on $\mathbb{R}^{d-1}$ equal to

$$\frac{1}{(2\pi)^{d-1}} \int_{\mathbb{R}^{d-1}} e^{-i(t_1x_1+\cdots+t_{d-1}x_{d-1})} \phi_{\mathbf{X}}(t_1, \ldots, t_{d-1}, 0) dt_1 \ldots dt_{d-1}. \quad (5.1.6)$$

We conclude from (5.1.4) that

$$g_r(x_1, \ldots, x_{d-1}) = f_{X_1,\ldots,X_{d-1}}(x_1, \ldots, x_{d-1}) \phi_{X_d|x_1,\ldots,x_{d-1}}(r).$$

Hence, for almost all realizations $X_1 = x_1, \ldots, X_{d-1} = x_{d-1}$ corresponding to $f_{x_1,\ldots,x_{d-1}}(x_1, \ldots, x_{d-1}) > 0$,

$$\begin{aligned} \phi_{X_d|x_1,\ldots,x_{d-1}}(r) &= \frac{g_r(x_1, \ldots, x_{d-1})}{f_{X_1,\ldots,X_{d-1}(x_1,\ldots,x_{d-1})}} \\ &= \frac{\int_{\mathbb{R}^{d-1}} e^{-i(t_1x_1+\cdots+t_{d-1}x_{d-1})} \phi_{\mathbf{X}}(t_1, \ldots, t_{d-1}, r) dt_1 \ldots dt_{d-1}}{\int_{\mathbb{R}^{d-1}} e^{-i(t_1x_1+\cdots+t_{d-1}x_{d-1})} \phi_{\mathbf{X}}(t_1, \ldots, t_{d-1}, 0) dt_1 \ldots dt_{d-1}}. \end{aligned} \quad (5.1.7)$$

Now that we have obtained an expression for the conditional characteristic function $\phi_{X_d|X_1,\ldots,X_{d-1}}$, we can exploit the well-known fact that finiteness of moments of a probability law is directly related to the behavior of its characteristic function at the origin.

Theorem 5.1.2 *Let F be a distribution function and ϕ the corresponding characteristic function.*

(*i*) *A necessary and sufficient condition for F to have the moment of order p, where $0 < p < 2$, is that, for some $c > 0$*

$$\int_0^c \frac{1 - \operatorname{Re} \phi(r)}{r^{1+p}} dr < \infty.$$

Moreover, F has a second moment if and only if $r^{-2}(1 - \operatorname{Re} \phi(r))$ is bounded for $0 < r < c$.

(*ii*) *A necessary and sufficient condition for F, having the moment of order $2n$, to have the moment of order $2n + p$, where $0 < p < 2$, is that, for some $c > 0$,*

$$\int_0^c \frac{\operatorname{Re} \phi^{(2n)}(0) - \operatorname{Re} \phi^{(2n)}(r)}{r^{1+p}} dr < \infty.$$

Moreover, F has the moment of order $2n+2$ if and only if $r^{-2}(\text{Re}\,\phi^{(2n)}(0) - \text{Re}\,\phi^{(2n)}(r))$ is bounded for $0 < r < c$.

PROOF: It is enough to prove part (i). Part (ii) follows from part (i) since $\phi^{(2n)}(r)/\phi^{(2n)}(0)$ is a characteristic function corresponding to the distribution G given by $G(dy) = y^{2n}F(dy)/EY^{2n}$.

Suppose that $\int_{-\infty}^{\infty} |x|^p F(dx) < \infty$ for $0 < p < 2$ and let $c > 0$. Now

$$\int_0^c \frac{1 - \text{Re}\,\phi(r)}{r^{1+p}} dr = \int_0^c \frac{1}{r^{1+p}} \int_{-\infty}^{\infty} (1 - \cos rx) F(dx) dr$$

$$= \int_{-\infty}^{\infty} \int_0^c \frac{1 - \cos rx}{r^{1+p}} dr\, F(dx) = \int_{-\infty}^{\infty} |x|^p \left(\int_0^{c|x|} \frac{1 - \cos t}{t^{1+p}} dt \right) F(dx) < \infty,$$

because $\int_0^{\infty} (1 - \cos t)/t^{1+p} dt < \infty$ for $0 < p < 2$.[1] Conversely, suppose that $\int_0^c (1 - \text{Re}\,\phi(r))/r^{1+p} dr < \infty$ for some $c > 0$. Then, as above,

$$\infty > \int_{-\infty}^{\infty} |x|^p \int_0^{c|x|} \frac{1 - \cos t}{t^{1+p}} dt\, F(dx) \geq \int_{|x|>1} |x|^p \int_0^c \frac{1 - \cos t}{t^{1+p}} dt\, F(dx)$$

$$= \text{const.} \int_{|x|>1} |x|^p F(dx),$$

which implies that F has a pth moment.

Now suppose that F has a second moment. Then ϕ is twice differentiable and

$$\phi''(0) = \lim_{r \to 0} \frac{\phi(r) + \phi(-r) - 2\phi(0)}{r^2} = -2 \lim_{r \to 0} \frac{1 - \text{Re}\,\phi(r)}{r^2}.$$

Hence, $r^{-2}(1 - \text{Re}\,\phi(r))$ is bounded in the neighborhood of zero. Conversely, if this quantity is bounded for $0 < r < c$,[2] then, by Fatou's lemma,

$$\infty > \liminf_{r \to 0} \frac{1 - \text{Re}\,\phi(r)}{r^2} = \liminf_{r \to 0} \int_{-\infty}^{\infty} \frac{1 - \cos rx}{r^2} F(dx)$$

$$\geq \int_{-\infty}^{\infty} \liminf_{r \to 0} \frac{1 - \cos rx}{r^2} F(dx) = \frac{1}{2} \int_{-\infty}^{\infty} x^2 F(dx),$$

and the theorem is proved. ■

---

[1] In fact, one can see that $\int_{-\infty}^{\infty} |x|^p F(dx) = \text{const.} \int_0^{\infty} (1 - \text{Re}\,\phi(r))/r^{1+p} dr$ (cf. also Theorem 11.4.3 in Kawata (1972)).

[2] The proof implies that it is enough to assume that $r^{-2}(1 - \text{Re}\,\phi(r))$ is bounded at some sequence of points tending to zero (cf. also Ramachandran and Rao (1968)).

Since we are mainly concerned with regression, i.e., the case $p = 1 < 2$ ($n = 0$), we want to study how fast $\phi_{X_d|x_1,\ldots,x_{d-1}}(r)$ approaches 1 as $r \to 0$. When $d = 2$, (5.1.7) reduces to

$$\phi_{X_2|x}(r) = 1 + \frac{1}{2\pi f_{X_1}(x)} \int_{-\infty}^{\infty} e^{-itx}(\phi_{\mathbf{X}}(t,r) - \phi_{\mathbf{X}}(t,0))dt, \qquad (5.1.8)$$

where we set $x_1 = x$. We shall now consider only the case $d = 2$ and study how fast the numerator in the right-hand side of (5.1.8) tends to 0 as $r \to 0$.

Let $\mathbf{X} = (X_1, X_2)$ be a $S\alpha S$ random vector in $\mathbb{R}^2$ with a spectral measure Γ satisfying (5.1.3). Substituting $d = 2$ in (2.4.2) we obtain

$$\phi_{\mathbf{X}}(t,0) = \exp\Big\{-\int_{S_2} |ts_1|^\alpha \Gamma(d\mathbf{s})\Big\} = e^{-|t|^\alpha \sigma_1^\alpha},$$

where, by (2.3.3), $\sigma_1 = (\int_{S_2} |s_1|^\alpha \Gamma(d\mathbf{s}))^{1/\alpha}$ is the scale parameter of the distribution of X_1. We therefore obtain the following expression for the conditional characteristic function of X_2 given $X_1 = x$:

$$\phi_{X_2|x}(r) = 1 + \frac{1}{2\pi f_{X_1}(x)} \int_{-\infty}^{\infty} e^{-itx} e^{-|t|^\alpha \sigma_1^\alpha}$$
$$\times \Big[\exp\{-\int_{S_2} (|ts_1 + rs_2|^\alpha - |ts_1|^\alpha)\Gamma(d\mathbf{s})\} - 1\Big] dt. \qquad (5.1.9)$$

The conditional moment $E(|X_2|^p | X_1 = x)$ is finite for almost every x if $p < \alpha$, but it can be infinite when $p \geq \alpha$. For example, let $(X_1, X_2) \stackrel{d}{=} (X_1, Y + Z)$, where (X_1, Y) is a $S\alpha S$ random vector, and Z is a $S\alpha S$ random variable independent of (X_1, Y). (The autoregressive model of order 1, i.e., $X_2 = aX_1 + Z$, is a special case of this.) The presence of the "independent increment" Z ensures, of course, that $E(|X_2|^p | X_1) = E(E(|Y + Z|^p | X_1, Y) | X_1) = \infty$ a.s. if $p \geq \alpha$, because for such p, $E(|Y + Z|^p | X_1 = x, Y = y) = E|y + Z|^p = \infty$. In terms of the spectral measure Γ of the vector (X_1, X_2), this "independent increment" Z is equivalent to the presence of atoms at the points (0,1) and $(0, -1)$ on the unit circle S_2. Therefore, heuristically, we would expect that the less the spectral measure Γ is concentrated around the points (0,1) and $(0, -1)$, the higher the conditional moments of X_2. The following theorem, which is the main result of this section, shows that these heuristics are essentially correct.

Theorem 5.1.3 *Let* $\mathbf{X} = (X_1, X_2)$ *be an* α*-stable random vector (not necessarily symmetric), with spectral measure* Γ*, satisfying*

$$\int_{S_2} |s_1|^{-\nu} \Gamma(d\mathbf{s}) < \infty \qquad (5.1.10)$$

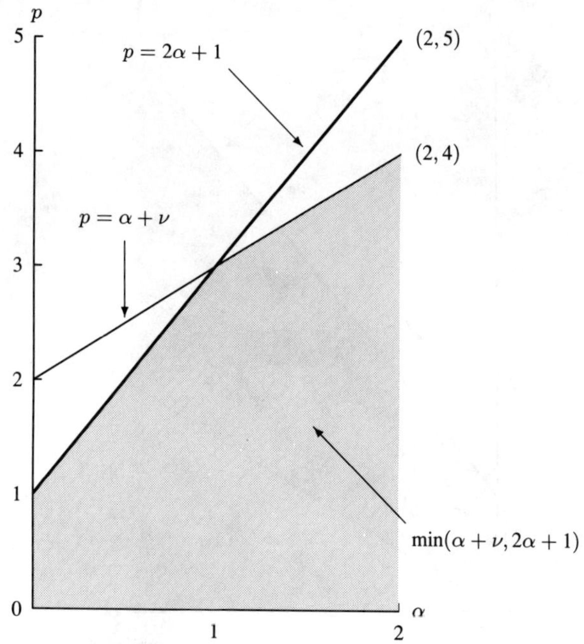

Figure 5.1: The conditional moment of order p is finite in the shaded region. The picture is drawn with $\nu = 2$.

for some $\nu \geq 0$. *Then* $E(|X_2|^p | X_1 = x) < \infty$ *for almost every* x *if*

$$p < \min\{\alpha + \nu,\ 2\alpha + 1\}. \tag{5.1.11}$$

Moreover, if $\alpha > 1/2$ *and* $\nu \geq 2 - \alpha$, *then* $E(X_2^2 | X_1 = x) < \infty$ *for almost every* x, *and if* $\alpha > 3/2$ *and* $\nu \geq 4 - \alpha$, *then* $E(X_2^4 | X_1 = x) < \infty$ *for almost every* x.

The region where (5.1.11) holds is displayed in Figures 5.1 and 5.2.

Remarks

1. "For almost every x" means for all x in the support of the probability density function of X_1.

2. Since $\int_{S_2} \Gamma(ds) < \infty$, the case $\nu = 0$ reduces to the trivial result $E(|X_2|^p | X_1 = x) < \infty$ for any $p < \alpha$.

3. Applying the theorem with $\alpha \leq 1$, we obtain $E(|X_2| | X_1 = x) < \infty$ a.e. if $\nu > 1 - \alpha$.

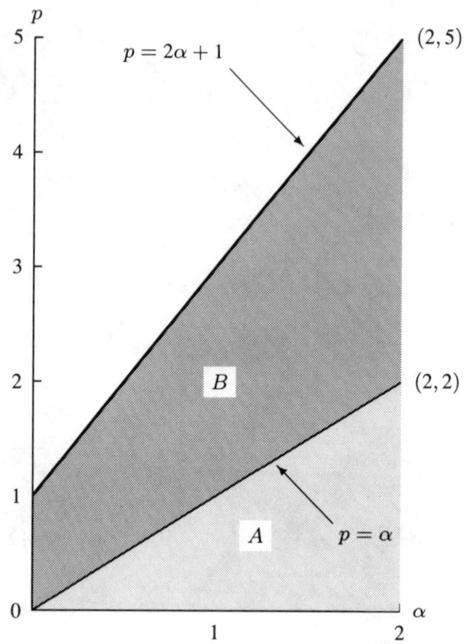

Figure 5.2: The shaded regions indicate where the conditional moment can be finite. Region A is for $\nu = 0$ (no improvement over the unconditional moment) and region $A \cup B$ is for $\nu \geq \alpha + 1$.

4. Cioczek-Georges and Taqqu (1994*a*, 1994*c*) prove that the condition $\nu > p - \alpha$ in Theorem 5.1.3 can be weakened to $\nu \geq p - \alpha$ not only for $p = 2$ or 4 but for all $\alpha < p < 2\alpha + 1$.

5. Cioczek-Georges and Taqqu (1994*b*) show that if $-\int_{S_2} \ln |s_1| \Gamma(ds) < \infty$, then $E[|X_2|^\alpha | X_1 = x] < \infty$ for almost every x.

The proof of Theorem 5.1.3 is quite technical. To obtain the existence of higher moments one may have to deal with the second and even the fourth derivative of $\phi_{X_2|x}(r)$. Formal computation of the derivatives can yield divergent terms, and hence special manipulations are needed. Moreover, one needs to bound the difference of derivatives (or the characteristic function itself) between points 0 and $r > 0$ by an adequate power of r.

To give a flavor of the proof and to demonstrate some of the basic techniques, we prove the theorem only in the two following special cases:

$$0 < \alpha \leq 1 \text{ and } 1 - \alpha < \nu < 1,$$

$$1 < \alpha < 2 \text{ and } 0 < \nu \leq 2 - \alpha.$$

The first case ensures the existence of $E(|X_2||X_1 = x)$ for almost every x and the second one (when $\nu = 2 - \alpha$) covers also the existence of $E[X_2^2|X_1 = x]$ for almost every x when $1 < \alpha < 2$. Other cases are considered in Samorodnitsky and Taqqu (1991*a*) ($0 < \nu \le 1 - \alpha$ for $0 < \alpha < 1$) and Cioczek-Georges and Taqqu (1994*a*, 1994*c*, 1994*b*).

Before proving Theorem 5.1.3 in the special cases indicated above, we need to collect certain technical results. Suppose that (5.1.10) holds for some $\nu \ge 0$, and let $C_\nu < \infty$ denote the corresponding value of the integral in (5.1.10).

Lemma 5.1.4 *(a) Let $0 < \alpha \le 1$. For any real, non-zero r and t,*

(i)

$$\left|\int_{S_2} (|ts_1 + rs_2|^\alpha - |ts_1|^\alpha)\Gamma(ds)\right| \le \Gamma(S_2)|r|^\alpha. \tag{5.1.12}$$

(ii)

$$\left|\int_{S_2} (|ts_1 + rs_2|^\alpha - |ts_1|^\alpha)\Gamma(ds)\right| \le \begin{cases} C_\nu 2^\nu (1+\alpha)|r|^{\alpha+\nu}|t|^{-\nu}, & \text{if } 0 < \nu < 1 - \alpha, \\ C_{1-\alpha} 2^{1-\alpha}(1+\alpha)|r||t|^{-(1-\alpha)}, & \text{if } \nu \ge 1 - \alpha. \end{cases} \tag{5.1.13}$$

(iii)

$$\int_{S_2} \left||ts_1 + rs_2|^\alpha + |ts_1 - rs_2|^\alpha - 2|ts_1|^\alpha\right|\Gamma(ds) \le \begin{cases} C_\nu 2^{1+\nu}(\alpha(1-\alpha)+1)|r|^{\alpha+\nu}|t|^{-\nu}, & \text{if } 0 < \nu < 2 - \alpha, \\ C_{2-\alpha} 2^{3-\alpha}(\alpha(1-\alpha)+1) r^2 |t|^{-(2-\alpha)}, & \text{if } \nu \ge 2 - \alpha. \end{cases} \tag{5.1.14}$$

(b) Let $1 < \alpha < 2$. For any real, non-zero r and t,

(i)

$$\left|\int_{S_2} (|ts_1 + rs_2|^\alpha - |ts_1|^\alpha)\Gamma(ds)\right| \le \alpha\Gamma(S_2)(|r|^\alpha + |r||t|^{\alpha-1}). \tag{5.1.15}$$

(ii)

$$\int_{S_2} \left||ts_1 + rs_2|^\alpha + |ts_1 - rs_2|^\alpha - 2|ts_1|^\alpha\right|\Gamma(ds) \le \begin{cases} C_\nu 2^{1+\nu}\alpha((\alpha-1)+3^{\alpha-1})|r|^{\alpha+\nu}|t|^{-\nu}, & \text{if } 0 < \nu < 2 - \alpha, \\ C_{2-\alpha} 2^{3-\alpha}\alpha((\alpha-1)+3^{\alpha-1}) r^2 |t|^{-(2-\alpha)}, & \text{if } \nu \ge 2 - \alpha. \end{cases} \tag{5.1.16}$$

PROOF: (a) Part (i) does not involve ν; it follows from $||ts_1+rs_2|^\alpha - |ts_1|^\alpha| \le |rs_2|^\alpha \le |r|^\alpha$. ((5.1.10) is not being used). For parts (ii) and (iii), decompose the whole integral into a sum of integrals J_1 and J_2 over $\{(s_1,s_2)\in S_2 : |ts_1| > 2|r|\}$ and $\{(s_1,s_2)\in S_2 : |ts_1|\le 2|r|\}$, respectively.

To obtain a bound for J_1 in part (ii), note that by the mean value theorem, there is a $0<\theta<1$

$$||ts_1+rs_2|^\alpha - |ts_1|^\alpha| = |rs_2|\alpha\,|ts_1+\theta rs_2|^{\alpha-1} \le |r|\alpha\,|ts_1/2|^{\alpha-1},$$

since $|ts_1| > 2|r|$ and $|\theta rs_2| < |r|$ imply $|ts_1+\theta rs_2| \ge |ts_1| - |\theta rs_2| \ge |ts_1|/2$. Hence $J_1 \le \alpha 2^{1-\alpha}|r||t|^{-(1-\alpha)}\int_{|ts_1|>2r}|s_1|^{-(1-\alpha)}\Gamma(ds)$ is bounded by $C_\nu\alpha 2^\nu |r|^{\nu+\alpha}|t|^{-\nu}$ if $0<\nu<1-\alpha$ and $C_{1-\alpha}\alpha 2^{1-\alpha}|r||t|^{-(1-\alpha)}$ if $\nu\ge 1-\alpha$. Moreover, J_2 is bounded above by

$$\int_{|ts_1|\le 2|r|}|rs_2|^\alpha\Gamma(ds) \le |r|^\alpha\int_{|ts_1|\le 2|r|}|s_1|^\nu|s_1|^{-\nu}\Gamma(ds) \le C_\nu 2^\nu|r|^{\alpha+\nu}|t|^{-\nu},$$

yielding the first inequality in (5.1.13). For the second inequality, take $\nu = 1-\alpha$.

For part (iii), in the estimate of J_1, use the fact that there exist constants $\theta_1,\theta_2,\gamma\in(0,1)$ such that

$$\begin{aligned}
&||ts_1+rs_2|^\alpha + |ts_1-rs_2|^\alpha - 2|ts_1|^\alpha| \\
&= \alpha|rs_2||(\alpha-1)(\theta_1+\theta_2)rs_2|ts_1+(\gamma\theta_1-(1-\gamma)\theta_2)rs_2|^{\alpha-2}| \\
&\le 2\alpha(1-\alpha)r^2|\frac{ts_1}{2}|^{\alpha-2},
\end{aligned}$$

and also use $J_2 \le \int_{|ts_1|\le 2|r|}2|rs_2|^\alpha\Gamma(ds)$.

(b) Part (i) follows from the mean value theorem and the proof of part (ii) is similar to the case $0<\alpha\le 1$. ∎

PROOF OF THEOREM 5.1.3 *(for $0<\alpha\le 1$, $1-\alpha<\nu<1$ and for $1<\alpha<2$, $0<\nu\le 2-\alpha$).* We first show that if the theorem holds for $S\alpha S$ random vectors, then it holds for α-stable vectors. Let (X_1,X_2) be α-stable, let ν and p be as in the theorem and let (Y_1,Y_2) be an independent copy of (X_1,X_2). Then $(Z_1,Z_2) = (X_1,X_2)-(Y_1,Y_2)$ is a $S\alpha S$ vector with spectral measure $\widetilde{\Gamma}$, defined by $\widetilde{\Gamma}(A) = \Gamma(A)+\Gamma(-A)$ for every Borel set A of S_2. Since $\int_{S_2}|s_1|^{-\nu}\widetilde{\Gamma}(ds) = 2\int_{S_2}|s_1|^{-\nu}\Gamma(ds) < \infty$, we have $E(E(|Z_2|^p|X_1,Y_1,Y_2)|Z_1) = E(|Z_2|^p|Z_1) < \infty$ a.s., since we are assuming the theorem holds for $S\alpha S$ random vectors. Consequently, $E(|Z_2|^p|X_1,Y_1,Y_2)<\infty$ a.s. by Fubini's theorem. Since (Y_1,Y_2) is independent of (X_1,X_2),

$$\begin{aligned}
E(|X_2|^p|X_1) &= E(|X_2-Y_2+Y_2|^p|\,X_1,Y_1,Y_2) \\
&\le 2^p(E(|X_2-Y_2|^p|\,X_1,Y_1,Y_2)+E(|Y_2|^p|\,X_1,Y_1,Y_2)) \\
&= 2^p(E(|Z_2|^p|\,X_1,Y_1,Y_2)+|Y_2|^p)
\end{aligned}$$

is a.s. finite, establishing the claim.

We suppose from now on that (X_1, X_2) is a $S\alpha S$ random vector and assume either $0 < \alpha \le 1$ with $1 - \alpha < \nu < 1$ or $1 < \alpha < 2$ with $0 < \nu \le 2 - \alpha$ in the sequel. Observe firstly that (5.1.10) implies (5.1.3), so (5.1.9) holds. Then decompose the integral in (5.1.9) as follows:

$$\begin{aligned}
&\int_{-\infty}^{\infty} e^{-itx} e^{-|t|^\alpha \sigma_1^\alpha} \left[\exp\left\{-\int_{S_2} (|ts_1 + rs_2|^\alpha - |ts_1|^\alpha)\Gamma(ds)\right\} - 1\right] dt \\
&= -\int_{-\infty}^{\infty} e^{-itx} e^{-|t|^\alpha \sigma_1^\alpha} \left[\int_{S_2} (|ts_1 + rs_2|^\alpha - |ts_1|^\alpha)\Gamma(ds)\right] dt \\
&\quad + \int_{-\infty}^{\infty} e^{-itx} e^{-|t|^\alpha \sigma_1^\alpha} \left[-1 + \int_{S_2} (|ts_1 + rs_2|^\alpha - |ts_1|^\alpha)\Gamma(ds)\right. \\
&\qquad\qquad \left. + \exp\left\{-\int_{S_2} (|ts_1 + rs_2|^\alpha - |ts_1|^\alpha)\Gamma(ds)\right\}\right] dt \\
&=: -I_1 + I_2.
\end{aligned} \tag{5.1.17}$$

We are now going to estimate I_1. We have

$$\begin{aligned}
I_1 &= \int_{-\infty}^{\infty} \cos tx\, e^{-|t|^\alpha \sigma_1^\alpha} \left[\int_{S_2} (|ts_1 + rs_2|^\alpha - |ts_1|^\alpha)\Gamma(ds)\right] dt \\
&\quad - i\int_{-\infty}^{\infty} \sin tx\, e^{-|t|^\alpha \sigma_1^\alpha} \left[\int_{S_2} (|ts_1 + rs_2|^\alpha - |ts_1|^\alpha)\Gamma(ds)\right] dt \\
&=: I_{11} - iI_{12}.
\end{aligned} \tag{5.1.18}$$

Lemma 5.1.4 (Relations (5.1.13) and (5.1.15)) implies that for all $0 < |r| \le 1$ (say), $|I_{12}| \le C|r|$. Here and in the sequel, C denotes a positive constant, which may change from line to line. Moreover, $|I_{11}|$ equals

$$\begin{aligned}
&\left|\int_0^\infty \cos tx\, e^{-t^\alpha \sigma_1^\alpha} \left[\int_{S_2} (|ts_1 + rs_2|^\alpha + |ts_1 - rs_2|^\alpha - 2|ts_1|^\alpha)\Gamma(ds)\right] dt\right| \\
&\le \int_0^\infty e^{-t^\alpha \sigma_1^\alpha} \left[\int_{S_2} \left||ts_1 + rs_2|^\alpha + |ts_1 - rs_2|^\alpha - 2|ts_1|^\alpha\right|\Gamma(ds)\right] dt.
\end{aligned} \tag{5.1.19}$$

Now use Lemma 5.1.4 to bound the inner integral in the right-hand side of (5.1.19). Specifically, use the first inequality of (5.1.14) for $0 < \alpha \le 1$ (note: $\nu < 1 \le 2 - \alpha$) and (5.1.16) for $1 < \alpha < 2$. We obtain for any $0 < |r| \le 1$,

$$|I_{11}| \le C|r|^{\alpha+\nu}. \tag{5.1.20}$$

We now turn to the term I_2 in (5.1.17). Using the elementary inequalities, valid for any $0 < y \le T$,

$$y - \frac{y^2}{2} \le 1 - e^{-y} \le y,$$

$$y \leq e^y - 1 \leq y + \frac{y^2}{2} e^T,$$

we conclude, using Relations (5.1.12) and (5.1.15) of Lemma 5.1.4, that $|I_2|$ is bounded above by

$$C \int_{-\infty}^{\infty} e^{-|t|^\alpha \sigma_1^\alpha} e^{\Gamma(S_2)|r|^\alpha} \left[\int_{S_2} (|ts_1 + rs_2|^\alpha - |ts_1|^\alpha) \Gamma(ds) \right]^2 dt$$

for $0 < \alpha \leq 1$, and by

$$C \int_{-\infty}^{\infty} e^{-|t|^\alpha \sigma_1^\alpha} e^{\alpha \Gamma(S_2)(|r|^\alpha + |r||t|^{\alpha-1})} \left[\int_{S_2} (|ts_1 + rs_2|^\alpha - |ts_1|^\alpha) \Gamma(ds) \right]^2 dt$$

for $1 < \alpha < 2$.

In the case $1 < \alpha \leq 2$, we may show that for all $0 < |r| \leq 1$

$$|I_2| \leq Cr^2 \tag{5.1.21}$$

by using (5.1.15). Consider, now, the case $0 < \alpha \leq 1$. If $1/2 < \alpha \leq 1$, then (5.1.13) implies that for all $0 < |r| \leq 1$,

$$|I_2| \leq Cr^2. \tag{5.1.22}$$

If $\alpha < 1/2$, then using both (5.1.12) and (5.1.13), we obtain, for $0 < |r| \leq 1$,

$$\begin{aligned} |I_2| &\leq C\Big(\int_{|t| \leq |r|} + \int_{|t| > |r|} \Big) \Big[\int_{S_2} (|ts_1 + rs_2|^\alpha - |ts_1|^\alpha) \Gamma(ds) \Big]^2 dt \\ &\leq C \int_{|t| \leq |r|} |r|^{2\alpha} dt + C \int_{|t| > r} r^2 |t|^{2\alpha - 2} dt \\ &\leq C|r|^{2\alpha+1}. \end{aligned} \tag{5.1.23}$$

If $\alpha = 1/2$, then consider $\nu' = (1-\epsilon)/2 < 1 - \alpha$ with $0 < \epsilon < 1$ and use the first inequality of (5.1.13) to obtain, for $0 \leq |r| \leq 1$,

$$|I_2| \leq C|r|^{2-\epsilon}. \tag{5.1.24}$$

We now gather the intermediate results. By (5.1.9), (5.1.17) and (5.1.18),

$$\begin{aligned} &1 - \text{Re}\, \phi_{X_2|x}(r) \\ &= -(2\pi f_{X_1}(x))^{-1}\, \text{Re}\,(-I_1 + I_2) \\ &= (2\pi f_{X_1}(x))^{-1}(I_{11} - \text{Re}\, I_2) \leq (2\pi f_{X_1}(x))^{-1}(|I_{11}| + |I_2|). \end{aligned}$$

Hence, by (5.1.21), (5.1.22), (5.1.23), (5.1.24) and (5.1.20),

$$1-\mathrm{Re}\,\phi_{X_2|x}(r) \le \begin{cases} C(f_{X_1}(x))^{-1}|r|^{\alpha+\nu}, & \text{if } 0<\alpha\le 1,\ 1-\alpha<\nu<1 \\ & \text{or if } 1<\alpha<2, 0<\nu<2-\alpha, \\ C(f_{X_1}(x))^{-1}|r|^2, & \text{if } 1<\alpha<2 \text{ and } \nu=2-\alpha, \end{cases} \tag{5.1.25}$$

for all $0\le|r|\le 1$.

Part (i) of Theorem 5.1.2 implies $E(|X_2|^p|X_1 = x) < \infty$ a.e. for any $p<\alpha+\nu$ in the cases listed in the first inequality of (5.1.25). Moreover, for $1<\alpha<2$, it implies $E(|X_2|^2|X_1=x)<\infty$ a.e. if $\nu=2-\alpha$. This completes the proof. ∎

Remark. Condition (5.1.10) involves the spectral measure Γ of the vector (X_1, X_2). Often, however, (X_1, X_2) is given by its integral representation

$$(X_1, X_2) \stackrel{d}{=} \Big(\int_E f_1(x)M(dx), \int_E f_2(x)M(dx)\Big), \tag{5.1.26}$$

where M is an α-stable random measure on the measure space $(E, \mathcal{E})$ with control measure m and skewness intensity $\beta(\cdot)$: $E\to[-1,1]$. It is easy to see by a change of variables that

$$\int_{S_2} \frac{|s_2|^{\alpha+\nu}}{|s_1|^{\nu}}\Gamma(ds) = \int_{E_+} \frac{|f_2(x)|^{\alpha+\nu}}{|f_1(x)|^{\nu}} m(dx), \tag{5.1.27}$$

where $E_+ = \{x\in E:\ f_1^2(x)+f_2^2(x)\ne 0\}$. (For a formal proof see Proposition 5.2.10 below.) Hence, Condition (5.1.10) is equivalent to the finiteness of the right-hand side of (5.1.27).

The following example shows that without additional assumptions (namely, $X_2 = aX_1$), it is not possible to obtain conditional moments of order $p\ge 2\alpha+1$ by merely supposing $\nu\ge\alpha+1$.

Example 5.1.5 Let Y_1 and Y_2 be i.i.d. $S\alpha S$ random variables and set $X_1 = Y_1+Y_2$ and $X_2 = Y_1-Y_2$. The spectral measure Γ of (X_1, X_2) is concentrated on the four points $(2^{1/2}, 2^{1/2})$, $(-2^{1/2}, -2^{1/2})$, $(2^{1/2}, -2^{1/2})$, $(-2^{1/2}, 2^{1/2})$ of the unit circle S_2 and hence, in this case, $\nu^*=\infty$, where $\nu^* = \sup\{\nu\ge 0:\ (5.1.10) \text{ holds}\}$.

On the other hand, let $g(x)$ be the density of Y_1 or Y_2 and let $f_{X_1}(x)$ be the density of X_1. Since $g(x)\sim C|x|^{-1-\alpha}$ as $|x|\to\infty$,

$$E(|X_2|^p|X_1=x_1) = \frac{1}{f_{X_1}(x_1)}\int_{-\infty}^{\infty} |x_2|^p \frac{1}{2} g\Big(\frac{x_1+x_2}{2}\Big) g\Big(\frac{x_1-x_2}{2}\Big) dx_2$$

is infinite for $p=1+2\alpha$ but is finite for all $p<1+2\alpha$.

The following proposition generalizes Example 5.1.5. It shows that one always has $E[|X_2|^{2\alpha+1}|X_1] = \infty$ for linearly independent $S\alpha S$ X_1, X_2, with a discrete spectral measure Γ.

Proposition 5.1.6 *Let (X_1, X_2) be an $S\alpha S$ random vector, $0 < \alpha < 2$. Suppose that there are points $\mathbf{s}^{(1)} \neq \pm\mathbf{s}^{(2)}$ such that the spectral measure assigns a positive mass to both. Then*

$$E\big(|X_2|^p|X_1\big) = \infty$$

for all $p \geq 1 + 2\alpha$.

PROOF: Assign the mass at the above two points with plus and minus signs to (Y_1, Y_2), all the rest to (Z_1, Z_2). Then (Y_1, Y_2) and (Z_1, Z_2) are independent (Proposition 2.3.7), and they sum to (X_1, X_2). By the argument of Example 5.1.5, $E(|Y_2|^p|Y_1) = \infty$ a.s. for $p \geq 1 + 2\alpha$ and, by independence, $E(|Y_2 + Z_2|^p|Y_1, Z_1, Z_2) = \infty$ a.s., and so

$$E\big(|X_2|^p|X_1\big) = E\big(E(|Y_2 + Z_2|^p|Y_1, Z_1, Z_2)\big|X_1\big) = \infty \text{ a.s.} \quad \blacksquare$$

Remark. It seems reasonable to conjecture that, even when Γ is not concentrated on a finite number of points, $E[|X_2|^{2\alpha+1}|X_1] = \infty$ unless X_1 and X_2 are linearly dependent. Exercise 5.2 illustrates what can happen in the skewed case.

5.2 Analytic representations of the non-linear regression functions

Let (X_1, X_2) be α-stable, $0 < \alpha < 2$, with spectral representation $(\Gamma, \boldsymbol{\mu}^0)$, i.e., characteristic function as in Theorem 2.3.1. Then $X_1 \sim S_\alpha(\sigma_1, \beta_1, \mu_1)$, where

$$\sigma_1 = \Big(\int_{S_2} |s_1|^\alpha \Gamma(d\mathbf{s})\Big)^{1/\alpha} \tag{5.2.1}$$

is the *scale parameter* of X_1,

$$\beta_1 = \frac{1}{\sigma_1^\alpha} \int_{S_2} s_1^{<\alpha>} \Gamma(d\mathbf{s}) \in [-1, 1] \tag{5.2.2}$$

is the *skewness parameter* of X_1, and

$$\mu_1 = \begin{cases} \mu_1^0 & \text{if } \alpha \neq 1, \\ \mu_1^0 - \frac{2}{\pi}\int_{S_2} s_1 \ln|s_1|\Gamma(d\mathbf{s}) & \text{if } \alpha = 1, \end{cases} \tag{5.2.3}$$

is the shift parameter of X_1 (see Example 2.3.4). In order to study the regression $E(X_2|X_1 = x)$, we must ensure that it is well defined. Clearly,

$$\alpha > 1 \Rightarrow E|X_2| < \infty \Rightarrow E\big(|X_2||X_1 = x\big) < \infty \text{ for a.e. } x.$$

When $\alpha \le 1$, we assume the following:

Condition 5.2.1 *If (X_1, X_2) is α-stable with spectral measure Γ and $\alpha \le 1$, then there is a number $\nu > 1 - \alpha$ such that*

$$\int_{S_2} \frac{\Gamma(ds)}{|s_1|^\nu} < \infty. \tag{5.2.4}$$

Theorem 5.1.3 ensures then $E(|X_2||X_1 = x) < \infty$ for a.e. x. Observe that the choice $\nu > 0$ is adequate when $\alpha = 1$ and the choice $\nu = 1$ is adequate for all $\alpha \le 1$.

Since we are interested in the regression $E(X_2|X_1 = x)$ as a function of x, we assume $\sigma_1 > 0$, because $\sigma_1 = 0$ implies that X_1 is degenerate and hence is $E(X_2|X_1) = EX_2$. By virtue of (5.2.1), $\sigma_1 > 0$ is equivalent to

$$\Gamma(S_2 \backslash \{(0,1) \cup (0,-1)\}) > 0. \tag{5.2.5}$$

We also assume, without loss of generality, $\boldsymbol{\mu}^0 = (\mu_1^0, \mu_2^0) = \mathbf{0}$, i.e., that (X_1, X_2) has representation $(\Gamma, \mathbf{0})$, because if $\boldsymbol{\mu}^0 \neq \mathbf{0}$, then setting $X_1 = \widetilde{X}_1 + \mu_1^0$ and $X_2 = \widetilde{X}_2 + \mu_2^0$ yields

$$E(X_2|X_1 = x) = E(\widetilde{X}_2|\widetilde{X}_1 = x - \mu_1^0) + \mu_2^0,$$

with $(\widetilde{X}_1, \widetilde{X}_2)$ having representation $(\widetilde{\Gamma}, \mathbf{0})$.

When $\boldsymbol{\mu}^0 = \mathbf{0}$, the density function of X_1, $f_{X_1}(x) = \frac{1}{2\pi}\int_{-\infty}^{\infty} e^{-itx}\phi_{X_1}(t)dt$, equals

$$\begin{aligned} f_{X_1}(x) &= \frac{1}{2\pi}\int_{-\infty}^{\infty} e^{-itx} \exp\{-\sigma_1^\alpha |t|^\alpha + ia\beta_1\sigma_1^\alpha t^{<\alpha>}\}dt \\ &= \frac{1}{\pi}\int_0^\infty e^{-\sigma_1^\alpha t^\alpha} \cos(tx - a\beta_1\sigma_1^\alpha t^\alpha)dt \end{aligned} \tag{5.2.6}$$

if $\alpha \neq 1$ (where $a = \tan(\alpha\pi/2)$), and

$$\begin{aligned} f_{X_1}(x) &= \frac{1}{2\pi}\int_{-\infty}^{\infty} e^{-itx} \exp\Big\{-\sigma_1|t| - i\beta_1\sigma_1\frac{2}{\pi}t\ln|t| + i\mu_1 t\Big\}dt \\ &= \frac{1}{\pi}\int_0^\infty e^{-\sigma_1 t} \cos\Big(t(x - \mu_1) + \frac{2}{\pi}\beta_1\sigma_1 t\ln t\Big)dt \end{aligned} \tag{5.2.7}$$

if $\alpha = 1$, where $\mu_1 = -\frac{2}{\pi}\int_{S_2} s_1 \ln|s_1|\Gamma(ds)$.

The following theorem provides an explicit formula for the regression in the case $\alpha \neq 1$.

Theorem 5.2.2 (case $\alpha \neq 1$). *Let (X_1, X_2) be α-stable, $\alpha \neq 1$, with spectral representation $(\Gamma, \mathbf{0})$. If $0 < \alpha < 1$, let (X_1, X_2) satisfy Condition 5.2.1.*

Then, for almost every x,

$$E(X_2|X_1 = x) = \kappa x + \frac{a(\lambda - \beta_1 \kappa)}{1 + a^2\beta_1^2}\left[a\beta_1 x + \frac{1 - xH(x)}{\pi f_{X_1}(x)}\right], \tag{5.2.8}$$

where

$$H(x) = \int_0^\infty e^{-\sigma_1^\alpha t^\alpha} \sin\left(tx - a\beta_1\sigma_1^\alpha t^\alpha\right) dt, \tag{5.2.9}$$

$$\kappa = \frac{[X_2, X_1]_\alpha}{\sigma_1^\alpha} = \frac{\int_{S_2} s_2 s_1^{<\alpha-1>}\Gamma(d\mathbf{s})}{\sigma_1^\alpha}, \tag{5.2.10}$$

$$\lambda = \frac{\int_{S_2} s_2|s_1|^{\alpha-1}\Gamma(d\mathbf{s})}{\sigma_1^\alpha}, \tag{5.2.11}$$

and where $a = \tan\frac{\pi\alpha}{2}$, and σ_1, β_1 and f_{X_1} are, respectively, the scale parameter (5.2.1), the skewness parameter (5.2.2) and the probability density function (5.2.6) of the random variable X_1.

If $\alpha < 1$ and $\beta_1 = 1$, Relation (5.2.8) is well defined only for $x \geq 0$, and if $\alpha < 1$ and $\beta_1 = -1$, it is well defined only for $x \leq 0$.

PROOF: Since (5.2.5) holds, the characteristic function $\phi_{\mathbf{X}}(t, r)$ of $\mathbf{X} = (X_1, X_2)$ is absolutely integrable with respect to t and therefore the conditional characteristic function $\phi_{X_2|x}(r)$ of X_2 given $X_1 = x$ equals

$$\phi_{X_2|x}(r) = 1 + \frac{1}{2\pi f_{X_1}(x)}\int_{-\infty}^\infty e^{-itx}(\phi_{\mathbf{X}}(t, r) - \phi_{X_1}(t))dt.$$

(See Section 5.1 and Relation (5.1.8).) Hence, for almost any x,

$$\begin{aligned} E(X_2|X_1 = x) &= -i\phi'_{X_2|x}(0) \\ &= -\frac{i}{2\pi f_{X_1}(x)}\frac{\partial}{\partial r}\int_{-\infty}^\infty e^{-itx}\phi_{\mathbf{X}}(t, r)dt\Big|_{r=0}. \end{aligned} \tag{5.2.12}$$

We start with the evaluation of $\frac{\partial}{\partial r}\phi_{\mathbf{X}}(t, r)|_{r=0}$. For any $t \neq 0$,

$$\frac{\partial}{\partial r}\phi_{\mathbf{X}}(t, r)\Big|_{r=0} = \phi_{\mathbf{X}}(t, 0)\lim_{r\to 0}\frac{1}{r}\left(\frac{\phi_{\mathbf{X}}(t, r)}{\phi_{\mathbf{X}}(t, 0)} - 1\right), \tag{5.2.13}$$

where $\phi_{\mathbf{X}}(t, r)/\phi_{\mathbf{X}}(t, 0) = e^{-u(t,r)}$ with

$$u(t, r) = \int_{S_2}[|ts_1 + rs_2|^\alpha(1 - ia\,\text{sign}\,(ts_1 + rs_2)) - |ts_1|^\alpha(1 - ia\,\text{sign}\,ts_1)]\Gamma(d\mathbf{s}).$$

Since $u(t,r) \to 0$ as $r \to 0$, we have

$$\begin{aligned}
&\lim_{r\to 0} \frac{1}{r}\left(\frac{\phi_{\mathbf{X}}(t,r)}{\phi_{\mathbf{X}}(t,0)} - 1\right) \\
&\quad = \lim_{r\to 0} \frac{\exp\{-u(t,r)\} - 1}{u(t,r)} \lim_{r\to 0} \frac{u(t,r)}{r} \\
&\quad = -\lim_{r\to 0} \frac{u(t,r)}{r} \\
&\quad = -\Big[\lim_{r\to 0} \int_{S_2} \frac{1}{r}(|ts_1 + rs_2|^\alpha - |ts_1|^\alpha)\Gamma(d\mathbf{s}) \\
&\qquad -ia \lim_{r\to 0} \int_{S_2} \frac{1}{r}((ts_1 + rs_2)^{<\alpha>} - (ts_1)^{<\alpha>})\Gamma(d\mathbf{s}))\Big] \\
&\quad =: -[L_1 - iaL_2].
\end{aligned}$$

To evaluate L_2, we express the corresponding integral as a sum of two integrals $Q_1(t,r)$ and $Q_2(t,r)$, the first over $S_2 \cap \{\mathbf{s} : |ts_1| \geq 2|r|\}$ and the second over $S_2 \cap \{s : |ts_1| < 2|r|\}$. The mean value theorem gives

$$(ts_1 + rs_2)^{<\alpha>} - (ts_1)^{<\alpha>} = \alpha s_2 r\, u^{\alpha-1},$$

where $u \in (|ts_1| \wedge |ts_1 + rs_2|,\ |ts_1| \vee |ts_1 + rs_2|)$, and so, for any $s_1 \neq 0$, the integrand of Q_1 converges to $s_2|ts_1|^{\alpha-1}$. The integrand is dominated by an integrable function because $|ts_1| \geq 2|r|$ implies $u \in \left[\frac{|ts_1|}{2},\ 2|t|\right]$ and therefore $|u|^{\alpha-1} \leq |ts_1/2|^{\alpha-1} + |2t|^{\alpha-1}$. This is certainly integrable with respect to Γ if $\alpha > 1$. If $\alpha < 1$, on S_2,

$$|u|^{\alpha-1} \leq C|t|^{\alpha-1}(|s_1|^{\alpha-1+\nu}|s_1|^{-\nu} + 1) \leq C|t|^{\alpha-1}(|s_1|^{-\nu} + 1),$$

which, by (5.2.4), is integrable with respect to Γ. Therefore, by the dominated convergence theorem,

$$\lim_{r\to 0} Q_1(t,r) = \alpha|t|^{\alpha-1} \int_{S_2} s_2|s_1|^{\alpha-1}\Gamma(d\mathbf{s}).$$

Now consider

$$Q_2(t,r) = \int_{S_2} \frac{1}{r} 1(|ts_1| < 2|r|)((ts_1 + rs_2)^{<\alpha>} - (ts_1)^{<\alpha>})\Gamma(d\mathbf{s}).$$

Suppose firstly $\alpha < 1$. When $|ts_1| < 2|r|$ and $(s_1, s_2) \in S_2$, the integrand is majorized by

$$\begin{aligned}
|r|^{-1}C|rs_2|^\alpha &\leq |r|^{-1}C|r|^\alpha|s_1|^\nu|s_1|^{-\nu} \leq |r|^{-1}C'|r|^{\alpha+\nu}|t|^{-\nu}|s_1|^{-\nu} \\
&\leq C'|t|^{-\nu}|s_1|^{-\nu},
\end{aligned}$$

for, say, $|r| \le 1$, as $\alpha + \nu - 1 > 0$. Since this is integrable with respect to Γ, we get $\lim_{r \to 0} Q_2(t,r) = 0$ by the dominated convergence theorem. The case $\alpha > 1$ is similar. Therefore, by (5.2.11),

$$L_2 = \alpha |t|^{\alpha-1} \int_{S_2} s_2 |s_1|^{\alpha-1} \Gamma(d\mathbf{s}) = \alpha \sigma_1^\alpha \lambda |t|^{\alpha-1}.$$

Turning to L_1, we have

$$L_1 = \alpha t^{<\alpha-1>} \int_{S_2} s_2 s_1^{<\alpha-1>} \Gamma(d\mathbf{s}) = \alpha \sigma_1^\alpha \kappa t^{<\alpha-1>}$$

by (5.2.10). We conclude by (5.2.13),

$$\left.\frac{\partial \phi_{\mathbf{X}}(t,r)}{\partial r}\right|_{r=0} = e^{-\sigma_1^\alpha |t|^\alpha} e^{ia\beta_1 \sigma_1^\alpha t^{<\alpha>}} \alpha \sigma_1^\alpha (\kappa t^{<\alpha-1>} - ia\lambda |t|^{\alpha-1}). \quad (5.2.14)$$

A similar argument applied to (5.2.12) yields

$$E(X_2|X_1 = x) = -\frac{i}{2\pi f_{X_1}(x)} \int_{-\infty}^{\infty} e^{-itx} \left(\left.\frac{\partial}{\partial r} \phi_{\mathbf{X}}(t,r)\right|_{r=0} \right) dt. \quad (5.2.15)$$

Substituting (5.2.14) in (5.2.15), we obtain

$$\begin{aligned} & E(X_2|X_1 = x) \\ &= \frac{i\alpha\sigma_1^\alpha}{2\pi f_{X_1}(x)} \int_{-\infty}^{\infty} e^{-itx} e^{ia\beta_1\sigma_1^\alpha t^{<\alpha>}} e^{-\sigma_1^\alpha |t|^\alpha} (\kappa t^{<\alpha-1>} - ia\lambda |t|^{\alpha-1}) dt \\ &= \frac{\alpha\sigma_1^\alpha}{\pi f_{X_1}(x)} [\kappa(I_{11} + I_{12}) + a\lambda(I_{21} + I_{22})], \end{aligned} \quad (5.2.16)$$

where

$$\begin{aligned} I_{11} &= \int_0^\infty \sin tx \cos(a\beta_1\sigma_1^\alpha t^\alpha) e^{-\sigma_1^\alpha t^\alpha} t^{\alpha-1} dt, \\ I_{12} &= -\int_0^\infty \cos tx \sin(a\beta_1\sigma_1^\alpha t^\alpha) e^{-\sigma_1^\alpha t^\alpha} t^{\alpha-1} dt, \\ I_{21} &= \int_0^\infty \cos tx \cos(a\beta_1\sigma_1^\alpha t^\alpha) e^{-\sigma_1^\alpha t^\alpha} t^{\alpha-1} dt, \\ I_{22} &= \int_0^\infty \sin tx \sin(a\beta_1\sigma_1^\alpha t^\alpha) e^{-\sigma_1^\alpha t^\alpha} t^{\alpha-1} dt. \end{aligned}$$

After integrating by parts,

$$I_{11} = -\frac{x}{K} \int_0^\infty e^{-\sigma_1^\alpha t^\alpha} [a\beta_1 \sin(a\beta_1\sigma_1^\alpha t^\alpha) - \cos(a\beta_1\sigma_1^\alpha t^\alpha)] \cos tx \, dt,$$

$$
\begin{aligned}
I_{12} &= -\frac{a\beta_1}{K} + \frac{x}{K}\int_0^\infty e^{-\sigma_1^\alpha t^\alpha}[a\beta_1 \cos(a\beta_1\sigma_1^\alpha t^\alpha) + \sin(a\beta_1\sigma_1^\alpha t^\alpha)] \sin tx\, dt,\\
I_{21} &= \frac{1}{K} + \frac{x}{K}\int_0^\infty e^{-\sigma_1^\alpha t^\alpha}[a\beta_1 \sin(a\beta_1\sigma_1^\alpha t^\alpha) - \cos(a\beta_1\sigma_1^\alpha t^\alpha)] \sin tx\, dt,\\
I_{22} &= -\frac{x}{K}\int_0^\infty e^{-\sigma_1^\alpha t^\alpha}[a\beta_1 \cos(a\beta_1\sigma_1^\alpha t^\alpha) + \sin(a\beta_1\sigma_1^\alpha t^\alpha)] \cos tx\, dt,
\end{aligned}
$$

where $K = \alpha\sigma_1^\alpha(1 + a^2\beta_1^2)$. Therefore, by (5.2.6) and (5.2.9),

$$
\begin{aligned}
I_{11} + I_{12} &= \frac{1}{\alpha\sigma_1^\alpha(1 + a^2\beta_1^2)}[-a\beta_1 + x\pi f_{X_1}(x) + a\beta_1 x H(x)],\\
I_{21} + I_{22} &= \frac{1}{\alpha\sigma_1^\alpha(1 + a^2\beta_1^2)}[1 + a\beta_1 x\pi f_{X_1}(x) - xH(x)].
\end{aligned}
$$

Substituting these expressions in (5.2.16) yields

$$
\begin{aligned}
E(X_2|X_1 = x) &= \frac{\kappa + a^2\beta_1\lambda}{1 + a^2\beta_1^2}x + \frac{a(\lambda - \beta_1\kappa)}{1 + a^2\beta_1^2}\left[\frac{1}{\pi f_{X_1}(x)} - \frac{xH(X)}{\pi f_{X_1}(x)}\right]\\
&= \kappa x + \frac{a(\lambda - \beta_1\kappa)}{1 + a^2\beta_1^2}\left[a\beta_1 x + \frac{1 - xH(x)}{\pi f_{X_1}(x)}\right]. \quad \blacksquare
\end{aligned}
$$

Remarks

1. To understand the reason for the last statement in the theorem, recall that when $\alpha < 1$ and $\beta_1 = 1$, the random variable X_1 is totally skewed to the right and when $\alpha < 1$ and $\beta_1 = -1$, X_1 is totally skewed to the left. The density function $f_{X_1}(x)$ vanishes for $x < 0$ when $\beta_1 = 1$ and it vanishes for $x > 0$ when $\beta_1 = -1$. Therefore, conditioning with respect to $X_1 = x$ makes no sense when $x < 0$ if $\beta_1 = 1$ or when $x > 0$ if $\beta_1 = -1$. When either $\alpha < 1$, $\beta_1 \neq \pm 1$ or $\alpha \geq 1$, the support of the density $f_{X_1}(x)$ is the whole real line.

2. The normalized covariation $[X_2, X_1]_\alpha/\sigma_1^\alpha$ is called here κ. If $\mathbf{X}$ is symmetric α-stable, $\alpha \neq 1$, we have Γ symmetric, $\beta_1 = 0$ and $\lambda = 0$ and hence,
$$E(X_2|X_1 = x) = \kappa x \quad \text{for a.e. } x.$$
We thus recover the statement of Theorem 4.1.2 in the case $1 < \alpha < 2$.

3. Observe that
$$\frac{H(x)}{\pi f_{X_1}(x)} = \frac{\text{Im} \int_0^\infty e^{-itx}\phi_{X_1}(t)dt}{\text{Re} \int_0^\infty e^{-itx}\phi_{X_1}(t)dt}.$$

4. As can easily be seen from the proof of the theorem, the following expression is equivalent to (5.2.8):

$$\begin{aligned} & E(X_2|X_1 = x) \\ = \quad & \kappa x + \alpha\sigma_1^\alpha a(\lambda - \beta_1\kappa) \frac{\int_0^\infty e^{-\sigma_1^\alpha t^\alpha} t^{\alpha-1} \cos(xt - a\beta_1\sigma_1^\alpha t^\alpha)dt}{\int_0^\infty e^{-\sigma_1^\alpha t^\alpha} \cos(xt - a\beta_1\sigma_1^\alpha t^\alpha)dt}. \end{aligned} \tag{5.2.17}$$

5. The constants κ and λ in the theorem are finite when $\alpha \leq 1$ because, by (5.2.4),

$$\left| \int_{S_2} s_2 |s_1|^{\alpha-1} \Gamma(ds) \right| \leq \int_{S_2} |s_2||s_1|^{\alpha-1+\nu} \frac{\Gamma(ds)}{|s_1|^\nu} \leq \int_{S_2} \frac{\Gamma(ds)}{|s_1|^\nu} < \infty$$

since $|s_1| \leq 1$, $|s_2| \leq 1$ and $\alpha - 1 + \nu > 0$.

6. If X_1 is (marginally) symmetric, then $\beta_1 = 0$ and

$$E(X_2|X_1 = x) = \kappa x + \left(\tan \frac{\pi\alpha}{2}\right)\lambda \frac{1}{\pi f_{X_1}(x)} \left(1 - x \int_0^\infty e^{-\sigma_1^\alpha t^\alpha} \sin txdt\right).$$

We now turn to the case $\alpha = 1$.

Theorem 5.2.3 (case $\alpha = 1$). *Let (X_1, X_2) be α-stable with $\alpha = 1$ and spectral representation $(\Gamma, \mathbf{0})$ satisfying Condition 5.2.1. Then, for almost every x,*

$$E(X_2|X_1 = x) = -\frac{2\sigma_1}{\pi} k_0 + \kappa(x - \mu_1) + \frac{\lambda - \beta_1\kappa}{\beta_1} \left[(x - \mu_1) - \sigma_1 \frac{U(x)}{\pi f_{X_1}(x)}\right] \tag{5.2.18}$$

if $\beta_1 \neq 0$, and

$$E(X_2|X_1 = x) = -\frac{2\sigma_1}{\pi} k_0 + \kappa(x - \mu_1) - \frac{2\sigma_1}{\pi} \lambda \frac{V(x)}{\pi f_{X_1}(x)} \tag{5.2.19}$$

if $\beta_1 = 0$. Here,

$$U(x) = \int_0^\infty e^{-\sigma_1 t} \sin\left(t(x - \mu_1) + \frac{2}{\pi}\beta_1\sigma_1 t \ln t\right) dt, \tag{5.2.20}$$

$$V(x) = \int_0^\infty e^{-\sigma_1 t}(1 + \ln t)(\cos t(x - \mu_1))dt, \tag{5.2.21}$$

$$\begin{aligned} k_0 &= \frac{1}{\sigma_1} \int_{S_2} s_2 \ln |s_1| \Gamma(ds), \\ \kappa &= \frac{[X_2, X_1]_1}{\sigma_1} = \frac{1}{\sigma_1} \int_{S_2} s_2 s_1^{<0>} \Gamma(ds) = \frac{1}{\sigma_1} \int_{S_2} s_2 \operatorname{sign}(s_1) \Gamma(ds), \\ \lambda &= \frac{1}{\sigma_1} \int_{S_2} s_2 \Gamma(ds) \end{aligned}$$

and

$$\mu_1 = -\frac{2}{\pi}\int_{S_2} s_1 \ln|s_1|\Gamma(ds). \tag{5.2.22}$$

PROOF: We have, as in (5.2.13),

$$\begin{aligned} \left.\frac{\partial\phi_{\mathbf{X}}(t,r)}{\partial r}\right|_{t=0} &= \phi_{\mathbf{X}}(t,0)\lim_{r\to 0}\frac{1}{r}\left(\frac{\phi_{\mathbf{X}}(t,r)}{\phi_{\mathbf{X}}(t,0)}-1\right) \\ &=: -\phi_{\mathbf{X}}(t,0)\left[L_1 + i\frac{2}{\pi}L_2\right], \end{aligned} \tag{5.2.23}$$

where

$$\begin{aligned} L_1 &= \lim_{r\to 0}\int_{S_2}\frac{1}{r}[|ts_1+rs_2|-|ts_1|]\Gamma(ds), \\ L_2 &= \lim_{r\to 0}\int_{S_2}\frac{1}{r}[(ts_1+rs_2)\ln|ts_1+rs_2| - ts_1\ln|ts_1|]\Gamma(ds). \end{aligned}$$

Clearly,

$$L_1 = \text{sign}\, t\int_{S_2} s_2\,\text{sign}\,(s_1)\Gamma(ds) = \sigma_1\kappa t^{<0>}. \tag{5.2.24}$$

As in the case $\alpha \neq 1$, to evaluate L_2, we write $L_2 = \lim_{r\to 0}(Q_1(t,r)+Q_2(t,r))$ where Q_1 involves integration over $S_2 \cap \{\mathbf{s} : |ts_1| \geq 2|r|\}$ and Q_2 over $S_2 \cap \{\mathbf{s} : |ts_1| < 2|r|\}$, and we apply the dominated convergence theorem.

We give details only for Q_2. Let $f : [0,\infty) \to [0,\infty)$ be defined by $f(r) = r|\ln r|$, $f(0) = 0$. For $|r|$ small enough ($0 < |r| < e^{-1}$), f is monotone increasing and therefore, when $|ts_1| < 2|r|$ and $0 < |r| < (3e)^{-1}$, one has $|r|^{-1}(f(|ts_1+rs_2|)+f(|ts_1|)) \leq |r|^{-1}(f(3|r|)+f(2|r|)) \leq 2|r|^{-1}f(3|r|)$ $\leq 6|\ln 3|r|| \leq 6|\ln 1/(\frac{3}{2}|ts_1|)| \leq 6(|\ln(\frac{2}{3}|t|)|+|s_1|^{-\nu/2})$, which is integrable by (5.2.4). Applying the dominated convergence theorem, we have $\lim_{r\to 0} Q_2(t,r) = 0$. Therefore,

$$\begin{aligned} L_2 &= \lim_{r\to 0} Q_1(t,r) + \lim_{r\to 0} Q_2(t,r) \\ &= \int_{S_2} s_2(1+\ln|ts_1|)\Gamma(ds) \\ &= (1+\ln|t|)\int_{S_2} s_2\Gamma(ds) + \int_{S_2} s_2\ln|s_1|\Gamma(ds) \\ &= \sigma_1\lambda(1+\ln|t|)+\sigma_1 k_0. \end{aligned} \tag{5.2.25}$$

Substituting (5.2.24) and (5.2.25) in (5.2.23) yields

$$\left.\frac{\partial\phi_{\mathbf{X}}(t,r)}{\partial r}\right|_{r=0} = -\phi_{\mathbf{X}}(t,0)\sigma_1\left[\kappa t^{<0>} + i\frac{2}{\pi}\lambda(1+\ln|t|)+i\frac{2}{\pi}k_0\right].$$

The dominated convergence theorem applied to (5.2.12), with μ_1 as in (5.2.22), gives

$$2\pi f_{X_1}(x)E(X_2|X_1=x) = -i\int_{-\infty}^{\infty} e^{-itx}\left(\frac{\partial\phi_{\mathbf{X}}(t,r)}{\partial r}\bigg|_{r=0}\right)dt$$

$$= i\sigma_1\int_{-\infty}^{\infty} e^{-it(x-\mu_1)}e^{-i\frac{2}{\pi}\beta_1\sigma_1 t\ln|t|}e^{-\sigma_1|t|}\left[\kappa t^{<0>} + i\frac{2}{\pi}\lambda(1+\ln|t|) + i\frac{2}{\pi}k_0\right]dt$$

$$= \frac{\sigma_1}{2}\left[\kappa U(x) - \frac{2}{\pi}\lambda(I_{21}+I_{22}) - \frac{2}{\pi}k_0\pi f_{X_1}(x)\right], \tag{5.2.26}$$

where

$$\begin{aligned} I_{21} &= \int_0^{\infty} e^{-\sigma_1 t}\cos t(x-\mu_1)(1+\ln t)\cos\left(\frac{2}{\pi}\beta_1\sigma_1 t\ln t\right)dt,\\ I_{22} &= -\int_0^{\infty} e^{-\sigma_1 t}\sin t(x-\mu_1)\,(1+\ln t)\sin\left(\frac{2}{\pi}\beta_1\sigma_1 t\ln t\right)dt. \end{aligned}$$

(i) *Case* $\beta_1 \neq 0$: After integrating by parts,

$$\begin{aligned} I_{21} &= \frac{\pi}{2\beta_1}\int_0^{\infty} e^{-\sigma_1 t}\cos t(x-\mu_1)\sin\left(\frac{2}{\pi}\beta_1\sigma_1 t\ln t\right)dt\\ &\quad+\frac{\pi}{2\beta_1\sigma_1}(x-\mu_1)\int_0^{\infty} e^{-\sigma_1 t}\sin t(x-\mu_1)\sin\left(\frac{2}{\pi}\beta_1\sigma_1 t\ln t\right)dt, \end{aligned}$$

$$\begin{aligned} I_{22} &= \frac{\pi}{2\beta_1}\int_0^{\infty} e^{-\sigma_1 t}\sin t(x-\mu_1)\cos\left(\frac{2}{\pi}\beta_1\sigma_1 t\ln t\right)dt\\ &\quad-\frac{\pi}{2\beta_1\sigma_1}(x-\mu_1)\int_0^{\infty} e^{-\sigma_1 t}\cos t(x-\mu_1)\cos\left(\frac{2}{\pi}\beta_1\sigma_1 t\ln t\right)dt, \end{aligned}$$

so that

$$I_{21}+I_{22} = -\frac{\pi}{2\beta_1\sigma_1}(x-\mu_1)\pi f_{X_1}(x) + \frac{\pi}{2\beta_1}U(x).$$

Substituting this expression in (5.2.26) and rearranging the terms, we obtain

$$\begin{aligned} &E(X_2|X_1=x)\\ &\quad= -\frac{2\sigma_1}{\pi}k_0 + \frac{\lambda}{\beta_1}(x-\mu_1) - \frac{\sigma_1}{\beta_1}(\lambda-\beta_1\kappa)\frac{U(x)}{\pi f_{X_1}(x)}\\ &\quad= -\frac{2\sigma_1}{\pi}k_0 + \kappa(x-\mu_1) + \frac{\lambda-\beta_1\kappa}{\beta_1}\left[(x-\mu_1) - \sigma_1\frac{U(x)}{\pi f_{X_1}(x)}\right]. \end{aligned}$$

(ii) *Case* $\beta_1 = 0$: After integrating by parts,

$$U(x) = \int_0^{\infty} e^{-\sigma_1 t}\sin t(x-\mu_1)dt = \frac{\pi f_{X_1}(x)}{\sigma_1}(x-\mu_1).$$

Since

$$I_{21} + I_{22} = \int_0^\infty e^{-\sigma_1 t} \cos t(x - \mu_1)(1 + \ln t) dt = V(x),$$

Relation (5.2.26) becomes

$$\begin{aligned} E(X_2|X_1 = x) \\ &= \frac{\sigma_1}{\pi f_{X_1}(x)} \left[\kappa \frac{\pi f_{X_1}(x)}{\sigma_1}(x - \mu_1) - \frac{2}{\pi}\lambda V(x) - \frac{2}{\pi} k_0 \pi f_{X_1}(x) \right] \\ &= -\frac{2\sigma_1}{\pi} k_0 + \kappa(x - \mu_1) - \frac{2\sigma_1}{\pi}\lambda \frac{V(x)}{\pi f_{X_1}(x)}. \quad \blacksquare \end{aligned}$$

Remarks

1. The shift parameter μ_1 also appears in expression (5.2.7) of $f_{X_1}(x)$.
2. The constants κ and λ are defined as in Theorem 5.2.2. When $\alpha = 1$, the constant λ is proportional to the skewness parameter of X_2.
3. If $\beta_1 \neq 0$,
$$\frac{U(x)}{\pi f_{X_1}(x)} = \frac{\operatorname{Im} \int_0^\infty e^{-itx} \phi_{X_1}(t) dt}{\operatorname{Re} \int_0^\infty e^{-itx} \phi_{X_1}(t) dt}.$$

The following corollary shows that the regression is linear when X_1 is totally skewed to the right ($\beta_1 = 1$) or when it is totally skewed to the left ($\beta_1 = -1$).

Corollary 5.2.4 (case $\beta_1 = \pm 1$). *Suppose that the conditions of Theorems 5.2.2 or 5.2.3 hold. If $\beta_1 = \pm 1$, then, for almost every x,*

$$E(X_2|X_1 = x) = \begin{cases} \kappa x, & \text{if } \alpha \neq 1, \\ -\frac{2\sigma_1}{\pi} k_0 - \kappa\mu_1 + \kappa x, & \text{if } \alpha = 1. \end{cases} \tag{5.2.27}$$

In the case $\alpha < 1$, this relation is well defined only for $x \geq 0$ when $\beta_1 = 1$, and for $x \leq 0$ when $\beta_1 = -1$.

PROOF: $\beta_1 = 1$ implies $s_1 \geq 0$, Γ-a.e., by (5.2.1) and (5.2.2) and therefore $\lambda = \kappa$ by (5.2.10) and (5.2.11). Similarly, $\beta_1 = -1$ implies $s_1 \leq 0$, Γ-a.e. and $\kappa = -\lambda$. In both cases, $\lambda - \beta_1\kappa = 0$, and the corollary follows from Theorems 5.2.2 and 5.2.3. ■

The next result shows that the regression $E(X_2|X_1 = x)$ is typically asymptotically linear as $x \to \pm\infty$. We know from Corollary 5.2.4 that the regression is linear when $\beta_1 = \pm 1$. For other values of β_1, one has:

Corollary 5.2.5 *(Asymptotic relations). Let (X_1, X_2) be α-stable, $0 < \alpha < 2$, with spectral representation $(\Gamma, \mathbf{0})$ and suppose that Condition 5.2.1 holds if $0 < \alpha \le 1$. Then, for $\beta_1 \ne \pm 1$,*

$$E(X_2|X_1 = x) \sim \frac{\kappa + \lambda}{1 + \beta_1} x \ , \ x \to \infty, \tag{5.2.28}$$

and

$$E(X_2|X_1 = x) \sim \frac{\kappa - \lambda}{1 - \beta_1} x \ , \ x \to -\infty. \tag{5.2.29}$$

PROOF: It is sufficient to consider the case $x \to \infty$. Indeed, consider the vector $(-X_1, X_2)$ whose parameters are $\widehat{\beta}_1 = -\beta_1$, $\widehat{\kappa} = -\kappa$ and $\widehat{\lambda} = \lambda$. Since $E(X_2|X_1 = x) = E(X_2| - X_1 = -x)$, one can obtain the asymptotic behavior of $E(X_2|X_1 = x)$ as $x \to -\infty$ from that of $E(X_2|X_1 = x)$ as $x \to \infty$ by replacing β_1 by $-\beta_1$, κ by $-\kappa$, and x by $-x = |x|$. Suppose $\beta_1 \ne \pm 1$.

We study first the cases with $\alpha \ne 1$ and then those with $\alpha = 1$.

(a) case $\alpha \ne 1$, $x \to \infty$: To obtain the asymptotic behavior of $E(X_2|X_1 = x)$, we use its expression in terms of I_{ij}, $i, j = 1, 2$, as given in (5.2.16). By (Titchmarsh 1986, Theorem 126), as $x \to \infty$,

$$\begin{aligned} I_{11} &\sim \Gamma(\alpha)(\sin \tfrac{\pi\alpha}{2}) x^{-\alpha}, \quad I_{12} = o(x^{-\alpha}), \\ I_{21} &\sim \Gamma(\alpha)(\cos \tfrac{\pi\alpha}{2}) x^{-\alpha}, \quad I_{22} = o(x^{-\alpha}), \end{aligned} \tag{5.2.30}$$

when $0 < \alpha < 1$ and also when $1 < \alpha < 2$. (If $1 < \alpha < 2$, use integration by parts to transform the terms $t^{\alpha-1}$ in I_{ij}, $i, j = 1, 2$, into $t^{\alpha-2}$.) Moreover,

$$f_{X_1}(x) \sim \frac{1}{\pi}(1 + \beta_1)\Big(\sin \frac{\pi\alpha}{2}\Big)\sigma_1^\alpha \Gamma(\alpha + 1) x^{-\alpha-1}$$

(Ibragimov & Linnik 1971, Theorem 2.4.2). Substituting in (5.2.16) and using $a = \tan \frac{\pi\alpha}{2}$, we obtain

$$\begin{aligned} E[X_2|X_1 = x] &\sim C\Big(\kappa\Gamma(\alpha)\Big(\sin \frac{\pi\alpha}{2}\Big)x^{-\alpha} + a\lambda\Gamma(\alpha)\Big(\cos \frac{\pi\alpha}{2}\Big)x^{-\alpha}\Big) \\ &\sim \frac{\kappa + \lambda}{1 + \beta_1} x, \end{aligned}$$

where

$$C = \frac{\alpha\sigma_1^\alpha}{\pi} \frac{\pi x^{\alpha+1}}{(1 + \beta_1)(\sin \frac{\pi\alpha}{2})\sigma_1^\alpha \Gamma(1 + \alpha)}.$$

(b) case $\alpha = 1$, $\beta_1 \ne 0$, $x \to \infty$: Since $x - \mu_1 \sim x$ as $x \to \infty$, we may assume $\mu_1 = 0$. Then,

$$U(x) = \int_0^\infty e^{-\sigma_1 t} \sin\Big(tx + \frac{2}{\pi}\beta_1\sigma_1 t \ln t\Big) dt = U_1(x) + U_2(x),$$

where

$$
\begin{aligned}
U_1(x) &= \int_0^\infty e^{-\sigma_1 t} \sin\Big(\frac{2}{\pi}\beta_1\sigma_1 t\ln t\Big)\cos tx\,dt,\\
U_2(x) &= \int_0^\infty e^{-\sigma_1 t} \cos\Big(\frac{2}{\pi}\beta_1\sigma_1 t\ln t\Big)\sin tx\,dt.
\end{aligned}
$$

After integrating by parts,

$$
\begin{aligned}
U_1(x) = -\frac{1}{x}\int_0^\infty & e^{-\sigma_1 t}\Big[-\sigma_1\sin\Big(\frac{2}{\pi}\sigma_1\beta_1 t\ln t\Big)\\
&+\frac{2}{\pi}\sigma_1\beta_1(1+\ln t)\cos\Big(\frac{2}{\pi}\sigma_1\beta_1 t\ln t\Big)\Big]\sin txdt,
\end{aligned}
$$

$$
\begin{aligned}
U_2(x) = \frac{1}{x}+\frac{1}{x}\int_0^\infty & e^{-\sigma_1 t}\Big[-\sigma_1\cos\Big(\frac{2}{\pi}\sigma_1\beta_1 t\ln t\Big)\\
&+\frac{2}{\pi}\sigma_1\beta_1(1+\ln t)\sin\Big(\frac{2}{\pi}\sigma_1\beta_1 t\ln t\Big)\Big]\cos tx\,dt.
\end{aligned}
$$

Since the factors of $\sin tx$ and $\cos tx$ are integrable, we see that $U_1(x) = o(x^{-1})$ and $U_2(x) \sim x^{-1}$ as $x \to \infty$ by the Riemann-Lebesgue lemma (Titchmarsh (1986), Theorem 1). Therefore,

$$U(x) \sim \frac{1}{x}. \tag{5.2.31}$$

Since, as is well known,

$$f_{X_1}(x) \sim \frac{\sigma_1(1+\beta_1)}{\pi}x^{-2}, \tag{5.2.32}$$

we obtain

$$\frac{U(x)}{\pi f_{X_1}(x)} \sim \frac{1}{\sigma_1(1+\beta_1)}x,$$

which, substituted in (5.2.18), gives

$$E(X_2|X_1=x) \sim \kappa x + \frac{\lambda-\beta_1\kappa}{\beta_1}\Big[x - \frac{x}{1+\beta_1}\Big] = \frac{\kappa+\lambda}{1+\beta_1}x.$$

This is the same result as in the case $\alpha \neq 1$.

(c) case $\alpha = 1,\ \beta_1 = 0,\ x \to \infty$: If Y is a $S\alpha S$ random variable with $\alpha = 1$ and scale parameter σ_1, then by the inversion formula (Lukacs (1970), Theorem 3.2.1), $P(0 < Y \le x) = \frac{1}{\pi}\int_0^\infty e^{-\sigma_1 t}t^{-1}\sin tx\,dt$. On the other

hand, because of the symmetry, $\lim_{x\to\infty} P(0 < Y < \infty) = 1/2$. Therefore, $\lim_{x\to\infty} \int_0^\infty e^{-\sigma_1 t} t^{-1} \sin tx \, dt = \frac{\pi}{2}$ and hence

$$\begin{aligned} V(x) &= \int_0^\infty e^{-\sigma_1 t}(1+\ln t)\cos tx \, dt \\ &= \frac{\sigma_1}{x}\int_0^\infty e^{-\sigma_1 t}(1+\ln t)\sin tx \, dt - \frac{1}{x}\int_0^\infty e^{-\sigma_1 t} t^{-1}\sin tx \, dt \\ &\sim -\frac{\pi}{2}\frac{1}{x}, \end{aligned} \tag{5.2.33}$$

since the first integral is $o(1)$ by the Riemann-Lebesgue lemma. Substituting (5.2.33) and (5.2.32) in (5.2.19) yields $E(X_2|X_1 = x) \sim (\kappa + \lambda)x$, i.e., (5.2.28) with $\beta_1 = 0$.

This completes the proof of the corollary. ■

Remark. If $\kappa \pm \lambda \neq 0$, then the regression is asymptotically linear. If either $\kappa + \lambda = 0$ or $\kappa - \lambda = 0$, then the respective tail of the regression does not grow in absolute value as fast as $|x|$.

The following corollary gives a necessary and sufficient condition for linearity of the regression.

Corollary 5.2.6 *(Linearity). Let (X_1, X_2) be α-stable, $0 < \alpha < 2$, with spectral representation $(\Gamma, \mathbf{0})$ and suppose that Condition 5.2.1 holds if $0 < \alpha \leq 1$. Then the regression $E(X_2|X_1 = x)$ is linear if and only if*

$$\lambda = \beta_1 \kappa. \tag{5.2.34}$$

If $\lambda = \beta_1\kappa$, then $E(X_2|X_1 = x)$ is given by (5.2.27) for a.e. x.

PROOF: If $\lambda = \beta_1\kappa$, then $E(X_2|X_1 = x) = \kappa x$ for a.e. x by Theorem 5.2.2 when $\alpha \neq 1$, and $E(X_2|X_1 = x) = -\frac{2\sigma_1}{\pi} k_0 + \kappa(x - \mu_1)$ for a.e. x by Theorem 5.2.3 when $\alpha = 1$.

Suppose, now, $E(X_2|X_1 = x) = Ax + B$ for a.e. x and also, to the contrary, that $\lambda \neq \beta_1\kappa$. In view of Corollary 5.2.4, this contradicts either (5.2.8) or (5.2.18) if $\beta_1 = \pm 1$. If $\beta_1 \neq \pm 1$, then, by linearity,

$$\lim_{x\to\infty} \frac{E(X_2|X_1 = x)}{x} = \lim_{x\to-\infty} \frac{E(X_2|X_1 = x)}{x}.$$

In view of (5.2.28) and (5.2.29), this implies

$$\frac{\kappa + \lambda}{1 + \beta_1} = \frac{\kappa - \lambda}{1 - \beta_1}$$

and hence $\lambda = \beta_1 \kappa$, a contradiction. ∎

In particular, in the $S\alpha S$ case, we obtain the following extension of Theorem 4.1.2:

Corollary 5.2.7 *If (X_1, X_2) is $S\alpha S$, $0 < \alpha \le 2$, with spectral measure Γ and if Condition 5.2.1 holds when $0 < \alpha \le 1$, then*

$$E(X_2|X_1 = x) = \kappa x \textit{ for a.e. } x.$$

Example 5.2.8 The regression can be linear even though (X_1, X_2) is not symmetric. For example, suppose $\alpha > 1$ and let $\mathbf{X} = (X_1, X_2)$ have spectral measure

$$\Gamma = \delta\Big(\Big(\frac{1}{\sqrt{2}}, \frac{1}{\sqrt{2}}\Big)\Big) + \delta\Big(\Big(-\frac{1}{\sqrt{2}}, -\frac{1}{\sqrt{2}}\Big)\Big) + \delta((0, -1)),$$

where $\delta((x_1, x_2))$ denotes a unit mass at the point (x_1, x_2). The vector $\mathbf{X}$ is not symmetric because Γ is not symmetric. However, $\beta_1 = 0$ (X_1 is $S\alpha S$) and $\lambda = 0$. Since (5.2.34) is satisfied in this case, the regression is linear. The slope is $\kappa = 1$ because both σ_1^α and the covariation $\int_{S_2} s_2 s_1^{<\alpha-1>} \Gamma(ds)$ equal $2(\frac{1}{\sqrt{2}})^\alpha$.

Example 5.2.9 The regression may be non-linear even though both components X_1 and X_2 are $S\alpha S$. For example, let $\alpha > 1$ and $\mathbf{X} = (X_1, X_2)$ have spectral measure

$$\Gamma = \delta((1,0)) + \delta((0,1)) + 2^{\alpha/2}\delta\Big(\Big(-\frac{1}{\sqrt{2}}, -\frac{1}{\sqrt{2}}\Big)\Big).$$

In this case, $\mathbf{X}$ is not symmetric, X_1 and X_2 are $S\alpha S$, and the regression is not linear because $\beta_1 = 0$ but $\sigma_1^\alpha \lambda = \int_{S_2} s_2 |s_1|^\alpha \Gamma(ds) = -2^{\alpha/2}(\frac{1}{\sqrt{2}})^{\alpha+1} \neq 0$.

Integral Representation

In applications, (X_1, X_2) is often given by its integral representation

$$(X_1, X_2) \stackrel{d}{=} \left(\int_E f_1(x) M(dx), \int_E f_2(x) M(dx) \right), \tag{5.2.35}$$

where M is an α-stable random measure on the measure space $(E, \mathcal{E})$ with control measure m and skewness intensity $\beta(\cdot) : E \to [-1, 1]$. The following proposition expresses Condition (5.2.4) and the constants of the regression in terms of $f_1, f_2, \beta(\cdot)$ and $m(dx)$.

Proposition 5.2.10

$$\int_{S_2} \frac{\Gamma(ds)}{|s_1|^\nu} < \infty \Leftrightarrow \int_{S_2} \frac{|s_2|^{\alpha+\nu}}{|s_1|^\nu} \Gamma(ds) = \int_{E_+} \frac{|f_2(x)|^{\alpha+\nu}}{|f_1(x)|^\nu} m(dx) < \infty, \tag{5.2.36}$$

where $E_+ = \{x \in E : f_1^2(x) + f_2^2(x) \neq 0\}$. *Moreover,*

$$
\begin{aligned}
\sigma_1^\alpha &= \int_E |f_1(x)|^\alpha m(dx), \\
\beta_1 &= \frac{1}{\sigma_1^\alpha} \int_E f_1(x)^{<\alpha>} \beta(x) m(dx), \\
\mu_1 &= -\frac{2}{\pi} \int_{E_+} f_1(x) \Big(\ln \frac{|f_1(x)|}{\sqrt{f_1^2(x) + f_2^2(x)}} \Big) \beta(x) m(dx) \qquad \text{(case } \alpha = 1\text{)}, \\
k_0 &= \frac{1}{\sigma_1} \int_{E_+} f_2(x) \Big(\ln \frac{|f_1(x)|}{\sqrt{f_1^2(x) + f_2^2(x)}} \Big) \beta(x) m(dx) \qquad (\textit{case } \alpha = 1), \\
\kappa &= \frac{[X_2, X_1]_\alpha}{\sigma_1^\alpha} = \frac{1}{\sigma_1^\alpha} \int_{E_+} f_2(x) f_1(x)^{<\alpha-1>} m(dx), \\
\lambda &= \frac{1}{\sigma_1^\alpha} \int_{E_+} f_2(x) |f_1(x)|^{\alpha-1} \beta(x) m(dx).
\end{aligned}
$$

PROOF: Suppose firstly $\alpha \neq 1$, and set $a = \tan \frac{\pi\alpha}{2}$ and $|\mathbf{f}| = \sqrt{f_1^2 + f_2^2}$. The characteristic function of (X_1, X_2) is

$$
\begin{aligned}
&E \exp\Big\{ i \sum_{j=1}^2 \theta_j \int_E f_j M(dx) \Big\} \\
&= \exp\Big\{ \int_E \Big[-\Big| \sum_{j=1}^2 \theta_j f_j \Big|^\alpha + ia \Big(\sum_{j=1}^2 \theta_j f_j \Big)^{<\alpha>} \beta(x) \Big] m(dx) \Big\} \\
&= \exp\Big\{ \int_{E_+} \Big[-\Big| \sum_{j=1}^2 \theta_j \frac{f_j}{|\mathbf{f}|} \Big|^\alpha + ia \Big(\sum_{j=1}^2 \theta_j \frac{f_j}{|\mathbf{f}|} \Big)^{<\alpha>} \Big] \frac{1+\beta(x)}{2} |\mathbf{f}|^\alpha m(dx) \\
&\quad + \int_{E_+} \Big[-\Big| \sum_{j=1}^2 \theta_j \frac{(-f_j)}{|\mathbf{f}|} \Big|^\alpha + ia \Big(\sum_{j=1}^2 \theta_j \frac{(-f_j)}{|\mathbf{f}|} \Big)^{<\alpha>} \Big] \frac{1-\beta(x)}{2} |\mathbf{f}|^\alpha m(dx) \Big\} \\
&= \exp\Big\{ \int_{S_2} \Big[-\Big| \sum_{j=1}^2 \theta_j s_j \Big|^\alpha + ia \Big(\sum_{j=1}^2 \theta_j s_j \Big)^{<\alpha>} \Big] \Gamma(d\mathbf{s}) \Big\}
\end{aligned}
$$

by making the change of variables

$$T : E_+ \to S_2, \;\; T : x \mapsto (s_1, s_2) = \Big(\frac{f_1(x)}{|\mathbf{f}(x)|}, \frac{f_2(x)}{|\mathbf{f}(x)|} \Big)$$

and setting $\Gamma(d\mathbf{s}) = \Gamma_+(d\mathbf{s}) + \Gamma_-(-d\mathbf{s})$, with

$$\Gamma_\pm(d\mathbf{s}) = \frac{1 \pm \beta(x)}{2} |\mathbf{f}(x)|^\alpha m(dx), \;\; x = T^{-1}(\mathbf{s}).$$

Hence, in order to transform an integral involving s_1, s_2 and Γ into one involving f_1, f_2 and m, one expresses it as a sum of two terms: the first is obtained by replacing s_j by $\frac{f_j(x)}{|\mathbf{f}(x)|}$, $j = 1, 2$, and $\Gamma(d\mathbf{s})$ by $\frac{1}{2}(1 + \beta(x))|\mathbf{f}(x)|^\alpha m(dx)$, and the second is obtained by replacing s_j by $\frac{-f_j(x)}{|\mathbf{f}(x)|}$, $j = 1, 2$, and $\Gamma(d\mathbf{s})$ by $\frac{1}{2}(1 - \beta(x))|\mathbf{f}(x)|^\alpha m(dx)$. (The same rule applies when $\alpha = 1$.) For example, when $\alpha = 1$,

$$\begin{aligned} k_0 &= \int_{S_2} s_1 \ln |s_2| \Gamma(d\mathbf{s}) \\ &= \int_{E_+} \frac{f_1}{|\mathbf{f}|} \left(\ln \frac{|f_2|}{|\mathbf{f}|} \right) \frac{1 + \beta(x)}{2} |\mathbf{f}| m(dx) \\ &\quad + \int_{E_+} \frac{-f_1}{|\mathbf{f}|} \left(\ln \frac{|-f_2|}{|\mathbf{f}|} \right) \frac{1 - \beta(x)}{2} |\mathbf{f}| m(dx) \\ &= \int_{E_+} f_1 \left(\ln \frac{|f_2|}{|\mathbf{f}|} \right) \beta(x) m(dx). \end{aligned}$$

Similarly,

$$\begin{aligned} &\int_{S_2} \frac{|s_2|^{\alpha+\nu}}{|s_1|^\nu} \Gamma(d\mathbf{s}) \\ &= \int_{E_+} \left(\frac{|f_2|}{|\mathbf{f}|} \right)^{\alpha+\nu} \left(\frac{|\mathbf{f}|}{|f_1|} \right)^\nu \left\{ \frac{1 + \beta(x)}{2} + \frac{1 - \beta(x)}{2} \right\} |\mathbf{f}|^\alpha m(dx) \\ &= \int_{E_+} \frac{|f_2|^{\alpha+\nu}}{|f_1|^\nu} m(dx). \quad \blacksquare \end{aligned}$$

5.3 Examples

Here, we apply the results proved in the preceding sections to various important classes of α-stable laws, namely those of moving averages, sub-Gaussian and harmonizable vectors.

Example 5.3.1 *Moving averages*

Often, $(X_1, X_2) = (X(t_1), X(t_2))$ where $t_1 \le t_2$ and $\{X(t), t \in T\}$ is some α-stable stochastic process. If the process $X(t)$ has a representation of the type $X(t) = \int_{-\infty}^t f(t - x) M(dx)$ and M has Lebesgue control measure, then there is no $\nu > 0$ satisfying Condition (5.1.27) and, clearly, if $0 < \alpha \le 1$, $E(|X(t_2)||X(t_1) = x) = \infty$. This is the case, for example, if $X(t) = \int_0^t M(dx)$ (α-stable Lévy motion) or if $X(t) = \int_{-\infty}^t e^{-\lambda(t-x)} M(dx)$ with $\lambda > 0$ (Ornstein-Uhlenbeck process).

Relation (5.2.36) however, can be satisfied by *two-sided moving averages*. Consider, for example, the two-sided α-stable Ornstein-Uhlenbeck process $X(t) = \int_{-\infty}^{\infty} e^{-\lambda|t-x|} M(dx)$ where M has Lebesgue control measure. Then any $\nu > 0$ satisfies (5.1.27) and, therefore,

(*i*) $E(|X(t_2)|^p | X(t_1) = x) < \infty$ a.e. for any $p < 2\alpha + 1$.
In particular, $E(X^2(t_2)|X(t_1) = x) < \infty$ a.e. if $\alpha > 1/2$.

(*ii*) If X is $S\alpha S$,

$$E(X(t_2)|X(t_1) = x)$$
$$= \left(\frac{1}{2}\lambda\alpha \int_{-\infty}^{\infty} \exp\{-\lambda(|t_2 - u| + (\alpha - 1)|t_1 - u|)\} du\right) x \text{ a.e.}$$

Example 5.3.2 *Sub-Gaussian vectors*

Let $0 < \alpha < 2$, let G_1, G_2 be zero mean jointly normal random variables and let A be a positive $\alpha/2$-stable random variable, independent of (G_1, G_2) with Laplace transform $Ee^{-\gamma A} = e^{-\gamma^{\alpha/2}}$, $\gamma > 0$. Then $(X_1, X_2) = (A^{1/2}G_1,\ A^{1/2}G_2)$ is a *sub-Gaussian* $S\alpha S$ random vector with underlying Gaussian vector $\mathbf{G} = (G_1, G_2)$ (see Sections 2.5 and 3.7).

Corollary 5.3.3 *Let* $\mathbf{X} = (X_1, X_2)$ *be a sub-Gaussian* α*-stable random vector with underlying Gaussian vector* (G_1, G_2). *Then*

(*i*) $E(|X_2|^p | X_1 = x) < \infty$ *a.e. for any* $p < \alpha + 1$.
In particular, $E(X_2^2 | X_1 = x) < \infty$ *a.e., if* $1 < \alpha \le 2$.

(*ii*) *If* $\mathbf{X}$ *is* $S\alpha S$,

$$E(X_2|X_1 = x) = \frac{\text{Cov}\,(G_1, G_2)}{\text{Var}\, G_1}\, x \ \text{a.e.} \tag{5.3.1}$$

The proof uses the following lemma which will also be useful in the sequel.

Lemma 5.3.4 *Let* $\mathbf{X} = (X_1, X_2)$ *be a sub-Gaussian* $S\alpha S$ *random vector with underlying Gaussian vector* $\mathbf{G}$ *having covariance matrix*

$$\begin{pmatrix} 1 & r \\ r & 1 \end{pmatrix}, \ -1 \le r \le 1.$$

Let Γ_r *denote the spectral measure of the sub-Gaussian* $S\alpha S$ *vector* $\mathbf{X}$. *Then for any* $0 < \nu < 1$, *there is a finite positive constant* K independent *of* r *such that*

$$\int_{S_2} \frac{|s_2|^{\alpha+\nu}}{|s_1|^{\nu}} \Gamma_r(ds) \le K. \tag{5.3.2}$$

PROOF: By Proposition 3.7.1,

$$(X_1, X_2) \stackrel{d}{=} C\left(\int_\Omega G_1(\omega)M(d\omega), \int_\Omega G_2(\omega)M(d\omega)\right), \tag{5.3.3}$$

where $(\Omega, \mathcal{F}, P)$ is the probability space on which the Gaussian vector (G_1, G_2) is defined, M is a $S\alpha S$ random measure on $(\Omega, \mathcal{F})$ with control measure P and $C = C(\alpha)$ is a finite positive constant. Using (5.1.27), we obtain

$$\int_{S_2} \frac{|s_2|^{\alpha+\nu}}{|s_1|^\nu}\Gamma_r(ds) = C(\alpha)E\left(\frac{|G_2|^{\alpha+\nu}}{|G_1|^\nu}\right). \tag{5.3.4}$$

Since $(G_1, G_2) \stackrel{d}{=} (G_1, rG_1 + (1-r^2)^{1/2}\epsilon)$ where ϵ is an $N(0,1)$ random variable independent of G_1, we have

$$E\frac{|G_2|^{\alpha+\nu}}{|G_1|^\nu} = E\frac{|rG_1 + (1-r^2)^{1/2}\epsilon|^{\alpha+\nu}}{|G_1|^\nu} \le 2^{\alpha+\nu}(E|G_1|^\alpha + E|\epsilon|^{\alpha+\nu}E|G_1|^{-\nu}),$$

proving the lemma. ■

PROOF OF COROLLARY 5.3.3: We assume, without loss of generality, that $EG_1^2 = EG_2^2 = 1$ and use the notation of the preceding lemma. That lemma, Relation (5.2.36) and Theorem 5.1.3 prove part (i). Corollary 5.2.7 shows that $E(X_2|X_1 = x) = \kappa x$ a.e. To identify κ, we use the representation (5.3.3) to obtain

$$\int_{S_2} |s_1|^\alpha \Gamma_r(ds) = C^\alpha \int_\Omega |G_1(\omega)|^\alpha P(d\omega) = C^\alpha E|G_1|^\alpha \tag{5.3.5}$$

and

$$\begin{aligned}
\int_{S_2} s_2 s_1^{<\alpha-1>}\Gamma_r(ds) &= C^\alpha \int_\Omega G_2(\omega)G_1(\omega)^{<\alpha-1>}P(d\omega) \\
&= C^\alpha E G_2 G_1^{<\alpha-1>} \\
&= C^\alpha E(rG_1 + \sqrt{1-r^2}\epsilon)G_1^{<\alpha-1>} \\
&= C^\alpha r E\, G_1\, G_1^{<\alpha-1>} \\
&= C^\alpha r E|G_1|^\alpha,
\end{aligned} \tag{5.3.6}$$

which gives the covariation . Hence, $\kappa = r$ when $a_1^2 := EG_1^2$ and $a_2^2 := EG_2^2$ are both equal to 1. In the general case,

$$E(X_2|X_1 = x) = a_2 E\left(\frac{X_2}{a_2}\middle|\frac{X_1}{a_1} = \frac{x}{a_1}\right) = \frac{a_2 r}{a_1}x = \frac{\text{Cov}\,(G_1, G_2)}{\text{Var}\,G_1}x \text{ a.e.} \quad ■$$

Statement (i) in Corollary 5.3.3 can be verified directly as follows. Let $p < \alpha + 1$ with $0 < \alpha < 2$ and write $G_2 = \epsilon + aG_1$ where ϵ is Gaussian, $E\epsilon G_1 = 0$ and a is a constant. Then

$$E(|A^{1/2}G_2|^p | A^{1/2}G_1 = x) = E(|A^{1/2}\epsilon + ax|^p | A^{1/2}G_1 = x) < \infty$$

because a simple computation shows that the conditional density $g(y|x)$ of $A^{1/2}\epsilon$ given $A^{1/2}G_1 = x$ satisfies $g(y|x) \sim C(x)y^{-2-\alpha}$ as $y \to \infty$, $C(x) > 0$. (Use Exercise 5.1.)

Example 5.3.5 *Real harmonizable vectors*

Let M be a (complex-valued) rotationally invariant $S\alpha S$ random measure on $(\mathbb{R}, \mathcal{B})$ with a *finite* control measure m, let t_1, t_2 be real numbers and let

$$X_j = \text{Re} \int_{-\infty}^{\infty} e^{it_j x} M(dx), \quad j = 1, 2.$$

Then (X_1, X_2) is a real harmonizable $S\alpha S$ random vector. (See Example 3.6.7.)

Corollary 5.3.6 *Let $\mathbf{X} = (X_1, X_2)$ be a real harmonizable $S\alpha S$ random vector as above. Then*

(i) *$E(|X_2|^p | X_1 = x) < \infty$ a.e. for any $p < \alpha + 1$.*

In particular, $E(X_2^2 | X_1 = x) < \infty$ a.e. if $1 < \alpha \le 2$.

(ii) *If $\mathbf{X}$ is $S\alpha S$,*

$$E(X_2|X_1 = x) = \left(\int_{-\infty}^{\infty} m(dy)\right)^{-1} \left(\int_{-\infty}^{\infty} \cos((t_2 - t_1)y) m(dy)\right) x \ \text{a.e.}$$

PROOF: By (3.6.6), the joint characteristic function $\phi_{\mathbf{X}}(\theta_1, \theta_2)$ of the $S\alpha S$ random vector $\mathbf{X} = (X_1, X_2)$ has the form

$$E \exp i(\theta_1 X_1 + \theta_2 X_2) = \exp\left\{-b_1(\alpha) \int_{-\infty}^{\infty} |\theta_1^2 + \theta_2^2 + 2\theta_1\theta_2 \cos \Delta y|^{\alpha/2} m(dy)\right\}, \tag{5.3.7}$$

where $b_1(\alpha)$ is a finite positive constant and $\Delta = t_2 - t_1$. By Proposition 2.5.2, for any $y \in \mathbb{R}$,

$$|\theta_1^2 + \theta_2^2 + 2\theta_1\theta_2 \cos \Delta y|^{\alpha/2} = b_2(\alpha) \int_{S_2} |\theta_1 s_1 + \theta_2 s_2|^\alpha \Gamma_{\cos \Delta y}(ds), \tag{5.3.8}$$

where $b_2(\alpha)$ is a finite positive constant and where we use the same notation as in Lemma 5.3.4. Therefore

$$\phi_{\mathbf{X}}(\theta_1, \theta_2) = \exp\left\{-b_1(\alpha)b_2(\alpha) \int_{S_2} |\theta_1 s_1 + \theta_2 s_2|^\alpha \int_{-\infty}^{\infty} \Gamma_{\cos \Delta y}(ds) m(dy)\right\}.$$

Comparing this expression with $\phi_{\mathbf{X}}(\theta_1, \theta_2) = \exp\{-\int_{S_2} |\theta_1 s_1 + \theta_2 s_2|^\alpha \Gamma(ds)\}$ and using the uniqueness of the spectral measure, we conclude that the spectral measure Γ of the real harmonizable $S\alpha S$ random vector (X_1, X_2) can be represented in the form

$$\Gamma(A) = b_1(\alpha)b_2(\alpha) \int_{-\infty}^{\infty} \Gamma_{\cos \Delta y}(A) m(dy) \qquad (5.3.9)$$

for any Borel set A on S_2.

For any $0 < \nu < 1$, we have, using Lemma 5.3.4,

$$\begin{aligned} \int_{S_2} \frac{|s_2|^{\alpha+\nu}}{|s_1|^{\nu}} \Gamma(ds) &= b_1(\alpha)b_2(\alpha) \int_{-\infty}^{\infty} \left[\int_{S_2} \frac{|s_2|^{\alpha+\nu}}{|s_1|^{\nu}} \Gamma_{\cos \Delta y}(ds) \right] m(dy) \\ &\leq b_1(\alpha)b_2(\alpha) \int_{-\infty}^{\infty} K m(dy) < \infty. \end{aligned}$$

Theorem 5.1.3 implies now that for almost every x, $E(|X_2|^p | X_1 = x) < \infty$ for any $p < \alpha + 1$. To prove (ii), we use Corollary 5.2.7 and (5.3.9). Computing the covariation , we obtain

$$\begin{aligned} \int_{S_2} s_2 s_1^{<\alpha-1>} \Gamma(ds) &= b_1(\alpha)b_2(\alpha) \int_{-\infty}^{\infty} \left[\int_{S_2} s_2 s_1^{<\alpha-1>} \Gamma_{\cos \Delta y}(ds) \right] m(dy) \\ &= b_1(\alpha)b_2(\alpha) C^{\alpha} E|G_1|^{\alpha} \int_{-\infty}^{\infty} \cos \Delta y \, m(dy) \end{aligned}$$

by (5.3.6). Moreover,

$$\int_{S_2} |s_1|^{\alpha} \Gamma(ds) = b_1(\alpha)b_2(\alpha) C^{\alpha} E|G_1|^{\alpha} \int_{-\infty}^{\infty} m(dy),$$

by (5.3.5). This establishes (ii). ∎

5.4 Graphical representations

In this section, we present graphical representations of the regression functions given analytically in Theorems 5.2.2 and 5.2.3. They were obtained by using the software package for computing bivariate regressions and probability density function of α-stable random variables given in Hardin, Samorodnitsky and Taqqu (1991*b*). That paper includes a description of the package and the listing of the source code, written in the C language.

In the case $\alpha \neq 1$, the five parameters $\alpha, \beta_1, \kappa, \lambda$, and σ_1 given in Proposition 5.2.10 are required to describe the regression function completely. In the case $\alpha = 1$, these five parameters, together with k_0 and μ_1, are required to describe the regression function completely. Thus, there is a myriad of possible choices for these parameters which are consistent with their definitions, giving rise to an unmanageably large family of regression functions. In producing the plots that follow, we have chosen to restrict the parameter space considerably, but in a way which we hope will give the reader some feeling for the general character of these functions.

To accomplish this, we use the stochastic integral representation (5.2.35) for (X_1, X_2). Here, the random measure M is taken to be totally right-skewed (i.e. , $\beta(\cdot) \equiv 1$), with domain $E = [0,1]$ and Lebesgue control measure m. The functions f_j, illustrated in Figure 5.3, are restricted to those of the form

$$f_j(t) = 1_{[0,c_j)}(t) - 1_{[c_j,1]}(t),$$

where 1_A is the indicator function of the set A. With these restrictions, all scale parameters equal unity and the three parameters α, c_1 and c_2 determine the regression function.

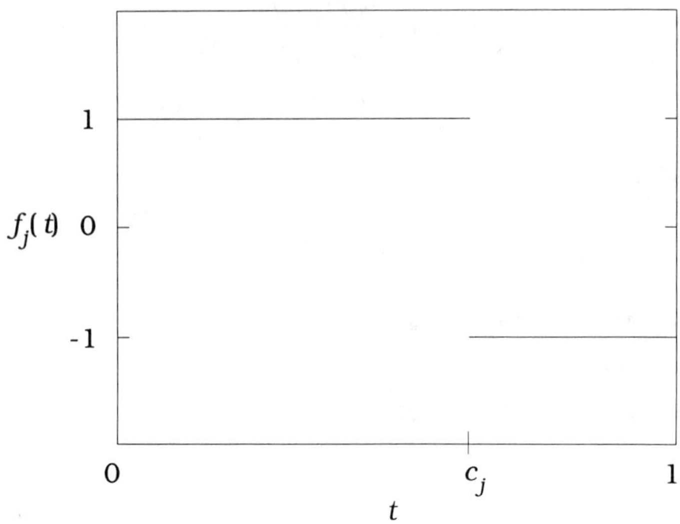

Figure 5.3: The function $f_j(t)$.

Regressions for (X_1, X_2) in this class may be interpreted as regressions involving three independent stable variables as follows. Define $c_{\min} = \min(c_1, c_2)$ and $c_{\max} = \max(c_1, c_2)$, and let Z_1, Z_2, and Z_3 be independent identically distributed totally right-skewed ($\beta = 1$) α-stable random variables with unit scale parameter. Then (X_1, X_2) is distributed as (Y_1, Y_2), where

$$Y_j = c_{\min}^{1/\alpha} Z_1 \pm (-1)^{j+1} (c_{\max} - c_{\min})^{1/\alpha} Z_2 - (1 - c_{\max})^{1/\alpha} Z_3$$

and the upper sign holds if $c_{max} = c_1$, and the lower sign prevails otherwise. For example, if $c_1 = 0.5$ and $c_2 = 1$, then the regression of X_2 on X_1 is the regression of $k(Z_1 + Z_2)$ on $k(Z_1 - Z_2)$ for the appropriate constant k.

From Proposition 5.2.10, the parameters defining the regressions in Theorem 5.2.3 are

$$\begin{aligned} \sigma_1 &= 1, \\ \beta_1 &= 2c_1 - 1. \\ \mu_1 &= \frac{2}{\pi}\ln(\sqrt{2})\beta_1, \\ k_0 &= -\ln(\sqrt{2})\lambda, \\ \kappa &= 1 - 2|c_1 - c_2|, \end{aligned}$$

and

$$\lambda = 2c_2 - 1.$$

We shall refer to the skewness parameter for X_2 as β_2. For the present class of distributions, β_2 is equal to λ.

The asymptotic results of Corollary 5.2.5 are illustrated in the graphs and translate to the present parametrization as follows:

$$\lim_{x\to+\infty} \frac{E(X_2|X_1 = x)}{x} = \begin{cases} 1 & \text{if } c_2 \geq c_1, \\ 2(c_2/c_1) - 1 & \text{if } c_1 > c_2 \end{cases}$$

and

$$\lim_{x\to-\infty} \frac{E(X_2|X_1 = x)}{x} = \begin{cases} 1 - 2(c_2 - c_1)/(1 - c_1) & \text{if } c_2 \geq c_1, \\ 1 & \text{if } c_1 > c_2. \end{cases}$$

The limits here do not depend on the value of α. Moreover, by Corollary 5.2.6, the regression is linear when $c_1 = c_2$ (here, this corresponds to $X_1 = X_2$).

The regression functions are computed by making use of Theorems 5.2.2 and 5.2.3. Unfortunately, the functions $H(x), U(x), V(x)$, and the density function $f_{X_1}(x)$ do not in general have representations in terms of elementary functions, and thus their values must be computed by means of numerical integration. Although the general problem of computing the integrals is straightforward in principle, the specific task is fraught with difficulties. The next section discusses the numerical techniques in detail.

In the graphs that follow, two of the parameters α, c_1 and c_2 are held constant while the third varies. Although not all possible variations are illustrated, much of the behavior for parameter choices not illustrated can be inferred from the graphs.

Figure 5.4 corresponds to $c_1 = 0.5$ and $c_2 = 1$ and, hence, to $\beta_1 = 0$ and $\beta_2 = \lambda = 1$. It shows the regression of a totally right-skewed variable X_2

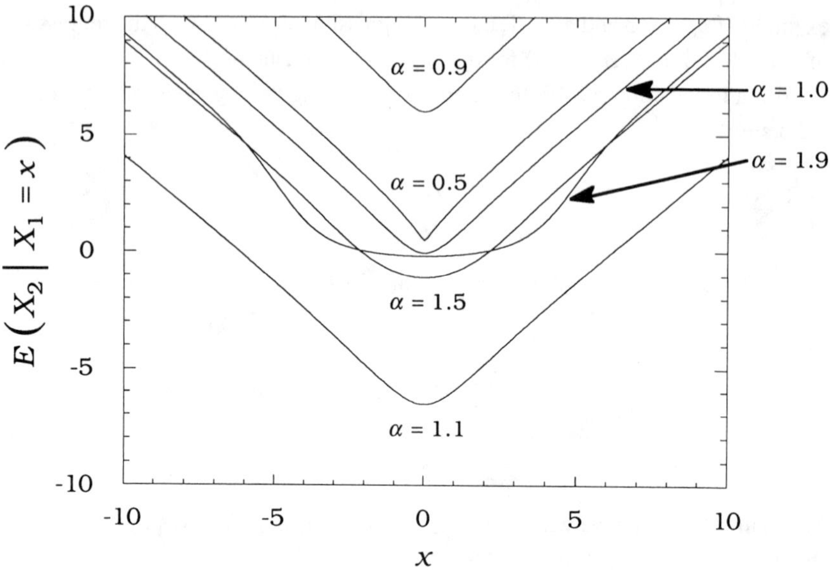

Figure 5.4: Regression functions for $\beta_1 = 0$, $\beta_2 = 1$, α varying.

upon a symmetric variable X_1 for selected α, or, equivalently, the regression of $k(Z_1 + Z_2)$ on $k(Z_1 - Z_2)$ as mentioned above. When $\alpha \neq 1$, the value of these regression functions at the origin has the same sign as $a = \tan \frac{\pi\alpha}{2}$ and hence is positive for $\alpha < 1$ and negative for $\alpha > 1$.

Figure 5.5 corresponds to $c_2 = 0.5$ and various values of c_1. It represents a regression of a symmetric variable X_2 upon variables X_1 with skewness β_1 varying from 0 to 1, for the case $\alpha = 1.9$. The value $c_1 = 0.5$ corresponds to X_1 symmetric, in which case $X_1 = X_2$ and the regression is linear. When $c_1 = 1$, X_1 is totally right-skewed and the regression is linear at zero.

In Figure 5.6, X_2 of varying skewness is regressed on a symmetric X_1 for the value $\alpha = 1.5$. Here, $c_1 = 0.5$. When $c_2 = 0.5$, the regression is linear since $X_2 = X_1$. The curious non-zero intersection of the regression lines occurs for all fixed values of $\alpha \geq 1$ at an x value depending on α but does not occur for values of $\alpha < 1$ (see Figure 5.12). Figure 5.7 corresponds also to $\alpha = 1.5$, but in this case it is X_2 that is symmetric ($\beta_2 = 0$).

In Figures 5.8 and 5.9, α equals 1.1. Figure 5.8 should be compared to Figure 5.7 because they both illustrate the regression of a symmetric random variable X_2 upon random variables X_1 of varying skewness. Figure 5.9 shows the regression of a totally right-skewed X_2 upon X_1 of varying skewness. Here $c_2 = 1$, and

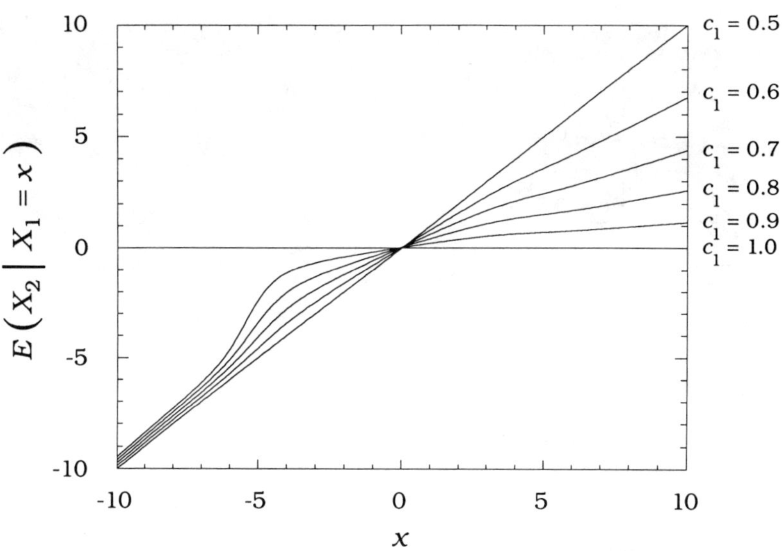

Figure 5.5: Regression functions for $\alpha = 1.9$, $\beta_1 = 2c_1 - 1$ and $\beta_2 = 0$.

hence $\beta_2 = \lambda = 1$. As the skewness of X_1 approaches that of X_2, the regression function approaches the identity, yet the left asymptote always has slope -1.

In Figure 5.10, $\alpha = 1$. The parameter c_1 is chosen to be 0.9, so that X_1 has skewness 0.8. The skewness of X_2 varies from -0.8 to 1. (For X_2 of skewness -1, the regression is the negative of that for X_2 of skewness 1.) The value $c_2 = 0.9$ corresponds to $X_2 = X_1$, in which case the regression is linear. This graph shows that a small change in skewness can result in a large change in the global shape of the regression function.

Figure 5.11 represents the regressions of variables of varying skewness upon a symmetric variable for the value $\alpha = 0.9$. The value $c_2 = 0.5$ corresponds to $X_2 = X_1$, in which case the regression is linear. This plot should be compared with the case $\alpha = 1.5$ illustrated in Figure 5.6.

Figures 5.12 and 5.13 both correspond to $\alpha = 0.5$, but $\beta_1 = 0$ in Figure 5.12, whereas $\beta_2 = 0$ in Figure 5.13. Observe that Figures 5.6 and 5.12 where $\beta_1 = 0$ have roughly the same shape. So do Figures 5.7, 5.8 and 5.13, where $\beta_2 = 0$. In Figure 5.13 the regression for $\beta_1 = 1$ ($c_1 = 1$) is defined only for $x \geq 0$, because the density of X_1 has support on $[0, \infty]$. It is linear with slope $\kappa = 0$.

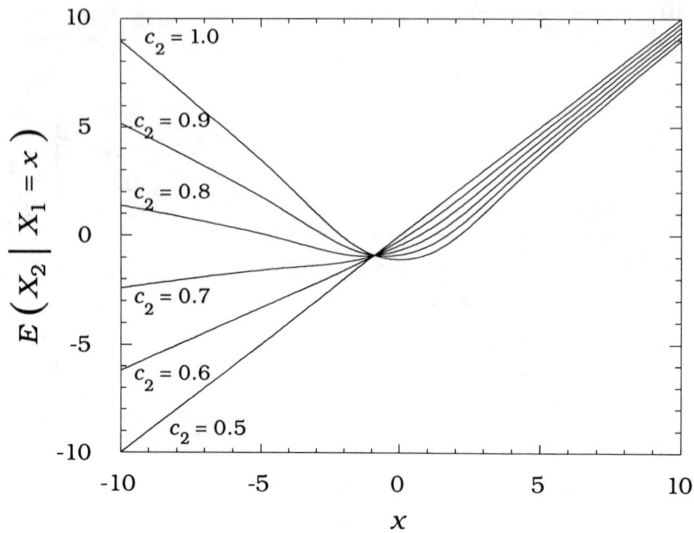

Figure 5.6: Regression functions for $\alpha = 1.5$, $\beta_1 = 0$ and $\beta_2 = 2c_2 - 1$.

5.5 Numerical techniques

We shall work with the integral representation since it is the one which is the most commonly used in practice. For convenience, we start by summarizing the main results of Sections 5.1 and 5.2.

The regression formulas

Let

$$(X_1, X_2) = \left(\int_E f_1(x)M(dx),\ \int_E f_2(x)M(dx)\right) \tag{5.5.1}$$

be α-stable, $0 < \alpha < 2$. A sufficient condition for the regression $E(X_2|X_1 = x)$ to be defined when $0 < \alpha \le 1$ is

$$\int_{E_+} \frac{|f_2(x)|^{\alpha+\nu}}{|f_1(x)|^{\nu}} m(dx) < \infty,$$

for some $\nu > 1 - \alpha$, where $E_+ = \{x \in E :\ f_1^2(x) + f_2^2(x) \neq 0\}$.

The following quantities enter in the expression of the regression:

$$\sigma_1^\alpha = \int_E |f_1(x)|^\alpha m(dx),$$

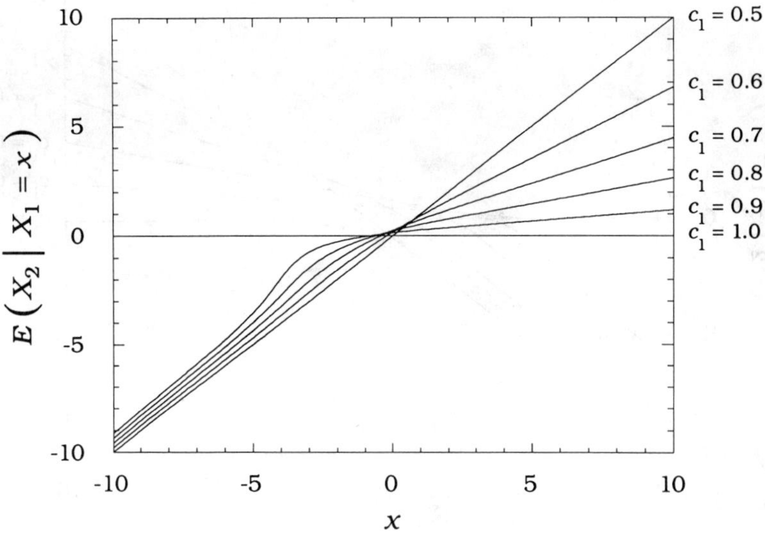

Figure 5.7: Regression functions for $\alpha = 1.5$, $\beta_1 = 2c_1 - 1$ and $\beta_2 = 0$.

$$
\begin{aligned}
\beta_1 &= \frac{1}{\sigma_1^\alpha}\int_E f_1(x)^{<\alpha>}\beta(x)m(dx),\\
\kappa &= \frac{[X_2, X_1]_\alpha}{\sigma_1^\alpha} = \frac{1}{\sigma_1^\alpha}\int_{E_+} f_2(x)f_1(x)^{<\alpha-1>}m(dx),\\
\lambda &= \frac{1}{\sigma_1^\alpha}\int_{E_+} f_2(x)|f_1(x)|^{\alpha-1}\beta(x)m(dx),\\
\mu_1 &= -\frac{2}{\pi}\int_{E_+} f_1(x)\Big(\ln\frac{|f_1(x)|}{\sqrt{f_1^2(x)+f_2^2(x)}}\Big)\beta(x)m(dx) \qquad (\text{case } \alpha = 1),\\
k_0 &= \frac{1}{\sigma_1}\int_{E_+} f_2(x)\Big(\ln\frac{|f_1(x)|}{\sqrt{f_1^2(x)+f_2^2(x)}}\Big)\beta(x)m(dx) \qquad (\text{case } \alpha = 1).
\end{aligned}
$$

σ_1 is the scale parameter of X_1, β_1 is the skewness parameter of X_1, κ is the normalized covariation of X_2 on X_1, λ is the normalized skewed covariation of X_2 on X_1, and μ_1 and k_0 are shift parameters that appear in the case $\alpha = 1$. When $\beta(\cdot) \equiv 0$, the regression is linear $E(X_2|X_1 = x) = \kappa x$ with slope equal to κ, the normalized covariation. When $\beta(\cdot) \not\equiv 0$, the formulas for the regression are different in the cases $\alpha \neq 1$ and $\alpha = 1$. In the case $\alpha \neq 1$,

$$E(X_2|X_1 = x) = \kappa x + \alpha\Big(\tan\frac{\pi\alpha}{2}\Big)(\lambda - \beta_1\kappa)\frac{r(x/\sigma_1)}{s(x/\sigma_1)}\sigma_1, \tag{5.5.2}$$

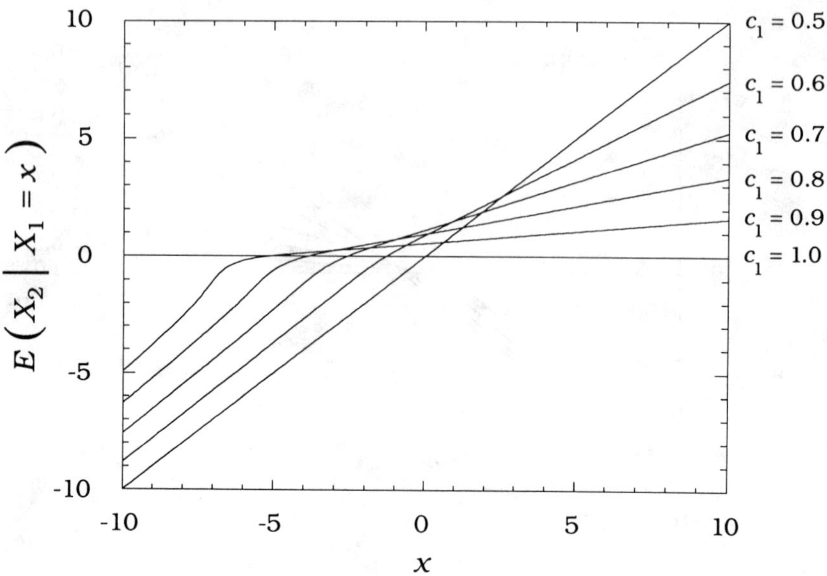

Figure 5.8: Regression functions for $\alpha = 1.1$, $\beta_1 = 2c_1 - 1$ and $\beta_2 = 0$.

where

$$r(x) = \int_0^\infty e^{-t^\alpha} t^{\alpha-1} \cos\Big(xt - \Big(\tan\frac{\pi\alpha}{2}\Big)\beta_1 t^\alpha\Big)\,dt, \tag{5.5.3}$$

$$s(x) = \int_0^\infty e^{-t^\alpha} \cos\Big(xt - \Big(\tan\frac{\pi\alpha}{2}\Big)\beta_1 t^\alpha\Big)\,dt. \tag{5.5.4}$$

(See (5.2.17).) The density function of X_1 equals $\frac{1}{\pi\sigma_1}s(\frac{x}{\sigma_1})$.

When $\alpha = 1$ and $\beta_1 \neq 0$,

$$E(X_2|X_1 = x) = -\frac{2\sigma_1}{\pi}k_0 + \kappa(x - \mu_1) + \frac{\lambda - \beta_1\kappa}{\beta_1}\Big((x - \mu_1) - \sigma_1\frac{U(x)}{W(x)}\Big), \tag{5.5.5}$$

where

$$U(x) = \int_0^\infty e^{-\sigma_1 t} \sin\Big(t(x - \mu_1) + \frac{2}{\pi}\beta_1\sigma_1 t \ln t\Big)\,dt, \tag{5.5.6}$$

$$W(x) = \int_0^\infty e^{-\sigma_1 t} \cos\Big(t(x - \mu_1) + \frac{2}{\pi}\beta_1\sigma_1 t \ln t\Big)\,dt. \tag{5.5.7}$$

When $\alpha = 1$ and $\beta_1 = 0$,

$$E(X_2|X_1 = x) = -\frac{2\sigma_1}{\pi}k_0 + \kappa(x - \mu_1) - \frac{2\sigma_1}{\pi}\lambda\frac{V(x)}{W(x)}, \tag{5.5.8}$$

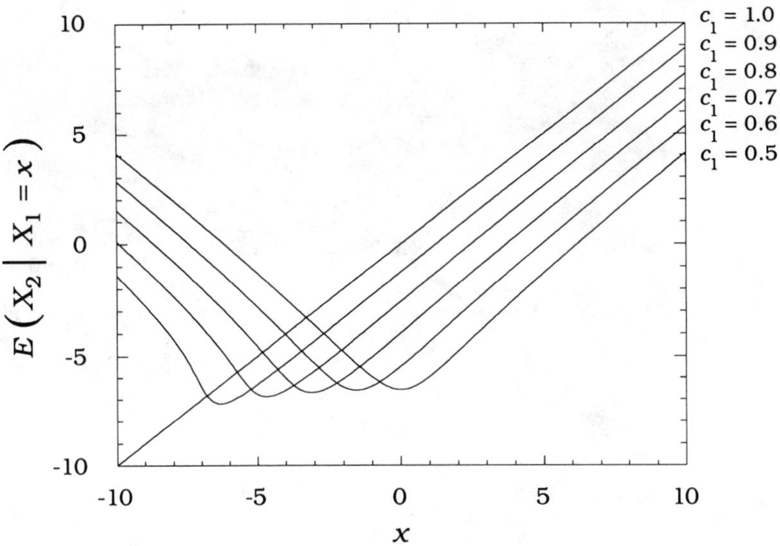

Figure 5.9: Regression functions for $\alpha = 1.1$, $\beta_1 = 2c_1 - 1$ and $\beta_2 = 1$.

where

$$V(x) = \int_0^\infty e^{-\sigma_1 t}(1 + \ln t)(\cos t(x - \mu_1))dt, \tag{5.5.9}$$

and $W(x)$ is as above. When $\alpha = 1$, the density of X_1 equals $\frac{1}{\pi}W(x)$.

Suppose $0 < \alpha < 2$. The regression is linear when $\beta_1 = \pm 1$, and more generally, when $\lambda = \beta_1 \kappa$. It is then equal to κx if $\alpha \neq 1$ and to $\kappa x - (\frac{2\sigma_1}{\pi} k_0 + \kappa \mu_1)$ if $\alpha = 1$.

The regression is typically asymptotically linear. When $\beta_1 \neq \pm 1$,

$$E(X_2 | X_1 = x) \sim \frac{\kappa + \lambda}{1 + \beta_1} x, \ x \to \infty,$$

and

$$E(X_2 | X_1 = x) \sim \frac{\kappa - \lambda}{1 - \beta_1} x, \ x \to -\infty.$$

Integrand characteristics

In the given form, the integrals defining $r(x), s(x), U(x), V(x)$ and $W(x)$ are improper integrals, and all have integrands which decay exponentially while oscillating, in many cases increasingly rapidly, about zero. The integrand of V and the integrand of r in the case $\alpha < 1$ have singularities at zero; otherwise the integrands are reasonably well behaved.

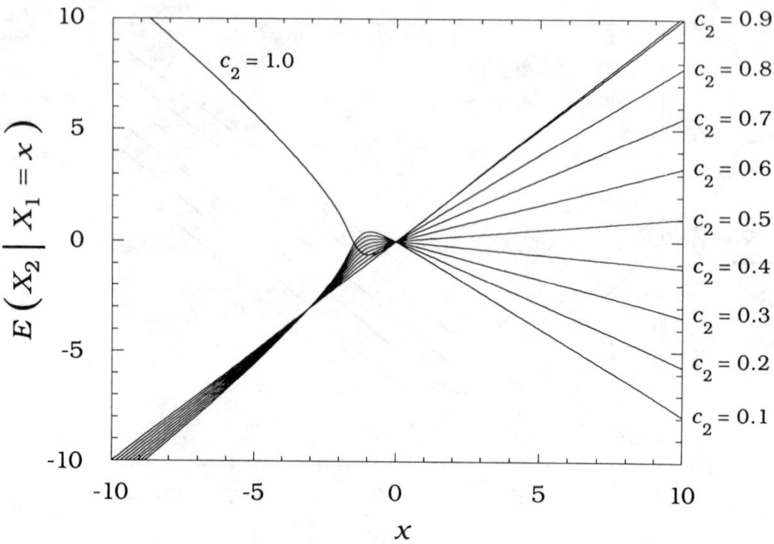

Figure 5.10: Regression functions for $\alpha = 1.0$, $\beta_1 = 0.8$ and $\beta_2 = 2c_2 - 1$.

It might be expected that with the exponential decay of the integrands, the improper integrals could be truncated at some adequately high cutoff and subjected to any one of the standard integration techniques, with acceptable results. One problem with this is, of course, the singularities. Another problem is that as the parameter x gets large, the sinusoidal oscillations become more rapid, and each integral decays to zero, imposing more stringent requirements on the accuracies of the individual integrals to ensure accuracy of the quotient. For example, the integral $r(x)$ decays as $x^{-\alpha}$ and $s(x)$ decays as $x^{-(\alpha+1)}$. (See the proof of Corollary 5.2.5.) For larger x, then, both more iterations and higher truncation levels may be required. The parameter α also affects the requirements for truncation due to the slower decay of the integrand for lower α. To illustrate some of the behavior mentioned above, Figure 5.14 shows a graph of the integrand for $r(x)$ for $\alpha = 0.5$ and $\alpha = 1.5$ when $x = 8$.

One can make the integrals more amenable to numerical techniques by making changes of variables in the integrands. For example, the change of variables $s = t^\alpha$ converts $r(x)$ to

$$r(x) = \frac{1}{\alpha} \int_0^\infty e^{-s} \cos\Big(xs^{1/\alpha} - \Big(\tan\frac{\pi\alpha}{2}\Big)\beta_1 s\Big)\, ds. \tag{5.5.10}$$

This eliminates the singularity in the integrand in (5.5.3) when $\alpha < 1$. The

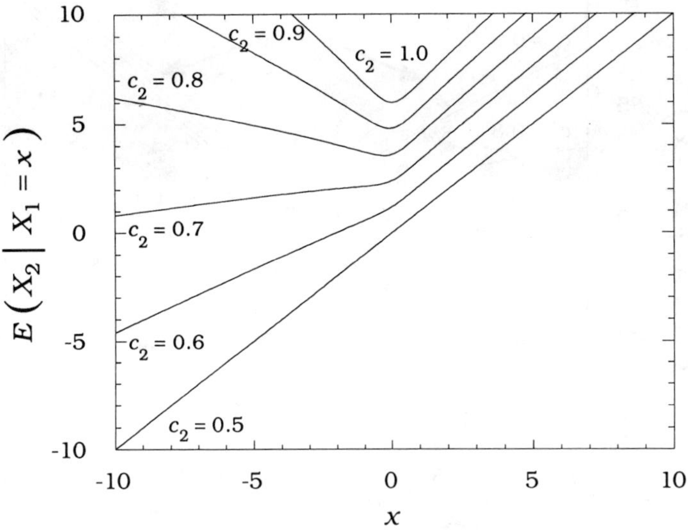

Figure 5.11: Regression functions for $\alpha = 0.9$, $\beta_1 = 0$ and $\beta_2 = 2c_2 - 1$.

same change of variables in the denominator integral $s(x)$, however, introduces a singularity at zero for values of α greater than one. To eliminate the need for truncation, the change of variables $s = e^{-t^\alpha}$ represents $r(x)$, after some fortuitous cancellations, as

$$r(x) = \frac{1}{\alpha} \int_0^1 \cos\Big(x(-\ln s)^{1/\alpha} + \Big(\tan \frac{\pi\alpha}{2}\Big)\beta_1 \ln s\Big) ds. \tag{5.5.11}$$

The same change of variables in the integrand for $s(x)$ introduces a singularity at zero for $\alpha < 1$, and at one for $\alpha > 1$. Although representation (5.5.11) results in a bounded integrand, the integrand oscillates increasingly rapidly about zero. Even though the contribution of the integrand from zero to ϵ is bounded by ϵ, the oscillations for s slightly greater than ϵ may be rapid enough to require an unacceptably small step size in the numerical procedure in order to obtain the necessary accuracy.

In the case $\alpha = 1$, the logarithms in the integrands do not cause singularities or discontinuities except in $V(x)$. That singularity may be removed at the expense of introducing an additional improper integral with the change of variables $t = e^{-s}$

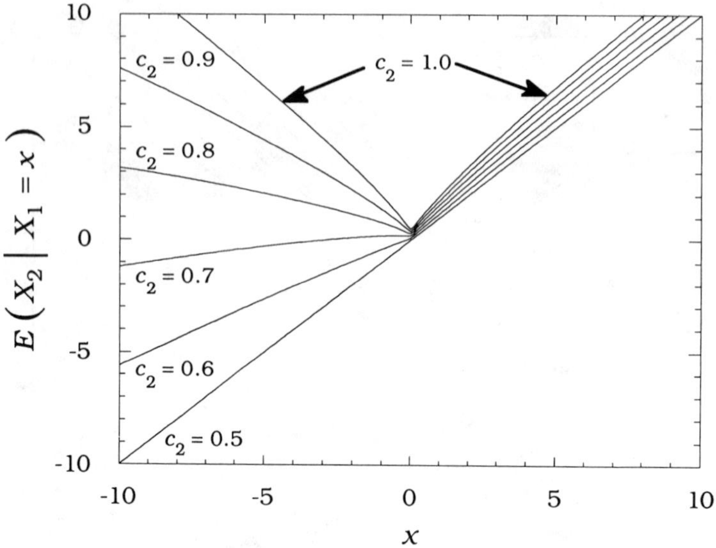

Figure 5.12: Regression functions for $\alpha = 0.5$, $\beta_1 = 0$ and $\beta_2 = 2c_2 - 1$.

for $0 < t \leq 1$:

$$\int_0^1 e^{-\sigma_1 t}(1+\ln t)\cos(t(x-\mu_1))dt = \int_0^\infty e^{-\sigma_1 e^{-s}-s}(1-s)\cos(e^{-s}(x-\mu_1))ds. \tag{5.5.12}$$

Numerical procedures

Because of the variation of integrand characteristics over the range of parameters considered, regardless of the representation of the integral, it is important for the integration scheme to allow truncation (if used) and step size to vary as the parameters vary. Bounding arguments can be applied to whatever form of the integrand one chooses in order to find a cutoff point for which the truncation error is small. Step size is somewhat more difficult to pick, as the usual error formulas involve derivatives which may be difficult to evaluate, or may have extremely high values not indicative of the actual errors.

A standard way of solving the step size selection process is to use one of the (by now standard) versions of the classical midpoint and trapezoidal integration formulas which allow the step size to be iteratively decreased. The integration sample point mesh is successively refined, and the corresponding numerical approximation is successively refined by updating the previous computations. When

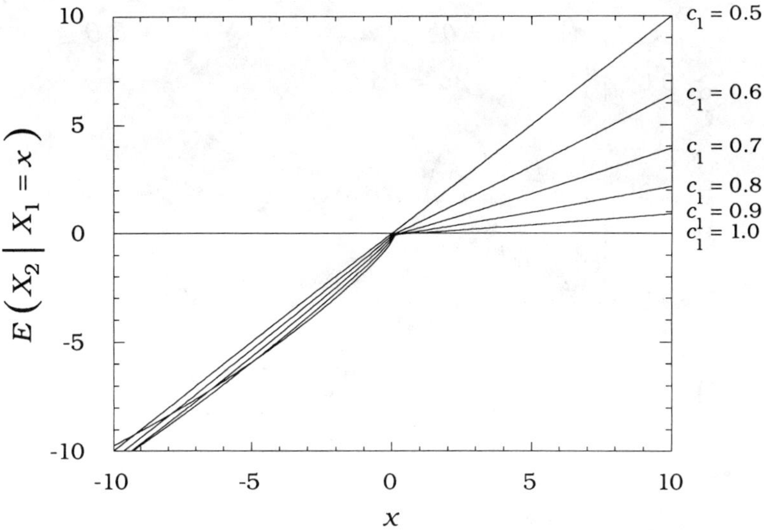

Figure 5.13: Regression functions for $\alpha = 0.5$, $\beta_1 = 2c_1 - 1$ and $\beta_2 = 0$.

successive approximations differ by some small prespecified amount, the iterations are terminated. Approaches such as these seem suited to the problem at hand since different parameter choices have different requirements. Unfortunately, these methods are slow to converge for our problem and tend to require more computation than other methods. This problem is not alleviated by any of the changes of variables mentioned. In fact, for some representations, the approximations did not converge.

Another promising possibility is to use a variable step size formula, since for all of the representations considered, the integrand varies much more rapidly in one part of the domain than in other parts. The bulk of the computation could take place in the regions of greatest change, enabling faster convergence and more control over the accuracy. Unfortunately, as for the iterative methods, this seemingly good idea fails the test of a practical implementation.

In the choice of integrand representation and integration technique one must consider computational expense, trade-offs involving number of iterations, number and complexity of function evaluations, and numerical accuracy. For example, in the case $\alpha \neq 1$, the integrands $r(x)$ and $s(x)$ in (5.5.3) and (5.5.4) are identical except for the factor $t^{\alpha-1}$. It is tempting to try an integration scheme which reuses the function evaluation for s in computing r in order to decrease the computational overhead. The savings realized in doing this, however, do not outweigh the costs

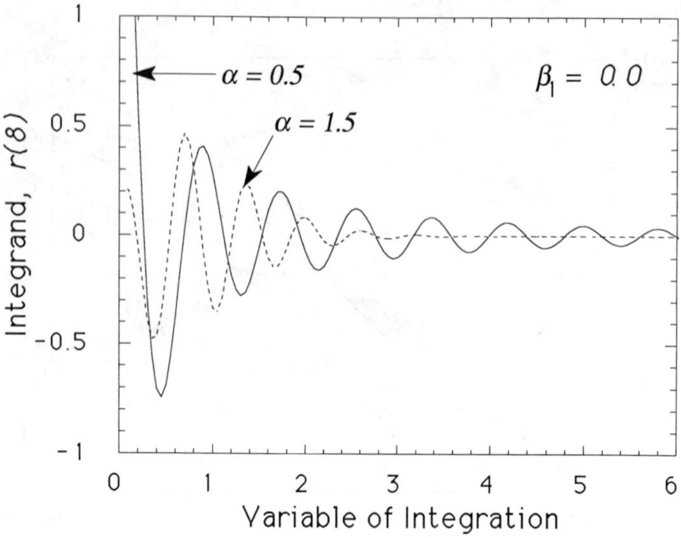

Figure 5.14: Integrand in (5.5.3) for $x = 8$.

incurred, compared with other methods, due to a larger number of iterations.

Many combinations of integration technique and integrand representation were tried. The combination that met with the most success was the version of Simpson's rule known as the "one-third" rule, coupled with the representation (5.5.10) for the function $r(x)$, the representation (5.5.4) for the function $s(x)$, the representations (5.5.6) and (5.5.7) for $U(x)$ and $W(x)$, and the representation for $V(x)$, obtained by writing the integral in (5.5.9) as the sum of two integrals, one from 0 to 1 and the other from 1 to ∞, and using the representation (5.5.12) for the first of these. The remainder of this section describes this approach in more detail.

To choose the step sizes, the following approach is taken. Require a certain minimum step size, but allow it to be chosen smaller, based on the frequency of the sinusoidal term. This frequency is, of course, not well defined since no unique frequency exists. An attempt is made, however, to approximate the frequency as effected by the regressor value, x, and to force the number of sample points per "period" to be bounded below by an input parameter. A parameter declaring the maximum number of samples is also provided which prevents the algorithm from taking the step size guidance to extremes.

Since all of the integrals are represented as improper integrals, they must be truncated. An input parameter, ϵ, specifies the greatest absolute truncation error to be suffered in either integral being computed. Analytically determined cutoff points, M, are then determined for each integral. By bounding the sinusoidal terms by unity, M is determined as follows. Solving

$$\epsilon = \int_M^\infty ke^{-ct} = \frac{k}{c}e^{-cM} \tag{5.5.13}$$

for M results in

$$M = -\frac{1}{c}\ln\left(\frac{c\epsilon}{k}\right) \tag{5.5.14}$$

for positive c and k. This implies that the truncation error is no greater than ϵ when the integral is truncated at M, where

- $M = -\ln(\alpha\epsilon)$ for $r(x)$
- $M = -\frac{1}{\sigma_1}\ln(\sigma_1\epsilon)$ for $U(x)$ and $W(x)$.

The cutoffs for $s(x)$ and $V(x)$ require the following lemma.

Lemma 5.5.1 *For $p > 0$ and $x \geq \max\{0, 4p(\ln(4p) - 1)\}$, $e^{-x}x^p \leq e^{-x/2}$.*

PROOF: The result follows if it is shown that for such x, $\ln x \leq \frac{x}{2p}$. Since $\ln x \leq x - 1$ for all $x > 0$, we have $\ln(cx) \leq cx - 1$ for all $x, c > 0$, in turn implying that $\ln x \leq cx - 1 - \ln c$ for such x and c. Now, for $0 < c < 1/(2p)$, we have $cx - 1 - \ln c \leq x/(2p)$ if and only if $x \geq (-1 - \ln c)/(1/(2p) - c)$. Choosing $c = 1/(4p)$ gives the result. ■

We consider firstly the cutoff for $s(x)$. Make the change of variables $u = t^\alpha$ and apply the lemma to obtain

$$\begin{aligned} \int_M^\infty e^{-t^\alpha}dt &= \frac{1}{\alpha}\int_{M^\alpha}^\infty e^{-u}u^{(1/\alpha)-1}du \\ &\leq \frac{1}{\alpha}\begin{cases} \int_{M^\alpha}^\infty e^{-u}du, & \alpha > 1,\ M > 1, \\ \int_{M^\alpha}^\infty e^{-u/2}du, & \alpha < 1,\ M^\alpha \geq B, \end{cases} \end{aligned}$$

where $B = 4(\frac{1}{\alpha} - 1)\ln[4(\frac{1}{\alpha} - 1) - 1]$. Consequently,

$$\int_M^\infty e^{-t^\alpha}dt \leq \begin{cases} \frac{1}{\alpha}e^{-M^\alpha}, & \alpha > 1,\ M > 1, \\ \frac{2}{\alpha}e^{-M^\alpha/2}, & \alpha < 1,\ M \geq B^{1/\alpha}. \end{cases}$$

This shows that truncating the integral defining $s(x)$ at M, where

$$M = \begin{cases} \{-\ln(\alpha\epsilon)\}^{1/\alpha} \vee 1, & \alpha > 1, \\ \{-2\ln(\alpha\epsilon/2)\}^{1/\alpha} \vee B^{1/\alpha}, & \alpha < 1 \end{cases}$$

results in truncation error not greater than ϵ.

We now turn to the cutoff for $V(x)$. Use the lemma to write for $M \geq 4\sigma_1^{-1}(\ln 4 - 1) \approx 1.545\sigma_1^{-1}$,

$$\begin{aligned}\int_M^\infty e^{-\sigma_1 t}(1+\ln t)dt &\leq \int_M^\infty e^{-\sigma_1 t} t dt \\ &\leq \frac{1}{\sigma_1}\int_M^\infty e^{-\sigma_1 t/2}dt \\ &= \frac{2}{\sigma_1^2}e^{-\sigma_1 M/2}.\end{aligned}$$

Thus, defining $M = \max\{4\sigma_1^{-1}(\ln 4 - 1), -2\sigma_1^{-1}\ln(\sigma_1^2\epsilon/2)\}$, the truncation error incurred by truncating the integral defining $V(x)$ at M is no greater than ϵ.

5.6 Exercises

Exercise 5.1 Let G_1 and G_2 be i.i.d. $N(0,1)$ and let $A \sim S_{1/2}(1,1,0)$ independent of G_1 and G_2. Let $X = A^{1/2}G_1$ and $Y = A^{1/2}G_2$. Prove that the joint probability density function $f_{(X,Y)}(x,y)$ of (X,Y) satisfies

$$f_{(X,Y)}(x,y) \sim C_1(x)y^{-2-\alpha} \text{ as } y \to \infty,$$

Hint: Use

$$f_{(X,Y)}(x,y) = \text{const.}\int_0^\infty \exp\Big\{-\frac{1}{2}(x^2+y^2)u\Big\}\frac{1}{u}f_A\Big(\frac{1}{u}\Big)du$$

and

$$\int_0^\infty e^{-\theta u}\frac{1}{u}f_A\Big(\frac{1}{u}\Big)du \sim \text{const. } \theta^{-1-\frac{\alpha}{2}} \text{ as } \theta \to \infty.$$

Exercise 5.2 Suppose in Example 5.1.5 that Y_1 and Y_2 are i.i.d. $S_\alpha(\sigma,1,0)$ with $\alpha < 1$. Their probability density function is supported on the positive real line. Show that $E(|X_2|^p|X_1 = x_1) < \infty$ a.s. for all $p > 0$.

Chapter 6

Complex stable stochastic integrals and harmonizable processes

A (real) harmonizable α-stable process is a stochastic process of the form

$$X(t) = \text{Re} \int_{-\infty}^{\infty} e^{itx} M(dx), \ -\infty < t < \infty,$$

where M is a complex-valued α-stable random measure on Borel subsets of the real line with a finite control measure m. A reader familiar with the spectral theory of stationary Gaussian processes will recall that every stationary Gaussian process, continuous in probability, can be so represented and hence is harmonizable. This is not true, however, for stationary α-stable processes with index $\alpha \in (0, 2)$. Nevertheless, the class of stationary harmonizable α-stable processes is of importance and it deserves a careful study. Because the definition involves the complex integral $\int_{-\infty}^{\infty} e^{itx} M(dx)$, we need to introduce complex stable random measures M and complex-valued integrals

$$I(f) = \int_{-\infty}^{\infty} f(x) M(dx),$$

which are of interest in their own right. For simplicity of notation and to make our formulas as short as possible, we restrict our discussion to the $S\alpha S$ case, although generalization to the skewed case should be possible by using similar arguments.

In Section 6.1 we introduce complex-valued $S\alpha S$ random measures and the related notions of "circular control measure" and "control measure." Integrals with respect to complex-valued $S\alpha S$ random measures are constructed in Section

6.2. In Section 6.3 we treat the important special case where M is a complex isotropic $S\alpha S$ random measure. In that case, $I(f)$ is a complex isotropic $S\alpha S$ random variable. Such random variables were first studied in Section 2.6. Their real and imaginary parts are identically distributed but, in general, not independent. They are independent if $\alpha = 2$.

Complex measures M and integrals $I(f)$ have also a series representation which is given in Section 6.4. In Section 6.5 we consider harmonizable processes, not only the real harmonizable process Re $\int_{-\infty}^{\infty} e^{itx} M(dx)$ but also its complex counterpart $\int_{-\infty}^{\infty} e^{itx} M(dx)$. The main theorem in the section gives necessary and sufficient conditions for these processes to be stationary. In Section 6.6 we consider stationary real harmonizable processes. Although these processes share properties with stationary sub-Gaussian processes, it is shown that the two classes are "almost" disjoint.

6.1 Complex-valued $S\alpha S$ random measures

Complex-valued $S\alpha S$ random measures are analogous to the real random measures defined in Section 3.3, but they are, naturally, a little more complicated.

Let $(\Omega, \mathcal{F}, P)$ be the underlying probability space, and let $L^0(\Omega)$ and $L^0_c(\Omega)$ be, respectively, the sets of all real and complex random variables defined on $(\Omega, \mathcal{F}, P)$. Every element of $L^0_c(\Omega)$ is of the form $X + iY$, where $X, Y \in L^0(\Omega)$. Let $(E, \mathcal{E})$ be a measurable space, and let $(S_2, \mathcal{B}_2)$ be the unit circle in $\mathbb{R}^2$ equipped with the Borel σ-algebra. Let k be a measure on the product space $(E \times S_2, \mathcal{E} \times \mathcal{B}_2)$ satisfying the following condition:

Condition 6.1.1 *For every $A \in \mathcal{E}$ such that $k(A \times S_2) < \infty$, $k(A \times \cdot)$ is a (finite) symmetric measure on $(S_2, \mathcal{B}_2)$.*

We shall consider

$$\mathcal{E}_0 = \{A \in \mathcal{E} : \ k(A \times S_2) < \infty\}. \tag{6.1.1}$$

Definition 6.1.2 A *complex-valued $S\alpha S$ random measure on $(E, \mathcal{E})$ with circular control measure* k is an independently scattered σ-additive complex-valued set function

$$M : \ \mathcal{E}_0 \to L^0_c(\Omega)$$

such that, for every $A \in \mathcal{E}_0$, $M^{(1)}(A) \equiv \text{Re}\, M(A)$ and $M^{(2)}(A) \equiv \text{Im}\, M(A)$ are *jointly* $S\alpha S$ with spectral measure $k(A \times \cdot\,)$.

The meaning of "independently scattered" and "σ-additive" is the same as in Section 3.1, but the reader should note that this time these notions are applied to complex-valued random variables, i.e., to the vectors formed by their real and imaginary parts. Hence, independence of the complex-valued random variables $M(A_1), M(A_2), \ldots, M(A_k)$ means independence of the random vectors

$$\begin{pmatrix} M^{(1)}(A_1) \\ M^{(2)}(A_1) \end{pmatrix}, \begin{pmatrix} M^{(1)}(A_2) \\ M^{(2)}(A_2) \end{pmatrix}, \ldots, \begin{pmatrix} M^{(1)}(A_k) \\ M^{(2)}(A_k) \end{pmatrix}.$$

One can apply Kolmogorov's existence theorem to show that the complex-valued $S\alpha S$ random measure M in Definition 6.1.2 exists, with finite-dimensional distributions

$$E \exp\Big\{ i \sum_{j=1}^{d} (\theta_j^{(1)} M^{(1)}(A_j) + \theta_j^{(2)} M^{(2)}(A_j)) \Big\}$$

$$= \exp\Big\{ - \int_E \int_{S_2} |\sum_{j=1}^{d} (s_1 \theta_j^{(1)} + s_2 \theta_j^{(2)}) 1_{A_j}(x)|^\alpha k(dx, ds) \Big\} \quad (6.1.2)$$

where $A_1, A_2, \ldots, A_d \in \mathcal{E}_0$ and $(\theta_1^{(1)}, \theta_1^{(2)}), (\theta_2^{(1)}, \theta_2^{(2)}), \ldots, (\theta_d^{(1)}, \theta_d^{(2)})$ are pairs of real numbers (Exercise 6.1).

Notation. $k(dx, ds)$ and $k(dx \times ds)$ are equivalent notations.

Remarks

1. We will often express the left-hand side of (6.1.2) as

$$E \exp\Big\{ i \,\mathrm{Re} \sum_{j=1}^{d} z_j \overline{M}_j(A_j) \Big\}, \quad (6.1.3)$$

   where $z_j = \theta_j^{(1)} + i\theta_j^{(2)}$, and $\overline{z}_j = \theta_j^{(1)} - i\theta_j^{(2)}$ denotes the complex conjugate of z_j.

2. A complex-valued $S\alpha S$ random measure is characterized by a circular control measure k defined on the *product space* $E \times S_2$, whereas a real-valued $S\alpha S$ random measure is characterized by a control measure m defined only on E.

3. To understand the need for k, think of $M(dx)$ as a $S\alpha S$ random variable measuring the "size" of a small set around point $x \in E$. Then, in the real-valued case, the distribution of the $S\alpha S$ random variable $M(dx)$ can be specified by a single number – its scale parameter $m(dx)$. Thus, in the real-valued case, M is defined by specifying m on E. In the complex-valued case, the distribution of the (complex-valued) $S\alpha S$ random variable

$M(dx)$ is specified by the spectral measure of its real and imaginary part, that is, by $k(dx, ds)$ where $x \in E$ and where the argument s belongs to the unit circle S_2. Hence, in the complex-valued case, M is defined by specifying k on $E \times S_2$.

4. The measure k is called "*circular* control measure" and not "control measure" because the latter term is reserved for the measure $k(\cdot \times S_2) = \int_{S_2} k(\cdot, ds)$ on E.

Definition 6.1.3 The measure on $(E, \mathcal{E})$ defined by $m(A) = k(A \times S_2)$ is called the *control measure* of complex-valued $S\alpha S$ random measure M.

Although, in the complex-valued case, the control measure m does not, in general, determine uniquely the (finite-dimensional distributions of) random measure M, it is a useful object because it is simpler than the circular control measure k and because there are properties of complex-valued $S\alpha S$ random measures M that can be conveniently stated in terms of m.

Example 6.1.4 The subset $\mathcal{E}_0 = \{A \in E : k(A \times S_2) < \infty\}$ of $\mathcal{E}$ introduced in (6.1.1) can be written $\mathcal{E}_0 = \{A \in \mathcal{E} : m(A) < \infty\}$.

Example 6.1.5 If the circular control measure k is concentrated on the two "lines" $E \times \{(1,0)\}$ and $E \times \{(-1,0)\}$, then the complex-valued $S\alpha S$ random measure M is, actually, real, and its control measure m (given, in this case, by $m(A) = 2k(A \times \{1,0\})$ does determine uniquely the random measure M.

Example 6.1.6 Suppose that the circular control measure k is a product measure of the form $k = m\gamma$, where m is the control measure of the random measure M and γ is the *uniform* probability measure on the unit circle S_2. Then the random measure M has the property

$$e^{i\phi} M \stackrel{d}{=} M, \quad \text{for any real } \phi \tag{6.1.4}$$

(see Section 2.6), where, as usual, $\stackrel{d}{=}$ denotes equality of the finite-dimensional distributions. We call such a measure *isotropic* (or *rotationally invariant*). Conversely, any complex-valued $S\alpha S$ random measure satisfying (6.1.4) must have a circular control measure of the form $k = m\gamma$ (see Exercise 6.2.) Note that γ must be a probability measure because the relations $k(A \times S_2) = m(A)$ and $k(A \times S_2) = m(A)\gamma(S_2)$ imply $\gamma(S_2) = 1$.

Of course, the finite-dimensional distributions of an isotropic $S\alpha S$ random measure M are determined uniquely by its control measure m.

It is obviously possible to view a complex-valued $S\alpha S$ random measure M as a pair of two dependent real $S\alpha S$ random measures $M^{(1)}$ and $M^{(2)}$, whose control measures $m^{(1)}$ and $m^{(2)}$ are given by

$$m^{(i)}(A) = \int_{S_2} |s_i|^\alpha k(A \times ds),\ i = 1, 2. \tag{6.1.5}$$

(See Exercise 6.3.) This perspective, however, is often not useful because the dependence between the random measures $M^{(1)}$ and $M^{(2)}$ can be very complicated.

Example 6.1.7 A complex-valued $S\alpha S$ random measure M has independent real and imaginary parts if and only if the circular control measure k is concentrated on the four "lines" $E \times \{0, 1\}$, $E \times \{0, -1\}$, $E \times \{1, 0\}$ and $E \times \{-1, 0\}$. (See Exercise 6.4.) Unlike the Gaussian case $\alpha = 2$, an isotropic measure M cannot, therefore, have independent real and imaginary parts, unless M is identically zero. (See Exercise 6.5.)

6.2 Integrals with respect to complex-valued $S\alpha S$ random measures

Having defined a complex-valued $S\alpha S$ random measure M, we shall now define the integral

$$I(f) = \int_E f(x) M(dx) \tag{6.2.1}$$

of a non-random complex-valued function f with respect to M.

Let $f = f^{(1)} + i f^{(2)} : E \to \mathbb{C}$ be a measurable function. Since we already know how to integrate real-valued functions against real-valued $S\alpha S$ random measures, it is tempting to write $\int f dM = \int (f^{(1)} + i f^{(2)}) d(M^{(1)} + i M^{(2)})$ and hence to define the integral in (6.2.1) by

$$\int_E f(x) M(dx) = \left(\int_E f^{(1)}(x) M^{(1)}(dx) - \int_E f^{(2)}(x) M^{(2)}(dx) \right)$$
$$+ i \left(\int_E f^{(1)}(x) M^{(2)}(dx) + \int_E f^{(2)}(x) M^{(1)}(dx) \right). \tag{6.2.2}$$

This relation should, of course, hold for any reasonable definition of the integral, but it is inconvenient to define the integral (6.2.1) by (6.2.2) because the nature of the dependence between the various integrals on the right-hand side of (6.2.2) is not readily apparent. A more convenient approach, it turns out, is to use one of the possible constructions, developed in Chapter 3, for the real-valued case. We choose the constructive definition discussed in Section 3.4 because it best demonstrates the Wiener-type nature of the integral. Throughout this section, unless explicitly stated otherwise, all $S\alpha S$ random measures M are complex-valued.

Let M be a $S\alpha S$ random measure on $(E, \mathcal{E})$ with a circular control measure k. For a simple function of the form $f(x) = \sum_{j=1}^{n} c_j 1_{A_j}(x)$ where the c_j are complex numbers and the A_js are disjoint sets in $\mathcal{E}_0$, we define

$$I(f) = \int_E f(x) M(dx) = \sum_{j=1}^{n} c_j M(A_j). \tag{6.2.3}$$

It is trivial to check that for simple functions, $I(f)$ is linear in f and satisfies (6.2.2) a.s. The real and imaginary parts of $I(f)$, denoted henceforth $I^{(1)}(f)$ and $I^{(2)}(f)$ respectively, are jointly $S\alpha S$ with joint characteristic function

$$\begin{aligned}
&E \exp\{i(\theta_1 I^{(1)}(f) + \theta_2 I^{(2)}(f))\} \\
&= E \exp\Big\{i\Big(\theta_1\Big[\int_E f^{(1)}(x) M^{(1)}(dx) - \int_E f^{(2)}(x) M^{(2)}(dx)\Big] \\
&\quad + \theta_2\Big[\int_E f^{(2)}(x) M^{(1)}(dx) + \int_E f^{(1)}(x) M^{(2)}(dx)\Big]\Big)\Big\} \\
&= E \exp\Big\{i\Big[\int_E (\theta_1 f^{(1)}(x) + \theta_2 f^{(2)}(x)) M^{(1)}(dx) \\
&\quad + \int_E (\theta_2 f^{(1)}(x) - \theta_1 f^{(2)}(x)) M^{(2)}(dx)\Big]\Big\} \\
&= \exp\Big\{-\int_E \int_{S_2} \Big|\theta_1(s_1 f^{(1)}(x) - s_2 f^{(2)}(x)) \\
&\quad + \theta_2(s_1 f^{(2)}(x) + s_2 f^{(1)}(x))\Big|^\alpha k(dx, ds)\Big\},
\end{aligned} \tag{6.2.4}$$

by Relation (6.1.2). Note also that

$$\begin{aligned}
&E \exp\{i(\theta_1 I^{(1)}(f) + \theta_2 I^{(2)}(f))\} \\
&= \exp\Big\{-\int_E \int_{S_2} |(s_1\theta_1 + s_2\theta_2) f^{(1)}(x) + (s_1\theta_2 - s_2\theta_1) f^{(2)}(x)|^\alpha k(dx, ds)\Big\}.
\end{aligned} \tag{6.2.5}$$

Having defined the integral for simple functions, we are ready to apply the "approximate-and-use-convergence-in-probability" argument. We will see below (Proposition 6.2.2) that, as in the real-valued case, the class of functions f for which the integral $I(f)$ can be defined is

$$L^\alpha(m) = \Big\{f : E \to \mathbb{C} : \int_E |f(x)|^\alpha m(dx) < \infty\Big\},$$

where m is the control measure of M and $|f|$ denotes the modulus of f.

Formally, for any $f = f^{(1)} + if^{(2)} \in L^\alpha(m)$, choose a sequence of simple functions $\{f_n\}_{n=1}^\infty$ such that

$$f_n(x) \to f(x) \text{ for almost every } x \in E, \tag{6.2.6}$$

$$|f_n(x)| \le \theta(x) \text{ for any } n, x \text{ and some } \theta \in L^\alpha(m). \tag{6.2.7}$$

Such a sequence always exists as we can take, for example, $f_n(x) = f_n^{(1)}(x) + if_n^{(2)}(x)$ with

$$f_n^{(j)}(x) = \begin{cases} \frac{i}{n} & \text{if } \frac{i}{n} \le f^{(j)}(x) < \frac{i+1}{n},\ i = 0, 1, \ldots, n^2 - 1, \\ -\frac{i}{n} & \text{if } -\frac{i+1}{n} < f^{(j)}(x) \le -\frac{i}{n},\ i = 0, 1, \ldots, n^2 - 1, \\ 0 & \text{if } |f^{(j)}(x)| \ge n, \end{cases}$$

for $j = 1, 2, \ldots,$ in which case $\theta = |f|$. The sequence of integrals $I(f_n)$, $n = 1, 2, \ldots,$ is well defined by (6.2.3). Note that by (6.2.4), for any n, m, $I^{(1)}(f_n) - I^{(1)}(f_m) \sim S_\alpha(\sigma_{n,m}^{(1)}, 0, 0)$, where

$$(\sigma_{m,n}^{(1)})^\alpha = \int_E \int_{S_2} |s_1(f_n^{(1)}(x) - f_m^{(1)}(x)) - s_2(f_n^{(2)}(x) - f_m^{(2)}(x))|^\alpha k(dx, ds).$$

Relations (6.2.6) and (6.2.7) imply $\sigma_{n,m}^{(1)} \to 0$ as $n, m \to \infty$ and hence the sequence $\{I^{(1)}(f_n)\}_{n=1}^\infty$ converges in probability. The sequence $\{I^{(2)}(f_n)\}_{n=1}^\infty$ converges in probability as well. Therefore, the sequence of integrals $\{I(f_n)\}_{n=1}^\infty$ converges in probability and *we define*

$$I(f) = \text{plim}_{n\to\infty} I(f_n),$$

and write $I(f) = \int_E f dM$. Arguing as in Section 3.4, the integral $I(f)$ does not depend on a particular choice of approximating sequence $\{f_n\}_{n=1}^\infty$. We summarize its properties in

Proposition 6.2.1 *(i) For any $f \in L^\alpha(m)$, the integral $I(f)$ is a complex-valued $S\alpha S$ random variable satisfying (6.2.2), with characteristic function given by Relation (6.2.4).*

(ii) For any $f, g \in L^\alpha(m)$, a, b complex constants,

$$I(af + bg) = aI(f) + bI(g) \ \ a.s.$$

(iii) For any $f_1, f_2, \ldots, f_d \in L^\alpha(m)$ and for any pairs of real numbers $(\theta_1^{(1)}, \theta_1^{(2)}), (\theta_2^{(1)}, \theta_2^{(2)}), \ldots, (\theta_d^{(1)}, \theta_d^{(2)})$,

$$-\ln E \exp\Big\{i \sum_{j=1}^d (\theta_j^{(1)} I^{(1)}(f_j) + \theta_j^{(2)} I^{(2)}(f_j))\Big\} \tag{6.2.8}$$

$$= \int_E \int_{S_2} \Big| \sum_{j=1}^d [\theta_j^{(1)}(s_1 f_j^{(1)}(x) - s_2 f_j^{(2)}(x)) + \theta_j^{(2)}(s_1 f_j^{(2)}(x) + s_2 f_j^{(1)}(x))] \Big|^\alpha k(dx, d\mathbf{s})$$

$$= \int_E \int_{S_2} \Big| \sum_{j=1}^d [(s_1\theta_j^{(1)} + s_2\theta_j^{(2)}) f_j^{(1)}(x) + (s_1\theta_j^{(2)} - s_2\theta_j^{(1)}) f_j^{(2)}(x)] \Big|^\alpha k(dx, d\mathbf{s}).$$

PROOF: (i) $I(f)$ is obviously a $S\alpha S$ random variable, being a limit in probability of a sequence of $S\alpha S$ random variables. The characteristic function of $I(f)$ is given by (6.2.4) because the characteristic functions of the $I(f_n)$ have this form, and convergence in probability implies convergence in distribution and, thus, pointwise convergence of characteristic functions. Finally, (6.2.2) holds for the $I(f_n)$, and every one of the five integrals involved converges in probability to the corresponding integral of f. Therefore (6.2.2) holds for $I(f)$.

(ii) The argument is identical to the one in the real-valued case and is left to the reader.

(iii) For $j = 1, 2, \ldots, d$ define

$$u_j = \frac{\theta_j^{(1)} + \theta_j^{(2)}}{2}, \quad v_j = \frac{\theta_j^{(1)} - \theta_j^{(2)}}{2}, \quad a_j = u_j + iv_j.$$

By part (ii) of this proposition,

$$I^{(1)}\Big(\sum_{j=1}^d a_j f_j\Big) = \sum_{j=1}^d (u_j I^{(1)}(f_j) - v_j I^{(2)}(f_j)),$$

$$I^{(2)}\Big(\sum_{j=1}^d a_j f_j\Big) = \sum_{j=1}^d (u_j I^{(2)}(f_j) + v_j I^{(1)}(f_j)).$$

Therefore,

$$E \exp\Big\{ i \sum_{j=1}^d (\theta_j^{(1)} I^{(1)}(f_j) + \theta_j^{(2)} I^{(2)}(f_j)) \Big\}$$

$$= E \exp\Big\{ i \Big(I^{(1)}\Big(\sum_{j=1}^d a_j f_j\Big) + I^{(2)}\Big(\sum_{j=1}^d a_j f_j\Big)\Big)\Big\}$$

$$= \exp\Big\{ - \int_E \int_{S_2} \Big| \Big[s_1 \,\mathrm{Re}\,\Big(\sum_{j=1}^d a_j f_j(x)\Big) - s_2 \,\mathrm{Im}\,\Big(\sum_{j=1}^d a_j f_j(x)\Big)\Big]$$

$$+ \Big[s_1 \,\mathrm{Im}\,\Big(\sum_{j=1}^d a_j f_j(x)\Big) + s_2 \,\mathrm{Re}\,\Big(\sum_{j=1}^d a_j f_j(x)\Big)\Big] \Big|^\alpha k(dx, d\mathbf{s}) \Big\},$$

by (6.2.4). Rearranging terms yields (6.2.8). ■

In the real-valued case, the form of the characteristic function of the stable integral (see Proposition 3.4.1) makes it apparent that the integral can only be defined for functions in $L^\alpha(m)$. In the complex-valued case, the form of characteristic function (see (6.2.4)) is more complicated, but it is still true that the integral cannot be extended outside of $L^\alpha(m)$. Indeed,

Proposition 6.2.2 *Let $f : E \to \mathbb{C}$ be a measurable function such that there is a sequence $\{f_n\}_{n=1}^\infty$ of measurable simple functions satisfying (6.2.6) and such that the sequence of integrals $\{I(f_n)\}_{n=1}^\infty$ defined by (6.2.3) converges in distribution. Then $f \in L^\alpha(m)$.*

PROOF: By (6.2.4), $I^{(1)}(f_n) \sim S_\alpha(\sigma_n, 0, 0)$, where $\sigma_n^\alpha = \int_E \int_{S_2} |s_1 f_n^{(1)}(x) - s_2 f_n^{(2)}(x)|^\alpha k(dx, ds)$. Since $\{I^{(1)}(f_n)\}_{n=1}^\infty$ converges in distribution, conclude that $\lim_{n\to\infty} \sigma_n^\alpha$ exists and is finite. By Fatou's lemma,

$$\int_E \int_{S_2} |s_1 f^{(1)}(x) - s_2 f^{(2)}(x)|^\alpha k(dx, ds)$$
$$\leq \lim_{n\to\infty} \int_E \int_{S_2} |s_1 f_n^{(1)}(x) - s_2 f_n^{(2)}(x)|^\alpha k(dx, ds) < \infty. \quad (6.2.9)$$

Similarly, $\{I^{(2)}(f_n)\}_{n=1}^\infty$ converges in distribution and, hence,

$$\int_E \int_{S_2} |s_1 f^{(2)}(x) + s_2 f^{(1)}(x)|^\alpha k(dx, ds) < \infty. \quad (6.2.10)$$

Now by (6.2.9) and (6.2.10), and Fubini's theorem,

$$\int_E |f(x)|^\alpha m(dx) = \int_E \int_{S_2} (f^{(1)}(x)^2 + f^{(2)}(x)^2)^{\alpha/2} k(dx, ds)$$
$$= \int_E \int_{S_2} [(s_1 f^{(1)}(x) - s_2 f^{(2)}(x))^2 + (s_1 f^{(2)}(x) + s_2 f^{(1)}(x)^2)]^{\alpha/2} k(dx, ds)$$
$$\leq \int_E \int_{S_2} [|s_1 f^{(1)}(x) - s_2 f^{(2)}(x)|^\alpha + |s_1 f^{(2)}(x) + s_2 f^{(1)}(x)|^\alpha] k(dx, ds),$$

which is bounded. ■

Many properties of stable integrals in the real-valued case have counterparts in the complex-valued case. The following result, for example, should be compared to Proposition 3.5.1 which gives the corresponding result in the real-valued case.

Proposition 6.2.3 *Let $X_j = \int_E f_j(x) M(dx)$, $j = 1, 2, \ldots,$ and $X = \int_E f(x) M(dx)$, where M is a $S\alpha S$ random measure with circular control measure k and control measure m, and f, f_j, $j = 1, 2, \ldots,$ are in $L^\alpha(m)$. Then*

$$\text{plim}_{j\to\infty} X_j = X,$$

if and only if

$$\lim_{j\to\infty}\int_E |f_j(x)-f(x)|^\alpha m(dx)=0. \tag{6.2.11}$$

PROOF: Note that $\operatorname{plim}_{j\to\infty}X_j = X$ is equivalent to $\operatorname{plim}_{j\to\infty}(X_j - X) = 0$, which is equivalent to $\operatorname{plim}_{j\to\infty}\operatorname{Re}(X_j - X) = 0$ and $\operatorname{plim}_{j\to\infty}\operatorname{Im}(X_j - X) = 0$. This, in turn, is equivalent to Re $(X_j - X) \stackrel{d}{\Rightarrow} 0$ and Im $(X_j - X) \stackrel{d}{\Rightarrow} 0$, where $\stackrel{d}{\Rightarrow}$ denotes convergence in distribution. By (6.2.4), $\operatorname{Re}(X_j - X) \sim S_\alpha(\sigma_j^{(1)}, 0, 0)$ and $\operatorname{Im}(X_j - X) \sim S_\alpha(\sigma_j^{(2)}, 0, 0)$ where

$$(\sigma_j^{(1)})^\alpha = \int_E\int_{S_2} |s_1(f_j^{(1)}(x)-f^{(1)}(x)) - s_2(f_j^{(2)}(x)-f^{(2)}(x))|^\alpha k(dx,ds),$$

$$(\sigma_j^{(2)})^\alpha = \int_E\int_{S_2} |s_1(f_j^{(2)}(x)-f^{(2)}(x)) + s_2(f_j^{(1)}(x)-f^{(1)}(x))|^\alpha k(dx,ds).$$

Therefore $\operatorname{plim}_{j\to\infty}X_j = X$ is equivalent to

$$\lim_{j\to\infty}\sigma_j^{(1)\alpha} = \lim_{j\to\infty}\sigma_j^{(2)\alpha} = 0. \tag{6.2.12}$$

Trivially, (6.2.11) implies (6.2.12).

On the other hand,

$$\begin{aligned}
&\int_E |f_j(x)-f(x)|^\alpha m(dx)\\
&= \int_E\int_{S_2} [(f_j^{(1)}(x)-f^{(1)}(x))^2 + (f_j^{(2)}(x)-f^{(2)}(x))^2]^{\alpha/2} k(dx,ds)\\
&= \int_E\int_{S_2} [(s_1(f_j^{(1)}(x)-f^{(1)}(x)) - s_2(f_j^{(2)}(x)-f^{(2)}(x)))^2\\
&\quad +(s_1(f_j^{(2)}(x)-f^{(2)}(x)) + s_2(f_j^{(1)}(x)-f^{(1)}(x)))^2]^{\alpha/2} k(dx,ds)\\
&\le (\sigma_j^{(1)})^\alpha + (\sigma_j^{(2)})^\alpha,
\end{aligned}$$

so that (6.2.12) implies (6.2.11) as well. ■

Let M be a $S\alpha S$ random measure on $(E,\mathcal{E})$ with circular control measure k, and let $f \in L^\alpha(m)$. Since this book deals mainly with real-valued stable laws, we will encounter $I^{(1)}(f) = \operatorname{Re} I(f)$ more frequently than $I(f)$. Of course, the properties of $I^{(1)}(f)$ can be deduced from those of $I(f)$. For future reference, we collect some useful facts about $I^{(1)}(f)$.

Proposition 6.2.4 *(i)* $I^{(1)}(f) \sim S_\alpha(\sigma_f, 0, 0)$, *where*

$$\sigma_f^\alpha = \int_E\int_{S_2} |s_1 f^{(1)}(x) - s_2 f^{(2)}(x)|^\alpha k(dx,ds).$$

(ii) For any $f,\ g \in L^\alpha(m)$ *and any* real *constants* a, b,

$$I^{(1)}(af + bg) = aI^{(1)}(f) + bI^{(1)}(g) \ a.s.$$

Example 6.2.5 The (real) *harmonizable* $S\alpha S$ *process* is defined as

$$X(t) = \text{Re} \int_{-\infty}^{\infty} e^{itx} M(dx), \ -\infty < t < \infty, \tag{6.2.13}$$

where M is a $S\alpha S$ random measure on $(\mathbb{R}, \mathcal{B})$ with a finite circular control measure k (equivalently, with a finite control measure m). Proposition 6.2.4 and Theorem 2.1.5 show that it is a $S\alpha S$ process. We shall investigate some of its properties in Sections 6.5 and 6.6.

6.3 The complex isotropic $S\alpha S$ case

We treat in this section the important special case where the complex random measure M is not only $S\alpha S$ but also *isotropic*, that is, its circular control measure is of the form

$$k = m\gamma,$$

where γ is the uniform probability measure on the unit circle S_2 and where m, the *control measure* of M, is a measure on $(E, \mathcal{E})$.

Let f be complex-valued with

$$\int_E |f(x)|^\alpha m(dx) < \infty$$

and consider the complex-valued integral

$$I(f) = \int_E f(x) M(dx),$$

where M is a complex isotropic $S\alpha S$ random measure with control measure m. Let $I^{(1)}(f)$ and $I^{(2)}(f)$ denote, respectively, the real and imaginary parts of $I(f)$. These are real-valued random variables. The following theorem gives the characteristic function of the complex-valued random variable $I(f)$, i.e., the characteristic function of the vector $(I^{(1)}(f),\ I^{(2)}(f))$.

Theorem 6.3.1 *Let* M *be complex isotropic* $S\alpha S$, $0 < \alpha \le 2$ *with control measure* m. *Then for any complex number* $z = \theta_1 + i\theta_2$,

$$E\exp\{i(\theta_1 I^{(1)}(f) + \theta_2 I^{(2)}(f))\} = E\exp\{i \,\text{Re}\, \overline{z} I(f)\} = E\exp\{i \,\text{Re}\, I(\overline{z} f)\}$$

$$= \exp\Big\{-|z|^\alpha c_0 \int_E |f(x)|^\alpha m(dx)\Big\}, \tag{6.3.1}$$

where

$$c_0 = \frac{1}{2\pi}\int_0^\pi |\cos\phi|^\alpha d\phi.$$

PROOF: Using Relation (6.2.4) and $k(dx, ds) = m(dx)\gamma(ds)$, we have

$$- \ln E \exp\{i(\theta_1 I^{(1)}(f) + \theta_2 I^{(2)}(f))\}$$

$$= \int_E\int_{S_2} |\theta_1(s_1 f^{(1)}(x) - s_2 f^{(2)}(x)) + \theta_2(s_1 f^{(2)}(x) + s_2 f^{(1)}(x))|^\alpha m(dx)\gamma(ds)$$

$$= \int_E\int_{S_2} |(s_1, s_2)\cdot(\theta_1 f^{(1)}(x)+\theta_2 f^{(2)}(x), -\theta_1 f^{(2)}(x)+\theta_2 f^{(1)}(x))|^\alpha m(dx)\gamma(ds).$$

Setting $g = g^{(1)} + ig^{(2)}$, $g^{(1)} = \theta_1 f^{(1)}(x) + \theta_2 f^{(2)}(x)$ and $g^{(2)} = -\theta_1 f^{(2)}(x) + \theta_2 f^{(1)}(x)$, we obtain

$$\begin{aligned} &E \exp\{i(\theta_1 I^{(1)}(f) + \theta_2 I^{(2)}(f))\} \\ &\quad = \exp\Big\{-\int_E\Big[\int_{S_2} |s_1 g^{(1)}(x) + s_2 g^{(2)}(x)|^\alpha \gamma(ds)\Big] m(dx)\Big\} \\ &\quad = \exp\Big\{-c_0\gamma(S_2)\int_E |g(x)|^\alpha m(dx)\Big\} \end{aligned}$$

as in the proof of Corollary 2.6.5. But $\gamma(S_2) = 1$ and $|g|^2 = (g^{(1)})^2 + (g^{(2)})^2$ equals

$$\theta_1^2(f^{(1)})^2 + \theta_2^2(f^{(2)})^2 + 2\theta_1\theta_2 f^{(1)}f^{(2)} + \theta_1^2(f^{(2)})^2 + \theta_2^2(f^{(1)})^2 - 2\theta_1\theta_2 f^{(1)}f^{(2)},$$

i.e., $|g|^2 = (\theta_1^2 + \theta_2^2)|f|^2$. Therefore

$$E \exp\{i(\theta_1 I^{(1)}(f) + \theta_2 I^{(2)}(f))\} = \exp\Big\{-(\theta_1^2 + \theta_2^2)^{\alpha/2} c_0 \int_E |f(x)|^\alpha m(dx)\Big\}$$

$$= \exp\Big\{-|z|^\alpha c_0 \int_E |f(x)|^\alpha m(dx)\Big\}. \quad \blacksquare$$

Remark. We normalized γ by $\gamma(S_2) = 1$. Had we normalized it by $\gamma(S_2) = c_0^{-1}$, the constant c_0 would not appear in (6.3.1).

Isotropic $S\alpha S$ random variables were introduced in Section 2.6. Comparing the statements of Theorem 6.3.1 and Corollary 2.6.5, we note that if M is isotropic $S\alpha S$, then the random variable $I(f)$ is isotropic $S\alpha S$ as well.

One advantage of dealing with isotropic $S\alpha S$ random measures or integrals is that we can ignore the circular control measure k and use exclusively the control measure m. Moreover, form (6.3.1) of the characteristic function is very similar to that of characteristic functions of real-valued $S\alpha S$ integrals. The only difference is the presence of the constant c_0 and that of $|z|^\alpha = (\theta_1^2 + \theta_2^2)^{\alpha/2}$ which appears in the exponent instead of $|\theta|^\alpha$. By analogy to the real case, we will sometimes express the characteristic function of $I(f) = \int_E f(x)M(dx)$ as

$$E\exp\{i(\theta_1 I^{(1)}(f) + \theta_2 I^{(2)}(f))\} = Ee^{i \text{ Re } I(\overline{z}f)} = e^{-\|\text{Re } I(\overline{z}f)\|_\alpha^\alpha},$$

where $z = \theta_1 + i\theta_2$. Relation (6.3.1) then becomes

$$\|\text{Re } I(\overline{z}f)\|_\alpha^\alpha = |z|^\alpha c_0 \int_E |f(x)|^\alpha m(dx) = |\overline{z}|^\alpha c_0 \|f\|_\alpha^\alpha, \tag{6.3.2}$$

Theorem 6.3.1 provides also information on the real and imaginary parts of $I(f)$.

Corollary 6.3.2 *For any real θ,*

$$\begin{aligned} E\exp\{i\theta I^{(1)}(f)\} &= E\exp\Big\{i\theta \text{ Re} \int_E f(x)M(dx)\Big\} \\ &= \exp\Big\{-c_0|\theta|^\alpha \int_E |f(x)|^\alpha m(dx)\Big\} \end{aligned} \tag{6.3.3}$$

and, moreover,

$$I^{(2)}(f) \stackrel{d}{=} I^{(1)}(f).$$

PROOF: In Theorem 6.3.1, set first $\theta_1 = \theta$ and $\theta_2 = 0$ and then set $\theta_1 = 0$ and $\theta_2 = \theta$. ■

The real and imaginary parts of $I(f)$ are identically distributed but they are, in general, not independent when $0 < \alpha < 2$. This follows from Section 2.6 because $I(f)$ is a complex isotropic $S\alpha S$ random variable. But it can also be directly observed here. Indeed, for $z = \theta_1 + i\theta_2 \neq 0$ and $\|f\|_\alpha \neq 0$,

$$E\exp\{i(\theta_1 I^{(1)}(f) + \theta_2 I^{(2)}(f))\} \neq (E\exp i\theta_1 I^{(1)}(f))(E\exp i\theta_2 I^{(2)}(f)),$$

because

$$(\theta_1^2 + \theta_2^2)^{\alpha/2} c_0 \|f\|_\alpha^\alpha \neq |\theta_1|^\alpha c_0 \|f\|_\alpha^\alpha + |\theta_2|^\alpha c_0 \|f\|_\alpha^\alpha.$$

There is equality, however, in the Gaussian case $\alpha = 2$, and in that case, the real and imaginary parts of $I(f)$ are independent.

The next corollary shows that a linear combination of the real and imaginary parts of $I(f)$ has, up to a multiplicative constant, the same distribution as the real part of $I(f)$.

Corollary 6.3.3 *Let A and B be real. Then*

$$AI^{(1)}(f) + BI^{(2)}(f) \stackrel{d}{=} (A^2+B^2)^{1/2} I^{(1)}(f).$$

PROOF: For any real θ,

$$\begin{aligned} E\exp\{i\theta(AI^{(1)}(f)+BI^{(2)}(f))\} &= \exp\{-c_0(\theta^2A^2+\theta^2B^2)^{\alpha/2}\|f\|_\alpha^\alpha\} \\ &= \exp\{-c_0|\theta|^\alpha\|(A^2+B^2)^{1/2}f\|_\alpha^\alpha\} \end{aligned}$$

by Theorem 6.3.1. Use now Relation (6.3.3). ∎

The following theorem is an extension of Theorem 6.3.1 to stochastic processes.

Theorem 6.3.4 *Let $\{I(f_t) = \int_E f_t(x)M(dx),\ t \in T\}$ be a complex-valued stochastic process where M is an isotropic $S\alpha S$ random measure with control measure m, and let $I^{(1)}$ and $I^{(2)}$ denote, respectively, the real and imaginary parts of I. Fix $t_1,\dots,t_d \in T$. Then for arbitrary complex numbers $z_j = \theta_j^{(1)} + i\theta_j^{(2)}$, $j=1,\dots,d$,*

$$E\exp\Big\{i\Big(\sum_{j=1}^d \theta_j^{(1)}I^{(1)}(f_{t_j}) + \theta_j^{(2)}I^{(2)}(f_{t_j})\Big)\Big\} = E\exp\Big\{i\sum_{j=1}^d \text{Re } I(\overline{z}_j f_{t_j})\Big\}$$

$$= \exp\Big\{-c_0\int_E |\sum_{j=1}^d \overline{z}_j f_{t_j}(x)|^\alpha m(dx)\Big\}. \tag{6.3.4}$$

PROOF: Apply Theorem 6.3.1 with $\theta_1 = 1,\ \theta_2 = 0$ and $f = \sum_{j=1}^d \overline{z}_j f_{t_j}$. ∎

Remark. Theorem 6.3.4 is often expressed as follows:

$$E\exp\Big\{i\,\text{Re}\sum_{j=1}^d I(\overline{z}_j f_{t_j})\Big\} = \exp\Big\{-\|\text{Re}\sum_{j=1}^d I(\overline{z}_j f_{t_j})\|_\alpha^\alpha\Big\},$$

where

$$\Big\|\text{Re}\sum_{j=1}^d I(\overline{z}_j f_{t_j})\Big\|_\alpha^\alpha = c_0\int_E |\sum_{j=1}^d \overline{z}_j f_{t_j}(x)|^\alpha m(dx) = c_0\Big\|\sum_{j=1}^d \overline{z}_j f_{t_j}(x)\Big\|_\alpha^\alpha. \tag{6.3.5}$$

Corollary 6.3.5 *Let* $\{I(f_t),\ t \in T\}$ *be defined as in Theorem 6.3.4. The joint characteristic function of* Re $I(f_t),\ t \in T$ *is*

$$E\exp\Big\{i\sum_{j=1}^{d}\theta_j \operatorname{Re} I(f_{t_j})\Big\} = \exp\Big\{-c_0\int_E |\sum_{j=1}^{d}\theta_j f_{t_j}(x)|^\alpha m(dx)\Big\} \quad (6.3.6)$$

and

$$\{\operatorname{Im} I(f_t),\ t \in T\} \stackrel{d}{=} \{\operatorname{Re} I(f_t),\ t \in T\}.$$

Remark. The major difference between the joint characteristic functions of the complex-valued process $\{I(f_t),\ t \in T\}$ and those of the real-valued process $\{\operatorname{Re} I(f_t),\ t \in T\}$ is the presence of the complex numbers $\overline{z}_j$ instead of θ_j in the characteristic functions.

Example 6.3.6 Let M be an isotropic $S\alpha S$ random measure on $\mathbb{R}$ with control measure m and consider the complex-valued process

$$H(t) = \int_{-\infty}^{\infty} e^{itx} M(dx),\ \ t \in \mathbb{R}.$$

Assume

$$\int_{-\infty}^{\infty} |e^{itx}|^\alpha m(dx) = m(\mathbb{R}) < \infty$$

in order to ensure that the process H is well defined. Its joint characteristic function is

$$E\exp\Big\{i\sum_{j=1}^{d}(\theta_j^{(1)}H^{(1)}(t_j) + \theta_j^{(2)}H^{(2)}(t_j))\Big\} \quad (6.3.7)$$

where $H^{(1)}(t)$ and $H^{(2)}(t)$ are, respectively, the real[1] and imaginary prats of $H(t)$. By Theorem 6.3.4, that characteristic function equals $\exp\{-c_0\int_{-\infty}^{\infty}|\sum_{j=1}^{d}\overline{z}_j e^{it_jx}|^\alpha m(dx)\}$ where $z_j = \theta_j^{(1)} + i\theta_j^{(2)}$, $j = 1,\ldots,d$. Now,

$$\Big|\sum_{j=1}^{d}\overline{z}_j e^{it_jx}\Big|^2$$

$$= \Big|\sum_{j=1}^{d}\theta_j^{(1)}\cos t_jx + \sum_{k=1}^{d}\theta_k^{(2)}\sin t_kx\Big|^2 + \Big|\sum_{j=1}^{d}\theta_j^{(1)}\sin t_jx - \sum_{k=1}^{d}\theta_k^{(2)}\cos t_kx\Big|^2$$

$$= \sum_{j=1}^{d}\sum_{k=1}^{d}(\theta_j^{(1)}\theta_k^{(1)} + \theta_j^{(2)}\theta_k^{(2)})\,(\cos t_jx\cos t_kx + \sin t_jx\sin t_kx)$$

[1] $H^{(1)}(t) = \operatorname{Re}\int_{-\infty}^{\infty} e^{itx}M(dx)$ was considered in Example 6.2.5, but we did not suppose that the random measure M is isotropic, as we do here.

$$+2\sum_{j=1}^{d}\sum_{k=1}^{d}(\theta_j^{(1)}\theta_k^{(2)}\cos t_jx\sin t_kx-\theta_j^{(1)}\theta_k^{(2)}\sin t_jx\cos t_kx)$$

$$=\sum_{j=1}^{d}\sum_{k=1}^{d}(\theta_j^{(1)}\theta_k^{(1)}+\theta_j^{(2)}\theta_k^{(2)})\cos(t_k-t_j)x$$

$$+2\sum_{j=1}^{d}\sum_{k=1}^{d}\theta_j^{(1)}\theta_k^{(2)}\sin(t_k-t_j)x.$$

Therefore the finite-dimensional characteristic function (6.3.7) of $\{H(t),\ t\in\mathbb{R}\}$ equals

$$\exp\Big\{-c_0\int_{-\infty}^{\infty}\Big|\sum_{j=1}^{d}\sum_{k=1}^{d}(\theta_j^{(1)}\theta_k^{(1)}+\theta_j^{(2)}\theta_k^{(2)})\cos(t_k-t_j)x$$
$$+2\sum_{j=1}^{d}\sum_{k=1}^{d}\theta_j^{(1)}\theta_k^{(2)}\sin(t_k-t_j)x\Big|^{\alpha/2}m(dx)\Big\} \tag{6.3.8}$$

The process H is the complex-valued *harmonizable* process and we will encounter it again in the sequel.

Setting $\theta_j^{(1)}=\theta_j$ and $\theta_j^{(2)}=0$, we obtain the finite-dimensional characteristic function of $\{\text{Re}\ H(t),\ t\in\mathbb{R}\}$:

$$E\exp\Big\{i\sum_{j=1}^{d}\theta_j\ \text{Re}\ H(t_j)\Big\}$$
$$=\exp\Big\{-c_0\int_{-\infty}^{\infty}\Big|\sum_{j=1}^{d}\sum_{k=1}^{d}\theta_j\theta_k\cos(t_j-t_k)x\Big|^{\alpha/2}m(dx)\Big\}. \tag{6.3.9}$$

6.4 Series representation of complex-valued $S\alpha S$ random measures and integrals

The series representation of real $S\alpha S$ random measures and stable stochastic integrals developed in Chapter 3 is extended here to the complex-valued $S\alpha S$ case.

Let M be a complex-valued $S\alpha S$ random measure on $(E,\mathcal{E})$ with *finite* circular control measure k and control measure m. Then $m(E)=k(E\times S_2)$ is finite and is both the total mass of k and the total mass of m. Thus $\widehat{k}=k/m(E)$ is a probability measure on $(E\times S_2,\ \mathcal{E}\times\mathcal{B}_2)$. Let $\{\Gamma_1,\Gamma_2,\ldots\}$ be a sequence of arrival times of a Poisson process with unit arrival rate and let $\{\epsilon_1, \epsilon_2, \ldots\}$ be a sequence of Rademacher random variables. Finally, let $\{V_1, V_2, \ldots\}$ be a sequence of i.i.d. $(E \times S_2)$-valued random variables with common law $\widehat{k}$. All three sequences, $\{\Gamma_1, \Gamma_2, \ldots\}$, $\{\epsilon_1, \epsilon_2, \ldots\}$ and $\{V_1, V_2, \ldots\}$, are assumed to be independent. We write $V_j = (Z_j, (S_j^{(1)}, S_j^{(2)}))$, $j = 1, 2, \ldots$, with $Z_j \in E$, $(S_j^{(1)}, S_j^{(2)}) \in S_2$. We have the following:

Theorem 6.4.1

$$\{M(A),\ A \in \mathcal{E}\}$$
$$\stackrel{d}{=} \left\{(C_\alpha m(E))^{1/\alpha} \sum_{j=1}^{\infty} \epsilon_j \Gamma_j^{-1/\alpha} 1(Z_j \in A)(S_j^{(1)} + iS_j^{(2)}),\ A \in \mathcal{E}\right\}, \quad (6.4.1)$$

where C_α is given in (1.2.9).

We will not prove this theorem, as it can be regarded as a particular case of Corollary 6.4.3 below, involving series representation for stochastic integrals with respect to complex-valued $S\alpha S$ random measures. We stated Theorem 6.4.1 because we feel it is the structure of a random measure which determines the structure of an integral with respect to that measure. But, in fact, a statement about a random measure follows from the corresponding statement about integrals with respect to that measure.

Theorem 6.4.2 *Let $f : E \to \mathbb{C}$ belong to $L^\alpha(m)$. Then*

$$\int_E f(x)M(dx) \stackrel{d}{=} (C_\alpha m(E))^{1/\alpha} \sum_{j=1}^{\infty} \epsilon_j \Gamma_j^{-1/\alpha} f(Z_j)(S_j^{(1)} + iS_j^{(2)}), \quad (6.4.2)$$

where C_α is given in (1.2.9). The series in the right-hand side of (6.4.2) converges a.s.

PROOF: The a.s. convergence of the series on the right-hand side of (6.4.2) is equivalent to a.s. convergence of its real and imaginary parts. We shall verify only the convergence of the real part of the series as the argument for the imaginary part is identical. Let $U_j = \mathrm{Re}(f(Z_j)(S_j^{(1)} + iS_j^{(2)}))$, $j = 1, 2, \ldots$. Then $|U_j| \le |f(Z_j)(S_j^{(1)} + iS_j^{(2)})| = |f(Z_j)|$, so that $E|U_j|^\alpha < \infty$. Applying Theorem 1.4.2, we conclude that the real part of the series on the right-hand side of (6.4.2) converges a.s.

Let $\theta^{(1)}$ and $\theta^{(2)}$ be arbitrary real numbers. From Proposition 6.2.1 we conclude

$$\theta^{(1)} \mathrm{Re} \int_E f(x)M(dx) + \theta^{(2)} \mathrm{Im} \int_E f(x)M(dx) \sim S_\alpha(\sigma, 0, 0),$$

where σ^α equals

$$\int_E \int_{S_2} |\theta^{(1)}(s_1 f^{(1)}(x) - s_2 f^{(2)}(x)) + \theta^{(2)}(s_1 f^{(2)}(x) + s_2 f^{(1)}(x))|^\alpha k(dx, d\mathbf{s}). \tag{6.4.3}$$

To prove (6.4.2) we need to show that the series

$$(C(\alpha)m(E))^{1/\alpha} \sum_{j=1}^{\infty} \epsilon_j \Gamma_j^{-1/\alpha} \Big[\theta^{(1)} \operatorname{Re}(f(Z_j)(S_j^{(1)} + iS_j^{(2)})) + \theta^{(2)} \operatorname{Im}(f(Z_j)(S_j^{(1)} + iS_j^{(2)}))\Big] \tag{6.4.4}$$

has the same distribution. Note that

$$\begin{aligned} &\theta^{(1)} \operatorname{Re}\, (f(Z_j)(S_j^{(1)} + iS_j^{(2)})) + \theta^{(2)} \operatorname{Im}\, (f(Z_j)(S_j^{(1)} + iS_j^{(2)})) \\ &= \theta^{(1)}(S_j^{(1)} f^{(1)}(Z_j) - S_j^{(2)} f^{(2)}(Z_j)) + \theta^{(2)}(S_j^{(1)} f^{(2)}(Z_j) + S_j^{(2)} f^{(1)}(Z_j)), \end{aligned}$$

and thus, applying Theorem 1.4.2, we conclude that the series in (6.4.4) has also a $S_\alpha(\sigma, 0, 0)$ distribution with σ given by (6.4.3). This proves (6.4.2). ∎

Because both the integral on the left-hand side of (6.4.2) and the series on its right-hand side are linear in f, we can, as in the real-valued case, extend immediately the result to $S\alpha S$ random vectors and to $S\alpha S$ random processes.

Corollary 6.4.3 *Let $\{f_t,\ t \in T\}$ be a family of functions from E into $\mathbb{C}$ such that $f_t \in L^\alpha(m)$ for every $t \in T$. Then*

$$\begin{aligned} &\Big\{\int_E f_t(x) M(dx),\ t \in T\Big\} \\ &\stackrel{d}{=} \Big\{(C_\alpha m(E))^{1/\alpha} \sum_{j=1}^{\infty} \epsilon_j \Gamma_j^{-1/\alpha} f_t(Z_j)(S_j^{(1)} + iS_j^{(2)}),\ t \in T\Big\}. \end{aligned} \tag{6.4.5}$$

PROOF: Use Theorem 6.4.2 and the linearity to show that every linear combination of finitely many of random variables on the left-hand side of (6.4.5) is equal in distribution to the corresponding linear combination of random variables on the right-hand side of (6.4.5). ∎

We now turn to a series representation for stochastic $S\alpha S$ integrals with respect to an isotropic complex-valued $S\alpha S$ random measure. The importance of this particular case will become evident in the next section in the context of $S\alpha S$ harmonizable processes.

Let M be an isotropic $S\alpha S$ random measure on $(E, \mathcal{E})$ with a finite control measure m. We remind the reader that the circular control measure of M is then given by $k = m\gamma$, where γ is the uniform probability measure on S_2. Let $f : E \to$

$\mathbb{C}$ belong to $L^\alpha(m)$. Theorem 6.4.2 of course applies. Note, however, that in this particular case the sequence of S_2-valued random variables $\{(S_j^{(1)}, S_j^{(2)}),\ j = 1, 2, \ldots\}$ is a sequence of i.i.d. random variables with common distribution γ and is *independent* of the sequence of E-valued i.i.d. random variables $\{Z_j,\ j = 1, 2, \ldots\}$ with common law $m/m(E)$. Let $\{G_j^{(i)},\ j = 1, 2, \ldots\}$, $i = 1, 2$, be two independent sequences of i.i.d. standard normal random variables. Then for each $j = 1, 2, \ldots,$

$$\left(\frac{G_j^{(1)}}{((G_j^{(1)})^2 + (G_j^{(2)})^2)^{1/2}}, \frac{G_j^{(2)}}{((G_j^{(1)})^2 + (G_j^{(2)})^2)^{1/2}}\right) \stackrel{d}{=} (S_j^{(1)}, S_j^{(2)})$$

(see Exercise 6.11). If we also assume that these normal sequences are independent of the sequences $\{\Gamma_1, \Gamma_2, \ldots\}$ and $\{Z_1, Z_2, \ldots\}$, then, by Theorem 6.4.2,

$$\int_E f(x) M(dx) \stackrel{d}{=} (C_\alpha m(E))^{1/\alpha} \sum_{j=1}^{\infty} \Gamma_j^{-1/\alpha} f(Z_j) \frac{G_j^{(1)} + iG_j^{(2)}}{((G_j^{(1)})^2 + (G_j^{(2)})^2)^{1/2}}. \tag{6.4.6}$$

Note that the Rademacher sequence $\{\epsilon_1, \epsilon_2, \ldots\}$ has been absorbed in the (symmetric) normal sequences $\{G_1^{(i)}, G_2^{(i)}, \ldots\}$, $i = 1, 2$, since $\epsilon G \stackrel{d}{=} G$ if ϵ and G are independent. How can one now eliminate the denominators $W_j = ((G_j^{(1)})^2 + (G_j^{(2)}))^{1/2}$? The sequence $\{W_j,\ j = 1, 2, \ldots\}$ is *independent* of the sequences $\{\Gamma_j,\ j = 1, 2, \ldots\}, \{Z_j,\ j = 1, 2, \ldots\}$ and $\{((G_j^{(1)})^2 + (G_j^{(2)})^2)^{-1/2}(G_j^{(1)}, G_j^{(2)}),\ j = 1, 2, \ldots\}$ (Exercise 6.11). Moreover, if $b_\alpha = EW_j^\alpha$, then $W_j / b_\alpha^{1/\alpha}$ has α^{th} moment equal to 1 and, therefore, by Theorem 1.4.2,

$$\int_E f(x) M(dx) \stackrel{d}{=} (C_\alpha m(E))^{1/\alpha} \sum_{j=1}^{\infty} \Gamma_j^{-1/\alpha} f(Z_j) \frac{W_j}{b_\alpha^{1/\alpha}} \frac{G_j + iG_j^{(2)}}{((G_j^{(1)})^2 + (G_j^{(2)})^2)^{1/2}}$$

$$\stackrel{d}{=} (C_\alpha b_\alpha^{-1} m(E))^{1/\alpha} \sum_{j=1}^{\infty} \Gamma_j^{-1/\alpha} f(Z_j)(G_j^{(1)} + iG_j^{(2)}).$$

(W_j has a *Rayleigh* distribution. Its density function is $xe^{-x^2/2} 1(x > 0)$ and therefore $b_\alpha = EW_j^\alpha$ is finite. It is easy to verify that $b_\alpha = 2^{\alpha/2}\Gamma(1 + \frac{\alpha}{2})$.) This establishes

Proposition 6.4.4 *Let M be an isotropic $S\alpha S$ random measure on $(E, \mathcal{E})$ with a finite control measure m. Let $\{G_j^{(i)},\ j = 1, 2, \ldots\}$, $i = 1, 2$, be two sequences of i.i.d. standard normal random variables, $\{\Gamma_j,\ j = 1, 2, \ldots\}$ be a sequence of arrival times of a unit rate Poisson process and $\{Z_1, Z_2 \ldots\}$ be i.i.d. E-valued* *random variables with the common law $m/m(E)$. These four sequences are assumed independent. Let $f : E \to \mathbb{C}$ belong to $L^\alpha(m)$. Then*

$$(i) \quad \int_E f(x)M(dx) \tag{6.4.7}$$

$$\stackrel{d}{=} (C_\alpha b_\alpha^{-1} m(E))^{1/\alpha} \sum_{j=1}^{\infty} \Gamma_j^{-1/\alpha} f(Z_j)(G_j^{(1)} + iG_j^{(2)}),$$

$$(ii) \quad \text{Re} \int_E f(x)M(dx) \tag{6.4.8}$$

$$\stackrel{d}{=} (C_\alpha b_\alpha^{-1} m(E))^{1/\alpha} \sum_{j=1}^{\infty} \Gamma_j^{-1/\alpha} (f^{(1)}(Z_j)G_j^{(1)} + f^{(2)}(Z_j)G_j^{(2)}),$$

where $b_\alpha = 2^{\alpha/2}\Gamma(1 + \frac{\alpha}{2})$ and C_α is given in (1.2.9).

Remarks

1. Since the series representations of the integrals in Proposition 6.4.4 are linear in the integrand f, they extend immediately to $S\alpha S$ processes represented as integrals with respect to an isotropic $S\alpha S$ random measure. See Corollary 6.4.3.

2. It seems at first glance that one could easily simplify the series representation (6.4.8) by observing that for any $j = 1, 2, \ldots,$

$$f^{(1)}(Z_j)G_j^{(1)} + f^{(2)}(Z_j)G_j^{(2)} \stackrel{d}{=} (f^{(1)}(Z_j)^2 + f^{(2)}(Z_j)^2)^{1/2} G_j^{(1)},$$

and thus

$$\text{Re} \int_E f(x)M(dx)$$

$$\stackrel{d}{=} (C_\alpha b_\alpha^{-1} m(E))^{1/\alpha} \sum_{j=1}^{\infty} \Gamma_j^{-1/\alpha} G_j^{(1)} (f^{(1)}(Z_j)^2 + f^{(2)}(Z_j)^2)^{1/2}.$$

Indisputably, this series representation requires only one sequence of i.i.d. standard normal random variables as compared to the two sequences required in the representation (6.4.8). But, unfortunately, it is not linear in the integrand f and therefore may not be directly extended to $S\alpha S$ processes that are represented as integrals with respect to an isotropic $S\alpha S$ random measure. Thus, in general,

$$\left\{ \text{Re} \int_E f_t(x)M(dx),\ t \in T \right\}$$

$$\stackrel{d}{\neq} \left\{ (C_\alpha b_\alpha^{-1} m(E))^{1/\alpha} \sum_{j=1}^{\infty} \Gamma_j^{-1/\alpha} G_j^{(1)} (f_t^{(1)}(Z_j)^2 + f_t^{(2)}(Z_j)^2)^{1/2},\ t \in T \right\}.$$

This is why we use the representation (6.4.8) instead.

6.5 Harmonizable process

The (real) $S\alpha S$ harmonizable process (6.2.13) has been defined in Example 6.2.5. Because of its importance in the theory of stationary $S\alpha S$ processes, we shall study its properties in more detail. It is convenient to work simultaneously with the real harmonizable process X defined in (6.2.13) and with its complex counterpart

$$Z(t) = \int_{-\infty}^{\infty} e^{itx} M(dx), \ -\infty < t < \infty. \tag{6.5.1}$$

We want to obtain conditions for stationarity. A real stochastic process $\{X(t),\ t \in \mathbb{R}\}$ is called (*strictly*) *stationary* if its finite-dimensional distributions are invariant under shift, i.e., for any $t_1, t_2, \ldots, t_d \in \mathbb{R}$ and any $h > 0$,

$$(X(t_1), X(t_2), \ldots, X(t_d)) \stackrel{d}{=} (X(t_1+h), X(t_2+h), \ldots, X(t_d+h)).$$

Similarly, a complex-valued stochastic process $\{Z(t),\ t \in \mathbb{R}\}$ is (*strictly*) *stationary* if all finite-dimensional distributions are invariant under shift, i.e., for any $t_1, t_2, \ldots, t_d \in \mathbb{R}$ and any $h > 0$

$$(\text{Re}\, Z(t_j), \text{Im}\, Z(t_j),\ j = 1, \ldots, d) \stackrel{d}{=} (\text{Re}\, Z(t_j+h), \text{Im}\, Z(t_j+h),\ j = 1, \ldots, d).$$

Obviously, if two harmonizable $S\alpha S$ processes, one real-valued, the other complex-valued, are defined through the same random measure M, then the real-valued process is stationary if the complex-valued process is stationary, but the converse is not necessarily true.

The goal of this section is to obtain conditions for the stationarity of the real- and complex-valued harmonizable $S\alpha S$ processes. Let us start with some heuristics and try to guess these conditions. Clearly, they must be expressed in terms of the circular control measure k of the random measure M. We begin with a simple observation: the behavior of the circular control measure k on $\{0\} \times S_2$ is of no importance as far as stationarity of either the real- or complex-valued harmonizable $S\alpha S$ processes is concerned. Indeed, let M_1 and M_2 be two $S\alpha S$ random measures on $\mathbb{R}$, with circular control measures $k_1 = (1 - 1_{\{0\}\times S_2})k$ and $k_2 = 1_{\{0\}\times S_2} k$, respectively. M_1 and M_2 are independent because their supports are disjoint. Let

$$Z_j(t) = \int_{-\infty}^{\infty} e^{itx} M_j(dx), \ -\infty < t < \infty, \ j = 1, 2.$$

Then Z_1 and Z_2 are independent complex-valued $S\alpha S$ harmonizable processes such that $\{Z_1(t)+Z_2(t),\ -\infty<t<\infty\} \stackrel{d}{=} \{Z(t),\ -\infty<t<\infty\}$. But Z_2 is a *constant* $S\alpha S$ process (see Exercise 6.9). Therefore, Z is stationary if and only if Z_1 is stationary, and Re Z is stationary if and only if Re Z_1 is stationary. Thus, we should not be concerned with the form of the circular control measure k on $\{0\}\times S_2$, and we may, and, indeed, shall assume throughout the rest of this section that the control measure m of the $S\alpha S$ random measure M in (6.2.13) or in (6.5.1) has no atom at the origin.

In the Gaussian case $\alpha=2$ the (complex-valued) harmonizable process (6.5.1) is stationary if and only if the two components $M^{(1)}$ and $M^{(2)}$ of the complex-valued Gaussian random measure M are independent and identically distributed. We may be tempted to guess that the same is true in the case $0<\alpha<2$, but this turns out to be false. This does not mean that we should not try to extrapolate from the Gaussian case; it only means that we considered the wrong property. In the Gaussian case, a complex-valued random measure has independent and identically distributed components if and only if it is isotropic in the sense of (6.1.4). It is the property of isotropy that should be carried over to the case $0<\alpha<2$. We know from Example 6.1.6 that a $S\alpha S$ random measure M with $0<\alpha<2$ is isotropic if and only if $k=m\gamma$, where γ is the uniform probability measure on S_2. This explains heuristically the conditions for stationarity of the complex-valued harmonizable $S\alpha S$ process that appear in the following theorem.

It should come as no surprise that M does not have to be fully isotropic in order for the *real* harmonizable $S\alpha S$ process to be stationary. To express the conditions for stationarity in the real case, we need to introduce some new notation.

Given a measure k on $(\mathbb{R}\times S_2, \mathcal{B}\times\mathcal{B}_2)$, define the *conjugate measure* $\overline{k}$ by

$$\overline{k}(A\times U) = k((-A)\times U^*), \tag{6.5.2}$$

$A\in\mathcal{B}$, $U\in\mathcal{B}_2$ and $U^* = \{(s_1,s_2)\in S_2 : (s_1,-s_2)\in U\}$. In other words, $\overline{k}$ is the image of k under the measurable map $h:\ \mathbb{R}\times S_2\to\mathbb{R}\times S_2$ defined by $h((x,s_1,s_2)) = (-x,s_1,-s_2)$. Similarly, for a measure m on $(\mathbb{R},\mathcal{B})$, we define the measure $\overline{m}$ as $\overline{m}(A) = m(-A)$, $A\in\mathcal{B}$. We are now ready to state the main theorem.

Theorem 6.5.1 *Let M be a $S\alpha S$ random measure on $(\mathbb{R},\mathcal{B})$ with a finite circular control measure k and a control measure m, and suppose that m has no atom at the origin. Then the complex-valued harmonizable $S\alpha S$ process Z in (6.5.1) is stationary if and only if the $S\alpha S$ random measure M is isotropic (rotationally invariant), i.e., if and only if it satisfies*

$$k = m\gamma, \tag{6.5.3}$$

where γ is the uniform probability measure on S_2.

The real harmonizable $S\alpha S$ process X in (6.2.13) is stationary if and only if the $S\alpha S$ random measure M satisfies the weaker condition

$$k + \overline{k} = (m + \overline{m})\gamma. \tag{6.5.4}$$

PROOF: Let $k_S = k + \overline{k}$ and let M_S be a $S\alpha S$ random measure on $(\mathbb{R}, \mathcal{B})$ with k_S as circular control measure. Then the control measure m_S of M_S satisfies $m_s = m + \overline{m}$. Define

$$Y(t) = \text{Re} \int_{-\infty}^{\infty} e^{itx} M_S(dx), \ -\infty < t < \infty. \tag{6.5.5}$$

It is trivial to check using Proposition 6.2.4 that

$$\{Y(t), \ -\infty < t < \infty\} \stackrel{d}{=} \{2^{1/\alpha} X(t), \ -\infty < t < \infty\}. \tag{6.5.6}$$

Therefore, X is stationary if and only if Y is stationary.

Proof of the necessity part of Theorem 6.5.1: We use two steps.

Step 1. We assume that the process Y (respectively, the process Z) is stationary, and we prove that the $S\alpha S$ random measure M_τ defined by

$$M_\tau(A) = \int_A e^{i\tau x} M_S(dx), \ \ A \in \mathcal{B}$$

(respectively, defined by $M_\tau(A) = \int_A e^{i\tau x} M(dx), \ A \in \mathcal{B}$) has the same distribution as M_S (respectively, M) for every real τ.

Step 2. We assume that M_τ has the same distribution as M, for every real τ, and prove that the $S\alpha S$ random measure M is isotropic.

These two steps imply the necessity part of the theorem. Indeed, if the process Z is stationary, then *Step 1* and *Step 2* imply that the $S\alpha S$ random measure M is isotropic, and then (6.5.3) follows from Example 6.2.5. Similarly, if X (and thus, Y) is stationary, then the same argument applies to M_S, and (6.5.4) follows. We start with the

Proof of Step 1. It is well known from Weierstrass' trigonometric approximation theorem that for any real continuous periodic function f with period 2π and for any $\epsilon > 0$, there is a trigonometric polynomial of the form $T(x) = \sum_{j=0}^{n}(a_j \cos jx + b_j \sin jx)$ such that $|f(x) - T(x)| \leq \epsilon$ for every real x. It follows immediately that if f is a continuous periodic function with period R, then there is a trigonometric polynomial of the form

$$T(x) = \sum_{j=0}^{n} \Big(a_j \cos\Big(\frac{2\pi}{R} jx\Big) + b_j \sin\Big(\frac{2\pi}{R} jx\Big)\Big) \tag{6.5.7}$$

such that $|f(x) - T(x)| \le \epsilon$ for every real x. Obviously, if f is even (respectively, odd), then we may always choose $b_j = 0,\ j = 1, \ldots, n$ (respectively, $a_j = 0,\ j = 1, \ldots, n$) in (6.5.7).

Let now g be a complex-valued function of the form $g = g^{(1)} + ig^{(2)}$, where $g^{(1)}$ (respectively, $g^{(2)}$) is an even (respectively, odd) continuous periodic function with period R, and let $\epsilon > 0$. It follows from the preceding remarks that there are two trigonometric polynomials $T^{(1)}(x) = \sum_{j=0}^{n} a_j \cos((2\pi/R)jx)$ and $T^{(2)}(x) = \sum_{j=0}^{n} b_j \sin((2\pi/R)jx)$ such that $|T^{(k)}(x) - g^{(k)}(x)| \le \frac{\epsilon}{2}$ for any $x \in R$, $k = 1, 2$. Choosing $\epsilon = 2^{-m},\ m = 1, 2, \ldots,$ we conclude that there is a sequence of functions of the form

$$g_m(x) = \sum_{j=0}^{n_m} [a_j^{(m)} (e^{i(2\pi/R)jx} + e^{-i(2\pi/R)jx}) + ib_j^{(m)} (e^{i(2\pi/R)jx} - e^{-i(2\pi R)jx})] \tag{6.5.8}$$

with real-valued coefficients $\{a_j^{(m)}, b_j^{(m)},\ j = 0, 1, \ldots, n_m\}$ such that for every $m = 1, 2, \ldots$ and for every real x we have $|g(x) - g_m(x)| \le 2^{-m}$. For any function g and its approximating sequence $\{g_m\}_{m=1}^{\infty}$ as above we define

$$Y = \text{Re} \int_{-\infty}^{\infty} g(x) M_S(dx),\ Y_m = \text{Re} \int_{-\infty}^{\infty} g_m(x) M_S(dx),\ m = 1, 2, \ldots.$$

Assume, now, that the real harmonizable $S\alpha S$ process $Y(\cdot)$ defined in (6.5.5) is stationary, and let τ be a real number. For M_τ given by $M_\tau(A) = \int_A e^{i\tau x} M_S(dx),\ A \in \mathcal{B}$, define

$$Y_\tau = \text{Re} \int_{-\infty}^{\infty} g(x) M_\tau(dx),\ Y_{\tau,m} = \text{Re} \int_{-\infty}^{\infty} g_m(x) M_\tau(dx),\ m = 1, 2, \ldots.$$

By the stationarity of $Y(\cdot)$, we have

$$\begin{aligned}
\{Y(t),\ -\infty < t < \infty\} &\stackrel{d}{=} \{Y(t+\tau),\ -\infty < t < \infty\} \\
&= \Big\{\text{Re} \int_{-\infty}^{\infty} e^{i(t+\tau)x} M_S(dx),\ -\infty < t < \infty\Big\} \\
&\stackrel{d}{=} \Big\{\text{Re} \int_{-\infty}^{\infty} e^{itx} M_\tau(dx),\ -\infty < t < \infty\Big\},
\end{aligned}$$

where Exercise 6.7 has been used in the last equality. Hence, by Proposition 6.2.4, $Y_m \stackrel{d}{=} Y_{\tau,m}$. Proposition 6.2.3 implies $\operatorname{plim}_{m\to\infty} Y_m = Y$, $\operatorname{plim}_{m\to\infty} Y_{\tau,m} = Y_\tau$. Therefore,

$$\operatorname{Re}\int_{-\infty}^{\infty} g(x) M_S(dx) \stackrel{d}{=} \operatorname{Re}\int_{-\infty}^{\infty} g(x) M_\tau(dx) \tag{6.5.9}$$

for any function $g : \mathbb{R} \to \mathbb{C}$ whose real and imaginary parts are, respectively, even and odd continuous periodic functions.

Our next goal is to extend the class of functions g for which (6.5.9) holds. Let g be a continuous function with bounded support whose real and imaginary parts are, respectively, even and odd. Choose K large enough so that $(-K, K)$ contains the support of g. Extend g periodically, with period $2K$, from $(-K, K]$ to the whole real line by setting $g_{(K)}(x) = g(x - 2Kj)$ if $x \in (2Kj - K,\ 2Kj + K]$, $j = \pm 1, \pm 2, \ldots$. Since $g_{(K)}$ is a continuous periodic function whose real and imaginary parts are, respectively, even and odd,

$$\operatorname{Re}\int_{-\infty}^{\infty} g_{(K)}(x) M_S(dx) \stackrel{d}{=} \operatorname{Re}\int_{-\infty}^{\infty} g_{(K)}(x) M_\tau(dx). \tag{6.5.10}$$

Since g is bounded,

$$\begin{aligned}\int_{-\infty}^{\infty} & |g(x) - g_{(K)}(x)|^\alpha m(dx) \\ &= \int_{-\infty}^{+\infty} |g(x) - g_{(K)}(x)|^\alpha 1_{(-K,K]^c}(x) m(dx) \to 0\end{aligned}$$

as $K \to \infty$ by the bounded convergence theorem. Applying Proposition 6.2.3, we see that

$$\operatorname{Re}\int_{-\infty}^{\infty} g(x) M_S(dx) = \operatorname{plim}_{K\to\infty} \operatorname{Re}\int_{\infty}^{\infty} g_{(K)}(x) M_S(dx),$$

$$\operatorname{Re}\int_{-\infty}^{\infty} g(x) M_\tau(dx) = \operatorname{plim}_{K\to\infty} \operatorname{Re}\int_{-\infty}^{\infty} g_{(K)}(x) M_\tau(dx).$$

From (6.5.10), we conclude that (6.5.9) holds for any continuous g with bounded support whose real and imaginary parts are, respectively, even and odd.

Let A be a Borel subset of the real line. We want to show that

$$k_S(A \times ds) = k_\tau(A \times ds), \tag{6.5.11}$$

where k_τ is the circular control measure of the $S\alpha S$ random measure M_τ. If (6.5.11) holds, then we will be able to conclude $M_S \stackrel{d}{=} M_\tau$.

Note that it is enough to prove (6.5.11) for A of the kind $A = (a, b)$, $0 < a < b$ or $a < b < 0$. Indeed, suppose (6.5.11) holds for every set A as above. Then, by the monotone class theorem (6.5.11) holds for every Borel set A such that either $A \subset (0, \infty)$ or $A \subset (-\infty, 0)$ and hence (6.5.11) follows from the decomposition $A = (A \cap (-\infty, 0)) \cup (A \cap \{0\}) \cup (A \cap (0, \infty))$ and the assumption that $m(\{0\}) = 0$. We will establish

$$k_S((a, b) \times ds) = k_\tau((a, b) \times ds) \tag{6.5.12}$$

for every $0 < a < b$ (the case $a < b < 0$ is identical).

Let θ_1 and θ_2 be arbitrary real numbers. Define

$$r(x) = \theta_1(1_{(a,b)}(x) + 1_{(-b,-a)}(x)) + i\theta_2(1_{(-b,-a)}(x) - 1_{(a,b)}(x)).$$

Let $\{h_n\}_{n=1}^{\infty}$ be a sequence of continuous real functions vanishing outside the interval *(a, b)* such that $h_n(x) \uparrow 1$ for every $x \in (a, b)$. Let

$$r_n(x) = \theta_1(h_n(x) + h_n(-x)) + i\theta_2(h_n(-x) - h_n(x)),\ n = 1, 2, \ldots.$$

Clearly, each r_n is a continuous function with bounded support whose real and imaginary parts are, respectively, even and odd. Applying (6.5.9), we have

$$\text{Re} \int_{-\infty}^{\infty} r_n(x) M_S(dx) \stackrel{d}{=} \text{Re} \int_{-\infty}^{\infty} r_n(x) M_\tau(dx),\ n = 1, 2, \ldots.$$

Moreover, $\text{Re} \int_{-\infty}^{\infty} r(x) M_S(dx) = \text{plim}_{n\to\infty} \text{Re} \int_{-\infty}^{+\infty} r_n(x) M_S(dx)$ and $\text{Re} \int_{-\infty}^{\infty} r(x) M_\tau(dx) = \text{plim}_{n\to\infty} \text{Re} \int_{-\infty}^{\infty} r_n(x) M_\tau(dx)$, by Proposition 6.2.3. Hence,

$$\text{Re} \int_{-\infty}^{\infty} r(x) M_S(dx) \stackrel{d}{=} \text{Re} \int_{-\infty}^{\infty} r(x) M_\tau(dx).$$

But this last equation states that two (real) $S\alpha S$ random variables have the same distribution. Applying Proposition 6.2.4, we have

$$\begin{aligned}
&\int_{S_2} |s_1\theta_1 + s_2\theta_2|^\alpha (k_S + \overline{k}_S)((a, b) \times ds) \\
&= \int_{-\infty}^{\infty} \int_{S_2} |s_1\theta_1 + s_2\theta_2|^\alpha 1_{(a,b)}(x) k_S(dx, ds) \\
&\quad + \int_{-\infty}^{\infty} \int_{S_2} |s_1\theta_1 - s_2\theta_2|^\alpha 1_{(-b,-a)}(x) k_S(dx, ds) \\
&= \int_{-\infty}^{\infty} \int_{S_2} |s_1 \,\text{Re}\, r(x) - s_2 \,\text{Im}\, r(x)|^\alpha k_S(dx, ds) \quad \text{(definition of } r(x)) \\
&= \int_{-\infty}^{\infty} \int_{S_2} |s_1 \,\text{Re}\, r(x) - s_2 \,\text{Im}\, r(x)|^\alpha k_\tau(dx, ds) \quad \text{(Proposition 6.2.4)} \\
&= \int_{S_2} |s_1\theta_1 + s_2\theta_2|^\alpha (k_\tau + \overline{k}_\tau)((a, b) \times ds).
\end{aligned}$$

We can now appeal to Theorem 2.3.1 (uniqueness of the spectral measure of a $S\alpha S$ random vector) and conclude

$$(k_S + \overline{k_S})((a,b) \times ds) = (k_\tau + \overline{k_\tau})((a,b) \times ds). \tag{6.5.13}$$

Courage, tired reader! We are very close to the end of the proof (of Step 1)! Relation (6.5.13) is very similar to Relation (6.5.12) which we want to establish. Since

$$\overline{k_S} = \overline{(k + \overline{k})} = \overline{k} + \overline{(\overline{k})} = \overline{k} + k = k_S,$$

the left-hand sides of (6.5.13) and (6.5.12) are, in fact, identical. Thus, to establish (6.5.12), it is sufficient to show that $\overline{k_\tau} = k_\tau$.

Using the definition of k_τ, we obtain for any real θ_1, θ_2 and any Borel set A

$$\begin{aligned}
&\int_{S_2} |\theta_1 s_1 + \theta_2 s_2|^\alpha \overline{k_\tau}(A \times ds) \\
&= \int_{S_2} |\theta_1 s_1 - \theta_2 s_2|^\alpha k_\tau((-A) \times ds) \\
&= \int_{-A}\int_{S_2} \left|\theta_1(s_1 \cos \tau x - s_2 \sin \tau x) - \theta_2(s_1 \sin \tau x + s_2 \cos \tau x)\right|^\alpha k_S(dx, ds) \\
&= \int_{A}\int_{S_2} \Big|\theta_1(s_1 \cos \tau(-x) - (-s_2) \sin \tau(-x)) \\
&\qquad -\theta_2(s_1 \sin \tau(-x) + (-s_2) \cos \tau(-x))\Big|^\alpha \overline{k_S}(dx, ds) \\
&= \int_{A}\int_{S_2} \left|\theta_1(s_1 \cos \tau x - s_2 \sin \tau x) + \theta_2(s_1 \sin \tau x + s_2 \cos \tau x)\right|^\alpha k_S(dx, ds) \\
&= \int_{S_2} |\theta_1 s_1 + \theta_2 s_2|^\alpha k_\tau(A \times ds),
\end{aligned}$$

where the second equality follows from Exercise 6.8, the fifth equality follows because $\overline{k_S} = k_S$, and the last equality follows once again from Exercise 6.8. Applying Theorem 2.3.1 once more, we have $\overline{k_S}(A \times ds) = k_\tau(A \times ds)$. This proves (6.5.12) and, therefore, completes the proof of Step 1 for the *real-valued* harmonizable $S\alpha S$ process.

The argument for the *complex-valued* harmonizable process Z defined by (6.5.1) is stationary is very similar and, in fact, simpler. The details are left to the reader (see Exercise 6.13). This completes the proof of Step 1.

Proof of Step 2. We must prove that if $M_\tau \stackrel{d}{=} M$ for every real τ, then $M \stackrel{d}{=} e^{i\phi} M$ for every real ϕ. In other words, we have to show that the circular control measures of M and $e^{i\phi}M$ coincide. Let $k^{(\phi)}$ denote the circular control measure of $e^{i\phi}M$. We must then prove that for every Borel set A, we have

$k^{(\phi)}(A \times ds) = k(A \times ds)$. By the monotone class theorem it is enough to prove that for any real $a < b$, $k^{(\phi)}((a, b] \times ds) = k((a, b] \times ds)$. Since $m(\{a\}) = 0$, it is sufficient to consider the case $0 < a < b$. That is, we will prove

$$e^{i\phi} M((a, b]) \stackrel{d}{=} M((a, b]) \tag{6.5.14}$$

for any $0 < a < b$. Let $\Delta = b - a > 0$, and for an $n \geq 1$, set $c_k(n) = a + \frac{k}{n}\Delta$, $k = 0, 1, \ldots, n$. Thus $c_0(n) = a$, $c_n(n) = b$. Let

$$J_k(n) = \int_{(c_{k-1}(n), c_k(n)]} e^{i\phi x / c_{k-1}(n)} M(dx), \ k = 1, \ldots, n,$$

and $J(n) = \sum_{k=1}^{n} J_k(n)$. The key observation is that for each $k = 1, \ldots, n$,

$$J_k(n) = M_{\phi / c_{k-1}(n)} \left((c_{k-1}(n), c_k(n)] \right) \stackrel{d}{=} M \left((c_{k-1}(n), c_k(n)] \right).$$

Hence,

$$J(n) = \sum_{k=1}^{n} J_k(n) \stackrel{d}{=} \sum_{k=1}^{n} M \left((c_{k-1}(n), c_k(n)] \right) = M((a, b]),$$

where the equality in distribution is between two sums of *independent* random variables whose corresponding terms are equal in distribution. Therefore, for each $n = 1, 2, \ldots,$ the distribution of $J(n)$ coincides with that of $M((a, b])$. In order to establish Relation (6.5.14), it is now sufficient to prove

$$\text{plim}_{n \to \infty} J(n) = e^{i\phi} M((a, b]). \tag{6.5.15}$$

Note that

$$J(n) = \int_{(a,b]} f^{(n)}(x) M(dx) \, , \ n = 1, 2, \ldots,$$

where

$$f_n(x) = e^{i\phi x / c_{k-1}(n)} \text{ if } c_{k-1}(n) < x \leq c_k(n), \ k = 1, \ldots, n.$$

By the Bounded Convergence Theorem, $\int_{(a,b]} |f_n(x) - e^{i\phi}|^\alpha m(dx) \to 0$ as $n \to \infty$ and, hence, Proposition 6.2.3 yields Relation (6.5.15). This completes the proof of Step 2 and also the proof of the necessity part of the theorem.

Proof of the sufficiency part of Theorem 6.5.1: Observe that (6.5.4) is equivalent to

$$k_S = m_S \gamma. \tag{6.5.16}$$

To prove sufficiency, we assume (6.5.16) and prove that the complex-valued harmonizable process

$$W(t) = \int_{-\infty}^{\infty} e^{itx} M_S(dx), \ -\infty < t < \infty \tag{6.5.17}$$

is stationary. This will establish the sufficiency part of the theorem for both the real- and complex-valued cases.

Take any $t_1, t_2, \dots, t_d \in \mathbb{R}$ and $u > 0$. By Proposition 6.2.1, for any real numbers $(\theta_1^{(1)}, \theta_1^{(2)}), (\theta_2^{(1)}, \theta_2^{(2)}), \dots, (\theta_d^{(1)}, \theta_d^{(2)})$,

$$E \exp\Big\{ i \sum_{j=1}^{d} (\theta_j^{(1)} \operatorname{Re} W(t_j + u) + \theta_j^{(2)} \operatorname{Im} W(t_j + u)) \Big\}$$

$$= \exp\Big\{ - \int_{-\infty}^{\infty} \int_{S_2} \Big| \sum_{j=1}^{d} [(s_1 \theta_j^{(1)} + s_2 \theta_j^{(2)}) \cos(x(t_j + u))$$

$$+ (s_1 \theta_j^{(2)} - s_2 \theta_j^{(1)}) \sin(x(t_j + u)) \Big|^{\alpha} k_S(dx, ds)] \Big\}$$

$$= \exp\Big\{ - \int_{-\infty}^{\infty} m_S(dx) \int_{S_2} \Big| s_1 \sum_{j=1}^{d} [\theta_j^{(1)} \cos(x(t_j + u)) + \theta_j^{(2)} \sin(x(t_j + u))]$$

$$+ s_2 \sum_{j=1}^{d} [\theta_j^{(2)} \cos(x(t_j + u)) - \theta_j^{(1)} \sin(x(t_j + u))] \Big|^{\alpha} \gamma(ds) \Big\}$$

$$= \exp\Big\{ - \int_{-\infty}^{\infty} m_S(dx) \int_{S_2} \Big| s_1 \operatorname{Re} \big(\sum_{j=1}^{d} R_j e^{i(\phi_j - (t_j + u)x)}\big)$$

$$+ s_2 \operatorname{Im} \big(\sum_{j=1}^{d} R_j e^{i(\phi_j - (t_j + u)x)}\big) \Big|^{\alpha} \gamma(ds) \Big\} \tag{6.5.18}$$

where $R_j e^{i\phi_j} = \theta_j^{(1)} + i\theta_j^{(2)}$, $j = 1, 2, \dots, d$. We now apply Proposition 2.5.5 to conclude that the integral $\int_{S_2}$ depends only on the magnitude

$$\Big| \sum_{j=1}^{d} R_j e^{i(\phi_j - (t_j + u)x)} \Big| = \Big| e^{-iux} \sum_{j=1}^{d} R_j e^{-i(\phi_j - t_j x)} \Big| = \Big| \sum_{j=1}^{d} R_j e^{-i(\phi_j - t_j x)} \Big|,$$

i.e., the characteristic function in (6.5.18) does not depend on u. This proves the stationarity of the process W defined in (6.5.17), and thus the proof of the theorem is complete. ∎

Example 6.5.2 The complex-valued harmonizable process considered in Example 6.3.6, $\{H(t) = \int_{-\infty}^{\infty} e^{itx} M(dx),\ t \in \mathbb{R}\}$, is stationary because the complex random measure M in that example is isotropic.

Remark. In the proof of Theorem 6.5.1, we encountered a uniqueness problem. Setting $X_1(t) = \int_{-\infty}^{\infty} e^{itx} M(dx)$, $-\infty < t < \infty$, and $X_2(t) = \int_{-\infty}^{\infty} e^{itx} M_\tau(dx)$, $-\infty < t < \infty$, we inquired whether $X_1 \stackrel{d}{=} X_2$ implies that M and M_τ have the same circular control measure. To solve this problem, we used a technique analogous to a standard one in the theory of characteristic functions of probability laws, namely that two laws with identical characteristic functions must coincide. (In our context, the "law" is the random measure M.) It is possible, however, to use the analog of another standard technique for establishing the uniqueness of characteristic functions of probability laws, that of inverse Fourier transforms. We will see in Section 11.6 that there is an "inversion formula" for harmonizable $S\alpha S$ processes, but we cannot use it at this point because it involves integrating the sample paths of α-stable processes. This technique will be developed in Chapter 11.

Proposition 6.5.3 *A harmonizable $S\alpha S$ process is continuous in probability.*

PROOF: Since $\int_{-\infty}^{\infty} |e^{itx} - e^{isx}|^\alpha m(dx) \to 0$ as $s \to t$, we conclude from Proposition 6.2.3 that $\text{plim}_{s \to t} X(s) = X(t)$. ∎

6.6 Stationary real harmonizable processes

We consider in this section some consequences of Theorem 6.5.1 concerning stationary real harmonizable processes.

Let M be a complex random measure with circular control measure k. If $\{Z(t) = \int_{-\infty}^{\infty} e^{itx} M(dx),\ t \in \mathbb{R}\}$ is stationary, then its real part $X = \text{Re}\, Z$ is also stationary. The converse is not necessarily true. However, given a stationary real harmonizable process X, there always exists *some* stationary complex-valued harmonizable process $\widetilde{Z}$ such that $X \stackrel{d}{=} \text{Re}\, \widetilde{Z}$. The proof of Theorem 6.5.1 shows that this special $\widetilde{Z}$ can be chosen as $2^{-1/\alpha} \int_{-\infty}^{\infty} e^{itx} M_S(dx)$ where M_S is isotropic with circular control measure $k_S = k + \overline{k}$, where $\overline{k}$ is the conjugate measure defined in (6.5.2). Indeed, $X = \text{Re}\, Z$ stationary $\Rightarrow k_S = (m + \overline{m})\gamma \Rightarrow \widetilde{Z}$ stationary. By (6.5.6), $X \stackrel{d}{=} \text{Re}\, \widetilde{Z}$. Therefore,

Proposition 6.6.1 *A real harmonizable $S\alpha S$ process X is stationary if and only if there is some complex-valued harmonizable $S\alpha S$ process Z such that $X \stackrel{d}{=} \text{Re}\, Z$.*

Example 6.6.2 Fix $u_0 > 0$ and let $m = \delta_{\{-u_0\}} + \delta_{\{u_0\}}$ be the control measure of a complex-valued $S\alpha S$ random measure M. To define the circular control measure k of M, we must specify $k\{\{-u_0\} \times \cdot\}$ and $k\{\{u_0\} \times \cdot\}$. Set $k\{\{-u_0\} \times \cdot\} = k\{\{u_0\} \times \cdot\} = \gamma^{13}$, where γ^{13} is the uniform probability measure concentrated on the first and third quadrant of the circle S_2. Then (6.5.3) does not hold because γ^{13} is not the uniform probability measure on the whole of S_2. Therefore the process $\{\int_{-\infty}^{\infty} e^{itx} M(dx),\ -\infty < t < \infty\}$ is not stationary.

Now let γ^{24} be the uniform probability measure concentrated on the second and fourth quadrant of S_2. Since $\overline{k}\{\{-u_0\} \times \cdot\} = \overline{k}\{\{u_0\} \times \cdot\} = \gamma^{24}$, we obtain $k + \overline{k} = (m + \overline{m})(\gamma^{13} + \gamma^{24})/2$ where $\gamma = (\gamma^{13} + \gamma^{24})/2$ is the uniform probability measure on S_2. Since (6.5.4) is satisfied, $X(t) = \text{Re} \int_{-\infty}^{\infty} e^{itx} M(dx)$ is stationary with control measure $m + \overline{m} = 2m$. Moreover, $X(t)$ can be represented as $X(t) \stackrel{d}{=} \text{Re} \int_{-\infty}^{\infty} e^{itx} M_S(dx)$, where M_S has circular control measure $k_S = (m + \overline{m})\gamma$. The measure M_S is isotropic and the process $\{\int_{-\infty}^{\infty} e^{itx} M_S(dx),\ t \in \mathbb{R}\}$ is stationary.

Proposition 6.6.3 *A stationary real harmonizable process $\{X(t),\ -\infty < t < \infty\}$ has finite-dimensional characteristic function*

$$E \exp\{i(\theta_1 X(t_1) + \cdots + \theta_d X(t_d))\}$$

$$= \exp\Big\{-c_0 \int_{-\infty}^{\infty} \Big| \Big(\sum_{j=1}^{d} \theta_j \cos t_j x\Big)^2 + \Big(\sum_{j=1}^{d} \theta_j \sin t_j x\Big)^2 \Big|^{\alpha/2} m(dx)\Big\} \quad (6.6.1)$$

$$= \exp\Big\{-c_0 \int_{-\infty}^{\infty} \Big| \sum_{j=1}^{d} \sum_{k=1}^{d} \theta_j \theta_k \cos((t_j - t_k)x) \Big|^{\alpha/2} m(dx)\Big\}, \quad (6.6.2)$$

where

$$c_0 = \frac{1}{2\pi} \int_0^{2\pi} (\cos \phi)^\alpha d\phi \quad (6.6.3)$$

and m is a finite measure. The measure m is the control measure of the isotropic complex-valued measure M in the representation $X(t) = \text{Re} \int_{-\infty}^{\infty} e^{itx} M(dx)$.

PROOF: See Relation (6.3.9). ■

Let $\{X(t),\ -\infty < t < \infty\}$ be a stationary real $S\alpha S$ harmonizable process. Like any other $S\alpha S$ process which is continuous in probability, it is, conditionally, a centered *Gaussian* process (Proposition 3.11.3). It is natural to ask whether it is, conditionally, a *stationary* centered *Gaussian* process. This question is both important and non-trivial. It is important because the structure of stationary Gaussian processes has been thoroughly studied and if a stationary real $S\alpha S$

harmonizable process is conditionally stationary Gaussian, it can have a number of important properties. The question is nontrivial because not every stationary $S\alpha S$ process is conditionally stationary Gaussian. We know from Corollary 2.5.4 that every stationary sub-Gaussian $S\alpha S$ process is conditionally stationary Gaussian but we will see later (Example 10.3.4) that there are stationary $S\alpha S$ processes that are not, conditionally, stationary Gaussian.

Proposition 6.6.4 *Let $\{X(t),\ -\infty < t < \infty\}$ be a stationary real $S\alpha S$ harmonizable process. Then X is conditionally stationary centered Gaussian. In fact,*

$$\{X(t),\ -\infty < t < \infty\} \stackrel{d}{=} \{(C_\alpha b_\alpha^{-1} m(E))^{1/\alpha} G(t),\ -\infty < t < \infty\}$$

where

$$G(t) = \sum_{j=1}^{\infty} \Gamma_j^{-1/\alpha} (G_j^{(1)} \cos tZ_j + G_j^{(2)} \sin tZ_j),\ -\infty < t < \infty, \tag{6.6.4}$$

and b_α and $\{\Gamma_j, Z_j\}$ are defined in Proposition 6.4.4. The process G is, conditionally on the Γ_js and Z_js, a stationary centered Gaussian process with autocovariance function

$$EG(t)G(s) = \sum_{j=1}^{\infty} \Gamma_j^{-2/\alpha} \cos Z_j(t-s). \tag{6.6.5}$$

PROOF: The series representation follows from (6.4.8). Conditionally on the Γ_js and Z_js, G is a centered Gaussian process with autocovariance function

$$\begin{aligned} EG(t)G(s) &= \sum_{j=1}^{\infty} \Gamma_j^{-2/\alpha} (\cos tZ_j \cos sZ_j + \sin tZ_j \sin sZ_j) \\ &= \sum_{j=1}^{\infty} \Gamma_j^{-2/\alpha} \cos Z_j(t-s). \end{aligned}$$

Thus, G is stationary. ∎

We have discussed in some detail two classes of stationary $S\alpha S$ processes: the sub-Gaussian processes and the real harmonizable processes. These two classes have certain similarities. For example, they are both conditionally stationary Gaussian. Are these classes really different? If so, are they disjoint? It turns out that they are different and "almost" disjoint. More precisely,

Theorem 6.6.5 *The only stationary $S\alpha S$ process which is both sub-Gaussian and harmonizable is*

$$X(t) = cA^{1/2}(G_1 \cos ut + G_2 \sin ut), \ -\infty < t < \infty, \tag{6.6.6}$$

where $c \geq 0$ is a real constant, $A \sim S_{\alpha/2}((\cos \frac{\pi\alpha}{4})^{2/\alpha}, 1, 0)$, $u \geq 0$, and G_1 and G_2 are i.i.d. standard normal random variables independent of A.

PROOF: Note that the $S\alpha S$ process given in (6.6.6) is stationary sub-Gaussian because it is of the form

$$X(t) = A^{1/2}G(t), \ -\infty < t < \infty, \tag{6.6.7}$$

where $\{G(t) = c(G_1 \cos ut + G_2 \sin ut), \ -\infty < t < \infty\}$ is a stationary centered Gaussian process with autocovariance function $c^2 \cos ut$. This implies

$$E \exp\Big\{ i \sum_{j=1}^{d} \theta_j X(t_j) \Big\} = \exp\Big\{ -2^{-\alpha/2} |c|^\alpha \Big| \sum_{j=1}^{d} \sum_{k=1}^{d} \theta_j \theta_k \cos(t_j - t_k) u \Big|^{\alpha/2} \Big\} \tag{6.6.8}$$

(see Proposition 2.5.2.)

On the other hand, $\{X(t), \ -\infty < t < \infty\}$ is also stationary harmonizable because it can be represented as

$$X(t) \stackrel{d}{=} \text{Re} \int_{-\infty}^{\infty} e^{itx} M(dx), \ -\infty < t < \infty, \tag{6.6.9}$$

where M is an isotropic $S\alpha S$ random measure with control measure $m = 2^{-\alpha/2}|c|^\alpha c_0^{-1} \delta_u$, where c_0 is the constant in (6.6.3) and δ_u is the unit mass at u. (Compare (6.6.8) with (6.6.2).)

This shows that the $S\alpha S$ process given in (6.6.6) is both sub-Gaussian and harmonizable.

To prove the converse, suppose that a stationary $S\alpha S$ process X can be represented (in distribution) both by (6.6.7), with a stationary centered Gaussian process G and autocovariance function R, and by (6.6.9), with an isotropic $S\alpha S$ random measure M with control measure m. Choose any $t_1 < t_2$, and let $\Delta = t_2 - t_1$. Then

$$(A^{1/2}G(t_1), A^{1/2}G(t_2)) \stackrel{d}{=} \Big(\text{Re} \int_{-\infty}^{+\infty} e^{it_1x} M(dx), \ \text{Re} \int_{-\infty}^{\infty} e^{it_2x} M(dx) \Big). \tag{6.6.10}$$

Proposition 2.5.2 and Proposition 6.6.3 give the characteristic functions of the $S\alpha S$ random vectors in the left- and right-hand sides of (6.6.10). Equating these characteristic functions, we see that, for any real θ_1 and θ_2,

$$\left(\frac{1}{2}(R(0)\theta_1^2 + 2R(\Delta)\theta_1\theta_2 + R(0)\theta_2^2)\right)^{\alpha/2}$$
$$= c_0 \int_{-\infty}^{\infty} |\theta_1^2 + 2\theta_1\theta_2 \cos \Delta x + \theta_2^2|^{\alpha/2} m(dx).$$

Setting $m_1 = (2^{\alpha/2} R(0)^{-\alpha/2} c_0) m$, we obtain

$$(\theta_1^2 + 2\rho\theta_1\theta_2 + \theta_2^2)^{\alpha/2} = \int_{-\infty}^{+\infty} (\theta_1^2 + 2\theta_1\theta_2 \cos \Delta x + \theta_2^2)^{\alpha/2} m_1(dx), \quad (6.6.11)$$

where $\rho = R(\Delta)/R(0)$. Setting $\theta_1 = 1$ and $\theta_2 = 0$ shows that m_1 is a probability measure. Suppose θ_1 and θ_2 are not both zero and let $u = 2\theta_1\theta_2/(\theta_1^2 + \theta_2^2)$. Then u takes values in $[-1, 1]$, and thus, for every u in this interval,

$$(1 + u\rho)^{\alpha/2} = \int_{-\infty}^{\infty} (1 + u \cos \Delta x)^{\alpha/2} m_1(dx). \quad (6.6.12)$$

Note that (6.6.12) equates two functions of u defined on $[-1, 1]$. It is easy to check (see Exercise 6.14) that these functions are infinitely differentiable on $(-1, 1)$; after successive differentiation, we obtain, for $k = 0, 1, 2, \ldots$.

$$\rho^k (1 + u\rho)^{(\alpha/2)-k} = \int_{-\infty}^{\infty} (\cos \Delta x)^k (1 + u \cos \Delta x)^{(\alpha/2)-k} m_1(dx).$$

Substituting $u = 0$ yields

$$\rho^k = \int_{-\infty}^{\infty} (\cos \Delta x)^k m_1(dx), \; k = 0, 1, 2, \ldots. \quad (6.6.13)$$

Note that (6.6.13) represents equality of all moments of two random variables: one is the constant ρ, the other is $\cos \Delta X$ where X is a random variable with distribution m_1. It is well known that if two *bounded* random variables have identical moments, then they must have identical distributions (Exercise 6.15.) Therefore $\cos \Delta X = \rho$ with probability 1. This implies that the probability measure m_1 and, therefore, the control measure m are concentrated on the set

$$V_\Delta = \{x \in (-\infty, +\infty) : \; \cos \Delta x = \rho\}, \quad (6.6.14)$$

where $\rho = R(\Delta)/R(0)$. Recall that the points t_1 and t_2 were chosen arbitrarily, so the control measure m must be concentrated on the set V_Δ for any $\Delta > 0$. Choosing, for example, $\Delta = 1, \sqrt{2}$ and $\sqrt{3}$ shows that the measure m is concentrated on a set $\{-u, u\}$ with $u \geq 0$, i.e., $m = a\delta_u + b\delta_{-u}$ for some $a, b \geq 0$ (see Exercise 6.16). From (6.6.11), we conclude $\rho = \cos \Delta u$ for every Δ and hence $\frac{R(\Delta)}{R(0)} = \cos \Delta u$. This proves that the process X must be representable in the form (6.6.6). The proof of the theorem is complete. ■

6.7 The codifference for stationary real harmonizable processes

The codifference $\tau(t)$ was defined in Section 4.7 (see also Section 2.10). For a real $S\alpha S$ harmonizable process X, it equals

$$\begin{aligned}\tau(t) &= 2\|X(0)\|_\alpha^\alpha - \|X(t)-X(0)\|_\alpha^\alpha \\ &= c_0\Big\{2m(\mathbb{R}) - \int_{-\infty}^{\infty} |2-2\cos tx|^{\alpha/2} m(dx)\Big\} \\ &= c_0\Big\{2m(\mathbb{R}) - 2^\alpha \int_{-\infty}^{\infty} \Big|\sin\frac{tx}{2}\Big|^\alpha m(dx)\Big\}.\end{aligned}$$

Although it may not converge as $t\to\infty$, its Cesaro mean $\frac{1}{T}\int_0^T \tau(t)dt$ converges as $T\to\infty$.

Proposition 6.7.1 *For a real harmonizable $S\alpha S$ process*

$$\lim_{T\to\infty}\frac{1}{T}\int_0^T \tau(t)dt = c_0\left\{2m(\mathbb{R}) - \left(2^\alpha \frac{1}{\pi}\int_0^\pi (\sin u)^\alpha du\right) m(\mathbb{R}\backslash\{0\})\right\}. \tag{6.7.1}$$

This limit is positive for $0<\alpha<2$.

PROOF: Let $K(x)$ denote the integer part of $x/(2\pi)$ and observe that $\lim_{T\to\infty} K(T)/T = 1/2\pi$:

$$\frac{1}{T}\int_0^T \tau(t)dt = c_0\left\{2m(\mathbb{R}) - 2^\alpha \frac{1}{T}\int_0^T dt \int_{-\infty}^{\infty}\Big|\sin\frac{tx}{2}\Big|^\alpha m(dx)\right\}.$$

Now

$$\begin{aligned}\frac{1}{T}\int_0^T dt\int_{-\infty}^{\infty}\Big|\sin\frac{tx}{2}\Big|^\alpha m(dx) &= \int_{x\neq 0}\frac{1}{Tx}\sum_{k=1}^{K(Tx)}\int_{2\pi(k-1)}^{2\pi k}\Big|\sin\frac{s}{2}\Big|^\alpha ds + o(1) \\ &\to m(\mathbb{R}\backslash\{0\})\frac{1}{2\pi}\int_0^{2\pi}\Big(\sin\frac{s}{2}\Big)^\alpha ds \\ &= m(\mathbb{R}\backslash\{0\})\frac{1}{\pi}\int_0^\pi (\sin s)^\alpha ds\end{aligned}$$

as $T\to\infty$, proving (6.7.1).

If $0<\alpha<2$, by Jensen's inequality,

$$\frac{1}{\pi}\int_0^\pi (\sin s)^\alpha ds = \frac{1}{\pi}\int_0^\pi (\sin^2 s)^{\alpha/2} ds < \left(\frac{1}{\pi}\int_0^\pi \sin^2 s\, ds\right)^{\alpha/2} = \left(\frac{1}{2}\right)^{\alpha/2},$$

so that

$$\begin{aligned}\lim_{T\to\infty}\frac{1}{T}\int_0^T \tau(t)dt &> c_0\{2m(\mathbb{R}) - 2^{\alpha/2}m(\mathbb{R}\backslash\{0\})\}\\ &= c_0\{2m(\{0\}) + (2-2^{\alpha/2})m(\mathbb{R}\backslash\{0\})\} > 0 \quad \blacksquare\end{aligned}$$

Now, $\lim \frac{1}{T}\int_0^T \tau(t)dt > 0$ implies that $\tau(t)$ cannot converge to zero. Since $\tau(t)$ converges to zero for moving average processes (Theorem 4.7.3), we obtain

Theorem 6.7.2 *Stationary $S\alpha S$ moving average processes are different from real harmonizable $S\alpha S$ processes for any $0 < \alpha < 2$.*

This theorem should be contrasted with the Gaussian case $\alpha = 2$. In that case, a stationary process may have both a moving average and a real harmonizable representation.

Remark. When $\alpha = 2$, the right-hand side of (6.7.1) equals $2c_0 m(\{0\})$. If $\tau(t) = \text{Cov}\,(X(t), X(0)) \to 0$, then $m(\{0\}) = 0$.

6.8 Exercises

Exercise 6.1 Use Kolmogorov's existence theorem to show that there is a stochastic process $\{(M^{(1)}(A), M^{(2)}(A)),\ A \in \mathcal{E}_0\}$ such that for every $A_1, A_2, \ldots, A_d$ in $\mathcal{E}_0$ and pairs of real numbers $(\theta_1^{(1)}, \theta_1^{(2)}), (\theta_2^{(1)}, \theta_2^{(2)}), \ldots, (\theta_d^{(1)}, \theta_d^{(2)})$, one has

$$\sum_{j=1}^{d}(\theta_j^{(1)}M^{(1)}(A_j) + \theta_j^{(2)}M^{(2)}(A_j)) \sim S_\alpha(\sigma, 0, 0), \tag{6.8.1}$$

where

$$\sigma^\alpha = \int_E\int_{S_2}|\sum_{j=1}^{d}(s_1\theta_j^{(1)} + s_2\theta_j^{(2)})1_{A_j}(x)|^\alpha k(dx, ds).$$

Show that this process is independently scattered and σ-additive, and that for any $A \in \mathcal{E}_0$, $M^{(1)}(A)$ and $M^{(2)}(A)$ are jointly $S\alpha S$ with spectral measure $k(A\times\cdot)$. *Hint:* To prove (6.8.1), express $A_1 \cup \cdots \cup A_d$ (and, hence, each A_j) as a union of disjoint sets.

Exercise 6.2 Prove that a complex-valued random measure M satisfies (6.1.4) if and only if its circular control measure has the form $k = m\gamma$, where m is the control measure of M, and γ is the uniform probability measure on S_2.

Exercise 6.3 Show that $M^{(1)}$ and $M^{(2)}$ are real-valued $S\alpha S$ random measures with control measures given by (6.1.5).

Exercise 6.4 Show that a complex-valued $S\alpha S$ random measure M has independent real and imaginary parts if and only its circular control measure k is concentrated on the four "lines" $E \times \{0,1\}$, $E \times \{0,-1\}$, $E \times \{1,0\}$ and $E \times \{-1,0\}$.

Exercise 6.5 Prove that the only isotropic complex-valued $S\alpha S$, $0 < \alpha < 2$, random measure M which has independent real and imaginary parts is the trivial measure $M \equiv 0$.

Exercise 6.6 Let M be a complex-valued $S\alpha S$ random measure on $(E, \mathcal{E})$ with control measure m, and let $f,\ g \in L^\alpha(E, \mathcal{E}, m)$. Show that $\int_E f(x)M(dx)$ and $\int_E g(x)M(dx)$ are independent iff $fg \equiv 0$, m-a.e.

Exercise 6.7 Let M be a complex-valued $S\alpha S$ random measure on $(E, \mathcal{E})$ with a control measure m and let $\phi:\ E \to \mathbb{C}$ be a measurable function. Let $\mathcal{E}_{0,\psi} = \{A \in \mathcal{E}: \int_A |\psi(x)|^\alpha m(dx) < \infty\}$. Show that the stochastic process $M_\psi(A) = \int_A \psi(x)M(dx)$, $A \in \mathcal{E}_{0,\psi}$, is a complex-valued $S\alpha S$ random measure on $(E, \mathcal{E})$. What is the control measure of M_ψ? Prove that for any measurable function $f:\ E \to \mathbb{C}$ such that $f \circ \psi \in L^\alpha(E, \mathcal{E}, m)$ the integral $\int_E f(x)M_\psi(dx)$ is well defined and equal in distribution to $\int_E f(x)\psi(x)M(dx)$.

Exercise 6.8 Under the assumptions of Exercise 6.7, show that the circular control measure k_ψ of the $S\alpha S$ random measure M_ψ, where $\psi = \psi^{(1)} + i\psi^{(2)}$ and $M_\psi(A) = \int_A \psi(x)M(dx)$, is

$$\int_{S_2} |\theta_1 s_1 + \theta_2 s_2|^\alpha k_\psi(A \times ds)$$

$$= \int_A \int_{S_2} |\theta_1(s_1\psi^{(1)}(x) - s_2\psi^{(2)}(x)) + \theta_2(s_1\psi^{(1)}(x) + s_2\psi^{(2)}(x))|^\alpha k(dx, ds)$$

for any $A \in \mathcal{E}_{0,\psi}$ and real θ_1, θ_2.

Hint: Use (6.1.2) and (6.2.4). The intermediate equality is then $-\ln E\exp\{i(\theta_1 M_\psi^{(1)}(A) + \theta_2 M_\psi^{(2)}(A))\}$.

Exercise 6.9 Let M be a (complex-valued) $S\alpha S$ random measure on $\mathbb{R}$ whose control measure m is concentrated at the origin. Let $f:\ \mathbb{R} \to \mathbb{C}$. Show that the process

$$X(t) = \int_{-\infty}^{\infty} f(xt)M(dx), \ -\infty < t < \infty,$$

is a constant $S\alpha S$ process, i.e., $P(X(t_1) = X(t_2)) = 1$ for every $t_1, t_2 \in \mathbb{R}$.

Exercise 6.10 Suppose that f and M in Theorem 6.4.2 are real-valued. What are the choices for $(S_j^{(1)}, S_j^{(2)})$ that reduce (6.4.2) to the real-valued series representation given in Corollary 3.10.4.

Exercise 6.11 Let G_1, G_2 be independent standard normal random variables. Then $\left(\frac{G_1}{(G_1^2+G_2^2)^{1/2}}, \frac{G_2}{(G_1^2+G_2^2)^{1/2}}\right)$ has the vector uniform distribution over the unit circle S_2 and is independent of $G_1^2 + G_2^2$.

Exercise 6.12 Prove (6.5.6).

Exercise 6.13 Give a proof of step 1 of the necessity part of Theorem 6.5.1 for the case when the complex-valued harmonizable process Z defined by (6.5.1) is stationary.

Hint: Apply the approximating procedure used in the text to the process Z instead of the process Y. Conclude that the following counterpart of (6.5.9) holds:

$$\int_{-\infty}^{\infty} g(x)M(dx) \stackrel{d}{=} \int_{-\infty}^{\infty} g(x)M_\tau(dx)$$

for every continuous periodic function $g : \mathbb{R} \to \mathbb{C}$. Deduce that $k(A \times ds) = k_\tau(A \times ds)$ for any Borel subset A of the real line.

Exercise 6.14 Show that the functions in the left- and right-hand sides of (6.6.12) have infinitely many derivatives at every point of $(-1, 1)$. Show that the functions in the right-hand side may be repeatedly differentiated inside the integral.

Exercise 6.15 Suppose that a sequence of real numbers $m_k,\ k = 1, 2, \ldots$, are moments of a bounded random variable with distribution F. Then there is no other distribution that has $m_k,\ k = 1, 2, \ldots$, as moments.

Hint: If $|X| \le C$, then $|m_k| \le C^k$. Use Carleman's condition (Feller (1971), VII (3.14)).

Exercise 6.16 Show that the intersection of the sets $V_1, V_{\sqrt{2}}$ and $V_{\sqrt{3}}$, defined in (6.6.14), is either empty, or of the type $\{-x, x\}, x \ge 0$.

Chapter 7

Self-similar processes

Self-similar processes are invariant in distribution under judicious scaling of time and space. They are important in probability because of their connection to limit theorems (e.g., Lamperti (1962)) and they are of great interest in modeling. Some aspects of self-similarity appear in geophysics, hydrology, turbulence, economics, communications and in relation to "1/*f* noises." Self-similar processes are also considered in physics, particularly in connection to the so-called "renormalization group theory" and "critical phenomena." Taqqu (1986) provides a bibliographical guide to many applications.

The scaling coefficient or index of self-similarity is a non-negative number denoted H. A process $X = \{X(t),\ t \in \mathbb{R}\}$ is *self-similar with index* H if, for any $a > 0$, the finite-dimensional distributions of $\{X(at),\ t \in \mathbb{R}\}$ are the same as those of $\{a^H X(t),\ t \in \mathbb{R}\}$. The process is H-*sssi* if it is self-similar with index H and has stationary increments.

Brownian motion, for example, is 1/2-sssi. Although the increments of Brownian motion are independent, those of other H-sssi processes can display *long-range dependence* or *long memory*. If the process is Gaussian, long-range dependence manifests itself by the presence of cycles of any order and, ultimately, by a spectral density that diverges at the origin like a power function. Equivalently, the covariances decay very slowly to zero, also like a power function. Mandelbrot and Wallis (1968) refer to this phenomenon as the *Joseph effect*, an allusion to the biblical figure Joseph who was faced with long periods of plenty followed by long periods of famine.

Although Gaussian processes can display long-range dependence, their marginal distributions, being normal, concentrate their mass around the mean. α-stable distributions with $0 < \alpha < 2$, on the other hand, allow for much greater variability. Mandelbrot refers to the variability of stable random variables as the

Noah effect, an allusion, we may recall, to the biblical figure Noah who lived through an exceptionally heavy flood. By considering in this chapter α-stable self-similar processes with $0 < \alpha < 2$, we are, in fact, dealing with models that can exhibit both the Joseph and the Noah effects.

In Section 7.1, we introduce general H-sssi processes (not necessarily Gaussian or α-stable) and describe some of their basic properties. We show that the existence of moments limits the possible values of H and, consequently, if an H-sssi process is α-stable, $0 < \alpha \leq 2$, then its self-similarity index H is restricted to the interval $(0, 1/\alpha)$ if $\alpha < 1$ and to the interval $(0, 1]$ if $\alpha \geq 1$.

The Gaussian case, because it is a paradigm, merits a separate section. For a given $H \in (0, 1)$, there is basically a single Gaussian H-sssi process, namely fractional Brownian motion. We provide two equivalent representations for that process, a "moving average" representation and a "harmonizable" one. The increments of fractional Brownian motion are called fractional Gaussian noise. They are stationary and if $1/2 < H < 1$, they exhibit long-range dependence; that is, their autocovariance function decreases slowly at large lags, like a power function, and their spectral density increases like a power function at low frequencies and explodes at the origin.

In Section 7.3, we describe the basic properties of α-stable H-sssi processes with $0 < \alpha < 2$. Whereas fractional Brownian motion is the only H-sssi Gaussian process with $0 < H < 1$, there are many different H-sssi processes when $0 < \alpha < 2$. The most common one is the linear fractional stable motion considered in Section 7.4. Its self-similarity index H belongs to the interval $(0, 1)$ and excludes the value $1/\alpha$. The α-stable Lévy motion introduced in Section 7.5, on the other hand, has self-similarity index $H = 1/\alpha$. It has independent increments and is the stable counterpart to Brownian motion. It is the unique $1/\alpha$-sssi process, if $0 < \alpha < 1$, but there exist others if $1 \leq \alpha < 2$. For example, if X is a 1-stable random variable, then $X(t) = tX$ is a 1-sssi process. If $1 < \alpha < 2$, then the log-fractional stable motion investigated in Section 7.6 is also $1/\alpha$-sssi.

The real harmonizable fractional stable motion considered in Section 7.7 is yet another H-sssi process. The kernel in its integral representation has the same form as the kernel of the harmonizable representation of fractional Brownian motion. Its properties, however, are quite different from those of the linear fractional stable motion. The complex harmonizable fractional stable motion defined in Section 7.8 is a complex-valued α-stable, H-sssi process. Finally in Section 7.9 we describe a large class of H-sssi processes that are subordinated to other processes, for example, sub-Gaussian or sub-stable processes.

In modeling, it is the increments of H-sssi processes that are of interest. Any H-sssi process $\{X(t),\ t \in \mathbb{R}\}$ induces a stationary sequence $\{Y_j = X(j+1) - X(j),\ j \in \mathbb{Z}\}$. The sequence $\{Y_j\}$ corresponding to a linear fractional stable motion is of particular interest because it displays both the "Joseph" and "Noah" effects of Mandelbrot. It is called linear fractional stable noise and is studied in Section 7.10. We analyze its asymptotic dependence structure by using the codifference $\tau(j)$. For most, but not all, values of α and H, $\tau(j)$ decreases as $j^{\alpha H-\alpha}$ for large j. This is analogous to the behavior of the autocovariance function in the Gaussian case $\alpha = 2$. Simulations of Gaussian and stable fractional noises and motions are presented in Section 7.11.

In Section 7.12, we discuss autoregressive-moving average (ARMA) sequences with stable innovations, and in Section 7.13, we extend to the stable case the family of models known as fractional ARIMA. A fractional ARIMA can also display long-range dependence, and because its definition involves more parameters than the linear fractional stable noise, it allows for greater modeling flexibility.

7.1 Self-similarity

Let T be either $\mathbb{R}$, $\mathbb{R}_+ = \{t\colon t \geq 0\}$ or $\{t\colon t > 0\}$.

Definition 7.1.1 The *real-valued* process $\{X(t),\ t \in T\}$ is *self-similar with index* $H > 0$ (H-ss) if for all $a > 0$, the finite-dimensional distributions of $\{X(at), t \in T\}$ are identical to the finite-dimensional distributions of $\{a^H X(t),\ t \in T\}$; i.e., if for any $d \geq 1, t_1, t_2, \ldots, t_d \in T$ and any $a > 0$,

$$(X(at_1),\ X(at_2), \ldots, X(at_d)) \stackrel{d}{=} (a^H X(t_1),\ a^H X(t_2), \ldots, a^H X(t_d)). \tag{7.1.1}$$

The *complex-valued* process $\{X(t),\ t \in T\}$ is *self-similar with index* H if for all $a > 0$, the finite-dimensional distributions of $\{\text{Re } X(at),\ \text{Im } X(at),\ t \in T\}$ are identical to those of $\{a^H \text{ Re } X(t),\ a^H \text{ Im } X(t),\ t \in T\}$.

We shall always suppose $H > 0$.

Notation. Relation (7.1.1) will be expressed succinctly as follows:

$$\{X(at),\ t \in T\} \stackrel{d}{=} \{a^H X(t),\ t \in T\}, \tag{7.1.2}$$

Sometimes, however, when we consider a *fixed* $t \in T$, we write $X(at) \stackrel{d}{=} a^H X(t)$ to mean that the *random variables* $X(at)$ and $a^H X(t)$ have identical (one-dimensional) distributions.

Example 7.1.2 *Brownian motion* $\{X(t),\ t \in \mathbb{R}\}$ is a Gaussian process with mean 0 and autocovariance function $EX(t_1)X(t_2) = \min(t_1, t_2)$. It is H-ss with $H = 1/2$ because for all $a > 0,\ -\infty < t_1, t_2 < \infty$,

$$EX(at_1)X(at_2) = \min(at_1, at_2) = a \min(t_1, t_2) = E(a^{1/2}X(t_1))(a^{1/2}X(t_2)).$$

Example 7.1.3 The *SαS Lévy motion*, $0 < \alpha < 2$ is H-ss with $H = 1/\alpha \in (1/2, \infty)$. This process was defined in Example 3.6.1 for $T = [0, \infty)$. (To define it, for $T = \mathbb{R}$, set $X(t) = X_1(t)$ for $t \geq 0$ and $X(t) = X_2(-t)$ for $t \leq 0$, where $\{X_1(t), t \geq 0\}$ and $\{X_2(t), t \geq 0\}$ are independent copies of a $S\alpha S$ Lévy motion). It is easy to verify that the $S\alpha S$ Lévy motion $X = \{X(t), t \in T\}$ is self-similar with $H = 1/\alpha$. Indeed, since X has independent increments, it is enough to show that for all $a > 0$ and $t, s \in T$, $X(at) - X(as) \stackrel{d}{=} a^{1/\alpha}(X(t) - X(s))$. But this follows from $X(t) - X(s) \sim S_\alpha(|t-s|^{1/\alpha}, 0, 0)$.

Remarks

1. Relation (7.1.1) states that a change of the time scale is equivalent to a change in the state space scale. Note, however, that it is the finite-dimensional distributions (the "odds") that are invariant under the transformation, not necessarily the sample paths. Lamperti (1962) uses the term *semi-stability* to describe the transformation in (7.1.1). The term *self-similarity*, coined by Mandelbrot, is now standard. That term is also used in the context of the scaling of non-random objects, such as fractals (Mandelbrot 1982).

2. If $T = \mathbb{R}$, then for each *fixed* $t \neq 0$,

$$X(t) \stackrel{d}{=} |t|^H X(\text{sign } t) = \begin{cases} t^H X(1) & \text{if } t > 0, \\ |t|^H X(-1) & \text{if } t < 0. \end{cases}$$

   This useful relation does not typically hold in the sense of the finite-dimensional distributions.

3. If $\{X(t), t \in T\}$ is H-ss, and $0 \in T$, then for each $a > 0$, $X(0) = X(a0) \stackrel{d}{=} a^H X(0)$ and hence

$$X(0) = 0 \text{ a.s.} \tag{7.1.3}$$

A non-degenerate H-ss process cannot be stationary because if it were, we would have for any $a > 0$ and $t > 0$, $X(t) \stackrel{d}{=} X(at) \stackrel{d}{=} a^H X(t)$ and we would obtain a contradiction because $a^H X(t) \to \infty$ as $a \to \infty$. There is, nevertheless, an important correspondence between self-similar and stationary processes.

Proposition 7.1.4 *If $\{X(t), 0 < t < \infty\}$ is H-ss, then*

$$Y(t) = e^{-tH} X(e^t), \quad -\infty < t < \infty,$$

is stationary. Conversely, if $\{Y(t), -\infty < t < \infty\}$ is stationary, then

$$X(t) = t^H Y(\ln t), \quad 0 < t < \infty,$$

is H-ss.

PROOF: Let $\theta_1, \ldots, \theta_d$ be real numbers. If $\{X(t),\ 0 < t < \infty\}$ is H-ss, then for any $t_1, \ldots, t_d \in \mathbb{R}^1$ and $h > 0$,

$$\begin{aligned}\sum_{j=1}^{d} \theta_j Y(t_j + h) &= \sum_{j=1}^{d} \theta_j e^{-t_j H} e^{-hH} X(e^h e^{t_j}) \\ &\overset{d}{=} \sum_{j=1}^{d} \theta_j e^{-t_j H} X(e^{t_j}) \\ &= \sum_{j=1}^{d} \theta_j Y(t_j),\end{aligned}$$

proving that $\{Y(t),\ -\infty < t < \infty\}$ is stationary.

Conversely, if $\{Y(t),\ -\infty < t < \infty\}$ is stationary, then for any $t_1, \ldots, t_d > 0$ and $a > 0$,

$$\begin{aligned}\sum_{j=1}^{d} \theta_j X(at_j) &= \sum_{j=1}^{d} \theta_j a^H t_j^H Y(\ln a + \ln t_j) \\ &\overset{d}{=} \sum_{j=1}^{d} \theta_j a^H t_j^H Y(\ln t_j) \\ &= \sum_{j=1}^{d} \theta_j a^H X(t_j),\end{aligned}$$

proving that $\{X(t),\ 0 < t < \infty\}$ is H-ss. ∎

Example 7.1.5 Let $\{X(t),\ t > 0\}$ be *Brownian motion.* It is thus Gaussian, has zero mean and is H-ss with $H = 1/2$. Therefore, $\{Y(t) = e^{-t/2} X(e^t),\ -\infty < t < \infty\}$ is a stationary Gaussian process with mean zero and autocovariance function

$$\begin{aligned}EY(t_1)Y(t_2) &= e^{-\frac{1}{2}(t_1+t_2)} EX(e^{t_1})X(e^{t_2}) \\ &= e^{-\frac{1}{2}(t_1+t_2)} \min(e^{t_1}, e^{t_2}) \\ &= e^{-\frac{1}{2}|t_2 - t_1|},\end{aligned}$$

i.e., $\{Y(t),\ -\infty < t < \infty\}$ is an Ornstein-Uhlenbeck process.

Proposition 7.1.4 shows that there are many different self-similar processes. We shall consider here those that have stationary increments. They are of great interest in applications because they give rise to stationary sequences with remarkable features.

Definition 7.1.6 A real-valued process $\{X(t),\ t \in T\}$ has *stationary increments* if

$$\{X(t+h) - X(h),\ t \in T\} \stackrel{d}{=} \{X(t) - X(0),\ t \in T\},\ \text{for all } h \in T.$$

A complex-valued process $\{X(t),\ t \in T\}$ has *stationary increments* if, for all $h \in T$,

$$\begin{aligned}&\{\mathrm{Re}\,(X(t+h) - X(h)),\ \mathrm{Im}\,(X(t+h) - X(h)),\ \ t \in T\}\\ &\quad \stackrel{d}{=} \{\mathrm{Re}\,(X(t) - X(0)),\ \mathrm{Im}\,(X(t) - X(0)),\ \ t \in T\}.\end{aligned}$$

Definition 7.1.7 The process $\{X(t),\ t \in T\}$ is H*-sssi* if it is *self-similar* with index H and has *stationary increments.*

Example 7.1.8 *Brownian motion* is H-sssi with $H = 1/2$. From the discussion in Example 7.1.3, we conclude that a *$S\alpha S$ Lévy motion* is H-sssi with $H = 1/\alpha$.

Remark. The following is a simple but useful fact: if $\{X(t),\ t \in \mathbb{R}\}$ is H-sssi, then for *fixed* t,

$$X(-t) \stackrel{d}{=} -X(t) \tag{7.1.4}$$

since $X(0) = 0$ and $X(-t) = X(-t) - X(0) \stackrel{d}{=} X(0) - X(t) = -X(t)$.

The next proposition shows that the existence of moments limits the possible values of H. It uses the following technical lemma:

Lemma 7.1.9 *Let $\{X(t),\ t \in \mathbb{R}\}$ be H-sssi. Then for any $s \neq 0$ and $t \neq 0$,*

$$P(X(s) = 0,\ X(t) = 0) = P(X(1) = 0)$$

and

$$P(X(s) \neq 0,\ X(t) \neq 0) = P(X(1) \neq 0).$$

PROOF: Let $p = P(X(1) = 0)$. By self-similarity, for any $s \neq 0$, we have

$$P(X(s) = 0) = P(X(\text{sign } s) = 0) = p \tag{7.1.5}$$

and, by the stationarity of the increments, for any $s \neq t$,

$$P(X(s) = X(t)) = P(X(s-t) = 0) = p. \tag{7.1.6}$$

Moreover, for any t and u and any $M > 0$,

$$\begin{aligned} &P(X(t) = X(u) \neq 0) \\ &\quad \leq P(|X(t)| \geq M,\ |X(u)| \geq M) + P(0 < |X(t)| < M,\ 0 < |X(u)| < M) \\ &\quad \leq P(|X(t)| \geq M) + P(0 < |X(u)| < M) \\ &\quad = P(|X(t)| \geq M) + P(0 < |X(1)| < Mu^{-H}), \end{aligned}$$

which is arbitrarily small for M and u sufficiently large. Thus,

$$\lim_{u \to \infty} P(X(t) = X(u) \neq 0) = 0. \tag{7.1.7}$$

Since, moreover,

$$\begin{aligned} &P(X(s) = X(u)) \\ &\quad = P(X(s) = X(u) \neq 0) + P(X(s) = 0,\ X(u) = 0) \\ &\quad = P(X(s) = X(u) \neq 0) + P(X(s) = 0) - P(X(s) = 0,\ X(u) \neq 0), \end{aligned}$$

we have by (7.1.5) and (7.1.6),

$$P(X(s) = 0,\ X(u) \neq 0) = P(X(s) = X(u) \neq 0). \tag{7.1.8}$$

Considering now s, t and u, we have

$$\begin{aligned} &P(X(s) = 0,\ X(t) \neq 0) \\ &\quad \leq P(X(s) = 0,\ X(u) \neq 0) + P(X(u) = 0,\ X(t) \neq 0) \\ &\quad = P(X(s) = X(u) \neq 0) + P(X(u) = X(t) \neq 0) \end{aligned}$$

by (7.1.8). We now use (7.1.7) to conclude that, for any s and t,

$$P(X(s) = 0,\ X(t) \neq 0) = 0. \tag{7.1.9}$$

Therefore, for any non-zero s and any t,

$$\begin{aligned} P(X(s) = X(t) = 0) &= P(X(s) = 0) - P(X(s) = 0,\ X(t) \neq 0) \\ &= p \\ &= P(X(1) = 0), \end{aligned}$$

by (7.1.5) and (7.1.9). This proves the first part of the lemma. From (7.1.5) and (7.1.9), we see for any non-zero s and t,

$$\begin{aligned} P(X(s) \neq 0,\ X(t) \neq 0) &= P(X(s) \neq 0) - P(X(s) \neq 0,\ X(t) = 0) \\ &= 1 - p \\ &= P(X(1) \neq 0). \end{aligned}$$

This proves the second part of the lemma. ■

Proposition 7.1.10 *If* $\{X(t),\ t \in \mathbb{R}\}$ *is* H*-sssi and*

$$P(X(1) \neq 0) > 0, \tag{7.1.10}$$

then the relation

$$E|X(1)|^{\gamma} < \infty$$

implies

$$0 < H < 1/\gamma \quad \textit{if} \quad 0 < \gamma < 1,$$

and it implies

$$0 < H \leq 1 \quad \textit{if} \quad \gamma \geq 1.$$

PROOF: Suppose first $0 < \gamma < 1$. Then $(a+b)^{\gamma} \leq a^{\gamma} + b^{\gamma}$ for any $a \geq 0,\ b \geq 0$ and $(a+b)^{\gamma} < a^{\gamma} + b^{\gamma}$ if $a > 0$, $b > 0$. Thus, $|X(2)|^{\gamma} \leq |X(2) - X(1)|^{\gamma} + |X(1)|^{\gamma}$, and this inequality is strict on the set $A = \{X(1) \neq 0,\ X(2) - X(1) \neq 0\}$. Since Lemma 7.1.9 and Relation (7.1.10) imply $P(A) > 0$, we have

$$E|X(2)|^{\gamma} < E|X(2) - X(1)|^{\gamma} + E|X(1)|^{\gamma}. \tag{7.1.11}$$

But $X(2) - X(1) \stackrel{d}{=} X(1) - X(0) \stackrel{d}{=} X(1)$ by the stationarity of the increments and self-similarity. Hence (7.1.11) reduces to $E|X(2)|^{\gamma} < 2E|X(1)|^{\gamma}$. On the other hand, by self-similarity, $E|X(2)|^{\gamma} = 2^{H\gamma} E|X(1)|^{\gamma}$. Therefore, $2^{H\gamma} < 2$, i.e., $H < 1/\gamma$.

If $\gamma \geq 1$, then $E|X(1)|^{p} < \infty$ for any $p < 1$ and hence $0 < H < 1/p$ for any $p < 1$, implying $0 < H \leq 1$. ■

Remark. Condition (7.1.10) is very weak. It is clearly satisfied if $\{X(t),\ t \in \mathbb{R}\}$ is an α-stable process $0 < \alpha \leq 2$ (not identically equal to zero).

Figure 7.1 displays the region of permissable values for (α, H). Since α-stable processes, $\alpha < 2$, have finite moments of order lower than α and Gaussian processes have finite moments of any order, we have

Corollary 7.1.11 *Let* $\{X(t),\ t \in \mathbb{R}\}$ *be a non-degenerate* α*-stable,* $0 < \alpha \leq 2$, H*-sssi process. Then*

$$\alpha < 1 \Rightarrow 0 < H \leq 1/\alpha,$$

$$\alpha \geq 1 \Rightarrow 0 < H \leq 1.$$

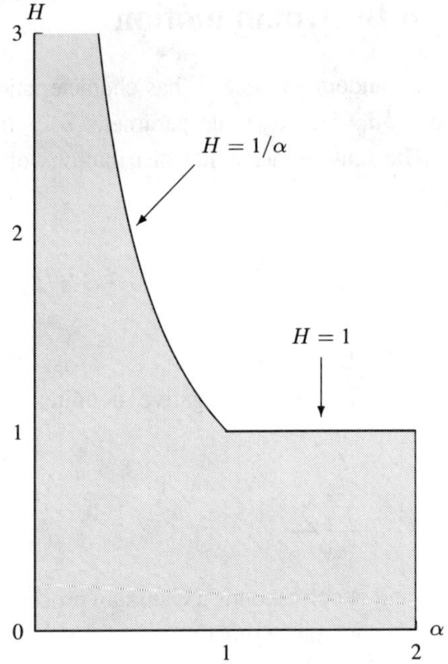

Figure 7.1: If the process is non-degenerate, α-stable and H-sssi, then α and H must lie in the shaded region.

The corollary is consistent with Example 7.1.3 where we noted that the $S\alpha S$ Lévy motion is H-ss with $H = 1/\alpha$. We will see in Section 12.4 that there is an α-stable H-sssi process for each permissible value of the pair (H, α).

Are there "different" α-stable H-sssi processes with a given α and H? In order to eliminate trivial answers, we adopt the following convention:

Convention. Stochastic processes that differ only by a multiplicative constant are identified. Hence, we say that $X_1 = \{X_1(t),\ t \in T\}$ and $X_2 = \{X_2(t),\ t \in T\}$ are *different* if there is no constant C such that X_1 and CX_2 have identical finite-dimensional distributions.

We will see later that for fixed $\alpha \in (0, 2)$ and H, there are often many *different* α-stable H-sssi processes. In the Gaussian case $\alpha = 2$, however, there is a unique H-sssi process, called *fractional Brownian motion.* Fractional Brownian motion is important not only because it is the unique H-sssi Gaussian process or because the various α-stable H-sssi processes reduce to it when setting $\alpha = 2$ but also because it has been widely applied, particularly in the context of *long-range dependence.* We shall now define it and study its properties.

7.2 Fractional Brownian motion

A mean zero Gaussian random variable X has characteristic function $Ee^{i\theta X} = \exp\{-\sigma^2\theta^2\} = \exp\{-\frac{1}{2}\sigma_0^2\theta^2\}$, i.e., scale parameter $\sigma > 0$ and standard deviation $\sigma_0 = \sqrt{2}\sigma$.[1] The finite-dimensional distributions of a Gaussian process $\{X(t),\ t \in \mathbb{R}\}$ satisfy

$$Ee^{i\sum_{j=1}^m \theta_j X(t_j)} = \exp\Big\{-\frac{1}{2}\sum_{j=1}^m \sum_{k=1}^m A(t_j,t_k)\theta_j\theta_j + \sum_{j=1}^m \mu(t_j)\theta_j\Big\},$$

$\theta_1,\ldots,\theta_m \in \mathbb{R},\ m \geq 1$, where $\mu(t),\ t \in \mathbb{R}$, is a real-valued function and $\{A(t_1,t_2),\ t_1,t_2 \in \mathbb{R}\}$ is non-negative definite; i.e., for any integer $m,\ t_1,\ldots,t_m \in \mathbb{R}, u_1,\ldots,u_m \in \mathbb{R}$,

$$\sum_{i=1}^m \sum_{j=1}^m A(t_i,t_j)u_i u_j \geq 0.$$

Conversely, to each μ and A corresponds a Gaussian process: μ is its mean value function and A is its autocovariance function.

Fix now $0 < H \leq 1$. Since the function $\{|t_1|^{2H} + |t_2|^{2H} - |t_1 - t_2|^{2H},\ t_1,t_2 \in \mathbb{R}\}$ is non-negative definite (see Lemma 2.10.8), there exists a Gaussian process $\{X(t),\ t \in \mathbb{R}\}$ with mean zero and autocovariance function

$$\begin{aligned} R_H(t_1,t_2) &:= \operatorname{Cov}(X(t_1),X(t_2)) \\ &= \frac{1}{2}\Big\{|t_1|^{2H} + |t_2|^{2H} - |t_1-t_2|^{2H}\Big\} \operatorname{Var} X(1). \quad (7.2.1) \end{aligned}$$

It is easy to check that this process is H-sssi. It is called fractional Brownian motion.[2]

Are there other Gaussian H-sssi processes? The following lemma implies that the answer is negative when $H < 1$.

Lemma 7.2.1 *Suppose that $\{X(t),\ t \in \mathbb{R}\}$ is a (non-degenerate) H-sssi finite variance process. Then*

$$0 < H \leq 1,$$

$$X(0) = 0 \ \ a.s.,$$

[1] There is an inconsistency in the literature between the definitions of the scale parameter for the Gaussian distribution ($\alpha = 2$) and for the stable non-Gaussian ones, $0 < \alpha < 2$. If $X \sim S\alpha S$ with $0 < \alpha < 2$, then the scale parameter of X is defined as the coefficient σ in $E\exp\{i\theta X\} = \exp\{-\sigma^\alpha|\theta|^\alpha\}$. If X is Gaussian, however, then its scale parameter is usually chosen to be the standard deviation σ_0. In order to be consistent, we define the scale parameter as the factor σ in $E\exp\{i\theta X\} = \exp\{-\sigma^\alpha|\theta|^\alpha\}$ in all cases $0 < \alpha \leq 2$ and hence distinguish between σ and the standard deviation $\sigma_0 = \sqrt{2}\sigma$ in the case $\alpha = 2$.

[2] The formal definition of fractional Brownian motion is given below.

and

$$\text{Cov}\,(X(t_1), X(t_2)) = \frac{1}{2}\left\{|t_1|^{2H} + |t_2|^{2H} - |t_1 - t_2|^{2H}\right\} \text{Var}\, X(1).$$

Moreover, for all $t \in \mathbb{R}$,

(i) in the case $0 < H < 1$*:*

$$EX(t) = 0;$$

(ii) in the case $H = 1$*:*

$$X(t) = tX(1) \quad a.s.$$

PROOF: By the stationarity of the increments and self-similarity,

$$\begin{aligned} EX(t_1)X(t_2) &= \frac{1}{2}\left\{EX^2(t_1) + EX^2(t_2) - E(X(t_1) - X(t_2))^2\right\} \\ &= \frac{1}{2}\left\{EX^2(t_1) + EX^2(t_2) - E(X(t_1 - t_2) - X(0))^2\right\} \\ &= \frac{1}{2}\left\{|t_1|^{2H} + |t_2|^{2H} - |t_1 - t_2|^{2H}\right\} EX^2(1) \qquad (7.2.2) \end{aligned}$$

since $X(0) = 0$ by (7.1.3), $X(1) \stackrel{d}{=} -X(-1)$ by (7.1.4) and $X(t) \stackrel{d}{=} |t|^H X(\text{sign}\, t)$ for fixed t. Corollary 7.1.11 and Lemma 2.10.8 yield $0 < H \leq 1$.

Suppose, now, $0 < H < 1$. Since $EX(1) = E(X(2) - X(1)) = 2^H EX(1) - EX(1) = (2^H - 1)EX(t)$, we obtain $EX(1) = 0$ and hence $EX(t) = 0$, because $EX(-1) = -EX(1)$ and $EX(t) = |t|^H EX(\text{sign}\, t)$. Form (7.2.1) of the autocovariance function follows from (7.2.2).

Suppose now $H = 1$. Then (7.2.2) yields $EX(t_1)X(t_2) = t_1 t_2 EX^2(1)$ and thus

$$\begin{aligned} E(X(t) - tX(1))^2 &= EX^2(t) - 2tEX(t)X(1) + t^2 EX^2(1) \\ &= (t^2 - 2t \cdot t + t^2)EX^2(1) \\ &= 0, \end{aligned}$$

that is, $X(t) = tX(1)$ a.s. for all t. Form (7.2.1) of the autocovariance function now follows. ■

Lemma 7.2.1 implies that all H-sssi Gaussian processes have the autocovariance function (7.2.1). For a given H, these processes only differ by a multiplicative constant (and by their mean, if $H = 1$). It is convenient, therefore, to define fractional Brownian motion as follows:

Definition 7.2.2 A Gaussian H-sssi process, $0 < H \leq 1$, is called *fractional Brownian motion* (FBM) and is denoted $\{B_H(t),\ t \in \mathbb{R}\}$. It is called *standard fractional Brownian motion* if $\sigma_0^2 \equiv \operatorname{Var} X(1) = 1$.

There are also *non-Gaussian finite variance* H-sssi processes (see Example 7.9.6; see also Taqqu (1981, 1988) and Major (1981)) but, here, we consider only the *Gaussian* ones. The following corollary is an immediate consequence of Definition 7.2.2 and Lemma 7.2.1.

Corollary 7.2.3 *Fix* $0 < H \leq 1$ *and* $\sigma_0^2 = EX^2(1)$. *The following statements are equivalent:*

(*i*) $\{X(t),\ t \in T\}$ *is Gaussian and* H*-sssi.*

(*ii*) $\{X(t),\ t \in T\}$ *is fractional Brownian motion with self-similarity index* H.

(*iii*) $\{X(t),\ t \in T\}$ *is Gaussian, has mean zero (if* $H < 1$*) and autocovariance function (7.2.1).*

Example 7.2.4 When $H = 1/2$, fractional Brownian motion $\{B_H(t),\ t \in \mathbb{R}\}$ is *Brownian motion* because (7.2.1) reduces to

$$EB_{1/2}(t_1)B_{1/2}(t_2) = \begin{cases} \sigma_0^2 \min(t_1, t_2) & \text{if } t_1 \text{ and } t_2 \text{ have the same signs,} \\ 0 & \text{if } t_1 \text{ and } t_2 \text{ have opposite signs.} \end{cases}$$

Example 7.2.5 For $H = 1$, we have $B_1(t) = t(Z + \mu)$ by Lemma 7.2.1, where $\mu \in \mathbb{R}$ and Z is a mean zero normal random variable.

Because the case $H = 1$ is degenerate, we shall suppose in the sequel that $0 < H < 1$.

Defining a process through its finite-dimensional distributions does not always give a clear picture of its underlying structure. An integral representation is often more helpful. The following subsections present the most common integral representations of fractional Brownian motion. The increment process, called fractional Gaussian noise, is introduced in the final subsection.

7.2.1 "Moving average" representations of fractional Brownian motion

We want to obtain integral representations for fractional Brownian motion of the form $\int_{-\infty}^{\infty} f_t(x)M(dx)$.

Let $(E, \mathcal{E}, m)$ be a measure space and M a $S\alpha S$ random measure, $0 < \alpha \leq 2$, with control measure m. The definition of the stable integrals $I(f_t) =$

$\int_E f_t(x)M(dx),\ t \in T$, given in Section 3.4 applies to any $0 < \alpha \leq 2$. When $\alpha = 2$, $I(f_t)$ is well defined if $\int_E |f_t(x)|^2 m(dx) < \infty$ and one has

$$EI(f_{t_1})I(f_{t_2}) = \text{Cov}\left(I(f_{t_1}), I(f_{t_2})\right) = 2\int_E f_{t_1}(x)f_{t_2}(x)m(dx)$$

by (3.5.4). In particular, $EM^2(A) = 2m(A)$ for any $A \in \mathcal{E}$ with finite measure m.

Fix $\alpha = 2$ and let

$$(E, \mathcal{E}, m) = (\mathbb{R}, \mathcal{B}, |\cdot|/2),$$

where $\mathcal{B}$ is the Borel σ-field and $|\cdot|$ is Lebesgue measure on $\mathbb{R}$. Then $EM^2(A) = |A|$ for any $A \in \mathcal{B}$ with a finite Lebesgue measure and

$$EI(f_{t_1})I(f_{t_2}) = \int_{-\infty}^{\infty} f_{t_1}(x)f_{t_2}(x)dx. \tag{7.2.3}$$

The following proposition gives a first integral representation for fractional Brownian motion. We shall use the function

$$u_+ = \begin{cases} u & \text{if } u \geq 0, \\ 0 & \text{if } u < 0, \end{cases}$$

and adopt the convention $0^0 = 0$.

Proposition 7.2.6 *Let $0 < H < 1$. Then standard fractional Brownian motion $\{B_H(t),\ t \in \mathbb{R}\}$ has the integral representation*

$$\frac{1}{C_1(H)} \int_{-\infty}^{\infty} \left(((t-x)_+)^{H-1/2} - ((-x)_+)^{H-1/2}\right) M(dx),\ \ t \in \mathbb{R}, \tag{7.2.4}$$

where

$$C_1(H) = \left\{ \int_0^{\infty} \left((1+x)^{H-1/2} - x^{H-1/2}\right)^2 dx + \frac{1}{2H} \right\}^{1/2}. \tag{7.2.5}$$

When $H = 1/2$, $C_1(1/2) = 1$ and the representation (7.2.4) is to be interpreted as $\int_0^t M(dx)$ if $t \geq 0$ and $-\int_t^0 M(dx)$ if $t < 0$, i.e., as an integral representation of Brownian motion.

PROOF: Let $X(t)$ denote the integral in (7.2.4) and let $f_t(x)$ denote its integrand. In order to verify that $X(t)$ is well defined, we show firstly that $\int_{-\infty}^{\infty} f_t^2(x)dx < \infty$. This relation is obvious when $H = 1/2$, because $\int_{-\infty}^{\infty} f_t^2(x)dx = \int_0^{|t|} dx < \infty$. Suppose now $0 < H < 1$, $H \neq 1/2$. Then as $x \to -\infty$, $f_t(x) \sim (H - 1/2)(-x)^{H-3/2}$ whose square is integrable around $-\infty$, $f_t(x) \sim (t-x)_+^{H-1/2}$

as $x \to t$ whose square is integrable around $x = t$, and similarly for $x = 0$ and $x = \infty$. Hence $\int_{-\infty}^{\infty} f_t^2(x)dx < \infty$ and (7.2.4) is well defined.

We now verify that $\{X(t),\ t \in \mathbb{R}\}$ has the autocovariance function (7.2.1) with $\mathrm{Var}X(1) = 1$. Notice that $X(0) = 0$ a.s. and, for every $t > 0$, $EX^2(t)$ equals

$$\begin{aligned}
&\frac{1}{C_1(H)^2}\int_{-\infty}^{\infty}((t-x)_+^{H-1/2} - (tx)_+^{H-1/2})^2 dx \\
&= \frac{1}{C_1(H)^2}\, t^{2H}\int_{-\infty}^{\infty}((1-u)_+^{H-1/2} - (-u)_+^{H-1/2})^2 du \\
&= \frac{1}{C_1(H)^2}\, t^{2H}\left[\int_{-\infty}^{0}((1-u)^{H-1/2} - (-u)^{H-1/2})^2 du + \int_0^1 (1-u)^{2H-1} du\right] \\
&= \frac{1}{C_1(H)^2}\, t^{2H}\left[\int_0^{\infty}((1+x)^{H-1/2} - x^{H-1/2})^2 dx + \frac{1}{2H}\right] = t^{2H},
\end{aligned}$$

and similarly $EX^2(t) = |t|^{2H}$ for $t < 0$. Further, for any $t, s \in \mathbb{R}$,

$$\begin{aligned}
E(X(t) - X(s))^2 &= \frac{1}{C_1(H)^2}\int_{-\infty}^{\infty}((t-x)_+^{H-1/2} - (s-x)_+^{H-1/2})^2 dx \\
&= \frac{1}{C_1(H)^2}\int_{-\infty}^{\infty}((t-s-x)_+^{H-1/2} - (-x)_+^{H-1/2})^2 dx \\
&= |t-s|^{2H}
\end{aligned}$$

by the previous calculation and hence (7.2.1) follows. ∎

The kernel

$$f_t(x) = ((t-x)_+)^{H-1/2} - ((-x)_+)^{H-1/2} \tag{7.2.6}$$

is plotted in Figures 7.2 and 7.3 for several values of t and several values of $d = H - 1/2$.

The representation (7.2.4) is far from unique. In fact, let

$$u_- = \begin{cases} 0 & \text{if } u \geq 0, \\ -u & \text{if } u < 0. \end{cases}$$

Then for any real a and b numbers,

$$\int_{-\infty}^{\infty}\Big[a\{(t-x)_+^{H-1/2} - (-x)_+^{H-1/2}\} + b\{(t-x)_-^{H-1/2} - (-x)_-^{H-1/2}\}\Big] M(dx), \quad t \in \mathbb{R}, \tag{7.2.7}$$

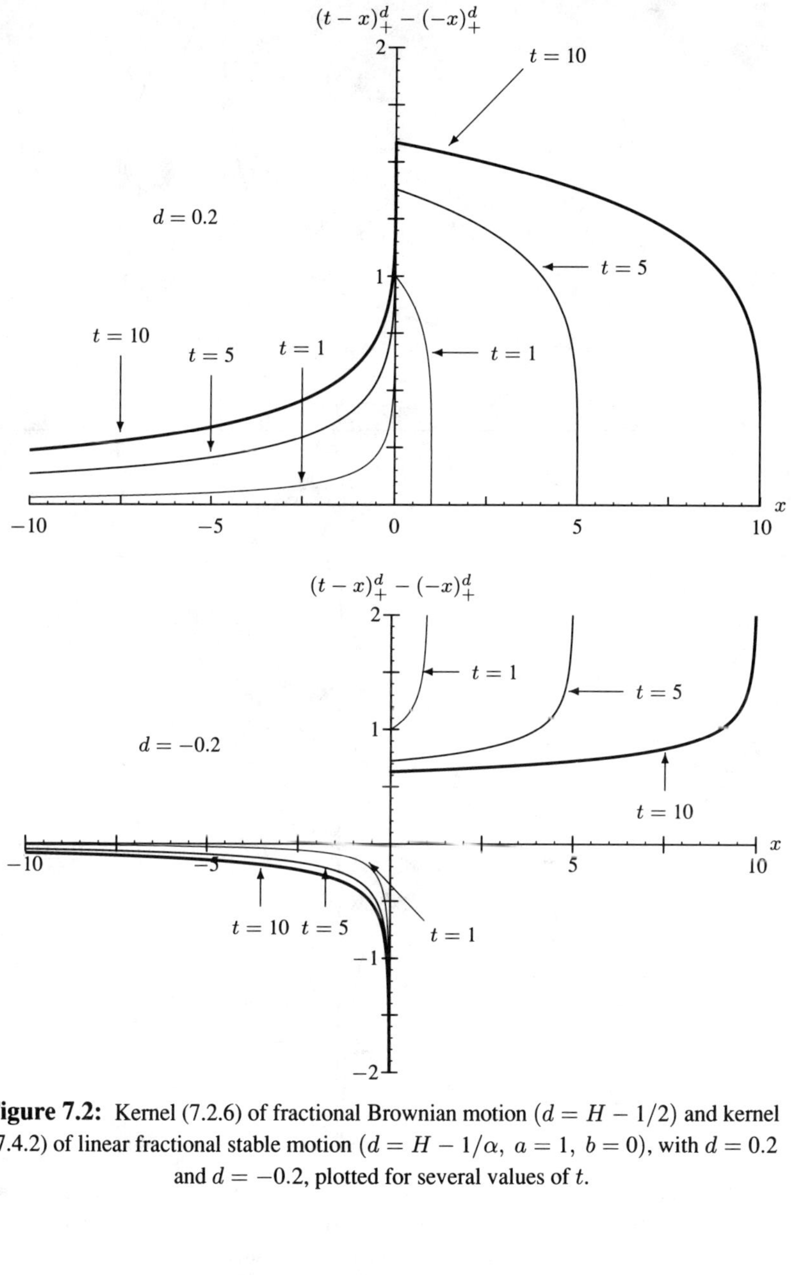

Figure 7.2: Kernel (7.2.6) of fractional Brownian motion $(d = H - 1/2)$ and kernel (7.4.2) of linear fractional stable motion $(d = H - 1/\alpha,\ a = 1,\ b = 0)$, with $d = 0.2$ and $d = -0.2$, plotted for several values of t.

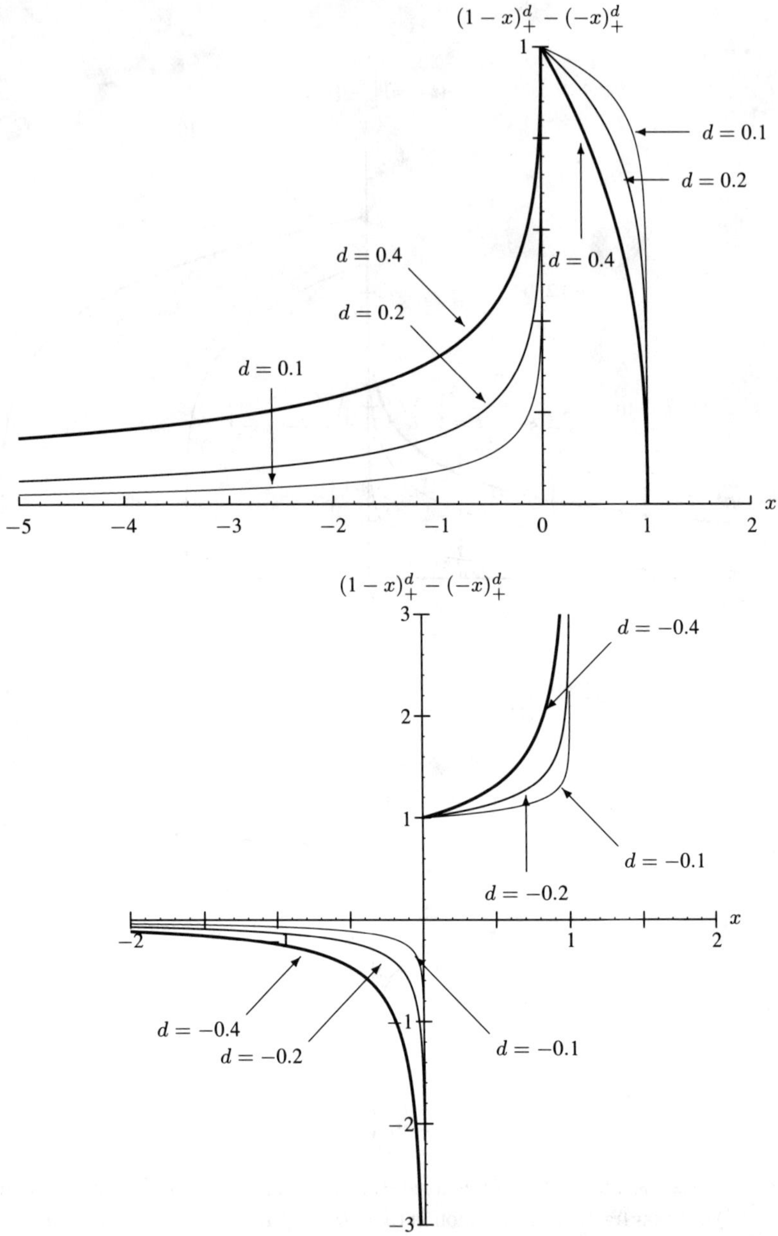

Figure 7.3: Kernel (7.2.6) of fractional Brownian motion ($d = H - 1/2$) and kernel (7.4.2) of linear fractional stable motion ($d = H - 1/\alpha,\ a = 1,\ b = 0$), with $t = 1$, plotted for several values of d.

is also (up to a multiplicative constant) a representation of fractional Brownian motion (Exercise 7.2).

The representation of fractional Brownian motion given in Proposition 7.2.6 corresponds to $a = 1/C_1(H)$ and $b = 0$. That representation is "non-anticipative" because it involves only integration on $x \in (-\infty, t]$. When $a = b = 1$, then (7.2.7) yields the "well-balanced" representation

$$\int_{-\infty}^{\infty} \left(|t-x|^{H-1/2} - |x|^{H-1/2}\right) M(dx).$$

7.2.2 "Harmonizable" representations of fractional Brownian motion

We now turn to representations of the "harmonizable" type, i.e., of the form $I(\widetilde{f}_t) = \int_{-\infty}^{\infty} \widetilde{f}_t(x)\widetilde{M}(dx)$ where $\widetilde{f}_t$ is complex and $\widetilde{M}$ is a complex isotropic random measure. Integrals ot his type were defined in Section 6.3 for the α-stable case, $0 < \alpha < 2$, but these definitions apply to the Gaussian case as well. [3] If we adopt this approach, $\int_{-\infty}^{\infty} \widetilde{f}_t(x)\widetilde{M}(dx)$ will be, in general, complex and we will have to consider Re $\int_{-\infty}^{\infty} \widetilde{f}_t(x)\widetilde{M}(dx)$, for example, if we want a real-valued process.[4]

Here, we present an alternative definition, one which has been traditionally followed in the context of Gaussian self-similar processes. In this approach, the real and imaginary components $M^{(1)}$ and $M^{(2)}$ of $\widetilde{M}$ are independent but they are not independently scattered, i.e., $M^{(j)}(A)$ may not be independent of $M^{(j)}(B)$ even for disjoint sets A and B. This second approach allows us to choose the complex $\widetilde{f}_t$ and $\widetilde{M}$ in such a way that the integral $\int_{-\infty}^{\infty} \widetilde{f}_t(x)\widetilde{M}(dx)$ is always real-valued.

Firstly, some heuristics. We would like to view the complex-valued $\widetilde{f}_t$ and $\widetilde{M}_t$ as "Fourier transforms" of real-valued f_t and M_t. We want them to be defined in such a way that a Parseval type relation $\int_{-\infty}^{\infty} f_t(x)M(dx) \stackrel{d}{=} \int_{-\infty}^{\infty} \widetilde{f}_t(x)\widetilde{M}(dx)$ holds. Since f_t and M are real, $\widetilde{f}_t$ and $\widetilde{M}_t$ should satisfy $\widetilde{f}_t(x) = \overline{\widetilde{f}_t(-x)}$ and $\widetilde{M}(A) = \overline{\widetilde{M}(-A)}$, where a bar denotes a "complex conjugate." The relation $\widetilde{M}(A) = \overline{\widetilde{M}(-A)}$ implies that $\widetilde{M}(A)$ and $\widetilde{M}(-A)$ are dependent even for disjoint sets A and $-A$; hence, the complex-valued random measure $\widetilde{M}$ will not be independently scattered on all Borel subsets of $\mathbb{R}$.

[3] A special feature of the Gaussian case is that the real and imaginary parts of $\widetilde{M}$ are independent.

[4] The reader who prefers this approach will find it in Section 7.7 where we define the real harmonizable fractional stable motion for $0 < \alpha \leq 2$. It is an harmonizable representation for fractional Brownian motion when $\alpha = 2$.

The following definitions of the random measure $\widetilde{M}$ and the integral $I(\widetilde{f}) = \int_{-\infty}^{\infty} \widetilde{f}(x)\widetilde{M}(dx)$ are motivated by the preceding heuristics. Define firstly $\widetilde{M} = M^{(1)} + iM^{(2)}$ where $M^{(1)}$ and $M^{(2)}$ are *independent* Gaussian random measures, independently scattered on $\mathbb{R}_+$ and satisfying $M^{(1)}(A) = M^{(1)}(-A)$, $M^{(2)}(A) = -M^{(2)}(-A)$ for any Borel set A of finite Lebesgue measure. Then $\widetilde{M}(A) = \overline{\widetilde{M}(-A)}$. Suppose also that $M^{(1)}$ and $M^{(2)}$ have control measure $(1/4)|\cdot|$ so that on $\mathbb{R}_+$, $E(M^{(1)}(A))^2 = E(M^{(2)}(A))^2 = \frac{1}{2}|A|$. Then the *variance* of $\widetilde{M}(A)$ equals

$$E|\widetilde{M}(A)|^2 = E(M^{(1)}(A))^2 + E(M^{(2)}(A))^2 = |A|$$

for any Borel set $A \in \mathbb{R}_+$. Since $\widetilde{M}$ is complex, the quantity $E(\widetilde{M}(A))^2$ has no special significance. In fact,

$$E(\widetilde{M}(A))^2 = E(M^{(1)}(A))^2 - E(M^{(2)}(A))^2 = 0.$$

We now describe the integrands. Let $\widetilde{f} = f^{(1)} + if^{(2)}$ where $f^{(1)}$ and $f^{(2)}$ are real-valued functions that satisfy $f^{(1)}(x) = f^{(1)}(-x)$, $f^{(2)}(x) = -f^{(2)}(-x)$ and $\int_{-\infty}^{\infty} (f^{(i)}(x))^2 dx < \infty$. Thus, $\widetilde{f}$ belongs to the set

$$\mathcal{F} = \left\{\widetilde{f}\colon \widetilde{f}(x) = \overline{\widetilde{f}(-x)} \text{ and } \int_{-\infty}^{\infty} |\widetilde{f}(x)|^2 dx < \infty\right\}.$$

For any $\widetilde{f} \in \mathcal{F}$, the integral

$$I(\widetilde{f}) = \int_{-\infty}^{\infty} \widetilde{f}(x)\widetilde{M}(dx)$$

is defined as

$$\begin{aligned} I(\widetilde{f}) &= \int_{-\infty}^{0} \widetilde{f}(x)\widetilde{M}(dx) + \int_{0}^{\infty} \widetilde{f}(x)\widetilde{M}(dx) \\ &= \int_{0}^{\infty} (f^{(1)}(x) - if^{(2)}(x))(M^{(1)}(dx) - iM^{(2)}(dx)) \\ &\quad + \int_{0}^{\infty} (f^{(1)}(x) + if^{(2)}(x))(M^{(1)}(dx) + iM^{(2)}(dx)) \\ &= 2\Big(\int_{0}^{\infty} f^{(1)}(x)M^{(1)}(dx) - \int_{0}^{\infty} f^{(2)}(x)M^{(2)}(dx)\Big), \end{aligned} \tag{7.2.8}$$

i.e.,

$$I(\widetilde{f}) = \int_{-\infty}^{\infty} f^{(1)}(dx)M^{(1)}(dx) - \int_{-\infty}^{\infty} f^{(2)}(x)M^{(2)}(dx).$$

$\{I(\widetilde{f}),\ \widetilde{f} \in \mathcal{F}\}$ is a Gaussian system in which each $I(\widetilde{f})$ is a *real-valued* Gaussian random variable with mean zero and variance

$$\begin{aligned} E\Big(\int_{-\infty}^{\infty} \widetilde{f}(x)\widetilde{M}(dx)\Big)^2 &= 4\Big\{\int_0^{\infty} (f^{(1)}(x))^2 \frac{dx}{2} + \int_0^{\infty} (f^{(2)}(x))^2 \frac{dx}{2}\Big\} \\ &= 2\int_0^{\infty} |\widetilde{f}(x)|^2 dx = \int_{-\infty}^{\infty} |\widetilde{f}(x)|^2 dx < \infty. \end{aligned}$$

Moreover, the covariance of $I(\widetilde{f}_1)$ and $I(\widetilde{f}_2)$, $\widetilde{f}_1, \widetilde{f}_2 \in \mathcal{F}$, equals

$$EI(f_1)I(\widetilde{f}_2) = \int_{-\infty}^{\infty} \widetilde{f}_1(x)\overline{\widetilde{f}_2(x)}dx. \tag{7.2.9}$$

With the preceding definitions, we can now start with a set of integrands

$$\mathcal{F}_0 = \Big\{f_t\colon \int_{-\infty}^{\infty} |f_t(x)|^2 dx < \infty,\ f_t \text{ real-valued},\ t \in T\Big\},$$

and define for each $f_t \in \mathcal{F}_0$ its L^2 Fourier transform $\widetilde{f}_t$ which we denote (with some abuse of notation) by

$$\widetilde{f}_t(x) = \frac{1}{\sqrt{2\pi}} \int_{-\infty}^{\infty} e^{ixu} f_t(u)du.$$

We shall now try to relate the Gaussian processes $\{I(f_t) = \int_{-\infty}^{\infty} f_t(x)M(dx),\ t \in T\}$ and $\{I(\widetilde{f}_t) - \int_{-\infty}^{\infty} \widetilde{f}_t(x)\widetilde{M}(dx),\ t \in T\}$. Here, M is a real-valued Gaussian random measure and $\widetilde{M}$ is the complex Gaussian random measure defined above. Observe that, since $f_t \in \mathcal{F}_0$ is real-valued, its Fourier transform $\widetilde{f}_t$ satisfies $\widetilde{f}_t(x) = \overline{\widetilde{f}_t(-x)}$.

Proposition 7.2.7

$$\Big\{I(f_t) - \int_{-\infty}^{\infty} f_t(x)M(dx),\ t \in T\Big\} \stackrel{d}{=} \Big\{I(\widetilde{f}_t) = \int_{-\infty}^{\infty} \widetilde{f}_t(x)\widetilde{M}(dx),\ t \in T\Big\}.$$

PROOF: Since both $\{I(f_t),\ t \in T\}$ and $\{I(\widetilde{f}_t),\ t \in T\}$ are Gaussian with mean zero, it is sufficient to prove that for any two functions $f_{t_1}, f_{t_2} \in \mathcal{F}_0$ one has $EI(f_{t_1})I(f_{t_2}) = EI(\widetilde{f}_{t_1})I(\widetilde{f}_{t_2})$. This, however, is an immediate consequence of (7.2.3), (7.2.9) and the Parseval identity:

$$\int_{-\infty}^{\infty} f_{t_1}(x)f_{t_2}(x)dx = \int_{-\infty}^{\infty} \widetilde{f}_{t_1}(x)\overline{\widetilde{f}_{t_2}(x)}dx. \quad \blacksquare$$

Thus, in order to obtain a harmonizable representation $\{\int_{-\infty}^{\infty} \widetilde{f}_t(x)\widetilde{M}(dx),\ t \in \mathbb{R}\}$ of fractional Brownian motion, one can start with any one of its representations $\{\int_{-\infty}^{\infty} f_t(x)M(dx),\ t \in \mathbb{R}\}$, and derive the Fourier transform $\widetilde{f}_t$ of f_t. This can be delicate at times and therefore it is easier to first guess the form of $\widetilde{f}_t$ and then verify that the resulting process $\{I(\widetilde{f}_t),\ t \in \mathbb{R}\}$ is H-sssi.

Consider, for example, the representation (7.2.4) of fractional Brownian motion. A heuristic computation[5] suggests the following representation (7.2.12).

Proposition 7.2.8 *Let $0 < H < 1$. Then standard fractional Brownian motion $\{B_H(t),\ t \in \mathbb{R}\}$ has the integral representation*

$$\frac{1}{C_2(H)} \int_{-\infty}^{\infty} \frac{e^{ixt}-1}{ix} |x|^{-(H-1/2)} \widetilde{M}(dx),\quad t \in \mathbb{R}, \tag{7.2.12}$$

where

$$C_2(H) = \left(\frac{\pi}{H\Gamma(2H)\sin H\pi}\right)^{1/2}. \tag{7.2.13}$$

PROOF: Let $X(t)$ denote the integral in (7.2.12) and $\widetilde{f}_t(x)$ its integrand. Since $|(e^{ixt}-1)/ix|$ is bounded for $x \to 0$ and behaves like $|x|^{-1}$ as $x \to \pm\infty$, one obtains $\int_{-\infty}^{\infty} |\widetilde{f}_t(x)|^2 dx < \infty$. Hence, $X(t)$ is well defined. It is easy to verify that the process is H-sssi (Exercise 7.6) and therefore by Corollary 7.2.3, $\{X(t),\ t \in \mathbb{R}\}$ is fractional Brownian motion.

---

[5]Here, we present the *heuristics* leading to (7.2.12). The computations are not precise because the integrals diverge! Suppose $H > 1/2$ and $t > 0$ and rewrite (7.2.4) as

$$C_1^{-1}(H)(H-1/2) \int_{-\infty}^{\infty} \left(\int_0^t (s-\xi)_+^{H-3/2} ds\right) M(d\xi) \tag{7.2.10}$$

$$= C_1^{-1}(H)(H-1/2) \int_{-\infty}^{\infty} \left\{\int_{-\infty}^{\infty} 1_{[0,t]}(s)(s-\xi)_+^{H-3/2} ds\right\} M(d\xi). \tag{7.2.11}$$

The inner integral in (7.2.11) is the convolution of the function $1_{[0,t]}(s)$ with $s_+^{H-3/2}$ and its Fourier transform should equal the product of the convolutions. Now the Fourier transform of $1_{[0,t]}(s)$ is

$$(\sqrt{2\pi})^{-1} \int_0^t e^{ixs} ds = (\sqrt{2\pi})^{-1}(ix)^{-1}(e^{ixt}-1).$$

Although the Fourier transform of $s_+^{H-3/2}$ does not exist (the function is not in L^1 nor L^2), let us write, nevertheless,

$$(\sqrt{2\pi})^{-1} \int_{-\infty}^{\infty} e^{ixs} s_+^{H-3/2} ds = |x|^{-(H-1/2)}(\sqrt{2\pi}) \int_0^{\infty} e^{iu} u^{H-3/2} du$$

by setting $u = xs$. Viewing the last integral as an improper Riemann integral, one can show that it equals $\exp\{i\pi\gamma/2\}\Gamma(H-3/2)$ (see Exercise 7.4). Use $(f \times g)\check{} = \sqrt{2\pi}\check{f}\check{g}$ and Proposition 7.2.7 to obtain (7.2.12).

This heuristic can be made rigorous (Exercise 7.5).

It remains to show that $EX(1)^2 = 1$, i.e.

$$\int_{-\infty}^{\infty} |\widetilde{f}_1(x)|^2 dx = C_2^2(H). \tag{7.2.14}$$

Since $|e^{ix} - 1|^2 = (\cos x - 1)^2 + \sin^2 x = 2(1 - \cos x) = 4\sin^2 \frac{x}{2}$, we have

$$\int_{-\infty}^{\infty} |\widetilde{f}_1(x)|^2 dx = 8 \int_0^{\infty} \frac{\sin^2 x/2}{x^{2H+1}} dx. \tag{7.2.15}$$

If $H = 1/2$, this equals $8\pi/4 = 2\pi = C_2^2(1/2)$ using Relation GR3.821 in Gradshteyn and Ryzhik (1980).

If $H \neq 1/2$, use Relation GR3.823 in Gradshteyn and Ryzhik (1980), $z\Gamma(z) = \Gamma(z+1)$ and $\Gamma(z)\Gamma(1-z) = \pi/\sin \pi z$ to show that (7.2.15) equals

$$-\frac{8}{2}\Gamma(-2\pi)\cos H\pi = -\frac{2\Gamma(2-2H)\cos H\pi}{H(2H-1)} = C_2^2(H).$$

($\Gamma(z)$ is the Gamma function.) This proves (7.2.14) and $EX(1)^2 = 1$. ■

Using (7.2.8), we can write the representation (7.2.12) as

$$\frac{2}{C_2(H)} \left\{ \int_0^{\infty} \frac{\sin xt}{x} x^{-(H-1/2)} M^{(1)}(dx) + \int_0^{\infty} \frac{1 - \cos xt}{x} x^{-(H-1/2)} M^{(2)}(dx) \right\}, \tag{7.2.16}$$

where $M^{(1)}$ and $M^{(2)}$ are two independent real-valued Gaussian random measures such that $E(M^{(i)}(A))^2 = |A|/2$ for any Borel set A of $\mathbb{R}_+$. The integrands of $M^{(1)}$ and $M^{(2)}$ are plotted in Figures 7.4 and 7.5, for $t = 1$ and $d = H - 1/2 = \pm 0.2$.

Remarks

1. When $H = 1/2$, $C_2(1/2) = \sqrt{2\pi}$ and (7.2.12) gives the following representation for Brownian motion,

$$\left\{ B(t) = \sqrt{2\pi} \int_{-\infty}^{\infty} \frac{e^{ixt} - 1}{ix} \widetilde{M}(dx),\ t \in \mathbb{R} \right\}.$$

2. For any real a and b, the process

$$\left\{ \int_{-\infty}^{\infty} \frac{e^{ixt} - 1}{ix} \left(a(x_+)^{-(H-1/2)} + b(x_-)^{-(H-1/2)} \right) \widetilde{M}(dx),\ t \in \mathbb{R} \right\}, \tag{7.2.17}$$

is also fractional Brownian motion (up to a multiplicative constant) because it is a Gaussian H-sssi process (Exercise 7.7). The representation (7.2.12) corresponds to the choice $a = b = C_2^{-1}(H)$.

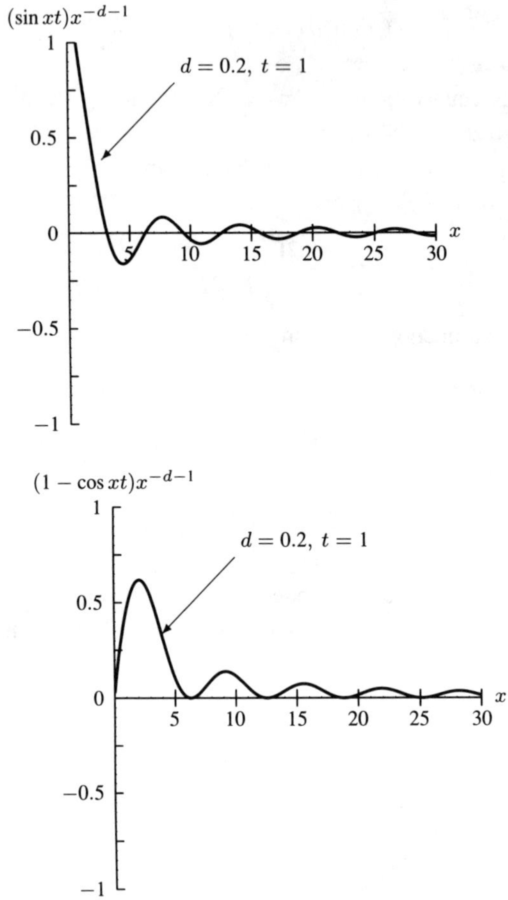

Figure 7.4: Integrands of $M^{(1)}$ and $M^{(2)}$, with $t = 1$, in the representation (7.2.16) of fractional Brownian motion ($d = H - 1/2$) and in the representation (7.7.4) of the real harmonizable fractional stable motion ($d = H - 1/\alpha$). Here, $d = 0.2$.

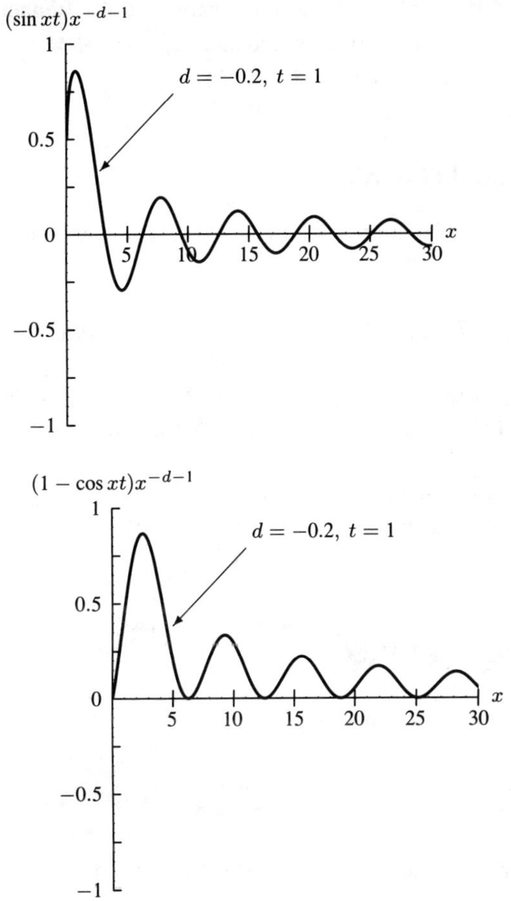

Figure 7.5: Integrands of $M^{(1)}$ and $M^{(2)}$, with $t = 1$, in the representation (7.2.16) of fractional Brownian motion ($d = H - 1/2$) and in the representation (7.7.4) of the real harmonizable fractional stable motion ($d = H - 1/\alpha$). Here, $d = -0.2$.

3. The representation (7.2.12) (or (7.2.17)) has a physical interpretation. The factor $(e^{ixt}-1)/ix$ is the Fourier transform of the indicator function of the interval $[0,t]$ and ensures stationarity of the increments. The term $|x|^{-(H-1/2)}$ colors the underlying Gaussian noise. In a sense that will be made precise in the following section, it gives rise to a spectral density $||x|^{-(H-1/2)}|^2 = x^{-(2H-1)}$, which diverges at low frequencies ($x \to 0$) when $1/2 < H < 1$, is colorless (equal to 1) when $H = 1/2$ and tends to 0 at low frequencies when $0 < H < 1/2$.

7.2.3 Fractional Gaussian noise

Since fractional Brownian motion $\{B_H(t),\ t \in \mathbb{R}\}$ has stationary increments, its increments

$$Y_j = B_H(j+1) - B_H(j),\ \ j = \dots, -1, 0, 1, \dots,$$

form a stationary sequence. The sequence $\{Y_j,\ j \in \mathbb{Z}\}$ is called *fractional Gaussian noise* (FGN).[6] It is called *standard fractional Gaussian noise* if $\sigma_0^2 \equiv \text{Var}\, Y_j^2 = 1$.

Any representation of $\{B_H(t),\ t \in \mathbb{R}\}$ induces a representation for $\{Y_j,\ j \in \mathbb{Z}\}$. For example, using (7.2.4) and (7.2.12) respectively, we obtain the moving average representation

$$Y_j \stackrel{d}{=} \frac{\sigma_0}{C_1(H)} \int_{-\infty}^{j+1} \left((j+1-x)_+^{H-1/2} - (j-x)_+^{H-1/2}\right) M(dx),\ j \in \mathbb{Z}, \quad (7.2.18)$$

and the harmonizable representation

$$Y_j \stackrel{d}{=} \frac{\sigma_0}{C_2(H)} \int_{-\infty}^{\infty} e^{ixj} \frac{e^{ix}-1}{ix} |x|^{-(H-1/2)} \widetilde{M}(dx),\ j \in \mathbb{Z}. \quad (7.2.19)$$

Fractional Gaussian noise has some remarkable properties. It is a stationary Gaussian sequence with mean zero and variance $EY_j^2 = EB_H^2(1) = \sigma_0^2$. Let $\{r(j) = EY_0Y_j,\ j \in \mathbb{Z}\}$, denote its autocovariance function and let $\{h(\lambda),\ -\pi \le \lambda \le \pi\}$ denote its spectral density, i.e.,

$$r(j) = \int_{-\pi}^{\pi} e^{i\lambda j} h(\lambda) d\lambda,\ j = 1, 2, \dots . \quad (7.2.20)$$

[6] One sometimes defines fractional Gaussian noise as

$$Y(u) = B_H(u+1) - B_H(u),\ u \in \mathbb{R}.$$

For most time series applications, however, one restricts u to $\mathbb{Z}$.

(It is possible to obtain h from r by using the relation

$$h(\lambda) = \frac{1}{2\pi} \sum_{j=-\infty}^{\infty} e^{-i\lambda j} r(j)$$

whenever the sum converges.) The correlation and spectral density are plotted in Figures 7.6 and 7.7 respectively.

Proposition 7.2.9 *Fractional Gaussian noise has autocovariance function*

$$r(j) = \frac{\sigma_0^2}{2}\left(|j+1|^{2H} - 2|j|^{2H} + |j-1|^{2H}\right), \quad j \in \mathbb{Z}, \tag{7.2.21}$$

and spectral density

$$h(\lambda) = \frac{\sigma_0^2}{C_2^2(H)} |e^{i\lambda} - 1|^2 \sum_{k=-\infty}^{\infty} \frac{1}{|\lambda + 2\pi k|^{2H+1}}, \quad -\pi \le \lambda \le \pi. \tag{7.2.22}$$

Equivalently,

$$h(\lambda) = \frac{\sigma_0^2 \int_0^\infty \cos x\lambda (\sin^2 \frac{x}{2}) x^{-2H-1} dx}{\int_0^\infty (\sin^2 \frac{x}{2}) x^{-2H-1} dx}, \quad -\pi \le \lambda \le \pi. \tag{7.2.23}$$

PROOF: Assume, without loss of generality, that $\sigma_0^2 = EY_j^2 = 1$. Then

$$\begin{aligned} r(j) = EY_0 Y_j &= EB_H(1)(B_H(j+1) - B_H(j)) \\ &= R_H(1, j+1) - R_H(1, j) \\ &= \frac{1}{2}\{|j+1|^{2H} - 2|j|^{2H} + |j-1|^{2H}\} \end{aligned} \tag{7.2.24}$$

by (7.2.1). An easy way to obtain the spectral density $h(\lambda)$ is to evaluate the autocovariance $r(j)$ by using the representation (7.2.19) of $\{Y_j,\ j \in \mathbb{Z}\}$ and Relation (7.2.9):

$$\begin{aligned} r(j) &= EY_0 Y_j \\ &= C_2^{-2}(H) \int_{-\infty}^{\infty} e^{ixj} \left| \frac{e^{ix} - 1}{ix} \right|^2 |x|^{-2(H-1/2)} d\lambda \qquad (7.2.25) \\ &= C_2^{-2}(H) \sum_{k=-\infty}^{\infty} \int_{-\pi+2\pi k}^{\pi+2\pi k} e^{ixj} |e^{ix} - 1|^2 |x|^{-2H-1} dx \\ &= C_2^{-2}(H) \sum_{k=-\infty}^{\infty} \int_{-\pi}^{\pi} e^{i\lambda j} |e^{i\lambda} - 1|^2 |\lambda + 2\pi k|^{-2H-1} d\lambda \\ &= C_2^{-2}(H) \int_{-\pi}^{\pi} e^{i\lambda j} |e^{i\lambda} - 1|^2 \sum_{k=-\infty}^{\infty} \frac{1}{|\lambda + 2\pi k|^{2H+1}} d\lambda, \end{aligned}$$

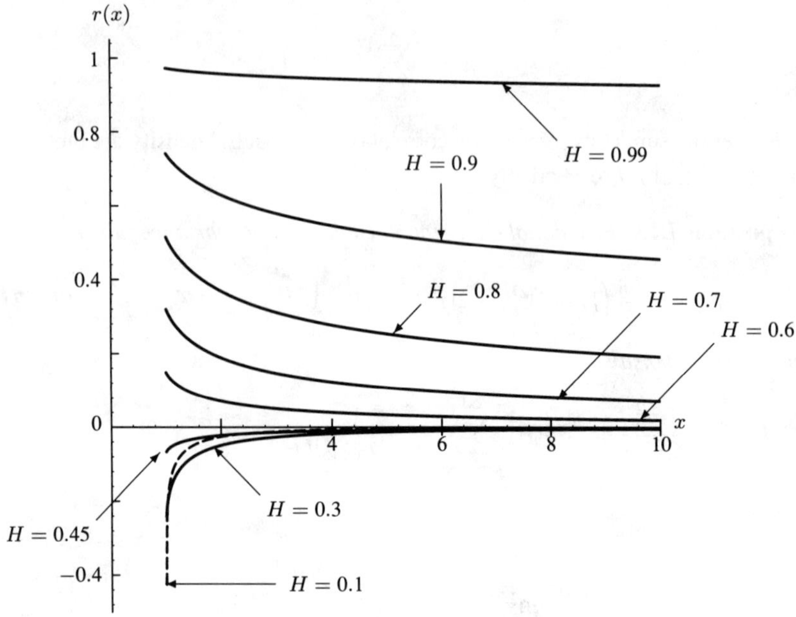

Figure 7.6: The correlation $r(x) = (|x+1|^{2H} - 2|x|^{2H} + |x-1|^{2H})/2$ of fractional Gaussian noise (see Relation (7.2.24)), plotted for several values of H. Note that it tends much faster to zero as $x \to \infty$ when $H < 1/2$ than when $H > 1/2$.

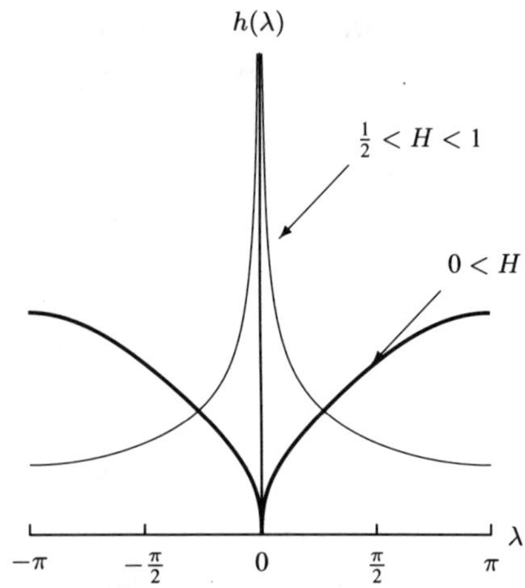

Figure 7.7: Spectral density $h(\lambda)$ of fractional Gaussian noise for $H = 0.7$ and $H = 0.3$ (see Relation (7.2.22)). $h(0)$ is infinite if $H > 1/2$ and is zero if $H < 1/2$.

and then deduce (7.2.22) by comparing this last expression with (7.2.20). To obtain (7.2.23), use $|e^{ix} - 1|^2 = 4\sin^2 \frac{x}{2}$, (7.2.14) and (7.2.15) in (7.2.25). ∎

If $H = 1/2$, then $r(j) = 0$ for $j \neq 0$ and hence $\{Y_j,\ j = \ldots, -1, 0, 1, \ldots\}$ is a sequence of i.i.d. Gaussian random variables. The Y_js are dependent when $H \neq 1/2$. The following result specifies the behavior of the autocovariance function $r(j)$ at large lags j, and, correspondingly, the behavior of the spectral density $h(\lambda)$ at low frequencies λ.

Proposition 7.2.10 *Let $\{Y_j,\ j = \ldots, -1, 0, 1, \ldots\}$ be fractional Gaussian noise. Then*

$$r(j) \sim \sigma_0^2 H(2H-1) j^{2H-2} \quad as \ j \to \infty \tag{7.2.26}$$

for $H \neq 1/2$, and

$$h(\lambda) \sim \sigma_0^2 C_2^{-2}(H) |\lambda|^{1-2H} \quad as \ \lambda \to 0. \tag{7.2.27}$$

PROOF: For large j,

$$r(j) = \frac{\sigma_0^2}{2} j^{2H-2} \left(j^2 \left[\left(1 + \frac{1}{j}\right)^{2H} - 2 + \left(1 - \frac{1}{j}\right)^{2H} \right] \right),$$

and the content of $\{\ \}$ tends to $2H(2H-1)$ as $j \to \infty$. As for h,

$$h(\lambda) = \sigma_0^2 C_2^{-2}(H) |e^{i\lambda} - 1|^2 \left\{ \frac{1}{|\lambda|^{2H+1}} + \sum_{\substack{k=-\infty \\ k \neq 0}}^{\infty} \frac{1}{|\lambda + 2\pi k|^{2H+1}} \right\}.$$

The result follows because $|e^{i\lambda} - 1|^2 \sim |\lambda|^2$ as $\lambda \to 0$, and the sum in the brackets is bounded for all $-\pi \leq \lambda \leq \pi$. ∎

The following are important points:

- Both $r(j)$ as $j \to \infty$ and $h(\lambda)$ as $|\lambda| \to 0$ behave like power functions.

- $r(j)$ tends to 0 as $j \to \infty$ for all $0 < H < 1$, but when $1/2 < H < 1$ it tends to zero so slowly that $\sum_{j=-\infty}^{\infty} r(j)$ diverges. We say that in this case $\{Y_j,\ j \in \mathbb{Z}\}$ exhibits *long-range dependence.*

  The divergence of $\sum_{j=-\infty}^{\infty} r(j)$ affects the behavior of the spectral density $h(\lambda) = \frac{1}{2\pi} \sum_{j=-\infty}^{\infty} e^{-i\lambda j} r(j)$ as $|\lambda| \to 0$: $h(\lambda)$ diverges at the origin when $1/2 < H < 1$ (see (7.2.27)).

- Suppose $0 < H < 1/2$. Then $\sum_{j=-\infty}^{\infty} |r(j)| < \infty$ and $\sum_{j=-\infty}^{\infty} r(j) = 0$. This last relation is due to the telescoping nature of $r(j)$ (see (7.2.21)) and is consistent with the fact that the spectral density $h(\lambda)$ tends to zero as $|\lambda| \to 0$ (see (7.2.27)). Although there is no long-range dependence, the case $0 < H < 1/2$ is a singular one. Because the coefficient $H(2H - 1)$ is negative, the $r(j)$s are negative for all large j, a behavior sometimes referred to as "negative dependence." It is easy to show that $r(j) < 0$ for all $j \neq 0$, and thus

$$0 = \sum_{j=-\infty}^{\infty} r(j) = r(0) + 2\sum_{j=1}^{\infty} r(j) = \sigma_0^2 + 2\sum_{j=1}^{\infty} r(j),$$

i.e., $\sum_{j=1}^{\infty} r(j) = -\sigma_0^2/2$.

- Fractional Gaussian noise $\{Y_j,\ j = \ldots, -1, 0, 1, \ldots\}$ with $H \neq 1/2$ provides a counterexample to the usual central limit theorem. Indeed, for $\frac{1}{d_N}\sum_{j=1}^{N} Y_j$ to converge in distribution as $N \to \infty$ to a non-trivial limit, one cannot choose $d_N \sim \sqrt{N}$ as $N \to \infty$, but one must choose $d_N \sim N^H$ as $N \to \infty$. In fact, $\frac{1}{N^H}\sum_{j=1}^{N} Y_j = \frac{1}{N^H} B_H(N) \stackrel{d}{=} B_H(1)$. More generally, the finite-dimensional distributions of $\frac{1}{N^H}\sum_{j=1}^{[Nt]} Y_j$, $0 \leq t \leq 1$, converge to those of $B_H(t)$, $0 \leq t \leq 1$, as $N \to \infty$. (Here [ ] means "integer part" and $\sum_{j=1}^{0} Y_j = 0$.)

- Because of long-range dependence, normalized sums of polynomials of the Y_js may not converge to a Gaussian limit when $1/2 < H < 1$ (Taqqu 1981).

The autocovariance $r(j)$ of fractional Gaussian noise is given by (7.2.21) for all lags j. In many modeling situations, however, one may want greater flexibility in the choice of the $r(j)$s at small lags j while retaining the asymptotic behavior $r(j) \sim Cj^{2H-2}$ as $j \to \infty$, which is characteristic of long-range dependence when $1/2 < H < 1$. This may be done in various ways. The simplest is to define the moving average

$$Z_j = \sum_{k=-\infty}^{j} a_{j-k}\epsilon_k,$$

where the ϵ_js are i.i.d. $N(0, 1)$ and

$$r(j) = EZ_0Z_j = \sum_{k=-\infty}^{j} a_{-k}a_{j-k} \sim Cj^{2H-2}$$

as $j \to \infty$. As the following theorem indicates, the Z_js satisfy the same type of non-central limit theorem as fractional Gaussian noise.

Theorem 7.2.11 *Fix $0 < H < 1$ and let $\{Z_j,\ j = \ldots,-1,0,1,\ldots\}$ be a stationary Gaussian sequence with mean zero and autocovariance function $r(j) = EZ_0Z_j$ satisfying:*
(i) Case $1/2 < H < 1$:

$$r(j) \sim cj^{2H-2} \text{ as } j \to \infty \text{ with } c > 0;$$

(ii) Case $H = 1/2$:

$$\sum_{j=1}^{\infty} |r(j)| < \infty\,, \quad \sum_{j=-\infty}^{\infty} r(j) = c > 0;$$

(iii) Case $0 < H < 1/2$:

$$r(j) \sim cj^{2H-2} \text{ as } j \to \infty \text{ with } c < 0 \text{ and } \sum_{j=-\infty}^{\infty} r(j) = 0.$$

Then the finite-dimensional distributions of $\{N^{-H}\sum_{j=1}^{[Nt]} Z_j,\ 0 \le t \le 1\}$ converge to those of $\{\sigma_0 B_H(t),\ 0 \le t \le 1\}$ where $\{B_H(t),\ 0 \le t \le 1\}$ is standard fractional Brownian motion and where

$$\sigma_0^2 = \begin{cases} H^{-1}(2H-1)^{-1}c & \text{if } 1/2 < H < 1, \\ c & \text{if } H = 1/2, \\ -H^{-1}(2H-1)^{-1}c & \text{if } 0 < H < 1/2. \end{cases}$$

PROOF: Since the Z_js are mean zero Gaussian random variables, it is sufficient to show that for any constants $\theta_1,\ldots,\theta_d$, $d \ge 1$, and $t_1,\ldots,t_d \in [0,1]$,

$$E\Big(\frac{1}{N^H}\sum_{u=1}^{d}\theta_u \sum_{j=1}^{[Nt_u]} Z_j\Big)^2 = \sum_{u=1}^{d}\sum_{v=1}^{d}\theta_u\theta_v \frac{1}{N^{2H}} \sum_{j=1}^{[Nt_u]}\sum_{k=1}^{[Nt_v]} r(j-k)$$

converges as $N \to \infty$ to $E(\sum_{u=1}^{d}\theta_u \sigma_0 B_H(t_u))^2$. Hence, it is enough to show that for any $0 < s \le t \le 1$,

$$\lim_{N\to\infty} \frac{1}{N^{2H}} \sum_{j=1}^{[Ns]}\sum_{k=1}^{[Nt]} r(j-k) = \frac{\sigma_0^2}{2}\{s^{2H} + t^{2H} - (t-s)^{2H}\}.$$

Since, moreover, $r(j) = r(-j)$, we have

$$\begin{aligned} \sum_{j=1}^{[Ns]}\sum_{k=1}^{[Nt]} r(j-k) &= \frac{1}{2}\Big\{\sum_{j=1}^{[Ns]}\sum_{k=1}^{[Ns]} r(j-k) + \sum_{j=1}^{[Nt]}\sum_{k=1}^{[Nt]} r(j-k) \\ &\quad - \sum_{j=1}^{[Nt]-[Ns]}\ \sum_{k=1}^{[Nt]-[Ns]} r(j-k)\Big\} \end{aligned}$$

and $[Nt]-[Ns]\sim N(t-s)$ as $N\to\infty$, it is enough to prove

$$\lim_{N\to\infty}\frac{1}{N^{2H}}\sum_{j=1}^{N}\sum_{k=1}^{N}r(j-k)=\sigma_0^2.$$

Now

$$A_N\equiv\sum_{j=1}^{N}\sum_{k=1}^{N}r(j-k)=r(0)+\sum_{j=1}^{N-1}\{r(0)+2\sum_{i=1}^{j}r(i)\}.$$

When $1/2<H<1$, $\sum_{i=1}^{j}r(i)\sim(2H-1)^{-1}cj^{2H-1}$ as $j\to\infty$ and thus, as $N\to\infty$,

$$A_N\sim 2(2H-1)^{-1}c\sum_{j=1}^{N}j^{2H-1}\sim 2(2H)^{-1}(2H-1)^{-1}cN^{2H}$$

(Feller (1971), p. 281).

When $H=1/2$, $r(0)+2\sum_{i=1}^{\infty}r(j)=\sum_{j=-\infty}^{\infty}r(j)=\sigma_0^2>0$ and we have $A_N\sim N\sigma_0^2$ as $N\to\infty$.

The case $0<H<1/2$ is more delicate. Since $\sum_{j=-\infty}^{\infty}r(j)=0$, we have

$$r(0)+2\sum_{i=1}^{j}r(j)=-2\sum_{i=j+1}^{\infty}r(j)\sim-2(2H-1)^{-1}cj^{2H-1}$$

(Feller (1971), p. 281), and hence as $N\to\infty$,

$$A_N\sim-2(2H-1)^{-1}c\sum_{j=1}^{N}j^{2H-1}\sim-2(2H)^{-1}(2H-1)^{-1}cN^{2H}.\quad\blacksquare$$

There is another way to formulate the result of Theorem 7.2.11. Consider the stationary sequence $Z=\{Z_j,\ j=\ldots,-1,0,1,\ldots\}$ of that theorem and define, for each $N\geq 1$, the transformation

$$T_N\colon Z\to T_NZ=\{(T_NZ)_i,\ i=\ldots,-1,0,1,\ldots\},$$

where

$$(T_NZ)_i=\frac{1}{N^H}\sum_{j=iN+1}^{(i+1)N}Z_j,\ \ i=\ldots,-1,0,1,\ldots.$$

T_N transforms the original sequence Z into a new sequence T_NZ obtained by summing the components of Z over successive blocks of size N and renormalizing by

$$N^H=(\text{size of the block})^H.$$

The sequence of transformations $\{T_N\}_{N\geq 1}$ forms a semi-group because $T_N T_M = T_{NM}$. It is called the *renormalization group with index H*. The following corollary is a consequence of Theorem 7.2.11.

Corollary 7.2.12 *Let $Z = \{Z_j,\ j = \ldots,-1,0,1,\ldots\}$ and σ_0 be as in Theorem 7.2.11 and let $Y = \{Y_j,\ j = \ldots,-1,0,1,\ldots\}$ be standard fractional Gaussian noise. Then, as $N \to \infty$,*

$$T_N Z \xrightarrow{d} \sigma_0 Y;$$

i.e., the finite-dimensional distributions of $\{(T_N Z)_j,\ j = \ldots,-1,0,1,\ldots\}$ converge to those of $\{\sigma_0 Y_j,\ j = \ldots,-1,0,1,\ldots\}$.

Any sequence $Y = \{Y_j,\ j = \ldots,-1,0,1,\ldots\}$ satisfying

$$T_N Y \stackrel{d}{=} Y \text{ for all } N \geq 1 \tag{7.2.28}$$

is called a *fixed point* of the renormalization group. Relation (7.2.28) is sometimes used as definition of H-self-similarity for stationary sequences (see, e.g., Major (1981)).

Corollary 7.2.13 *Fractional Gaussian noise is the only Gaussian fixed point of the renormalization group.*

PROOF: That fractional Gaussian noise is a fixed point follows from Corollary 7.2.12. This fact can also be readily verified by noting that for any $\theta_1,\ldots,\theta_d,\ d \geq 1$, and $N \geq 1$,

$$\begin{aligned}
\sum_{i=1}^{d} \theta_i (T_N Y)_i &= \sum_{i=1}^{d} \theta_i \frac{1}{N^H} \sum_{j=iN+1}^{(i+1)N} Y_j \\
&= \sum_{i=1}^{d} \theta_i \frac{1}{N^H} (B_H((i+1)N) - B_H(iN)) \\
&\stackrel{d}{=} \sum_{i=1}^{d} \theta_i (B_H(i+1) - B_H(i)) \\
&= \sum_{i=1}^{d} \theta_i Y_i.
\end{aligned}$$

Fractional Gaussian noise is the unique fixed point because fractional Brownian motion is the unique Gaussian H-sssi process. ■

Although fractional Gaussian noise is not the only Gaussian sequence with long-range dependence, it is the only one which is a Gaussian fixed point of the renormalization group, and as such, it serves as an important paradigm.

7.3 General characteristics of processes that are α-stable and H-sssi

We now turn to α-stable H-sssi processes $\{X(t),\ t \in \mathbb{R}\}$ with $0 < \alpha < 2$. The admissible values of H are $H \in (0, 1/\alpha]$ if $0 < \alpha < 1$ and $H \in (0, 1]$ if $1 < \alpha < 2$ (Corollary 7.1.11). We will see in the next sections that, in contrast to the Gaussian case $\alpha = 2$, the property of "H-sssi" does not typically determine the finite-dimensional distributions. It imposes, however, a number of restrictions.

Lemma 7.3.1 *Let $\{X(t),\ t \in \mathbb{R}\}$ be α-stable, $0 < \alpha < 2$, and H-ss, and let Γ denote the spectral measure of the random vector $(X(t_1), \ldots, X(t_d))$ in $\mathbb{R}^d$. Then, for any $a > 0$, the spectral measure of the random vector $(X(at_1), \ldots, X(at_d))$ is $a^{\alpha H}\Gamma$.*

PROOF: Use Theorem 2.3.1. ∎

Lemma 7.3.2 *Let $\{X(t),\ t \in \mathbb{R}\}$ be α-stable with $0 < \alpha < 2$ and H-ss. Then $X(0) = 0$ and for fixed $t \in \mathbb{R}$, $X(t) \sim S_\alpha(\sigma_{X(t)}, \beta_{X(t)}, \mu_{X(t)})$ where*

$$\begin{aligned} \sigma_{X(t)} &= |t|^H \sigma_{X(1)}, \\ \beta_{X(t)} &= (\text{sign } t)\beta_{X(1)}, \\ \mu_{X(t)} &= \begin{cases} |t|^H (\text{sign } t)\mu_{X(1)} & \text{if } \alpha \neq 1, \\ |t|^H (\text{sign } t)(\mu_{X(1)} - \frac{2}{\pi}|t|^H (\ln |t|^H)\sigma_{X(1)}\beta_{X(1)}) & \text{if } \alpha = 1. \end{cases} \end{aligned}$$

PROOF: Use (1.1.6) and the relation $X(t) \stackrel{d}{=} |t|^H X(\text{sign } t)$, valid for fixed t. ∎

Proposition 7.3.3 *Let $X = \{X(t),\ t \in \mathbb{R}\}$ be an α-stable H-sssi process with $0 < \alpha < 2$.*

(i) If $H \neq 1$, then X is strictly stable.

(ii) If $H = 1$ and $\alpha \neq 1$, then there is a constant μ such that $\{X(t) - \mu t,\ t \in \mathbb{R}\}$ is strictly stable.

PROOF: Set $\sigma_t = \sigma_{X(t)}$, $\beta_t = \beta_{X(t)}$ and $\mu_t = \mu_{X(t)}$ and recall that the strict stability of the finite-dimensional distributions is equivalent to the strict stability of all the marginals (Corollary 2.4.2). We shall consider separately the three cases (i) $H \neq 1,\ \alpha \neq 1$, (ii) $H \neq 1\ \alpha = 1$, and (iii) $H = 1,\ \alpha \neq 1$.

(i) *Case* $H \neq 1,\ \alpha \neq 1$. Since $\alpha \neq 1$, we need to show that $\mu_t = 0$ for all t. The stationarity of the increments implies $X(2) - X(1) \stackrel{d}{=} X(1)$ and hence by

Example 2.3.4, $\mu_2 - \mu_1 = \mu_1$, i.e., $\mu_2 = 2\mu_1$. On the other hand, H-ss implies $\mu_2 = 2^H\mu_1$ by Lemma 7.3.2. Equating the last two relations yields $\mu_1 = 0$ and hence $\mu_t = 0$ for all $t \in \mathbb{R}$ by Lemma 7.3.2.

(ii) *Case* $H \neq 1$, $\alpha = 1$. Since $\alpha = 1$, it is enough to show that $\beta_t = 0$ for all t. We firstly show that $\beta_1 = 0$. Let Γ denote the spectral measure of the vector $(X(2), X(1))$. The relation $X(2) - X(1) \stackrel{d}{=} X(1)$ implies $\sigma_{X(2)-X(1)}\beta_{X(2)-X(1)} = \sigma_1\beta_1$. Since, on the other hand (see Example 2.3.4),

$$\begin{aligned}\sigma_{X(2)-X(1)}\beta_{X(2)-X(1)} &= \int_{S_2} |s_2 - s_1| \operatorname{sign}(s_2 - s_1)\Gamma(ds) \\ &= \int_{S_2} (s_2 - s_1)\Gamma(ds) \\ &= \sigma_2\beta_2 - \sigma_2\beta_1,\end{aligned}$$

we have $\sigma_2\beta_2 = 2\sigma_1\beta_1$. By H-ss, $\sigma_2\beta_2 = 2^H\sigma_1\beta_1$ and hence $2\sigma_1\beta_1 = 2^H\sigma_1\beta_1$, which is possible only if $\beta_1 = 0$, and thus $\beta_t = 0$ for all t by Lemma 7.3.2.

(iii) *Case* $H = 1$, $\alpha \neq 1$. Then $\{X(t) - \mu_t,\ t \in \mathbb{R}\}$ is strictly stable (Theorem 2.4.1) and $\mu_t = t\mu_1,\ t \in \mathbb{R}$ (Lemma 7.3.2). ∎

One can make no general statement in the case $H = 1$ and $\alpha = 1$. For example, if $X(1) \sim X_1(\sigma_1, \beta_1, \mu_1)$ with $\beta_1 \neq 0$, then the process $\{X(t) = tX(1),\ t \in \mathbb{R}\}$, is 1-stable, H-sssi but is not strictly stable, even if $\mu_1 = 0$, because $\beta_t = \beta_1 \neq 0$.

Corollary 7.3.4 *Let $\{X(t),\ t \in \mathbb{R}\}$ be α-stable, $0 < \alpha < 2$, and H-sssi. Fix $t_1, t_2,\ t \in \mathbb{R}$. Then*

(i) if $\alpha \neq 1$:

$$\begin{aligned}\sigma_{X(t_2)-X(t_1)} &= |t_2 - t_1|^H\sigma_{X(1)}, \\ \beta_{X(t)} &= (\operatorname{sign} t)\beta_{X(1)}, \\ \mu_{X(t)} &= \begin{cases} 0 & \text{if } H \neq 1, \\ t\mu_{X(1)} & \text{if } H = 1; \end{cases}\end{aligned}$$

(ii) if $\alpha = 1$ and $H \neq 1$:

$$\begin{aligned}\sigma_{X(t_2)-X(t_1)} &= |t_2 - t_1|^H\sigma_{X(1)}, \\ \beta_{X(t)} &= 0, \\ \mu_{X(t)} &= |t|^H(\operatorname{sign} t)\mu_{X(1)},\end{aligned}$$

that is,

$$\mu_{X(t)} = -|t|^H (\text{sign } t)\frac{1}{2-2^H}\frac{2}{\pi}\int_{S_2}(s_2 - s_1)\ln|s_2 - s_1|\Gamma_{X(2),X(1)}(ds).$$

PROOF: The only new statement concerns $\mu_{X(t)}$ when $\alpha = 1$ and $H \neq 1$, which we now verify. Since $\beta_{X(1)} = 0$ (Proposition 7.3.3), we have $\mu_{X(t)} = |t|^H(\text{sign } t)\mu_{X(1)}$ by Lemma 7.3.2. To devise an expression for $\mu_{X(1)}$, use

$$\mu_{X(2)-X(1)} = \mu_{X(2)} - \mu_{X(1)} - \frac{2}{\pi}\int_{S_2}(s_2 - s_1)\ln|s_2 - s_1|\Gamma_{X(2),X(1)}(ds),$$

Relation (2.3.5), $\mu_{X(2)-X(1)} = \mu_{X(1)}$ and $\mu_{X(2)} = 2^H\mu_{X(1)}$. Combining these expressions yields the desired result. ∎

When $H \neq 1$, the parameter $\mu_{X(t)}$ depends on α and H, and also on $\Gamma_{X(2),X(1)}$ if $\alpha = 1$. What is important, however, is the following implication:

Corollary 7.3.5 *If $\{X(t),\ t \in \mathbb{R}\}$ is α-stable H-sssi with $H \neq 1$, then $\{X(t) + \mu_t^0,\ t \in T\}$ cannot be H-sssi unless the "shift" μ_t^0 is identically equal to zero.*

H-sssi processes are typically defined by their integral representation, a more "physical" quantity than the joint characteristic function. Suppose, then, that the α-stable H-sssi process $\{X(t),\ t \in \mathbb{R}\}$ has the integral representation

$$X(t) = \int_E f_t(x)M(dx),\ t \in \mathbb{R},$$

where M is an α-stable measure with skewness intensity $\beta(x)$, $x \in E$, and control measure m and $\int_E |f_t(x)|^\alpha m(dx) < \infty$ for all $t \in \mathbb{R}$. [7] What restrictions on f_t, $t \in \mathbb{R}$, does the condition of "H-sssi" impose?

Proposition 7.3.6 *Suppose $\beta(\cdot) = 0$ if $\alpha = 1$. Then the α-stable process $\{X(t),\ t \in \mathbb{R}\}$ is H-sssi if and only if for any integer $d \geq 1$ and real numbers θ_j, t_j, $j = 1, \ldots, d$, the integrals*

$$a^{-\alpha H}\int_E\left|\sum_{j=1}^{d}\theta_j(f_{at_j+h}(x) - f_h(x))\right|^\alpha m(dx)$$

[7] When $\alpha = 1$, we must also suppose

$$\int_E |f(x)\beta(x)\ln|f(x)||m(dx) < \infty,$$

but this will not be necessary because, in the sequel, we will always set $\beta(\cdot) = 0$ when $\alpha = 1$.

and

$$a^{-\alpha H}\int_E\Big[\sum_{j=1}^d \theta_j(f_{at_j+h}(x)-f_h(x))\Big]^{<\alpha>}\beta(x)m(dx)$$

are independent of $a>0$ *and* $h\in\mathbb{R}$.

PROOF: H-ss requires $\sum_{j=1}^d \theta_j X(at_j)\stackrel{d}{=}a^H\sum_{j=1}^d\theta_j X(t_j)$ and the stationarity of the increments requires the distribution of $\sum_{j=1}^d\theta_j(X(t_j+h)-X(h))$ to be independent of h. Now use the expression for the characteristic function given in (3.2.2). ■

Remarks

1. Different representations may define the same process and, hence, in the sequel it will be important to know when this happens.

2. In the case $\alpha=1$, we choose the skewness intensity $\beta(\cdot)=0$. Then the process X is, in fact, $S1S$.

7.4 Linear fractional stable motion

There are many different extensions of fractional Brownian motion to the α-stable case. The one which is most commonly used is the *linear fractional stable motion* (also called *linear fractional Lévy motion*). It is defined as follows.

Definition 7.4.1 The *linear fractional stable motion* (LFSM) is the stochastic process $\{L_{\alpha,H}(a,b;t),-\infty<t<\infty\}$ given by

$$L_{\alpha,H}(a,b;t)=\int_{-\infty}^{\infty}f_{\alpha,H}(a,b;t,x)M(dx), \tag{7.4.1}$$

where

$$\begin{aligned} f_{\alpha,H}(a,b;t,x) &= a\Big(((t-x)_+)^{H-1/\alpha}-((-x)_+)^{H-1/\alpha}\Big)\\ &\quad +b\Big(((t-x)_-)^{H-1/\alpha}-((-x)_-)^{H-1/\alpha}\Big), \end{aligned} \tag{7.4.2}$$

and where a,b are real constants, $|a|+|b|>0, 0<\alpha<2$, $0<H<1, H\neq 1/\alpha$, and M is an α-stable random measure on $\mathbb{R}$ with Lebesgue control measure and skewness intensity $\beta(x)$, $-\infty<x<\infty$, satisfying:

(i) $\beta(\cdot)=0$ if $\alpha=1$,

(ii) for all integers $d \geq 1$ and real $\theta_j, t_j,\ j = 1, \ldots, d$,

$$\int_{-\infty}^{\infty} \Big(\sum_{j=1}^{d} \theta_j \big(f_{\alpha,H}(a,b;t_j,x) - f_{\alpha,H}(a,b;0,x) \big) \Big)^{<\alpha>} \beta(cx+h)dx \quad (7.4.3)$$

is independent of $c > 0$ and $-\infty < h < \infty$.

The representation of LFSM is based on the representation (7.2.7) of fractional Brownian motion. Here, the exponent is $H - 1/\alpha$ instead of $H - 1/2$. Therefore LFSM reduces to fractional Brownian motion if one sets $\alpha = 2$. Figures 7.2 and 7.3 show the kernel $f_{\alpha,H}(a,b;t,x)$ for $a = 1,\ b = 0$ and various values of t and $d = H - 1/\alpha$; and Figure 7.10 shows it for $a = b = 1,\ t = 1$ and $d = H - 1/\alpha = \pm 0.05$.

Remarks

1. The process $\{L_{\alpha,H}(a,b;t),\ -\infty < t < \infty\}$ is well defined because $\int_{-\infty}^{\infty} |f_{\alpha,H}(a,b;t,x)|^{\alpha} dx < \infty$, as in Example 3.6.5.

2. $L_{\alpha,H}(a,b;t)$ is not defined for $H = 1$ because $\int_{-\infty}^{\infty} |f_{\alpha,1}(a;b;t,x)|^{\alpha} dx = \infty,\ t \neq 0$. Hence $0 < H < 1$ is a necessary condition for $L_{\alpha,H}$ to be well defined.

3. The *well-balanced linear fractional stable motion*

$$L_{\alpha,H}(1,1;t) = \int_{-\infty}^{\infty} \left(|t-x|^{H-1/\alpha} - |x|^{H-1/\alpha} \right) M(dx) \qquad (7.4.4)$$

introduced in Example 3.6.5 corresponds to the choice $a = b = 1$.

4. The $S\alpha S$ case corresponds to $\beta(\cdot) = 0$.

5. We suppose $\beta(\cdot) = 0$ when $\alpha = 1$ in order to provide a simple condition to ensure that the process is strictly stable.

6. Condition (7.4.3) is, in general, very difficult to verify. In practice, we take β constant, typically $\beta(\cdot) = 1$. The condition then holds trivially with such a choice of β.

Proposition 7.4.2 *The LFSM is H-sssi.*

PROOF: It is sufficient to verify the conditions of Proposition 7.3.6. For example, for any $c > 0$ and $-\infty < h < \infty$,

$$c^{-\alpha H} \int_{-\infty}^{\infty} \Big(\sum_{j=1}^{d} \theta_j (f_{\alpha,H}(a,b;ct_j+h,x) - f_{\alpha,H}(a,b;h,x))\Big)^{<\alpha>} \beta(x)dx$$

$$= c^{-\alpha H} \int_{-\infty}^{\infty} \left(\sum_{j=1}^{d} \theta_j (f_{\alpha,H}(a,b;ct_j,cx) - f_{\alpha,H}(a,b;0,cx))\right)^{<\alpha>} \beta(cx+h)dcx$$

$$= \int_{-\infty}^{\infty} \left(\sum_{j=1}^{d} \theta_j (f_{\alpha,H}(a,b;t_j,x) - f_{\alpha,H}(a,b;0,x))\right)^{<\alpha>} \beta(cx+h)dx$$

which does not depend on c, by Assumption (7.4.3) if $\alpha \neq 1$ and by the choice $\beta = 0$ when $\alpha = 1$. ■

Remark. By analogy to the case $\alpha = 2$, we say that the increments of LFSM have *long-range dependence* when $H > 1/\alpha$ and *negative dependence* when $H < 1/\alpha$. There is no long-range dependence when $0 < \alpha \leq 1$ because H is constrained to lie in the interval $(0,1)$. The value $H = 1/\alpha$ lies on the boundary between long-range and negative dependence. Processes with $H = 1/\alpha$ will be considered in Sections 7.5 and 7.6.

Proposition 7.4.3 *If $\{X(t),\ -\infty < t < \infty\}$ is LFSM, then for each fixed $t \in \mathbb{R}$, $X(t)$ has a $S_\alpha(\sigma_t, \beta_t, 0)$ distribution, where*

$$\sigma_t = K_{\alpha,H}(a,b)|t|^H$$

and

$$K_{\alpha,H}^{\alpha}(a,b) = (|a|^\alpha + |b|^\alpha) \int_0^\infty \left((1+x)^{H-1/\alpha} - x^{H-1/\alpha}\right)^\alpha dx$$
$$+ \int_0^1 |a(1-x)^{H-1/\alpha} - b\,x^{H-1/\alpha}|^\alpha dx. \qquad (7.4.5)$$

If, moreover, the skewness intensity $\beta(\cdot)$ equals a constant β, then

$$\beta_t = \frac{\beta}{K_{\alpha,H}^{\alpha}(a,b)} (a^{<\alpha>} - b^{<\alpha>}) \int_0^\infty \left((1+x)^{H-1/\alpha} - x^{H-1/\alpha}\right)^\alpha dx$$
$$+ \int_0^1 \left(a(1-x)^{H-1/\alpha} - bx^{H-1/\alpha}\right)^{<\alpha>} dx.$$

PROOF: We establish the expression for β_t. (The expression for σ_t is obtained in a similar fashion.) Assume, without loss of generality, that $t > 0$. If $\beta(x) = \beta$, then, by Property 3.2.2,

$$\begin{aligned}\beta_t &= \frac{1}{\sigma_t^\alpha}\int_{-\infty}^{\infty}(f_{\alpha,H}(a,b;t,x))^{<\alpha>}\beta(x)dx \\ &= \frac{\beta}{\sigma_t^\alpha}\int_{-\infty}^{\infty}(f_{\alpha,H}(a,b;t,x))^{<\alpha>}dx.\end{aligned}$$

Now use the expression (7.4.2) for $f_{\alpha,H}(a,b;t,x)$. If $\beta \neq 0$,

$$\begin{aligned}\sigma_t^\alpha\frac{\beta_t}{\beta} &= a^{<\alpha>}\int_{-\infty}^{0}\left((t-x)^{H-1/\alpha}-(-x)^{H-1/\alpha}\right)^\alpha dx \\ &\quad+\int_0^t\left(a(t-x)^{H-1/\alpha}-b\,x^{H-1/\alpha}\right)^{<\alpha>}dx \\ &\quad+b^{<\alpha>}\int_t^\infty\left((x-t)^{H-1/\alpha}-x^{H-1/\alpha}\right)^{<\alpha>}dx.\end{aligned}$$

The third integral can be written as

$$-\int_t^\infty\left(x^{H-1/\alpha}-(x-t)^{H-1/\alpha}\right)^\alpha dx = -\int_0^\infty\left((t+u)^{H-1/\alpha}-u^{H-1/\alpha}\right)du,$$

so that

$$\begin{aligned}\sigma_t^\alpha\frac{\beta_t}{\beta} = |t|^{\alpha H}\Big\{&\left(a^{<\alpha>}-b^{<\alpha>}\right)\int_0^\infty\left((1+x)^{H-1/\alpha}-x^{H-1/\alpha}\right)^\alpha dx \\ &+\int_0^1\left(a(1-x)^{H-1/\alpha}-b\,x^{H-1/\alpha}\right)^{<\alpha>}dx\Big\}. \quad \blacksquare\end{aligned}$$

The process

$$\left\{X(t)=\frac{1}{K_{\alpha,H}(a,b)}L_{\alpha,H}(a,b,t),\ -\infty<t<\infty\right\},$$

where the constant $K_{\alpha,H}(a,b)$ is given in (7.4.5), is called a *standard linear fractional stable motion* (standard LFSM). It is normalized in the sense that the scaling parameter of $X(1)$ equals 1.

We want to show now that the (standard) linear fractional stable motions $\{L_{\alpha,H}(a,b;t),\ t\in\mathbb{R}\}$ are typically different for different a and b. The proof uses the following result (Lemma 4 of Kanter (1973)).

Lemma 7.4.4 *(Kanter). Let $f, g \in L^\alpha(\mathbb{R})$. If for some $0 < \alpha < 2$, any integer d, and any real numbers $\theta_1, \ldots, \theta_d,\ t_1, \ldots, t_d$,*

$$\int_{-\infty}^{\infty} \Big|\sum_{j=1}^{d} \theta_j f(t_j + x)\Big|^\alpha dx = \int_{-\infty}^{\infty} \Big|\sum_{j=1}^{d} \theta_j g(t_j + x)\Big|^\alpha dx,$$

then there exists a number ϵ equal to -1 or $+1$ and a real number τ such that

$$f(x) = \epsilon g(x + \tau) \text{ a.e. } x.$$

Theorem 7.4.5 *Let $0 < \alpha < 2, 0 < H < 1,\ H \neq 1/\alpha$, and let a, a', b, b' be real numbers satisfying $|a| + |b| > 0$ and $|a'| + |b'| > 0$. Then the standard LFSMs*

$$\frac{1}{K_{\alpha,H}(a,b)} L_{\alpha,H}(a, b; t)\ ,\ -\infty < t < \infty,$$

and

$$\frac{1}{K_{\alpha,H}(a',b')} L_{\alpha,H}(a', b'; t)\ ,\ -\infty < t < \infty,$$

have identical finite-dimensional distributions if and only if one of the following conditions holds:

(i) $a = a' = 0$,

(ii) $b = b' = 0$,

(iii) a, a', b, b' are non-zero and $a/b = a'/b'$.

PROOF: Suppose that one of the three conditions holds, for example, (iii). Then

$$\frac{1}{K_{\alpha,H}(a,b)} L_{\alpha,H}(a, b; t) \stackrel{d}{=} \frac{b}{K_{\alpha,H}(a,b)} L_{\alpha,H}\left(\frac{a}{b}, 1; t\right)$$

has same finite-dimensional distributions as

$$\frac{1}{K_{\alpha,H}(a',b')} L_{\alpha,H}(a', b'; t) \stackrel{d}{=} \frac{b'}{K_{\alpha,H}(a',b')} L_{\alpha,H}\left(\frac{a}{b}, 1; t\right)$$

because they differ by a multiplicative constant which must be equal to 1 since both processes have been normalized.

To prove the converse, suppose

$$K_{\alpha,H}^{-1}(a,b) L_{\alpha,H}(a, b; t) \quad \text{and} \quad K_{\alpha,H}^{-1}(a',b') L_{\alpha,H}(a', b'; t)$$

have the same finite-dimensional distributions. Then the stationary processes

$$Y(a, b, t) = K_{\alpha,H}^{-1}(a,b)[L_{\alpha,H}(a, b; t + h) - L_{\alpha,H}(a, b; t)],\ -\infty < t < \infty,$$

and

$$Y(a',b',t) = K_{\alpha,H}^{-1}(a',b')[L_{\alpha,H}(a',b';t+h) - L_{\alpha,H}(a',b';t)],\ -\infty < t < \infty,$$

also have identical finite-dimensional distributions, for any given $h > 0$. We want to show that this implies that one of the conditions (i)-(iii) holds. The process $Y(a,b;t)$ can be represented as

$$Y(a,b;t) = K_{\alpha,H}^{-1}(a,b)\int_{-\infty}^{\infty} g(a,b;t-x)M(dx),$$

where

$$g(a,b;t) = a[(t+h)_+^{H-1/\alpha} - t_+^{H-1/\alpha}] + b[(t+h)_-^{H-1/\alpha} - t_-^{H-1/\alpha}].$$

Since $Y(a,b;t)$ and $Y(a',b';t)$ have the same finite-dimensional distributions,

$$\left|E\exp\Big\{i\sum_{j=1}^{d}\theta_j Y(a,b;t_j)\Big\}\right| = \left|E\exp\Big\{i\sum_{j=1}^{d}\theta_j Y(a',b';t_j)\Big\}\right|,$$

i.e.,

$$\exp\Big\{-\int_{-\infty}^{\infty}\Big|\sum_{j=1}^{d}\theta_j K_{\alpha,H}^{-1}(a,b)g(a,b;t_j-x)\Big|^{\alpha}dx\Big\}$$
$$= \exp\Big\{-\int_{-\infty}^{\infty}\Big|\sum_{j=1}^{d}\theta_j K_{\alpha,H}^{-1}(a',b')g(a',b';t_j-x)\Big|^{\alpha}dx\Big\},$$

for all integers d and real numbers $\theta_1,\ldots,\theta_d,t_1,\ldots,t_d$. Hence, by Kanter's Lemma 7.4.4, there exists a real number τ and an $\epsilon \in \{+1,-1\}$ depending on h,a,b,a',b' such that

$$K_{\alpha,H}^{-1}(a,b)g(a,b;t) = \epsilon K_{\alpha,H}^{-1}(a',b')g(a',b';t-\tau) \text{ a.e. } t,$$

i.e.,

$$K_{\alpha,H}^{-1}(a,b)\Big[a[(t+h)_+^{H-1/\alpha} - t_+^{H-1/\alpha}] + b[(t+h)_-^{H-1/\alpha} - t_-^{H-1/\alpha}]\Big]$$
$$= \epsilon K_{\alpha,H}^{-1}(a',b')\Big[a'[(t+h-\tau)_+^{H-1/\alpha} - (t-\tau)_+^{H-1/\alpha}]$$
$$+ b'[(t+h-\tau)_-^{H-1/\alpha} - (t-\tau)_-^{H-1/\alpha}]\Big] \text{ a.e. } t. \tag{7.4.6}$$

Since both sides of (7.4.6) represent functions continuous everywhere except at the four points $t = -h, 0, \tau-h, \tau$, Relation (7.4.6) can be extended by continuity to all t not equal to these four points.

We claim that this implies

$$\frac{a}{K_{\alpha,H}(a,b)} = \epsilon\frac{a'}{K_{\alpha,H}(a',b')}, \quad \frac{b}{K_{\alpha,H}(a,b)} = \epsilon\frac{b'}{K_{\alpha,H}(a',b')}. \tag{7.4.7}$$

Indeed, differentiating (7.4.6) with respect to $t \neq -h, 0, \tau - h, \tau$, we obtain

$$K^{-1}_{\alpha,H}(a,b)\Big[a[(t+h)_+^{H-1/\alpha-1} - t_+^{H-1/\alpha-1}] - b[(t+h)_-^{H-1/\alpha-1} - t_-^{H-1/\alpha-1}]\Big]$$

$$= \epsilon K^{-1}_{\alpha,H}(a',b')\Big[a'[(t+h-\tau)_+^{H-1/\alpha-1} - (t-\tau)_+^{H-1/\alpha-1}]$$

$$- b'[(t+h-\tau)_-^{H-1/\alpha-1} - (t-\tau)_-^{H-1/\alpha-1}]\Big] \tag{7.4.8}$$

for all t except $t = -h, 0, \tau - h, \tau$. Suppose, without loss of generality, that $a \neq 0$ and let $t \downarrow 0$. Then the left-hand side of (7.4.8) tends to $\pm\infty$. For the right-hand side to do the same it is necessary that $\tau = 0$ or $\tau = h$. To see that $\tau \neq h$, set $\tau = h$ and let $t \downarrow h$. This makes the left-hand side of (7.4.8) tend to a finite number and the right-hand side tend to $\pm\infty$, which is not possible. Hence $\tau = 0$ and (7.4.7) holds. Relation (7.4.7) implies that one of the following

(i) $a = a' = 0$
(ii) $b = b' = 0$
(iii) a, a', b, b' are non-zero and

$$\epsilon\frac{K_{\alpha,H}(a,b)}{K_{\alpha,H}(a',b')} = \frac{a}{a'} = \frac{b}{b'}. \quad \blacksquare$$

7.5 α-stable Lévy motion

The linear fractional stable motion (LFSM) was defined for $H \neq 1/\alpha$. Its definition can be extended to the case $H = 1/\alpha$ in two different ways. One leads to the *α-stable Lévy motion* (discussed here), the other to the *log-fractional stable motion* (discussed in the following section).

Definition 7.5.1 An *α-stable Lévy motion*, $0 < \alpha < 2$, is a process $\{X(t),\ t \in \mathbb{R}\}$ with stationary independent increments having a strictly α-stable distribution.

By independent increments, we mean, as usual, that the random variables $X(t_j) - X(t_{j-1})$, $j = 2, \ldots, d$, are mutually independent for any $-\infty < t_1 < t_2 < \cdots < t_d < \infty$, $d \geq 3$.

Theorem 7.5.2 *Let $0 < \alpha < 2$ and let M be an α-stable random measure on $\mathbb{R}$ with Lebesgue control measure and constant skewness intensity $\beta(\cdot)$. (Assume $\beta(\cdot) = 0$ if $\alpha = 1$.) Then*

$$X(t) = \begin{cases} \int_0^t M(dx) & t \geq 0, \\ \int_{-t}^0 M(dx) & t < 0, \end{cases} \tag{7.5.1}$$

is α-stable Lévy motion.

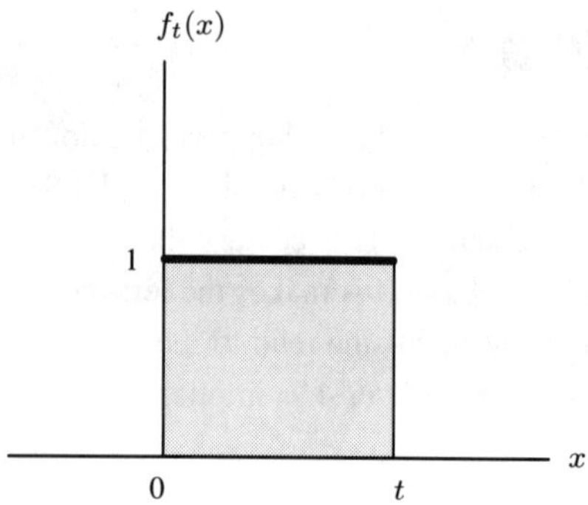

Figure 7.8: Integration kernel for Brownian motion ($\alpha = 2$) and α-stable Lévy motion ($0 < \alpha < 2$) when $t \geq 0$.

Remarks

1. The choice $\beta(\cdot)$ constant and $\beta(\cdot) = 0$ if $\alpha = 1$ is made for practical reasons. X, with $\beta(\cdot) = 0$, is the $S\alpha S$ Lévy motion introduced in Example 7.1.3.

2. The definition extends to $\alpha = 2$. The 2-stable Lévy motion is obviously Brownian motion.

3. For $t \geq 0$, Brownian motion ($\alpha = 2$) and α-stable Lévy motion ($0 < \alpha < 2$) can be represented as $\int_{-\infty}^{\infty} 1_{[0,t]}(x) M(dx) = M([0,t])$ (see Figure 7.8).

The proofs of the previous theorem and of the following proposition are straightforward.

Proposition 7.5.3 *The α-stable Lévy motion is H-sssi with*

$$H = \frac{1}{\alpha} \in \left(\frac{1}{2}, \infty\right).$$

By allowing the value $H = 1/\alpha$ and using the convention $0^\gamma = 0$ for all values of γ, Definition 7.4.1 of LFSM can be extended to include the α-stable motion. Indeed, for $t > 0$,

$$\begin{aligned} &f_{\alpha,1/\alpha}(a, b; t, x) \\ &\quad = a(1_{(-\infty,t)}(x) - 1_{(-\infty,0)}(x)) + b(1_{(t,\infty)}(x) - 1_{(0,\infty)}(x)) \\ &\quad = a1_{[0,t)}(x) - b\, 1_{[0,t]}(x). \end{aligned}$$

Integrating this kernel with respect to the random measure M, we obtain $(a - b) \int_0^t M(dx)$. This differs from (7.5.1) only by the multiplicative factor $a - b$.

Setting, in the same way, $H = 1/2$ in the representation (7.2.4) of fractional Brownian motion yields Brownian motion.

Observe that although the α-stable Lévy motion is H-sssi with $H = 1/\alpha \in (1/2, \infty)$, the self-similarity parameter H of the LFSM restricted to the interval $(0, 1)$. Observe also that Theorem 7.4.5 is not valid for α-stable Lévy motion: the parameters a and b of that process yield here only a multiplicative factor $a - b$.

Are there other α-stable $(1/\alpha)$-sssi processes, besides α-stable Lévy motion? The following theorem shows that the answer is in the negative when $0 < \alpha < 1$.

Theorem 7.5.4 *The only non-degenerate α-stable $1/\alpha$-sssi processes with $0 < \alpha < 1$ are the α-stable Lévy motions.*

PROOF: Let $\{X(t),\ t \in \mathbb{R}\}$ be a non-degenerate (i.e., $X(1) \neq 0$ a.s.) α-stable $1/\alpha$-sssi process with $0 < \alpha < 1$. Corollary 7.3.4 implies that $\{X(t),\ t \in \mathbb{R}\}$ is strictly α-stable. Let σ_t denote the scaling parameter of the α-stable random variable $X(t)$. Fix arbitrary $s_1 < s_2 \le t_1 < t_2$. The random variables $X(s_1),\ X(s_2),\ X(t_1)$ and $X(t_2)$ are jointly strictly α-stable, and thus, by the representation theorem 3.5.6, there are functions $f_{s_1}, f_{s_2}, f_{t_1}$ and f_{t_2} in $L^\alpha([0, 1])$ such that

$$(X(s_1), X(s_2), X(t_1), X(t_2)) \stackrel{d}{=} \left(\int_0^1 f_j(x) M(dx),\ j = s_1, s_2, t_1, t_2 \right),$$

where M is an independently scattered α-stable measure on $([0, 1], \mathcal{B})$ with Lebesgue control measure and skewness intensity $\beta(\cdot) = 1$. Then

$$(t_2 - s_1)\sigma_1^\alpha = \sigma_{t_2 - s_1}^\alpha = \int_0^1 |f_{t_2}(x) - f_{s_1}(x)|^\alpha dx$$

$$\le \int_0^1 |f_{s_2}(x) - f_{s_1}(x)|^\alpha dx + \int_0^1 |f_{t_1}(x) - f_{s_2}(x)|^\alpha dx + \int_0^1 |f_{t_2}(x) - f_{t_1}(x)|^\alpha dx \tag{7.5.2}$$

$$= \sigma^\alpha_{s_2-s_1} + \sigma^\alpha_{t_1-s_2} + \sigma^\alpha_{t_2-t_1} = (s_2 - s_1)\sigma_1^\alpha + (t_1 - s_2)\sigma_1^\alpha + (t_2 - t_1)\sigma_1^\alpha$$
$$= (t_2 - s_1)\sigma_1^\alpha.$$

Here we have used the self-similarity and the stationarity of the increments of $\{X(t),\ t \in \mathbb{R}\}$ (Corollary 7.3.4). Thus, the inequality in (7.5.2) is actually an equality, implying

$$(f_{s_2}(x) - f_{s_1}(x))(f_{t_2}(x) - f_{t_1}(x)) = 0 \text{ a.e.}$$

It follows from Theorem 3.5.3 that $X(s_2) - X(s_1)$ and $X(t_2) - X(t_1)$ are independent for any $s_1 < s_2 \le t_1 < t_2$ and, since for jointly stable random variables pairwise independence is equivalent to total independence (Corollary 3.5.7), we conclude that $\{X(t),\ t \in \mathbb{R}\}$ has independent increments. Thus $\{X(t),\ t \in \mathbb{R}\}$ is an α-stable Lévy motion. ∎

When $1 \le \alpha < 2$, the α-stable Lévy motion is not the only $1/\alpha$-sssi process. If $\alpha = 1$, the process $X(t) = tX$, where X is a 1-stable random variable, is 1-sssi. If $1 < \alpha < 2$, there is the log-fractional stable motion defined in the following section and also the processes defined in Relation (7.14.1) (see Exercise 7.12).

7.6 Log-fractional stable motion

We indicated at the beginning of Section 7.5 that there are several ways to extend Definition 7.4.1 of LFSM to the case $H = 1/\alpha$. One way yields the α-stable Lévy motion. Here is a different extension: set $\alpha > 1$ and $a = b = 1$ and consider

$$\frac{1}{H - 1/\alpha} L_{\alpha,H}(1,1,t) = \int_{-\infty}^{\infty} \Big\{ \frac{|t-x|^{H-1/\alpha} - 1}{H - 1/\alpha} - \frac{|x|^{H-1/\alpha} - 1}{H - 1/\alpha} \Big\} M(dx),$$

The fact that for any real $u \ne 0$,

$$\lim_{H \to 1/\alpha} \frac{|u|^{H-1/\alpha} - 1}{H - 1/\alpha} = \lim_{H \to 1/\alpha} \frac{e^{(H-1/\alpha)\ln|u|} - 1}{H - 1/\alpha} = \ln|u|$$

suggests the following definition:

Definition 7.6.1 Set $1 < \alpha < 2$ and let M be defined as in Definition 7.5.1 (with $\beta(\cdot) = \beta$ constant). The process

$$\Lambda_{\alpha,1/\alpha}(t) = \int_{-\infty}^{\infty} \Big(\ln|t - x| - \ln|x| \Big) M(dx), \quad -\infty < t < \infty, \tag{7.6.1}$$

is called *log-fractional stable motion* (log-FSM).[8]

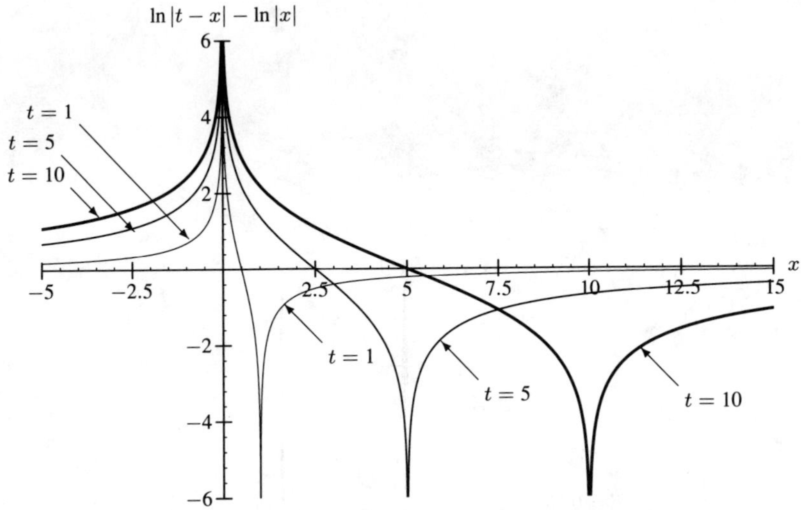

Figure 7.9: Kernel of the log-fractional stable motion (see (7.6.1)) plotted for several values of t.

The kernel is illustrated in Figure 7.9 and it is compared in Figure 7.10 to that of the well-balanced fractional stable motion. To verify that the process $\Lambda_{\alpha,1/\alpha}(t)$ is well defined, we will check that its integrand belongs to L^α. The function $\ln^\alpha |x|$ is integrable around $x = 0$, and, as $x \to \infty$, the integrand behaves like $(1/x)^\alpha$, which is an integrable function for $\alpha > 1$. The process $\{\Lambda_{\alpha,1/\alpha}(t),\ t \in \mathbb{R}\}$ is therefore well defined.

Remark. The log-fractional stable motion is not defined for $\alpha \le 1$ because $x^{-\alpha}$ is not integrable as $x \to \infty$.

Proposition 7.6.2 *The log-fractional stable motion is H-sssi with $H = 1/\alpha$.*

PROOF: Since $H = 1/\alpha$ and β is constant, the conditions of Proposition 7.3.6 are satisfied. For example for any $c > 0$ and $-\infty < h < \infty$,

$$c^{-\alpha H} \int_{-\infty}^{\infty} \Big[\sum_{j=1}^{d} \theta_j (\ln |ct_j + h - x| - \ln |h - x|)\Big]^{<\alpha>} \beta dx$$

$$= \int_{-\infty}^{\infty} \Big[\sum_{j=1}^{d} \theta_j (\ln |ct_j - cx| - \ln |cx|)\Big]^{<\alpha>} \beta dx$$

$$= \int_{-\infty}^{\infty} \Big[\sum_{j=1}^{d} \theta_j (\ln |t_j - x| - \ln |x|)\Big]^{<\alpha>} \beta dx$$

[8] It is sometimes called *log-fractional Lévy motion.*

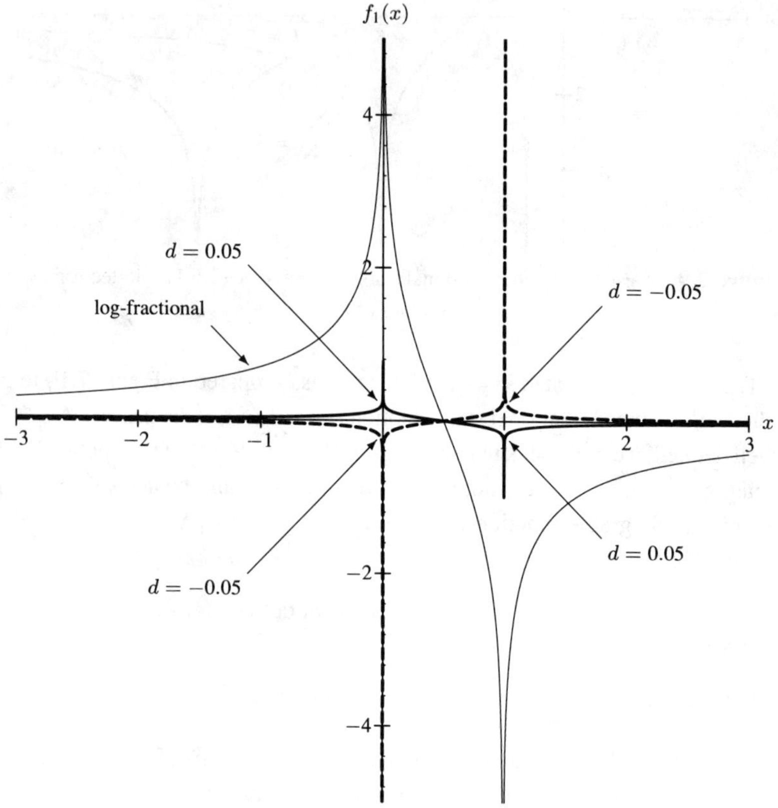

Figure 7.10: Comparison, at $t = 1$, of the kernel $\ln|t-x| - \ln|x|$ of the log-fractional stable motion and the kernels $|t-x|^d - |x|^d$, $d = H - 1/\alpha$, of the well-balanced fractional stable motion for small values of d.

is independent of c and h. ∎

We noted that the definition of the log-fractional stable motion cannot be extended to $\alpha \le 1$. It can be extended, however, to $\alpha = 2$, but in that case, one does not obtain anything new. Since the resulting process is Gaussian and H-sssi with $H = 1/2$, it must be Brownian motion.

When $1 < \alpha < 2$, the log-fractional stable motion does not have independent increments and therefore is not the α-stable Lévy motion. To see that the increments are dependent, set $\Lambda_{\alpha,1/\alpha}(1) = \int_{-\infty}^{\infty} f(x)M(dx)$ and $\Lambda_{\alpha,1/\alpha}(2) - \Lambda_{\alpha,1/\alpha}(1) = \int_{-\infty}^{\infty} g(x)M(dx)$ and observe that one does not have $fg = 0$ a.e. (Theorem 3.5.3).

Finally, note that Definition 7.6.1 of log-fractional stable motion involves a "well-balanced" representation analogous to that of the linear fractional stable motion (7.4.2) with $a = b = 1$. A representation that is not "well-balanced" would not yield a self-similar process. Indeed, set $\ln_0 x$ equal $\ln x$ if $x > 0$ and 0 if $x \le 0$. Let a and b be two real numbers. Then the process

$$\int_{-\infty}^{\infty} \Big(a\big[\ln_0(t-x)_+ - \ln_0(-x)_+\big] + b\big[\ln_0(t-x)_- - \ln_0(-x)_-\big]\Big) M(dx)$$

is not self-similar unless $a \equiv b$, in which case we have $a\Lambda_{\alpha,1/\alpha}(t)$. For example, if $a = 1$ and $b = 0$, the scale parameter σ_t of the process at a fixed $t > 0$ satisfies

$$\begin{aligned}
\sigma_t^\alpha &= \int_0^\infty [\ln(t+x) - \ln x]^\alpha dx + \int_0^t |\ln x|^\alpha dx \\
&= t\int_0^\infty [\ln(1+x) - \ln x]^\alpha dx + t\int_0^1 |\ln x + \ln t|^\alpha dx \\
&\neq t\sigma_1^\alpha.
\end{aligned}$$

7.7 The real harmonizable fractional stable motion

We saw in Proposition 7.2.8 that fractional Brownian motion has the "harmonizable representation" $\int_{-\infty}^{\infty} \frac{e^{ixt}-1}{ix}|x|^{-(H-1/2)}\widetilde{M}(dx)$ where $\widetilde{M} = M^{(1)} + iM^{(2)}$ is a Gaussian complex measure whose real and imaginary parts $M^{(1)}$ and $M^{(2)}$ are independent with Lebesgue control measure.

The corresponding representation in the case $\alpha < 2$ involves a random measure $\widetilde{M}$ whose real and imaginary parts $M^{(1)}$ and $M^{(2)}$ are dependent. The random measure $\widetilde{M}$, is, however, isotropic, i.e., its circular control measure k on $\mathbb{R} \times S_2$ equals $m\gamma$, where γ is the uniform probability measure on S_2. (Refer to Section 6.3 for details.) Here, m is the Lebesgue measure on $\mathbb{R}$.

Definition 7.7.1 The *real harmonizable fractional stable motion* (RHFSM) is the process

$$\Big\{\mathcal{H}_{\alpha,H}(t) = \text{Re}\int_{-\infty}^{\infty}\frac{e^{ixt}-1}{ix}|x|^{-H+1-1/\alpha}\widetilde{M}(dx),\ t\in\mathbb{R}\Big\}, \tag{7.7.1}$$

where $0<\alpha\le 2,\ 0<H<1$ and where $\widetilde{M}$ is a complex isotropic $S\alpha S$ random measure with Lebesgue control measure.

We must first verify that the integral in (7.7.1) is well defined. Recall that an integral $I(f) = \text{Re}\int_{-\infty}^{\infty} f(x)\widetilde{M}(dx)$, where $f = f^{(1)} + i f^{(2)}$ (f_1, f_2 real-valued) and where $\widetilde{M}$ is isotropic (with control measure m) has characteristic function

$$E\exp i\theta I(f) = \exp\Big\{-|\theta|^{\alpha}c_0\int_{-\infty}^{\infty}|f(x)|^{\alpha}m(dx)\Big\} \tag{7.7.2}$$

where $c_0 = \frac{1}{2\pi}\int_0^{2\pi}(\cos\phi)^{\alpha}d\phi$ (Corollary 6.3.2). Applying this to $\mathcal{H}_{\alpha,H}(t)$, we obtain its finite-dimensional characteristic function

$$\begin{aligned} & E\exp\Big\{i\sum_{j=1}^{d}\theta_j\mathcal{H}_{\alpha,H}(t_j)\Big\} \\ &= \exp\Big\{-c_0\int_{-\infty}^{\infty}\Big|\sum_{j=1}^{d}\theta_j\frac{e^{ixt_j}-1}{ix}x^{-H+1-1/\alpha}\Big|^{\alpha}dx\Big\} \\ &= \exp\Big\{-c_0\int_{-\infty}^{\infty}\Big|\sum_{j=1}\theta_j(e^{ixt_j}-1)\Big|^{\alpha}x^{-\alpha H-1}dx\Big\}. \end{aligned} \tag{7.7.3}$$

As $|x|\to\infty$, the integrand behaves like $|x|^{-\alpha H-1}$, which is integrable at infinity because $H>0$; as $x\to 0$, the integrand behaves like $|x|^{\alpha-\alpha H-1}$, which is integrable at zero because $H<1$. Therefore the process $\{\mathcal{H}_{\alpha,H}(t),\ -\infty<t<\infty\}$ is well defined.

Remarks

1. Writing

$$\begin{aligned} I(f) &= \text{Re}\int_{-\infty}^{\infty} f(x)\widetilde{M}(dx) \\ &= \int_{-\infty}^{\infty} f^{(1)}(x)M^{(1)}(dx) - \int_E f^{(2)}(x)M^{(2)}(dx), \end{aligned}$$

we obtain the following alternative representation for the process $\mathcal{H}_{\alpha,H}$:

$$\mathcal{H}_{\alpha,H}(t) = \int_{-\infty}^{\infty}\frac{\sin xt}{x}|x|^{-H+1-1/\alpha}M^{(1)}(dx)$$

$$+\int_{-\infty}^{\infty} \frac{1-\cos xt}{x} |x|^{-H+1-1/\alpha} M^{(2)}(dx), \quad (7.7.4)$$

where $M^{(1)}$ and $M^{(2)}$ are the real and imaginary parts of the isotropic random measure M. The factors of $M^{(1)}$ and $M^{(2)}$ were plotted in Figures 7.4 and 7.5 for $t = 1$ and $d = H - 1/\alpha = \pm 0.2$. The measures $M^{(1)}$ and $M^{(2)}$ are dependent, but they become independent when $\alpha = 2$ (Section 2.6).

2. Comparing Relation (7.7.4) with (7.2.16), we see that by setting $\alpha = 2$, the process $\mathcal{H}_{\alpha,H}(t)$ becomes $CB_H(t)$ where C is a constant and where the fractional Brownian motion $B_H(t)$ is given by its harmonizable representation (7.2.12). (See also Exercise 7.14.)

Proposition 7.7.2
The real harmonizable fractional stable motion $\{\mathcal{H}_{\alpha,H}(t),\ -\infty < t < \infty\}$ *is H-sssi.*

PROOF: Exercise 7.13. ∎

The following proposition shows that there is nothing to gain by considering the process

$$\mathcal{H}_{\alpha,H}(a,b;t) = \text{Re}\int_{-\infty}^{\infty} \frac{e^{ixt}-1}{ix}\Big(a(x_+)^{-H+1-1/\alpha} + b(x_-)^{-H+1-1/\alpha}\Big)\widetilde{M}(dx),$$

where a, b are real numbers.

Proposition 7.7.3 *The finite-dimensional distributions of* $\{\mathcal{H}_{\alpha,H}(a,b;t),\ -\infty < t < \infty\}$ *are equal to those of*

$$\Big\{\Big(\frac{|a|^\alpha + |b|^\alpha}{2}\Big)^{1/\alpha}\mathcal{H}_{\alpha,H}(t),\ -\infty < t < \infty\Big\}.$$

PROOF: Fix $d \ge 1$, $\theta_1, \ldots, \theta_d$, $t_1, \ldots, t_d$, and let c_0 be the constant introduced in (7.7.2). Writing $E\exp\{iI(f)\} = \exp\{-\|I(f)\|_\alpha^\alpha\}$, we have

$$\begin{aligned}
&\|\sum_{j=1}^{d}\theta_j\mathcal{H}_{\alpha,H}(a,b;t_j)\|_\alpha^\alpha \\
&= c_0\int_{-\infty}^{\infty}\Big|\sum_{j=1}^{d}\theta_j\frac{e^{ixt_j}-1}{ix}\Big|^\alpha |a(x_+)^{-H+1-1/\alpha} + b(x_-)^{-H+1-1/\alpha}|^\alpha dx \\
&= |a|^\alpha c_0\int_{-\infty}^{\infty}\Big|\sum_{j=1}^{d}\theta_j(e^{ixt_j}-1)\Big|^\alpha |x|^{-\alpha}(x_+)^{-\alpha H+\alpha-1}dx
\end{aligned}$$

$$+ |b|^\alpha c_0 \int_{-\infty}^{\infty} \Big| \sum_{j=1}^{d} \theta_j (e^{ixt_j} - 1) \Big|^\alpha |x|^{-\alpha} (x_-)^{-\alpha H + \alpha - 1} dx$$

since $x_+ x_- = 0$. Clearly, each of these last two integrals equals $\frac{1}{2} \int_{-\infty}^{\infty} | \sum_{j=1}^{n} \theta_j (e^{ixt_j} - 1)|^\alpha |x|^{-\alpha H - 1} dx$ and therefore, by (7.7.3),

$$\| \sum_{j=1}^{d} \theta_j \mathcal{H}_{\alpha,H}(a,b;t_j) \|_\alpha^\alpha = \frac{|a|^\alpha + |b|^\alpha}{2} \| \sum_{j=1}^{d} \theta_j \mathcal{H}_{\alpha,H}(t_j) \|_\alpha^\alpha . \quad \blacksquare$$

We have been considering the real harmonizable fractional stable motion. Would a linear combination of the real and imaginary part, namely

$$A \, \mathrm{Re} \int_{-\infty}^{\infty} \frac{e^{ixt} - 1}{ix} |x|^{-H+1-1/\alpha} \widetilde{M}(dx)$$
$$+ B \, \mathrm{Im} \int_{-\infty}^{\infty} \frac{e^{ixt-1}}{ix} |x|^{-H-1-1/\alpha} \widetilde{M}(dx)$$

yield new H-sssi processes? The answer is in the negative (Collorary 6.3.3).

Finally, in sharp contrast to the Gaussian case, we have the following consequence of Theorem 6.7.2.

Theorem 7.7.4 *The linear fractional stable motion defined in (7.4.1) and the real harmonizable fractional stable motion defined in (7.7.1) are different processes.*

PROOF: If the two processes have the same finite-dimensional distributions (up to a multiplicative factor), then their increments (which are stationary processes) would also have the same finite-dimensional distrubutions. This would contradict Theorem 6.7.2. $\blacksquare$

7.8 Complex harmonizable fractional stable motion

Here, we consider the *complex harmonizable fractional stable motion* (CHFSM)

$$\Big\{ \mathcal{C}_{\alpha,H}(a,b;t)$$

$$= \int_{-\infty}^{\infty} \frac{e^{ixt} - 1}{ix} \Big(a(x_+)^{-H+1-1/\alpha} + b(x_-)^{-H+1-1/\alpha} \Big) \widetilde{M}(dx), \;\; t \in \mathbb{R} \Big\}, \tag{7.8.1}$$

where $0 < \alpha < 2$, $0 < H < 1$, $a \geq 0$, $b \geq 0$, $a + b > 0$, and where $\widetilde{M}$ is a complex isotropic random measure $S\alpha S$ with Lebesgue control measure as in Definition 7.7.1. We studied in Section 7.7 the real harmonizable fractional stable motion $\mathcal{H}_{\alpha,H}(t) = \text{Re } \mathcal{C}_{\alpha,H}(1,1;t)$, and we noted that Re $\mathcal{C}_{\alpha,H}(a,b;t)$, or for that matter Im $\mathcal{C}_{\alpha,H}(a,b;t)$, is again $\mathcal{H}_{\alpha,H}(t)$ up to a multiplicative constant. In contrast, different choices of a and b typically yield different *complex* harmonizable fractional stable motions. To show this, we need firstly to derive the finite-dimensional characteristic function of $\{\mathcal{C}_{\alpha,H}(a,b;t),\ -\infty < t < \infty\}$.

Fix $t_1, \ldots, t_d$ and let $z_j = \theta_j^{(1)} + i\theta_j^{(2)},\ j = 1, \ldots, d$ be arbitrary complex numbers. Applying Theorem 6.3.4, we obtain

$$E\exp\Big\{i\Big(\sum_{j=1}^d \theta_j^{(1)} \text{Re } \mathcal{C}_{\alpha,H}(a,b;t_j)\Big) + \sum_{j=1}^d \theta_j^{(2)} \text{Im } \mathcal{C}_{\alpha,H}(a,b;t_j)\Big\}$$

$$= \exp\Big\{-\|\sum_{j=1}^d \overline{z}_j \mathcal{C}_{\alpha,H}(a,b;t_j)\|_\alpha^\alpha\Big\}, \tag{7.8.2}$$

where

$$\|\sum_{j=1}^d \overline{z}_j \mathcal{C}_{\alpha,H}(a,b;t_j)\|_\alpha^\alpha$$

$$= |a|^\alpha c_0 \int_{-\infty}^{\infty} |\sum_{j=1}^d \overline{z}_j (e^{ixt_j} - 1)|^\alpha \big(|a|^\alpha (x_+)^{-\alpha H - 1} dx + |b|^\alpha (x_-)^{-\alpha H - 1}\big) dx$$

$$= c_0 \int_0^\infty \Big[|a \sum_{j=1}^d \overline{z}_j (e^{ixt_j} - 1)|^\alpha + |b \sum_{j=1}^d \overline{z}_j (e^{-ixt_j} - 1)|^\alpha\Big] x^{-H\alpha - 1} dx \tag{7.8.3}$$

as in the proof of Proposition 7.7.3, with $c_0 = \frac{1}{2\pi}\int_0^{2\pi} (\cos\phi)^\alpha d\phi$.

Proposition 7.8.1 *The complex harmonizable fractional stable motion*

$$\{\mathcal{C}_{\alpha,H}(a,b;t),\ t \in \mathbb{R}\}$$

is H-sssi.

PROOF: It is sufficient to show that for any real c_1 and c_2, $\{c_1 \text{ Re } C_{\alpha,H}(a,b;t) + c_2 \text{ Im } C_{\alpha,H}(a,b;t),\ t \in \mathbb{R}\}$ is H-sssi. This, however, follows from (7.8.2). ■

The normalization factor used to standardize the process $\mathcal{C}_{\alpha,H}(a,b;\cdot)$ is $\|\mathcal{C}_{\alpha,H}(a,b;1)\|_\alpha$. Setting $d = 1,\ z_1 = 1$ and $t_1 = 1$ in (7.8.3), we obtain

$$\|\mathcal{C}_{\alpha,H}(a,b;1)\|_\alpha^\alpha = c_0 \int_0^\infty (a^\alpha |e^{ix} - 1|^\alpha + b^\alpha |e^{-ix} - 1|^\alpha) x^{-\alpha H - 1} dx$$

$$= (a^\alpha + b^\alpha)c_0 \int_0^\infty \left|2\sin\frac{x}{2}\right|^\alpha x^{-\alpha H-1}dx$$

$$= (a^\alpha + b^\alpha)c_0 2^{\alpha(1-H)} \int_0^\infty |\sin x|^\alpha x^{-\alpha H-1}dx. \quad (7.8.4)$$

The *standard* CHFSM is the process

$$\frac{\mathcal{C}_{\alpha,H}(a,b;t)}{\|\mathcal{C}_{\alpha,H}(a,b;1)\|_\alpha} = \frac{\mathcal{C}_{\alpha,H}(a,b;t)}{(a^\alpha+b^\alpha)^{1/\alpha}2^{(1-H)}(c_0\int_0^\infty |\sin x|^\alpha x^{-\alpha H-1}dx)^{1/\alpha}}.$$

Note also that the parameters a and b in the definition (7.8.1) of the complex harmonizable fractional stable motion were restricted to $\mathbb{R}_+$. This is because choosing them in $\mathbb{R}$ would not have yielded new processes as Relation (7.8.3) demonstrates.

We shall now show that each ray through the origin of the parameter space $(a,b) \in \mathbb{R}_+^2$ defines a distinct complex harmonizable fractional stable motion.

Theorem 7.8.2 *Let $0 < \alpha < 2$, $0 < H < 1$, and let a, a', b, b' be non-negative real numbers satisfying $a + b > 0$ and $a' + b' > 0$. Then the processes*

$$\left\{\frac{1}{(a^\alpha+b^\alpha)^{1/\alpha}}\mathcal{C}_{\alpha,H}(a,b;t),\ -\infty < t < \infty\right\},$$

and

$$\left\{\frac{1}{(a'^\alpha+b'^\alpha)^{1/\alpha}}\mathcal{C}_{\alpha,H}(a',b';t),\ -\infty < t < \infty\right\},$$

have identical finite-dimensional distributions if and only if one of the following conditions holds:

(i) $a = a' = 0$,

(ii) $b = b' = 0$,

(iii) a, a', b, b' *are non-zero and* $a/b = a'/b'$.

PROOF: The "if" part follows immediately from (7.8.3). Consider now the "only if" part and, in particular, case (iii) (the others are the same). Setting, without loss of generality, $a = a' = 1$, suppose

$$(1+b^\alpha)^{-1/\alpha}\mathcal{C}_{\alpha,H}(1,b;\cdot) \stackrel{d}{=} (1+b'^\alpha)^{-1/\alpha}\mathcal{C}_{\alpha,H}(1,b';\cdot) \quad (7.8.5)$$

for some b and b'. We want to show that this implies $b = b'$.

We will work not with $\mathcal{C}_{\alpha,H}(1,b,\cdot)$ but with

$$Y(1,b;t) = \int_{-\infty}^0 [\mathcal{C}_{\alpha,H}(1,b;t) - \mathcal{C}_{\alpha,H}(1,b';t+u)]e^u du,\ -\infty < t < \infty, \quad (7.8.6)$$

instead. By Theorems 11.3.2 and 11.4.1 below, the process $\{Y(1,b;t),\ -\infty < t < \infty\}$ is well defined and has the following integral representation:

$$\int_{-\infty}^{\infty} e^{itx}\Big[\int_{-\infty}^{0} \frac{e^u - e^{u+iux}}{ix} du\Big](x_+^{-H+1-1/\alpha} + bx_-^{-H+1-1/\alpha})\widetilde{M}(dx)$$

$$= \int_{-\infty}^{\infty} e^{itx} \frac{1}{1+ix}(x_+^{-H+1-1/\alpha} + bx_-^{-H+1-1/\alpha})\widetilde{M}(dx).$$

Its bivariate characteristic function

$$E\exp\{i(\overline{z}_1 Y(1,b;t_1) + \overline{z}_2 Y(1,b;t_2))\}$$

at $t_1 = 0,\ t_2 = t > 0,\ z_1 = 1,\ z_2 = -i$ equals $\exp\{-c_0 B(b;t)\}$, where

$$B(b;t) = \|Y(1,b;0) + iY(1,b;t)\|_\alpha^\alpha$$

$$= \int_0^\infty \{|1+ie^{ixt}|^\alpha + b^\alpha|1+ie^{-ixt}|^\alpha\} x^{-\alpha H+\alpha-1}(1+x^2)^{-\alpha/2}dx$$

$$= 2^{\alpha/2}\int_0^\infty \{(1-\sin xt)^{\alpha/2} + b^\alpha(1+\sin xt)^{\alpha/2}\} x^{-\alpha H+\alpha-1}(1+x^2)^{-\alpha/2}dx.$$

The relation (7.8.5) implies

$$(1+b^\alpha)^{-1}B(b;t) = (1+b'^\alpha)^{-1}B(b';t)$$

or

$$\frac{b'^\alpha - b^\alpha}{(1+b^\alpha)(1+b'^\alpha)} f(t) = 0$$

for all $t > 0$, where

$$f(t) = \int_0^\infty \{(1+\sin xt)^{\alpha/2} - (1-\sin xt)^{\alpha/2}\} x^{-\alpha H+\alpha-1}(1+x^2)^{-\alpha/2}dx. \tag{7.8.7}$$

Since the following lemma shows that this is possible only if $b = b'$, the proof is complete. ■

Lemma 7.8.3 *Let $0 < \alpha < 2,\ 0 < H < 1$ and let $f(t)$ be defined as in (7.8.7). Then $f(t) \neq 0$ for some $t > 0$.*

PROOF: Suppose $t \neq 0$ and consider firstly the case $1 \leq \alpha H < 2$. Then $f(t)$ equals

$$\Big(\int_0^{\pi/2t} + \int_{\pi/2t}^\infty\Big)\{(1+\sin xt)^{\alpha/2} - (1-\sin xt)^{\alpha/2}\} x^{-\alpha H+\alpha-1}(1+x^2)^{-\alpha/2}dx$$

$$> \int_0^{\pi/2t} \left\{ \left(1 + \frac{2}{\pi}xt\right)^{\alpha/2} - \left(1 - \frac{2}{\pi}xt\right)^{\alpha/2} \right\} x^{-\alpha H+\alpha-1}(1+x^2)^{-\alpha/2}dx$$
$$- 2\int_{\pi/2t}^{\infty} x^{-\alpha H+\alpha-1}(1+x^2)^{-\alpha/2}dx$$

since $\sin x \geq \frac{2}{\pi}x$ for $0 \leq x \leq \pi/2$ and since $(1+\sin tx)^{\alpha/2} - (1-\sin tx)^{\alpha/2} \geq 0 - 2^{\alpha/2} > -2$. Now, for $tx > 0$, $(1+\frac{2}{\pi}xt)^{\alpha/2} - (1-\frac{2}{\pi}xt)^{\alpha/2}$ has a Taylor series expansion with only positive terms and hence it is strictly greater than $\frac{2}{\pi}\alpha tx$. Using also the inequality $-(1+x^2)^{-\alpha/2} > -x^{\alpha}$ for $x \geq 1$, we obtain, for small enough t,

$$f(t) > \frac{2}{\pi}\alpha t \int_0^{\pi/2t} x^{-\alpha H+\alpha}(1+x^2)^{-\alpha/2}dx - 2\int_{\pi/2t}^{\infty} x^{-\alpha H-1}dx.$$

Note that the integrand in the first integral is integrable as $x \to \infty$ if $\alpha H > 1$ because it behaves like $x^{-\alpha H}$ for large x. If $\alpha H = 1$, the integral behaves like $\ln \frac{\pi}{2t}$ for small enough t. The second integral, on the other hand, is proportional to $t^{\alpha H}$. Letting $c_j(\cdot,\cdot)$, $j \geq 1$, denote positive constants depending only on the parameters in the parentheses, we have for sufficiently small $t > 0$,

$$f(t) > c_1(\alpha,H)t - c_2(\alpha,H)t^{\alpha H} > 0 \quad \text{if } 1 < \alpha H < 2,$$

and

$$f(t) > c_3(\alpha)t\ln(1/t) - c_4 t > 0 \qquad \text{if } \alpha H = 1.$$

This proves the lemma in the case $1 \leq \alpha H < 2$.

Suppose now $0 < \alpha H < 1$. By changing variables, we obtain

$$f(t) = t^{\alpha H}\int_0^{\infty} \{(1+\sin x)^{\alpha/2} - (1-\sin x)^{\alpha/2}\} x^{-\alpha H+\alpha-1}(t^2+x^2)^{-\alpha/2}dx$$

$$= t^{\alpha H}\int_0^{\infty} \{(1+\sin x)^{\alpha/2} - (1-\sin x)^{\alpha/2}\} x^{-\alpha H-1}dx$$

$$+ t^{\alpha H}\int_0^{\infty} \{(1+\sin x)^{\alpha/2} - (1-\sin x)^{\alpha/2}\} \left[\left(1+\frac{t^2}{x^2}\right)^{-\alpha/2} - 1\right] x^{-\alpha H-1}dx \tag{7.8.8}$$

$$:= t^{\alpha H}(c_5(\alpha,H) + g(t)).$$

It is easy to verify that the first integral in (7.8.8) converges. Indeed, its integrand is integrable as $x \to 0$ because it is asymptotically proportional to $x^{1-\alpha H-1}$ and $\alpha H < 1$; it is also integrable as $x \to \infty$ because it is bounded by $4x^{-\alpha H-1}$ and $\alpha H > 0$. The integral is therefore equal to a constant, which we denote $c_5(\alpha,H)$. We now want to show that

$$c_5(\alpha,H) > 0.$$

Write $\int_0^\infty = \sum_{k=0}^\infty \int_{2\pi k}^{2\pi+2\pi k}$ and change variables to obtain

$$c_5(\alpha, H) = \int_0^{2\pi} \{(1+\sin u)^{\alpha/2} - (1-\sin u)^{\alpha/2}\} \sum_{k=0}^\infty (u+2\pi k)^{-\alpha H-1} du =$$

$$\int_0^\pi \{(1+\sin u)^{\alpha/2} - (1-\sin u)^{\alpha/2}\} \sum_{k=0}^\infty ((u+2\pi k)^{-\alpha H-1} - (u+2\pi k+\pi)^{-\alpha H-1}) du.$$

Since $x^{-\alpha H-1}$ is strictly decreasing on $(0, \infty)$, we conclude that $c_5(\alpha, H) > 0$.

We now turn to the second integral in (7.8.8) which we denoted $g(t)$. Choosing $\epsilon > 0$ and $t > 0$ sufficiently small, we see that $|g(t)|$ is bounded above by

$$\int_0^\infty |(1+\sin x)^{\alpha/2} - (1-\sin x)^{\alpha/2}|\, |1-(1+t^2/x^2)^{-\alpha/2}| x^{-\alpha H-1} dx$$

$$\leq 2\int_0^\epsilon x^{-\alpha H} dx + 2\int_\epsilon^\infty (t^2/x^2) x^{-\alpha H-1} dx$$

$$= \frac{2}{1-\alpha H}\epsilon^{1-\alpha H} + \frac{2}{\alpha H+2}\epsilon^{-\alpha H-2} t^2$$

$$< c_5(\alpha, H).$$

Using (7.8.8), we conclude that $f(t) > 0$ for sufficiently small $t > 0$. ■

7.9 Subordinated processes

A *subordinated* process $X(t)$ is an α-stable process with representation

$$X(t) = \int_\Omega Y(t, x) M(dx), \tag{7.9.1}$$

where $Y = \{Y(t, x),\ t \in T,\ x \in \Omega\}$ is a stochastic process defined on $(\Omega, \mathcal{F}, P)$ and M is an α-stable random measure with control measure P.

For example, suppose that M is $S\alpha S$, $0 < \alpha < 2$, and Y is Gaussian. Then X is the $S\alpha S$ sub-Gaussian process introduced in Section 3.7. Suppose, now, that Y is $S\alpha' S$ with $0 < \alpha < \alpha' < 2$. Then X is the $S\alpha S$ sub-stable process introduced in Section 3.8.

The measure M in (7.9.1) does not have to be symmetric nor the process Y be Gaussian or stable. In order to ensure that the process X is well defined, it is only necessary to suppose $E|Y(t)|^\alpha < \infty$ for all $t \in T$.

For example, if M is symmetric, then

$$E\Big\{\exp i\Big(\sum_{j=1}^d \theta_j X(t_j)\Big)\Big\} = \exp\Big\{-\int_\Omega |\sum_{j=1}^d \theta_j Y(t_j, x)|^\alpha P(dx)\Big\},$$

that is,

$$E\Big\{\exp i\Big(\sum_{j=1}^{d}\theta_j X(t_j)\Big)\Big\} = \exp\Big\{-E_P|\sum_{j=1}^{d}\theta_j Y(t_j)|^\alpha\Big\}, \tag{7.9.2}$$

where E_P is the expectation with respect to the probability measure P.

If M is not symmetric, then one has also to take into account the skewness intensity $\beta(\cdot)$. In this context, $\beta(\cdot)$ is a random variable because it is a measurable function on $(\Omega, \mathcal{F}, P)$. For convenience, suppose $\alpha \neq 1$. Then the finite-dimensional characteristic function of $X(t)$ is

$$\exp\Big\{-\int_\Omega |\sum_{j=1}^{d}\theta_j Y(t_j,x)|^\alpha\Big(1 - i\beta(x)\,\text{sign}\,\Big(\sum_{j=1}^{d}\theta_j Y(t_j,x)\tan\frac{\pi\alpha}{2}\Big)\Big)P(dx)\Big\}.$$

If β and Y, which are both defined on $(\Omega, \mathcal{F}, P)$, are independent, then the finite-dimensional characteristic function of $X(t)$ simplifies to

$$\exp\Big\{-E_P|\sum_{j=1}^{d}\theta_j Y(t_j)|^\alpha - i(E_P\beta)E_P\Big(\sum_{j=1}^{d}\theta_j Y(t_j)\Big)^{<\alpha>}\tan\frac{\pi\alpha}{2}\Big\},$$

so assuming β and Y independent is equivalent to assuming that β is constant.

For convenience, we will suppose in the sequel that M is symmetric, so that the finite-dimensional characteristic function of $X(t)$ is given by (7.9.2). Observe that the finite-dimensional distributions of X depend on those of Y only through the α-moments of linear combinations of $Y(t)$s. Thus, if two different processes $\{Y_1(t),\ t \in T\}$ and $\{Y_2(t),\ t \in T\}$, the first defined on $(\Omega_1, \mathcal{F}_1, P_1)$, the second on $(\Omega_2, \mathcal{F}_2, P_2)$, satisfy

$$E_{P_1}\Big|\sum_{j=1}^{d}\theta_j Y_1(t_j)\Big|^\alpha = E_{P_2}\Big|\sum_{j=1}^{d}\theta_j Y_2(t_j)\Big|^\alpha$$

for all $d \geq 1$, $\theta_1, \ldots, \theta_d$ real, $t_1, \ldots, t_d \in T$, then the processes

$$\Big\{X_1(t) = \int_{\Omega_1} Y_1(t,x)M_{P_1}(dx),\ t \in T\Big\}$$

and

$$\Big\{X_2(t) = \int_{\Omega_2} Y_2(t,x)M_{P_2}(dx),\ t \in T\Big\}$$

have identical finite-dimensional distributions, where M_{P_1} and M_{P_2} are $S\alpha S$ random measures with control measures P_1 and P_2, respectively.

We now examine what happens if Y is an H-sssi process.

Theorem 7.9.1 *If $\{Y(t),\ -\infty < t < \infty\}$ is H-sssi with finite α-moments defined on $(\Omega, \mathcal{F}, P)$ and if M is a $S\alpha S$ random measure with control measure P, then*

$$X(t) = \int_\Omega Y(t,x)M(dx)$$

is $S\alpha S$ and also H-sssi.

PROOF: To verify self-similarity, let $a > 0$, and observe that

$$E\exp\Big\{i\sum_{j=1}^{d}\theta_j X(at_j)\Big\} = \exp\Big\{-E|\sum_{j=1}^{d}\theta_j Y(at_j)|^\alpha\Big\}$$

$$= \exp\Big\{-a^{\alpha H}E|\sum_{j=1}^{d}\theta_j Y(t_j)|^\alpha\Big\} = E\exp\Big\{i\sum_{j=1}^{d}\theta_j a^H X(t_j)\Big\}.$$

Verifying stationarity of the increments is also straightforward. ■

Observe that if $\alpha = 2$, then whatever the choice of the H-sssi process Y, the resulting subordinated process X is always fractional Brownian motion (see Definition 7.2.2). We now present several examples with $0 < \alpha < 2$ and M $S\alpha S$.

Example 7.9.2 Suppose Y is fractional Brownian motion ($\alpha' = 2$). The corresponding subordinated process X is called *sub-FBM*. Its parameters are

$$\alpha \in (0,2), \quad H \in (0,1).$$

Because the process X is sub-Gaussian, it is different from the real harmonizable fractional stable motion (Theorem 6.6.5) and from the linear fractional stable motion (Theorem 4.7.5).

Example 7.9.3 Suppose Y is the linear fractional stable motion (LFSM) with $\alpha' < 2$ and take $\alpha < \alpha'$. The corresponding process X is sub-stable with parameters

$$\alpha \in (0,2), \quad \alpha' \in (\alpha, 2), \quad H \in (0,1), \quad a, b \in \mathbb{R}.$$

Example 7.9.4 Suppose Y is the log-fractional stable motion (logFSM) with $1 < \alpha' < 2$ (and $H = 1/\alpha'$). The corresponding process X is sub-stable with parameters

$$\alpha \in (0,2), \quad \alpha' \in (\max(1,\alpha), 2), \quad H = 1/\alpha'.$$

Example 7.9.5 Suppose Y is the real harmonizable fractional stable motion (RHFSM) with $\alpha' < 2$ and take $\alpha < \alpha'$. The corresponding process X is sub-stable with parameters

$$\alpha \in (0,2), \quad \alpha' \in (\alpha, 2), \quad H \in (0,1).$$

Example 7.9.6 Suppose that $Y(t)$ is a Hermite process (see Taqqu (1979), where the process is denoted $\overline{Z}_m(t)$). Formally,

$$\Big\{Y(t) = \int_{-\infty}^{\infty} dB(\xi_1) \int_{-\infty}^{\xi_1} dB(\xi_2) \cdots \int_{-\infty}^{\xi_{m-1}} dB(\xi_m)$$

$$\times \int_0^t \prod_{i=1}^{m} (s-\xi_i)^{H_0-3/2} 1(\xi_i < s) ds,\ t \geq 0\Big\},$$

where B is a Gaussian random measure with Lebesgue control measure, m is a positive integer and $1 - 1/(2m) < H_0 < 1$. The process $\{Y(t),\ t \geq 0\}$ is H-sssi with $H = m(H_0 - 1) + 1$. It has moments of all order. When $m = 1$, $Y(t)$ becomes fractional Brownian motion with $1/2 < H = H_0 < 1$. When $m = 2, 3, \ldots, Y(t)$ is non-Gaussian.

The parameters of the corresponding process X are

$$\alpha \in (0,2), \quad m = 1,2,3,\ldots, \quad H \in (0,1).$$

See Section 12.4 for more examples.

7.10 Fractional stable noises

If $\{X(t),\ t \in \mathbb{R}\}$ is a process with stationary increments, then its increments $Y_j = X(j+1) - X(j)$, $j = \ldots, -1, 0, 1, \ldots$, form a stationary sequence. The sequence $\{Y_j\}$ is called a *noise*. "Fractional noises" are obtained by taking the increments of H-sssi processes. For example, the fractional Gaussian noise defined in Section 7.2.3 is the increment of the H-sssi Gaussian process fractional Brownian motion. But although there is only one H-sssi Gaussian process, there are many different H-sssi α-stable processes with a given $0 < \alpha < 2$, and each of these processes gives rise to a different fractional α-stable noise.

Consider, for example, the linear fractional stable motion $L_{\alpha,H}(a,b;t)$ introduced in Section 7.4. The *linear fractional stable noise* is the stationary sequence

$$\begin{aligned} Y_j &= L_{\alpha,H}(a,b;j+1) - L_{\alpha,H}(a,b;j) \\ &= \int_{-\infty}^{\infty} \Big(a[(j+1-x)_+^{H-\frac{1}{\alpha}} - (j-x)_+^{H-\frac{1}{\alpha}}] \\ &\quad + b[(j+1-x)_-^{H-\frac{1}{\alpha}} - (j-x)_-^{H-\frac{1}{\alpha}}]\Big) M(dx), \qquad (7.10.1) \end{aligned}$$

where M is an α-stable random measure with Lebesgue control measure.

The integrand in (7.10.1) is similar to that in (7.2.18) with $H - 1/\alpha$ replacing $H - 1/2$. By analogy, we will say that the Y_js display long-range dependence when $H > 1/\alpha$ and negative dependence when $H < 1/\alpha$. Because H is still constrained to lie in the interval $(0, 1)$, long-range dependence is only possible when $\alpha > 1$. This fact can be understood heuristically as follows. When $\alpha \leq 1$, M generates many large values. If, in addition, $H > 1/\alpha$, then their effect is long lasting because the integrand in (7.10.1) changes very slowly with x. As a result, the integral diverges.

We now list, for reference purposes, some other important fractional stable noises. The *log-fractional stable noise* is

$$\begin{aligned} Y_j &= \Lambda_{\alpha,1/\alpha}(j+1) - \Lambda_{\alpha,1/\alpha}(j) \\ &= \int_{-\infty}^{\infty} (\ln|j+1-x| - \ln|j-x|) M(dx), \end{aligned}$$

where M is α-stable with Lebesgue control measure. It is defined for $1 < \alpha < 2$.

The *real harmonizable fractional stable noise* is

$$\begin{aligned} Y_j &= \mathcal{H}_{\alpha,H}(j+1) - \mathcal{H}_{\alpha,H}(j) \\ &= \text{Re} \int_{-\infty}^{\infty} e^{ixj} \frac{e^{ix}-1}{ix} |x|^{-H+1-1/\alpha} \widetilde{M}(dx), \end{aligned}$$

where $0 < \alpha \leq 2$, $0 < H < 1$ and where $\widetilde{M}$ is a complex isotropic $S\alpha S$ random measure with Lebesgue control measure.

Finally, the *sub-Gaussian fractional stable noise* is

$$Y_j = A^{1/2}(B_H(j+1) - B_H(j)) = A^{1/2} Y_{j,2},$$

where B_H is fractional Brownian motion, $Y_{j,2}$ fractional Gaussian noise and A is an independent $\frac{\alpha}{2}$-stable random variable defined in (2.5.1).

Additional fractional stable noises can be obtained by taking increments of other H-sssi subordinated process; for example, those defined in see Section 7.9.

The asymptotic dependence structure of the fractional Gaussian noise $\{Y(j)\}$ was investigated in Proposition 7.2.10 where we noted that the autocovariance function $EY(j)Y(0)$ is asymptotically proportional to j^{2H-2} as $j \to \infty$. Here, we shall use the codifference $\tau(j)$ to analyze the asymptotic dependence of fractional stable noises.

We consider only the linear fractional stable noise and the log-fractional stable noise because the codifference does not tend to zero for either the real harmonizable fractional stable noise or for the sub-Gaussian fractional stable noise (see Propositions 6.7.1 and 4.7.4). The first two types of noises are stationary moving averages and, therefore, their codifference $\tau(j)$ tends to 0 as the lag $j \to \infty$ (Theorem 4.7.3).

The following theorems specify the asymptotic behavior of $I(\theta_1,\theta_2;j)$ defined in Relation (4.7.5) for any $\theta_1,\theta_2 \in \mathbb{R}$ and, hence, of $\tau(j) = -I(1,-1;j)$. The proofs, which are very technical, are omitted. They can be found in Astrauskas, Levy and Taqqu (1991).

Theorem 7.10.1 *Consider the linear fractional stable noise defined in Relation (7.10.1) and suppose that it is $S\alpha S$. If either*

(i) $0 < \alpha \le 1, 0 < H < 1$

or

(ii) $1 < \alpha < 2,\ 1 - \frac{1}{\alpha(\alpha-1)} < H < 1,\ H \ne 1/\alpha$

then,[9] *as* $j \to \infty$,

$$\begin{aligned} I(\theta_1,\theta_2;j) &\sim B(\theta_1,\theta_2) j^{\alpha H - \alpha}, \\ \tau(j) &\sim -B(1,-1) j^{\alpha H - \alpha}. \end{aligned}$$

If

(iii) $1 < \alpha < 2,\ 0 < H < 1 - \frac{1}{\alpha(\alpha-1)}$,

then, as $t \to \infty$,

$$\begin{aligned} I(\theta_1,\theta_2;j) &\sim F(\theta_1,\theta_2) j^{H-(1/\alpha)-1}, \\ \tau(j) &\sim -F(1,-1) j^{H-(1/\alpha)-1}. \end{aligned}$$

The constant $B(\theta_1,\theta_2)$ *equals*

$$\begin{aligned} \left|H - \frac{1}{\alpha}\right|^{\alpha} \Bigg[|a|^{\alpha} \int_{-\infty}^{0} \Big\{ & |\theta_1(1-x)^{H-\frac{1}{\alpha}-1} + \theta_2(-x)^{H-\frac{1}{\alpha}-1}|^{\alpha} \\ & - |\theta_1(1-x)^{H-\frac{1}{\alpha}-1}|^{\alpha} - |\theta_2(-x)^{H-\frac{1}{\alpha}-1}|^{\alpha} \Big\} dx \\ + \int_0^1 \Big\{ & |a\theta_1(1-x)^{H-\frac{1}{\alpha}-1} + b\theta_2 x^{H-\frac{1}{\alpha}-1}|^{\alpha} \\ & - |a\theta_1(1-x)^{H-\frac{1}{\alpha}-1}|^{\alpha} - |b\theta_2 x^{H-\frac{1}{\alpha}-1}|^{\alpha} \Big\} dx \\ + |b|^{\alpha} \int_0^{\infty} \Big\{ & |\theta_2(1+x)^{H-\frac{1}{\alpha}-1} + \theta_1 x^{H-\frac{1}{\alpha}-1}|^{\alpha} \\ & - |\theta_2(1+x)^{H-\frac{1}{\alpha}-1}|^{\alpha} - |\theta_1 x^{H-\frac{1}{\alpha}-1}|^{\alpha} \Big\} dx \Bigg]. \end{aligned}$$

---

[9] If $\alpha = 1$, the asymptotic behavior of I is valid under the following additional conditions: either sign $ab = 1$ or sign $\theta_1\theta_2 = -1$. If $\alpha = 1$ and both sign $ab \ne 1$ and sign $\theta_1\theta_2 \ne -1$, then $B = 0$. In that case one has to consider the next term in the asymptotic expansion and one finds that

$$I(\theta_1,\theta_2;j) \sim C(\theta_1,\theta_2) j^{H-2}$$

as $j \to \infty$, where

$$C(\theta_1,\theta_2) = -2(1-H)(|a\theta_1| + |b\theta_2|).$$

(Recall: sign $u \ne 1$ means $u \le 0$.)

The constant $F(\theta_1, \theta_2)$ equals

$$
\begin{aligned}
&(H\alpha - 1)\Bigg[a\theta_1\Bigg\{\int_{-\infty}^{0}\left(a\theta_2[(1-x)^{H-\frac{1}{\alpha}} - (-x)^{H-\frac{1}{\alpha}}]\right)^{<\alpha-1>} dx \\
&+ \int_0^1 \left(\theta_2[a(1-x)^{H-\frac{1}{\alpha}} - bx^{H-\frac{1}{\alpha}}]\right)^{<\alpha-1>} dx \\
&+ \int_1^{\infty} \left(b\theta_2[(x-1)^{H-\frac{1}{\alpha}} - x^{H-\frac{1}{\alpha}}]\right)^{<\alpha-1>} dx\Bigg\} \\
&+ b\theta_2\Bigg\{\int_{-\infty}^{0}\left(b\theta_1[(1-x)^{H-\frac{1}{\alpha}} - (-x)^{H-\frac{1}{\alpha}}]\right)^{<\alpha-1>} dx \\
&+ \int_0^1 \left(\theta_1[b(1-x)^{H-\frac{1}{\alpha}} - ax^{H-\frac{1}{\alpha}}]\right)^{<\alpha-1>} dx \\
&+ \int_1^{\infty} \left(a\theta_1[(x-1)^{H-\frac{1}{\alpha}} - x^{H-\frac{1}{\alpha}}]\right)^{<\alpha-1>} dx\Bigg\}\Bigg].
\end{aligned}
$$

PROOF: See Astrauskas, Levy and Taqqu (1991), Theorems 2.1, 2.2 and 2.3. ■

Theorem 7.10.2 *For the log-fractional noise, as $j \to \infty$,*

$$
\begin{aligned}
I(\theta_1, \theta_2; j) &\sim G(\theta_1, \theta_2) j^{1-\alpha}, \\
\tau(j) &\sim -G(1, -1) j^{1-\alpha},
\end{aligned}
$$

where

$$
\begin{aligned}
G(\theta_1, \theta_2) = &\int_{\infty}^{1} \left\{\left|\frac{\theta_1}{1+x} - \frac{\theta_2}{x}\right|^{\alpha} - \left|\frac{\theta_1}{1+x}\right|^{\alpha} - \left|\frac{\theta_2}{x}\right|^{\alpha}\right\} dx \\
&+ \int_0^{\infty} \left|\frac{\theta_2}{1+x} + \frac{\theta_1}{x}\right|^{\alpha} - \left|\frac{\theta_2}{1+x}\right|^{\alpha} - \left|\frac{\theta_1}{x}\right|^{\alpha} dx
\end{aligned}
$$

PROOF: See Astrauskas, Levy and Taqqu (1991), Theorem 2.4. ■

The codifference τ equals the covariance when $\alpha = 2$ (Property 2.10.3). Thus, for fractional Gaussian noise, $\tau(j) \approx j^{2H-2}$ as $j \to \infty$, where $\approx$ denotes asymptotic proportionality (compare with Theorem 7.2.11). Theorem 7.10.1 states that in most cases $\tau(j) \approx j^{\alpha H - \alpha}$ when $0 < \alpha < 2$, with the following exception: $\tau(j) \approx j^{H-(1/\alpha)-1}$, when $\alpha > 1$ and $0 < H < \overline{H}$, where

$$
\overline{H} = \overline{H}(\alpha) = 1 - \frac{1}{\alpha(\alpha - 1)}.
$$

This phase transition [10] at $H = \overline{H}$ is peculiar. Since $\overline{H}(\alpha) < 1/\alpha$, it occurs only when there is negative dependence.

Note that $\overline{H} \geq 0$ if and only if $\alpha \geq \alpha_0$ where

$$\alpha_0 = \frac{1+\sqrt{5}}{2} = 1.618\ldots$$

equals the golden ratio.[11] Thus, $\tau(j) \approx j^{H-(1/\alpha)-1}$ takes place only when

$$1 < \alpha_0 < \alpha < 2 \ \text{ and } \ 0 < H < \overline{H}(\alpha),$$

that is, for large α and small H. In this region,

$$\alpha H - \alpha < H - \frac{1}{\alpha} - 1 < 0.$$

Finally, observe that the two exponents $\alpha H - \alpha$ and $H - (1/\alpha) - 1$ are equal on the boundary $H = \overline{H}(\alpha)$.

Remarks

1. We have assumed that the noises are $S\alpha S$. If they are skewed α-stable, the asymptotic behavior of I and τ is the same but the constants B, F and G are more complicated. They are given in Astrauskas, Levy and Taqqu (1991).

2. If X is an H-sssi process, one can also define $Y(t) = X(t+1) - X(t)$ for all $t \in \mathbb{R}$ and not just for $t = \ldots, -1, 0, 1, \ldots,$ as we have done in this section. Theorems 7.10.1 and 7.10.2 continue to hold with j replaced by $t \in \mathbb{R}$.

3. The parameters θ_1 and θ_2 affect only the multiplicative constants in the limit and not the exponents.

7.11 Simulation of fractional noises and motions

To simulate fractional Brownian or stable noise (or motion) one approximates the integral representation. This creates two types of errors, a "low frequency" one due to the truncation of the limits of integration and a "high frequency" error caused by replacing the integral by a sum. Consider then the fractional Gaussian

[10] It does not occur, for example, when the process is the projection of a linear fractional random field of dimension greater than one (Kokoszka & Taqqu 1993a).

[11] Private communication of Jonathan Taqqu. The golden ratio a/b satisfies the relation $a/b = (a+b)/a$.

noise as defined in (7.2.18) and linear fractional stable noise with $a = 1,\ b = 0$, given in (7.10.1). We simulate them as follows:

$$Y_j = \sum_{u=1}^{Mm} K_d(\frac{u}{m})\epsilon_{j-(u/m)},\ j = 1, \dots, T \tag{7.11.1}$$

where the ϵs are i.i.d. α-stable random variables, $0 < \alpha \leq 2$. M and m are integers. M defines the memory (cutoff in the limits of integration), $1/m$ is the mesh (discretization of the integral) and $T \leq M$ is the length of the sequence. The kernel

$$K_d(x) = \begin{cases} x^d - (x-1)^d,\ x > 1, \\ x^d,\ \ 0 < x \leq 1 \end{cases}$$

with $d = H - \frac{1}{\alpha}$ is plotted in Figure 7.11 for $d = \pm 0.4$.

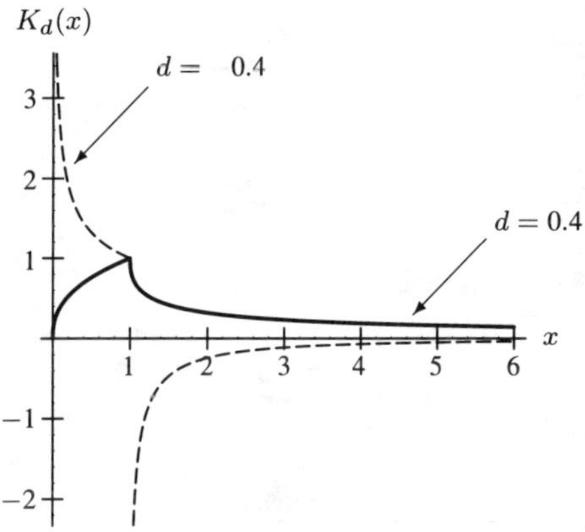

Figure 7.11: The kernel $K_d(x)$

This procedure requires generating $m(T + M - 1)$ i.i.d. ϵ_js. We avoided infinite values for $d < 0$ by starting the summation in (7.11.1) at $1/m$ and taking $K_d(1)$ equal to 1. It is best to choose a large value of m when d is negative. It is also convenient to renormalize the sequence $\{Y_j,\ j = 1, \dots, T\}$ by subtracting the sample mean and dividing by the sample standard deviation. To approximate fractional Brownian or stable motion, one can use $T^{-H} \sum_{j=1}^{[tT]} Y_j,\ 1/T \leq t \leq 1$, where [ ] denotes the integer part. For display purposes, however, it is often sufficient to plot $\sum_{j=1}^{t} Y_j,\ t = 1, \dots, T$ with an adequate aspect ratio. The following figures illustrate this procedure in the case $\alpha = 2$ and $\alpha = 1.7$. The random variables ϵ_js are symmetric. We chose $T = M = 1000$ and $m = 30$ and focused on $d = 0.4,\ 0,\ -0.4$. Since $H = d + 1/\alpha$, this choice of d means $H = 0.9,\ 0.5,\ 0.1$ in the Gaussian case and $H = 0.99,\ 1/1.7 = 0.59,\ 0.19$

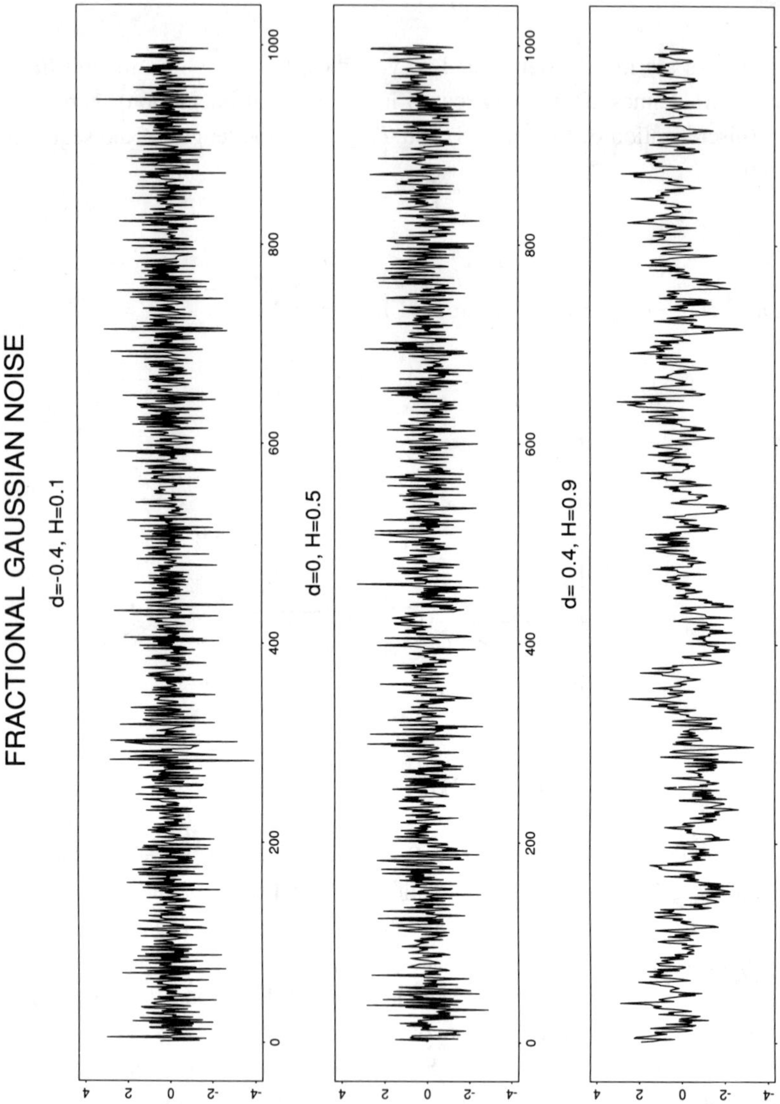
FRACTIONAL GAUSSIAN NOISE
d=-0.4, H=0.1
d=0, H=0.5
d= 0.4, H=0.9
0
200
400
600
800
1000
-4
-2
0
2
4

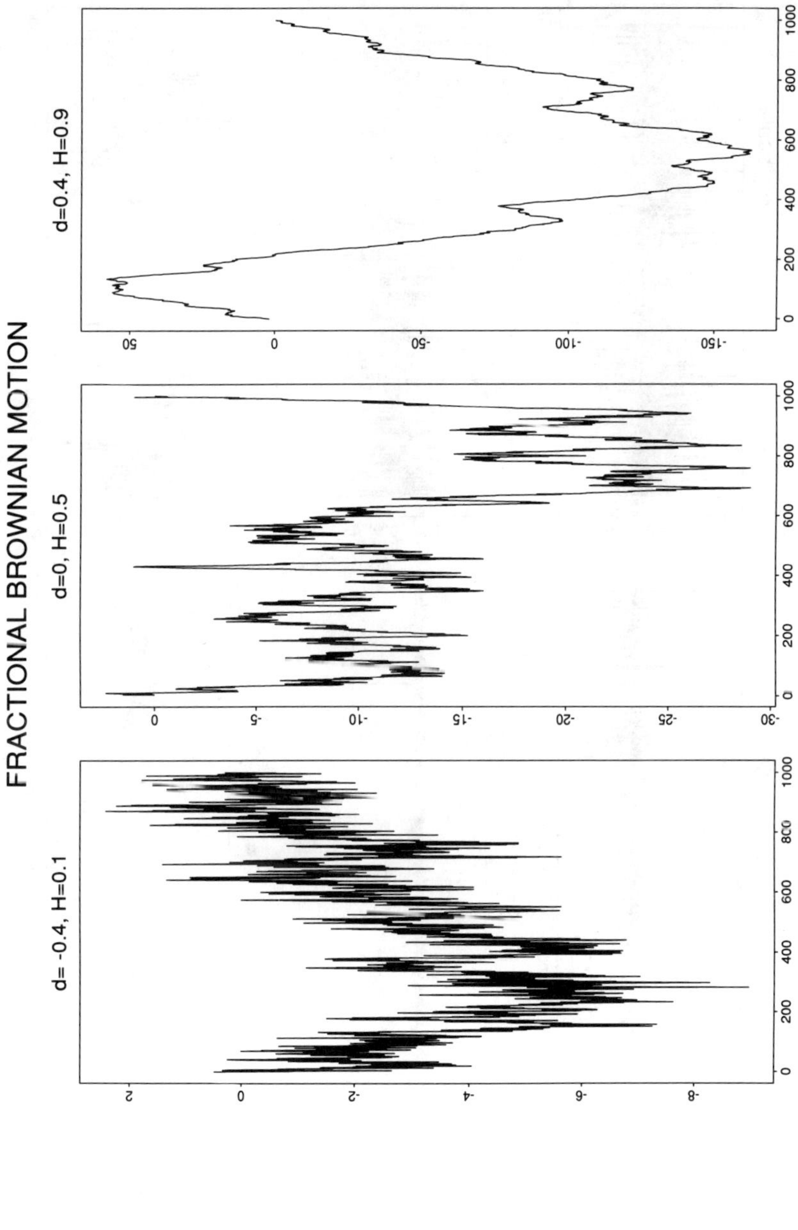
FRACTIONAL BROWNIAN MOTION
d= -0.4, H=0.1
d=0, H=0.5
d=0.4, H=0.9

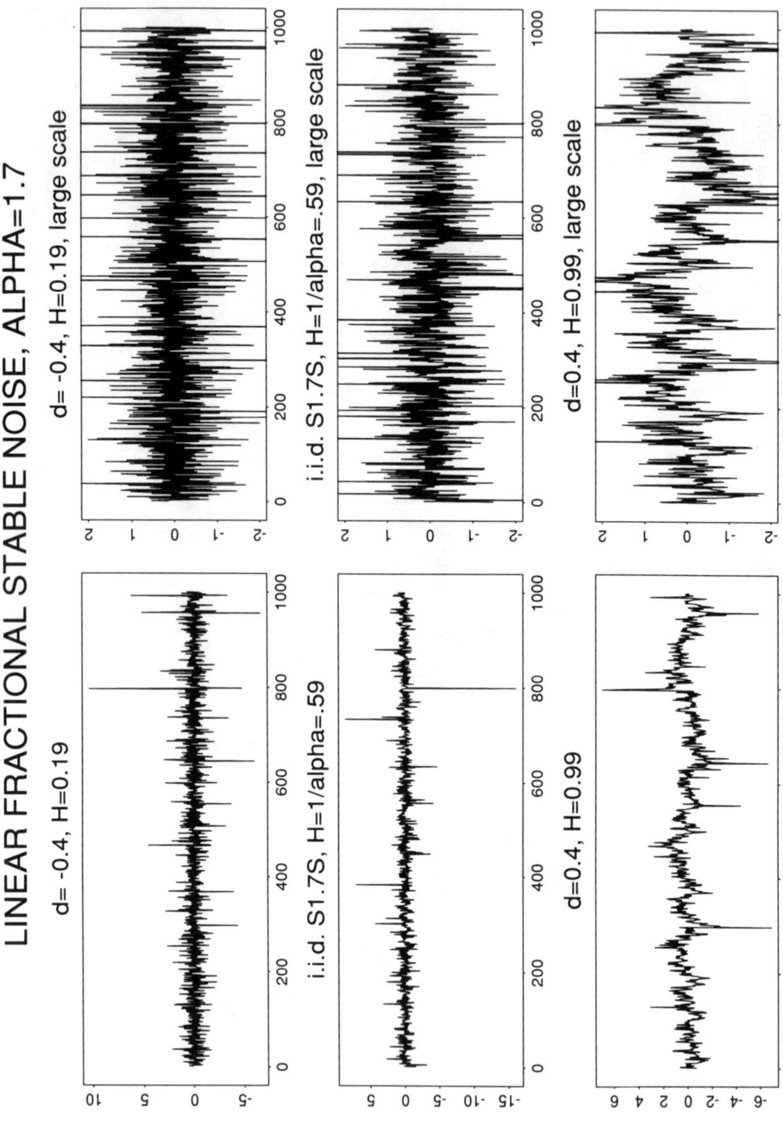
LINEAR FRACTIONAL STABLE NOISE, ALPHA=1.7
d= -0.4, H=0.19
d= -0.4, H=0.19, large scale
i.i.d. S1.7S, H=1/alpha=.59
i.i.d. S1.7S, H=1/alpha=.59, large scale
d=0.4, H=0.99
d=0.4, H=0.99, large scale

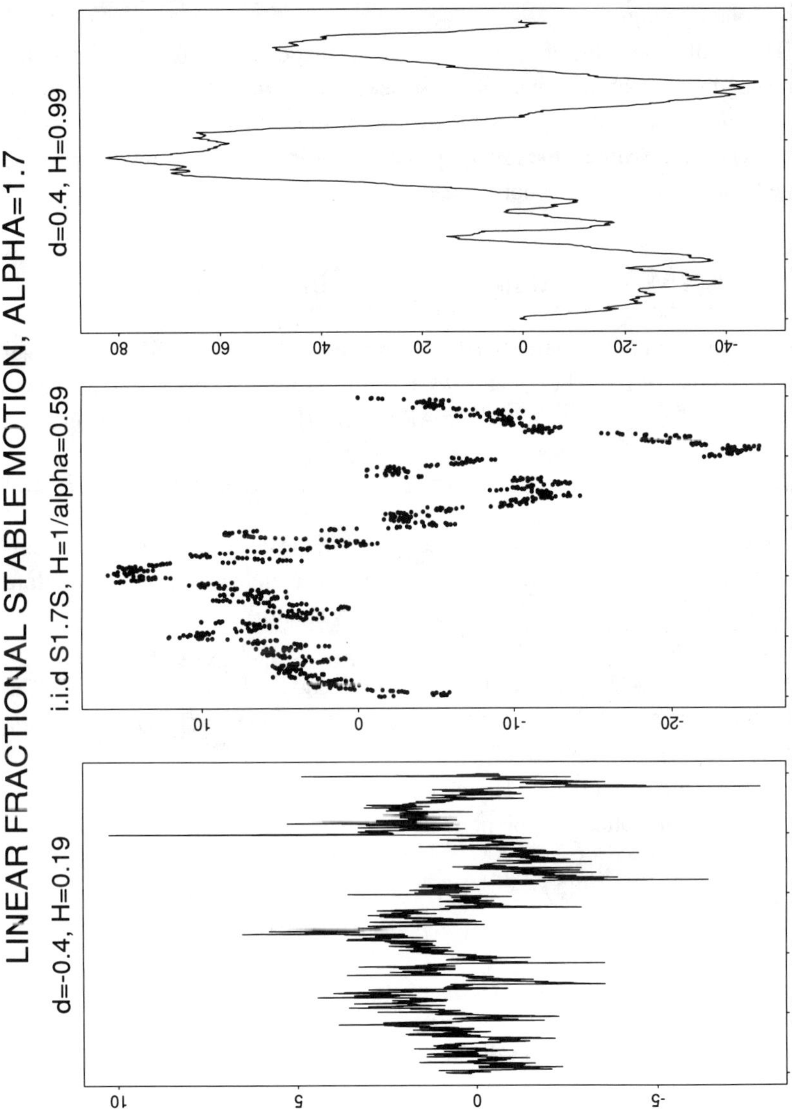
LINEAR FRACTIONAL STABLE MOTION, ALPHA=1.7
d=0.4, H=0.99
i.i.d S1.7S, H=1/alpha=0.59
d=-0.4, H=0.19

in the $S1.7S$ case. The case $d = 0$ is included for reference and is obtained by simply setting $Y_j = \epsilon_j$. The corresponding cumulative sum simulates Brownian motion ($\alpha = 2$) and stable Lévy motion ($\alpha = 1.7$). Since the sample paths of the stable Lévy motion are discontinuous, we have not connected the points in the corresponding figure.

The smaller the value of d, the higher the fluctuations (high frequencies) of the corresponding noises. Because of the scale, the high frequencies are not visible in the figures displaying fractional Brownian or stable motion at $d = 0.4$. Long-range dependence affects the low frequencies. Their presence is quite noticeable in the figures displaying the noise when $d = 0.4$.

7.12 ARMA sequences with stable innovations

Autoregressive-moving average (ARMA) processes are often used for modeling empirical time series. Let p and q be non-negative integers. The sequence $\{X_n,\ n = \ldots, -1, 0, 1, \ldots\}$ is called ARMA (p, q) if it satisfies the equations

$$X_n - \phi_1 X_{n-1} - \cdots - \phi_p X_{n-p} = \epsilon_n + \theta_1 \epsilon_{n-1} + \cdots + \theta_q \epsilon_{n-q}. \qquad (7.12.1)$$

The *innovations* ϵ_n are i.i.d. random variables. In the classical time series literature, the ϵ_ns are either Gaussian or non-Gaussian with finite variance and therefore the probability that they take large values is very small.

Here, we shall suppose that the ϵ_ns are i.i.d. α-stable with $0 < \alpha \leq 2$; in other words, $\epsilon_n \sim S_\alpha(\sigma, \beta, \mu)$ if $0 < \alpha < 2$ and $N(\mu, 2\sigma^2)$ if $\alpha = 2$. The finite-dimensional distributions of the X_ns depend on the coefficients $\theta_1, \ldots, \theta_p$ and $\phi_1, \ldots, \phi_q$.

Consider the system (7.12.1) with real coefficients $\phi_0 = 1,\ \phi_1, \ldots, \phi_p,\ \theta_0 = 1,\ \theta_1, \ldots, \theta_q$ and define the polynomials

$$\begin{aligned} \Phi(z) &= 1 - \phi_1 z - \cdots - \phi_p z^p, \\ \Theta(z) &= 1 + \theta_1 z + \cdots + \theta_q z^q, \end{aligned}$$

where z is a complex variable. One can write (7.12.1) symbolically as

$$\Phi(B) X_n = \Theta(B) \epsilon_n,\ n = \ldots, -1, 0, 1, \ldots, \qquad (7.12.2)$$

where B is the *backward operator*, formally, $B(X_n) = X_{n-1},\ B^2(X_n) = X_{n-2}, \ldots$. As in the Gaussian case, one solves (7.12.1) by showing that $X_n = \Phi(B)^{-1}\Theta(B)\epsilon_n$ is well defined. It is natural to suppose

Condition 7.12.1 *The polynomials $\Theta(z)$ and $\Phi(z)$ do not have common roots.*

The following theorem shows that, as in the Gaussian case, a non-anticipating solution exists if and only if $\Phi(z)$ has no roots in the closed unit disk $\{z : |z| \le 1\}$.

Theorem 7.12.2 *The system (7.12.1) has a unique solution of the form*

$$X_n = \sum_{j=0}^{\infty} c_j \epsilon_{n-j}, \; n \in \mathbb{Z}, \; \text{a.s.} \tag{7.12.3}$$

with real c_js satisfying $|c_j| < Q^{-j}$ eventually,[12] *$Q > 1$, if and only if $\Phi(z)$ has no roots in the closed unit disk $\{z\colon |z| \le 1\}$. The sequence $\{X_n, \; n \in \mathbb{Z}\}$ is then stationary and α-stable.*

The c_js are the coefficients in the series expansion of $\Theta(z)/\Phi(z)$, $|z| < 1$.

PROOF: Suppose $\Phi(z)$ has no roots in $\{z\colon |z| \le 1\}$. The function

$$C(z) = \frac{\Theta(z)}{\Phi(z)}$$

is, therefore, analytic in the disk $\{z\colon |z| < R\}$, where $R > 1$ is the radius of convergence of the series $C(z) = \sum_{j=0}^{\infty} c_j z^j$. Since $1/R = \limsup_{j\to\infty} |c_j|^{1/j}$, for any $1 < Q < R$, $|c_j| < Q^{-j}$ eventually. Using the relation $\Phi(z)C(z) = \Theta(z)$, which holds for $|z| \le 1$, and the fact that the series $\sum_{j=0}^{\infty} c_j z^j$ converges absolutely for $|z| \le 1$, we obtain the following system of equations:

$$\begin{cases} c_0 = 1, \\ c_1 - \phi_1 c_0 = \theta_1, \\ c_2 - \phi_1 c_1 - \phi_2 c_0 = \theta_2, \\ \vdots \\ c_q - \phi_1 c_{q-1} - \phi_2 c_{q-2} - \ldots - \phi_q c_0 = \theta_q, \\ c_s - \phi_1 c_{s-1} - \phi_2 c_{s-2} - \ldots - \phi_s c_0 = 0, \; s > q, \end{cases} \tag{7.12.4}$$

with the understanding that $\phi_i = 0$ if $i > p$. It follows from (7.12.4) that the c_js are real and since $|c_j| < Q^{-j}$ eventually, the series (7.12.3) is well defined and, in fact, converges *absolutely* a.s. (Exercise 1.26).

To see that the process (7.12.3) with the c_js uniquely defined by (7.12.4) satisfies (7.12.1), just use Relation (7.12.4) and the fact that the series (7.12.3) converges absolutely a.s.. Rearranging the terms in

$$\sum_{j=0}^{\infty} c_j \epsilon_{n-j} - \phi_1 \sum_{j=0}^{\infty} c_j \epsilon_{n-1-j} - \cdots - \phi_p \sum_{j=0}^{\infty} c_j \epsilon_{n-p-j}$$

yields (7.12.1).

[12] "$a_j < b_j$ eventually" means that there is a j_0 such that $a_j < b_j$ for all $j > j_0$.

To prove the converse suppose that the system of equations (7.12.1) has a solution of the form (7.12.3) with the c_js satisfying $|c_j| < Q^{-j}$ eventually for some $Q > 1$. We want to show that $\Phi(z) \neq 0$ for $|z| \leq 1$.

Consider the series $C(z) = \sum_{j=0}^{\infty} c_j z^j$ which, under our assumptions, converges absolutely and uniformly in the closed unit disk $\{z\colon |z| \leq 1\}$. Setting

$$\widetilde{\Theta}(z) := \Phi(z)C(z) =: \sum_{j=0}^{\infty} \widetilde{\theta}_j z^j \,, \quad |z| \leq 1, \tag{7.12.5}$$

we obtain

$$\begin{cases} \widetilde{\theta}_0 = c_0, \\ \widetilde{\theta}_s = c_s - \phi_1 c_{s-1} - \cdots - \phi_s c_0 \,, \ s \geq 1. \end{cases} \tag{7.12.6}$$

Since, for any n, the series $\sum_{j=0}^{\infty} c_j \epsilon_{n-j}$ converges absolutely a.s., (7.12.1) and (7.12.6) imply

$$\sum_{j=0}^{q} \theta_j \epsilon_{n-j} = X_n - \phi_1 X_{n-1} - \cdots - \phi_p X_{n-p} = \sum_{j=0}^{\infty} \widetilde{\theta}_j \epsilon_{n-j} \ \ a.s., \tag{7.12.7}$$

which, in turn, yields $\widetilde{\theta}_j = \theta_j$ for $j = 0, 1, \ldots, q$ and $\widetilde{\theta}_j = 0$ for $j > q$. Thus, $\widetilde{\Theta}(z) = \Theta(z)$ and, by (7.12.5),

$$\Phi(z) = \frac{\Theta(z)}{C(z)}, \quad |z| \leq 1.$$

As $C(z)$ is bounded on $\{z\colon |z| \leq 1\}$, $\Phi(z) = 0$ implies $\Theta(z) = 0$. But $\Phi(z)$ and $\Theta(z)$ do not have common roots, so $\Phi(z) \neq 0$, for all $|z| \leq 1$, proving the converse.

The solution (7.12.3) is α-stable because it is a linear combination of α-stable random variables. It is clearly stationary. ∎

The condition that $\Phi(z)$ has no roots in the closed unit disk $\{z\colon |z| \leq 1\}$ is a natural one for it ensures that the system of equations (7.12.1) has the *stationary non-anticipating* solution (7.12.3). We shall suppose from now on that this condition holds.

The c_js are obtained by identifying the coefficients of $C(z) = \sum_{j=0}^{\infty} c_j z^j$ with those in the power series expansion of $\Theta(z)/\Phi(z)$. Since this is the same procedure as in the Gaussian case, the explicit form of the c_js for specific ARMA (p, q) models can be readily found in the time series literature.

Example 7.12.3 Consider the autoregressive process $\{X_n\}$ of order 2 defined by

$$X_n - \phi_1 X_{n-1} - \phi_2 X_2 = \epsilon_n.$$

If $\Phi(z) = 1 - \phi_1 z - \phi_2 z^2$ has two different roots z_1 and z_2, satisfying $|z_i| > 1,\ i = 1, 2$, then

$$\Phi(z) = \frac{1}{z_1 z_2}(z - z_1)(z - z_2)$$

and

$$\frac{1}{\Phi(z)} = \frac{z_1 z_2}{z_2 - z_1}\left(\frac{1}{z_1 - z} - \frac{1}{z_2 - z}\right) = \frac{z_1 z_2}{z_2 - z_1}\left(\frac{z_1^{-1}}{1 - (z/z_1)} - \frac{z_2^{-1}}{1 - (z/z_2)}\right).$$

The coefficients c_j in the series expansion of $1/\Phi(z)$ are, therefore,

$$c_j = \frac{z_1 z_2}{z_2 - z_1}\left(z_1^{-j-1} - z_2^{-j-1}\right),\ j \geq 0.$$

If $\Phi(z)$ has complex conjugate roots $\rho e^{\pm i\mu}$, $\mu \neq k\pi$, then it is not difficult to see that

$$c_j = \frac{\sin \mu(j+1)}{\sin \mu}\rho^{-j}.$$

The ARMA (p, q) time series is *invertible* if there exists a sequence of constants $\{\widetilde{c}_j\}$ such that $\sum_{j=0}^{\infty} |\widetilde{c}_j| < \infty$ and $\sum_{j=0}^{\infty} \widetilde{c}_j X_{n-j} = \epsilon_n,\ n \in \mathbb{Z}$, where convergence holds in probability. Invertibility is particularly useful for prediction because it allows X_n to be expressed in terms of the previous observations $X_j,\ j \leq n$. The following theorem provides a condition for invertibility.

Theorem 7.12.4 *Suppose that $\Theta(z)$ has no roots in the closed unit disk $\{z\colon |z| \leq 1\}$. Then ARMA (p, q) is invertible, i.e.,*

$$\sum_{j=0}^{\infty} \widetilde{c}_j X_{n-j} = \epsilon_n,\ n \in \mathbb{Z},\ \text{a.s..}$$

The $\widetilde{c}_j$s are the coefficients in the series expansion of $\Theta^{-1}(z)\Phi(z),\ |z| < 1$.

PROOF: Because $\Theta^{-1}(z)\Phi(z)$ is analytic in the disk $\{z\colon |z| < R\},\ R > 1$, we have $\sum_{j=0}^{\infty} |\widetilde{c}_j| < \infty$. If $\alpha > 1$, we have $E|X_n| < \infty$ and the result follows from Exercise 7.17. In the general case $0 < \alpha \leq 2$, it follows from Exercise 7.18. ∎

Since the coefficients c_j of the moving average (7.12.3) satisfy $|c_j| < Q^{-j}$ eventually with $Q > 1$, they lie within two exponentially decreasing functions. This behavior is quite different from that of either fractional stable noise or of the fractional ARIMA considered in the next section. In these cases, c_j decreases like a power function.

As for the codifference $\tau(n) = \tau_{X_n, X_0}$, we have:

Theorem 7.12.5 *Suppose that $\Phi(z)$ has no roots in the closed unit disk $\{z\colon |z| \le 1\}$ and let $Q > 1$ be as in Theorem 7.12.2. Then there are constants K_1 and K_2 depending on α, Q and the c_js such that*

$$\limsup_{n\to\infty} Q^n|\tau(n)| \le K_1 \quad \textit{for } 1 \le \alpha \le 2,$$

$$\limsup_{n\to\infty} Q^{\alpha n}|\tau(n)| < K_2 \quad \textit{for } 0 < \alpha < 1.$$

PROOF: See Kokoszka and Taqqu (1993*d*). ■

If $\alpha > 1$, a similar result holds for the covariation, namely

$$\limsup_{n\to\infty} Q^n|[X_n, X_0]_\alpha| < K_3,$$

because

$$Q^n|[X_n, X_0]_\alpha| \le \sum_{i=0}^{\infty} Q^n|c_{j+n}||c_j|^{\alpha-1}$$
$$\le \text{const.} \sum_{j=0}^{\infty} Q^n Q^{-(j+n)} Q^{-(\alpha-1)j} = \text{const.} \sum_{j=0}^{\infty} Q^{-\alpha j} < \infty.$$

7.13 Fractional ARIMA with stable innovations

We now turn to the fractional ARIMA time series, also called FARIMA or fractional ARMA. "ARIMA" stands for "autoregressive integrated moving average." Let Δ be the difference operator, defined by $\Delta X_n = X_n - X_{n-1} = (I - B)X_n$ and let Δ^d, for $d = 1, 2, \dots,$ be Δ iterated d times. The fractional ARIMA model is based on the classical ARIMA (p, d, q) model

$$\Phi(B)\Delta^d X_n = \Theta(B)\epsilon_n,$$

but the parameter d, instead of being a non-negative integer, is allowed to take fractional values, either positive or negative. To simplify the discussion, set $p = q = 0$. If d is a non-negative integer, then $\Delta^d X_n = \epsilon_n$ describes a model where X_n, differenced d times, yields a sequence of i.i.d. random variables (the random walk, for example, is ARIMA $(0, 1, 0)$).

Now define $\Delta^d X_n = \epsilon_n$ for d fractional as $X_n = \Delta^{-d}\epsilon_n$ and interpret $\Delta^{-d} = (I - B)^{-d}$ by using the formal power series expansion $(1 - z)^{-d} = \sum_{j=0}^{\infty} b_j(-d) z^j$ as follows:

$$\Delta^{-d} = (I - B)^{-d} = \sum_{j=0}^{\infty} b_j(-d) B^j,$$

where B^j denotes the backward operator B iterated j times. The coefficients $b_j(-d)$ in the expansion are $b_0(-d) = 1$ and

$$b_j(-d) = \prod_{k=1}^{j} \frac{k+d-1}{k} = \frac{\Gamma(j+d)}{\Gamma(d)\Gamma(j+1)}, \; j = 1, 2, \ldots, \tag{7.13.1}$$

where Γ denotes the gamma function: $\Gamma(x+1) = x\Gamma(x)$ and $\Gamma(j+1) = j!$ for j integer.

The formal definition of fractional ARIMA $(0, d, 0)$ is, then, as follows: it is the moving average

$$X_n = \sum_{j=0}^{\infty} b_j(-d)\epsilon_{n-j}, \; n = \ldots, -1, 0, 1, \ldots, \tag{7.13.2}$$

where the b_js are given by (7.13.1).

Note that for $d = 0, -1, -2, \ldots$, the operator Δ^{-d} is merely the difference operator Δ iterated $|d|$ times, and in this case,

$$b_j(-d) = 0 \;\text{ for } j \geq d \;\text{ if } d = 0, -1, -2, \ldots.$$

For example, $X_n = \epsilon_n$ when $d = 0$, $X_n = \epsilon_n - \epsilon_{n-1}$ when $d = -1$ and $X_n = \epsilon_n - 2\epsilon_{n-1} + \epsilon_{n-2}$ when $d = -2$.

Applying Stirling's formula to (7.13.1), we obtain

$$b_j(-d) \sim \frac{1}{\Gamma(d)} j^{d-1} \;\text{ as } j \to \infty \;\text{ if } d \neq 0, -1, -2, \ldots. \tag{7.13.3}$$

Now suppose $\epsilon_n \sim S_\alpha(\sigma, \beta, \mu)$. The following theorem specifies the conditions on α, d and μ under which fractional ARIMA $(0, d, 0)$ is defined.

Theorem 7.13.1 *Let $\epsilon_n \sim S_\alpha(\sigma, \beta, \mu)$ be i.i.d. and suppose*

$$-\infty < d < 1 - \frac{1}{\alpha}. \tag{7.13.4}$$

Condition (7.13.4) is necessary for the series (7.13.2) to converge. When it holds, the series (7.13.2) converges in the following sense:

(1) $0 < \alpha \leq 1$: absolutely a.s.;

(2) $1 < \alpha \leq 2$: absolutely a.s. if $d \leq 0$, and a.s. if $d > 0$ and $\mu = 0$.

PROOF: Observe firstly that $\sum_{j=j_0}^{\infty} (j^{d-1})^\alpha < \infty$, $j_0 \geq 1$, if and only if $d < 1 - \frac{1}{\alpha}$. Relation (7.13.3) and Exercise 1.26 imply that the series (7.13.2) converges absolutely a.s. in the cases indicated. When $1 < \alpha \leq 2$ and $d > 0$, one does not

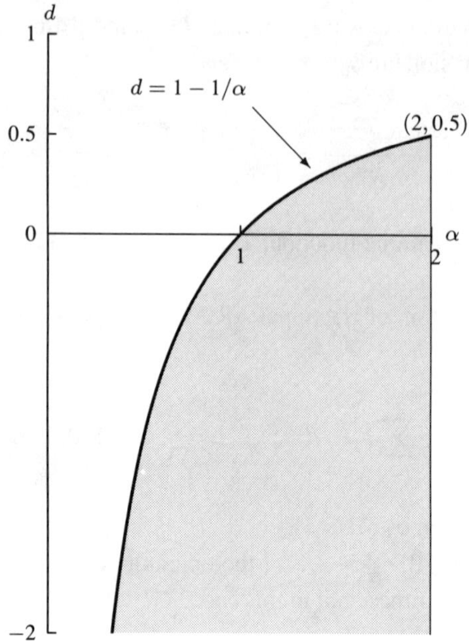

Figure 7.12: The shaded region indicates the allowable values of (d, α) for fractional ARIMA.

have $\sum_{j=0}^{\infty} |b_j(-d)| < \infty$. However, $\sum_{j=0}^{\infty} |b_j(-d)|^{\alpha} < \infty$ and $\mu = 0$ imply that the series (7.13.2) converges a.s. ■

Figure 7.12 illustrates the region of the (d, α) plane where Condition (7.13.4) is satisfied.

Both fractional stable noise and fractional ARIMA $(0, d, 0)$ are moving averages with coefficients that decrease like a power function. It is tempting to compare the asymptotic rate of decay of the kernels. For fractional stable noise $(j+1-x)^{H-1/\alpha} - (j-x)^{H-1/\alpha} \approx (j-x)^{H-(1/\alpha)-1}$ as $j - x \to \infty$ and for fractional ARIMA $(0, d, 0)$, $b_j(-d) \approx j^{d-1}$ as $j \to \infty$. This suggests the following relation between H and d:

$$d = H - \frac{1}{\alpha}. \tag{7.13.5}$$

As for fractional stable noise we will say that $d > 0$ corresponds to *long-range dependence* and $d < 0$ corresponds to *negative dependence*. Since $d < 1 - 1/\alpha$, long-range dependence can occur only if $\alpha > 1$.

Recall that fractional stable noise is defined only for $0 < H < 1$, whereas fractional ARIMA $(0, d, 0)$ is defined for all $d < 1 - 1/\alpha$. The condition $H < 1$ is equivalent to $d < 1 - 1/\alpha$. The lower bound $H > 0$, expressed in terms of

$d = H - 1/\alpha$, becomes $d > -\frac{1}{\alpha}$. Thus $0 < H < 1$ becomes

$$-\frac{1}{\alpha} < d < 1 - \frac{1}{\alpha},$$

an important range for fractional ARIMA.

Indeed, suppose $d < 1 - 1/\alpha$. Since $\Delta^{-d} = \Delta^m \Delta^{-d'}$, where m is a non-negative integer and $-1/\alpha < d' < 1 - 1/\alpha$, the operator Δ^{-d} with $d < 1 - 1/\alpha$ can be interpreted as $\Delta^{-d'}$ followed by a number of full differences (Exercise 7.16).

We have been supposing until now $p = q = 0$. It is straightforward to define a general fractional ARIMA (p, d, q) with $d < 1 - \frac{1}{\alpha}$, by

$$\Phi(B)X_n = \Theta(B)\Delta^{-d}\epsilon_n, \tag{7.13.6}$$

or, formally,

$$X_n = \Phi(B)^{-1}\Theta(B)\Delta^{-d}\epsilon_n, \tag{7.13.7}$$

under the assumptions given in Section 7.12, namely Condition 7.12.1 and that $\Phi(z)$ has no roots in the closed unit disk $\{z\colon |z| \le 1\}$. Relations (7.13.6) and (7.13.7) are to be interpreted as

$$X_n = \sum_{j=0}^{\infty} u_j \epsilon_{n-j}, \tag{7.13.8}$$

where

$$u_j = (c * b(-d))_j = \sum_{i=0}^{j} c_i b_{j-i}(-d), \tag{7.13.9}$$

and where the coefficients c_i are those in (7.12.3). The u_js are thus the coefficients in the power series expansion of $\Phi^{-1}(z)\Theta(z)(1-z)^{-d}$, $|z| < 1$.

The effect of the operator $\Phi^{-1}(B)\Theta(B)$ is to provide modeling flexibility. It modifies the coefficients of the moving average without modifying their asymptotic behavior.

Combining Theorem 7.12.2 with the proof of Theorem 7.13.1, we obtain:

Theorem 7.13.2 *Suppose that the polynomials $\Phi(z)$ and $\Theta(z)$ satisfy Condition 7.12.1 and that $\Phi(z)$ has no roots in the closed unit disk $\{z\colon |z| \le 1\}$. Suppose also that*

$$-\infty < d < 1 - \frac{1}{\alpha},$$

and $\epsilon_n \sim S_\alpha(\sigma, \beta, \mu)$ i.i.d. Then the series (7.13.8) defining fractional ARIMA (p, d, q) converges in the following sense:

(1) $0 < \alpha \le 1$: absolutely a.s.;

(2) $1 < \alpha \le 2$: absolutely a.s. if $d \le 0$, and a.s. if $d > 0$ and $\mu = 0$.

It is a solution of the system of equations (7.13.6).

PROOF: Use Exercises 7.15 and 7.18. ∎

We now address the question of invertibility of fractional ARIMA (p, d, q) when $\alpha > 1$ (see Figure 7.13).

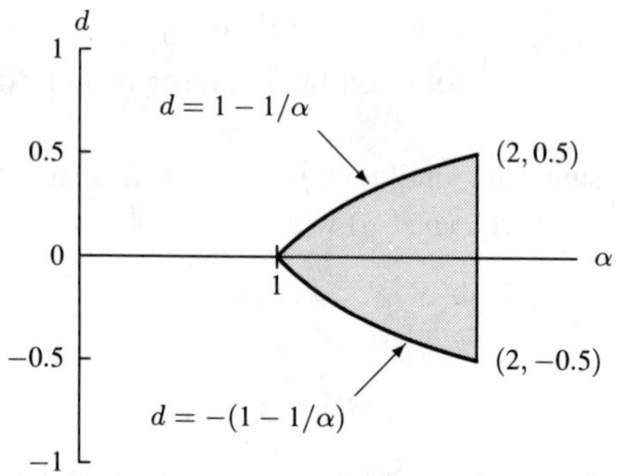

Figure 7.13: Fractional ARIMA is invertible if (d, α) lies in the shaded region.

Theorem 7.13.3 *Suppose that $\Theta(z)$ has no roots in the closed unit disk $\{z\colon |z| \le 1\}$ and let $\widetilde{u}_j$ be the coefficients in the series expansion of $\Theta^{-1}(z)\Phi(z)(1-z)^d$. If*

$$|d| < 1 - \frac{1}{\alpha},$$

then

$$\sum_{j=0}^{\infty} \widetilde{u}_j X_{n-j} = \epsilon_n,\ n \in \mathbb{Z},$$

where the convergence is in L^p, $0 < p < \alpha$. For $0 < d < 1 - 1/\alpha$, the partial sums $\sum_{j=0}^{m} \widetilde{u}_j X_{n-j}$ converges to ϵ_n absolutely a.s.

PROOF: Use Exercise 7.18. ∎

The following theorem gives the asymptotic behavior of the codifference $\tau(n) = \tau_{X_{t+n}, X_t}$.

Theorem 7.13.4 *Suppose $X_n = \sum_{j=0}^{\infty} u_j \epsilon_{n-j}$ is a fractional ARIMA (p, d, q) process with $S\alpha S$ innovations ϵ_n defined in (7.13.8). Suppose $0 < \alpha \le 2$ and d is not an integer.*

(a) If either (i) $\alpha \le 1$ or (ii) $\alpha > 1$ and $(\alpha - 1)(d - 1) > -1$, then

$$\lim_{n\to\infty} \frac{\tau(n)}{n^{\alpha(d-1)+1}} = \left|\frac{\Theta(1)}{\Gamma(d)\Phi(1)}\right|^{\alpha} \int_0^{\infty} g(x)dx, \tag{7.13.10}$$

where

$$g(x) = x^{(d-1)\alpha} + (1+x)^{(d-1)\alpha} - (x^{d-1} - (1+x)^{d-1})^{\alpha}. \qquad (7.13.11)$$

(b) If $\alpha > 1$ and $(\alpha - 1)(d - 1) < -1$, then

$$\lim_{n\to\infty} \frac{\tau(n)}{n^{d-1}} = \frac{\alpha\Theta(1)}{\Gamma(d)\Phi(1)} \sum_{j=0}^{\infty} u_j^{<\alpha-1>}. \qquad (7.13.12)$$

(If d is an integer, then $\{X_n\}$ is a finite moving average and, consequently, $\tau_n = 0$ for large n.)

PROOF: See Kokoszka and Taqqu (1993*c*). ■

Remarks. The condition $(\alpha-1)(d-1) > -1$ can be rewritten as $d > 1-1/(\alpha-1)$. Figure 7.14 shows the regions of the (d, α) plane where the different rates are in effect.[13] The rates are the same as those for the linear fractional stable noise (Theorem 7.10.1).[14]

Instead of the codifference one may consider the covariation $[X_n, X_0]_\alpha$ of X_n and X_0. This covariation, however, is defined only for $\alpha > 1$. Since $X_n = \sum_{j=0}^{\infty} u_j \epsilon_{n-j}$, we have, by Proposition 3.5.2,

$$[X_n, X_0]_\alpha = \sum_{j=0}^{\infty} u_{j+n} u_j^{<\alpha-1>}. \qquad (7.13.13)$$

The next theorem shows that the asymptotic behavior of this covariation is similar to that of the codifference.

Theorem 7.13.5 *Suppose that $\{X_n\}$ is a $S\alpha S$ fractional ARIMA(p, d, q) process defined in (7.13.8), and that $1 < \alpha \le 2$ and that d is not an integer.*
(a) If $(\alpha - 1)(d - 1) > -1$, then

$$\lim_{n\to\infty} \frac{[X_n, X_0]_\alpha}{n^{\alpha(d-1)+1}} = \left|\frac{\Theta(1)}{\Gamma(d)\Phi(1)}\right|^{\alpha} \int_0^{\infty} (y+1)^{d-1} y^{(\alpha-1)(d-1)} dy. \qquad (7.13.14)$$

(b) If $(\alpha - 1)(d - 1) < -1$, then

$$\lim_{n\to\infty} \frac{[X_n, X_0]_\alpha}{n^{(d-1)}} = \frac{\Theta(1)}{\Gamma(d)\,\Phi(1)} \sum_{j=0}^{\infty} u_j^{<\alpha-1>}. \qquad (7.13.15)$$

---

[13] See the remark at the end of this section for the case $\alpha = 2$.

[14] To compare the rates, assume that $d > -1/\alpha$ and set $H = d - 1/\alpha$, $\theta_1 = 1$ and $\theta_2 = -1$ in Theorem 7.10.1. The assumption $H \ne 1/\alpha$ corresponds to $d \ne 0$. Cases (i) and (ii) of Theorem 7.10.1 correspond to case (a) of Theorem 7.13.4 and case (iii) corresponds to case (b).

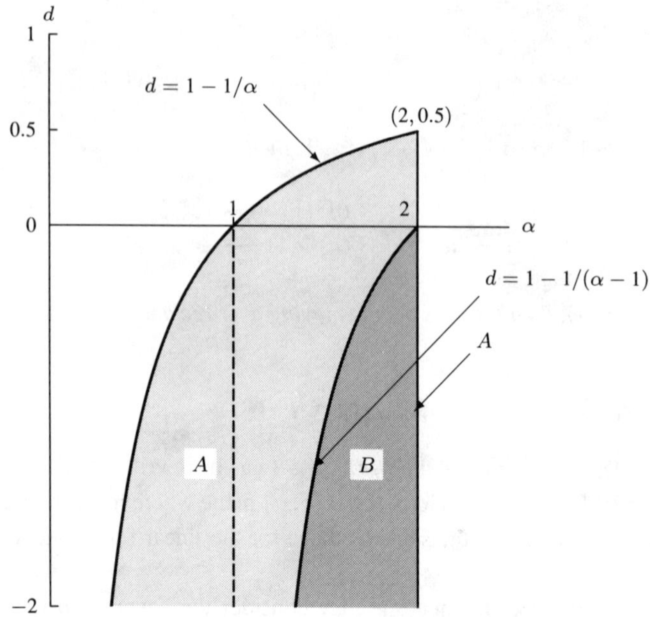

Figure 7.14: Regions in the (d, α) plane where the various rates for the codifference (and for the covariation, if $\alpha > 1$) are in effect. (See Theorems 7.13.4 and 7.13.5.) Regions A and B correspond, respectively, to the rates $n^{\alpha(d-1)+1}$ and n^{d-1}.

PROOF: See Kokoszka and Taqqu (1993*c*). ■

We conclude from these results that, for $\alpha > 1$, the covariation $[X_n, X_0]_\alpha$ and codifference $\tau(n)$ are asymptotically proportional. The constant of proportionality depends only on α and d.

Remark. Theorems 7.13.4 and 7.13.5 and their proofs remain valid in the Gaussian case $\alpha = 2$. Notice that if $\alpha = 2$ and $d < 0$, Relations (7.13.12) and (7.13.15) become, respectively,

$$\lim_{n\to\infty} \frac{\tau(n)}{n^{d-1}} = 0$$

and

$$\lim_{n\to\infty} \frac{[X_n, X_0]_\alpha}{n^{d-1}} = 0$$

because for $d < 0$,

$$\sum_{j=0}^{\infty} u_j = \Phi^{-1}(1)\Theta(1)(1-1)^{-d} = 0.$$

This does not contradict the fact that for $\alpha = 2$ the actual rate of decay is n^{2d-1}. It is the special relation $\sum_{j=0}^{\infty} u_j^{<\alpha-1>} = 0$ valid for $\alpha = 2$, which accounts for the discontinuity at $\alpha = 2$ (and $d < 0$) in the rate of decay of the codifference and the covariation of the fractional ARIMA (p, d, q) time series. Observe finally that $2d - 1 = \alpha(d-1) + 1$ when $\alpha = 2$. Thus, in the Gaussian case $\alpha = 2$, the rate in part (a) of Theorem 7.13.4 is in effect not only for $d > 0$ but for all $-\infty < d < 1/2$, as Figure 7.14 illustrates.

For more details on fractional ARIMA with stable innovations, see Kokoszka and Taqqu (1993*c*).

7.14 Exercises

Exercise 7.1 Let X_1 and X_2 be two independent stochastic processes on $\mathbb{R}$ and let

$$X_1 \circ X_2 = \{X_1(X_2(t)),\ t \in \mathbb{R}\}$$

be the composition process. (Assume X_1 measurable, i.e., the map $(t, \omega) \mapsto X_1(t, \omega)$ is measurable, so that the composition process $X_1 \circ X_2$ is measurable.) Show that if X_1 and X_2 are H_j-sssi, $j = 1, 2$, then $X_1 \circ X_2$ is $H_1 H_2$-sssi.
Hint: Vervaat (1985).

Exercise 7.2 Prove that (7.2.7) represents standard fractional Brownian motion up to a multiplicative constant. Determine that multiplicative constant.

Exercise 7.3 Prove that $\{\int_{-\infty}^{\infty} (\ln|t - x| - \ln|x|) M(dx),\ t \in \mathbb{R}\}$, is well defined and that it represents standard Brownian motion up to a multiplicative constant. Determine that multiplicative constant.

Exercise 7.4 Prove that for every $\gamma > 0$ and complex $z = x + iy,\ x > 0$,

$$\int_0^{\infty} e^{-zu} u^{\gamma-1} du = \frac{\Gamma(\gamma)}{z^{\gamma}},$$

where Γ is the Gamma function and where z^{γ} denotes the branch obtained by continuity for $x > 0$ starting from the positive real branch for z positive. Show by continuity that this formula is still valid for $x \geq 0$, $z \neq 0$ if $0 < \gamma < 1$. Thus, in particular

$$\int_0^{\infty} e^{iu} u^{\gamma-1} du = e^{i\pi\gamma/2} \Gamma(\gamma).$$

Hint: See Choquet-Bruhat (1967).

Exercise 7.5 Make the passage from the representations (7.2.10) to (7.2.12) rigorous.
Hint: See Section 6 of Taqqu (1979).

Exercise 7.6 Show that the representation (7.2.12) defines an H-sssi process.

Exercise 7.7 Show that (7.2.17) represents standard fractional Brownian motion up to a multiplicative constant. Determine that multiplicative constant.

Exercise 7.8 Let $\{Y_j,\ j = \ldots, -1, 0, 1 \ldots\}$ be fractional Gaussian noise. Use its moving average representation (7.2.18) to compute its autocovariance $r(j) = EY_0Y_j$.

Exercise 7.9 Prove Corollary 7.2.12.

Exercise 7.10 Let $\|\cdot\|$ be the Euclidean norm on $\mathbb{R}^n$. For $n \geq 2, 0 < \alpha < 2,\ 0 < H < 1$, set

$$X_{n,\alpha,H}(t) = \int_{\mathbb{R}^n} (\|\mathbf{x} - t\mathbf{1}\|^{H-(n/\alpha)} - \|\mathbf{x}\|^{H-(n/\alpha)})M(d\mathbf{x}),\ t \in \mathbb{R}, \quad (7.14.1)$$

where $\mathbf{x} = (x_1, \ldots, x_n),\ \mathbf{1} = (1, \ldots, 1) \in \mathbb{R}^n$ and M is a $S\alpha S$ random measure on $\mathbb{R}^n$, $0 < \alpha < 2$, with Lebesgue control measure. Show that the processes $X_{n,\alpha,H}$ are well defined and H-sssi.

Exercise 7.11 Show that for any $m,\ n \geq 2,\ m \neq n, 0 < \alpha < 2,\ 0 < H < 1$, the processes $X_{n,\alpha,H}$ and $X_{m,\alpha H}$ of Exercise 7.10 are different.
Hint: See Samorodnitsky and Taqqu (1990*a*), Theorem 3.1. In addition, make the following corrections to page 312 of that paper: on line 7, replace $n = 2$ by $n = 2, 3, 4, 5$, and on line 9, replace $n \geq 3$ by $n \geq 6$.

Exercise 7.12 Show that the α-stable Lévy motion in (7.5.1) and the log-fractional α-stable motion defined in (7.6.1) are different from the process $X_{n,\alpha,H}$ defined in Exercise 7.10 for any $1 < \alpha < 2,\ n \geq 2,\ H = 1/\alpha$. Observe that $X_{n,\alpha,1/\alpha},\ n \geq 2,\ 1 < \alpha < 2$ provide new examples of $1/\alpha$-sssi processes.

Exercise 7.13 Prove that the real harmonizable fractional stable motion (7.7.1) is H-sssi.

Exercise 7.14 Let

$$X_1(t) = \text{Re} \int_{-\infty}^{\infty} e^{ixt} f(x) \widetilde{M}(dx),\ \ t \in \mathbb{R},$$

where $\widetilde{M}$ is a complex $S\alpha S,\ 0 < \alpha < 2$, random measure on $\mathbb{R} \times S_2$ with circular control measure $m\gamma$ where m is the Lebesgue measure on $\mathbb{R}$ and γ is the uniform probability measure on S_2. Show that if one sets $\alpha = 2$ in its joint characteristic functions, the latter become that of

$$X_2(t) = C \int_{-\infty}^{\infty} e^{ixt} f(x) M(dx),\ \ t \in \mathbb{R},$$

where C is a constant and M is a complex Gaussian random measure with control measure m defined as in Section 7.2.

Exercise 7.15 Prove that in Theorem 7.13.2, for d a non-negative integer,

$$\lim_{j\to\infty} j^{-(d-1)} \sum_{i=0}^{j} c_j b_{j-i} = \Gamma(d) \sum_{i=0}^{\infty} c_i.$$

Hint: Divide the inner summation in two: $1 \le i < j/2$ and $j/2 \le i \le j$ and bound the terms accordingly.

Exercise 7.16 Let $0 < \alpha < 2$, $d < 1 - 1/\alpha$, ϵ_j i.i.d. $S\alpha S$ and $X_t^{(d)} = \sum_{j=0}^{\infty} b_j(-d)\epsilon_{t-j}$, $t \in \mathbb{Z}$, where $b_j(-d)$ is defined in (7.13.1). Then for any $m = 0, 1, 2, \ldots$,

$$X_t^{(d-m)} = (1-B)^m X_t^{(d)},\ t \in \mathbb{Z}. \tag{7.14.2}$$

Hint: Let $m = 1$. Show $\lim_{s\to\infty} b_s(-d)\epsilon_{t-1-s} = 0$ a.s. (Borel-Cantelli lemma) and $b_j(-d) - b_{j-1}(-d) = b_j(-d-1)$, $j \ge 1$. Then consider $(1-B)\lim_{s\to\infty} \sum_{j=0}^{s} b_j(-d)\epsilon_{t-j}$. (You could use Exercise 7.18 below.)

Exercise 7.17 Consider an arbitrary sequence of random variables $\{X_n\}_{n=-\infty}^{\infty}$ such that $\sup_n E|X_n| < \infty$. Show that if $\sum_{j=-\infty}^{\infty} |a_j| < \infty$, then the series $\sum_{j=-\infty}^{\infty} a_j X_{n-j}$ converges absolutely a.s.
Hint: Proposition 3.1.1 of Brockwell and Davis (1991).

Exercise 7.18 Prove the following:

Suppose $\{\epsilon_n,\ n \in \mathbb{Z}\}$ is a sequence of i.i.d. $S\alpha S$ random variables with $0 < \alpha \le 2$. Let $\{c_j,\ j = 0, 1, 2, \ldots\}$ be a sequence of real numbers satisfying $\sum_{j=0}^{\infty} |c_j|^\alpha < \infty$ and let $\{\psi_j,\ j = 0, 1, 2, \ldots\}$ be a sequence of real numbers such that $\sum_{j=0}^{\infty} |\psi_j| < \infty$ if $\alpha > 1$, and $\sum_{j=0}^{\infty} |\psi_j|^\alpha < \infty$ if $\alpha \le 1$. Let

$$X_n := C(B)\epsilon_n = \sum_{j=0}^{\infty} c_j \epsilon_{n-j} \text{ and } Y_n := \Psi(B)\epsilon_n = \sum_{j=0}^{\infty} \psi_j \epsilon_{n-j},$$

and let $A(B)\epsilon_n := \sum_{j=0}^{\infty} a_j \epsilon_{n-j}$, where $a_j = \sum_{k=0}^{j} \psi_k c_{j-k}$, $j = 0, 1, 2, \ldots$. Then

$$\lim_{m\to\infty} \sum_{k=0}^{m} \psi_k X_{n-k} = \sum_{j=0}^{\infty} a_j \epsilon_{n-j} \tag{7.14.3}$$

and

$$\lim_{s\to\infty} \sum_{j=0}^{s} c_j Y_{n-j} = \sum_{j=0}^{\infty} a_j \epsilon_{n-j}, \tag{7.14.4}$$

where convergence is in the L^p-norm for any $0 < p < \alpha$. Consequently,

$$\Psi(B)[C(B)\epsilon_n] = A(B)\epsilon_n = C(B)[\Psi(B)\epsilon_n] \text{ a.s.} \tag{7.14.5}$$

Moreover, the left-hand side of (7.14.3) converges absolutely a.s. for $\alpha > 1$. If $\alpha \leq 1$, the absolute a.s. convergence in (7.14.3) takes place under the additional assumption that $\sum_{j=0}^{\infty} |\psi_j|^r < \infty$ for some $0 < r < \alpha$.
Hint: See Kokoszka and Taqqu (1993*c*).

Chapter 8

Chentsov random fields

Stochastic processes $\{X(t),\ t \in \mathbb{R}^n\}$, $n \geq 1$, whose parameter space is the Euclidean space $\mathbb{R}^n$, $n \geq 1$, are called *random fields*.[1] We shall be interested in H-*sssis fields*, that is, in random fields that are self-similar with index H and have stationary increments in the strong sense. Although the definition of "self-similarity" in $\mathbb{R}^n$ is analogous to that in $\mathbb{R}^1$, there are several possible definitions of "stationary increments" in $\mathbb{R}^n$. Here, we suppose that the increments are invariant under all Euclidean rigid body motions.

In Section 8.1 we introduce the H-sssis fields. A well-known example is the Lévy Brownian motion whose autocovariance function is $EX(t)X(s) = \frac{1}{2}\{\|t\| + \|s\| - \|t-s\|\}$. It was introduced by Paul Lévy in 1948 (Lévy 1965) and given a geometric construction by Chentsov (1957). Chentsov's construction allows the field to be defined as $M(V_t)$, $t \in \mathbb{R}^n$, where M is a Gaussian random measure and V_t is the set of all hyperplanes separating the origin zero from the point t. We define the Chentsov fields in Section 8.2 by generalizing Chentsov's construction: we let the measure M be $S\alpha S$, $0 < \alpha \leq 2$, and consider arbitrary measurable sets V_t. We then consider the important subclass of H-sssis Chentsov fields, defined only for $H \leq 1/\alpha$. A first example, the Lévy-Chentsov random field, is given in Section 8.3. That field is $\frac{1}{\alpha}$-sssis and is the stable analog of the Lévy Brownian motion with the same V_ts.

A second example, Takenaka fields, is obtained in Section 8.4 by letting the V_ts be all spheres that separate 0 from t. By choosing the control measure m judiciously, the Takenaka fields become H-sssis with $H < 1/\alpha$, and are then called (α, H)-Takenaka. They are the stable counterpart the Lévy fractional Brownian field, defined for $H < 1/2$, whose autocovariance function equals $EX(t)X(s) = \frac{1}{2}\{\|t\|^{2H} + \|s\|^{2H} - \|t-s\|^{2H}\}$.

[1] To lighten the notation of the chapter we do not use bold letters to denote elements of $\mathbb{R}^n$.

In Sections 8.5 and 8.6 we give properties of the finite-dimensional distributions of, respectively, Chentsov fields and H-sssis Chentsov fields. We show that the spectral measure is concentrated on a finite number of points and that normalized $S\alpha S$ H-sssis Chentsov fields with the same α and H have the same two-dimensional distributions.

In Section 8.7 we consider the codifference, which provides information on the two-dimensional distributions. Because the codifference is defined for stationary stochastic processes, we first consider the H-sssi process $\{X(ue),\ u \in \mathbb{R}\}$, which is the projection of an H-sssis Chentsov field $\{X(t),\ t \in \mathbb{R}^n\}$ in a direction $e \in \mathbb{R}^n$, and then consider the corresponding increment process $\{Y(u) = X((u+1)e) - X(ue),\ u \in \mathbb{R}\}$. We show, using the codifference, that any projection process $\{X(ue),\ u \in \mathbb{R}\}$ obtained from a Chentsov field is different from the linear fractional stable motion and from other H-sssi processes considered in Chapter 7.

The last section deals with Takenaka processes on $[0, \infty)$. In this case, the two-dimensional distributions determine all the higher-dimensional distributions.

8.1 Self-similar fields with stationary increments in the strong sense

The definition of self-similarity for $T = \mathbb{R}^n$ is analogous to the one for $T = \mathbb{R}^1$.

Definition 8.1.1 A random field $\{X(t),\ t \in \mathbb{R}^n\}$ is *self-similar with index* $H > 0$ (H-ss) if $\{X(at),\ t \in \mathbb{R}^n\} \stackrel{d}{=} \{a^H X(t),\ t \in \mathbb{R}^n\}$ for all $a > 0$, where, as usual, $\stackrel{d}{=}$ denotes equality of the finite-dimensional distributions.

The extension of the notion of stationary increments to $\mathbb{R}^n$, $n > 1$, is more delicate. Recall that a process $\{X(t),\ t \in \mathbb{R}\}$ has stationary increments if

$$\{X(t+s) - X(s),\ t \in \mathbb{R}\} \stackrel{d}{=} \{X(t) - X(0),\ t \in \mathbb{R}\},\ \text{for all } s \in \mathbb{R},$$

i.e., if the finite-dimensional distributions of the increments are invariant under translation. Translations are the only Euclidean rigid body motions in $\mathbb{R}$, but in $\mathbb{R}^n$, the Euclidean rigid body motions include all rotations and translations. Let $\mathcal{G}(\mathbb{R}^n)$ denote the group of Euclidean rigid body motions in $\mathbb{R}^n$.

Definition 8.1.2 The random field $\{X(t),\ t \in \mathbb{R}^n\}$ has *stationary increments in the strong sense* (sis) if

$$\{X(g(t)) - X(g(0)),\ t \in \mathbb{R}^n\} \stackrel{d}{=} \{X(t) - X(0),\ t \in \mathbb{R}^n\},$$

for all Euclidean rigid body motions $g \in \mathcal{G}(\mathbb{R}^n)$.

Notation. A random field which is both H-self-similar and has stationary increments in the strong sense is denoted H*-sssis*.

Example 8.1.3 *Lévy fractional Brownian field.* Let $0 < H < 1$ and $\sigma > 0$. The Gaussian field $X = \{X(t),\ t \in \mathbb{R}^n\}$ with mean 0 and autocovariance function

$$EX(t)X(s) = \frac{\sigma^2}{2}\{\|t\|^{2H} + \|s\|^{2H} - \|t-s\|^{2H}\},\ t, s \in \mathbb{R}^n, \tag{8.1.1}$$

is called the *Lévy fractional Brownian field.* There are several ways to verify that X is well defined. One could note, for example, that the right-hand side of (8.1.1) is non-negative-definite (Lemma 2.10.8) and hence defines an autocovariance function. One can also construct X as the integral

$$\sigma_0 \int_{\mathbb{R}^n} (\|t-x\|^{H-\frac{n}{2}} - \|x\|^{H-\frac{n}{2}})M(dx), \tag{8.1.2}$$

where M is a Gaussian random measure on $\mathbb{R}^n$ with Lebesgue control measure and σ_0 is a constant proportional to σ (Exercise 8.1).

The Lévy fractional Brownian field X is an extension of fractional Brownian motion to $\mathbb{R}^n$. In fact,

Proposition 8.1.4 *Let $0 < H < 1$. Then the Lévy fractional Brownian fields are the only H-sssis Gaussian fields on $\mathbb{R}^n$.*

PROOF: Let $\{X(t),\ t \in \mathbb{R}^n\}$ be an H-sssis Gaussian field. H-self-similarity implies that for each $a > 0$, $X(0) = X(a0) \stackrel{d}{=} a^H X(0)$ and hence $X(0) = 0$.

Now any unit vector $e \in \mathbb{R}^n$ can be expressed as $e = g_e(e_0)$ where $e_0 = (1, 0, \dots, 0) \in \mathbb{R}^n$ and where $g_e \in \mathcal{G}(\mathbb{R}^n)$ is the rotation that maps e_0 onto e. Since

$$EX(e) = EX(g_e(e_0)) = E(X(g_e(e_0)) - X(g_e(0))) = EX(e_0),$$

we obtain

$$E(X(s+t) - X(s)) = E(X(t) - X(0)) = EX(t) = \|t\|^H EX(e_0). \tag{8.1.3}$$

By H-self-similarity, on the other hand,

$$E(X(s+t) - X(s)) = (\|s+t\|^H - \|s\|^H)EX(e_0). \tag{8.1.4}$$

Equating (8.1.4) and (8.1.3) proves that $EX(t) = 0$.

Similarly, $EX(t)^2 = \|t\|^{2H}\sigma^2$ for all $t \in \mathbb{R}^n$ where $\sigma^2 = EX^2(e_0)$. Therefore for all $t, s \in \mathbb{R}^n$,

$$E(X(t) - X(s))^2 = E(X(t-s) - X(0))^2 = EX^2(t-s) = \|t-s\|^{2H}\sigma^2$$

and

$$\begin{aligned} EX(t)X(s) &= \frac{1}{2}\{EX^2(t) + EX^2(s) - E(X(t) - X(s))^2\} \\ &= \frac{\sigma^2}{2}\{\|t\|^{2H} + \|s\|^{2H} - \|t-s\|^{2H}\}, \end{aligned}$$

which is the autocovariance function of a Lévy fractional Brownian field. ■

8.2 Chentsov random fields

We want to study $S\alpha S$ random fields of the form $X(t) = \int_E 1_{V_t}(x)M(dx),\ t \in \mathbb{R}^n$, where M is a $S\alpha S$ random measure with control measure m and the V_ts are sets parametrized by $t \in \mathbb{R}^n$.

Definition 8.2.1 Let $0 < \alpha \le 2$, $(E, \mathcal{E}, m)$ be a measure space, $\mathcal{E}$ be non-trivial, M be a $S\alpha S$ random measure with control measure m and $\{V_t,\ t \in \mathbb{R}^n\}$ be a family of measurable subsets satisfying

$$m(V_t) < \infty \quad \text{for all} \quad t \in \mathbb{R}^n.$$

The random field

$$X(t) = M(V_t),\ \ t \in \mathbb{R}^n,$$

is called a $S\alpha S$ *Chentsov field*. The measure m is called the *associated measure*.

Example 8.2.2 Suppose $\alpha = 2$ and $n = 2$ and let M be a Gaussian random measure on $(\mathbb{R}_+)^2$ with Lebesgue control measure. The so-called *Brownian sheet* is the Chentsov field $\{X(t) = M(V_t),\ t \in (\mathbb{R}_+)^2\}$ with

$$V_t = \{(x_1, x_2)\colon 0 \le x_1 \le t,\ 0 \le x_2 \le t\},\ t \in (\mathbb{R}_+)^2.$$

X has mean zero and autocovariance function $EX(s)X(t) = (s_1 \wedge t_1)\,(s_2 \wedge t_2)$ where $s = (s_1, s_2),\ t = (t_1, t_2)$. It is self-similar with $H = 1$ but it does not have stationary increments in the strong sense.

We will be interested in Chentsov fields that are self-similar and have stationary increments in the strong sense. Before showing that such fields exist, let us determine the restrictions that the additional assumption of H-sssis imposes on the associated measure. Fix $0 < \alpha \le 2$, $H > 0$ and let $\{X(t),\ t \in \mathbb{R}^n\}$ be a $S\alpha S$, H-sssis Chentsov field with associated measure m.

Proposition 8.2.3 *H-ss implies*

$$m(V_{at}) = a^{\alpha H} m(V_t), \quad \text{for all } a > 0, \quad t \in \mathbb{R}^n. \tag{8.2.1}$$

The stationarity of increments in the strong sense implies

$$m(V_{g(t)} \,\Delta\, V_{g(0)}) = m(V_t \,\Delta\, V_0), \quad \text{for all } g \in \mathcal{G}(\mathbb{R}^n), \quad t \in \mathbb{R}^n, \tag{8.2.2}$$

where Δ denotes the symmetric difference.

PROOF: H-ss implies $M(V_{at}) \stackrel{d}{=} a^H M(V_t)$ for any $a > 0$ and any fixed $t \in \mathbb{R}^n$. Equating the characteristic functions yields $m(V_{at}) = a^{\alpha H} m(V_t)$.

Stationarity of the increments in the strong sense implies

$$M(V_{g(t)}) - M(V_{g(0)}) \stackrel{d}{=} M(V_t) - M(V_0) \tag{8.2.3}$$

for any fixed t and $g \in \mathcal{G}(\mathbb{R}^n)$. Letting V^c denote the complement of V, we have

$$\begin{aligned} &M(V_{g(t)}) - M(V_{g(0)}) \\ &\quad = M(V_{g(t)} \cap V_{g(0)}) + M(V_{g(t)} \cap V_{g(0)}^c) \\ &\qquad - M(V_{g(0)} \cap V_{g(t)}) - M(V_{g(0)} \cap V_{g(t)}^c) \\ &\quad = M(V_{g(t)} \cap V_{g(0)}^c) - M(V_{g(0)} \cap V_{g(t)}^c), \end{aligned} \tag{8.2.4}$$

Since these last two terms are independent, we obtain

$$\begin{aligned} &-\ln E \exp\{i(M(V_{g(t)}) - M(V_{g(0)}))\} \\ &\quad = m(V_{g(t)} \cap V_{g(0)}^c) + m(V_{g(0)} \cap V_{g(t)}^c) \\ &\quad = m\Big((V_{g(t)} \cap V_{g(0)}^c) \cup (V_{g(0)} \cap V_{g(t)}^c)\Big) \\ &\quad - m(V_{g(t)} \,\Delta\, V_{g(0)}). \end{aligned} \tag{8.2.5}$$

Relation (8.2.3) establishes (8.2.2). ∎

Remark. Setting $t = 0$ in (8.2.1) gives

$$m(V_0) = 0,$$

i.e., $X(0) = M(V_0) = 0$ a.s.. Then (8.2.2) becomes

$$m(V_{g(t)} \,\Delta\, V_{g(0)}) = m(V_t), \quad \forall g \in \mathcal{G}(\mathbb{R}^n), \ t \in \mathbb{R}^n.$$

The measure m and the family $\{V_t, \ t \in \mathbb{R}^n\}$ induce a *pseudo metric*

$$d(t, s) = m(V_t \,\Delta\, V_s)$$

in $\mathbb{R}^n$. Indeed, $d(t,t) = 0,\ d(t,s) = d(s,t)$ and $d(t,s) \le d(t,u) + d(u,s)$. To establish the last relation, note that

$$\begin{aligned} V_t \,\Delta\, V_s &= (V_t \,\Delta\, V_u) \,\Delta\, (V_u \,\Delta\, V_s) \\ &\subset (V_t \,\Delta\, V_u) \cup (V_u \,\Delta\, V_s). \end{aligned}$$

The function d does not define a metric because $d(t,s) = 0$ does not imply $t = s$ in general. If X is a H-sssis process, however, then d is a metric. Indeed,

Proposition 8.2.4 *H-sssis implies*

$$d(t,s) = m(V_t \,\Delta\, V_s) = c\|t - s\|^{\alpha H},$$

where $c = d(e_0, 0)$ *and* $e_0 = (1, 0, \ldots, 0)$.

PROOF: Relation (8.2.2) implies $d(g(v),\ g(0)) = d(v,0)$ for any Euclidean rigid body motion g in $\mathbb{R}^n$. Setting $g(v) = v - u$ gives $d(v-u, -u) = d(v,0)$ and setting $t = v - u,\ s = -u$ yields

$$d(t,s) = d(t - s, 0). \tag{8.2.6}$$

Now $t - s = \|t - s\| e$ where e is the unit vector in the direction $t - s$. However, relation (8.2.1) implies

$$d(t - s, 0) = d(\|t - s\| e, 0) = \|t - s\|^{\alpha H}\, d(e, 0). \tag{8.2.7}$$

Now let g be the rotation around the origin that maps the unit vector e into the unit vector $e_0 = (1, 0, \ldots, 0)$. Since $g(0) = 0$, we obtain by using (8.2.1),

$$\begin{aligned} d(e,0) &= m(V_e \,\Delta\, V_0) = m(V_{g(e)} \,\Delta\, V_{g(0)}) \\ &= m(V_{e_0} \,\Delta\, V_0) = d(e_0, 0). \end{aligned} \tag{8.2.8}$$

Relations (8.2.6), (8.2.7) and (8.2.8) prove the proposition. ∎

Corollary 8.2.5 *If $\{X(t),\ t \in \mathbb{R}^n\}$ is an H-sssis $S\alpha S$ Chentsov field, then*

$$H \le 1/\alpha.$$

PROOF: Proposition 8.2.4 and the triangle inequality imply $c|2t|^{\alpha H} = d(2t, 0) \le d(2t, t) + d(t, 0) = 2c|t|^{\alpha H}$, i.e., $2^{\alpha H} \le 2$ or $H \le 1/\alpha$. ∎

Corollary 8.2.5 implies that the Lévy fractional Brownian field cannot be represented as a Chentsov field when $1/2 < H < 1$. We will see that it can be so represented when $0 < H \le 1/2$ (Examples 8.3.3 and 8.4.5).

The Lévy-Chentsov and (α, H)-Takenaka fields defined below provide examples of $S\alpha S$ H-sssis Chentsov fields with $H = 1/\alpha$ and $0 < H < 1/\alpha$, respectively.

Propositions 8.2.3 and 8.2.4 were established using only the marginal distributions of the field X. They provide *necessary conditions* for H-sssis. The next result gives *sufficient conditions* for H-sssis and involves the finite-dimensional distributions of X.

The following *notation* will now be used:

For $d \geq 1$, consider the set of all functions

$$\Sigma_d = \{ \Delta : \{1, 2, \ldots, d\} \to \{0,1\}^d \backslash (0, 0, \ldots, 0)\} \tag{8.2.9}$$

and let $\delta = (\delta(1), \delta(2), \ldots, \delta(d))$. Each component of $\delta \in \Sigma_d$ is either 0 or 1, but the components are not all zero. Moreover, if B is a set, let

$$\begin{cases} B^1 = B, \\ B^0 = B^c = \text{the complement of } B. \end{cases}$$

Thus, $(B^c)^1 = B^0$ and $(B^c)^0 = B^1 = B$.

Theorem 8.2.6 *Set $0 < \alpha \leq 2$, $0 < H \leq 1/\alpha$. Let $(E, \mathcal{E}, m)$ be a measure space, M be a $S\alpha S$ random measure on $(E, \mathcal{E})$ with control measure m and $\{V_t,\ t \in \mathbb{R}^n\}$ be a family of m-measurable sets. Then*

$$X(t) = M(V_t),\ t \in \mathbb{R}^n,$$

is an H-sssis $S\alpha S$ Chentsov process if the following three conditions hold:

(i) $m(V_t) < \infty$ for any $t \in \mathbb{R}^n$;

(ii)

$$m\left(\bigcap_{k=1}^{d} V_{at_k}^{\delta(k)}\right) = a^{\alpha H} m\left(\bigcap_{k=1}^{d} V_{t_k}^{\delta(k)}\right) \tag{8.2.10}$$

for any $t_1, t_2, \ldots, t_d \in \mathbb{R}^n$, $\delta \in \Sigma_d$, $d \geq 1$, $a > 0$;

(iii)

$$m\left(\bigcap_{k=1}^{d}\left(V_{g(t_k)}^{\delta(k)} \Delta V_{g(0)}\right)\right) = m\left(\bigcap_{k=1}^{d} V_{t_k}^{\delta(k)}\right), \tag{8.2.11}$$

for any $t_1, \ldots, t_d \in \mathbb{R}^n$, $\delta \in \Sigma_d$, $d \geq 1$ and for any $g \in \mathcal{G}(\mathbb{R}^n)$.

PROOF: Condition (i) ensures that $\{X(t),\, t \in \mathbb{R}^n\}$ is a $S\alpha S$ Chentsov field. To prove that it is H-ss, let $t_1, \ldots, t_d \in \mathbb{R}^n$ and notice that each set $V_{t_1}, \ldots, V_{t_d}$ can be expressed as a union of disjoint sets as follows:

$$V_{t_j} = \bigcup_{\delta:\, \delta(j)=1} \bigcap_{k=1}^{d} V_{t_k}^{\delta(k)},$$

where the union is over all functions $\delta \in \Delta_\alpha$ such that $\delta(j) = 1$ (see Exercise 8.2).

We shall now prove that the characteristic function of the vector $(X(at_1), \ldots, X(at_d))$, $a > 0$, equals that of $(a^H X(t_1), \ldots, a^H X(t_d))$. We have

$$\begin{aligned}
&E\exp\Big\{i\sum_{j=1}^{d}\theta_j X(at_j)\Big\}\\
&= E\exp\Big\{i\sum_{j=1}^{d}\theta_j M(V_{at_j})\Big\}\\
&= E\exp\Big\{i\sum_{j=1}^{d}\theta_j M\Big(\bigcup_{\delta:\,\delta(j)=1}\bigcap_{k=1}^{d} V_{at_k}^{\delta(k)}\Big)\Big\}\\
&= E\exp\Big\{i\sum_{j=1}^{d}\theta_j \sum_{\delta:\,\delta(j)=1} M\Big(\bigcap_{k=1}^{d} V_{at_k}^{\delta(k)}\Big)\Big\}\\
&= E\exp\Big\{i\sum_{\delta\in\Sigma_d} M\Big(\bigcap_{k=1}^{d} V_{at_k}^{\delta(k)}\Big)\sum_{j:\,\delta(j)=1}\theta_j\Big\}\\
&= \prod_{\delta\in\Sigma_d} E\exp\Big\{i\Big(\sum_{j=1}^{d}\theta_j\delta(j)\Big) M\Big(\bigcap_{k=1}^{d} V_{at_k}^{\delta(k)}\Big)\Big\}\\
&= \prod_{\delta\in\Sigma_d} \exp\Big\{-|\sum_{j=1}^{d}\theta_j\delta(j)|^\alpha m\Big(\bigcap_{k=1}^{d} V_{at_k}^{\delta(k)}\Big)\Big\} \qquad (8.2.12)\\
&= \prod_{\delta\in\Sigma_d} \exp\Big\{-|\sum_{j=1}^{d}\theta_j\delta(j)a^H|^\alpha m\Big(\bigcap_{k=1}^{d} V_{t_k}^{\delta(k)}\Big)\Big\}\\
&= \cdots = E\exp\Big\{i\sum_{j=1}^{d}\theta_j a^H X(t_j)\Big\},
\end{aligned}$$

by retracing the preceding steps.

We establish the stationarity of the increments by showing that for any $g \in \mathcal{G}(\mathbb{R}^n)$, the characteristic function of $(X(g(t_1)) - X(g(0)), \ldots, X(g(t_d) -$

$X(g(0)))$ equals that of $(X(t_1), \ldots, X(t_d))$. (Notice that (ii) implies $m(V_0) = 0$.) We have

$$
\begin{aligned}
&E \exp\Big\{ i \sum_{j=1}^{d} \theta_j (X(g(t_j)) - X(g(0))) \Big\} \\
&= E \exp\Big\{ i \sum_{j=1}^{d} \theta_j (M(V_{g(t_j)}) - M(V_{g(0)})) \Big\} \\
&= E \exp\Big\{ i \sum_{j=1}^{d} \theta_j (M(V_{g(t_j)} \cap V_{g(0)}^c) - M(V_{g(t_j)}^c \cap V_{g(0)})) \Big\} \quad \text{as in (8.2.4)} \\
&= E \exp\Big\{ i \sum_{j=1}^{d} \theta_j \Big(M\Big(\bigcup_{\delta:\, \delta(j)=1} \bigcap_{k=1}^{d} \Big(V_{g(t_k)}^{\delta(k)} \cap V_{g(0)}^c \Big)\Big) \\
&\qquad - M\Big(\bigcup_{\delta:\, \delta(j)=1} \bigcap_{k=1}^{d} \Big(\Big(V_{g(t_k)}^c \Big)^{\delta(k)} \cap V_{g(0)} \Big)\Big)\Big) \Big\} \\
&= \prod_{\delta \in \Sigma_d} \exp\Big\{ -\Big| \sum_{j:\, \delta(j)=1} \theta_j \Big|^\alpha m\Big(\bigcap_{k=1}^{d} \Big(V_{g(t_k)}^{\delta(k)} \cap V_{g(0)}^c \Big)\Big) \Big\} \\
&\qquad \times \exp\Big\{ -\Big| \sum_{j:\, \delta(j)=1} \theta_j \Big|^\alpha m\Big(\bigcap_{k=1}^{d} \Big(\Big(V_{g(t_k)}^c \Big)^{\delta(k)} \cap V_{g(0)} \Big)\Big) \Big\} \quad \text{as in (8.2.12)} \\
&= \prod_{\delta \in \Sigma_d} \exp\Big\{ -\Big| \sum_{j=1}^{d} \theta_j \delta(j) \Big|^\alpha m\Big(\bigcap_{k=1}^{d} V_{g(t_k)}^{\delta(k)} \,\Delta\, V_{g(0)} \Big) \Big\} \quad \text{as in (8.2.5)} \\
&= \prod_{\delta \in \Sigma_d} \exp\Big\{ -\Big| \sum_{j=1}^{d} \theta_j \delta(j) \Big|^\alpha m\Big(\bigcap_{k=1}^{d} V_{t_k}^{\delta(k)} \Big) \Big\} \quad \text{by (iii)} \qquad (8.2.13) \\
&= \cdots = E \exp\Big\{ i \sum_{j=1}^{d} \theta_j X(t_j) \Big\}, \qquad (8.2.14)
\end{aligned}
$$

after retracing our steps. This concludes the proof. ∎

The preceding proof also establishes:

Corollary 8.2.7 *Let $\{X(t),\ t \in \mathbb{R}^n\}$ be a Chentsov field with associated measure m. The characteristic function of the vector $(X(t_1), \ldots, X(t_d))$, $d \geq 1$, equals*

$$
E \exp\Big\{ i \sum_{j=1}^{d} \theta_j X(t_j) \Big\} = \prod_{\delta \in \Sigma_d} \exp\Big\{ -\Big| \sum_{j:\, \delta(j)=1} \theta_j \Big|^\alpha m\Big(\bigcap_{k=1}^{d} V_{t_k}^{\delta(k)} \Big) \Big\}, \qquad (8.2.15)
$$

and is therefore determined by the numbers $m(\bigcap_{k=1}^{d} V_{t_k}^{\delta(k)})$, $\delta \in \Sigma_d$.

8.3 Example: the Lévy-Chentsov random field

As a first example, we construct the Lévy-Chentsov random field, which is an extension to $\alpha \leq 2$ of the Lévy fractional Brownian field defined in Example 8.1.3.

Let S_n denote as usual the ($(n-1)$-dimensional) unit sphere in $\mathbb{R}^n$ and let s be a point on S_n. A hyperplane h in $\mathbb{R}^n$ which does not contain the origin is a set of dimension $n-1$ parameterized by $(s, r) \in S_n \times \mathbb{R}_+$ where $\mathbb{R}_+ = \{r\colon r > 0\}$.

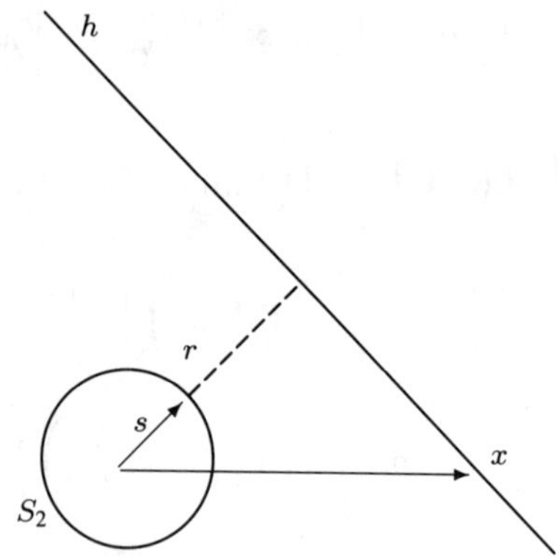

Figure 8.1: The hyperplane h is defined by the parameters s and r.

Here, it will be more convenient to regard the hyperplane h not as the set $\{x \in \mathbb{R}^n\colon <x, s>= r\}$, where $<\ ,\ >$ denotes the scalar product in $\mathbb{R}^n$, but as the pair of parameters $(s, r) \in S_n \times \mathbb{R}_+$. We are now in position to define the measure space $(E, \mathcal{E}, m)$ and the sets $V_t,\ t \in \mathbb{R}^n$. Let

$$
\begin{aligned}
E &= \text{all hyperplanes } h \text{ in } \mathbb{R}^n \text{ of dimension } n-1 \\
&\quad\ \text{that do not contain the origin} \\
&= \{(s, r)\colon s \in S_n,\ 0 < r < \infty\} \\
&= S_n \times \mathbb{R}_+, \\
\mathcal{E} &= \text{Borel } \sigma\text{-field on } E,
\end{aligned}
$$

and let m be the measure on $(E, \mathcal{E})$ given by

$$m(ds, dr) = ds\, dr$$

where ds denotes the Lebesgue (i.e., uniform) measure on S_n $(m(S_n) =$ area of $S_n)$. The measure m is σ-finite since

$$\{s \in S_n,\ r > 0\} = \bigcup_{i=0}^{\infty} \{s \in S_n,\ i < r \leq i+1\}.$$

If B is a set of hyperplanes in $\mathcal{E}$, then

$$m(B) = \int_B ds dr. \tag{8.3.1}$$

Hence:

(i) $m(\{h\}) = 0$ (the measure of a single hyperplane is 0).

(ii) A set of hyperplanes, all containing a given point has m-measure 0.

(iii) A set of parallel hyperplanes has m-measure 0.

A rigid body motion $g \in \mathcal{G}(\mathbb{R}^n)$ induces naturally a map g' on $S_n \times \mathbb{R}_+$. Since

$$m(g'(B)) = m(B), \tag{8.3.2}$$

we say, simply, that the measure m is invariant with respect to Euclidean rigid body motions of hyperplanes. Thus, in particular, the measure of all hyperplanes that intersect segments of equal length is the same.

Finally, for each $t \in \mathbb{R}^n$, let

$$\begin{aligned} V_t &= \text{all hyperplanes } h \text{ separating the origin 0 and the point } t \\ &= \{(s, r) \colon 0 < r << s, t >\} \end{aligned}$$

and $V_0 = \emptyset$. Observe that

$$\begin{aligned} m(V_t) &= \int_{S_n} \text{Leb}_1(r \colon 0 < r << s, t >) ds \\ &= \frac{1}{2} \int_{S_n} | < s, t > | ds = c\|t\|, \end{aligned} \tag{8.3.3}$$

where $c = (1/2) \int_{S_n} < s, e_0 > ds$ and $e_0 = (1, 0, \ldots, 0) \in S_n$.

Definition 8.3.1 A Lévy-Chentsov field $\{X(t) = M(V_t),\ t \in \mathbb{R}^n\}$ is a $S\alpha S$, $0 < \alpha \leq 2$, Chentsov random field with the associated measure m and the sets $\{V_t,\ t \in \mathbb{R}^n\}$ defined above.

Theorem 8.3.2 *A $S\alpha S$, $0 < \alpha \leq 2$, Lévy-Chentsov field $\{X(t) = M(V_t),\ t \in \mathbb{R}^n\}$ has the following properties:*

- *It is $\frac{1}{\alpha}$-sssis.*
- *There is a constant $c > 0$ such that*

$$d(t, s) = m(V_t \,\Delta\, V_0) = c\|t - s\|.$$

- *If $n = 1$, non-overlapping increments are independent.*

PROOF: Conditions (i), (ii) and (iii) of Theorem 8.2.6 hold with $H = 1/\alpha$. (The verification of condition (ii) is similar to that of (8.3.3), and condition (iii) holds because $m(V_0) = 0$ and m is invariant under rigid body motions of hyperplanes.) Hence X is an $\frac{1}{\alpha}$-sssis random field. The relation $d(t,s) = c\|t-s\|$ follows from Proposition 8.2.4. Finally, set $n = 1$ and let $0 < t_1 < t_2 < t_3 < t_4$. Hyperplanes that pass through an interval can be identified here with that interval, and since the intervals $[t_1, t_2]$ and $[t_3, t_4]$ are disjoint, we conclude from Theorem 3.5.3 that $X(t_2) - X(t_1)$ is independent of $X(t_2) - X(t_1)$. ■

Example 8.3.3 When $\alpha = 2$, the Lévy-Chentsov field $\{X(t) = M(V_t),\ t \in \mathbb{R}^n\}$ is Gaussian with mean 0 and autocovariance

$$EX(t)X(s) = \frac{\sigma^2}{2}\{\|t\| + \|s\| - \|t-s\|\},\ t, s \in \mathbb{R}^n,$$

by Proposition 8.1.4. It is the Lévy fractional Brownian field with $H = 1/2$, and is usually known as *Lévy Brownian motion.* If, in addition, $n = 1$, then $EX(t)X(s)$ equals either $\min(t,s)$, if t and s have the same signs, and equals 0 if they have opposite signs. Hence, $\{X(t),\ t \in \mathbb{R}^1\}$ is *Brownian motion.*

8.4 Example: Takenaka random fields

Although, in the preceding section, we defined V_t as the set of *hyperplanes* separating 0 and t, we shall now take V_t to be the set of *hyperspheres* that separate 0 and t. Our goal is to construct a family of $S\alpha S$ Chentsov fields that are H-sssis with $H < 1/\alpha$.

A (hyper) sphere in $\mathbb{R}^n$ of dimension $n-1$ can be represented by a pair (x, λ) where $x \in \mathbb{R}^n$ is its center and $\lambda \in \mathbb{R}_+$ is its radius. It will be more convenient here to regard the sphere not as the subset $\{y\colon \|y - x\| = \lambda\}$ of $\mathbb{R}^n$ but as the pair of parameters $(x, \lambda) \in \mathbb{R}^n \times \mathbb{R}_+$.

Define

$$\begin{aligned} E &= \text{all spheres in } \mathbb{R}^n \text{ of dimension } n-1 \\ &= \{(x,\lambda)\colon x \in \mathbb{R}^n,\ 0 < \lambda < \infty\} \qquad (8.4.1) \\ &= \mathbb{R}^n \times \mathbb{R}_+, \\ \mathcal{E} &= \text{Borel } \sigma\text{-field on } E, \end{aligned}$$

and for each $t \in \mathbb{R}^n$, let

$$\begin{aligned} V_t &= \text{all spheres in } E \text{ separating the origin 0 and the point } t \\ &= \{(x,\lambda)\colon \|x\| \le \lambda\} \,\Delta\, \{(x,\lambda)\colon \|x - t\| \le \lambda\}. \qquad (8.4.2) \end{aligned}$$

The set V_t is illustrated in Figure 8.2 in the case $n = 1$.

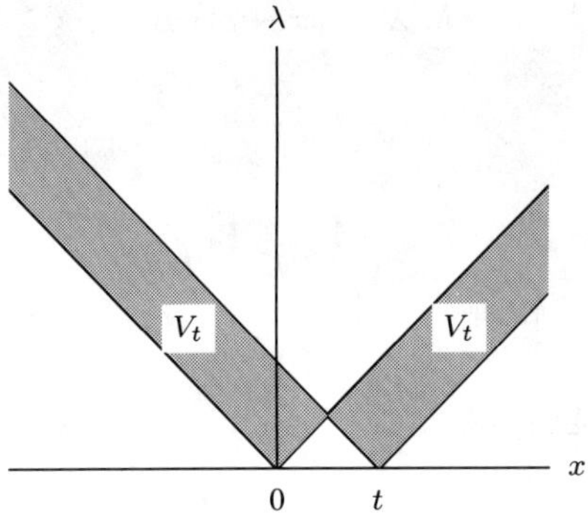

Figure 8.2: The set V_t for $t \in \mathbb{R}^1$.

To understand the analytic definition of V_t, note that the set V_t excludes spheres that contain both 0 and t or that contain neither 0 nor t. Thus, for a sphere (x, λ) to be in V_t, we must have either

$$\|0 - x\| \leq \lambda \text{ and } \|t - x\| > \lambda$$

(0 is in the sphere but t is not) or

$$\|t - x\| \leq \lambda \text{ and } \|0 - x\| > \lambda$$

(t is in the sphere but 0 is not).

Definition 8.4.1 Let $0 < \alpha \leq 2$. If $(E, \mathcal{E})$ and $\{V_t,\ t \in \mathbb{R}^d\}$ are defined as in (8.4.1) and (8.4.2), respectively, then the $S\alpha S$ Chentsov field $\{X(t) = M(V_t),\ t \in \mathbb{R}^n\}$ is called a ($S\alpha S$) *Takenaka field.*

A Takenaka field can have any associated measure m, as long as $m(V_t) < \infty$, $t \in \mathbb{R}^n$. Because we are interested in self-similar fields, we shall consider the family of associated measures $m = m_\beta$, $0 < \beta < 1$, where

$$m_\beta(dx, d\lambda) = \lambda^{\beta-n-1} dx\, d\lambda. \tag{8.4.3}$$

Lemma 8.4.2 *(a) Fix $0 < \beta < 1$. Then $m_\beta(V_t) < \infty$ for all $t \in \mathbb{R}^n$ and, moreover,*

$$m_\beta(V_t) = \frac{2}{n-\beta} \int_{\{x \in \mathbb{R}^n:\, \|x\| < \|x-t\|\}} (\|x\|^{\beta-n} - \|x-t\|^{\beta-n}) dx \tag{8.4.4}$$

which equals

$$\frac{\|t\|^\beta}{n-\beta} \int_{\mathbb{R}^n} \left| \left((z_1 + \frac{1}{2})^2 + \sum_{k=2}^n z_k^2 \right)^{(\beta-n)/2} - \left((z_1 - \frac{1}{2})^2 + \sum_{k=2}^n z_k^2 \right)^{(\beta-n)/2} \right| dz. \tag{8.4.5}$$

(b) If $m_\beta(V_t) < \infty$ for some $t \neq 0$ and some $-\infty < \beta < \infty$, then $0 < \beta < 1$.

PROOF: Set $\mu_\beta(d\lambda) = \lambda^{\beta-n-1}d\lambda$ so that $m_\beta(dx, d\lambda) = \mu_\beta(d\lambda)dx$ and define, for each $x \in \mathbb{R}^n$, $V_t^{(x)} = \{\lambda\colon \lambda \geq \|x - t\|\} \,\Delta\, \{\lambda\colon \lambda \geq \|x\|\} \subset \mathbb{R}^1$. Then $m_\beta(V_t)$ equals

$$\begin{aligned}
&\int_{\mathbb{R}^n} \mu_\beta(V_t^{(x)})dx \\
&= \int_{\{x\in\mathbb{R}^n\colon \|x\|<\|x-t\|\}} \mu_\beta(V_t^{(x)})dx + \int_{\{x\in\mathbb{R}^n\colon \|x\|>\|x-t\|\}} \mu_\beta(V_t^{(x)})dx \\
&= 2\int_{\{x\in\mathbb{R}^n\colon \|x\|<\|x-t\|\}} \mu_\beta(V_t^{(x)})dx \\
&= 2\int_{\{x\in\mathbb{R}^n\colon \|x\|<\|x-t\|\}} \Big(\int_{\|x\|}^{\|x-t\|} \lambda^{\beta-n-1}d\lambda\Big)dx \\
&= \frac{2}{n-\beta}\int_{\{x\in\mathbb{R}^n\colon \|x\|<\|x-t\|\}} (\|x\|^{\beta-n} - \|x-t\|^{\beta-n})dx,
\end{aligned}$$

establishing (8.4.4).

Notice that the integrand in (8.4.4) diverges as $x \to 0$ or $x \to \infty$. As $x \to 0$, the integrand behaves like $\|x\|^{\beta-n}$, so the integral converges in a neighborhood of 0 if and only if $\beta - n + (n-1) + 1 > 0$, i.e., $\beta > 0$. As $x \to \infty$, the integrand behaves like $\|x\|^{\beta-n-1}$, so convergence takes place if and only if $\beta - n - 1 + (n-1) + 1 < 0$, i.e., $\beta < 1$.

Finally, we prove (8.4.5). Let $< , >$ denote the scalar product in $\mathbb{R}^n$ and notice that the inequality $\|x\| < \|t - x\|$ is equivalent to $< x, x > \; < \; < x-t,\, x-t > \; = \; < x, x > -2 < t, x > + < t, t >$, that is, to $< t, x > \; < \; \|t\|^2/2$. Let $t = \|t\|e$ where e is a unit vector in $\mathbb{R}^n$.

If $t \neq 0$, the change of variable $x \to x/|t|$ in (8.4.4) yields

$$m_\beta(V_t) = \frac{2\|t\|^\beta}{n-\beta}\int_{\{x\in\mathbb{R}^n\colon <e,x> \,<\, 1/2\}} (\|x\|^{\beta-n} - \|x-e\|^{\beta-n})dx.$$

Since Lebesgue measure in $\mathbb{R}^n$ is invariant under rotations, the integral does not depend on e. Replacing e by $(1, 0, \ldots, 0)$, we see that $m_\beta(V_t)$ equals

$$\frac{2\|t\|^\beta}{n-\beta}\int_{\{x\in\mathbb{R}^n\colon x_1<1/2\}} \Big[\Big(\sum_{k=1}^n x_k^2\Big)^{(\beta-n)/2} - \Big((x_1-1)^2 + \sum_{k=2}^n x_k^2\Big)^{(\beta-n)/2}\Big]dx.$$

The change of variable $z_1 = x_1 - \frac{1}{2}$, $z_k = x_k$, $k \geq 2$, yields (8.4.5). ■

When $n = 1$, we have (see Exercise 8.3)

$$m_\beta(V_t) = \frac{1}{\beta(1-\beta)}2^{2-\beta}|t|^\beta. \tag{8.4.6}$$

Definition 8.4.3 Let $0 < \alpha \leq 2$ and $0 < \beta < 1$. The $S\alpha S$ Takenaka field $\{X(t) = M(V_t),\ t \in \mathbb{R}^n\}$ with associated measure m_β is called an (α, H)-*Takenaka field,* where $H = \beta/\alpha \in (0, 1/\alpha)$.

Theorem 8.4.4 *An* (α, H)*-Takenaka field, with* $0 < \alpha \leq 2$ *and* $0 < H < 1/\alpha$*, is* H*-sssis.*

PROOF: We must verify conditions (i), (ii) and (iii) of Theorem 8.2.6. Condition (i) follows from Lemma 8.4.2. To verify condition (ii), one could return to the proof of Lemma 8.4.2, but it is easier to note that

$$\begin{aligned} V_{at} &= \{(x,\lambda)\colon \|x\| \leq \lambda\}\,\Delta\,\{(x,\lambda)\colon \|x - at\| \leq \lambda\} \\ &= \{(ax, a\lambda)\colon \|x\| \leq \lambda\}\,\Delta\,\{(ax, a\lambda)\colon \|x - t\| \leq \lambda\}, \end{aligned}$$

and thus $V_{at} = aV_t$, where aV_t means that every element of V_t is multiplied by a. Similarly, $V_{at}^c = aV_t^c$ and $\widetilde{V}_a = a\widetilde{V}_1$ where $\widetilde{V}_a = \bigcap_{k=1}^d V_{at_k}^{\delta(k)}$.

Therefore,

$$m_\beta(\widetilde{V}_a) = \int_{\widetilde{V}_a} \lambda^{\beta-n-1} dx d\lambda = a^\beta m_\beta(\widetilde{V}_1)$$

since $d(ax) = a^n dx$ and $d(a\lambda) = a d\lambda$.

We now verify Condition (iii) of Theorem 8.2.6. Since the transformation $(x, \lambda) \to (g^{-1}(x), \lambda)$ is an isometry, its Jacobian equals 1 and therefore

$$m_\beta\left(\bigcap_{k=1}^d \left(V_{g(t_k)}^{\delta(k)}\,\Delta\,V_{g(0)}\right)\right) = m_\beta\left(\bigcap_{k=1}^d \left(V_{t_k}^{\delta(k)}\,\Delta\,V_0\right)\right) = m_\beta\left(\bigcap_{k=1}^d V_{t_k}^{\delta(k)}\right)$$

using $V_0 = \emptyset$.

The three conditions of Theorem 8.2.6 hold with $H = \beta/\alpha$ and, therefore, the $S\alpha S$ Takenaka field $\{X(t) = M(V_t),\ t \in \mathbb{R}^n\}$ with associated measure m_β is β/α-sssis. ■

Example 8.4.5 A $(2, H)$-Takenaka field is a Lévy fractional Brownian field (Proposition 8.1.4).

8.5 Properties of Chentsov random fields

In this section we investigate the properties of the finite-dimensional distributions of Chentsov fields on $\mathbb{R}^n$. In Theorem 8.5.1 we show that the spectral measure is discrete and in Theorem 8.5.3 we investigate what happens when α varies.

Theorem 8.5.1 *Fix $0 < \alpha < 2$ and let $\{X(t),\ t \in \mathbb{R}^n\}$ be a $S\alpha S$ Chentsov random field with associated measure m. Then the spectral measure Γ of the vector $(X(t_1), \dots, X(t_d))$, $d \geq 1$, is discrete and concentrated on at most $2(2^d - 1)$ points of the unit sphere S_d. These points are $\pm\zeta^\delta$, where*

$$\zeta^\delta = \frac{1}{(\sum_{j=1}^d \delta(j))^{1/2}}(\delta(1), \dots, \delta(d)),\quad \delta \in \Sigma_d,$$

and where Σ_d is defined in (8.2.9). The point masses of the measure Γ are

$$\Gamma(-\zeta^\delta) = \Gamma(\zeta^\delta) = \frac{1}{2}\Big(\sum_{j=1}^d \delta(j)\Big)^{\alpha/2} m\Big(\bigcap_{k=1}^d V_{t_k}^{\delta(k)}\Big). \tag{8.5.1}$$

PROOF: The characteristic function of $(X(t_1), \dots, X(t_d))$ is

$$\begin{aligned} &E\exp\Big\{i\sum_{j=1}^d \theta_j X(t_j)\Big\} \\ &= \exp\Big\{-\sum_{\delta\in\Sigma_d} | \sum_{j:\,\delta(j)=1} \theta_j|^\alpha m\Big(\bigcap_{k=1}^d V_{t_k}^{\delta(k)}\Big)\Big\} \quad \text{(as in (8.2.12))} \end{aligned}$$

by Corollary 8.2.7. The theorem follows by identifying this expression with

$$\exp\Big\{-\int_{S_d} |\theta_1 s_1 + \cdots + \theta_d s_d|^\alpha\Big\}\Gamma(ds). \quad \blacksquare$$

The distribution of a $S\alpha S$ Chentsov field $\{X(t) = M(V_t),\ t \in \mathbb{R}^n\}$ clearly depends not only on the control measure m of M but also on the value of α. To make the dependence on α explicit, we now write $X_\alpha(t) = M_\alpha(t),\ t \in \mathbb{R}^n$, denoting by M_α a $S\alpha S$ random measure with control measure m. What happens when we fix m and let α vary?

Definition 8.5.2 Fix $(E, \mathcal{E}, m)$ and a family of sets $\{V_t,\ t \in \mathbb{R}^n\}$. The family of random fields

$$\{\{X_\alpha(t) = M_\alpha(V_t),\ t \in \mathbb{R}^n\},\ 0 < \alpha \leq 2\}$$

is called a *conjugate class* of Chentsov fields. The field $\{X_2(t),\ t \in \mathbb{R}^n\}$ is the Gaussian element of the class.

Theorem 8.5.3 *Let $\{\{X_\alpha(t),\ t \in \mathbb{R}^n\},\ 0 < \alpha \leq 2\}$ and $\{\{X'_\alpha(t),\ t \in \mathbb{R}^n\},\ 0 < \alpha \leq 2\}$ be two conjugate classes defined by $((E, \mathcal{E}, m),\ \{V_t\})$ and*

$\{(E',\mathcal{E}',m'),\ \{V'_t\}\}$, *respectively. If the Gaussian elements* X_2 *and* X'_2 *are identically distributed, then for any* $0<\alpha<2$, *the fields* X_α *and* X'_α *have the same two-dimensional distributions.*

Moreover, if $\{X_{\alpha_0}(t),\ t\in\mathbb{R}^n\}\overset{d}{=}\{X'_{\alpha_0}(t),\ t\in\mathbb{R}^n\}$ *for some* $0<\alpha_0<2$, *then* $\{X_\alpha(t),\ t\in\mathbb{R}^n\}\overset{d}{=}\{X'_\alpha(t),\ t\in\mathbb{R}^n\}$ *for all* $0<\alpha\le 2$.

PROOF: X_α has two-dimensional characteristic function

$$E\exp\{\theta_1X_\alpha(t_1)+\theta_2X_\alpha(t_2)\}$$
$$=\exp\{-\{|\theta_1|^\alpha m(V_{t_1}\cap V_{t_2}^c)+|\theta_2|^\alpha m(V_{t_1}^c\cap V_{t_2})+|\theta_1+\theta_2|^\alpha m(V_{t_1}\cap V_{t_2})\}\}. \tag{8.5.2}$$

The Gaussian element has characteristic function

$$E\exp\{\theta_1X_2(t_1)+\theta_2X_2(t_2)\}=\exp\Big\{-\Big\{\frac{1}{2}\theta_1^2\sigma_{t_1}^2+\theta_1\theta_2\sigma_{t_1,t_2}+\frac{1}{2}\sigma_{t_2}^2\Big\}\Big\}, \tag{8.5.3}$$

where $\sigma_t^2=E(X_2(t))^2$ and $\sigma_{t_1,t_2}=EX(t_1)X(t_2)$. Identifying (8.5.2) with $\alpha=2$ and (8.5.3), we obtain

$$\begin{aligned}m(V_{t_1}\cap V_{t_2}^c)+m(V_{t_1}\cap V_{t_2}) &= \frac{1}{2}\sigma_{t_1}^2,\\ m(V_{t_1}^c\cap V_{t_2})+m(V_{t_1}\cap V_{t_2}) &= \frac{1}{2}\sigma_{t_2}^2,\\ 2m(V_{t_1}\cap V_{t_2}) &= \sigma_{t_1,t_2},\end{aligned}$$

i.e.,

$$\begin{cases} m(V_{t_1}\cap V_{t_2}^c)=\frac{1}{2}(\sigma_{t_1}^2-\sigma_{t_1,t_2}),\\ m(V_{t_1}^c\cap V_{t_2})=\frac{1}{2}(\sigma_{t_2}^2-\sigma_{t_1,t_2}),\\ m(V_{t_1}\cap V_{t_2})=\frac{1}{2}\sigma_{t_1,t_2}.\end{cases} \tag{8.5.4}$$

Any two conjugate families with equally distributed Gaussian elements have the same σs and, hence, the same values for $m(V_{t_1}\cap V_{t_2}^c)$, $m(V_{t_1}^c\cap V_{t_1})$, $m(V_{t_1}\cap V_{t_2})$. By virtue of (8.5.2), they have the same two-dimensional characteristic functions.

The last part of the theorem follows from Theorem 8.5.1. ∎

8.6 Properties of H-sssis Chentsov random fields

Whereas the preceding section covered arbitrary $S\alpha S$ Chentsov fields, we consider in this section $S\alpha S$ Chentsov fields that are also H-sssis, for example, (α,H)-Takenaka fields.

Proposition 8.6.1 *Fix $\alpha_0 \in (0,2]$ and $H_0 \in (0, 1/\alpha_0]$ and let $\{X(t),\ t \in \mathbb{R}^n\}$ be a $S\alpha_0S$, H_0-sssis Chentsov random field. Then the conjugate class $\{\{X_\alpha(t),\ t \in \mathbb{R}^n\},\ 0 < \alpha \le 2\}$ with $X_{\alpha_0} = X$ is well defined and, moreover, each member $X_\alpha,\ 0 < \alpha \le 2$, is a $S\alpha S$, H_α-sssis Chentsov random field, with $H_\alpha = \alpha_0 H_0/\alpha$.*

PROOF: Fix $0 < \alpha < 2$ and consider $\{X_\alpha(t) = M_\alpha(V_t),\ t \in \mathbb{R}^n\}$ where M_α has control measure m. Since $\alpha H_\alpha = \alpha_0 H_0$, the control measure m satisfies the conditions of Theorem 8.2.6 and therefore X_α is a $S\alpha S$ H_α-sssis Chentsov random field. ∎

Normalized $S\alpha S$ random fields have identical one-dimensional distributions. The next result states that if they are H-sssis Chentsov with the same H, then they also have identical two-dimensional distributions.

Theorem 8.6.2 *Fix $0 < \alpha \le 2$ and $0 < H \le 1/\alpha$. Then any two $S\alpha S$, H-sssis Chentsov random fields have, up to a multiplicative constant, the same two-dimensional distributions.*

PROOF: To avoid confusion with previous notation, let α_0 and H_0 denote, respectively, the α and H in the statement of the theorem. Let $\{X(t) = M(V_t),\ t \in \mathbb{R}^n\}$ and $\{X'(t) = M'(V'_t),\ t \in \mathbb{R}^n\}$ be the two random fields with associated measures m and m', respectively. Consider, as in Proposition 8.6.1, the conjugate class $\{\{X_\alpha(t),\ t \in \mathbb{R}^n\},\ 0 < \alpha \le 2\}$ with $X_{\alpha_0} = X$, and also the conjugate class $\{\{X'_\alpha(t),\ t \in \mathbb{R}^n\},\ 0 < \alpha \le 2\}$ with $X'_{\alpha_0} = X'$. Proposition 8.6.1 states that the Gaussian elements X_2 and X'_2 are both H_2-sssis with $H_2 = \alpha_0 H_0/2$ and hence they are both Lévy fractional Brownian fields. Since the Lévy fractional Brownian fields with same self-similarity index differ only by a multiplicative constant C, we have $X_2 \stackrel{d}{=} CX'_2$. Therefore, by Theorem 8.5.3, the relation $(X_\alpha(t_1), X_\alpha(t_2)) \stackrel{d}{=} (CX_\alpha(t_1),\ CX_\alpha(t_2))$ holds for all $0 < \alpha \le 2$ and, in particular, for $\alpha = \alpha_0$. ∎

What happens if we restrict the argument $t \in \mathbb{R}^n$ in (α, H)-Takenaka or Lévy-Chentsov fields to a linear subspace $\mathbb{R}^m \subset \mathbb{R}^n$ with $m < n$? Do we obtain a new type of random field? The answer is in the negative.

Theorem 8.6.3 *Fix $0 < \alpha \le 2,\ 0 < \beta < 1,\ m < n$, set $H = \beta/\alpha$, and let $\{X^n(\tilde{t}),\ \tilde{t} \in \mathbb{R}^n\}$ denote an (α, H)-Takenaka field with associated measure $dm^n_\beta(x_n, \lambda) = \lambda^{\beta-n-1}dx_n d\lambda,\ x_n \in \mathbb{R}^n$. Then the random field*

$$Y^m(t) = X^m(t, \underbrace{0, \dots, 0}_{n-m}),\ t \in \mathbb{R}^m,$$

has the same finite-dimensional distributions as $\{CX^m(t),\ t \in \mathbb{R}^m\}$, *where* C *is a constant and* X^m *is an* (α, H)*-Takenaka field with associated measure* $dm_\beta^m(x_m, \lambda) = \lambda^{\beta-m-1}dx_m d\lambda$, $x_m \in \mathbb{R}^m$.

A similar statement is true if X^n *and* X^m *are Lévy-Chentsov fields.*

PROOF: We shall consider Takenaka fields (the argument is similar for a Lévy-Chentsov field). We may assume $m = n - 1$ (the general result follows by induction). Let $x_n = (x_m, x_1) \in \mathbb{R}^n$, where $x_m \in \mathbb{R}^m$, $x_1 \in \mathbb{R}^1$ and let $\tilde{t} = (t, 0) \in \mathbb{R}^n$, where $t \in \mathbb{R}^m$. Then $Y^m(t) = M^n(V_{\tilde{t}}^n)$, $t \in \mathbb{R}^m$, with

$$\begin{aligned} V_t^n &= \{(x_m, x_1, \lambda)\colon \|(x_m, x_1)\| \le \lambda\} \\ &\quad \Delta\, \{(x_m, x_1, \lambda)\colon \|(x_m, x_1) - (t, 0)\| \le \lambda\} \\ &= \{(x_m, x_1, \lambda)\colon \|x_m\|^2 \le \lambda^2 - x_1^2\} \\ &\quad \Delta\, \{(x_m, x_1, \lambda)\colon \|x_m - t\|^2 \le \lambda^2 - x_1^2\} \\ &= \{(x_m, \theta, \eta)\colon \|x_m\| < \eta,\ |\theta| \le \pi/2\} \\ &\quad \Delta\, \{(x_m, \theta, \eta)\colon \|x_m - t\| < \eta,\ |\theta| \le \pi/2\}, \end{aligned}$$

where we set $\eta = (\lambda^2 - x_1^2)^{1/2}$, $\sin\theta = x_1/\lambda$ and, thus, $\cos\theta = \eta/\lambda$. Therefore

$$V_t^n = V_t^m \times [-\pi/2, \pi/2].$$

Since the Jacobian of the transformation $(x_1, \lambda) \mapsto (\theta, \eta)$ equals $\eta/\cos\theta$, we have

$$\begin{aligned} m_\beta^n(V_t^n) &= \int_{\mathbb{R}^n \times \mathbb{R}_+} 1_{V_t^n}(x_m, x_1, \lambda)\lambda^{\beta-n-1} dx_m dx_1 d\lambda \\ &= \int_{-\pi/2}^{\pi/2} \int_{\mathbb{R}^m \times \mathbb{R}_+} 1_{V_t^m}(x_m, \eta) \left(\frac{\eta}{\cos\theta}\right)^{\beta-n-1} \frac{\eta}{\cos\theta} dx_m d\eta d\theta \\ &= \int_{\mathbb{R}^m \times \mathbb{R}_+} 1_{V_t^m}(x_m, \eta)\eta^{\beta-m-1} dx_m d\eta \int_{-\pi/2}^{\pi/2} (\cos\theta)^{m+1-\beta} d\theta \\ &= C m_\beta^m(V_t^m), \end{aligned}$$

where $C = \int_{-\pi/2}^{\pi/2} (\cos\theta)^{m+1-\beta} d\theta$.

But the finite-dimensional distributions of X^n are determined by the measure of sets of the form $\bigcap_{k=1}^d (V_{t_k}^n)^{\delta(k)}$ (see (8.2.12)). Since T transforms these sets into $\bigcap_{k=1}^d (V_{t_k}^m)^{\delta(k)} \times [-\frac{\pi}{2}, \frac{\pi}{2}]$, the preceding argument also shows that

$$m_\beta^n\left(\bigcap_{k=1}^d (V_{t_k}^n)^{\delta(k)}\right) = C m_\beta^m\left(\bigcap_{k=1}^d (V_{t_k}^m)^{\delta(k)}\right).$$

The result now follows from Theorem 8.5.1. ∎

8.7 Codifference induced by (α, H)-Takenaka fields

We now turn to the codifference $\tau(\cdot)$ and to the function $I(\theta_1, \theta_2; \cdot)$ defined in (4.7.1) and (4.7.5), respectively. These functions provide information on the two-dimensional distributions. Since they are defined for *stationary* stochastic *processes* only and not for non-stationary random fields, we want to apply them to stationary stochastic processes *induced* by an (α, H)-Takenaka random field $\{X(t),\ t \in \mathbb{R}^n\}$.

Choose an arbitrary unit vector $e \in \mathbb{R}^n$ centered at the origin and consider the *projection process* $\{X(ue),\ u \in \mathbb{R}^1\}$ which is defined on the line spanned by the vector e. Since the random field X is H-sssis, the projection process $\{X(ue),\ u \in \mathbb{R}^1\}$ is H-sssi. Therefore, the *increment process*

$$Y(u) = X((u+1)e) - X(ue),\ u \in \mathbb{R}^1$$

is stationary.

The following analysis centers on Y. Our goal is to derive the asymptotic behavior of the function

$$\begin{aligned} &I(\theta_1, \theta_2; u) \\ &\quad = -\ln Ee^{i(\theta_1 Y(u) + \theta_2 Y(0))} + \ln Ee^{i\theta_1 Y(u)} + \ln Ee^{i\theta_2 Y(0)} \\ &\quad = \|\theta_1 Y(u) + \theta_2 Y(0)\|_\alpha^\alpha - \|\theta_1 Y(u)\|_\alpha^\alpha - \|\theta_2 Y(0)\|_\alpha^\alpha \end{aligned} \tag{8.7.1}$$

as $u \to \infty$ and hence that of the codifference

$$\tau(u) = -I(1, -1; u)$$

when X is an (α, H)-Takenaka field. To simplify the notation, we let m denote its associated measure (m is the m_β of (8.4.3)). The normalization constant

$$\|Y(1)\|_\alpha^\alpha = \|X((u+1)e) - X(ue)\|_\alpha^\alpha = \|M(V_e)\|_\alpha^\alpha = m(V_e)$$

which appears in the results does not depend on the chosen direction e. Its numerical value is given by (8.4.5) (with $\beta = \alpha H$ and $\|t\|^\beta = 1$).

Theorem 8.7.1 *Let $\{X(t),\ t \in \mathbb{R}^n\}$ be an (α, H)-Takenaka field, $0 < \alpha \le 2,\ 0 < H < 1/\alpha$ with associated m, let $e \in \mathbb{R}^n$ be an arbitrary unit vector centered at the origin and let I and τ refer to the increment process*

$$Y(u) = X((u+1)e) - X(ue),\ \ u \in \mathbb{R}^1.$$

Then

$$I(\theta_1, \theta_2; u) = m(V_e) F_\alpha(\theta_1, \theta_2)\{|u+1|^{\alpha H} - 2|u|^{\alpha H} + |u-1|^{\alpha H}\},\ \ u > 1,$$

where

$$F_\alpha(\theta_1, \theta_2) = |\theta_1|^\alpha + |\theta_2|^\alpha - |\theta_1 - \theta_2|^\alpha. \tag{8.7.2}$$

If, in addition, $\theta_1, \theta_2 \neq 0$, then, as $u \to \infty$,

$$I(\theta_1, \theta_2; u) \sim -m(V_e)F_\alpha(\theta_1, \theta_2)\frac{\alpha H(1-\alpha H)}{2}u^{\alpha H - 2}$$

and

$$\tau(u) \sim -m(V_e)(2^\alpha - 2)\frac{\alpha H(1-\alpha H)}{2}u^{\alpha H - 2}.$$

PROOF: Suppose, without loss of generality, that θ_1 and θ_2 are not equal to zero and assume firstly that $n = 1$. Since the process Y is $S\alpha S$, it is sufficient to take $e = 1$. Since $Y(u) = M(V_{u+1}) - M(V_u)$ and $V_0 = \emptyset$, we have

$$\begin{aligned}
\|\theta_1 Y(u) + \theta_2 Y(0)\|_\alpha^\alpha &= \|\theta_1 M(V_{u+1}) - \theta_1 M(V_u) + \theta_2 M(V_1)\|_\alpha^\alpha \\
&= |\theta_1|^\alpha (m(V_u \,\Delta\, V_{u+1}) - m(B_u \cup B_u')) \\
&\quad + |\theta_2|^\alpha (m(V) - m(B_u \cup B_u')) \\
&\quad + |\theta_1 + \theta_2|^\alpha m(B_u') + |\theta_1 - \theta_2|^\alpha m(B_u),
\end{aligned}$$

where $B_u = (V_u \backslash V_{u+1}) \cap V_1$ and $B_u' = (V_{u+1} \backslash V_u) \cap V_1$. By stationarity, $(V_u \,\Delta\, V_{u+1}) = m(V_u \,\Delta\, V_0) = m(V_1)$. Since Lemma 8.7.2 below shows that for $u > 1$, $B_u' = \emptyset$, we obtain

$$\begin{aligned}
\|\theta_1 Y(u) + \theta_2 Y(0)\|_\alpha^\alpha &= |\theta_1|^\alpha (m(V_1) - m(B_u)) \\
+ \quad & |\theta_2|^\alpha (m(V_1) - m(B_u)) + |\theta_1 - \theta_2|^\alpha m(B_u), \ u > 1,
\end{aligned}$$

and hence using (8.7.1), we obtain

$$I_\alpha(\theta_1, \theta_2; u) = -F_\alpha(\theta_1, \theta_2) m(B_u), \ u > 1, \tag{8.7.3}$$

where I_α and F_α are given by (8.7.1) and (8.7.2), respectively. (We have written I_α instead of I to avoid confusion.) We shall evaluate $m(B_u)$ by considering the analog of (8.7.3) for $\alpha = 2$.

The field X is a member of a conjugate class that includes a Gaussian element X_2. By Proposition 8.6.1, X_2 is H_2-sssis with $H_2 = \alpha H/2$. The induced projection process $\{X_2(ue), \ u \in \mathbb{R}^1\}$ is H_2-sssi and, hence, is Brownian motion if $H = 1/\alpha$ (i.e., if $H_2 = 1/2$) and fractional Brownian motion (see Definition 7.2.2) if $H < 1/\alpha$ (i.e., if $H_2 < 1/2$). Let $\{Y_2(u) = X_2((u+1)e) - X_2(ue), \ u \in \mathbb{R}^1\}$ denote its increment process. Letting $I_2(\theta_1, \theta_2; u)$ be defined as in (8.7.1) (with $\alpha = 2$ and $Y = Y_2$), we obtain

$$\begin{aligned}
I_2(\theta_1, \theta_2; u) &= \frac{1}{2}\{\mathrm{Var}(\theta_1 Y_2(u) + \theta_2 Y_2(0)) - \mathrm{Var}\,(\theta_1 Y_2(u)) - \mathrm{Var}\,(\theta_2 Y_2(0))\} \\
&= \theta_1 \theta_2 E Y_2(u) Y_2(0).
\end{aligned}$$

The process $\{Y_2(u),\ u \in \mathbb{R}^1\}$ is fractional Gaussian noise,[2] with variance

$$\sigma_0^2 = 2\|Y_2(1)\|_2^2 = 2\|M_2(V_e)\|_2^2 = 2m(V_e).$$

On the other hand, by (8.7.3), we have, for $u > 1$,

$$I_2(\theta_1, \theta_2; u) = -F_2(\theta_1, \theta_2)m(B_u),\ u > 1.$$

Therefore, for $u > 1$,

$$\begin{aligned} I_\alpha(\theta_1, \theta_2; u) &= \frac{F_\alpha(\theta_1, \theta_2)}{F_2(\theta_1, \theta_2)} I_2(\theta_1, \theta_2; u) \\ &= \frac{F_\alpha(\theta_1, \theta_2)}{2\theta_1\theta_2} \theta_1\theta_2 EY_2(u)Y_2(0) \\ &= \frac{1}{2} F_\alpha(\theta_1, \theta_2) EY_2(u)Y_2(0). \end{aligned} \tag{8.7.4}$$

It remains to evaluate $EY_2(u)Y_2(0)$. Since $2H_2 = \alpha H$, this covariance equals

$$EY_2(u)Y_2(0) = m(V_1)\{|u+1|^{\alpha H} - 2|u|^{\alpha H} + |u-1|^{\alpha H}\}$$

by Proposition 7.2.9 and, as $u \to \infty$,

$$EY_2(u)Y(0) \sim m(V_1)\alpha H(\alpha H - 1)u^{\alpha H - 2},$$

as in Proposition 7.2.10. Substituting these expressions in (8.7.4) concludes the proof in the case $n = 1$.

Suppose, now, $n > 1$. Let $\{X^1(u) = M(\widetilde{V}_u),\ u \in \mathbb{R}\}$ be the (α, H)-Takenaka process on $\mathbb{R}$ with associated measure m^1. The theorem, as we have shown, holds for X^1, with $m(V_e)$ equal to $m^1(\widetilde{V}_1)$. If $e_0 = (1, 0, \ldots, 0) \in \mathbb{R}^n$, then by Theorem 8.6.3,

$$\{X(ue_0),\ u \in \mathbb{R}\} \stackrel{d}{=} \{CX^1(u),\ u \in \mathbb{R}\}$$

in the sense of equality of the finite-dimensional distributions. To determine the constant C, notice that

$$m(V_{e_0}) = \|X(e_0)\|_\alpha^\alpha = |C|^\alpha \|X^1(u)\|_\alpha^\alpha = |C|^\alpha m^1(\widetilde{V}_1),$$

i.e.,

$$|C|^\alpha = \frac{m(V_{e_0})}{m^1(\widetilde{V}_1)}.$$

[2] In Section 7.2.3, fractional Gaussian noise was defined for $u = \ldots, -1, 0, 1, \ldots,$ but its definition can be extended naturally to all of $u \in \mathbb{R}^1$.

Therefore the theorem holds for the field X when $e = e_0$. But, because of rotation invariance, for any vector $e \in \mathbb{R}^n$ centered at the origin, we have

$$\{X(ue),\, u \in \mathbb{R}\} \stackrel{d}{=} \{X(ue_0),\, u \in \mathbb{R}\},$$

in the sense of finite-dimensional distributions. Hence our theorem holds for the field X and arbitrary e. ■

The proof used the following:

Lemma 8.7.2 *For $u > 1$, $(V_{u+1} \backslash V_u) \cap V_1 = \emptyset$.*

PROOF: Recall that $V_u = C_u \,\Delta\, C_0$ and $C_u = \{(\lambda, x) \colon |x - u| \le \lambda\}$. Write

$$\begin{aligned}
(V_{u+1}\backslash V_u) \cap V_1 &= (V_{u+1} \cap V_u^c) \cap V_1 \\
&= [(C_{u+1} \,\Delta\, C_0) \cap (C_u \,\Delta\, C_0)^c] \cap [C_1 \,\Delta\, C_0] \\
&= [\{(C_{u+1} \cap C_0^c) \cup (C_{u+1}^c \cap C_0)\} \cap \{(C_u^0 \cup C_0) \cap (C_u \cup C_0^c)\}] \\
&\quad \cap\ [(C_1 \cap C_0^c) \cup (C_1^c \cap C_0)] \\
&= [\{(C_{u+1} \cap C_0^c) \cup (C_{u+1}^c \cap C_0)\} \cap \{(C_u^0 \cap C_0^c) \cup (C_u \cap C_0)\}] \\
&\quad \cap\ [(C_1 \cap C_0^c) \cup (C_1^c \cap C_0)] \\
&= [(C_{u+1} \cap C_0^c \cap C_u^c) \cup (C_{u+1}^c \cap C_0 \cap C_u)] \\
&\quad \cap\ [(C_1 \cap C_0^c) \cup (C_1^c \cap C_0)] \\
&= (C_0^c \cap C_1 \cap C_u^c \cap C_{u+1}) \cup (C_0 \cap C_1^c \cap C_u \cap C_{u+1}^c).
\end{aligned}$$

Now $(\lambda, x) \in C_0^c \cap C_1 \cap C_u^c \cap C_{u+1}$ iff simultaneously

$$|x| > \lambda,\ |x - 1| \le \lambda,\ |x - u| > \lambda,\ |x - u - 1| \le \lambda. \tag{8.7.5}$$

Notice that inequalities (8.7.5) hold iff

$$\lambda < x \le \lambda + 1 \ \text{ and } \ \lambda + u < x \le \lambda + 1 + u. \tag{8.7.6}$$

If $u > 1$, there are no λ and x satisfying (8.7.6), so $C_0^c \cap C_1 \cap C_u^c \cap C_{u+1} = \emptyset$. In a similar way, one can show that $C_0 \cap C_1^0 \cap C_u \cap C_{u+1} = \emptyset$. ■

We have evaluated the asymptotic behavior of the codifference of a number of increment processes and found the following. For the projection processes of (α, H)-Takenaka random fields, the codifference behaves like $u^{\alpha H - 2}$. For the *linear fractional stable noise*, it behaves either like $u^{\alpha H - \alpha}$ or like $u^{H-(1/\alpha)-1}$ (see Theorem 7.10.1 and a remark following the theorem). For the *log-fractional noise*, where $H = 1/\alpha$, it behaves like $u^{1-\alpha}$ (Theorem 7.10.2). It does not tend to zero for either the *real harmonizable* or the *sub-Gaussian fractional stable noise* (Propositions 6.7.1 and 4.7.4 respectively). Therefore we have:

Corollary 8.7.3 *Projection processes of* (α, H)*-Takenaka random fields,* $0 < \alpha < 2, 0 < H < 1/\alpha$*, are different from the linear fractional stable motion, from real harmonizable fractional stable motion and from the sub-Gaussian fractional stable motion.*

8.8 Takenaka processes on $[0, \infty)$

We shall now consider Takenaka processes on $[0, \infty)$, i.e., Chentsov fields $\{X(t) = M(V_t),\ t \geq 0\}$ where the V_ts are defined in (8.4.2). We have seen that any d-dimensional distribution of a $S\alpha S$ Chentsov field on $\mathbb{R}^n$ has a spectral measure concentrated on at most $2(2^d - 1)$ points of the unit sphere S_d. One can say more if that field is a Takenaka process on $[0, \infty)$.

Lemma 8.8.1 *Let* $0 < \alpha < 2$*. If* $\{X(t) = M(V_t),\ t \geq 0\}$ *is a* $S\alpha S$ *Takenaka process, then the spectral measure* Γ *of* $(X(t_1), \ldots, X(t_d))$ *is concentrated on at most* $d(d+1)$ *points of* S_d*.*

PROOF: The spectral measure of a Chentsov process given in (8.5.1) involves the sets $\bigcap_{k=1}^{d} V_{t_k}^{\delta(k)}$, $\delta \in \Sigma_d$. It is easy to see that for any $0 < t_1 < t_2 < t_3 < \infty$,

$$V_{t_1} \cap V_{t_2}^c \cap V_{t_3} = \emptyset \tag{8.8.1}$$

and hence it suffices to consider sets of the form

$$R_{p,q} = \left(\bigcap_{j=1}^{p-1} V_{t_j}^c\right) \cap \left(\bigcap_{j=p}^{q} V_{t_j}\right) \cap \left(\bigcap_{j=q+1}^{d} V_{t_j}^c\right), \quad 1 \leq p \leq q \leq d. \tag{8.8.2}$$

These sets are disjoint $(R_{p,q} \cap R_{p',q'}) = \emptyset$, if $(p, q) \neq (p', q')$, and there are $d(d+1)/2$ of them. The (symmetric) spectral measure Γ is therefore concentrated on at most $d(d+1)$ points. ■

Remark. The proof fails if the process X is defined on $\mathbb{R}^1$ instead of $\mathbb{R}_+$ because Relation (8.8.1) does not hold, for example, when $t_1 < t_2 < 0 < t_3$. However, $V_{t_1} \cap V_{t_2}^c \cap V_{t_3} \cap V_{t_4}^c = \emptyset$ and $V_{t_1}^c \cap V_{t_2} \cap V_{t_3}^c \cap V_{t_4} = \emptyset$ for any $t_1 < t_2 < t_3 < t_4$, and so the measure is concentrated on at most $2d(d-1) + 2$ points in this case.

Let $0 < \alpha < 2,\ 0 < \beta < 1,\ H = \beta/\alpha$. The proof of the preceding lemma suggests how to construct $S\alpha S$, $H = \beta/\alpha$-sssis Chentsov processes which are *not* (α, H)-Takenaka.

Example 8.8.2 Let $n = 1$ and define a $S\alpha S$ Chentsov process $\{X(t) = M(V_t),\ t \geq 0\}$ with

$$V_t = C_t \,\Delta\, C_0,$$

where

$$C_t = \{(x, \lambda) \in \mathbb{R}^1 \times \mathbb{R}_+ \colon -\lambda < x - t < -\lambda/3 \text{ or } \lambda/3 < x - t < \lambda\} \quad (8.8.3)$$

and where M has control measure $m_\beta,\ 0 < \beta < 1$, given in (8.4.3). It is easy to verify (Exercise 8.7) that $m_\beta(V_t) < \infty$ for all $t \geq 0$ and that the process $X(t) = M(V_t),\ t \geq 0$, is H-sssis with $H = \beta/\alpha$. Note also that $m_\beta(V_{t_1} \cap V_{t_2}^c \cap V_{t_3}) > 0$ for $0 < t_1 < t_2 < t_3 < \infty$; so in view of (8.8.1), this process cannot have the same finite-dimensional distributions as an (α, H)-Takenaka process.

Remarks

1. The argument presented in the preceding example shows that an (α, H)-Takenaka process and the process defined in the example are different even if these processes are defined for all $t \in \mathbb{R}^1$.

2. One can construct other examples by setting $V_t = C_t \,\Delta\, C_0$ where

$$\begin{aligned} C_t &= C_t(a_1, b_1, a_2, b_2, \ldots, a_N, b_N) \\ &= \{(x, \lambda) \in \mathbb{R}^1 \times \mathbb{R}_+ \colon a_i\lambda < x - t < b_i\lambda,\ i = 1, \ldots, N\}, \end{aligned}$$

   where $-\infty < a_1 < b_1 < a_2 < b_2 < \cdots < a_N < b_N < \infty$ (see Sato and Takenaka (1991)).

3. While the spectral measure of any d-dimensional distribution of a general Chentsov field is concentrated on $2(2^d - 1)$ points (Theorem 8.5.1), that of a Takenaka field defined on $[0, \infty)$ is concentrated on $d(d+1)$ points (Lemma 8.8.1). Observe that while $2(2^d - 1) \geq d(d+1)$, there is equality when $d = 2$.

4. An (α, H)-Takenaka process as well as the process defined in Example 8.8.2 are H-sssis Chentsov. They therefore have proportional two-dimensional distributions (Theorem 8.6.2). Example 8.8.2 shows that their three-dimensional distributions are different.

The following theorem states that the two-dimensional distributions of Takenaka processes on $[0, \infty)$, in particular (α, H)-Takenaka processes, determine all the higher-dimensional distributions.

Theorem 8.8.3 *Let $0 < \alpha < 2$ and let $\{X(t),\ t \geq 0\}$ be a $S\alpha S$ Takenaka process. Then for any $d \geq 3$ and distinct $t_1, t_2, \ldots, t_d$, the distribution of $(X(t_1), X(t_2), \ldots, X(t_d))$ is determined by its two-dimensional marginal distributions.*

PROOF: Let m be the measure associated with X. Theorem 8.5.1 states that the distribution of $(X(t_i), X(t_j))$ is characterized by the three numbers $m(V_{t_i} \cap V_{t_j}^c)$, $m(V_{t_i}^c \cap V_{t_j})$ and $m(V_{t_i} \cap V_{t_j})$. The proof of Lemma 8.8.1 shows that the distribution of $(X(t_1), X(t_2), \cdots, X(t_d))$, $0 < t_1 < t_2 < \ldots < t_d$, is determined by $m(R_{p,q})$, $1 \leq p \leq q \leq d$, where the $R_{p,q}$ are disjoint sets defined in (8.8.2). It is therefore sufficient to prove that any $r_{p,q} = m(R_{p,q})$, $1 \leq p \leq q \leq d$, can be obtained from

$$a_{i,j} = m(V_{t_i} \cap V_{t_j}^c), \quad b_{i,j} = m(V_{t_i} \cap V_{t_j}), \quad 1 \leq i < j \leq d.$$

Relation (8.8.2) implies

$$V_{t_i} \cap V_{t_j}^c = \bigcup_{p=1}^{i} \bigcup_{q=i}^{j-1} R_{p,q}, \quad 1 \leq i < j \leq d,$$

and thus

$$a_{i,j} = m(V_{t_i} \cap V_{t_j}^c) = \sum_{p=1}^{i} \sum_{q=i}^{j-1} r_{p,q}, \quad 1 \leq i < j \leq d, \tag{8.8.4}$$

since the $R_{p,q}$s are disjoint. For $1 < i < j < d$, we obtain

$$a_{i,j+1} - a_{i,j} = \sum_{p=1}^{i} r_{p,j}, \text{ and } a_{i-1,j+1} - a_{i-1,j} = \sum_{p=1}^{i-1} r_{p,j},$$

i.e.,

$$r_{i,j} = (a_{i,j+1} - a_{i,j}) - (a_{i-1,j+1} - a_{i-1,j}), \ 1 < i < j < d.$$

When $i = 1$, Relation (8.8.4) yields

$$r_{1,j} = a_{1,j+1} - a_{1,j}, \quad 1 < j < d.$$

Similarly, when $j = d$,

$$V_{t_i} \cap V_{t_d} = \bigcup_{p=1}^{i} R_{p,d}, \quad 1 \leq i \leq d,$$

and so $b_{i,d} = m(V_{t_i} \cap V_{t_d}) = \sum_{p=1}^{i} r_{p,d}$, that is, $r_{i,d} = b_{i,d} - b_{i-1,d}$, $1 < i \leq d$, and $r_{1,d} = b_{1,d}$. ∎

The extension of the preceding result to non-Gaussian $S\alpha S$ Takenaka fields on $\mathbb{R}^1$ and on $\mathbb{R}^n$, $n \geq 2$, is more delicate.

Theorem 8.8.4 *Let* $0 < \alpha < 2$, $n \geq 1$ *and let* $\{X(t),\ t \in \mathbb{R}^n\}$ *be a* $S\alpha S$ *Takenaka field. Then, for any* $d \geq n+2$ *and distinct* $t_1, t_2, \ldots, t_d \in \mathbb{R}^n$, *the distribution of* $(X_{t_1}, X_{t_2}, \ldots, X_{t_d})$ *is determined by its* $(n+1)$*-dimensional marginal distributions.*

PROOF: See Sato (1992*b*) for $n = 1, 2$ and Sato (1991) for general n. ∎

8.9 Exercises

Exercise 8.1 Prove that the integral (8.1.2) is well defined and that the corresponding process has autocovariance (8.1.1). Express σ in terms of σ_0.

Exercise 8.2 Consider d sets $B_1, \ldots, B_d$, let Σ_d be defined as in (8.2.9) and, for $\delta \in \Sigma_d$, write $B_k^{\delta(k)} = B_k$ if $\delta(k) = 1$ and $B_k^{\delta(k)} =$ the complement of B_k if $\delta(k) = 0$. Show firstly that

$$B_1 \cup \cdots \cup B_d = \bigcup_{\delta \in \Sigma_d} \bigcap_{k=1}^{d} B_k^{\delta(k)}.$$

Then prove

$$B_j = \bigcup_{\substack{\delta \in \Sigma_d \\ \delta(j)=1}} B_k^{\delta(k)}$$

for $j = 1, \ldots, d$.

Exercise 8.3 Evaluate $m_\beta(V_t)$ when $n = 1$ (see (8.4.6)).

Exercise 8.4 Let $X^n_{\alpha,H}$ and $X^m_{\alpha,H}$ be (α, H)-Takenaka fields on $\mathbb{R}^n$ and $\mathbb{R}^m$, respectively. Fix unit vectors $e_n \in \mathbb{R}^n$ and $e_m \in \mathbb{R}^m$ and define the corresponding projection processes $\{X^n_{\alpha,H,e_n}(u) = X^n_{\alpha,H}(ue_n),\ u \in \mathbb{R}^1\}$ and $\{X^m_{\alpha,H,e_m}(u) = X^m_{\alpha,H}(ue_m),\ u \in \mathbb{R}^1\}$. Show that these processes are $S\alpha S$, H-sssis Chentsov and that there is a constant C such that $X^n_{\alpha,H,e_n} \stackrel{d}{=} CX^m_{\alpha,H,e_m}$ in the sense of equality of the *two*-dimensional distributions.

Exercise 8.5 Let $\{X(t) = M(V_t),\ t \in \mathbb{R}^n\}$ be a $S\alpha S$ Chentsov field with associated measure m. Fix $\mathbf{t} = (t_1, \ldots, t_d) \in (\mathbb{R}^n)^d$ and define Σ_d as in Exercise 8.2. For $\delta \in \Sigma_d$, let

$$V_{t_k}^{\delta(k)} = \begin{cases} V_{t_k} & \text{if } \delta(k) = 1, \\ V_{t_k}^c & \text{if } \delta(k) = 0, \end{cases} \qquad \widetilde{V}_{t_k}^{\delta(k)} = \begin{cases} V_{t_k} & \text{if } \delta(k) = 1, \\ \mathbb{R}^d \times \mathbb{R}^+ & \text{if } \delta(k) = 0, \end{cases}$$

$$V(\mathbf{t},\delta) = \bigcap_{k=1}^{d} V_{t_k}^{\delta(k)}, \quad \widetilde{V}(\mathbf{t},\delta) = \bigcap_{k=1}^{d} \widetilde{V}_{t_k}^{\delta(k)},$$

$$\Sigma_d(\delta) = \{\delta' = (\delta'(1),\ldots,\delta'(d)) \in \Sigma_d\colon \delta'(k) \geq \delta(k) \text{ for } k = 1,\ldots,d\}.$$

Show that the associated measure m satisfies the following $2^d - 2$ equation

$$\sum_{\delta' \in \Sigma_d(\delta)} m(V(\mathbf{t},\delta')) = m(\widetilde{V}(t,\delta)) \text{ for } \delta \in \Sigma_d \backslash \{(1,\ldots,1)\}. \tag{8.9.1}$$

This can be viewed as a system of $2^d - 2$ linear equations in $2^d - 1$ unknowns $m(V(\mathbf{t},\delta))$, $\delta \in \Sigma_d$. The right-hand sides are the values of $m(\widetilde{V}(\mathbf{t},\delta))$, $\delta \in \Sigma_d \backslash \{1,\ldots,1\}$, which are determined by the $(d-1)$-dimensional marginal distributions of $(X(t_1),\ldots,X(t_d))$.
Hint: See Sato (1991), p. 124.

Exercise 8.6 Write (8.9.1) in matrix form as $M_\alpha \mathbf{x} = \mathbf{b}$ where M_d is a $(2^d-2) \times (2^d-1)$ matrix of 0's and 1's. Let $M_d(k)$ be the $(2^d-2) \times (2^d-2)$ matrix obtained from M_d by deleting the kth column. Prove that $M_d(k)$ is invertible for any $k = 1,2,\ldots,2^d-1$.
Hint: See Sato (1991).

Exercise 8.7 Verify the statements in Example 8.8.2.

Exercise 8.8 Consider the H-sssi process $\{X(t),\ t \geq 0\}$ defined in Example 8.8.2. This process is not (α, H)-Takenaka. Show, however, that the codifference $\tau(j)$ for the increment sequence $Y(j) = X(j+1) - X(j),\ j \geq 1$, is still asymptotically proportional to $j^{\alpha H - 2}$ as $j \to \infty$.

Chapter 9

Introduction to sample path properties

We have defined a stochastic process $\{X(t),\ t \in T\}$ by its finite-dimensional distributions. We now want to view it as a collection of random variables. Our goal is to study its sample paths, i.e., the functions $\{X(t,\omega),\ t \in T\}, \omega \in \Omega$. We shall do this in several chapters. In this chapter we introduce a number of basic notions such as "versions," "separability" and "measurability," that are used in the study of sample path behavior. We shall not assume at first that $\{X(t),\ t \in T\}$ is stable because these notions apply to all stochastic processes. Some of the more technical proofs in this chapter and the following ones can be omitted in the first reading.

Two stochastic processes are *versions* of each other if they have the same finite-dimensional distributions. We will say, for example, that $\{X(t),\ t \in T\}$ is sample bounded if there is a version whose sample paths are bounded. *Separable* versions, which are introduced in Section 9.2, allow the use of finite-dimensional distributions to compute the probability of events involving uncountably many ts. We use separability in Section 9.3 to obtain criteria for sample boundedness and sample continuity in terms of finite-dimensional distributions. In Section 9.4 we discuss measurability. Whereas separable versions of a stochastic process always exist, there may be no *measurable* versions. We provide in this section necessary and sufficient conditions for the existence of a measurable version. The conditions involve the Condition S introduced in Section 3.11.

Section 9.5 is devoted to *zero-one* laws. Is it possible for a proportion $0 < p < 1$ of the sample paths to have a given property or must p be either 0 or 1? Theorem 9.5.4 presents a very general zero-one law for stable processes that will be used on several occasions in the sequel.

9.1 Versions

In the preceding chapters, we characterized a stochastic process by its finite-dimensional distributions. We now view a stochastic process $X = \{X(t),\ t \in T\}$ as a collection of random variables defined on a probability space $(\Omega, \mathcal{F}, P)$. Each element $\omega \in \Omega$ gives rise to a "sample path" or "realization," which is denoted $\{X(t, \omega),\ t \in T\}$.

Definition 9.1.1 Two stochastic processes $\{X(t),\ t \in T\}$ and $\{Y(t),\ t \in T\}$ are said to be *versions* of each other if they have the same finite-dimensional distributions, i.e.,

$$\{X(t_n),\ n = 1, \ldots, N\} \stackrel{d}{=} \{Y(t_n),\ n = 1, \ldots, N\}$$

for any N and $t_1, \ldots, t_N \in T$.

Hence a version of a stochastic process is a representative of the equivalence class of all stochastic processes with a given set of finite-dimensional distributions. Even if two versions of the same process are defined on the same probability space they represent, in general, two different collections of measurable functions on that space; their sample paths, in particular, may be quite different.

Example 9.1.2 Let $e, e_1, e_2, \ldots$ be a sequence of i.i.d. exponential random variables defined on a probability space $(\Omega, \mathcal{F}, P)$. Define three stochastic processes on $(\Omega, \mathcal{F}, P)$ indexed by $t \in [0, \infty)$ as follows:

$$X(0) = 0,\ \ X(t) = n \text{ if } e_1 + \cdots + e_n \le t < e_1 + \cdots + e_{n+1},\ n = 0, 1, 2, \ldots,$$

$$Y(0) = 0,\ \ Y(t) = n \text{ if } e_1 + \cdots + e_n < t \le e_1 + \cdots + e_{n+1},\ n = 0, 1, 2, \ldots,$$

$$Z(t) = \begin{cases} X(t) & \text{if } t \neq e, \\ X(t) + 1 & \text{if } t = e. \end{cases}$$

Observe that X is left-continuous, Y is right-continuous and Z is neither left- nor right-continuous. But $\{X(t),\ t \ge 0\}$, $\{Y(t),\ t \ge 0\}$ and $\{Z(t),\ t \ge 0\}$ are all versions of the same process, the standard time-homogeneous Poisson process on $[0, \infty)$. Of course, only $\{X(t),\ t \ge 0\}$ is a Poisson process as it is usually defined, i.e., with right-continuous sample paths. It is useful to draw a typical sample path of each of the three processes and to note the difference between the pictures.

If $\{X(t),\ t \in T\}$ and $\{Y(t),\ t \in T\}$ are defined on the same probability space and if for every $t \in T$,

$$P(X(t) = Y(t)) = 1, \tag{9.1.1}$$

then obviously the two processes are versions of each other. Processes satisfying (9.1.1) are typically called *indistinguishable*, but even indistinguishable processes can have very different sample paths (the three processes in Example 9.1.2 are indistinguishable but they have different sample paths).

9.2 Separability

Some versions of a stochastic process $\{X(t),\ t \in T\}$ have properties that facilitate the study of sample path behavior. One such property is "separability."

Definition 9.2.1 A stochastic process $\{X(t),\ t \in T\}$ on a probability space $(\Omega, \mathcal{F}, P)$ is called *separable* if there is a countable subset $T^* \subseteq T$ and an event $\Omega_0 \in \mathcal{F}$ with $P(\Omega_0) = 0$ such that for any closed set $F \subseteq \mathbb{R}$ we have

$$\{\omega: X(t) \in F,\ \forall t \in T^*\} \backslash \{\omega: X(t) \in F,\ \forall t \in T\} \subseteq \Omega_0.$$

T^* is called a *separant* for $\{X(t),\ t \in T\}$.

This definition is useful if the parameter set T is uncountable. If the process is separable, we may consider a countable subset T^* of T and use finite-dimensional distributions to compute probability of events, even those involving uncountably many ts. Note that separability does not imply "smoothness" of the paths. The following example of a process consisting of i.i.d. random variables may be the best illustration of this point. Such a process is, in a sense, the "least smooth" one can think of, yet not only does it have a separable version but, in fact, *every version* of this process is separable provided that the distribution of $X(t)$ has full support.

Example 9.2.2 Let $X = \{X(t),\ t \in T\}$ be a stochastic process consisting of i.i.d. random variables, i.e., for any $t_1, \ldots, t_n \in T$, the random variables $X(t_1), \ldots, X(t_n)$ are i.i.d.. Assume that the distribution of $X(t)$ is supported on the whole of $\mathbb{R}$, i.e., $P(X(t) \in (a, b)) > 0$ for any $-\infty < a < b < \infty$. If T is countable, take $T^* = T$. If T is uncountable, take T^* to be any infinite countable subset of T. Then *any* version of X is separable with separant T^*. To verify this, let

$$\begin{aligned}\Omega_0^c &= \{\omega\colon \{X(t,\omega),\ t \in T^*\} \text{ is dense in } \mathbb{R}\} \\ &= \bigcap_{r \in \mathbb{Q}} \bigcap_{m=1}^{\infty} \bigcup_{t \in T^*} \left\{\omega\colon |X(t,\omega) - r| \le \frac{1}{m}\right\},\end{aligned}$$

where $\mathbb{Q}$ denotes the rationals. Clearly, $P(\Omega_0) = 0$, since the support of X is the whole of $\mathbb{R}$. Choose, now, $\omega \in \Omega_0^c$, and let $A = \{X(t,\omega),\ t \in T^*\} \subset F$, where F is a closed subset in $\mathbb{R}$. Since the set A is dense in $\mathbb{R}$ and its closure $\overline{A}$ is also in F, it follows that $F = \mathbb{R}$. Therefore, for any closed set $F \subseteq \mathbb{R}$, the set $\{\omega\colon X(t) \in F,\ \forall t \in T^*\} \backslash \{\omega\colon X(t) \in F,\ \forall t \in T\}$ has probability zero because it belongs to Ω_0 if $F \neq \mathbb{R}$ and it is the empty set if $F = \mathbb{R}$. This proves that X is separable with separant T^*.

The definition of separability did not use any special structure of the parameter set T. The set T, however, may have a special structure of which one can take advantage. We will often suppose that (T, ρ) is a metric space with metric ρ; for example, T can be the closed interval $[0, 1]$, the open interval (0,1), the positive real line or a Euclidean space with the usual Euclidean distance. This permits us to talk of (open) balls centered at $y \in T$, i.e., $B_y(r) = \{x\colon \rho(x, y) < r\}$, and, more generally, of open sets (a subset U of a metric space is open if for any $y \in U$, it contains a ball centered at y).

Sometimes, however, what is important are the open sets themselves and not how they are generated. In these cases, it is not necessary to suppose that T is metrizable, i.e., that T is a space whose open sets are generated by a distance ρ. It is sufficient to suppose that (T, τ) is a topological space. Recall that a topology τ on T is a collection of subsets of T, called "open," such that T and $\emptyset$ are open, the intersection of two open sets is open and the union (not necessarily countable) of open sets is open. Any topology on T generates $\mathcal{B}(T)$, the Borel σ-algebra on T, which is the smallest σ-algebra generated by the open sets.

A *base* for the topology τ is a subset of τ such that every open set in τ is the union of open sets in the base. In a metric space, the (open) balls form a base for the topology generated by the metric. Some bases are countable. For example, in $\mathbb{R}$, the set of open intervals (a, b) with a and b rational forms a countable base for the topology generated by the Euclidean metric. In the following, we always suppose that (T, τ) is a topological space with a countable base.

Definition 9.2.3 Let (T, τ) be a topological space with a countable base. A stochastic process $\{X(t),\ t \in T\}$ on a probability space $(\Omega, \mathcal{F}, P)$ is called *strongly separable* in the topology τ if there is a countable subset $T^* \subseteq T$ and an event $\Omega_0 \in \mathcal{F}$ with $P(\Omega_0) = 0$ such that for every $I \in \tau$, every closed set $F \subseteq \mathbb{R}$

$$\{\omega\colon X(t) \in F,\ \forall t \in I \cap T^*\} \backslash \{\omega\colon X(t) \in F,\ \forall t \in I\} \subseteq \Omega_0.$$

T^* is called a *strong separant* for $\{X(t),\ t \in T\}$.

Remarks

1. Strong separability implies separability because $T \in \tau$. Since Ω_0 has probability zero, separability allows us to replace events involving all ts in

T by events involving all ts in the countable subset T^*. Strong separability allows us to go one step further and replace events involving all ts in any open set I by events involving all ts in the corresponding countable subset $I \cap T^*$.

2. The notion of separability is due to Doob (1953). Doob's "separability" is called here "strong separability."

3. Separability and strong separability are defined here relative to the class of closed sets because this is what we need, but these notions can also be defined relative to other classes of sets, for example, the class of closed intervals. Obviously, the larger the class, the more useful the (strong) separability property (but fewer stochastic processes possess it).

The following proposition gives a useful criterion for strong separability.

Proposition 9.2.4 *Let (T, τ) be a topological space with a countable base. A stochastic process $\{X(t),\ t \in T\}$ is strongly separable in the topology τ if and only if there is a countable dense subset $T^* \subseteq T$ and an event $\Omega_0 \in \mathcal{F}$ with $P(\Omega_0) = 0$ such that for every $\omega \notin \Omega_0$ and every $t \in T$, there is a sequence $\{t_n\}_{n=1}^{\infty} \subseteq T^*$ such that $t_n \to t$ and $X(t_n, \omega) \to X(t, \omega)$ as $n \to \infty$. T^* is a strong separant for $\{X(t),\ t \in T\}$.*

PROOF: The sufficiency is obvious because if $X(t_n, \omega) \to X(t, \omega)$ and $X(t_n, \omega)$ is in a closed set, then so is $X(t, \omega)$. To establish the necessity, assume that X is strongly separable, and let T^* be a strong separant for X and Ω_0 the corresponding exceptional set. Fix $t \in T$ and $\omega \notin \Omega_0$. Since τ has a countable base, let B_n, $n = 1, 2, \ldots$, be the base sets containing t. Then $I_n = \bigcap_{i=1}^{n} B_i$, $n = 1, 2, \ldots$, are open, ordered by inclusion and contain t. If $F_n = \{y\colon |y - X(t, \omega)| \geq \frac{1}{n}\}$, then, by strong separability, we cannot have $X(s, \omega) \in F_n$ for all $s \in I_n \cap T^*$. There must therefore be a $t_n \in I_n \cap T^*$ such that $|X(t_n, \omega) - X(t, \omega)| < \frac{1}{n}$. Thus, $t_n \to t$ and $X(t_n, \omega) \to X(t, \omega)$. ∎

In Exercises 9.1 and 9.3 we consider the existence of strongly separable versions in two very important cases: (i) When $\{X(t),\ t \in T\}$ consists of i.i.d. random variables and (ii) when $\{X(t),\ t \in T\}$ is continuous in probability.

As it turns out, *every* stochastic process has a separable version.

Theorem 9.2.5 *Let $\{X(t),\ t \in T\}$ be a stochastic process defined on an arbitrary probability space $(\Omega, \mathcal{F}, P)$.*

(i) *A separable version $\{Y(t),\ t \in T\}$ of $\{X(t),\ t \in T\}$ exists. Moreover, $\{Y(t),\ t \in T\}$ can be defined on the same probability space $(\Omega, \mathcal{F}, P)$ in such a way that $P(X(t) = Y(t)) = 1$ for every $t \in T$.*

(*ii*) *Assume, in addition, that* (T, τ) *is a topological space with countable base. Then there is a version* $\{Y(t),\ t \in T\}$ *of* $\{X(t),\ t \in T\}$, *strongly separable in the topology* τ, *that takes values in the extended real line* $[-\infty, +\infty]$. *Moreover,* $\{Y(t),\ t \in T\}$ *can be defined on the same probability space* $(\Omega, \mathcal{F}, P)$ *in such a way that* $P(X(t) = Y(t)) = 1$ *for every* $t \in T$.

The proof of this theorem parallels the corresponding argument in Doob (1953), but in a more general setting. Firstly, we present a few remarks.

Remarks

1. The theorem states that a strongly separable version of a given stochastic process may take the infinite values $\pm\infty$. We will see in the proof why such infinite values appear.

2. The proof of part (ii) of Theorem 9.2.5 provides a construction of a strongly separable version. Is there a different construction which provides a version that takes only finite values? In general, the answer is in the negative (see Exercise 9.4).

3. One should not be unduly concerned about the infinite values. Firstly, for each t, $P(|Y(t)| < \infty) = 1$ anyway. Secondly, for "smooth" stochastic processes (which are our main concern here) there is often a strongly separable version that takes no infinite values. In fact, the "most irregular" of stochastic process, the one consisting of i.i.d. random variables, admits a strongly separable version that takes only finite values!

Lemma 9.2.6 *Let* $\{X(t),\ t \in T\}$ *be a stochastic process on a probability space* $(\Omega, \mathcal{F}, P)$. *Let* $\mathcal{A}_0$ *be a countable collection of Borel subsets of* $\mathbb{R}$ *and let* $\mathcal{A}$ *be the collection of intersections of subfamilies of* $\mathcal{A}_0$. *Then there is a countable subset* T^* *of* T *such that, for every* $t \in T$, *there is an event* $\Omega_t \in \mathcal{F}$ *with* $P(\Omega_t) = 0$ *such that*

$$\{\omega \colon X(s) \in A,\ \forall s \in T^*,\ X(t) \notin A\} \subseteq \Omega_t$$

for every $A \in \mathcal{A}$.

PROOF: We start with the particular case where $\mathcal{A}_0$ consists of a single Borel set A. Then, of course, $\mathcal{A} = \{A\}$ as well. For every countable subset C of T, define

$$P_C(A) = P(X(t) \in A,\ \forall t \in C) \tag{9.2.1}$$

and let $\alpha(A) = \inf_C P_C(A)$. This infimum is achieved because there is a sequence of countable subsets C_m of T, $m = 1, 2, \ldots,$ such that $P_{C_m}(A) \leq \alpha(A) + \frac{1}{m}$.

Since $P_{C_m}(A) \downarrow \alpha(A)$ as $m \uparrow \infty$ and $T_A^* = \bigcup_{m=1}^{\infty} C_m$, we have $P_{T_A^*}(A) = \alpha(A)$, i.e., $C = T_A^*$ achieves the infimum. In the sequel, T_A^* continues to denote a countable subset of T for which this infimum is achieved.

Now choose any $t \in T$ and let

$$\Omega_t(A) = \{\omega\colon X(s) \in A,\ \forall s \in T_A^*,\ X(t) \notin A\}.$$

Since (9.2.1) with $C = T_A^* \cup \{t\}$ yields

$$P(X(s) \in A,\ \forall s \in T_A^*,\ X(t) \in A) \geq \alpha(A) = P_{T_A^*}(A),$$

we conclude

$$P(\Omega_t(A)) = P_{T_A^*}(A) - P(X(s) \in A,\ \forall s \in T_A^*,\ X(t) \in A) = 0.$$

This proves the statement of the lemma in the particular case of $\mathcal{A}_0$ consisting of a single Borel set. To prove the general case where $\mathcal{A}_0$ is (at most) countable, fix $t \in T$ and construct T_A^* and $\Omega_t(A)$ for each $A \in \mathcal{A}_0$. If $T^* = \bigcup_{A \in \mathcal{A}_0} T_A^*$ and $\Omega_t = \bigcup_{A \in \mathcal{A}_0} \Omega_t(A)$, then, clearly, $P(\Omega_t) = 0$ for every $t \in T$. Now, if $B \in \mathcal{A}$, then $B = \bigcap_k A_k$ for some sequence $A_k \in \mathcal{A}_0$, $k = 1, 2, \ldots,$ and

$$\begin{aligned}
&\Big\{\omega\colon X(s) \in B,\ \forall s \in T^*,\ X(t) \notin B\Big\} \\
&\quad = \bigcup_k \Big\{\omega\colon X(s) \in B,\ \forall s \in T^*,\ X(t) \notin A_k\Big\} \\
&\quad \subseteq \bigcup_k \Big\{\omega\colon X(s) \in A_k,\ \forall s \in T_{A_k}^*,\ X(t) \notin A_k\Big\} \\
&\quad \subseteq \bigcup_k \Omega_t(A_k) \subseteq \Omega_t,
\end{aligned}$$

as required. ■

PROOF OF THEOREM 9.2.5: Let $\mathcal{A}_0$ be the collection of all finite unions of closed intervals with rational or infinite endpoints, and let $\mathcal{A}$ be defined as in Lemma 9.2.6. Observe that $\mathcal{A}$ is the collection of all closed sets in $\mathbb{R}$.

(i) To establish the first part of the theorem, choose T^* as in Lemma 9.2.6 and define, for each $t \in T$,

$$Y(t) = \begin{cases} X(t) & \text{if } \omega \notin \Omega_t, \\ X(t_0) & \text{if } \omega \in \Omega_t, \end{cases}$$

where Ω_t is the null event described in Lemma 9.2.6 and t_0 is any fixed point in T^*. Then $\{Y(t),\ t \in T\}$ is a separable process, T^* is a separant and $\Omega_0 = \emptyset$. Moreover, $P(X(t) = Y(t)) = 1$ for every $t \in T$.

(ii) To establish the second part of the theorem, let β denote a countable base of (T, τ). Apply Lemma 9.2.6 to conclude that for each $B \in \beta$, there is a countable subset $T^*(B)$ of B with the following property: for every $t \in B$ there is an event Ω_t^B with $P(\Omega_t^B) = 0$ such that

$$\{X(t) \in F, \forall t \in T^*(B), X(t) \notin F\} \subseteq \Omega_t^B$$

for every closed subset F of $\mathbb{R}$. Let $T^* = \bigcup_{B \in \beta} T^*(B)$, and for every $t \in T$, set $\Omega_t = \emptyset$ if $t \in T^*$ and $\Omega_t = \bigcup_{\substack{B \in \beta \\ B \ni t}} \Omega_t^B$ if $t \notin T^*$. Since the base β is countable, $P(\Omega_t) = 0$.

Now define the process $\{Y(t), t \in T\}$ as follows: for each $t \in T$,

$$Y(t) = \begin{cases} X(t) & \text{if } \omega \notin \Omega_t, \\ \overline{\lim}_{\substack{s \to t \\ s \in T^*}} X(s) & \text{if } \omega \in \Omega_t. \end{cases}$$

Obviously, $Y(t)$ is a well-defined random variable (taking values in $[-\infty, +\infty]$), and $P(Y(t) \neq X(t)) \leq P(\Omega_t) = 0$ for every $t \in T$. Hence Y is a version of the process X defined on the same probability space as X.

To prove that $\{Y(t), t \in T\}$ is strongly separable, fix any $I \in \tau$ and any closed subset F of $[-\infty, \infty]$. It is enough to show

$$\{Y(t) \in F, \forall t \in I \cap T^*\} = \{Y(t) \in F, \forall t \in I\}. \tag{9.2.2}$$

We will prove firstly that for $t \in I$, the set

$$\Delta_t(I, F) = \{Y(t) \in F, \forall t \in I \cap T^*, Y(t) \notin F\}$$

is empty. We only need to consider the case $t \notin T^*$. Choose any $B \in \beta$ such that $t \in B \subseteq I$. Then $\Delta_t(I; F) \subseteq \Delta_t(B; F)$. If $\omega \notin \Omega_t$, then $\omega \notin \Omega_t^B$, so that $\omega \notin \Delta_t(B; F)$. On the other hand, if $\omega \in \Omega_t$, then $X(t) = Y(t) \in F$ for every $t \in B \cap T^*$ implies

$$Y(t) = \overline{\lim}_{\substack{s \to t \\ s \in B \cap T^*}} X(s) = \overline{\lim}_{\substack{s \to t \\ s \in B \cap T^*}} Y(s) \in F,$$

since F is closed. Therefore, again $\omega \notin \Delta_t(B; F)$. This means that $\Delta_t(B; F) = \emptyset$ and thus $\Delta_t(I; F) = \emptyset$. Since

$$\{Y(t) \in F, \forall t \in I \cap T^*\} \backslash \{Y(t) \in F, \forall t \in I\} = \bigcup_{t \in I} \Delta_t(I; F) = \emptyset,$$

Relation (9.2.2) follows and we conclude that the process Y is strongly separable with T^* as a separant and the exceptional set $\Omega_0 = \emptyset$. ∎

9.3 Applications

Here, we consider sample boundedness, sample continuity and pointwise sample continuity.

Definition 9.3.1 A stochastic process is said to possess a sample path property $\mathcal{P}$ if there is a version of this process in which *each* sample path has the property $\mathcal{P}$.

Example 9.3.2 A stochastic process $\{X(t),\ t \in T\}$ is said to be *sample bounded* if there is a version $\{Y(t),\ t \in T\}$ of $\{X(t),\ t \in T\}$ such that

$$\{\omega\colon \sup_{t \in T} |Y(t)| = \infty\} = \emptyset.$$

A stochastic process $\{X(t),\ t \in T\}$ on a topological space (T, τ) is called *sample continuous* if there is a version $\{Y(t),\ t \in T\}$ of $\{X(t),\ t \in T\}$ such that

$$\{\omega\colon \text{ the map } Y\colon T \to \mathbb{R} \text{ is not continuous }\} = \emptyset.$$

It is important to note that not every version of a sample continuous process has all (or even "almost" all) continuous sample paths. Indeed, using the methods of Example 9.1.2 it is easy to construct a version of Brownian motion $\{B(t),\ 0 \leq t \leq 1\}$ where all sample paths are discontinuous, i.e., $\{\omega\colon$ the map $B\colon [0,1] \to \mathbb{R}$ is continuous $\} = \emptyset$! Note that according to Definition 9.3.1, Brownian motion is then both sample continuous and sample discontinuous.

The study of sample continuity and other sample path properties requiring a topology on T is often simpler when there is a metric ρ on T, that is, when (T, ρ) is a metric space. The assumption "the topological space has a countable base" is then equivalent to "(T, ρ) is a *separable* metric space." The meaning, here, of the term "separable" is different from that in Definition 9.2.1.

Definition 9.3.3 A metric space (T, ρ) is *separable* if T has a countable dense subset.

For example, the real line $\mathbb{R}$ with the usual Euclidean metric is separable because the rational numbers are dense in $\mathbb{R}$. If T is separable with a countable dense subset $\{t_1, t_2, \ldots\}$, then the balls $B_{t_i}(r) = \{x\colon \rho(x, t_i) < r\}$, $i = 1, 2, \ldots, r$ rational, form a countable base. Because most of the stochastic processes in which we are interested are defined on separable metric spaces, we will frequently assume that (T, ρ) is such a space.

We will also be interested in *compact* metric spaces (T, ρ), i.e., metric spaces where every (infinite) sequence in T contains an (infinite) subsequence which converges to a point in T. (The qualifier "infinite" is always subsumed.) The real line is obviously not compact (no subsequence of the integers converges in $\mathbb{R}$), but the extended real line $[-\infty, +\infty]$ is compact and so is any closed interval $[a, b]$ in $\mathbb{R}$.

The following proposition can simplify the search for a version in which all sample paths are bounded and/or continuous.

Proposition 9.3.4 *Let* $\{X(t),\ t \in T\}$ *be a stochastic process defined on a probability space* $(\Omega, \mathcal{F}, P)$.

If T^* *is a separant for a separable version of* $\{X(t),\ t \in T\}$, *then* $\{X(t),\ t \in T\}$ *is sample bounded if and only if*

$$P(\sup_{t \in T^*} |X(t)| < \infty) = 1. \tag{9.3.1}$$

If (T, ρ) *is a compact separable metric space and* T^* *is a strong separant for a strongly separable version of* $\{X(t),\ t \in T\}$, *then* $\{X(t),\ t \in T\}$ *is sample continuous if and only if*

$$P(X(t) \textit{ is uniformly continuous on } T^*) = 1. \tag{9.3.2}$$

Remarks

1. The statements (9.3.1) and (9.3.2) are well defined in the sense that they both involve collections $\omega \in \Omega$ which are events (i.e., belong to $\mathcal{F}$). The probabilities of these events, moreover, depend only on the finite-dimensional distributions of $\{X(t),\ t \in T\}$. (See Exercise 9.7.)

2. The version where all sample paths are bounded can be constructed as a modification of the original separable version. It can be made separable with the same T^* as separant.

PROOF OF PROPOSITION 9.3.4: It follows from the first remark above that sample boundedness (respectively, continuity) of $\{X(t),\ t \in T\}$ implies (9.3.1) (respectively, (9.3.2)). On the other hand, suppose (9.3.1) holds, and let $\{Y(t),\ t \in T\}$, defined on a probability space $(\Omega', \mathcal{F}', P')$, be a separable version of $\{X(t),\ t \in T\}$ with T^* as separant. Let $\Delta_0' = \{\omega' \in \Omega' \colon \sup_{t \in T^*} |Y(t)| = \infty\}$, and let Ω_0' be the null subset of the sample space Ω' which enters the definition of separability of $\{Y(t),\ t \in T\}$. Set $\Sigma_0' = \Delta_0' \cup \Omega_0'$. From the first remark above and (9.3.1), conclude that $P'(\Sigma_0') = 0$. Notice that for any $\omega' \notin \Sigma_0'$, one has $\sup_{t \in T} |Y(t)| < \infty$. Now define for every $t \in T$,

$$Z(t) = \begin{cases} Y(t) & \text{if } \omega' \notin \Sigma_0', \\ 0 & \text{if } \omega' \in \Sigma_0'. \end{cases}$$

Every sample path of $\{Z(t),\ t \in T\}$ is bounded and, obviously, $\{Z(t),\ t \in T\}$ is a version of $\{X(t),\ t \in T\}$. This proves that $\{X(t),\ t \in T\}$ is sample bounded. Similarly, (9.3.2) implies the sample continuity of $\{X(t),\ t \in T\}$. ■

Continuity in probability is a pointwise property, whereas sample continuity is a global one. Because a process which is not continuous in probability cannot be sample continuous, we may want to assume from the outset that $\{X(t),\ t \in T\}$ is continuous in probability. In this case, we can choose T^* to be any countable dense subset of T (see Exercise 9.3). We then obtain the following modification of the second part of Proposition 9.3.4. Here, the separable metric space (T, ρ) can be assumed to be locally compact, i.e., every point of T is contained in some open ball with compact closure. This allows (T, ρ), for example, to be a Euclidean space.

Proposition 9.3.5 *Let (T, ρ) be a locally compact separable metric space and let T^* be a countable dense subset of T. Let $\{X(t),\ t \in T\}$ be a stochastic process defined on a probability space $(\Omega, \mathcal{F}, P)$ which is continuous in probability. Then $\{X(t),\ t \in T\}$ is sample continuous if and only if for every compact subset C of T,*

$$P(X(t) \text{ is uniformly continuous on } T^* \cap C) = 1. \tag{9.3.3}$$

PROOF: Obviously, the sample continuity of $\{X(t),\ t \in T\}$ implies (9.3.3). On the other hand, suppose that (9.3.3) holds for every compact subset C of T. We know from Proposition 9.3.4 that for each compact C there is a sample continuous version defined on C. Our aim is to construct a sample continuous version on T.

The Lindelöf theorem (Kelley 1955) states that there is a countable subcover of each open cover of a space whose topology has a countable base. Therefore one can express T as a countable union of compact subsets such that each point t of T belongs to the interior of one of these subsets.

Now, let $\{Y(t),\ t \in T\}$ be a strongly separable version of $\{X(t),\ t \in T\}$ with T^* as separant (Exercise 9.3), defined on (say) the same probability space $(\Omega, \mathcal{F}, P)$. Conclude as in the proof of Proposition 9.3.4 that there are events $\Sigma_0^{(n)} \in \mathcal{F}$ with $P(\Sigma_0^{(n)}) = 0$ for every $n = 1, 2, \ldots$ such that for every $\omega \notin \Sigma_0^{(n)}$, $\{Y(t),\ t \in B_n\}$ is uniformly continuous. Denote $\Sigma_0 = \bigcup_{n=1}^{\infty} \Sigma_0^{(n)}$. Then $P(\Sigma_0) = 0$. Define

$$Z(t) = \begin{cases} Y(t) & \text{if } \omega \notin \Sigma_0, \\ 0 & \text{if } \omega \in \Sigma_0, \end{cases}$$

$t \in T$. It is obvious that $\{Z(t),\ t \in T\}$ is a version of $\{X(t),\ t \in T\}$ and that all realizations of $\{Z(t),\ t \in T\}$ corresponding to $\omega \in \Sigma_0$ are continuous. Fix $\omega \notin \Sigma_0$. Since every point $t_0 \in T$ belongs to the interior of some B_n, the non-random function $Z(t,\omega)$, $t \in T$, is continuous at t_0 and hence is a continuous function on T. Thus $\{Z(t),\ t \in T\}$ is a version of $\{X(t),\ t \in T\}$ with all sample paths continuous, proving that $\{X(t),\ t \in T\}$ is sample continuous. ■

A weaker property than sample path continuity is pointwise sample path continuity.

Definition 9.3.6 A stochastic process $\{X(t),\ t \in T\}$ on a topological space (T,τ) is *pointwise sample continuous* if, for every $t_0 \in T$, there is a version $\{Y(t),\ t \in T\}$ of $\{X(t),\ t \in T\}$ such that every sample path of $\{Y(t),\ t \in T\}$ is continuous at t_0. We say that $\{X(t),\ t \in T\}$ is *sample continuous at each* $t_0 \in T$.

A sample continuous stochastic process is obviously pointwise sample continuous and if $\{X(t),\ t \in T\}$ is a non-random function on (T,τ), then the two notions are equivalent. When $\{X(t),\ t \in T\}$ is random, the implication "pointwise sample continuity $\Rightarrow$ sample continuity" is false. Nevertheless, the implication holds if we restrict ourselves to a certain class of stochastic processes. For example, it holds for Gaussian processes on a compact metric space. Unfortunately, it does not hold for stable processes on a compact metric space (Exercise 9.5). But, as we will see later, pointwise sample continuity implies sample continuity for a certain subclass of stable processes.

Here is a version of Proposition 9.3.4 tailored to pointwise sample continuity. The proof is left as an exercise (Exercise 9.9).

Proposition 9.3.7 *Let $\{X(t),\ t \in T\}$ be a stochastic process on a separable metric space (T,ρ). Let $t_0 \in T$, and let T^* be a strong separant for a strongly separable version of $\{X(t),\ t \in T\}$. Then $\{X(t),\ t \in T\}$ is sample continuous at t_0 if and only if*

$$P\Big(\lim_{t\to t_0,\ t\in T^*} X(t) = X(t_0)\Big) = 1. \tag{9.3.4}$$

9.4 Measurability

We have introduced separable versions of stochastic processes and proved in Theorem 9.2.5 that every stochastic process possesses one. The situation is quite different with *measurable versions*. We assume again that (T,ρ) is a separable metric space.

Definition 9.4.1 A stochastic process $\{X(t),\ t \in T\}$ on a probability space $(\Omega,\mathcal{F},P)$ is called *measurable* if $X(t,\omega): T \times \Omega \to \mathbb{R}$ is a (jointly) measurable function.

It is often preferable to work with a measurable version of a given stochastic process $\{X(t),\ t \in T\}$. Unfortunately, not every stochastic process has a measurable version. Necessary and sufficient conditions on the finite-dimensional distributions of the process which ensure the existence of a measurable version are given below. They are due to Hoffmann-Jørgensen (1973) and involve the Condition S introduced in Section 3.11. Recall that $\{X(t),\ t \in T\}$ satisfies Condition S if there is a countable subset $T_0 \subseteq T$ such that for every $t \in T$, $X(t) = \text{plim}_{k\to\infty} X(t_k)$ where $t_k \in T_0,\ k = 1, 2, \ldots$.

Theorem 9.4.2 *Let $\{X(t),\ t \in T\}$ be a stochastic process. Then $\{X(t),\ t \in T\}$ has a measurable version if and only if the following conditions hold:*

(1) $\{X(t),\ t \in T\}$ satisfies Condition S ;

(2) for every two Borel sets G and H and every $s \in T$ the map

$$t \to P\big(X(s) \in G,\ X(t) \in H\big)$$

from T to $[0, 1]$ is measurable. Moreover, if (1) and (2) hold, then a measurable version $\{Y(t),\ t \in T\}$ can be chosen to be strongly separable, defined on the same space $(\Omega, \mathcal{F}, P)$ as $\{X(t),\ t \in T\}$ and satisfy $P(X(t) = Y(t)) = 1$ for every $t \in T$.

Example 9.4.3 Consider the process $\{X(t),\ t \in [0, 1]\}$ comprising i.i.d. non-degenerate random variables. As indicated in Section 3.11, this process does not satisfy Condition S and hence by Theorem 9.4.2, does not have a measurable version.

Remark. Condition S and condition (2) of Theorem 9.4.2 involve only two-dimensional distributions. Hence only the two-dimensional distributions of $\{X(t),\ t \in T\}$ play a role in determining whether the process has a measurable version. The one-dimensional distributions of $\{X(t),\ t \in T\}$ are not sufficient (Exercise 9.10).

The proof of Theorem 9.4.2 is based on the following proposition due to Cohn (1972). To formulate it, we first need to introduce the Ky Fan distance. Let X and Y be random variables defined on the same probability space. The *Ky Fan distance* between X and Y is

$$\alpha(X, Y) := \inf\{\epsilon \geq 0\colon P(|X - Y| > \epsilon) < \epsilon\}. \tag{9.4.1}$$

This distance metrizes convergence in probability (i.e., $X_n \to X$ in probability as $n \to \infty$ if and only if $\alpha(X_n, X) \to 0$ as $n \to \infty$), and it converts the space $L^0(\Omega)$ of random variables into a metric space (Exercise 9.11).

Proposition 9.4.4 *Let $\{X(t),\ t \in T\}$ be a stochastic process. Then $\{X(t),\ t \in T\}$ has a measurable version if and only if the map $t \to X(t)$, from the metric space (T, ρ) to the metric space $(L^0(\Omega), \alpha)$, is (Borel) measurable and has separable range, i.e., the range has a countable dense subset. In the latter case, a measurable version $\{Y(t),\ t \in T\}$ can be chosen to be strongly separable, be defined on the same probability space $(\Omega, \mathcal{F}, P)$ as $\{X(t),\ t \in T\}$ and satisfy $P(X(t) = Y(t)) = 1$ for each $t \in T$.*

PROOF: (i) Suppose firstly that the map $t \to X(t)$ is measurable and has separable range. Our goal is to construct a measurable version $\{Y(t),\ t \in T\}$.

Fix $n \geq 1$. Because the range is separable, we can choose a collection $C_i(n)$, $i = 1, 2, \ldots$, of disjoint Borel sets in $L^0(\Omega)$ of α-diameter less than 2^{-n} that covers the range $\{X(t),\ t \in T\}$. The sets $B_i(n) = \{t \in T\colon X(t) \in C_i(n)\}$, $i = 1, 2, \ldots$, constitute a Borel partition of T. For any two points $s, t \in B_i(n)$, we have $\alpha(X(s), X(t)) < 2^{-n}$, implying

$$P(|X(s) - X(t)| > 2^{-n}) < 2^{-n} \tag{9.4.2}$$

by the definition (9.4.1) of the Ky Fan distance α.

For each n and i choose $t_{n,i} \in B_i(n)$. Let $X_n(t, \omega) = X(t_{n,i}, \omega)$ if $t \in B_i(n)$, $i = 1, 2, \ldots$, $n = 1, 2, \ldots$. Each $\{X_n(t),\ t \in T\}$, $n = 1, 2, \ldots$, is a measurable process because for every Borel set $A \subseteq \mathbb{R}$, we have $\{(t, \omega)\colon X_n(t, \omega) \in A\} = \bigcup_{i=1}^{\infty}(B_i(n) \times \{\omega \in \Omega\colon X(t_{n,i}, \omega) \in A\})$. It follows from (9.4.2) that $P(|X_n(t) - X(t)| > 2^{-n}) < 2^{-n}$ for every $t \in T$ and $n \geq 1$. Therefore, by the Borel-Cantelli lemma,

$$P(\lim_{n \to \infty} X_n(t) = X(t)) = 1 \quad \text{for every } t \in T. \tag{9.4.3}$$

Now consider the measurable set $E = \{(t, \omega)\colon X_n(t, \omega)$ converges as $n \to \infty\}$. The stochastic process $\widetilde{X}_n$, defined as $\widetilde{X}_n(t, \omega) = X_n(t, \omega)$ if $(t, \omega) \in E$ and $\widetilde{X}_n(t, \omega) = 0$ otherwise, is measurable for every $n = 1, 2, \ldots,$ and $\widetilde{X}_n(t, \omega)$ converges as $n \to \infty$ for every $t \in T$, $\omega \in \Omega$. Let $Y(t) = \lim_{n \to \infty} \widetilde{X}_n(t)$, $t \in T$. Then $\{Y(t),\ t \in T\}$ is measurable as the limit of a sequence of measurable functions and $P(X(t) = Y(t)) = 1$ for every $t \in T$ in view of (9.4.3). To show that the version $\{Y(t),\ t \in T\}$ can be chosen to be strongly separable, one has to modify slightly the last step above and use the measurable choice of limit points lemma of Cohn (1972). We omit the details.

(ii) Suppose, now, that $\{X(t),\ t \in T\}$ has a measurable version $\{Y(t),\ t \in T\}$ defined on some probability space $(\Omega_1, \mathcal{F}_1, P_1)$. We want to show that the map $t \to X(t)$ is measurable and has a separable range. Without loss of generality, we may assume that the probability space $(\Omega_1, \mathcal{F}_1, P_1)$ is rich enough so that the following is true: for every random variable $X_0 \in L^0(\Omega)$ there is a random variable $Y_0 \in L^0(\Omega_1)$ such that $\{\{X(t),\ t \in T\}, X_0\}$ has the same finite-dimensional distributions as $\{\{Y(t),\ t \in T\}, Y_0\}$. Then for every $\epsilon > 0$, $\{t \in T\colon \alpha(X(t), X_0) < \epsilon\} = \{t \in T\colon \alpha(Y(t), Y_0) < \epsilon\}$, and so it is enough to prove that the map $t \to Y(t)$, from (T, ρ) to $(L^0(\Omega_1), \alpha)$, is measurable and has separable range.

Let $\mathcal{M}$ be the family of all *measurable* processes $\{Y(t),\ t \in T\}$ on $(\Omega_1, \mathcal{F}_1, P_1)$ such that the map $t \to Y(t)$ from (T, ρ) to $(L^0(\Omega_1), \alpha)$ is measurable and has separable range. We want to show that $\mathcal{M}$ is closed under pointwise convergence. More precisely, let Y_k, $k = 1, 2, \dots$, be a sequence of processes in $\mathcal{M}$ such that the sequence $\{Y_k(t, \omega_1)\}$ converges as $k \to \infty$ for each $t \in T$ and $\omega_1 \in \Omega_1$. Let $Y(t, \omega_1) = \lim_{k\to\infty} Y_k(t, \omega_1)$. We must show that $Y\colon T \times \Omega_1 \to \mathbb{R}$ is in $\mathcal{M}$. Clearly, Y is measurable. Moreover, for each $t \in T$, as $k \to \infty$, $Y_k(t) \to Y(t)$ a.s. and thus in probability. The map $t \to Y(t)$ from (T, ρ) to $(L^0(\Omega_1), \alpha)$ is therefore the limit of a sequence of measurable maps and is thus measurable itself. Moreover, denoting by S_j a countable subset of $(L^0(\Omega_1), \alpha)$ separating the range of $t \to Y_j(t)$, $j = 1, 2, \dots$, we conclude that $\bigcup_{j=1}^{\infty} S_j$ separates the range of $t \to Y(t)$. Therefore $Y \in \mathcal{M}$, and thus $\mathcal{M}$ is closed under pointwise convergence.

Now let $\mathcal{M}_S$ be the family of "simple" maps Y from $T \times \Omega_1 \to \mathbb{R}$ of the form

$$Y(t, \omega_1) = \sum_{i=1}^{n} a_i \mathbf{1}\Big((t, \omega_1) \in \bigcup_{j=1}^{m_i} (A_{i,j} \times B_{i,j})\Big),$$

where, respectively, $A_{i,j}$, $B_{i,j}$, $j = 1, \dots, m_i$, $i = 1, \dots, n$, $n = 1, 2, \dots$, are measurable sets in $\mathcal{B}(T)$ and $\mathcal{F}_1$. Clearly, $\mathcal{M}_S \subset \mathcal{M}$.

Since $\mathcal{M}$ contains $\mathcal{M}_S$ and is closed under pointwise convergence, we can use a monotone class argument (see Exercise 9.13) to conclude that every measurable process $\{Y(t),\ t \in T\}$ on $(\Omega_1, \mathcal{F}_1, P_1)$ belongs to $\mathcal{M}$. This completes the proof of the proposition. ■

PROOF OF THEOREM 9.4.2: Suppose that $\{X(t),\ t \in T\}$ has a measurable version $\{Y(t),\ t \in T\}$ defined on a probability space $(\Omega_1, \mathcal{F}_1, P_1)$. By Proposition 9.4.4, the map $t \to X(t)$ from T to $(L^0(\Omega), \alpha)$ has separable range, which implies that $\{X(t),\ t \in T\}$ satisfies Condition S (as noted in the remarks following Definition 3.11.2 and established in Exercise 3.21). Moreover, for every two Borel sets G and H and every set $s \in T$,

$$\begin{aligned} P\big(X(s) \in G,\ X(t) \in H\big) &= P\big(Y(s) \in G,\ Y(t) \in H\big) \\ &= \int_{\Omega_1} \mathbf{1}\big(Y(s, \omega_1) \in G\big)\mathbf{1}\big(Y(t, \omega_1) \in H\big) P_1(d\omega_1) \end{aligned}$$

is a measurable function of t by Fubini's theorem.

Suppose, on the other hand, that (1) and (2) of Theorem 9.4.2 hold. Let $\mathcal{M}$ be the class of all bounded measurable functions f: $\mathbb{R} \times \mathbb{R} \to \mathbb{R}$ such that for any $s \in T$, the map $t \to Ef(X(s), X(t))$ from T to $\mathbb{R}$ is measurable. Clearly, $\mathcal{M}$ is closed under the pointwise convergence of uniformly bounded sequences of functions. Let $\mathcal{M}_S$ be the family of "simple" functions f: $\mathbb{R} \times \mathbb{R} \to \mathbb{R}$ of the form

$$f(x,y) = \sum_{i=1}^{n} a_i 1\Big((x,y) \in \bigcup_{j=1}^{m_i} (A_{i,j} \times B_{i,j})\Big),$$

where $A_{i,j}, B_{i,j},\ j = 1, \ldots, m_i,\ i = 1, \ldots, n,\ n = 1, 2, \ldots,$ are Borel sets. It follows from (2) that $\mathcal{M}_S \subseteq \mathcal{M}$. We use Exercise 9.13 to conclude that $\mathcal{M}$ contains every bounded measurable function f: $\mathbb{R} \times \mathbb{R} \to \mathbb{R}$. Taking, in particular, $f(x,y) = |x - y| \wedge 1$, we see that the map $t \to \widetilde{\alpha}(X(s), X(t))$ from T to $\mathbb{R}$ is measurable for any $s \in T$ where $\widetilde{\alpha}(X,Y) = E(|X - Y| \wedge 1),\ X, Y \in L^0(\Omega)$. Note that $\widetilde{\alpha}$ is a metric on $L^0(\Omega)$ equivalent to the Ky Fan metric α (see Exercise 9.12). Therefore, by (1), $(R(\mathbf{X}), \widetilde{\alpha})$ is separable, where $\mathbf{X} = \{X(t),\ t \in T\}$ and $R(\mathbf{X})$ is the range of the map $t \to X(t)$ in $L^0(\Omega)$.

Let R^* be a countable dense subset of $(R(\mathbf{X}), \widetilde{\alpha})$ and let V be an open subset of $L^0(\Omega)$. For every random variable $Y \in R^* \cap V$, let

$$r_Y = \sup\big\{r > 0\colon \forall Z \in L^0(\Omega)\colon \widetilde{\alpha}(Y, Z) \le r \Rightarrow Z \in V\big\}.$$

Since

$$R(\mathbf{X}) \cap V = \bigcup_{Y \in R^* \cap V} \{Z \in R(\mathbf{X})\colon \widetilde{\alpha}(Y, Z) < r_Y\},$$

we conclude that

$$\{t \in T\colon X(t) \in V\} = \bigcup_{Y \in R^* \cap V} \{t \in T\colon \widetilde{\alpha}(X(s_Y), X(t)) < r_Y\}$$

is a $\mathcal{B}(T)$-measurable set. Here, s_Y is an element in T such that $X(s_Y) = Y$, $Y \in R^* \cap V$. Therefore the map $t \to X(t)$ from (T, ρ) to $(L^0(\Omega), \widetilde{\alpha})$ is measurable with separable range. Applying firstly Exercise 9.12 and then Proposition 9.4.4, we conclude that the process $\{X(t),\ t \in T\}$ has a measurable version, which can also be chosen to have the required properties. This completes the proof of Theorem 9.4.2 ■

9.5 Zero-one laws

In a previous section we discussed the meaning of some sample path properties such as sample boundedness and sample continuity. The careful reader may have noticed that we have not discussed the possibility that such properties hold only with some probability $0 < p < 1$, for example, that only a proportion p of the sample paths is sample bounded. This is because, as we will see below, for stable processes, properties such as sample boundedness and sample continuity satisfy a zero-one law; i.e., they hold either with probability 0 or with probability 1.

Definition 9.5.1 A stochastic process $\{X(t),\ t \in T\}$ is said to possess a sample path property $\mathcal{P}$ with probability $p > 0$ if

$$p = \sup \Big\{ q > 0: \begin{array}{l} \text{there is a version } \{Y(t),\ t \in T\} \text{ of } \{X(t),\ t \in T\} \\ \text{defined on a probability space } (\Omega, \mathcal{F}, P) \text{ and an event} \\ \Omega_0 \in \mathcal{F} \text{ with } P(\Omega_0) \leq 1 - q \text{ such that, for } \textit{any } \omega \notin \Omega_0, \\ \text{the sample path } Y(t, \omega),\ t \in T \text{ has the property } \mathcal{P} \end{array} \Big\}.$$

It is obvious that Propositions 9.3.4 and 9.3.5 extend immediately to the following:

Proposition 9.5.2 *Let $\{X(t),\ t \in T\}$ be a stochastic process defined on a probability space $(\Omega, \mathcal{F}, P)$ and let T^* be a separant for a separable version of $\{X(t),\ t \in T\}$. Then $\{X(t),\ t \in T\}$ is sample bounded with probability p if and only if*

$$P(\sup_{t \in T^*} |X(t)| < \infty) = p. \tag{9.5.1}$$

Let (T, ρ) be a compact separable metric space and let T^* be a strong separant for a strongly separable version of $\{X(t),\ t \in T\}$. Then $\{X(t),\ t \in T\}$ is sample continuous with probability p if and only if

$$P(X(t) \textit{ is uniformly continuous on } T^*) = p. \tag{9.5.2}$$

This proposition, whose proof is left to the reader, implies in particular that the supremum over q in Definition 9.5.1 is achieved. We can now relate the terminology introduced in the previous sections with the one we use here, in the context of zero-one laws. Recall that by "$\{X(t),\ t \in T\}$ is sample bounded" we mean "there is a version such that all paths $\{\omega\colon\ X(t, \omega),\ t \in T\}$ are bounded". As a consequence of the previous proposition we obtain

Corollary 9.5.3 *(i) A stochastic process $\{X(t),\ t \in T\}$ is sample bounded if and only if it is sample bounded with probability 1.*

(ii) Let (T, ρ) be a locally compact separable metric space. Then the process $\{X(t),\ t \in T\}$ is sample continuous if and only if it is sample continuous with probability 1.

The next theorem presents a basic zero-one law for α-stable processes. Because it is used on several occasions in the sequel, it is formulated in greater generality than is required for sample boundedness and sample continuity. The theorem involves a family $\mathcal{M}$ of stochastic processes and a property $\mathcal{P}$ of the sample paths. Because of the interplay between $\mathcal{M}$ and $\mathcal{P}$, it is convenient to regard explicitly each stochastic process as a function $X: T \times \Omega \to \mathbb{R}$ of two variables $t \in T$ and $\omega \in \Omega$. $X(t, \cdot)$ is, as usual, Borel-measurable for each $t \in T$. We shall have to distinguish explicitly between the process $X = \{X(t,\omega),\ t \in T,\ \omega \in \Omega\}$, the function $X_{\omega_0}: T \times \Omega \to \mathbb{R}$ given by $X(t,\omega) = X(t,\omega_0)$ for a fixed $\omega_0 \in \Omega$ and the realization $X(\cdot, \omega_0): T \to \mathbb{R}$ where $\omega_0 \in \Omega$.

Theorem 9.5.4 *Let $(\Omega, \mathcal{F}, P)$ be a probability space. Let $\mathcal{M}$ be a family of stochastic processes*

$$X = \{X(t,\widetilde{\omega}),\ t \in T,\ \widetilde{\omega} = (\omega_1, \omega_2, \ldots) \in \Omega^{\mathbb{N}}\}$$

defined on the product space $(\widetilde{\Omega}, \widetilde{\mathcal{F}}, \widetilde{P}) = (\Omega, \mathcal{F}, P)^{\mathbb{N}}$ and depending on finitely many coordinates[1] *of the product space, with the following properties:*

(i) *For every X and Y in $\mathcal{M}$ and any two real numbers a, b, the process $aX + bY$ is also in $\mathcal{M}$.*

(ii) *If the stochastic process X is in $\mathcal{M}$, then for every $\widetilde{\omega}_0 \in \Omega^{\mathbb{N}}$, $X_{\widetilde{\omega}_0}$ defined above is also in $\mathcal{M}$.*

(iii) *If the stochastic process $X = \{X(t,\widetilde{\omega}),\ t \in T,\ \widetilde{\omega} \in \Omega^{\mathbb{N}}\}$ is in $\mathcal{M}$, then the stochastic process $X^{(i)} = \{X(t,\widetilde{\omega}^{(i)}),\ t \in T,\ \widetilde{\omega} \in \Omega^{\mathbb{N}}\}$ is also in $\mathcal{M}$ for every $i = 1, 2, \ldots$, where for $\widetilde{\omega} = (\omega_1, \omega_2, \ldots) \in \Omega^{\mathbb{N}}$,*

$$\widetilde{\omega}^{(i)} := (\omega_{i+1}, \omega_{i+2}, \ldots).$$

Moreover, let $\mathcal{P}$ be a sample path property such that:

(iv) *For every X in $\mathcal{M}$ the set $\{\widetilde{\omega}\colon X(\cdot,\widetilde{\omega})$ has property $\mathcal{P}\}$ is in $\widetilde{\mathcal{F}}$.*

(v) *If X and Y are in $\mathcal{M}$ and $X \stackrel{d}{=} Y$, then*

$$\widetilde{P}\big(\widetilde{\omega}\colon X(\cdot,\widetilde{\omega}) \text{ has property } \mathcal{P}\big) = \widetilde{P}\big(\widetilde{\omega}\colon Y(\cdot,\widetilde{\omega}) \text{ has property } \mathcal{P}\big).$$

(vi) *For every two functions $f, g\colon T \to \mathbb{R}$ with property $\mathcal{P}$ and any two real numbers a, b, the function $af + bg\colon T \to \mathbb{R}$ has property $\mathcal{P}$.*

[1] $\mathcal{M}$ may also include processes depending on zero coordinates, i.e., non-random processes.

Then for any α-stable stochastic process X in $\mathcal{M}$,

$$\widetilde{P}\big(\widetilde{\omega}\colon X(\cdot,\widetilde{\omega}) \text{ has property } \mathcal{P}\big) = 0 \text{ or } 1.$$

Remark. The family $\mathcal{M}$ is defined independently of $\mathcal{P}$. In some applications, one may choose $\mathcal{M}$ to include *all* processes depending on *any* number of finitely many coordinates, including zero coordinates. In that case, conditions (ii) and (iii) are automatically satisfied. Moreover, because the probability space $(\widetilde{\Omega}, \widetilde{\mathcal{F}}, \widetilde{P})$ is rich enough and because X depends only on a finite number of coordinates, it is possible, using the algorithm in (iii), to construct as many independent copies of X as we wish. These copies will all be in $\mathcal{M}$.

PROOF OF THEOREM 9.5.4: Assume, to the contrary, that

$$p := \widetilde{P}\big(\widetilde{\omega}\colon X(\cdot,\widetilde{\omega}) \text{ has property } \mathcal{P}\big) \in (0,1). \tag{9.5.3}$$

Let $\widetilde{X}$ be an independent copy of X belonging to $\mathcal{M}$ and depending on a different set of coordinates. Then $Y = X - \widetilde{X}$ is a $S\alpha S$ process and is in $\mathcal{M}$. Moreover,

$$q := \widetilde{P}\big(\widetilde{\omega}\colon Y(\cdot,\widetilde{\omega}) \text{ has property } \mathcal{P}\big) \in (0,1) \tag{9.5.4}$$

since $q \geq p^2 > 0$ and $1-q \geq 2p(1-p) > 0$.

Our aim is to show that (9.5.4) cannot hold. Let $\widetilde{Y}$ be an independent copy of Y belonging to $\mathcal{M}$ and dependent on a still different set of coordinates. Then using the fact that $Y + \widetilde{Y} \stackrel{d}{=} 2^{1/\alpha} Y$ and the properties of $\mathcal{M}$ and $\mathcal{P}$, we obtain

$$\begin{aligned} q &= \widetilde{P}\big(\widetilde{\omega}\colon Y(\cdot,\widetilde{\omega}) \text{ has property } \mathcal{P}\big) \\ &= \widetilde{P}\big(\widetilde{\omega}\colon Y(\cdot,\widetilde{\omega}) + \widetilde{Y}(\cdot,\widetilde{\omega}) \text{ has property } \mathcal{P}\big) \\ &= \widetilde{P}\big(\widetilde{\omega}\colon Y(\cdot,\widetilde{\omega}) + \widetilde{Y}(\cdot,\widetilde{\omega}) \text{ has property } \mathcal{P},\ \widetilde{Y}(\cdot,\widetilde{\omega}) \text{ does not have it}\big) \\ &\qquad + \widetilde{P}\big(\widetilde{\omega}\colon Y(\cdot,\widetilde{\omega}) \text{ has property } \mathcal{P},\ \widetilde{Y}(\cdot,\widetilde{\omega}) \text{ has property } \mathcal{P}\big) \\ &= \widetilde{P}\big(\widetilde{\omega}\colon Y(\cdot,\widetilde{\omega}) + \widetilde{Y}(\cdot,\widetilde{\omega}) \text{ has property } \mathcal{P},\ \widetilde{Y}(\cdot,\widetilde{\omega}) \text{ does not have it}\big) \\ &\qquad + q^2. \end{aligned} \tag{9.5.5}$$

It is enough to show

$$\widetilde{P}\big(\widetilde{\omega}\colon Y(\cdot,\widetilde{\omega}) + \widetilde{Y}(\cdot,\widetilde{\omega}) \text{ has property } \mathcal{P}, \widetilde{Y}(\cdot,\widetilde{\omega}) \text{ does not have property } \mathcal{P}\big) = 0 \tag{9.5.6}$$

since (9.5.5) and (9.5.6) imply $q = q^2$, i.e., $q = 0$ or 1, contradicting (9.5.4).

Suppose that (9.5.6) does not hold. Then, by Fubini's theorem, there is a function $f\colon T \to \mathbb{R}$ that *does not* have property $\mathcal{P}$ such that

$$\widetilde{P}\big(\widetilde{\omega}\colon Y(\cdot,\widetilde{\omega}) + f \text{ has property } \mathcal{P}\big) > 0. \tag{9.5.7}$$

Take any $\theta \in (0,1)$, and let

$$\widetilde{\Omega}_\theta = \{\widetilde{\omega}\colon Y(\cdot,\widetilde{\omega}) + \theta f \text{ has property } \mathcal{P}\}.$$

The events $\widetilde{\Omega}_\theta,\ \theta \in (0,1)$, are all disjoint. If this were not so, there would be $\theta_1 \neq \theta_2$ such that $\widetilde{\Omega}_{\theta_1} \cap \widetilde{\Omega}_{\theta_2} \neq \emptyset$. Then, taking $\widetilde{\omega} \in \widetilde{\Omega}_{\theta_1} \cap \widetilde{\Omega}_{\theta_2}$, we obtain a contradiction: on one hand $(\theta_1 - \theta_2)f = (Y(\cdot,\widetilde{\omega}) + \theta_1 f) - (Y(\cdot,\widetilde{\omega}) + \theta_2 f)$ has property $\mathcal{P}$ by the properties of $\mathcal{M}$ and $\mathcal{P}$, whereas on the other hand, the function f does not have property $\mathcal{P}$. This shows that the events $\widetilde{\Omega}_\theta,\ \theta \in (0,1)$, are all disjoint.

Let $\widetilde{Y}$ be once again an independent copy of Y, in $\mathcal{M}$, and dependent on a different set of coordinates. Using the fact $Y + \theta f \stackrel{d}{=} \theta Y + (1-\theta^\alpha)^{1/\alpha}\widetilde{Y} + \theta f$ and the properties of $\mathcal{M}$ and $\mathcal{P}$, we obtain

$$\begin{aligned}\widetilde{P}(\widetilde{\Omega}_\theta) &= P(\widetilde{\omega}\colon \theta Y(\cdot,\widetilde{\omega}) + (1-\theta^\alpha)^{1/\alpha}\widetilde{Y}(\cdot,\widetilde{\omega}) + \theta f \text{ has property } \mathcal{P}) \\ &\geq \widetilde{P}(\widetilde{\omega}\colon Y(\cdot,\widetilde{\omega}) + f \text{ has property } \mathcal{P}, \widetilde{Y}(\cdot,\widetilde{\omega}) \text{ has property } \mathcal{P}) \\ &= \widetilde{P}(\widetilde{\omega}\colon Y(\cdot,\widetilde{\omega}) + f \text{ has property } \mathcal{P}) \cdot \widetilde{P}(\widetilde{\omega}\colon Y(\cdot,\widetilde{\omega}) \text{ has property } \mathcal{P}),\end{aligned}$$

which is positive by (9.5.4) and (9.5.7), implying that the probability $\widetilde{P}(\widetilde{\Omega}_\theta)$ is positive and bounded away from 0. This is, of course, impossible, since there are uncountably many of such $\widetilde{\Omega}_\theta$s, all disjoint. Hence (9.5.6) must hold. This completes the proof of the theorem. ■

Corollary 9.5.5 *(i) An α-stable process $\{X(t),\ t \in T\}$ cannot be sample bounded with probability p, if $0 < p < 1$.*

(ii) If (T,ρ) is a locally compact separable metric space, then an α-stable process $\{X(t),\ t \in T\}$ cannot be sample continuous with probability p if $0 < p < 1$.

PROOF: (i) Assume that X is sample bounded with probability p. Then Proposition 9.5.2 implies that for some countable subset T^* of T, $p = P(\sup_{t\in T^*} |X(t)| < \infty)$. Choose $\mathcal{M}$ to be the family of all stochastic processes $\widetilde{X}$ indexed by T^*, defined on the probability space $(\Omega, \mathcal{F}, P)^{\mathbb{N}}$ and depending only on finitely many coordinates of the product space, including zero coordinates. Observe that our process X belongs to $\mathcal{M}$ since it depends on one coordinate. Choose $\mathcal{P}$ to be the property of boundedness of functions on T^*. Conditions (i), (iii) and (vi) of Theorem 9.5.4 are obviously satisfied. Condition (ii) holds because $\widetilde{X}_{\widetilde{\omega}_0}$ depends on zero coordinates. Conditions (iv) and (vi) follow from the fact that T^* is countable. Applying Theorem 9.5.4, we conclude that $p = 0$ or 1.

(ii) To prove the second part of the corollary, suppose, to the contrary, that $\{X(t),\ t \in T\}$ is sample continuous with probability $p \in (0,1)$. Let C be a compact subset of T. Then $\{X(t),\ t \in C\}$ is sample continuous with probability $q \geq p > 0$. Our aim is to prove that $q = 1$.

To this end, we apply firstly Proposition 9.5.2 to conclude that there is a countable subset C^* of C such that

$$P\big(X(t) \text{ is uniformly continuous on } C^*\big) = q.$$

Then we take $\mathcal{M}$ to be the family of all stochastic processes indexed by C^* defined on the probability space $(\Omega, \mathcal{F}, P)^{\mathbb{N}}$ and depending only on finitely many coordinates of the probability space. Choosing $\mathcal{P}$ to be the property of uniform continuity for functions on C^*, we can apply Theorem 9.5.4 and conclude that $q = 0$ or 1. Since $q \geq p > 0$, we can only have $q = 1$.

To see that $q = 1$ leads to a contradiction, observe that if $\{X(t),\ t \in C\}$ is sample continuous with probability 1 for *any* compact $C \subseteq T$, then it is continuous in probability on T (continuity in probability is a pointwise property). Letting T^* be a countable dense subset of T and using Propositions 9.5.2 and 9.3.5 we conclude that the process $\{X(t),\ t \in T\}$ is sample continuous, contradicting our assumption. ∎

Remarks

1. There is *no* zero-one law that states that $P(\sup_{n\geq 1} X_n < \infty) = 0$ or 1 for a family of jointly α-stable random variables $X_1, X_2, \dots$. (Take X_1 to be $S_\alpha(1, 0, 0)$ and set $X_n = nX_1,\ n \geq 2$. Then $P(\sup_{n\geq 1} X_n < \infty) = P(X_1 \leq 0) = 1/2$.) Therefore, there is *no* zero-one law for *one-sided* sample boundedness of an α-stable process!

2. An argument similar to that of Corollary 9.5.5 can be applied to derive a zero-one law for pointwise sample continuity; see Exercise 9.17.

9.6 Exercises

Exercise 9.1 As in Example 9.2.2, let $\{X(t),\ t \in T\}$ be a stochastic processes consisting of i.i.d. random variables and assume that the support of the distribution of $X(t)$ is the whole of $\mathbb{R}$. Let (T, τ) be a topological space with a countable base. Prove that *every version* of $\{X(t),\ t \in T\}$ is strongly separable in the topology τ.

Hint: If $I \in \tau$ is uncountable, choose as T^* any countable dense subset of T such that $\#(I \cap T^*) = \infty$, where # denotes cardinality.

Exercise 9.2 Give an example to show that the statements in Example 9.2.2 and in Exercise 9.1 are not generally true if the support of the distribution of $X(t)$ is a proper subset of $\mathbb{R}$.

Exercise 9.3 Let (T, τ) be a topological space with a countable base and let $\{X(t),\ t \in T\}$ be a stochastic process continuous in probability. Show directly (i.e., without appealing to Theorem 9.2.5) how to construct a version of the process $\{X(t),\ t \in T\}$ which is strongly separable in the topology τ and whose separant T^* is a given countable dense subset of T.
Hint: For each $t \in T$, there is a subsequence $\{t_{n_k}\}_{k=1}^{\infty} \subseteq T^*$ such that $X(t_{n_k}) \to X(t)$ a.s. Define for *each* $\omega \in \Omega$, $Y(t) = \limsup_{k\to\infty} X(t_{n_k})$.

Exercise 9.4 Let $\Omega = [0,1]$, $\mathcal{F} = \mathcal{B}$, $P =$ Lebesgue measure and define

$$X(t,\omega) = \begin{cases} \frac{1}{t-\omega} & \text{if } \omega \neq t, \\ 0 & \omega = t. \end{cases}$$

Show that every strongly separable version of the process $\{X(t),\ 0 \le t \le 1\}$ has to admit infinite values.
Hint: Let T^* be the countable dense subset of $[0,1]$ appearing in Proposition 9.2.4 and suppose without loss of generality that $1/2 \in T^*$. Let $\{Y(t),\ 0 \le t \le 1\}$ be any strongly separable version of $\{X(t),\ 0 \le t \le 1\}$ and let $(\Omega', \mathcal{F}', P')$ be the probability space on which it is defined. Set

$$A' = \left\{\omega' \in \Omega' \colon \frac{1}{Y(t)} - \frac{1}{Y(s)} = t - s \ \ \forall t, s \in T^*,\ \frac{1}{2} - \frac{1}{Y(\frac{1}{2})} \notin (0,1) \backslash T^* \right\},$$

take $\omega' \in \Omega_0'^c \cap A'$ where $P'(\Omega_0') = 0$ and let $t \to \frac{1}{2} - \frac{1}{Y(1/2)}$. Then use Proposition 9.2.4.

Exercise 9.5 The purpose of this exercise and of the following one is to give a feeling for the difference between continuity in probability, sample continuity and pointwise sample continuity. At the same time, it provides a direct treatment of the most important stable process–the $S\alpha S$ Lévy motion–without using the general theory developed later in this chapter.

Let $\{X(t),\ t \ge 0\}$ be a $S\alpha S$ Lévy motion.

(a) Show that $\{X(t),\ t \ge 0\}$ is continuous in probability.

(b) Let $\{r_n\}_{n=1}^{N} \subseteq [0,t]$. Show, using a proof similar to that of Kolmogorov's inequality, that for any $\lambda > 0$,

$$P(\max_{n=1,\dots,N} X(r_n) > \lambda) \le 2P(X(t) > \lambda).$$

Conclude that for each $t_0 \ge 0$, $\{X(t),\ t \ge 0\}$ is sample continuous at t_0.

Hint: Let $\mathbb{Q}$ be the set of rational numbers, fix $t_0 \in (0,1)$ and show that for any $\lambda > 0$ and $0 < \epsilon < \min(t_0,\ 1-t_0)$,

$$P\Big(\sup_{r \in \mathbb{Q} \cap [t_0-\epsilon, t_0+\epsilon]} |X(t_0) - X(r)| > \lambda\Big) \le 8P(X(\epsilon) > \lambda).$$

(c) Use (b) to conclude that the sample paths of $\{X(t),\ t \in [a,b]\}$ are bounded for any $0 \le a < b < \infty$.

(d) For $n = 1, 2, \dots,\ i = 1, \dots, 2^n$, define

$$Z_i^{(n)} = \sup_{r_1, r_2 \in \mathbb{Q} \cap (2^{-n}(i-1),\ 2^{-n}i)} |X(r_1) - X(r_2)|,$$

and let $Z_n^* = \max_{i=1,\dots,2^n} Z_i^{(n)}$. Show that for any $\lambda > 0$,

$$\liminf_{n \to \infty} P(Z_n^* \le \lambda) \le e^{-C\lambda^{-\alpha}}$$

for some $C > 0$. Conclude that $P(\lim_{n\to\infty} Z_n^* = 0) = 0$ and not 1, so that the $S\alpha S$ Lévy motion $\{X(t),\ t \ge 0\}$ is not sample continuous.

(e) Show that $\{X(t), 0 \le t \le 1\}$ has a version in $D([0,1])$, the space of functions on [0,1] that are right-continuous and have left limits.

Hint: See Breiman (1968), Theorem 14.27 and Corollary 14.29.

Exercise 9.6 Let $Z_1, Z_2, \dots$ be i.i.d. $S\alpha S$ random variables, and let $a_n \downarrow 0$ as $n \to \infty$. Define $\{X(t),\ t \in [0,1]\}$ by $X(\frac{1}{n}) = a_n Z_n$, $n = 1, 2, \dots,\ X(0) = 0$, and

$$X(t) = n(n+1)t\Big(X\Big(\frac{1}{n}\Big) - X\Big(\frac{1}{n+1}\Big)\Big) + (n+1)X\Big(\frac{1}{n+1}\Big) - nX\Big(\frac{1}{n}\Big)$$

if $1/(n+1) < t \le 1/n$.

Show that $\{X(t),\ t \in [0,1]\}$ is a $S\alpha S$ process and that it is continuous in probability. Show that $\{X(t),\ t \in [0,1]\}$ is pointwise sample continuous if and only if $\sum_{n=1}^{\infty} a_n^\alpha < \infty$. Thus, continuity in probability does not imply pointwise sample continuity.

Exercise 9.7 Under the assumptions of Proposition 9.3.4, let

$$\Delta_{\text{bound}} = \{\omega \in \Omega\colon \sup_{t \in T^*} |X(t)| < \infty\},$$

$$\Delta_{\text{cont}} = \{\omega \in \Omega\colon X(t) \text{ is uniformly continuous on } T^*\}.$$

Show that Δ_{bound} and Δ_{cont} belong to $\mathcal{F}$ and that $P(\Delta_{\text{bound}})$ and $P(\Delta_{\text{cont}})$ are determined by the finite-dimensional distributions of $\{X(t),\ t \in T\}$.

Exercise 9.8 (i) Let $\{X(t),\ t \in T\}$ be a sample bounded stochastic process, and let $\{Y(t),\ t \in T\}$ be a separable version of $\{X(t),\ t \in T\}$, defined on a probability space $(\Omega, \mathcal{F}, P)$. Then there is an event $\Sigma_0 \in \mathcal{F}$ with $P(\Sigma_0) = 0$ such that for every $\omega \notin \Sigma_0$, $\sup_{t \in T} |Y(t, \omega)| < \infty$.

(ii) Let (T, ρ) be a compact separable metric space, and let $\{X(t),\ t \in T\}$ be a sample continuous stochastic process. Let $\{Y(t),\ t \in T\}$ be a version of $\{X(t),\ t \in T\}$ strongly separable in the topology τ and defined on a probability space $(\Omega, \mathcal{F}, P)$. Then there is an event $\Sigma_0 \in \mathcal{F}$ with $P(\Sigma_0) = 0$ such that, for every $\omega \notin \Sigma_0$, $Y\colon T \to \mathbb{R}$ is continuous.

Exercise 9.9 Prove Proposition 9.3.7.

Exercise 9.10 Let μ be an arbitrary non-degenerate probability measure on the real line. Let $X_1 = \{X_1(t),\ 0 \le t \le 1\}$ and $X_2 = \{X_2(t),\ 0 \le t \le 1\}$ be two stochastic processes. Assume that $P(X_1(t) = X_1(s)) = 1$ for all $s,\ t \in [0, 1]$, whereas $(X_2(t_1), \ldots, X_2(t_n))$ are i.i.d. for any $t_1, \ldots, t_n \in [0, 1]$. Assume also that the one-dimensional distributions of the processes $\{X_1(t),\ t \in [0, 1]\}$ and $\{X_2(t),\ t \in [0, 1]\}$ are identical and equal to μ. Show that $\{X_1(t),\ t \in [0, 1]\}$ has a measurable version, whereas $\{X_2(t),\ 0 \le t \le 1\}$ does not.

Exercise 9.11 Show that the Ky Fan distance (9.4.1) is a metric on $L^0(\Omega)$ which metrizes convergence in probability.

Exercise 9.12 Show that the functional

$$\tilde{\alpha}(X, Y) = E(|X - Y| \wedge 1)$$

on $L^0(\Omega) \times L^0(\Omega)$ is a metric equivalent to the Ky Fan metric.

Exercise 9.13 Let $(E, \mathcal{E})$ be a measurable space, and let $\mathcal{E}_0$ be an algebra generating $\mathcal{E}$. Let $\mathcal{M}_S$ be the collection of all functions from E to $\mathbb{R}$ of the type

$$f(x) = \sum_{i=1}^{n} a_i 1(x \in A_i),$$

$a_i \in \mathbb{R},\ A_i \in \mathcal{E}_0,\ i = 1, \ldots, n,\ n = 1, 2, \ldots$. Let $\mathcal{M}_1$ be a family of functions from E to $\mathbb{R}$ that is closed under pointwise convergence and contains $\mathcal{M}_S$.

(a) Let $n \ge 1,\ a_1, \ldots, a_n \in \mathbb{R}$ and $A_2, \ldots, A_n$ be a partition of E into $\mathcal{E}_0$-sets. Let $\mathcal{H}$ be the family of all subsets H of E such that the function $f(x) = a_1 1(x \in H) + \sum_{i=2}^{n} a_i 1(x \in A_i \backslash H)$ is in $\mathcal{M}_1$. Show that $\mathcal{H}$ is a monotone class that contains $\mathcal{E}_0$. Conclude that $\mathcal{H} = \mathcal{E}$.

(b) Show that every function of the form

$$f(x) = \sum_{i=1}^{n} a_i 1(x \in A_i),\ x \in E,$$

$a_i \in \mathbb{R},\ A_i \in \mathcal{E},\ i = 1, \ldots, n,\ n \geq 1$, belongs to $\mathcal{M}_1$.

(c) Conclude that $\mathcal{M}_1$ contains every measurable function $f\colon\ E \to \mathbb{R}$.

(d) Show that if we assume only that $\mathcal{M}_1$ is closed under pointwise convergence of *uniformly bounded* sequences of functions (as opposed to assuming that $\mathcal{M}_1$ is closed under pointwise convergence), then $\mathcal{M}_1$ contains all *bounded* measurable functions $f\colon\ E \to \mathbb{R}$.

Exercise 9.14 Let $\{X(t),\ t \in T\}$ be a *deterministic* process on a separable metric space (T, ρ), i.e., there is a deterministic function $\eta\colon\ T \to \mathbb{R}$ such that for every $t \in T$,

$$P(X(t) = \eta(t)) = 1.$$

Show that $\{X(t),\ t \in T\}$ has a measurable version if and only if the function $\eta\colon T \to \mathbb{R}$ is measurable.

Exercise 9.15 (i) Show that every stochastic process $\{X(t),\ t \in T\}$ continuous in probability has a measurable version.
(ii) Show that every stochastic process $\{X(t),\ t \in T\}$ with all sample paths continuous is measurable.

Exercise 9.16 Prove Proposition 9.5.2.

Exercise 9.17 Let $\{X(t),\ t \in T\}$ be an α-stable process on a separable metric space (T, ρ). Let $t_0 \in T$. Prove that $\{X(t),\ t \in T\}$ cannot be sample continuous at t_0 with probability $p,\ 0 < p < 1$.

Exercise 9.18 In the course of the proof of Theorem 9.5.4 we have shown that (9.5.4) and (9.5.7) cannot be simultaneously true for a jointly $S\alpha S$ sequence $\{Y_n\}_{n=1}^{\infty}$. Show that the following even stronger statement is true: (9.5.7) cannot hold *even by itself* for a *marginally* $S\alpha S$ sequence $\{Y_n\}_{n=1}^{\infty}$.

Chapter 10

Boundedness, continuity and oscillations

Stable processes display several types of sample path behavior. The α-stable Lévy motion, for example, has versions with right-continuous bounded paths. Depending on the value of the self-similarity parameter, the linear fractional stable motion has either versions with continuous paths or all its versions have paths that are unbounded in any finite interval. We want to find convenient criteria for such properties.

Let $\{X(t),\ t \in T\}$ be an α-stable process with integral representation $X(t) = \int_E f(t,x)M(dx),\ t \in T$. We will see that the sample path properties of $\{X(t),\ t \in T\}$ are related to the smoothness of the non-random functions $\{f(t,x),\ t \in T\},\ x \in E$.

In the first part of this chapter we find conditions on f for sample boundedness and sample continuity of α-stable processes. Applying the general criteria stated in Proposition 9.3.4, we obtain in Sections 10.2 and 10.3 *necessary* conditions for boundedness, continuity and pointwise continuity valid for any $0 < \alpha < 2$. In Section 10.4 we show that these conditions are *sufficient* as well when $0 < \alpha < 1$ and also provide the precise asymptotic behavior of the distribution function of the supremum (Theorem 10.4.2). The situation is more complicated when $1 \leq \alpha < 2$. The necessary conditions for boundedness are no longer sufficient (Example 10.4.1), but the formula for the asymptotic behavior of the distribution function of the supremum still holds for $\alpha \geq 1$ *provided* that the process is a.s. bounded (Theorem 10.5.1).

In the second part of the chapter we consider the oscillation process. This process describes more delicate properties of sample paths than sample boundedness or sample continuity. If the non-random functions $\{f(t,x),\ t \in T\},\ x \in E$, in the integral representation of the stable process are uniformly continuous (the precise assumption is stated in Condition 10.6.4), then the oscillation process is a non-random upper-semi-continuous function with values in the extended interval $[0, \infty]$. That function takes values in the two-point set $\{0, \infty\}$ when $0 < \alpha < 1$.

We use the oscillation process to relate pointwise continuity to sample continuity (Corollary 10.6.9). In Section 10.9, for example, we show that under Condition 10.6.4, the sample paths of a stationary α-stable process or those of a self-similar α-stable process are either a.s. continuous or a.s. unbounded.

The linear fractional stable motion however, does not satisfy Condition 10.6.4. How general then is that condition? We prove in Section 10.11 that it is always satisfied when the oscillation process is non-random and finite.

10.1 Introduction

Consider an α-stable process given in the form

$$X(t) = \int_E f(t, x) M(dx) + \eta(t), \ t \in T, \tag{10.1.1}$$

where M is an α-stable random measure on a measurable space $(E, \mathcal{E})$ with control measure m and skewness intensity β, $f(t, \cdot)$, $t \in T$, is a family of functions in F (i.e., measurable functions satisfying (3.4.1) and (3.4.2)), and $\eta(t)$, $t \in T$ are constants.

This representation is very general as we will see in Chapter 13. We shall use it because it allows us to describe the conditions for sample path regularity in a convenient way. We will see that the "regularity" properties of the family of functions $f(t, \cdot)$, $t \in T$, and of the family of constants $\eta(t)$, $t \in T$, are strongly related to the sample path regularity of the process $\{X(t),\ t \in T\}$ given in (10.1.1). This may be surprising at first because a given α-stable process $\{X(t),\ t \in T\}$ admits many different representations of the type (10.1.1).

We begin our discussion with an example illustrating the type of results we hope to obtain.

Example 10.1.1 Consider the well-balanced symmetric linear fractional stable motion

$$X(t) = \int_{-\infty}^{\infty} (|t - x|^{H - 1/\alpha} - |x|^{H - 1/\alpha}) M(dx), \ -\infty < t < \infty,$$

defined in Example 3.6.5. Here, M is a $S\alpha S$ random measure on $(\mathbb{R}, \mathcal{B})$ with Lebesgue control measure, $0 < \alpha < 2$, and $0 < H < 1$, $H \neq 1/\alpha$. Let us try to *guess* the type of sample paths the process $\{X(t),\ t \in \mathbb{R}\}$ posesses. We shall consider how the value of H affects the sample continuity of $\{X(t),\ t \in \mathbb{R}\}$ for a given α. Is $\{X(t),\ t \in \mathbb{R}\}$ sample continuous for all possible choices of H? Is it never sample continuous? Or, perhaps, is there a *critical value* of H, such that the crossing of this critical value implies a radical change in the character of the sample paths of our process (a kind of phase transition)?

Let us concentrate on the kernel

$$f(t,x) = |t-x|^{H-1/\alpha} - |x|^{H-1/\alpha},\ -\infty < t < \infty,\ -\infty < x < \infty. \quad (10.1.2)$$

Previously, we viewed this kernel as a family of functions in x, i.e., as $f(t,\cdot)$, parameterized by $-\infty < t < \infty$. Let us now change perspective and regard it as a family of functions in t, i.e., $f(\cdot, x)$, parameterized by $-\infty < x < \infty$. Choose any "typical" x, and thus regard $f(t,x)$ in (10.1.2) as a function of one variable t. It is a "nice" continuous function if $H - 1/\alpha > 0$, but it becomes "less nice" if $H - 1/\alpha < 0$: it "explodes" at $t = x$. In the former case, the well-balanced $S\alpha S$ linear fractional stable motion is an integral of a "nice" continuous function, whereas in the latter case the process is an integral of a rather "badly behaved" one. It turns out that this difference in the properties of the family of functions in t, $f(\cdot, x)$, $-\infty < x < \infty$, is reflected in the properties of our linear fractional stable motion $\{X(t),\ -\infty < t < \infty\}$. We will see in the sequel that the process is sample continuous if $H > 1/\alpha$ and not sample continuous (in fact, much "worse" than that) if $H < 1/\alpha$. Taking into account the range of our parameters, $0 < H < 1,\ 0 < \alpha < 2$, we observe that the case $H > 1/\alpha$ is possible only if $\alpha > 1$. Therefore in the case $1 < \alpha < 2$, we will have a "phase transition" (an abrupt change in the structure of sample paths) as H crosses the critical value of $1/\alpha$, whereas no such abrupt change occurs for $0 < \alpha \le 1$, in which case our process has "bad" sample paths for all possible values of H. (See Example 10.2.5, Exercise 10.21, and Example 12.2.3.)

It should be noted that another "phase transition" occurs as $\alpha \to 2$. Our fractional process is perfectly well defined for $\alpha = 2$ and $0 < H < 1,\ H \ne 1/2$ (it is called fractional Brownian motion). But as can easily be shown, fractional Brownian motion is sample continuous in all cases, whether $H - 1/2$ is positive or negative! (Exercise 10.1).

10.2 Necessary conditions for sample boundedness

We start with the following result which, although subsumed by the more precise estimates that are to follow, is useful in itself.

Proposition 10.2.1 *Let $\{X_n,\ n \geq 1\}$ be an α-stable process. If $Z := \sup_{n\geq 1} |X_n| < \infty$ a.s., then*

$$\overline{\lim}_{\lambda\to\infty} \lambda^\alpha P(Z > \lambda) < \infty. \tag{10.2.1}$$

PROOF: Let $\{\widetilde{X}_n,\ n \geq 1\}$ be an independent copy of $\{X_n,\ n \geq 1\}$ and let $Y_n = X_n - \widetilde{X}_n,\ n \geq 1$. Then $\{Y_n,\ n \geq 1\}$ is a $S\alpha S$ process such that $W = \sup_{n\geq 1} |Y_n| < \infty$ a.s.. Since, for arbitrary $M > 0$,

$$\begin{aligned} P(W > \lambda) &\leq P(\sup_{n\geq 1} |X_n| > \lambda + M,\ \sup_{n\geq 1} |\widetilde{X}_n| \leq M) \\ &= P(Z > \lambda + M) P(Z \leq M), \end{aligned}$$

we have $\overline{\lim}_{\lambda\to\infty} \lambda^\alpha P(W > \lambda) \geq \overline{\lim}_{\lambda\to\infty} \lambda^\alpha P(Z > \lambda) P(Z \leq M)$. Letting $M \to \infty$ we obtain

$$\overline{\lim}_{\lambda\to\infty} \lambda^\alpha P(W > \lambda) \geq \overline{\lim}_{\lambda\to\infty} \lambda^\alpha P(Z > \lambda)$$

since Z is bounded. It is therefore sufficient to prove

$$\overline{\lim}_{\lambda\to\infty} \lambda^\alpha P(W > \lambda) < \infty. \tag{10.2.2}$$

Let $\{\widetilde{Y}_n,\ n \geq 1\}$ be an independent copy of $\{Y_n,\ n \geq 1\}$. For any $\lambda > 0$,

$$\begin{aligned} P(W > \lambda) &= P(\sup_{n\geq 1} |Y_n - \widetilde{Y}_n| > 2^{1/\alpha}\lambda) \\ &\geq 2P(W_1 - W_2 > 2^{1/\alpha}\lambda), \end{aligned} \tag{10.2.3}$$

where W_1 and W_2 are independent copies of W. As we will see, (10.2.3) is the only property of W that is needed to prove (10.2.2).

Let λ_0 be any number such that $P(W \leq 2^{1/\alpha}\lambda_0) \geq 3/4$. For any $\lambda \geq \lambda_0$, by (10.2.3),

$$\begin{aligned} P(W > \lambda) &\geq 2P(W_1 > 2^{1+1/\alpha}\lambda) \cdot P(W_2 \leq 2^{1/\alpha}\lambda) \\ &\geq \frac{3}{2} P(W > 2^b \lambda), \end{aligned}$$

where $b = 1 + 1/\alpha$. Thus, for every $m \geq 1$, $(\frac{3}{2})^m P(W > 2^{mb}\lambda_0) \leq P(W > \lambda_0)$, which implies

$$\overline{\lim}_{\lambda\to\infty} \lambda^{b_1} P(W > \lambda) < \infty, \tag{10.2.4}$$

where $b_1 = (\log_2 3 - 1)/(1 + 1/\alpha)$. Since $b_1 < \alpha$, (10.2.4) is weaker than (10.2.2) but it is an important first step. It implies $\sum_{j=1}^\infty P(W > C2^{j\theta/\alpha}) < \infty$ for any $\theta \in (0, 1)$ and any constant $C > 0$, or equivalently,

$$\prod_{j=k}^\infty P(W \leq C2^{j\theta/\alpha}) > 0 \tag{10.2.5}$$

In particular, if $\{X(t),\ t \in T\}$ *is sample bounded, then, necessarily,*

$$\sup_{T^* \subseteq T} \int_E f^*(T^*; x)^\alpha m(dx) < \infty \tag{10.2.14}$$

and

$$\begin{cases} \sup_{t \in T} |\eta(t)| < \infty & \text{if } \alpha \neq 1, \\ \sup_{t \in T} \left|\eta(t) - \frac{2}{\pi} \int_E f(t,x) \ln |f(t,x)| \beta(x) m(dx)\right| < \infty & \text{if } \alpha = 1. \end{cases} \tag{10.2.15}$$

PROOF: Note firstly that by the separability of $\{Y(t),\ t \in T\}$, the left-hand sides of (10.2.10) and (10.2.12) are well defined because $\sup_{t \in T} Y(t)$ and $\sup_{t \in T} |Y(t)|$ are well-defined random variables. Moreover,

$$\sup_{t \in T} Y(t) \geq \sup_{t \in T^*} Y(t) \ \text{ and } \ \sup_{t \in T} |Y(t)| \geq \sup_{t \in T^*} |Y(t)|.$$

Hence (10.2.10) and (10.2.12) follow immediately from Theorem 4.4.5 and Corollary 4.4.6, respectively. Of course, (10.2.14) now follows from Propositions 9.3.4 and 10.2.1 and from (10.2.12).

We now prove (10.2.15). The case $\alpha \neq 1$ is easy. Let $\{\widetilde{X}(t),\ t \in T\}$ be an independent copy of $\{X(t),\ t \in T\}$ and let

$$Z(t) = 2^{-1/\alpha}(X(t) + \widetilde{X}(t)) + (1 - 2^{1-1/\alpha})\eta(t),\ t \in T.$$

Then $\{Z(t),\ t \in T\}$ is a version of $\{X(t),\ t \in T\}$, so it is sample bounded. Since $\{X(t),\ t \in T\}$ and $\{\widetilde{X}(t),\ t \in T\}$ are also sample bounded, the above definition of $\{Z(t),\ t \in T\}$ shows that $\sup_{t \in T} |\eta(t)| < \infty$. Unfortunately, this "trick" does not work for $\alpha = 1$. We will consider this case separately.

Let $\alpha = 1$ and suppose, conversely, that $\{X(t),\ t \in T\}$ is sample bounded and (10.2.15) does not hold. Then there is a sequence of points $\{t_n,\ n = 1, 2, \ldots\} \subseteq T$ such that

$$\lim_{n \to \infty} \left|\eta(t_n) - \frac{2}{\pi} \int_E f(t_n, x) \ln |f(t_n, x)| \beta(x) m(dx)\right| = \infty.$$

Of course, we may assume, without loss of generality, that

$$\lim_{n \to \infty} a_n = \infty, \tag{10.2.16}$$

where

$$a_n = \eta(t_n) - \frac{2}{\pi} \int_E f(t_n, x) \ln \left|f(t_n, x)\right| \beta(x) m(dx),\ n = 1, 2, \ldots.$$

Let $\widehat{X}(t_n) = X(t_n) - a_n$, $n = 1, 2, \ldots$. Then

$$P\big(\sup_{n\geq 1} |\widehat{X}(t_n) + a_n| < \infty\big) = 1 \tag{10.2.17}$$

because $\{X(t),\ t \in T\}$ is sample bounded. Note that each $\widehat{X}(t_n)$ has a distribution of the type $S_1(\sigma_n, \beta_n, 0)$ with

$$\sigma_n = \int_E |f(t_n, x)| m(dx) \leq \int_E \sup_{n\geq 1} |f(t_n, x)| m(dx) := K < \infty$$

by (10.2.14). Therefore there is a finite constant γ_K and a $p > 0$ such that $P(\widehat{X}(t_n) > \gamma_K) \geq p$ for every $n = 1, 2, \ldots$ (see Exercise 1.18). By (10.2.16), there is for any $\lambda > 0$ an $n(\lambda)$ such that $a_{n(\lambda)} > \lambda - \gamma_K$. Then

$$\begin{aligned} P(\sup_{n\geq 1} |\widehat{X}(t_n) + a_n| > \lambda) &\geq P(\widehat{X}(t_{n(\lambda)}) + a_{n(\lambda)} > \lambda) \\ &\geq P(\widehat{X}(t_{n(\lambda)}) > \gamma_K) \\ &\geq p. \end{aligned}$$

Since $\lambda > 0$ is arbitrary, we obtain $P(\sup_{n\geq 1} |\widehat{X}(t_n) + a_n| = \infty) \geq p$, contradicting (10.2.17) and thus proving (10.2.15) for $\alpha = 1$. This completes the proof of the theorem. ■

Corollary 10.2.4 *Let $\{X(t),\ t \in \mathbb{R}\}$ be an α-stable process, $0 < \alpha < 2$. Suppose there is a a countable subset T^* of $\mathbb{R}$ such that for every $a < b$ we have*

$$\int_a^b f^*(T^*; x) m(dx) = \infty. \tag{10.2.18}$$

Then for every version $\{Y(t), t \in \mathbb{R}\}$ of $\{X(t), t \in \mathbb{R}\}$ there is an event Ω_0 of probability 1 such that

$$\sup_{a<t<b} |Y(t)| = \infty \tag{10.2.19}$$

for every $\omega \in \Omega_0$ and every $a < b$.

We use the expression "the process $\{X(t),\ t \in \mathbb{R}\}$ is *unbounded on every interval of positive length*" to summarize the conclusion of the corollary

Example 10.2.5 Applying Corollary 10.2.4 to the symmetric well-balanced linear fractional stable motion, we conclude that if $H < 1/\alpha$, the process is unbounded on every interval of positive length. Indeed, for a given pair (a, b) take $T^* = \{r \text{ rational: } a \leq r \leq b\}$. Since for every $x \in [a, b]$,

$$f^*(T^*; x) = \sup_{t\in T^*} \left| |t - x|^{H-1/\alpha} - |x|^{H-1/\alpha} \right| = \infty,$$

the necessary condition for sample boundedness (10.2.14) fails.

Example 10.2.6 Applying the argument in Example 10.2.5 to the log-fractional stable motion, we note that

$$f^*(T^*;x) = \sup_{t\in T^*} \big| \ln|t-x| - \ln|x| \big| = \infty.$$

Therefore, the log-fractional stable motion is also unbounded on every interval of positive length.

Example 10.2.7 Let $0 < \alpha < 2$, $0 < H < 1/\alpha$ and consider an (α, H)-Takenaka field with $n = 1$ (see Definition 8.4.3), i.e., the H-sssi process $\{X(t) = M(V_t),\ t \in \mathbb{R}^1\}$, where M is a $S\alpha S$ random measure with control measure

$$m_{\alpha H}(dx, d\lambda) = \lambda^{\alpha H - 2} dx d\lambda,\ x \in \mathbb{R}^1,\ \lambda > 0,$$

and where V_t is given in (8.4.2) and illustrated in Figure 8.2. As in the proof of Lemma 8.4.2, for any $T = \{a \le t \le b\}$ and any countable $T^* \subset T$,

$$\begin{aligned}
&\int_{-\infty}^{\infty}\int_0^{\infty} \big(\sup_{t\in T^*} 1_{V_t}(x,\lambda)\big)^{\alpha} m_{\alpha H}(dx, d\lambda) \\
&\ge \int_{-\infty}^{\infty} \sup_{t\in T^*} \int_0^{\infty} 1_{V_t}(x,\lambda) m_{\alpha H}(dx, d\lambda) \\
&= \frac{1}{1-\alpha H} \int_{-\infty}^{\infty} \sup_{t\in T^*} \big||x|^{\alpha H - 1} - |x-t|^{\alpha H - 1}\big| 1_{\{|x| \ne |x-t|\}} dx \ = \ \infty.
\end{aligned}$$

By Corollary 10.2.4, the Takenaka process $\{X(t),\ t \in \mathbb{R}^1\}$ is unbounded on every interval of positive length.

It is obvious that there is always a countable $T^* \subseteq T$ for which the supremum in (10.2.10), (10.2.12) and (10.2.14) is achieved, and in many cases this T^* can be identified as follows. Suppose that there is a probability measure λ on $(E, \mathcal{E})$ *equivalent* to the control measure m of the α-stable random measure M in (10.1.1). (This is always the case if the control measure m is σ-finite; see Exercise 10.4.) Regard $(E, \mathcal{E}, \lambda)$ as a probability space on which random variables $f(t,\cdot)$s are defined. Then $\{f(t,\cdot),\ t \in T\}$ is a stochastic process defined on that probability space, and by Theorem 9.2.5 this process has a separable version $\{g(t,\cdot),\ t \in T\}$ defined on the same probability space $(E, \mathcal{E}, \lambda)$. We will call $\{g(t,\cdot),\ t \in T\}$ *a separable kernel* of the process $\{X(t),\ t \in T\}$ in (10.1.1). Let T^* be a separant for $\{g(t,\cdot),\ t \in T\}$. We claim that T^* is the countable subset for which we are looking. Indeed, for any other countable subset $T_1^* \subseteq T$, $\lambda\{x \in E\colon f^*(T^*;x) < f^*(T_1^*;x)\} = 0$, so that

$$m\big(x \in E\colon f^*(T^*;x) < f^*(T_1^*;x)\big) = 0$$

as well. Hence,

$$\int_E f^*(T^*;x)^\alpha m(dx) \geq \int_E f^*(T_1^*;x)^\alpha m(dx),$$

as required. Similarly, the supremum in (10.2.10) and (10.2.12) is also achieved on T^*.

The preceding result also yields a necessary condition for sample boundedness of complex-valued $S\alpha S$ processes given in the form

$$X(t) = \int_E f_c(t,x) M_c(dx), \tag{10.2.20}$$

where M_c is a complex-valued $S\alpha S$ random measure on $(E, \mathcal{E})$ with circular control measure k, and $f_c(t,\cdot)$, $t \in T$, are complex-valued and in $L^\alpha(E, \mathcal{E}, m)$, where m is the control measure of M (i.e., $m(A) = k(A \times S_2),\ A \in \mathcal{E}$).

Corollary 10.2.8 *Let $\{X(t),\ t \in T\}$ be a complex-valued $S\alpha S$ process given in the form (10.2.20). If $\{X(t),\ t \in T\}$ is sample bounded, then*

$$\sup_{T^* \subseteq T} \int_E f_c^*(T^*;x)^\alpha m(dx) < \infty, \tag{10.2.21}$$

where

$$f_c^*(T^*;x) = \sup_{t \in T^*} |f_c(t,x)| = \sup_{t \in T^*} |f_c^{(1)}(t,x)^2 + f_c^{(2)}(t,x)^2|^{1/2}, \tag{10.2.22}$$

and $f_c^{(1)}$ and $f_c^{(2)}$ are, respectively, the real and the imaginary parts of f_c.

PROOF: Let $Y_1(t) = \text{Re } X(t),\ Y_2(t) = \text{Im } X(t)$. Then $\{Y_1(t),\ t \in T\}$ and $\{Y_2(t),\ t \in T\}$ are real-valued $S\alpha S$ processes, which are both sample-bounded if $\{X(t),\ t \in T\}$ is. By Proposition 6.2.4,

$$\{Y_1(t),\ t \in T\} \stackrel{d}{=} \left\{ \int_{E \times S_2} g(t,(x,\mathbf{s})) M(dx, d\mathbf{s}),\ t \in T \right\}, \tag{10.2.23}$$

where

$$g(t,(x,\mathbf{s})) = f_c^{(1)}(t,x)s_1 - f_c^{(2)}(t,x)s_2, \quad (x,\mathbf{s}) \in E \times S_2,$$

and M is a (real-valued) $S\alpha S$ random measure on $(E \times S_2,\ \mathcal{E} \times \mathcal{B})$ with control measure k. The sample boundedness of $\{Y_1(t),\ t \in T\}$ implies, by Theorem 10.2.3,

$$\sup_{T^* \subseteq T} \int_E \int_{S_2} \sup_{t \in T^*} \left| f_c^{(1)}(t,x)s_1 - f_c^{(2)}(t,x)s_2 \right|^\alpha k(dx, d\mathbf{s}) < \infty, \tag{10.2.24}$$

and, similarly, the sample boundedness of $\{Y_2(t),\ t \in T\}$ implies

$$\sup_{T^* \subseteq T} \int_E \int_{S_2} \sup_{t \in T^*} \left| f_c^{(1)}(t,x)s_2 + f_c^{(2)}(t,x)s_1 \right|^\alpha k(dx,ds) < \infty. \tag{10.2.25}$$

Hence,

$$\begin{aligned}
&\sup_{T^* \subseteq T} \int_E f_c^*(T^*;x)^\alpha m(dx) \\
&\quad = \sup_{T^* \subseteq T} \int_E \int_{S_2} \sup_{t \in T^*} \left| f_c^{(1)}(t,x)^2 + f_c^{(2)}(t,x)^2 \right|^{\alpha/2} k(dx,ds) \\
&\quad \le \sup_{T^* \subseteq T} \int_E \int_{S_2} \sup_{t \in T^*} \left| f_c^{(1)}(t,x)s_1 - f_c^{(2)}(t,x)s_2 \right|^\alpha k(dx,ds) \\
&\qquad + \sup_{T^* \subseteq T} \int_E \int_{S_2} \sup_{t \in T^*} \left| f_c^{(1)}(t,x)s_2 + f_c^{(2)}(t,x)s_1 \right|^\alpha k(dx,ds) \\
&\quad < \infty.
\end{aligned}$$

This concludes the proof. ∎

10.3 Necessary conditions for sample continuity

Having obtained necessary conditions for sample boundedness, we now turn to necessary conditions for sample continuity.

Theorem 10.3.1 *Let (T,ρ) be a locally compact separable metric space, and let $\{X(t),\ t \in T\}$ be an α-stable process, $0 < \alpha < 2$, given in the standard form (10.1.1). Let T^* be a countable dense subset of T. If $\{X(t),\ t \in T\}$ is sample continuous, then for every compact subset C of T, we have*

$$m\Big(x\text{: } f(t,x),\ t \in T^* \cap C \text{ is not uniformly continuous}\Big) = 0, \tag{10.3.1}$$

$$\int_E \sup_{t \in T^* \cap C} |f(t,x)|^\alpha m(dx) < \infty \tag{10.3.2}$$

and

$$\begin{cases} \{\eta(t),\ t \in T\} \text{ is continuous} & \text{if } \alpha \neq 1, \\ \{\eta(t) - \frac{2}{\pi}\int_E f(t,x)\ln|f(t,x)|\beta(x)m(dx),\ t \in T\} \text{ is continuous} & \text{if } \alpha = 1. \end{cases} \tag{10.3.3}$$

PROOF: We start with our standard symmetrization argument. Let $\{\widetilde{X}(t),\ t \in T\}$ be an independent copy of $\{X(t),\ t \in T\}$ and define $Y(t) = X(t) - \widetilde{X}(t),\ t \in T$. Then $\{Y(t),\ t \in T\}$ is a $S\alpha S$ process, is sample continuous since $\{X(t),\ t \in T\}$ is sample continuous and has the integral representation

$$Y(t) = 2^{1/\alpha} \int_E f(t,x) M_1(dx),\ t \in T,$$

where M_1 is a $S\alpha S$ random measure on $(E, \mathcal{E})$ with control measure m.

Note that sample continuity implies continuity in probability. Hence, if C is a compact subset of T, then, by Proposition 9.3.5,

$$P\big(Y(t),\ t \in T^* \cap C \text{ is uniformly continuous}\big) = 1. \tag{10.3.4}$$

Observe that the control measure m is σ-finite on the support of each function $\{f(t,x),\ x \in E\},\ t \in T^* \cap C$, and, therefore, it is σ-finite on the set $E_1 = \{x \in E\colon \sum_{t \in T^* \cap C} f(t,x)^2 > 0\}$. Thus (10.3.1) is equivalent to

$$m\big(x \in E_1\colon\ f(t,x),\ t \in T^* \cap C, \text{ is not uniformly continuous}\big) = 0, \tag{10.3.5}$$

and also

$$Y(t) = 2^{1/\alpha} \int_{E_1} f(t,x) M_1(dx),\ t \in T^* \cap C.$$

We shall show that (10.3.4) implies (10.3.5).

Let λ be a probability measure on $(E_1, \mathcal{E}_1)$ (where $\mathcal{E}_1$ is the restriction of $\mathcal{E}$ to E_1), equivalent to the σ-finite measure m on $(E_1, \mathcal{E}_1)$. Let $g(x) = \frac{dm}{d\lambda}(x)$ be the corresponding Radon-Nikodym derivative, which can be chosen to take values in $(0, \infty)$. Then on changing the variable of integration (Proposition 3.5.5), we conclude that

$$\{Y(t),\ t \in T^* \cap C\} \stackrel{d}{=} \Big\{2^{1/\alpha} \int_{E_1} f(t,x) g(x)^{1/\alpha} M_2(dx),\ t \in T^* \cap C\Big\}, \tag{10.3.6}$$

where M_2 is a $S\alpha S$ random measure on $(E_1, \mathcal{E}_1)$ with control measure λ. Now we use the series representation of the integral in the right-hand side of (10.3.6) (Theorem 3.10.1) to conclude that

$$\{Y(t),\ t \in T^* \cap C\} \stackrel{d}{=} \Big\{d_\alpha \sum_{i=1}^{\infty} \epsilon_i \Gamma_i^{-1/\alpha} f(t, V_i) g(V_i)^{1/\alpha},\ t \in T^* \cap C\Big\}, \tag{10.3.7}$$

where d_α is a constant, and the three independent series of random variables are, as usual, a Rademacher sequence $\{\epsilon_1, \epsilon_2, \ldots\}$, a sequence of arrival times of a Poisson process with unit arrival rate $\{\Gamma_1, \Gamma_2, \ldots\}$ and a sequence of i.i.d.

E_1-valued random variables $\{V_1, V_2, \ldots\}$ with common distribution λ. Now let $\{Z(t),\ t \in T^* \cap C\}$ denote the stochastic process in the right-hand side of (10.3.7). Then, by (10.3.4) and (10.3.7),

$$P\big(Z(t),\ t \in T^* \cap C, \text{ is uniformly continuous}\big) = 1.$$

Let $\{\widehat{Z}(t),\ t \in T^* \cap C\}$ be the stochastic process obtained by replacing ϵ_i by $-\epsilon_i$ for $i = 2, 3, \ldots$, on the right-hand side of (10.3.7). Since $\epsilon_1, -\epsilon_2, -\epsilon_3, \ldots$ is again a Rademacher sequence, we have $\{\widehat{Z}(t),\ t \in T^* \cap C\} \stackrel{d}{=} \{Z(t),\ t \in T^* \cap C\}$ and thus

$$P\big(\widehat{Z}(t),\ t \in T^* \cap C, \text{ is uniformly continuous}\big) = 1$$

so

$$P\big(Z(t) + \widehat{Z}(t),\ t \in T^* \cap C, \text{ is uniformly continuous}\big) = 1.$$

But $Z(t) + \widehat{Z}(t) = 2d_\alpha \epsilon_1 \Gamma_1^{1/\alpha} g(V_1)^{1/\alpha} f(t, V_1)$. Since $g(V_1) \in (0, \infty)$, we obtain

$$P\big(f(t, V_1),\ t \in T^* \cap C \text{ is uniformly continuous}\big) = 1$$

or, equivalently,

$$\lambda\big(x \in E_1\colon\ f(t, x),\ t \in T^* \cap C, \text{ is not uniformly continuous}\big) = 0.$$

This proves (10.3.5) because m and λ are equivalent measures (on E_1).

Relation (10.3.2) follows directly from the fact that sample continuity of $\{X(t),\ t \in C\}$ implies sample boundedness of $\{X(t),\ t \in C\}$ and from Theorem 10.2.3.

It remains to prove (10.3.3). Once again, the argument for $\alpha \neq 1$ is simple, and is identical to the proof of the corresponding part of (10.2.15) in Theorem 10.2.3. This leaves us with the case $\alpha = 1$. Denote

$$\mu(t) = \eta(t) - \frac{2}{\pi} \int_E f(t, x) \ln \big|f(t, x)\big| \beta(x) m(dx),\ t \in T,$$

and suppose, to the contrary, that $\{X(t),\ t \in T\}$ is sample continuous but $\{\mu(t),\ t \in T\}$ is discontinuous. Then there is a point $t_0 \in T$, an $\epsilon > 0$ and a sequence of points $\{t_n,\ n = 1, 2, \ldots\} \subseteq T$ such that $\rho(t_n, t_0) \to 0$ as $n \to \infty$ and $|\mu(t_n) - \mu(t_0)| \geq \epsilon$ for every $n \geq 1$. Clearly, sample continuity of $\{X(t),\ t \in T\}$ implies

$$P\big(\lim_{n\to\infty} X(t_n) = X(t_0)\big) = 1 \tag{10.3.8}$$

(Proposition 9.3.7). Let

$$Y_n = \big(X(t_n) - \mu(t_n)\big) - \big(X(t_0) - \mu(t_0)\big),$$

$a_n = \mu(t_n) - \mu(t_0)$, $n = 1, 2, \ldots$. Then, by (10.3.8),

$$\begin{aligned} 1 &= P\left(\lim_{n\to\infty} X(t_n) = X(t_0)\right) \\ &= P\left(\lim_{n\to\infty} (Y_n + a_n) = 0\right). \end{aligned} \tag{10.3.9}$$

The sequence $\{\mu(t_n),\ n = 1, 2, \ldots\}$ is bounded by Theorem 10.2.3. Therefore we may (and will) assume that there is an $L \in [\epsilon, \infty)$ such that $0 < a_n \to L$ as $n \to \infty$. Note also that each Y_n has the distribution $S_1(\sigma_n, \beta_n, 0)$ for some $\sigma_n \geq 0$ and $-1 \leq \beta_n \leq 1$. If $\lim_{n\to\infty} \sigma_n = 0$, then, as $n \to \infty$, Y_n converges in distribution to 0 and $Y_n + a_n$ to L, contradicting (10.3.9). Therefore all limit points of the sequence $\{\sigma_n,\ n \geq 1\}$ must be non-zero. On the other hand, let $\{\widetilde{Y}_n,\ n = 1, 2, \ldots\}$ be an independent copy of $\{Y_n,\ n = 1, 2, \ldots\}$. Then $Z_n = Y_n - \widetilde{Y}_n$ is a $S_1(2\sigma_n, 0, 0)$ random variable, $n \geq 1$, and

$$P\left(\lim_{n\to\infty} Z_n = 0\right) \geq P\left(\lim_{n\to\infty} (Y_n + a_n) = 0,\ \lim_{n\to\infty} (\widetilde{Y}_n + a_n) = 0\right) = 1$$

by independence and (10.3.9). Hence, as $n \to \infty$, Z_n converges in distribution to 0, implying $\lim_{n\to\infty} \sigma_n = 0$. This contradiction proves that $\{\mu(t),\ t \in T\}$ must be continuous, and the proof of Theorem 10.3.1 is now complete. ∎

Theorem 10.3.1 can be extended to complex-valued $S\alpha S$ processes given in the form (10.2.20). The proof of the following corollary is similar to the proof of Corollary 10.2.8 and is left to the reader (Exercise 10.5).

Corollary 10.3.2 *Let (T, ρ) be a locally compact separable metric space and let $\{X(t),\ t \in T\}$ be a complex-valued $S\alpha S$ process given in the form (10.2.20). Let T^* be a countable dense subset of T. If $\{X(t),\ t \in T\}$ is sample continuous, then for every compact subset C of T we have*

$$m\big(x\colon\ f_c(t, x),\ t \in T^* \cap C \text{ is not uniformly continuous}\big) = 0 \tag{10.3.10}$$

and

$$\int_E \sup_{t \in T^* \cap C} |f_c(t, x)|^\alpha m(dx) < \infty. \tag{10.3.11}$$

Let us now discuss the intuitive meaning of the results we have obtained, namely Theorems 10.2.3 and 10.3.1. Assume that our α-stable process $\{X(t),\ t \in T\}$ is given in the standard form (10.1.1) (take $\eta(t) = 0$ for simplicity) and that there is a probability measure λ on $(E, \mathcal{E})$ equivalent to the control measure m. We can view $\{f(t, \cdot),\ t \in T\}$ as a stochastic process defined on the probability space $(E, \mathcal{E}, \lambda)$. Let (T, ρ) be a locally compact separable metric space. We know from Theorem 9.2.5 that there is a strongly separable version $\{g(t, \cdot),\ t \in T\}$ of the stochastic process $\{f(t,\cdot),\ t \in T\}$ which is defined on the same measurable space $(E, \mathcal{E})$ and such that $\lambda\{x \in E: g(t,x) \neq f(t,x)\} = 0$ for every $t \in T$. Since λ is equivalent to m, we obtain

$$\int_E |f(t,x) - g(t,x)|^\alpha m(dx) = 0$$

and therefore

$$X(t) = \int_E g(t,x) M(dx) \text{ a.s., } t \in T. \tag{10.3.12}$$

The representation (10.3.12) is called a *strongly separable integral representation* of $\{X(t),\ t \in T\}$ and $\{g(t,\cdot),\ t \in T\}$ is called a *strongly separable kernel.*

Suppose, now, that $\{X(t),\ t \in T\}$ is sample continuous. Let T^* be a countable dense subset of T which is, at the same time, a strong separant for $\{g(t,\cdot),\ t \in T\}$. Applying Theorem 10.3.1 and Proposition 9.3.4, we conclude that for every compact subset C of T, $\{f(t,\cdot),\ t \in C\}$ regarded as stochastic process on the probability space $(E, \mathcal{E}, \lambda)$ is sample continuous. It is therefore continuous in measure λ and, thus, by Proposition 9.3.5, $\{f(t,\cdot),\ t \in T\}$ is sample continuous.

We thus arrive at the following conclusion. *Let $\{X(t),\ t \in T\}$ be an α-stable process in a standard form (10.1.1). If $\{X(t),\ t \in T\}$ is sample continuous, then $\{f(t,\cdot),\ t \in T\}$ is sample continuous as well.* We remind the reader that the sample path properties of the kernel in an integral representation are always understood to hold with respect to a probability measure equivalent to the control measure.

Of course, a similar argument can be used to show that an α-stable process with a separable (not necessarily strongly separable) integral representation is sample bounded only if the kernel $\{g(t,\cdot),\ t \in T\}$ of this representation is sample bounded. However, Theorem 10.2.3 yields actually a stronger result, namely not only

$$m\big(x\colon \sup_{t \in T} |g(t,x)| = \infty\big) = 0$$

but also

$$\int_E (\sup_{t \in T} |g(t,x)|)^\alpha m(dx) < \infty.$$

Needless to say, none of the above results hold for Gaussian processes, i.e., when $\alpha = 2$ (see Example 10.1.1).

The following theorem gives necessary conditions for *pointwise* sample continuity of α-stable processes.

Theorem 10.3.3 *Let (T, ρ) be a separable metric space, $\{X(t),\ t \in T\}$ be an α-stable process, $0 < \alpha < 2$, and let T^* be a strong separant for a strongly* *separable version of* $\{X(t),\ t \in T\}$. *Suppose that* $\{X(t),\ t \in T\}$ *is sample continuous at* $t_0 \in t$. *Then*

$$m\Big(x\text{: either } \lim_{\substack{t\to t_0\\ t\in T^*}} f(t,x) \text{ does not exist or } \lim_{\substack{t\to t_0\\ t\in T^*}} f(t,x) \neq f(t_0,x)\Big) = 0 \tag{10.3.13}$$

and in the cases $\alpha \neq 1$ *and* $\alpha = 1$, *respectively,*

$$\begin{cases} \{\eta(t),\ t \in T\} \text{ is continuous at } t_0, \\ \{\eta(t) - \frac{2}{\pi}\int_E f(t,x)\ln|f(t,x)|\beta(x)m(dx),\ t \in T\} \text{ is continuous at } t_0. \end{cases} \tag{10.3.14}$$

Suppose, in addition, that the point t_0 *has a relatively compact neighborbood, i.e., a neighborhood whose closure is compact. Then for some open set* A *with compact closure such that* $t_0 \in A \subseteq T$,

$$\int_E \sup_{t\in T^*\cap A} |f(t,x)|^\alpha m(dx) < \infty. \tag{10.3.15}$$

The proof is very similar to the proof of Theorem 10.3.1 and is left to the reader (Exercise 10.6).

The following example is a curious application of our results. It answers the question posed in Section 6.6 by providing an example of a stationary $S\alpha S$ process which is not conditionally stationary Gaussian.

Example 10.3.4 Let $X(t) = L(t) - L(t-1),\ 1 \le t \le 2$, where $\{L(t),\ 0 \le t \le 2\}$ is the $S\alpha S$ motion. Clearly, $X = \{X(t),\ 1 \le t \le 2\}$ is a stationary $S\alpha S$ process. We claim that X cannot be conditionally stationary Gaussian.

Indeed, suppose to the contrary, that X is conditionally stationary Gaussian. We know that $\{X(t),\ 0 \le t \le 2\}$ is a.s. bounded (Exercise 9.5) and it follows from Theorem 10.3.1 that $\{X(t),\ 0 \le t \le 2\}$ is not sample continuous. Therefore the conditional stationary Gaussian process must be sample bounded as well but not sample continuous. But this violates Belyaev's theorem (1961) that every stationary Gaussian process is either a.s. continuous or else a.s. unbounded on every interval of positive length. Thus X cannot be conditionally stationary Gaussian.

10.4 Necessary and sufficient conditions for sample boundedness and continuity when $0 < \alpha < 1$

In the preceding section, we established necessary conditions for sample boundedness and sample continuity of α-stable processes. The reader may wonder whether these conditions are sufficient as well. We will show that the conditions are sufficient if $0 < \alpha < 1$, but they are not if $1 \le \alpha < 2$.

The following example demonstrates that the necessary condition (10.2.14) for sample boundedness is not sufficient when $1 \le \alpha < 2$.

Example 10.4.1 Let $1 \le \alpha < 2$ and define a $S\alpha S$ process on the countable parameter space

$$T = \Big\{(N, A)\colon\ N = 1, 2, \ldots,\ A \subseteq \{1, 2, \ldots, N\}\Big\}$$

as follows: start with an arbitrary sequence $\{a_n,\ n \ge 1\}$ of positive numbers satisfying the following two conditions:

$$\sum_{n=1}^{\infty} a_n^{\alpha} < \infty,\ 1 \le \alpha < 2, \tag{10.4.1}$$

and

$$\begin{cases} \sum_{n=1}^{\infty} a_n |\ln a_n| = \infty & \text{if } \alpha = 1, \\ \sum_{n=1}^{\infty} a_n = \infty & \text{if } 1 < \alpha < 2. \end{cases} \tag{10.4.2}$$

(A possible choice is $a_n = n^{-1}$ if $\alpha > 1$ and $a_n = (n \log^2(n+1))^{-1}$ if $\alpha = 1$.) Set $b_n = a_n (n(n+1))^{1/\alpha}$, $n = 1, 2, \ldots,$ and define the family of functions $\{f((N, A), x),\ 0 < x < 1\}$, $(N, A) \in T$, as

$$f((N, A), x) = \begin{cases} b_n & \text{if } x \in (\frac{1}{n+1}, \frac{1}{n}] \text{ and } n \in A, \\ -b_n & \text{if } x \in (\frac{1}{n+1}, \frac{1}{n}],\ n \le N \text{ and } n \notin A, \\ 0 & \text{otherwise.} \end{cases}$$

Consider the stochastic process

$$\Big\{X((N, A)) = \int_0^1 f((N, A), x) M(dx),\ (N, A) \in T\Big\},$$

where M is a $S\alpha S$ random measure on $((0, 1), \mathcal{B})$ with Lebesgue control measure. Clearly,

$$\int_0^1 |f((N, A), x)|^{\alpha} dx = \sum_{n=1}^{N} b_n^{\alpha} \Big(\frac{1}{n} - \frac{1}{n+1}\Big) = \sum_{n=1}^{N} a_n^{\alpha} < \infty$$

for every $(N, A) \in T$, so $\{X((N, A)), (N, A) \in T\}$ is a well-defined $S\alpha S$ process.

The necessary condition for sample boundedness (10.2.14) holds for this process because, by taking $T^* = T$, we obtain

$$f^*(T;x) = \sup_{(N,A)\in T} |f((N,A),x)| = b_n \text{ if } x \in \Big(\frac{1}{n+1}, \frac{1}{n}\Big]$$

and therefore

$$\int_0^1 f^*(T;x)^\alpha dx = \sum_{n=1}^\infty b_n^\alpha \Big(\frac{1}{n} - \frac{1}{n+1}\Big) = \sum_{n=1}^\infty a_n^\alpha < \infty$$

by (10.4.1). We will show nevertheless that

$$P\Big(\sup_{(N,A)\in T} |X((N,A))| = \infty\Big) = 1, \tag{10.4.3}$$

i.e., $\{X((N,A)),\ (N,A) \in T\}$ is not sample bounded.

To prove (10.4.3), define

$$Z_n = [n(n+1)]^{-1/\alpha} M\Big(\Big(\frac{1}{n+1}, \frac{1}{n}\Big]\Big),\ n = 1, 2, \ldots.$$

Then $Z_1, Z_2, \ldots$ is a sequence of i.i.d. $S\alpha S$ random variables. Fix any $\omega \in \Omega$, let $K \geq 1$ and $A_\omega = \{n \in \{1, 2, \ldots, K\}\colon Z_n(\omega) \geq 0\}$. Clearly,

$$\begin{aligned} X((K, A_\omega))(\omega) &= \sum_{n=1}^K b_n \Big| M\Big(\Big(\frac{1}{n+1}, \frac{1}{n}\Big]\Big)\Big| \\ &= \sum_{n=1}^K a_n |Z_n(\omega)|. \end{aligned}$$

Therefore, for any $K \geq 1$,

$$\sup_{(N,A)\in T} |X((N,A))(\omega)| \geq \sum_{n=1}^K a_n |Z_n(\omega)|,$$

and thus

$$\sup_{(N,A)\in T} |X((N,A))| \geq \sum_{n=1}^\infty a_n |Z_n|.$$

But Condition (10.4.2) implies $\sum_{n=1}^\infty a_n |Z_n| = \infty$ a.s. (see Exercise 1.26), establishing (10.4.3).

Exercise 10.7 describes a similar construction for the parameter space $T = [0, 1]$, and also gives an example of a $S\alpha S$ process, $1 \leq \alpha < 2$, which satisfies both (10.3.1) and (10.3.2) but is not sample continuous. We will encounter more examples of this type later in the chapter.

The following is the main "positive" result of this section. It shows that in the case $0 < \alpha < 1$, sample boundedness and sample continuity of α-stable processes are completely determined by the corresponding properties of the kernel that appears in their integral representation (see the discussion after Theorem 10.3.1).

Theorem 10.4.2 *Let $0 < \alpha < 1$, and let $\{X(t),\ t \in T\}$ be an α-stable process given in the standard form (10.1.1).*

(i) Suppose that $\eta(t) = 0$ for every $t \in T$ in (10.1.1) and let $T^* \subseteq T$ be a separant for a separable version $\{Y(t),\ t \in T\}$ of $\{X(t),\ t \in T\}$. Suppose that $\{Y(t),\ t \in T\}$ is defined on a probability space $(\Omega, \mathcal{F}, P)$. Then

$$\lim_{\lambda\to\infty} \lambda^\alpha P(\sup_{t\in T} Y(t) > \lambda)$$

$$= \frac{C_\alpha}{2}\Big[\int_E h_+(T^*;x)^\alpha(1+\beta(x))m(dx) + \int_E h_-(T^*;x)^\alpha(1-\beta(x))m(dx)\Big] \tag{10.4.4}$$

and

$$\lim_{\lambda\to\infty} \lambda^\alpha P(\sup_{t\in T} |Y(t)| > \lambda) = C_\alpha \int_E f^*(T^*;x)^\alpha m(dx). \tag{10.4.5}$$

In particular, $\{X(t),\ t \in T\}$, with a general $\{\eta(t),\ t \in T\}$ in (10.1.1), is sample bounded if and only if

$$\int_E f^*(T^*;x)^\alpha m(dx) < \infty \tag{10.4.6}$$

and (10.2.15) holds.

(ii) Let (T, ρ) be a locally compact separable metric space. Then the process $\{X(t),\ t \in T\}$ is sample continuous if and only if (10.3.3) holds and for every compact subset C of T, for any countable dense subset C^* of C, we have

$$m\big(x \in E\colon\ f(t,x),\ t \in C^*,\ \text{is not uniformly continuous}\big) = 0 \tag{10.4.7}$$

and

$$\int_E \sup_{t\in C^*} |f(t,x)|^\alpha m(dx) < \infty. \tag{10.4.8}$$

PROOF: (i) In view of (10.2.10) and (10.2.12), the statements

$$\overline{\lim}_{\lambda\to\infty} \lambda^\alpha P\big(\sup_{t\in T} Y(t) > \lambda\big)$$

$$\leq \frac{C_\alpha}{2}\Big[\int_E h_+(T^*;x)^\alpha(1+\beta(x))m(dx) + \int_E h_-(T^*;x)^\alpha(1-\beta(x))m(dx)\Big] \tag{10.4.9}$$

and

$$\overline{\lim}_{\lambda\to\infty}\lambda^\alpha P\big(\sup_{t\in T}|Y(t)|>\lambda\big) \leq C_\alpha \int_E f^*(T^*;x)^\alpha m(dx) < \infty \tag{10.4.10}$$

will imply (10.4.4) and (10.4.5), respectively. We prove (10.4.9). (The proof of (10.4.10) is similar; moreover, (10.4.5) can be deduced directly from (10.4.4); see Exercise 10.8.)

Since there is nothing to prove if either one of the two integrals in the right-hand side of (10.4.9) is infinite, assume

$$\begin{cases} \int_E h_+(T^*;x)^\alpha(1+\beta(x))m(dx) < \infty, \\ \int_E h_-(T^*;x)^\alpha(1-\beta(x))m(dx) < \infty. \end{cases} \tag{10.4.11}$$

Let M_1 and M_2 be independent α-stable random measures on $(E,\mathcal{E})$ with the same control measure m (which is the control measure of M) and with the same skewness intensity $\beta \equiv 1$. Setting

$$W(t) = \int_E f(t,x)\big(1+\beta(x)\big)^{1/\alpha}M_1(dx),\ t\in T^*,$$

$$U(t) = \int_E f(t,x)\big(1-\beta(x)\big)^{1/\alpha}M_2(dx),\ t\in T^*,$$

we obtain

$$\{Y(t),\ t\in T^*\} \stackrel{d}{=} \{2^{-1/\alpha}(W(t)-U(t)),\ t\in T^*\},$$

and, since $\{Y(t),\ t\in T\}$ is separable, we conclude

$$\begin{aligned} P\big(\sup_{t\in T} Y(t)>\lambda\big) &= P\big(\sup_{t\in T^*} Y(t)>\lambda\big) \\ &= P\big(\sup_{t\in T^*}(W(t)-U(t)) > 2^{1/\alpha}\lambda\big) \\ &\leq P\big(\sup_{t\in T^*} W(t) + \sup_{t\in T^*}(-U(t)) > 2^{1/\alpha}\lambda\big). \end{aligned} \tag{10.4.12}$$

Set also

$$\begin{aligned} W_* &= \int_E h_+(T^*;x)\big(1+\beta(x)\big)^{1/\alpha}M_1(dx), \\ U_* &= \int_E h_-(T^*;x)\big(1-\beta(x)\big)^{1/\alpha}M_2(dx). \end{aligned}$$

By virtue of (10.4.11), W_* and U_* are well-defined α-stable random variables. They are also independent, and it follows from Proposition 3.4.1 and Proposition 1.2.11 that for each $t \in T^*$,

$$W_* - W(t) = \int_E \big(h_+(T^*;x) - f(t,x)\big)\big(1+\beta(x)\big)^{1/\alpha} M_1(dx) \geq 0 \text{ a.s.}$$

Therefore $W_* \geq \sup_{t\in T^*} W(t)$ a.s., and, similarly, $U_* \geq \sup_{t\in T^*}(-U(t))$ a.s. We conclude from (10.4.12) that

$$\overline{\lim}_{\lambda\to\infty}\lambda^\alpha P\big(Y(t) > \lambda\big) \leq \lim_{\lambda\to\infty} \lambda^\alpha P\big(W_* + U_* > 2^{1/\alpha}\lambda\big)$$

$$= \frac{1}{2}C_\alpha\Big[\int_E h_+(T^*;x)^\alpha(1+\beta(x))m(dx) + \int_E h_-(T^*;x)^\alpha(1-\beta(x))m(dx)\Big],$$

by Property 1.2.15 and the independence of W_* and U_*. This proves (10.4.9).

The necessity of (10.4.6) and (10.2.15) follows from Theorem 10.2.3. On the other hand, if (10.4.6) and (10.2.15) hold, then (10.4.5) implies

$$\lim_{\lambda\to\infty} \lambda^\alpha P\big(\sup_{t\in T}|Y(t)| > \lambda\big) < \infty,$$

and thus $P(\sup_{t\in T}|Y(t)| < \infty) = 1$ since $0 < \alpha < 1$. This proves that $\{X(t),\ t\in T\}$ is sample bounded and completes the proof of the first part of the theorem.

(ii) The necessity of (10.3.3), (10.4.7) and (10.4.8) for sample continuity of $\{X(t),\ t\in T\}$ was established in Theorem 10.3.1. It remains to prove the sufficiency of these conditions. We may and will assume $\eta(t) = 0$ for every $t \in T$.

We firstly show that the process $\{X(t),\ t\in T\}$ is continuous in probability under (10.4.7) and (10.4.8). Let $t_0 \in T$ and let B be a compact neighborhood of t_0. Let $\{t_n,\ n = 1,2,\ldots\} \subseteq T$ be such that $\rho(t_n, t_0) \to 0$ as $n\to\infty$. We may assume that $\{t_n,\ n=1,2,\ldots\}\subseteq B$. By (10.4.7), $\lim_{n\to\infty} f(t_n,x) = f(t_0,x)$ m-almost everywhere, and so by (10.4.8) and the dominated convergence theorem

$$\lim_{n\to\infty}\int_E \Big|f(t_n,x) - f(t_0,x)\Big|^\alpha m(dx) = 0.$$

Proposition 3.5.1 implies that $X(t_n) \to X(t_0)$ in probability as $n\to\infty$ and, since t_0 was chosen arbitrarily, we conclude that the process $\{X(t),\ t\in T\}$ is continuous in probability.

In view of Propositions 9.3.4 and 9.3.5, it is enough to prove that $\{X(t),\ t\in C\}$ is sample continuous for every compact subset $C\subseteq T$; we may therefore assume without loss of generality that T itself is compact.

Let T^* be a countable dense subset of T. According to Proposition 9.3.5, we must prove

$$P\big(\{X(t),\ t \in T^*\} \text{ is uniformly continuous}\big) = 1. \tag{10.4.13}$$

Set, as usual, $f^*(T^*;x) = \sup_{t\in T^*} |f(t,x)|$ and define a new measure m_1 on $(E,\mathcal{E})$ by

$$m_1(dx) = f^*(T^*;x)^\alpha m(dx).$$

Because of (10.4.8), m_1 is a finite measure. Let M_1 be an α-stable random measure on $(E,\mathcal{E})$ with control measure m_1 and the same skewness intensity β as the random measure M. By Proposition 3.5.5,

$$\{X(t),\ t \in T^*\} \stackrel{d}{=} \{Y(t),\ t \in T^*\},$$

where

$$Y(t) = \int_E f(t,x) f^*(T^*;x)^{-1} M_1(dx),\ t \in T^*,$$

(we adopt the convention $0/0 = 0$.) Moreover, by Theorem 3.10.1,

$$\{Y(t),\ t \in T^*\} \stackrel{d}{=} \{CZ(t),\ t \in T^*\},$$

where C is a positive constant and

$$Z(t) = \sum_{i=1}^\infty \gamma_i \Gamma_i^{-1/\alpha} f(t,V_i) f^*(T^*;V_i)^{-1},\ t \in T^*.$$

In the preceding expression, $\Gamma_1, \Gamma_2, \ldots$ is a sequence of arrival times of a Poisson process with unit arrival rate; $\binom{V_1}{\gamma_1}, \binom{V_2}{\gamma_2}, \ldots$ is a sequence of i.i.d. random vectors, independent of the sequence $\Gamma_1, \Gamma_2, \ldots,$ taking values in $E \times \{-1,1\}$ and such that V_1 has distribution $\widehat{m} = m_1/m_1(E)$ on E and $P(\gamma_1 = 1|V_i) = (1+\beta(V_1))/2$. We must therefore prove

$$P\big(\{Z(t),\ t \in T^*\} \text{ is uniformly continuous}\big) = 1. \tag{10.4.14}$$

Set $g(t,x) = f(t,x) f^*(T^*;x)^{-1}$, $t \in T^*$, $x \in E$, and note that $|g(t,x)| \le 1$ for every t and x. For $t,s \in T^*$,

$$\big|Z(t) - Z(s)\big| \le \sum_{i=1}^\infty \Gamma_i^{-1/\alpha} \big|g(t,V_i) - g(s,V_i)\big|. \tag{10.4.15}$$

Slightly abusing notation, denote by $(\Omega, \mathcal{F}, P)$ the probability space in which the stochastic process $\{Z(t),\ t \in T^*\}$ is defined (even though it is $\{X(t),\ t \in T\}$ that was originally supposed to be defined on that space).

Introduce the subset $\Omega_1 = \{\omega \in \Omega\colon \lim_{i\to\infty} \Gamma_i/i = 1$ and $\{g(t, V_i),\ t \in T^*\}$ is uniformly continuous for each $i = 1, 2, \ldots\}$ and note that $P(\Omega_1) = 1$ by the strong law of large numbers and (10.4.7). In order to prove (10.4.14), it is sufficient to show that $\{Z(t),\ t \in T^*\}$ is uniformly continuous for every $\omega \in \Omega_1$. But the latter statement follows directly from (10.4.15) and the dominated convergence theorem (the dominating factor is $ci^{-1/\alpha}$ for some $c \in (0, \infty)$; note that $\sum_{i=1}^{\infty} i^{-1/\alpha} < \infty$ since $0 < \alpha < 1$). This proves (10.4.14). The proof of Theorem 10.4.2 is now complete. ■

Remark. Theorem 10.4.2 solves completely the problems of sample boundedness and sample continuity for α-stable processes with $0 < \alpha < 1$ and gives the precise asymptotic behavior of the distribution function of the supremum (of its separable version). As we shall see, it is much more difficult to obtain analogous results in the case $1 \le \alpha < 2$, and there, different ideas are needed. This may seem surprising at first, at least as far as sample boundedness and sample continuity are concerned, because of the following argument.

Let $\{X(t),\ t \in T\}$ be, say, a $S\alpha S$ process with $1 \le \alpha < 2$ and suppose that we want to determine whether this process is sample bounded (continuous). Let A be a $S_{1/(2\alpha)}(1, 1, 0)$ random variable, independent of the process $\{X(t),\ t \in T\}$. Then

$$Y(t) = A^{1/\alpha} X(t),\ t \in T,$$

is a $S\frac{1}{2}S$ process (see Proposition 3.8.1). Clearly, $\{X(t),\ t \in T\}$ is sample bounded (continuous) if and only if $\{Y(t), t \in T\}$ is. Since $\{Y(t),\ t \in T\}$ is 1/2-stable, we can apply Theorem 10.4.2 and determine precisely when it is sample bounded (continuous). Hence, we know precisely what happens to $\{X(t),\ t \in T\}$ as well!

Unfortunately, this procedure is not practical. The necessary and sufficient conditions for sample boundedness (continuity) of $\{Y(t),\ t \in T\}$ are in terms of its integral representation. We would be in good shape if we could obtain an integral representation of $\{Y(t),\ t \in T\}$ which is explicitly related to "basic data" involving the original process $\{X(t),\ t \in T\}$ (say, to an integral representation of $\{X(t),\ t \in T\}$). But this seems to be very hard to obtain in general. We are only aware of one general integral representation of the sub-stable process $\{Y(t),\ t \in T\}$, the one given in Proposition 3.8.2,

$$\{Y(t),\ t \in T\} \stackrel{d}{=} \left\{c \int_{\Omega} X(t, \omega) M(d\omega),\ t \in T\right\}, \tag{10.4.16}$$

where M is a $S\frac{1}{2}S$ random measure on the probability space $(\Omega, \mathcal{F}, P)$ – the probability space where the $S\alpha S$ process $\{X(t),\ t \in T\}$ is defined – and $X(t, \omega)$ is X evaluated at a fixed $t \in T$ and $\omega \in \Omega$. Applying Theorem 10.4.2 to (10.4.16) gives no information on sample path properties of $\{X(t),\ t \in T\}$ because all we can conclude is that $\{X(t),\ t \in T\}$ is sample bounded (continuous) if and only if $\{Y(t),\ t \in T\}$ is sample bounded (continuous) – a complete tautology!

Theorem 10.4.2 has a counterpart for complex-valued $S\alpha S$ processes.

Corollary 10.4.3 *Let $0 < \alpha < 1$, and let $\{X(t),\ t \in T\}$ be a complex-valued $S\alpha S$ process given in the form (10.2.20).*

(i) Let T^* be a separant for a separable version of $\{X(t),\ t \in T\}$. Then the process $\{X(t),\ t \in T\}$ is sample bounded if and only if

$$\int_E f_c^*(T^*; x)^\alpha m(dx) < \infty, \tag{10.4.17}$$

where $f_c^*(T^*; x)$ is defined by (10.2.22).

(ii) Let (T, ρ) be a locally compact separable metric space. Then the process $\{X(t),\ t \in T\}$ is sample continuous if and only if for every compact subset C of T and for any countable dense subset C^* of C we have

$$m\big(x \in E\colon\ f_c(t,x),\ t \in C^* \text{ is not uniformly continuous}\big) = 0 \tag{10.4.18}$$

and

$$\int_E \sup_{t \in C^*} |f_c(t,x)|^\alpha m(dx) < \infty. \tag{10.4.19}$$

PROOF: (i) The necessity of (10.4.17) has been proven in Corollary 10.3.2. Suppose, now, that (10.4.17) holds. Let $Y_1(t) = \text{Re}\, X(t)$, $Y_2(t) = \text{Im}\, X(t)$, $t \in T$. The processes $\{Y_1(t),\ t \in T\}$ and $\{Y_2(t),\ t \in T\}$ are sample bounded by (10.2.23) and Theorem 10.4.2. Therefore $\{X(t),\ t \in T\}$ is sample bounded. The proof of (ii) is similar and is left to the reader (Exercise 10.9). ∎

Example 10.4.4 An immediate consequence of Corollary 10.4.3 is that, *for $0 < \alpha < 1$, every harmonizable $S\alpha S$ process (6.5.1) is sample continuous.*

The situation in the case $1 \le \alpha < 2$ is more complicated, as we will see in the sequel.

Example 10.4.5 Consider the α-stable moving average

$$X(t) = \int_{-\infty}^{\infty} f(t-x)\, M(dx),\ t \in \mathbb{R}, \tag{10.4.20}$$

where M is an α-stable random measure on $(\mathbb{R}, \mathcal{B})$ with Lebesgue control measure and skewness intensity β, $0 < \alpha < 1$, and f is a measurable function satisfying

$$\int_{-\infty}^{\infty} |f(x)|^\alpha dx < \infty. \tag{10.4.21}$$

We will use Theorem 10.4.2 to show that the following is true:

The α-stable moving average defined in (10.4.20) with $0 < \alpha < 1$ is sample bounded on an interval $[a, b]$, $-\infty < a < b < \infty$, if and only if for every countable set $T^* \subseteq [0, 1]$,

$$\int_{-\infty}^{\infty} \sup_{t \in T^*} |f(t+x)|^\alpha dx < \infty. \tag{10.4.22}$$

It is sample continuous if and only if there is a continuous function g: $\mathbb{R} \to \mathbb{R}$ such that

$$\text{Leb}\big(x\colon\ f(x) \neq g(x)\big) = 0 \tag{10.4.23}$$

and

$$\int_{-\infty}^{\infty} \sup_{0 \le t \le 1} |g(t+x)|^\alpha dx < \infty. \tag{10.4.24}$$

Applying the first part of Theorem 10.4.2, we note that $\{X(t),\ a \le t \le b\}$ is sample bounded if and only if for every countable subset S^* of $[a, b]$,

$$\int_{-\infty}^{\infty} \sup_{s \in S^*} |f(s-x)|^\alpha < \infty. \tag{10.4.25}$$

(See Theorem 10.2.3 and the subsequent discussion to reconcile (10.4.25) with (10.4.6).) It is easy to show that (10.4.25) is equivalent to (10.4.22). Indeed, suppose, for example, that $b - a \le 1$ and let S^* be a countable subset of $[a, b]$. Then $S^* - a$ is a countable subset of $[0, 1]$ and, if (10.4.22) holds,

$$\int_{-\infty}^{\infty} \sup_{s \in S^*} |f(s-x)|^\alpha dx = \int_{-\infty}^{\infty} \sup_{t \in S^* - a} |f(t+x)|^\alpha dx < \infty.$$

Suppose, on the other hand, that (10.4.25) holds. Then, as above,

$$\int_{-\infty}^{\infty} \sup_{t \in T^*} |f(t+x)|^\alpha dx < \infty$$

for every countable subset $T^* \subseteq [0, b-a]$. Thus,

$$\int_{-\infty}^{\infty} \sup_{t \in T^* + \theta} |f(t+x)|^\alpha dx < \infty$$

for every $\theta > 0$ and therefore (10.4.22) follows. The case $b - a > 1$ is similar.

For the sample continuity part of our claim, we apply the second part of Theorem 10.4.2 and conclude that $\{X(t),\ t \in T\}$ is sample continuous if and only if for every closed interval $[a, b]$, $-\infty < a < b < +\infty$, and for every countable dense subset S^* of $[a, b]$, Relation (10.4.25) holds and

$$\text{Leb}\big(x \in \mathbb{R}\colon\ f(t-x),\ t \in S^* \text{ is not uniformly continuous}\big) = 0. \tag{10.4.26}$$

We have already proved that (10.4.25) is equivalent to (10.4.22). It follows from (10.4.26) that there exists a continuous function $g\colon \mathbb{R} \to \mathbb{R}$ such that (10.4.23) holds. (The proof of this fact is an exercise in real analysis and is left to the reader – see Exercise 10.10.) It follows immediately from (10.4.25) and (10.4.23) that g satisfies (10.4.24).

On the other hand, suppose that a continuous g satisfying (10.4.23) and (10.4.24) exists. Then (10.4.25) and (10.4.26) holds, and so $\{X(t),\ t \in T\}$ is sample continuous.

The problem of pointwise sample continuity can be also fully solved in the case $0 < \alpha < 1$.

Theorem 10.4.6 *Let $0 < \alpha < 1$ and suppose that $\{X(t),\ t \in T\}$ is an α-stable process on a separable metric space (T, ρ). Let T^* be a strong separant for a strongly separable version of $\{X(t),\ t \in T\}$. Let $t_0 \in T$ have a relatively compact neighborhood. Then $\{X(t),\ t \in T\}$ is sample continuous at t_0 if and only if (10.3.13) holds, $\{\eta(t),\ t \in T\}$ is continuous at t_0 and for some open set A with compact closure such that $t_0 \in A \subseteq T$, (10.3.15) holds.*

The proof is left to the reader (Exercise 10.11).

10.5 Probability tails of suprema of bounded α-stable processes, with index $0 < \alpha < 2$

Let $\{X(t),\ t \in T\}$ be an α-stable process given in the standard form (10.1.1), and let $\{Y(t),\ t \in T\}$ be a separable version of $\{X(t),\ t \in T\}$. For $\alpha \in (0,1)$, the asymptotic behavior of $P(\sup_{t\in T} Y(t) > \lambda)$ as $\lambda \to \infty$ is described in Theorem 10.4.2. For $\alpha \in [1,2)$, however, we have only the asymptotic lower bound given in Theorem 10.2.3. One may wonder whether either Relations (10.4.4) or (10.4.5) continue to hold when $\alpha \in [1,2)$. Example 10.4.1 shows that this is not the case in general. Indeed, the α-stable process constructed in that example is a.s. unbounded. It turns out that the difficulties are caused precisely by the a.s. unboundedness of the paths. The following theorem shows that (10.4.4) and (10.4.5) do hold when the paths are a.s. bounded. Its proof uses some results of the theory of probability on Banach spaces.

Theorem 10.5.1 *Let $\{X(t),\ t \in T\}$ be an a.s. bounded α-stable process, $0 < \alpha < 2$, given in the form (10.1.1). Let T^* be a separant for a separable version $\{Y(t),\ t \in T\}$ of $\{X(t),\ t \in T\}$, defined on a probability space $(\Omega, \mathcal{F}, P)$. Then (10.4.4) and (10.4.5) hold, i.e.,*

$$\lim_{\lambda\to\infty} \lambda^\alpha P\big(\sup_{t\in T} Y(t) > \lambda\big)$$

$$= \frac{C_\alpha}{2}\Big[\int_E h_+(T^*;x)^\alpha(1+\beta(x))m(dx) + \int_E h_-(T^*;x)^\alpha(1-\beta(x))m(dx)\Big],$$

and

$$\lim_{\lambda\to\infty} \lambda^\alpha P\big(\sup_{t\in T}|Y(t)| > \lambda\big) = C_\alpha \int_E f^*(T^*;x)^\alpha m(dx),$$

where C_α is defined in (1.2.9), h_+ and h_- in (10.2.11) and f^* in (10.2.13).

PROOF: We start with the symmetric case, i.e., $\beta \equiv 0$ and $\eta \equiv 0$. Let $\widetilde{M}$ be a $S\alpha S$ random measure on $(E,\mathcal{E})$ with control measure $\widetilde{m}$ given by $\widetilde{m}(A) = \int_A f^*(T^*;x)^\alpha m(dx)$. Theorem 10.2.3 implies that $\widetilde{m}$ is a finite measure and hence $\widetilde{m}/\widetilde{m}(E)$ is a probability measure on E. Set $g(t,x) = f(t,x)/f^*(T^*;x)$ with the convention 0/0 = 0. Then

$$\{Y(t),\ t\in T^*\} \stackrel{d}{=} \Big\{\int_E g(t,x)\widetilde{M}(dx),\ t\in T^*\Big\}$$

by Proposition 3.5.5 and $\{Y(t),\ t\in T^*\} \stackrel{d}{=} \{S(t),\ t\in T^*\}$ by Corollary 3.10.4 where, in the notation of the corollary,

$$S(t) = (C_\alpha \widetilde{m}(E))^{1/\alpha}\sum_{i=1}^\infty \epsilon_i\Gamma_i^{-1/\alpha} g(t,V_i),\ t\in T^*,$$

and $V_1,, V_2,\ldots$ are i.i.d. E-valued random variables with distribution $\widetilde{m}(\cdot)/\widetilde{m}(E)$.

Define the following stochastic processes:

$$S_1(t) = (C_\alpha \widetilde{m}(E))^{1/\alpha}\sum_{i=1}^\infty \epsilon_i\Gamma_i^{-1/\alpha} 1(\Gamma_i > 1) g(t,V_i),\ t\in T^*,$$

$$S_2(t) = (C_\alpha \widetilde{m}(E))^{1/\alpha}\sum_{i=1}^\infty \epsilon_i\Gamma_i^{-1/\alpha} 1(\Gamma_i \le 1) g(t,V_i),\ t\in T^*.$$

Obviously, $S(t) = S_1(t) + S_2(t)$ for any $t\in T^*$. The processes $\{S_1(t),\ t\in T^*\}$ and $\{S_2(t),\ t\in T^*\}$ are independent because the properties of Poisson random measures ensure that $\{\{\epsilon_i\Gamma_i^{-1/\alpha}1(\Gamma_i > 1)g(t,V_i),\ t\in T^*\},\ i = 1,2,\ldots\}$ and $\{\{\epsilon_i\Gamma_i^{-1/\alpha}1(\Gamma_i \le 1)g(t,V_i),\ t\in T^*\},\ i = 1,2,\ldots\}$ are independent Poisson random measures on $\mathbb{R}^{T^*}$ (see, e.g., Resnick (1987), Section 3.3.2). Note also that $\{S_2(t),\ t\in T^*\}$ is the sum of an a.s. finite number of bounded functions. Therefore $\{S_2(t),\ t\in T^*\}$ is an a.s. bounded stochastic process and, consequently, $\{S_1(t),\ t\in T^*\}$ is also a.s. bounded.

It follows from Proposition 2.7 of Rosinski (1990) that $\{S_1(t),\ t\in T^*\}$ is an a.s. bounded symmetric infinitely divisible process whose characteristic function satisfies

$$-\ln E \exp i \sum_{t\in T^*} \theta(t)S_1(t)$$

$$= \tilde{m}(E)^{-1} \int_1^\infty \int_E \Big[1 - \cos\Big((C_\alpha \tilde{m}(E))^{1/\alpha} u^{-1/\alpha} \sum_{t\in T^*} \theta(t) g(t,x)\Big)\Big] \tilde{m}(dx)du \tag{10.5.1}$$

for every $\{\theta(t),\ t \in T^*\}$ with finitely many non-zero coordinates. We claim that there is an $\epsilon > 0$ such that

$$E\exp\Big(\epsilon \sup_{t\in T^*} |S_1(t)|\Big) < \infty. \tag{10.5.2}$$

To show this, identify T^* with $\mathbb{N} = \{1, 2, \ldots\}$ for simplicity and take an arbitrary sequence $\{a_j,\ j \geq 1\}$ of real numbers converging to zero. Let $Z_j = a_j S_1(j)$, $j = 1, 2, \ldots$. Since $Z_j \to 0$ a.s. as $j \to \infty$, the vector $\mathbf{Z} = (Z_1, Z_2, \ldots)$ can be viewed as a symmetric infinitely divisible random vector taking values in the separable Banach space c_0 of sequences of real numbers converging to zero with the supremum norm $\|\cdot\|_\infty$, i.e., $\|\mathbf{Z}\|_\infty = \sup_{j\geq 1} |Z_j|$. The dual $(c_0)'$ of c_0 is the space l^1 of absolutely summable sequences (Reed and Simon (1972), Section III.2). It follows from (10.5.1) and the lebesgue dominated convergence theorem that for any $\boldsymbol{\gamma} = (\gamma_1, \gamma_2, \ldots) \in l^1 = (c_0)'$,

$$-\ln E \exp i(\boldsymbol{\gamma}, \mathbf{Z}) = -\ln E \exp\Big\{i \sum_{j=1}^\infty \gamma_j Z_j\Big\}$$

$$= \tilde{m}(E)^{-1} \int_1^\infty \int_E \Big[1 - \cos\Big((C_\alpha \tilde{m}(E))^{1/\alpha} u^{-1/\alpha} \sum_{j=1}^\infty \gamma_j a_j g(j,x)\Big)\Big] \tilde{m}(dx)du$$

$$= \int_{c_0} \Big(1 - \cos(\boldsymbol{\gamma}, \mathbf{y})\Big) F(d\mathbf{y}), \tag{10.5.3}$$

where F is the image of the measure $\tilde{m}(E)^{-1}$ (Leb$\times\tilde{m}$) on $[1,\infty) \times E$ under the transformation ϕ: $[1,\infty) \times E \to c_0$ given by

$$\phi(u,x) = \Big((C_\alpha \tilde{m}(E))^{1/\alpha} u^{-1/\alpha} (a_1 g(1,x), a_2 g(2,x), \ldots)\Big).$$

By virtue of (10.5.3), F is the Lévy measure of the symmetric infinitely divisible random vector $\mathbf{Z}$ with values in c_0. Since $|g(j,x)| \leq 1$, we have for $u \geq 1$ and $x \in E$,

$$\|\phi(u,x)\|_\infty = (C_\alpha \tilde{m}(E))^{1/\alpha} u^{-1/\alpha} \sup_{j\geq 1} |a_j g(j,x)| \leq (C_\alpha \tilde{m}(E)^{1/\alpha} \sup_{j\geq 1} |a_j|).$$

Therefore

$$F\Big(\{\mathbf{y} \in c_0: \|\mathbf{y}\|_\infty \geq (C_\alpha \tilde{m}(E))^{1/\alpha} \sup_{j\geq 1} |a_j|\}\Big) = 0,$$

i.e., the Lévy measure F has bounded support. But de Acosta (1980), Corollary 3.3, shows that an infinitely divisible random random vector $\mathbf{Z}$ on a separable Banach space with norm $\|\cdot\|$, whose Lévy measure has bounded support, must have all exponential moments finite, i.e., for any $\delta > 0$, $E\exp(\delta\|\mathbf{Z}\|) < \infty$. Since here $\mathbf{Z} = \{a_j S_1(j),\ j \geq 1\}$ is a vector in c_0, we have, for every $\delta > 0$,

$$E\exp\Big(\delta \sup_{j\geq 1} |a_j||S_1(j)|\Big) < \infty. \tag{10.5.4}$$

Since (10.5.4) holds for every sequence $\{a_j,\ j = 1, 2, \ldots\}$ converging to zero, then (10.5.2) holds, because if (10.5.2) fails for every $\epsilon > 0$, we may choose a sequence of positive integers $M_m \uparrow \infty$ as $m \to \infty$ such that $E\exp\{\frac{1}{m}\max_{1\leq j\leq M_m} |S_1(j)|\} \geq m$. Then, setting $a_j = 1/m$ for $M_{m-1} < j \leq M_m$, $m = 1, 2, \ldots$; $M_0 = 0$, we obtain a sequence converging to zero for which (10.5.4) fails with $\delta = 1$. This contradiction proves that (10.5.2) holds for some $\epsilon > 0$.

We now turn to $\{S_2(t),\ t \in T^*\}$. The number of Γ_is with values in the interval $(0, 1]$ is a Poisson random variable N with mean 1, and given N, the Γ_is have the same distribution as N i.i.d. $U(0, 1)$ random variables arranged in increasing order. Reasoning as in the proof of Theorem 1.4.2, we have the representation

$$\{S_2(t),\ t \in T^*\} \stackrel{d}{=} (C_\alpha \tilde{m}(E))^{1/\alpha} \sum_{i=1}^{N} \epsilon_i U_i^{-1/\alpha} g(t, V_i),\ t \in T^*, \tag{10.5.5}$$

where N is Poisson with mean 1, independent of the sequences $\{\epsilon_1, \epsilon_2, \ldots\}$ and $\{V_1, V_2, \ldots\}$, and where $\{U_1, U_2, \ldots\}$ is a sequence of i.i.d. $U(0, 1)$ random variables independent of the other random quantities appearing in the right-hand side of (10.5.5).

For any $\lambda > 0$, we have

$$\begin{aligned} P(\sup_{t\in T^*} S_2(t) > \lambda) &= P\Big(\sup_{t\in T^*} (C_\alpha \tilde{m}(E))^{1/\alpha} \sum_{i=1}^{N} \epsilon_i U_i^{-1/\alpha} g(t, V_i) > \lambda\Big) \\ &\leq P\Big((C_\alpha \tilde{m}(E))^{1/\alpha} \sum_{i=1}^{N} \sup_{t\in T^*} \epsilon_i U_i^{-1/\alpha} g(t, V_i) > \lambda\Big) \\ &= P\Big(\sum_{i=1}^{N} W_i > \lambda\Big), \end{aligned}$$

where $W_i = (C_\alpha \tilde{m}(E))^{1/\alpha} \sup_{t\in T^*} \epsilon_i U_i^{-1/\alpha} g(t, V_i)$, $i = 1, 2, \ldots$. Set

$$H^* = \frac{1}{2} C_\alpha \Big[\int_E h_+(T^*; x)^\alpha m(dx) + \int_E h_-(T^*; x)^\alpha m(dx)\Big], \tag{10.5.6}$$

and observe that

$$\begin{aligned} P(W_i > \lambda) &= \frac{1}{2}P\left((C_\alpha \widetilde{m}(E))^{1/\alpha}U_i^{-1/\alpha}\frac{h_+(T^*;V_i)}{f^*(T^*;V_i)} > \lambda\right) \\ &\quad + \frac{1}{2}P\left((C_\alpha \widetilde{m}(E))^{1/\alpha}U_i^{-1/\alpha}\frac{h_-(T^*;V_i)}{f^*(T^*;V_i)} > \lambda\right) \\ &= \lambda^{-\alpha}H^*, \end{aligned}$$

for $\lambda > (C_\alpha \widetilde{m}(E))^{1/\alpha}$, because for such λ,

$$\begin{aligned} P\left(U_i < \lambda^{-\alpha}C_\alpha \widetilde{m}(E)\frac{h_\pm(T^*;V_i)^\alpha}{f^*(T^*;V_i)^\alpha}\right) &= \lambda^{-\alpha}C_\alpha \widetilde{m}(E)E\frac{h_\pm(T^*;V_i)^\alpha}{f^*(T^*;V_i)^\alpha} \\ &= \lambda^{-\alpha}C_\alpha \int_E h_\pm(T^*;x)^\alpha m(dx), \end{aligned}$$

since V_i has distribution $\widetilde{m}/\widetilde{m}(E)$ with $\widetilde{m}(dx) = f^*(T^*;x)^\alpha m(dx)$. Use now Exercise 10.12 to conclude

$$\overline{\lim}_{\lambda\to\infty}\lambda^\alpha P(\sup_{t\in T^*} S_2(t) > \lambda) \le H^*. \tag{10.5.7}$$

On the other hand, for any $\lambda > 0$ and $0 < \theta < 1/2$,

$$\begin{aligned} &P(\sup_{t\in T^*} S_2(t) > \lambda) \\ &= \sum_{n=0}^\infty e^{-1}\frac{1}{n!}P\left(\sup_{t\in T^*}(C_\alpha \widetilde{m}(E))^{1/\alpha}\sum_{i=1}^n \epsilon_i U_i^{-1/\alpha}g(t,V_i) > \lambda\right) \\ &\ge \sum_{n=0}^\infty e^{-1}\frac{1}{n!}P\left(\bigcup_{i=1}^n\Big\{\sup_{t\in T^*}(C_\alpha \widetilde{m}(E))^{1/\alpha}\epsilon_i U_i^{-1/\alpha}g(t,V_i) > \lambda(1+\theta),\right. \\ &\quad \left.\sup_{t\in T^*}|(C_\alpha \widetilde{m}(E))^{1/\alpha}\epsilon_j U_j^{-1/\alpha}g(t,V_j)| \le \frac{\lambda\theta}{n+1}\ \forall j=1,\dots,n,\ j\ne i\Big\}\right) \\ &= \sum_{n=0}^\infty e^{-1}\frac{n}{n!}\Big[\sigma\Big(\frac{\lambda\theta}{n+1}\Big)\Big]^{n-1} \\ &\quad \times P\Big(\sup_{t\in T^*}(C_\alpha \widetilde{m}(E))^{1/\alpha}\epsilon_1 U_1^{-1/\alpha}g(t,V_1) > \lambda(1+\theta)\Big), \end{aligned}$$

where $\sigma(\lambda) = P(\sup_{t\in T^*}|(C_\alpha \widetilde{m}(E))^{1/\alpha}\epsilon_1 U_1^{-1/\alpha}g(t,V_1)| \le \lambda)$. For $\lambda > (C_\alpha \widetilde{m}(E))^{1/\alpha}$,

$$P(\sup_{t\in T^*} S_2(t) > \lambda) \ge \sum_{n=1}^\infty e^{-1}\frac{n}{n!}H^*\lambda^{-\alpha}(1+\theta)^{-\alpha}\Big[\sigma\Big(\frac{\lambda\theta}{n+1}\Big)\Big]^{n-1}.$$

Therefore, by Fatou's lemma, $\underline{\lim}_{\lambda\to\infty}\lambda^\alpha P(\sup_{t\in T^*} S_2(t) > \lambda) \geq H^*(1+\theta)^{-\alpha}$. Since we may take θ arbitrarily close to 0, we obtain, using (10.5.7),

$$\lim_{\lambda\to\infty} \lambda^\alpha P(\sup_{t\in T^*} S_2(t) > \lambda) = H^*. \tag{10.5.8}$$

Our claim (10.4.4), in the symmetric case, follows now from (10.5.2), (10.5.6), (10.5.8) and Lemma 4.4.2.

To prove (10.4.4) in general, let $\{Y_i(t),\ t\in T\}$, $i = 1,2$, be independent copies of $\{Y(t),\ t\in T\}$, and let $Z(t) = Y_1(t) - Y_2(t)$, $t\in T$. Then the process $\{Z(t),\ t\in T\}$ is $S\alpha S$ and applying (10.4.4), we obtain

$$\lim_{\lambda\to\infty} \lambda^\alpha P(\sup_{t\in T}(Y_1(t) - Y_2(t)) > \lambda) = 2H^*. \tag{10.5.9}$$

From Theorem 10.2.3, we need only prove that

$$\overline{\lim}_{\lambda\to\infty}\lambda^\alpha P(\sup_{t\in T} Y(t) > \lambda)$$

$$\leq \frac{C_\alpha}{2}\Big[\int_E h_+(T^*;x)^\alpha(1+\beta(x))m(dx) + \int_E h_-(T^*;x)^\alpha(1-\beta(x))m(dx)\Big]. \tag{10.5.10}$$

Suppose, to the contrary, that there is a sequence $\lambda_k \uparrow \infty$ as $k\to\infty$ such that for some $\epsilon > 0$,

$$\lim_{k\to\infty} \lambda_k^\alpha P(\sup_{t\subset T} Y(t) > \lambda_k)$$

$$\geq \frac{C_\alpha}{2}\Big[\int_E h_+(T^*;x)^\alpha(1+\beta(x))m(dx) + \int_E h_-(T^*;x)^\alpha(1-\beta(x))m(dx)\Big] + \epsilon.$$

Then for every $M > 0$,

$$\begin{aligned}
&\underline{\lim}_{k\to\infty}(\lambda_k - M)^\alpha P\big(\sup_{t\in T}(Y_1(t) - Y_2(t)) > \lambda_k - M\big) \\
&\quad > \underline{\lim}_{k\to\infty}(\lambda_k - M)^\alpha P\big(\sup_{t\in T} Y_1(t) > \lambda_k,\ \sup_{t\in T}|Y_2(t)| \leq M\big) \\
&\quad\ + \underline{\lim}_{k\to\infty}(\lambda_k - M)^\alpha P\big(\sup_{t\in T}(-Y_2(t)) > \lambda_k,\ \sup_{t\in T}|Y_1(t)| \leq M\big) \\
&\quad \geq (2H^* + \epsilon)P\big(\sup_{t\in T}|Y(t)| \leq M\big).
\end{aligned}$$

Since M can be taken arbitrarily large, we obtain a contradiction to (10.5.9). This proves (10.5.10), and thus establishes (10.4.4). As we have seen before, this also implies (10.4.5). The proof of the theorem is now complete. ∎

Remark. Theorem 10.5.1 follows from Theorem 4.4.5 when $T = \{1,\ldots,n\}$.

10.6 The oscillation process

We consider in this section a more delicate characteristic of the sample paths, namely their oscillations.

Let $\{X(t),\ t \in T\}$ be a stochastic process on a locally compact separable metric space (T, ρ) and let T_0 be a strong separant for a strongly separable version of $\{X(t),\ t \in T\}$.

Definition 10.6.1 The stochastic process

$$W_X(t) = \overline{\lim}_{t_1,t_2 \to t,\ t_1,t_2 \in T_0} |X(t_2) - X(t_1)|,\ t \in T,$$

that takes values in $[0, \infty]$ is called the oscillation process of $\{X(t),\ t \in T\}$.

We leave it to the reader to verify that different choices of T_0 yield different versions of the same oscillation process. In fact, much more is true (see the remark following Property 10.6.10).

The following is an immediate consequence of the definition of the oscillation, the properties of a strong separant (Proposition 9.2.4) and Propositions 9.3.4 and 9.3.7.

Proposition 10.6.2 *For any fixed* $t_0 \in T$, $\{X(t),\ t \in T\}$ *is*
(i) sample bounded at t_0 *if and only if* $W(t_0)$ *is finite a.s.,*
(ii) sample continuous at t_0 *if and only if* $W(t_0) = 0$ *a.s.*

Itô and Nisio (1968) proved that the oscillation process of any *Gaussian* process continuous in probability is deterministic (non-random) in the following sense. Let $(\Omega, \mathcal{F}, P)$ be the probability space on which such a process $\{X(t),\ t \in T\}$ is defined. Then there is an event $\Omega_0 \subseteq \Omega$ with $P(\Omega_0) = 0$ and a (non-random) function $\alpha_X(t),\ t \in T$, such that $W_X(t, \omega) = \alpha_X(t)$ for every $\omega \in \Omega_0^c$ and every $t \in T$. This turns out to be false, in general, for stable processes.

Example 10.6.3 Let $\{L(t),\ 0 \le t \le 1\}$ be the $S\alpha S$ motion. We know (Exercise 9.5) that this process is pointwise sample continuous and thus, by Proposition 10.6.2, $W_X(t) = 0$ a.s. for every $t \in [0, 1]$. W_X is not deterministic in the above sense because if it were, we would have $\alpha_X(t) = 0$ for all $t \in [0, 1]$. It would then follow from Proposition 9.3.4 that the $S\alpha S$ motion is sample continuous which, as we know, is not the case.

To characterize the α-stable processes for which the oscillation process is deterministic, we will use their integral representation (10.1.1) rather than continuity in probability.

Recall that, here, (T, ρ) is a locally compact separable metric space. We consider the subclass of α-stable processes on T satisfying the following condition:

Condition 10.6.4 *$\{X(t),\ t \in T\}$ has a representation (10.1.1) with η: $T \to \mathbb{R}$ continuous and such that for every countable dense subset T^* of T and every compact subset C of T,*

$$m\big(x\colon \text{the function } f(\cdot, x)\colon T^* \cap C \to \mathbb{R} \text{ is not uniformly continuous}\big) = 0. \tag{10.6.1}$$

We will use the following technical result.

Lemma 10.6.5 *An α-stable process $\{X(t),\ t \in T\}$ satisfies Condition 10.6.4 if and only if $\{X(t),\ t \in T\}$ has an integral representation (10.1.1) with the following properties:*

- *η: $T \to \mathbb{R}$ continuous,*
- *$f(\cdot, x)$: $T \to \mathbb{R}$ continuous for every $x \in E$,*
- *there is a probability measure λ on $(E, \mathcal{E})$ equivalent to the control measure m.*

PROOF: The sufficiency part is trivial. To prove the necessity, suppose that $\{X(t),\ t \in T\}$ satisfies (10.6.1). Let T^* be a countable dense subset of T, and let

$$\begin{aligned} E_0 = \{x \in E : \quad & \text{for some } t \in T^* \text{ and for some rational } r > 0 \\ & \text{such that the closed ball } \overline{B(t,r)} \text{ is compact ,} \\ & \text{the map } f(\cdot, x)\colon T^* \cap B(t,r) \to \mathbb{R} \\ & \text{is not uniformly continuous}\}. \end{aligned}$$

Observe that $m(E_0) = 0$ and the limit $\lim_{s \to t,\ s \in T^*} f(s, x)$ exists for every $t \in T$ and $x \in E_0^c$. Therefore, the function $g(\cdot, x)$: $T \to \mathbb{R}$, given by

$$g(t, x) = \begin{cases} \lim_{s \to t,\ s \in T^*} f(s, x) & \text{if } x \in E_0^c, \\ 0 & \text{if } x \in E_0, \end{cases}$$

is well defined and continuous for every $x \in E$.

Fix any $t \in T$ and let $E_0(t)$ be defined as E_0 but with $T^* \cup \{t\}$ replacing T^*. Then $m(E_0(t)) = 0$. Clearly, for each $x \in E_0(t)^c$,

$$f(t, x) = \lim_{s \to t,\ s \in T^*} f(s, x) = g(t, x),$$

and so for every $t \in T$,

$$m\big(x \in E\colon f(t, x) \neq g(t, x)\big) = 0.$$

Thus, a.s., for every $t \in T$,

$$X(t) = \int_E g(t, x) M(dx) + \eta(t) = \int_{E_+} g(t, x) M(dx) + \eta(t),$$

where

$$E_+ = \bigcup_{t \in T} \{x \in E \colon |g(t,x)| \neq 0\}.$$

Note that $E_+ \in \mathcal{E}$ since $g(\cdot, x)\colon T \to \mathbb{R}$ is continuous for every $x \in E$, and thus $E_+ = \bigcup_{t \in T^*} \{x \in E \colon |g(t,x)| \neq 0\}$. Moreover, m is σ-finite on E_+ (Exercise 10.14). Using Exercise 10.4, conclude that there is a probability measure λ on $(E_+, \mathcal{E} \cap E_+)$ equivalent to m. ∎

$W_X(t)$ is the oscillation of X at a point t. We now define $W_X(C)$, the oscillation of X on a set C. This will allow us to overcome measurability difficulties when dealing with $\{W_X(t),\ t \in T\}$, which is generally an uncountable set of random variables.

Definition 10.6.6 The oscillation of a process $\{X(t),\ t \in T\}$ on a set $C \subseteq T$ is

$$W_X(C) = \overline{\lim}\, \big|X(t_1) - X(t_2)\big|,$$

where the limit superior is taken over all $t_1, t_2 \in T_0$ such that $d(t_1, t_2) \to 0$ and $d(t_1, C) \to 0$.

It is easy to verify the following properties of the oscillation on a set. (If U and V are random variables, we write $U \overset{\text{st}}{\geq} V$ to indicate that U is stochastically greater than V, i.e., $P(U \geq \lambda) \geq P(V \geq \lambda)$ for all λ.)

Proposition 10.6.7 *(i) $W_X(\{t\}) = W_X(t)$ a.s. for every $t \in T$.*

(ii) $W_{aX}(C) = |a| W_X(C)$ a.s. for every real a, $C \subseteq T$.

(iii) $W_{X+Y}(C) \leq W_X(C) + W_Y(C)$ a.s. for every $C \subseteq T$, where $\{X(t),\ t \in T\}$ and $\{Y(t),\ t \in T\}$ are stochastic processes defined on the same probability space.

(iv) Suppose additionally in (iii) that the set C is compact and that $\{Y(t),\ t \in T\}$ is sample continuous. Then $W_{X+Y}(C) = W_X(C)$ a.s.

(v) Suppose additionally in (iii) that the set C is compact, and that $\{Y(t),\ t \in T\}$ is independent of $\{X(t),\ t \in T\}$ and continuous in probability. Then $W_{X+Y}(C) \overset{\text{st}}{\geq} W_X(C)$.

PROOF: We prove only (v) the rest is left to the reader. See Exercise 10.15.

We can assume that $\{X(t),\ t \in T\}$ is defined on the probability space $(\Omega_1, \mathcal{F}_1, P_1)$, and $\{Y(t),\ t \in T\}$ is defined on a different probability space $(\Omega_2, \mathcal{F}_2, P_2)$, where $\mathcal{C}^T$ is the σ-algebra of cylinder sets. We view the three stochastic processes $\{X(t),\ t \in T\}$, $\{Y(t),\ t \in T\}$ and $\{X(t) + Y(t),\ t \in T\}$ as defined on the product probability space $(\Omega_1 \times \Omega_2, \mathcal{F}_1 \times \mathcal{F}_2, P_1 \times P_2)$. Let T_0 be a common strong separant for a strongly separable version of $\{X(t),\ t \in T\}$ and a strongly separable version of $\{Y(t),\ t \in T\}$. Let $W_X(C)$ and $W_{X+Y}(C)$ be defined by the same T_0 and regard the random vector $(W_X(C), W_{X+Y}(C))$ as an element of the product space. The set $A = \{(\omega_1, \omega_2) \in \Omega_1 \times \Omega_2 \colon W_{X+Y}(C) \geq W_X(C)\}$ is then in $\mathcal{F}_1 \times \mathcal{F}_2$.

In order to establish the result, we prove $(P_1 \times P_2)(A) = 1$ by using Fubini's theorem as follows: for a fixed $\omega_1 \in \Omega_1$, there are two sequences depending on ω_1, $\{t_n^{(i)},\ n = 1, 2, \ldots\} \subseteq T_0$, $i = 1, 2$, and a point $t_0 \in C$ such that as $n \to \infty$, $t_n^{(i)} \to t_0$, $i = 1, 2$, and

$$\lim_{n\to\infty} \left| X(t_n^{(1)}) - X(t_n^{(2)}) \right| = W_X(C).$$

Since $\{Y(t),\ t \in T\}$ is continuous in probability at t_0, there are subsequences $\{t_{n_k}^{(i)},\ k = 1, 2, \ldots\}$, $i = 1, 2$, and an event $A_{\omega_1} \in \mathcal{F}_1$ with $P_2(A_{\omega_1}) = 1$ such that $\lim_{k\to\infty} Y(t_{n_k}^{(i)}, \omega_2) = Y(t_0, \omega_2)$, $i = 1, 2$, for every $\omega_2 \in A_{\omega_1}$. Clearly, $(\omega_1, \omega_2) \in A$ for any $\omega_2 \in A_{\omega_1}$. Thus $\int_{\Omega_2} 1_A(\omega_1, \omega_2) P_2(d\omega_2) = 1$ for all $\omega_1 \in \Omega_1$, which implies $(P_1 \times P_2)(A) = 1$. ∎

We shall now use the notion of oscillation on sets to derive sufficient conditions for the oscillation process $\{W_X(t),\ t \in T\}$ to be non-random.

Theorem 10.6.8 *Let (T, ρ) be a locally compact separable metric space and let $\{X(t),\ t \in T\}$ be an α-stable process satisfying Condition 10.6.4. Then the oscillation process $\{W_X(t),\ t \in T\}$ is non-random, i.e., there is an event $\Omega_0 \subseteq \Omega$ with $P(\Omega_0) = 0$ and a deterministic function $\alpha_X \colon T \to \mathbb{R}_+ \cup \{\infty\}$ such that $W_X(t) = \alpha_X(t)$ for every $\omega \notin \Omega_0$ and every $t \in T$.*

PROOF: Our argument depends only on the distributional properties of the oscillations $W_X(C)$ on compact sets C, which are the same for all versions of $\{X(t),\ t \in T\}$. We may therefore assume, without loss of generality that the representation (10.1.1) is of the form prescribed by Lemma 10.6.5 and that $\eta(t) \equiv 0$. Let λ be a probability measure on $(E, \mathcal{E})$ equivalent to the control measure m and let

$$(r(x))^\alpha = \frac{m(dx)}{\lambda(dx)}, \quad x \in E.$$

We suppose that $\alpha \neq 1$ as the case $\alpha = 1$ is similar to the case $1 < \alpha < 2$ considered below. Using Proposition 3.5.5 and Theorem 3.10.1, we obtain $\{X(t),\ t \in T\} \stackrel{d}{=} \{cY(t),\ t \in T\}$ where $c > 0$ and

$$Y(t) = \sum_{i=1}^{\infty} \left[\gamma_i \Gamma_i^{-1/\alpha} r(V_i) f(t, V_i) - b_i^{(\alpha)} \int_E r(x) f(t, x) \beta(x) \lambda(dx) \right].$$

As usual, $\Gamma_1, \Gamma_2, \ldots$ is the sequence of arrival times of a Poisson process with unit rate, $\binom{V_1}{\gamma_1}, \binom{V_2}{\gamma_2}, \ldots$ is a sequence of i.i.d. random vectors such that V_i has the distribution λ on E,

$$P\Big(\gamma_i = 1\Big|V_i\Big) = 1 - P\Big(\gamma_i = -1\Big|V_i\Big) = \frac{1+\beta(V_i)}{2}$$

($\beta(\cdot)$ is the skewness intensity of the random measure M in (10.1.1)) and the two random sequences are independent. The constants $b_i^{(\alpha)}$ are given in Theorem 3.9.1.

Let $C \subseteq T$ be a compact set. We firstly show that the oscillation $W_X(C)$ is non-random.

The case $0 < \alpha < 1$ is particularly straightforward because $b_i^{(\alpha)} = 0$ for every i and Proposition 10.6.7 shows that for every $N \geq 1$, $W_Y(C) = W_{Y_N}(C)$ a.s., where

$$Y_N(t) = \sum_{i=N+1}^{\infty} \gamma_i \Gamma_i^{-1/\alpha} r(V_i) f(t, V_i),\ t \in T.$$

Let $e_1, e_2, \ldots$ be i.i.d. standard exponential random variables such that $\Gamma_i = e_1 + \cdots + e_i$, $i = 1, 2, \cdots$. Let π be an arbitrary permutation of the indices $\{1, 2, \cdots, N\}$ and

$$Y^\pi(t) = \sum_{i=1}^{N} \gamma_{\pi(i)} \big(e_{\pi(1)} + \cdots + e_{\pi(i)}\big)^{-1/\alpha} r(V_{\pi(i)}) f(t, V_{\pi(i)}) + Y_N(t),\ t \in T.$$

Applying Proposition 10.6.7, we obtain $W_{Y^\pi}(C) = W_{Y_N}(C)$ a.s. Hence, $W_{Y^\pi}(C) = W_Y(C)$ a.s. for any $N \geq 1$ and any permutation π of $\{1, 2, \ldots, N\}$. The Hewitt-Savage zero-one law applies and yields

$$W_X(C) = |c| W_Y(C) = \text{constant} =: \alpha_X(C) \text{ a.s.} \tag{10.6.2}$$

We now turn to the case $1 < \alpha < 2$. Let $g(t) = \int_E r(x) f(t, x) \beta(x) \lambda(dx)$, $t \in T$. If for each $t \in C$,

$$g(t) = \lim_{t_n \to t,\ t_n \in T_0} g(t_n), \tag{10.6.3}$$

then the same argument as in the case $0 < \alpha < 1$ applies, and (10.6.2) still holds.

Suppose, now, that (10.6.3) is violated at a point $t_0 \in C$. Consider a closed ball $B(t_0, r), r > 0$, and the integral

$$J_r(t_0) = \int_E \sup_{t \in T_0 \cap B(t_0, r)} |f(t, x)|^\alpha m(dx).$$

If $J_r(t_0) < \infty$ for some $r > 0$, then (10.6.3) holds at t_0 by the dominated convergence theorem. Thus $J_r(t_0) = \infty$ for all $r > 0$. Applying Theorem 10.2.3, we conclude that $\{Y(t),\ t \in T_0 \cap B(t_0, r)\}$ is not a.s. bounded for any $r > 0$ and hence, by the zero-one law (Theorem 9.5.4),

$$P\Big(\omega\colon \sup_{t \in T_0 \cap B(t_0,r)} |Y(t)| = \infty \text{ for all } r > 0\Big) = 1.$$

This implies $W_Y(t_0) = \infty$ a.s., so (10.6.2) still holds (with an infinite constant).

We shall use (10.6.2) to conclude the proof of the theorem.

Let $\mathcal{A}$ be the collection of finite intersections of closed balls of the type $\overline{B(t,q)}$, where $t \in T_0,\ q \geq 0$ is rational and such that $\overline{B(t,q)}$ is compact. Since $\mathcal{A}$ is a countable collection of compact sets,

$$P\big(W_X(C) = \alpha_X(C) \text{ for every } C \in \mathcal{A}\big) = 1. \tag{10.6.4}$$

Let Ω_0^c denote the event in (10.6.4). Fixing $\omega \in \Omega_0^c$ and $t \in T$, choose $r > 0$ such that $\overline{B(t,r)}$ is compact, and choose a sequence $\{t_n,\ n = 1, 2, \ldots\} \subseteq T_0$ such that $\rho(t_n, t) \leq r2^{-(n+1)},\ n = 1, 2, \ldots$. Choose also rational numbers $q_n,\ n = 1, 2, \ldots,$ such that $\rho(t_n, t) \leq q_n \leq 2\rho(t_n, t)$. Then

$$C_n = \bigcap_{j=1}^{n} B(t_j, q_j),\ n = 1, 2, \ldots$$

is a decreasing sequence of $\mathcal{A}$-sets, and $\{t\} = \bigcap_{n=1}^{\infty} C_n$. Hence (Exercise 10.16),

$$W_X(t) = \lim_{n\to\infty} W_X(C_n), \tag{10.6.5}$$

which, of course, does not involve $\omega \in \Omega_0^c$. This completes the proof of the theorem. ∎

The following is an immediate and important corollary of Theorem 10.6.8.

Corollary 10.6.9 *Let (T, ρ) be a locally compact separable metric space, and let $\{X(t),\ t \in T\}$ be an α-stable process satisfying Condition 10.6.4. If the process $\{X(t),\ t \in T\}$ is pointwise sample continuous, then it is sample continuous.*

PROOF: By pointwise sample continuity and Proposition 10.6.2 we conclude that in Theorem 10.6.8, $\alpha_X(t) = 0$ for every $t \in T$. It now follows from Proposition 9.3.5 that $\{X(t),\ t \in T\}$ is sample continuous. ∎

We have seen that an α-stable process $\{X(t),\ t \in T\}$ satisfying Condition 10.6.4 has a non-random oscillation function $\{\alpha_X(t),\ t \in T\}$. What are possible shapes for $\alpha_X(\cdot)$? Recall that a function $f\colon T \to [-\infty, \infty]$ is said to be *upper semi-continuous* if for every $t \in T$, $f(t) \geq \overline{\lim}_{s\to t} f(s)$.

Property 10.6.10 *The oscillation function* $\{\alpha_X(t),\ t \in T\}$ *is upper semi-continuous.*

The proof is left as an exercise.

Remarks

1. This is an important example of a situation where the knowledge of the finite-dimensional distributions is clearly insufficient. If we are told that the oscillation process $\{W_X(t),\ t \in T\}$ is 0 in the sense of the *finite-dimensional distributions*, we cannot conclude that $\{X(t),\ t \in T\}$ is sample continuous because the $S\alpha S$ motion, $0 < \alpha < 2$, which is not sample continuous, also has oscillations 0 (Example 10.6.3). In order to reach the desired conclusion we would need to know additionally that the oscillation process is *non-random*, and this fact can certainly not be determined by the finite-dimensional distributions of the oscillation process.

2. Choosing different T_0s provides oscillation processes which are not only versions of each other but also share some additional properties: they are all non-random under Condition 10.6.4 (Theorem 10.6.8) or all random under some other condition (Theorem 10.11.1).

3. The oscillation process $\{W_X(t),\ 0 \le t \le 1\}$ of a Poisson process $\{X(t),\ 0 \le t \le 1\}$ is random and takes values in $\{0, 1\}$; the value 1 at $t = t_0$ indicates that X has a jump at t_0.

4. We noted that the oscillation process of a (symmetric) stable motion is also random. Now a stable process can be viewed as an integral with respect to stable motion. Theorem 10.6.8 can then heuristically be understood as follows: if the integrand is not uniformly continuous, then we may notice the individidual jumps of the underlying stable motion and the resulting oscillation may be expected to be random. Condition 10.6.4, therefore, requires the integrand to be uniformly continuous. The condition is very close to being necessary (Theorem 10.11.1).

5. The same heuristic argument explains why Condition 10.6.4 is also a necessary condition for sample continuity (Theorem 10.3.1.)

10.7 The case $0 < \alpha < 1$

Property 10.7.1 *Let* $\{X(t),\ t \in T\}$ *be an* α*-stable process,* $0 < \alpha < 1$*, satisfying Condition 10.6.4. Then its oscillation function* $\{\alpha_X(t),\ t \in T\}$ *takes values in the two-point set* $\{0, \infty\}$.

PROOF: Assume, without loss of generality, that $\eta(t) = 0$ for all $t \in T$. Then $\{X(t),\ t \in T\}$ is strictly α-stable, and by Proposition 10.6.7, for every $t \in T$,

$$\begin{aligned} \alpha_X(t) &= W_X(t) = W_{2^{-1/\alpha}(X_1+X_2)}(t) = 2^{-1/\alpha} W_{X_1+X_2}(t) \\ &\leq 2^{-1/\alpha}(W_{X_1}(t) + W_{X_2}(t)) = 2^{1-1/\alpha}\alpha_X(t) \text{ a.s.,} \end{aligned}$$

where $\{X_i(t),\ t \in T\},\ i = 1, 2$, are independent copies of $\{X(t),\ t \in T\}$. The last relation implies $\alpha_X(t) \in \{0, \infty\}$. ∎

We are now able to describe fully the sample path behavior of α-stable processes, $0 < \alpha < 1$, satisfying Condition 10.6.4.

Theorem 10.7.2 *Let $\{X(t),\ t \in T\}$ be an α-stable process, $0 < \alpha < 1$, satisfying Condition 10.6.4.*

(i) *There is an open set $T_{\text{cont}} \subseteq T$ such that $\{X(t),\ t \in T\}$ is sample continuous on T_{cont} and is not sample bounded on any open set A such that $A \cap T^c_{\text{cont}} \neq \emptyset$.*

(ii) *For every $t_0 \in T$, $\{X(t),\ t \in T\}$ is either sample continuous at t_0, or is not sample bounded in any neighborhood of t_0.*

(iii) *If $\{X(t),\ t \in T\}$ is sample continuous at $t_0 \in T$, then it is sample continuous in a neighborhood of t_0.*

PROOF: (i) The set $T_{\text{cont}} = \{t \in T: \alpha_X(t) = 0\}$ is open by Property 10.6.10 and has the required properties.

(ii) The statement follows from (i) since either $t_0 \in T_{\text{cont}}$ or $t_0 \in T^c_{\text{cont}}$.

(iii) Sample continuity at t_0 implies $t_0 \subset T_{\text{cont}}$. The claim follows because T_{cont} is open. ∎

10.8 The case $1 \leq \alpha < 2$

What happens in the case $1 \leq \alpha < 2$? Can the oscillation function take values other than 0 or ∞? The answer to the latter question is "yes," as the following example demonstrates:

Example 10.8.1 Let $1 < \alpha < 2$, let $V_1, V_2 \ldots$ be a sequence of i.i.d. $S\alpha S$ random variables, and let $W_1 = V_1, W_2 = -V_1,\ W_3 = 2^{-1}(V_1 + V_2), W_4 = 2^{-1}(V_1 - V_2), W_5 = 2^{-1}(-V_1 + V_2), W_6 = 2^{-1}(-V_1 - V_2), W_7 = 3^{-1}(V_1 + V_2 + V_3)$, etc. By the strong law of large numbers,

$$\overline{\lim}_{n\to\infty} W_n = \lim_{n\to\infty} n^{-1}\left(|V_1| + \ldots + |V_n|\right) = E|V_1| \text{ a.s.}$$

Now let $X(0) = 0$, $X(1/n) = W_n$, $n = 1, 2, \ldots$, and define $X(t)$ for all other $t \in (0, 1)$ by linear interpolation. The process $\{X(t),\ 0 \le t \le 1\}$ is $S\alpha S$ and satisfies Condition 10.6.4. However, $\alpha_X(t) = 0$ for $t \in (0, 1]$ and $\alpha_X(0) = E|V_1| \in (0, \infty)$. For a similar construction in the case $\alpha = 1$ see Exercise 10.18.

The following properties of the oscillation function hold for all $0 < \alpha < 2$. (They follow trivially from Property 10.7.1 in the case $0 < \alpha < 1$.)

Proposition 10.8.2 *Let $\{X(t),\ t \in T\}$ be a $S\alpha S$ process satisfying Condition 10.6.4. Then, for each $t_0 \in T$,*

$$P\Big(\overline{\lim}_{t \to t_0,\ t \in T_0} X(t) = X(t_0) + \frac{1}{2}\alpha_X(t_0),$$
$$\underline{\lim}_{t \to t_0,\ t \in T_0} X(t) = X(t_0) - \frac{1}{2}\alpha_X(t_0)\Big) = 1. \tag{10.8.1}$$

PROOF: Let

$$U_X(t_0) = \overline{\lim}_{t \to t_0,\ t \in T_0} (X(t) - X(t_0)). \tag{10.8.2}$$

Since $\{X(t),\ t \in T\}$ satisfies Condition 10.6.4, conclude that $U_X(t_0)$ is, in fact, a constant, i.e., there is a $\beta(t_0) \in [0, \infty]$ such that the event $\Omega_1 = \{\omega\colon U_X(t_0) = \beta(t_0)\}$ occurs with probability 1 (see Exercise 10.19). By the symmetry of $\{X(t),\ t \in T\}$, we have $U_X(t_0) \stackrel{d}{=} V_X(t_0)$, where

$$V_X(t_0) = \overline{\lim}_{t \to t_0,\ t \in T_0} (X(t_0) - X(t)).$$

Therefore, the event $\Omega_2 = \{\omega\colon V_X(t_0) = \beta(t_0)\}$ occurs also with probability 1. Let Ω_0 be the null event of Theorem 10.6.8. For $\omega \in \Omega_0^c \cap \Omega_1 \cap \Omega_2$, we have

$$\begin{aligned} \alpha_X(t_0) &= \overline{\lim}_{t_1, t_2 \to t_0,\ t_1, t_2 \in T_0} |X(t_1) - X(t_2)| \\ &= \overline{\lim}_{t_1 \to t_0,\ t_1 \in T_0} (X(t_1) - X(t_0)) + \overline{\lim}_{t_2 \to t_0,\ t_2 \in T_0} (X(t_0) - X(t_2)) \\ &= 2\beta(t_0), \end{aligned}$$

i.e., $\beta(t_0) = \frac{1}{2}\alpha_X(t_0)$. This proves the result. ∎

10.9 The level sets of the oscillation function

Condition 10.6.4 neither implies nor is implied by continuity in probability (Exercise 10.20). Nevertheless, the following result holds:

Proposition 10.9.1 *Let $\{X(t),\ t \in T\}$ be an α-stable, $\alpha \ne 1$, or $S1S$ process satisfying Condition 10.6.4. If there is a $t_0 \in T$ such that $\alpha_X(t_0) < \infty$, then there is an $r > 0$ such that $\{X(t),\ t \in B(t_0, r)\}$ is continuous in probability.*

PROOF: Let $\{t_n,\ n \geq 1\} \subseteq T$ be such that $t_n \to t_0$ as $n \to \infty$. We shall prove that $X(t_0) = \text{plim}_{n\to\infty} X(t_n)$. Let T_0 be, as usual, a strong separant for X. We can assume, without loss of generality, that $\{t_n,\ n \geq 1\} \subseteq T_0$, since this will not affect α_X. The assumption $\alpha_X(t_0) < \infty$ implies

$$P\Big(\sup_{t\in T_0\cap B(t_0,r)} |X(t)| < \infty\Big) = 1$$

for some $r > 0$, and so

$$\int_E \sup_{t\in T_0\cap B(t_0,r)} |f(t,x)|^\alpha m(dx) < \infty \tag{10.9.1}$$

by Theorem 10.2.3. Relations (10.6.1), (10.9.1), and the dominated convergence theorem then yield

$$\lim_{n\to\infty} \int_E |f(t_0,x) - f(t_n,x)|^\alpha m(dx) = 0.$$

Applying Proposition 3.5.1, we conclude that $X(t_0) = \text{plim}_{n\to\infty} X(t_n)$, i.e., the process $\{X(t),\ t \in T\}$ is continuous in probability at t_0. Obviously, $\alpha_X(t) < \infty$ for every $t \in B(t_0, r)$, and the same argument shows that the process is continuous in probability at every point of $B(t_0, r)$. ∎

We have now the tools needed to prove

Property 10.9.2 *Let $\{X(t),\ t \in T\}$ be an α-stable, $\alpha \neq 1$, or $S1S$ process satisfying Condition 10.6.4. Then for every $\epsilon > 0$, the set*

$$T_\epsilon = \{t \in T\colon \alpha_X(t) \in [\epsilon, \infty)\}$$

is nowhere dense.

PROOF: Suppose firstly that the α-stable process $\{X(t),\ t \in T\}$ is symmetric, and assume to the contrary that there is a non-empty open set $A \subseteq T$ such that T_ϵ is dense in A. Property 10.6.10 implies $\alpha_X(t) \geq \epsilon$ for every $t \in A$. Now take a $t_0 \in T_\epsilon \cap A$ and observe that there is an $r > 0$ such that $\epsilon \leq \alpha_X(t) < \infty$ for all $t \in B(t_0, r)$. Let Ω_0 be the null set in Theorem 10.6.8 and

$$\begin{aligned}\Omega_1 = \Big\{\omega \in \Omega_0^c\colon X(t) &= \overline{\lim}_{u\to t,\ u\in T_0} X(u) - \frac{1}{2}\alpha_X(t)\ \forall t \in T_0 \cap B(t_0, r),\\ X(t_0) &= \overline{\lim}_{u\to t_0,\ u\in T_0} X(u) - \frac{1}{2}\alpha_X(t_0)\Big\}\end{aligned}$$

By Proposition 10.8.2, $P(\Omega_1) = 1$. Fixing an $\omega \in \Omega_1$, we have

$$\begin{aligned} X(t_0) &= \overline{\lim}_{u \to t_0,\ u \in T_0} X(u) - \frac{1}{2}\alpha_X(t_0) \\ &= \overline{\lim}_{u \to t_0,\ u \in T_0} \left(\overline{\lim}_{s \to u,\ s \in T_0} X(s) - \frac{1}{2}\alpha_X(u) \right) - \frac{1}{2}\alpha_X(t_0) \\ &\le -\frac{1}{2}\epsilon + \overline{\lim}_{u \to t_0,\ u \in T_0} \overline{\lim}_{s \to u,\ s \in T_0} X(s) - \frac{1}{2}\alpha_X(t_0) \\ &\le -\frac{1}{2}\epsilon + \overline{\lim}_{u \to t_0,\ u \in T_0} X(u) - \frac{1}{2}\alpha_X(t_0) \\ &= -\frac{1}{2}\epsilon + X(t_0). \end{aligned}$$

This contradiction proves our claim in the symmetric case.

In the general case $\alpha \neq 1$, let $\{X_i(t),\ t \in T\}$ be two independent copies of $\{X(t),\ t \in T\}$, and let $Y(t) = X_1(t) - X_2(t),\ t \in T$. Then $\{Y(t),\ t \in T\}$ is a $S\alpha S$ process satisfying Condition 10.6.4. Let α_X and α_Y denote the corresponding oscillation functions. Suppose firstly $\alpha_X(t_0) < \infty$ for some $t_0 \in T$. Then $\{X(t),\ t \in T\}$ is continuous in probability in a neighborhood of t_0 by Proposition 10.9.1. Proposition 10.6.7 (v) implies $\alpha_Y(t_0) \ge \alpha_X(t_0)$. If $\alpha_X(t_0) = \infty$, then $\alpha_Y(t_0) = \infty$ as well (see the proof of Theorem 9.5.4). Using Proposition 10.6.7 (iii), we conclude that, in all cases, $\alpha_X(t) \le \alpha_Y(t) \le 2\alpha_X(t)$ for every $t \in T$. Hence T_ϵ is a subset of the set $T_\epsilon^Y = \{t \in T\colon \alpha_Y(t) \in [\epsilon, \infty)\}$. Since we have proved that T_ϵ^Y is nowhere dense, the set T_ϵ is also nowhere dense, and the proof is now complete. ∎

Remark. We supposed that the process is symmetric in the case $\alpha = 1$. The above argument shows, however, that if the process is 1-stable, satisfies Condition 10.6.4 *and* is continuous in probability, then Property 10.9.2 holds as well.

10.10 A sample path alternative

Recall that the Belyaev's alternative states that the sample paths of a stationary Gaussian process are either a.s. continuous or a.s. unbounded. We now state a corresponding alternative for the sample paths of either a stationary α-stable process or a self-similar α-stable process.

Theorem 10.10.1 *Let $\{X(t),\ t \in T\}$ be an α-stable process satisfying Condition 10.6.4. If $\alpha = 1$, assume additionally that $\{X(t),\ t \in T\}$ is either symmetric or continuous in probability.*

(i) *Suppose that* (T, ρ) *is a locally compact separable metric group*[1], *and* $\{X(t),\ t \in T\}$ *is stationary. Then either* $\{X(t),\ t \in T\}$ *is sample continuous or* $\{X(t),\ t \in T\}$ *is not sample bounded in any non-empty open subset of* T.

(ii) *Suppose* $T = (0, \infty)$, *and that* $\{X(t),\ t \in (0, \infty)\}$ *is* H*-self-similar (i.e.,* $\{X(ct),\ t \in (0, \infty)\} \stackrel{d}{=} \{c^H X(t)\ t \in (0, \infty)\}$ *for every* $c > 0$*). Then the conclusion of (i) holds as well.*

PROOF: (i) The stationarity of $\{X(t),\ t \in T\}$ implies that $\{\alpha_X(t),\ t \in T\}$ is a constant function and it follows from Property 10.9.2 that either $\alpha_X(t) = 0$ for every $t \in T$, or $\alpha_X(t) = \infty$ for all $t \in T$. In the former case, $\{X(t),\ t \in T\}$ is sample continuous (Proposition 10.6.2 and Corollary 10.6.9), whereas in the latter, $\{X(t),\ t \in T\}$ cannot be sample bounded in an open ball around any given point in T (Proposition 10.6.2).

(ii) It is easy to check that the α-stable process $\{Y(t) = e^{-Ht}X(e^t),\ t \in \mathbb{R}\}$ is stationary. But $\{X(t),\ t \in (0, \infty)\}$ is sample continuous if and only if $\{Y(t),\ t \in \mathbb{R}\}$ is, and there is an open set A in $(0, \infty)$ such that $\{X(t),\ t \in A\}$ is sample bounded if and only if there is an open set $\widehat{A}$ in $\mathbb{R}$ such that $\{Y(t),\ t \in \widehat{A}\}$ is sample bounded. Part (i) now implies the result. ∎

Let us summarize our findings. We proved that any α-stable process $\{X(t),\ t \in T\}$ satisfying Condition 10.6.4 has a non-random oscillation function $\{\alpha_X(t),\ t \in T\}$ which is upper semi-continuous. If $0 < \alpha < 1$ then, this function is, in fact, a function from T to the two-point set $\{0, \infty\}$, whereas if $1 < \alpha < 2$, the function must have nowhere dense level sets of the form $\{t \in T\colon \epsilon \leq \alpha_X(t) < \infty\}$. The same is true in the case $\alpha = 1$ if $\{X(t),\ t \in T\}$ is, additionally, either symmetric or continuous in probability.

It turns out, at least in the case $T = \mathbb{R}^d$, that the oscillation function has, in general, no additional properties, i.e., the following is true:

Proposition 10.10.2 *Let* $d \geq 1$ *and let* $\{\alpha(t),\ t \in \mathbb{R}^d\}$ *be an upper semi-continuous function* $\mathbb{R}^d \to [0, \infty]$ *such that for every* $\epsilon > 0$ *the set* $\{t \in \mathbb{R}^d\colon \alpha(t) \in [\epsilon, \infty)\}$ *is nowhere dense. Then there is a* $S\alpha S$ *process* $\{X(t),\ t \in \mathbb{R}^d\}$, $1 \leq \alpha < 2$, *continuous in probability and satisfying Condition 10.6.4, whose oscillation function is precisely* $\{\alpha(t),\ t \in \mathbb{R}^d\}$. *If* $\{\alpha(t),\ t \in \mathbb{R}^d\}$ *takes values in* $\{0, \infty\}$, *then one can choose* $\{X(t),\ t \in \mathbb{R}^d\}$ *to be a* $S\alpha S$ *process with* $0 < \alpha < 1$.

The proof of Proposition 10.10.2 follows closely the corresponding proofs in the Gaussian case $\alpha = 2$ given by Itô and Nisio (1968) and Jain and Kallianpur

[1]The parameter space T of a stationary process must be a group, e.g., $T = \mathbb{R}$.

(1972). We explain in Exercise 10.23 how to adapt the Gaussian construction to the stable case.

10.11 How strong is the basic assumption?

We have been assuming that Condition 10.6.4 holds. But must any integral representation of an α-stable process $\{X(t),\ t \in T\}$ with non-random oscillation satisfy that condition? Of course, the requirement that $\{\eta(t),\ t \in T\}$ be continuous is not necessary for $\{X(t),\ t \in T\}$ to have a non-random oscillation. What about the second part of Condition 10.6.4, namely (10.6.1)?

Observe that the well-balanced symmetric linear fractional stable motion (Example 3.6.5) with $H < 1/\alpha$ is not sample bounded on any interval of positive length (see Example 10.2.5) and thus its oscillation process $\{W_X(t),\ t \in \mathbb{R}\}$ is non-random; in fact, $\alpha_X(t) = \infty$ for every $t \in \mathbb{R}$. Since, clearly, the representation (3.6.4) of the well-balanced symmetric linear fractional stable motion does not satisfy Condition 10.6.4 when $H < 1/\alpha$, we have a counterexample.

It turns out that the above counterexample points to the only way that we can have a non-random oscillation when the integral representation (10.1.1) does not satisfy the second part of Condition 10.6.4: this non-random oscillation must be infinite. More precisely, the following holds:

Theorem 10.11.1 *Let $\{X(t),\ t \in T\}$ be an α-stable process given in the form (10.1.1). If $\alpha = 1$, suppose, additionally, that $\{X(t),\ t \in T\}$ is continuous in probability. Suppose that the oscillation process $\{W_X(t),\ t \in T\}$ is non-random and finite, i.e., there is an event $\Omega_0 \in \Omega$ with $P(\Omega_0) = 0$ and a function $\alpha_X \colon T \to \mathbb{R}_+$ such that for every $\omega \in \Omega_0^c$, every $t \in T$, $W_X(t) = \alpha_X(t)$. Then for every countable dense subset T^* of T, every compact subset C of T,*

$$m\big(x\colon \text{ the function } f(\cdot, x)\colon T^* \cap C \to \mathbb{R} \text{ is not uniformly continuous}\big) = 0.$$

Remark. The conclusion of Theorem 10.11.1 holds if $\{X(t),\ t \in T\}$ is $S1S$; in this case, one does not need to assume that X is continuous in probability because continuity in probability follows from the argument in the proof of Proposition 10.9.1.

PROOF OF THEOREM 10.11.1: The assumptions of the theorem and the argument in the proof of Proposition 10.9.1 imply that $\{X(t),\ t \in T\}$ is continuous in probability. Let T^* be a countable dense subset of T and let C be a compact subset of T. Since C is compact,

$$W_X(C) = \sup_{t \in C} \alpha_X(t) =: \alpha_X(C) < \infty \text{ a.s.}$$

as α_X is, obviously, upper semi-continuous. In particular, $W_X(C)$ is non-random. Let $\{X_i(t),\ t \in T\}$, $i = 1, 2$, be two independent copies of $\{X(t),\ t \in T\}$, and let $Y(t) = X_1(t) - X_2(t)$, $t \in T$. Then $\{Y(t),\ t \in T\}$ is a $S\alpha S$ process with an integral representation

$$Y(t) = 2^{1/\alpha} \int_E f(t,x) \widetilde{M}(dx),\ t \in T,$$

where $\widetilde{M}$ is a $S\alpha S$ random measure on $(E, \mathcal{E})$ with control measure m. Note that by Proposition 10.6.7,

$$W_Y(C) \leq 2\alpha_X(C) \text{ a.s.}$$

and, in particular, $\{Y(t),\ t \in C\}$ is sample bounded. Applying Theorem 10.2.3 we obtain

$$\int_E f^*(T^* \cap C; x)^\alpha m(dx) < \infty,$$

where $f^*(T^* \cap C; x) = \sup_{t \in T^* \cap C} |f(t,x)|$. Therefore $\{Y(t),\ t \in T^* \cap C\}$ has the following integral representation:

$$Y(t) = \int_{E_+} \frac{f(t,x)}{f^*(T^* \cap C; x)} M^*(dx),\ t \in T^* \cap C,$$

where $E_+ = \{x \in E : f^*(T^* \cap C; x) \neq 0\}$ and M^* is a $S\alpha S$ random measure on $(E, \mathcal{E})$ with a *finite* control measure $m^*(dx) = 2f^*(T^* \cap C; x)^\alpha m(dx)$. Applying Theorem 3.10.1, we obtain $\{Y(t),\ t \in T^* \cap C\} \stackrel{d}{=} \{Z(t),\ t \in T^* \cap C\}$, where, in the notation of that theorem,

$$Z(t) = (C_\alpha m^*(E))^{1/\alpha} \sum_{i=1}^{\infty} \epsilon_i \Gamma_i^{-1/\alpha} \frac{f(t, V_i)}{f^*(T \cap C; V_i)},\ t \in T^* \cap C.$$

In particular,

$$\overline{\lim}_{d(t_1,t_2)\to 0,\ t_1,t_2 \in T^* \cap C} |Z(t_1) - Z(t_2)| \leq 2\alpha_X(C) \text{ a.s.} \tag{10.11.1}$$

We now use a "trick" similar to the one we applied in the proof of Theorem 10.2.3. Let $\tilde{\epsilon}_1 = \epsilon_1$, $\tilde{\epsilon}_i = -\epsilon_i$, $i \geq 2$, and define

$$\widetilde{Z}(t) = (C_\alpha m^*(E))^{1/\alpha} \sum_{i=1}^{\infty} \tilde{\epsilon}_i \Gamma_i^{-1/\alpha} \frac{f(t, V_i)}{f^*(T^* \cap C; V_i)},\ t \in T^* \cap C.$$

Then $\{\widetilde{Z}(t),\ t \in T^* \cap C\} \stackrel{d}{=} \{Z(t),\ t \in T^* \cap C\}$. If $U(t) = Z(t) + \widetilde{Z}(t)$, $t \in T^* \cap C$, then by (10.11.1),

$$\begin{aligned} &\Gamma_1^{-1/\alpha} f^*(T^* \cap C; V_1)^{-1} \overline{\lim}_{d(t_1,t_2)\to 0,\ t_1,t_2 \in T^* \cap C} |f(t_1, V_1) - f(t_2, V_1)| \\ &= \overline{\lim}_{d(t_1,t_2)\to 0,\ t_1,t_2 \in T^* \cap C} (C_\alpha m^*(E))^{-1/\alpha} |U(t_1) - U(t_2)| \\ &\leq 4(C_\alpha m^*(E))^{-1/\alpha} \alpha_X(C) \text{ a.s.} \end{aligned} \tag{10.11.2}$$

This implies

$$m^*\left(x \in E\colon\ \overline{\lim}_{d(t_1,t_2)\to 0,\ t_1,t_2\in T^*\cap C} |f(t_1,x) - f(t_2,x)| \neq 0\right) = 0. \quad (10.11.3)$$

Indeed, suppose that (10.11.3) does not hold. For any V_1 belonging to the set on the left-hand side of (10.11.3), there is a positive probability that the left-hand side of (10.11.2) exceeds the constant on its right-hand side (this happens when Γ_1 is small), and thus, by Fubini's theorem, (10.11.2) is violated.

Observe lastly that (10.11.3) is equivalent to the statement of the theorem. ∎

10.12 Exercises

Exercise 10.1 Prove that fractional Brownian motion is sample continuous.
Hint: Prove that $E(X(t) - X(s))^{2n} = \sigma^{2n}|t-s|^{2Hn}$, $n \geq 1$, and use Kolmogorov's criterion for sample continuity.

Exercise 10.2 Let $u_0 > 0$ and define $u_m = 2^{1/\alpha}u_{m-1} + u_{m-1}^\theta$, $m \geq 1$, for some $\alpha > 0$ and $\theta \in (0,1)$. Show that the limit $l = \lim_{m\to\infty} 2^{-m/\alpha}u_m$ exists and is finite.
Hint: Use $0 < 2^{-1/\alpha}u_1 \leq 2^{-2/\alpha}u_2 \leq \cdots$ to show that the limit exists and is non-negative. To prove that it is finite show that for $k \geq 1$ and m satisfying $2^{-m/\alpha}u_m > 1$,

$$\begin{aligned} 2^{-(m+k)}u_{m+k} &\leq 2^{-m/\alpha}u_m \prod_{j=1}^{k}\left(1 + 2^{-j(1-\theta)/\alpha}\right) \\ &\leq 2^{-m/\alpha}u_m \prod_{j=1}^{\infty} \exp\left\{2^{-j(1-\theta)/\alpha}\right\}. \end{aligned}$$

Exercise 10.3 Let $a_n \in (0,1)$, $n \geq 1$. Show that $\prod_{n=1}^\infty a_n > 0$ if and only if $\sum_{n=1}^\infty (1 - a_n) < \infty$.

Exercise 10.4 (i) Let $(E, \mathcal{E}, m)$ be a σ-finite measure space. Construct a probability measure P on $(E, \mathcal{E})$ equivalent to the measure m.

(ii) Give an example of a measurable space $(E, \mathcal{E})$ on which there are two equivalent measures, m and P, where P is a probability measure and m is not σ-finite.

Exercise 10.5 Prove Corollary 10.3.2.

Exercise 10.6 Prove Theorem 10.3.3.
Hint: You may find it useful to use the zero-one law of Theorem 9.5.4 (i) to prove that $\{X(t),\ t \in T\}$ is sample bounded in a neighborhood of t_0.

Exercise 10.7 (i) Let $1 \le \alpha < 2$, $\{a_n,\ n \ge 1\}$ be a sequence of positive numbers satisfying (10.4.1) and (10.4.2) and let $b_n = a_n(n(n+1))^{1/\alpha}$. Let

$$X(t) = \int_0^1 f(t,x)M(dx),\ 0 \le t \le 1,$$

where M is a $S\alpha S$ random measure on $((0,1), \mathcal{B})$ with Lebesgue control measure and define the functions $\{f(t,x),\ x \in (0,1)\}$, $0 \le t \le 1$, as follows:

Let $t = 0.d_1d_2\ldots$ be the binary expansion of a $t \in (0,1)$. For $N = 1, 2, \ldots$, define

$$T_N = \Big\{ t \in (0,1)\colon t = 0.d_1d_2\ldots,\ d_i = 0 \ \ \forall i \le N-1, \\ \text{where } d_N = 1, d_{2N+1} = 1, d_i = 0 \ \ \forall i \ge 2N+2 \Big\}.$$

For a $t \in T_N$ let $A(t) = \{i \in \{N+1, \ldots, 2N\}\colon d_i = 1\}$ and set

$$f(t,x) = \begin{cases} b_n & \text{if } x \in (\frac{1}{n+1}, \frac{1}{n}] \text{ and } n+N \in A(t), \\ -b_n & \text{if } x \in (\frac{1}{n+1}, \frac{1}{n}] \text{ and } n \le N \text{ but } n+N \notin A(t), \\ 0 & \text{otherwise.} \end{cases}$$

To complete the definition of f, set $f(0,x) = f(1,x) = 0\ \forall x \in (0,1)$ and define $f(t,x)$ for the remaining ts in $(0,1)$ by interpolating linearly between the ts in $\bigcup_{N=1}^{\infty} T_N \cup \{0,1\}$.

Show that $\{X(t),\ 0 \le t \le 1\}$ is a well-defined $S\alpha S$ process. Show also that the kernel $f(t,x)$ is strongly separable, that

$$\int_0^1 \sup_{0 \le t \le 1} |f(t,x)|^\alpha dx < \infty,$$

but $\{X(t),\ 0 \le t \le 1\}$ is not sample bounded.

Hint: For the last question, define $Z_n = \{(n(n+1))^{1/\alpha} M((\frac{1}{n+1}, \frac{1}{n}])\}$ and show, as in the text, that $\sum_{n=1}^{\infty} a_n|Z_n| = \infty$ a.s.

(ii) Of course, the $S\alpha S$ process in (i) is also not sample continuous either, but this is not a good counterexample for sample continuity because the functions $\{f(t,x),\ t \in [0,1]\}$ are not uniformly continuous on rational numbers for any $x \in (0,1)$. We can easily overcome this point. Let Z_n be defined as above and let $S_N = \sum_{n=1}^{N} a_n|Z_n|$. You showed in (i) that $S_N \uparrow \infty$ a.s. as $N \to \infty$.

Show firstly that there is a non-decreasing sequence of integers $\{k_n,\ n = 1, 2, \ldots\}$ such that $k_n \uparrow \infty$ as $n \to \infty$, and $\lim_{N\to\infty} S_N - S_{k_N} = \infty$ a.s.

For $t \in T_N$, redefine $f(t,x)$ as follows:

$$f(t,x) = \begin{cases} b_n & \text{if } x \in (\frac{1}{n+1}, \frac{1}{n}] \text{ and } k_N < n \le N \text{ and } n+N \in A(t), \\ -b_n & \text{if } x \in (\frac{1}{n+1}, \frac{1}{n}] \text{ and } k_N < n \le N \text{ for } n+N \notin A(t), \\ 0 & \text{otherwise,} \end{cases}$$

and, as before, define $f(t,x)$ for the rest of the ts by linear interpolation.

Show that the resulting kernel $f(t,x)$ is strongly separable, that $\{f(t,x),\ 0 \le t \le 1\}$ is continuous for every $x \in (0,1)$ and that

$$\int_0^1 \sup_{t\in[0,1]} |f(t,x)|^\alpha dx < \infty.$$

Show that $\{X(t),\ 0 \le t \le 1\}$ is not a.s. bounded and therefore not sample continuous.

Exercise 10.8 Give a direct argument showing that (10.4.10) implies (10.4.5). *Hint:* Define a new α-stable process $W((t,d)) = (-1)^d X(t),\ t \in T,\ d \in \{0,1\}$.

Exercise 10.9 Prove the second part of Corollary 10.4.3.

Exercise 10.10 Let f be a measurable function $\mathbb{R} \to \mathbb{R}$ satisfying

$$\text{Leb}\big(x \in \mathbb{R}\colon f(\tau + x),\ \tau \in \mathbb{Q} \text{ is not uniformly continuous}\big) = 0,$$

where $\mathbb{Q}$ denotes the set of rationals.

(i) Show that for any $x_0 \in \mathbb{R}$ there cannot exist two numbers $a > b$ such that for every $\delta > 0$,

$$\text{Leb}\big(x \in \mathbb{R}\colon f(x) \ge a,\ |x - x_0| \le \delta\big) > 0,$$

$$\text{Leb}\big(x \in \mathbb{R}\colon f(x) \le b,\ |x - x_0| \le \delta\big) > 0.$$

Hint: You may find it useful to prove first that for any two measurable sets A, B with $\text{Leb}(A) > 0$, $\text{Leb}(B) > 0$ there are points $x \in A$ and $y \in B$ such that $x - y \in \mathbb{Q}$.

(ii) For an $x_0 \in \mathbb{R}$, define

$$a(x_0) = \sup\Big\{a \in \mathbb{R}\colon \forall\delta > 0,\ \text{Leb}\big(x\colon f(x) \ge a,\ |x - x_0| \le \delta\big) > 0\Big\},$$

$$b(x_0) = \inf\Big\{b \in \mathbb{R}\colon \forall\delta > 0,\ \text{Leb}\big(x\colon f(x) \le b,\ |x - x_0| \le \delta\big) > 0\Big\}.$$

Show that $a(x_0) = b(x_0)$.

(iii) Show that the function

$$g(x) = a(x) \ \ (= b(x)), \ \ x \in \mathbb{R},$$

is continuous.

(iv) Prove that Leb $\big(x\colon g(x) \neq f(x)\big) = 0$.

Exercise 10.11 Prove Theorem 10.4.6.

Exercise 10.12 Let $W_1, W_2, \ldots$ be i.i.d. random variables such that

$$\lim_{\lambda \to \infty} \lambda^\alpha P(W_1 > \lambda) = H$$

for some $\alpha > 0$. Let N be a Poisson random variable with mean m, independent of the sequence $W_1, W_2, \ldots$. Show that

$$\lim_{\lambda \to \infty} \lambda^\alpha P\Big(\sum_{i=1}^{N} W_i > \lambda\Big) = mH.$$

Exercise 10.13 Let (T, d) be a separable metric space and let T_0 and T_1 be the respective strong separants of two strongly separable versions of a stochastic process $\{X(t),\ t \in T\}$. Set

$$W_i(t) = \overline{\lim}_{t_1, t_2 \to t,\ t_1, t_2 \in T_i} |X(t_2) - X(t_1)|,\ t \in T,\ \ i = 0, 1.$$

Show that $\{W_0(t),\ t \in T\} \stackrel{d}{=} \{W_1(t),\ t \in T\}$.

Exercise 10.14 Let $(E, \mathcal{E}, m)$ be a measure space and suppose that there is a never-vanishing function $g : E \to \mathbb{R}$ such that $g \in L^\alpha(m)$ for some $\alpha > 0$. Show that m must be σ-finite.

Hint: Let $B_0 = \{x : |g(x)| > 1\}$, $B_n = \{x : (n+1)^{-1} < |g(x)| \le n^{-1}\}$, $n = 1, 2, \ldots,$ and note that $m(B_n) < \infty$ for each n.

Exercise 10.15 Prove the rest of Proposition 10.6.7.

Exercise 10.16 Prove Relation (10.6.5).

Exercise 10.17 Prove Property 10.6.10 of the oscillation function.

Exercise 10.18 Let $V_1, V_2, \ldots$ be a sequence of i.i.d. $S1S$ (Cauchy) random variables. Use the weak law of large numbers (Theorem VII.8 II of Feller (1971)) to conclude that there are numbers $\{a_n,\ n \ge 1\}$ such that $\overline{\lim}_{n \to \infty} a_n^{-1} \sum_{i=1}^{n} |V_i| = 1$ a.s. Use this to construct, as in Example 10.8.1, a $S1S$ process $\{X(t),\ 0 \le t \le 1\}$ satisfying Condition 10.6.4 such that its oscillation function $\{\alpha(t),\ 0 \le t \le 1\}$ takes values outside the set $\{0, \infty\}$.

Exercise 10.19 Let $\{X(t),\ t \in T\}$ be an α-stable process satisfying Condition 10.6.4, and let $\{U_X(t),\ t \in T\}$ be defined by (10.8.2). Show that there is a function β: $T \to [0, \infty]$ such that for every $t \in T$, $P(U_X(t) = \beta(t)) = 1$.

Exercise 10.20 (i) Construct an example of a $S\alpha S$ process satisfying Condition 10.6.4 but not continuous in probability.

(ii) Construct an example of a $S\alpha S$ process continuous in probability which does not satisfy Condition 10.6.4.

Hint for (i): Let $T = [0, 1]$ and $m =$ Leb. Let $f(\frac{1}{n}, x) = n^{2/\alpha} 1_{(\frac{1}{n+1}, \frac{1}{n}]}(x)$, $n = 1, 2, \ldots$, $f(0, x) \equiv 0$, and define the rest of $f(t, x)$ by linear interpolation in t.
Hint for (ii): The $S\alpha S$ Lévy motion.

Exercise 10.21 Let $d \geq 2$, and let M be a $S\alpha S$ random measure on $(\mathbb{R}^d, \mathcal{B}^d)$ with d-dimensional Lebesgue control measure. Let $0 < H < 1$, and define for $t \in \mathbb{R}$

$$X(t) = \int_{\mathbb{R}^d} \left(\|\mathbf{x} - t\mathbf{1}\|^{H-d/\alpha} - \|\mathbf{x}\|^{H-d/\alpha} \right) M(d\mathbf{x}),$$

where $\mathbf{x} = (x_1, \ldots, x_d), \mathbf{1} = (1, \ldots, 1)$ and $\|\mathbf{x}\| = (\sum_{i=1}^d x_i^2)^{1/2}$.
(i) Show that $\{X(t),\ t \in \mathbb{R}\}$ is a well-defined $S\alpha S$ process, continuous in probability and satisfying Condition 10.6.4.
(ii) Show that $\{X(t),\ t \in \mathbb{R}\}$ is H-self-similar and has stationary increments.
(iii) Show that $\{X(t),\ t \in \mathbb{R}\}$ is sample continuous (and hence bounded on bounded intervals) if $H > 1/\alpha$, $\alpha > 1$. Recall that it is not sample bounded on any interval of positive length if $H \leq 1/\alpha$ (Example 10.2.5).

Exercise 10.22 Modify the construction of Example 10.8.1 to construct a $S\alpha S$ process $\{X(t),\ 0 \leq t \leq 1\}$, $1 \leq \alpha < 2$, satisfying Condition 10.6.4 and having the following property: for given $0 \leq u < v \leq 1$, $a > 0$,

(i) $X(t) = 0$ a.s., $0 \leq t \leq u$,
(ii) $\{X(t),\ u < t \leq v\}$ is sample continuous,
(iii) $\overline{\lim}_{t \downarrow u} X(t) = a$ a.s.,
(iv) $X(t) = X(v)$ a.s. for all $t \in [v, 1]$.

Exercise 10.23 Follow the proof of Theorem 4(b) of Itô and Nisio (1968) in order to construct a $S\alpha S$ process $\{X(t),\ 0 \leq t \leq 1\}$ with a given oscillation function. Go through the following steps:

(i) Given $0 < \alpha < 2$, $0 \leq u < v \leq 1$ and $\epsilon > 0$ construct a $S\alpha S$ process $\{X(t),\ 0 \leq t \leq 1\}$ satisfying Condition 10.6.4, continuous in probability, and such that

(a) $X(t) = 0$ a.s. for every $t \notin (u, v)$,
(b) $\alpha_X(t) = \infty$ for every $t \in [u, v]$,

(c) $\sigma_{X(t)} \leq \epsilon$ for every $t \in [0, 1]$.

Here $\sigma_{X(t)}$ is the scale parameter of the $S\alpha S$ random variable $X(t)$.

Hint: Use Exercise 10.21 and follow the proof of Lemma 6.1 of Itô and Nisio (1968).

(ii) Given $1 \leq \alpha < 2,\ a > 0,\ \epsilon > 0$ and $0 \leq u < v \leq 1$, construct a $S\alpha S$ process $\{X(t),\ 0 \leq t \leq 1\}$ satisfying Condition 10.6.4, continuous in probability and such that

(a) $\{X(t),\ 0 \leq t \leq 1\}$ is sample continuous,

(b) $X(t) = 0$ a.s. if $t \notin (u, v)$,

(c) $\sigma_{X(t)} \leq \epsilon$ for every $t \in [0, 1]$,

(d) $P(|\sup_{u \leq t \leq v} X(t) - a| > \epsilon) < \epsilon$.

Hint: Use Exercise 10.22 and follow the proof of Lemma 6.2 of Itô and Nisio (1968).

(iii) Let $\{\alpha_i(t),\ 0 \leq t \leq 1\}$ be two upper semi-continuous functions such that the sets $\{t \in [0, 1]\colon \alpha_i(t) > 0\}, i = 1, 2$, are nowhere dense. Let $\{X_1(t),\ 0 \leq t \leq 1\}$ be a $S\alpha S$ process, $1 \leq \alpha < 2$, continuous in probability, satisfying Condition 10.6.4, such that $\alpha_{X_1}(t) = \alpha_1(t)$ for all $t \in [0, 1]$. Show that for every $\epsilon > 0$ one can construct a $S\alpha S$ process $\{X_2(t),\ 0 \leq t \leq 1\}$, continuous in probability, satisfying Condition 10.6.4 and independent of the process $\{X_1(t),\ 0 \leq t \leq 1\}$, such that

(a) $\alpha_{X_1+X_2}(t) = \alpha_1(t) + \alpha_2(t)$ for every $t \in [0, 1]$,

(b) $\sigma_{X_2(t)} \leq \epsilon$ for every $t \in [0, 1]$,

(c) for a separable version $\{Y_2(t),\ 0 \leq t \leq 1\}$ of $\{X_2(t),\ 0 \leq t \leq 1\}$ we have

$$P\Big(\sup_{t\in[0,1]} |X_2(t)| > \sup_{t\in[0,1]} \alpha_2(t)\Big) \leq \epsilon.$$

Hint: Follow the proof of Lemma 6.3 of Itô and Nisio (1968). In the case $\alpha = 1$ you will need a slightly different argument to show that $\{X_2(t),\ 0 \leq t \leq 1\}$ is continuous in probability, and you will have to show additionally that $\{X_2(t),\ 0 \leq t \leq 1\}$ satisfies Condition 10.6.4.

(iv) Let $\{\alpha(t),\ 0 \leq t \leq 1\}$ be an upper semi-continuous function $[0, 1] \to [0, \infty]$ such that, for any $\epsilon > 0$, the set $\{t \in [0, 1]\colon \epsilon \leq \alpha(t) < \infty\}$ is nowhere dense. Prove that there is a $S\alpha S$ process $\{X(t),\ 0 \leq t \leq 1\}$, $1 \leq \alpha < 2$, continous in probability, satisfying Condition 10.6.4 and such that $\alpha_X(t) = \alpha(t)$ for all $t \in [0, 1]$.

Hint: Follow the proof of Theorem 4(b) of Itô and Nisio (1968).

(v) Let $\{\alpha(t),\ 0 \leq t \leq 1\}$ be an upper semi-continuous function $[0, 1] \to \{0, \infty\}$ and let $0 < \alpha < 1$. Construct a $S\alpha S$ process $\{X(t),\ 0 \leq t \leq 1\}$ continuous probability, satisfying Condition 10.6.4 and such that $\alpha_X(t) = \alpha(t)$ for every $t \in [0, 1]$.

Hint: The set $T_0 = \{t \in [0,1]\colon \alpha(t) = 0\}$ is open and hence is a countable union of open intervals. Write $T_0 = \bigcup_{u=1}^{\infty}(a_n, b_n)$, where (a_n, b_n) are disjoint open intervals . For $n = 1, 2, \dots,$ define

$$X(t) = \int_{a_n}^{b_n} f(t,x) M(dx), \; a_n < t < b_n,$$

in such a way that $\{X(t),\ a_n < t < b_n\}$ is sample continuous, $\lim_{t\to a_n} |X(t)| = \lim_{t\to b_n} |X(t)| = \infty$, $\sigma_{X(t)} \to 0$ as $t \to a_n,\ t \to b_n$, and for any $x \in (a_n, b_n)$, $\lim_{t\to a_n} f(t,x) = \lim_{t\to b_n} f(t,x) = 0$.

Then define $X(t)$ on the closed set $T_\infty = \{t \in (0,1)\colon \alpha(t) = \infty\}$ as follows: if t is a boundary point of T_∞, set $X(t) = 0$ a.s.. If $I = [a,b]$ is a maximal subinterval of T_∞, then define on (a,b)

$$X(t) = \int_a^b f(t,x) M(dx), \; a < t < b,$$

in such a way that $\{X(t),\ t \in (a,b)\}$ is nowhere bounded, continuous in probability, $\lim_{t\to a} \sigma_{X(t)} = \lim_{t\to b} \sigma_{X(t)} = 0$, and for every $x \in (a,b)$

$$\lim_{t\to a} f(t,x) = \lim_{t\to b} f(t,x) = 0.$$

(vi) Generalize the above construction to get $S\alpha S$ processes on $[0,1]^d$ with a given oscillation function.

Hint: Follow the proof of Theorem 4 of Jain and Kallianpur (1972).

(vii) Prove Proposition 10.10.2.

Chapter 11

Measurability, integrability and absolute continuity

In this chapter, we consider the integrability of α-stable processes. For a process to be integrable, it must be measurable, and we have already seen in Chapter 9 that not all stochastic processes have a measurable version. We give in Section 11.1 necessary and sufficient conditions for measurability in terms of the integral representation of the process.

Suppose that the α-stable process is given as $\{X(t) = \int_E f(t,x)M(dx), t \in T\}$. In Section 11.2 we consider the integrability of that process with respect to t. More precisely, we investigate whether $J_p(X) = \int_T |X(t)|^p \nu(dt)$ converges or diverges with probability 1, for a given $p > 0$ and σ-finite measure ν. Necessary and sufficient conditions for convergence are given in Section 11.3.

In Section 11.4 we determine conditions under which we can change the order of integration. Theorem 11.4.1 states that if $\int |X(t)|\nu(dt) < \infty$ a.s., then the order of integration can be changed, i.e., $\int_T X(t)\nu(dt) \stackrel{d}{=} \int_E(\int_T f(t,x)\nu(dt))M(dx)$.

In Section 11.5 we investigate properties of the random "L^p-norm" $(J_p(X))^{1/p} = (\int_T |X(t)|^p\nu(dt))^{1/p}$. The random variable $(J_p(X))^{1/p}$ is a norm on the sample paths of $\{X(t), t \in T\}$ when $p \geq 1$. Theorem 11.5.1 specifies the asymptotic behavior of $P\{(J_p(X))^{1/p} > \lambda\}$, $p > 0$, as $\lambda \to \infty$ if $J_p(X) < \infty$ a.s.

This last result is used in Section 11.6 to obtain an inversion formula for the harmonizable $S\alpha S$ process $Z(t) = \int_{-\infty}^{\infty} e^{itx}M(dx)$, $-\infty < t < \infty$. The process Z can be viewed as a Fourier transform of the *random* measure M. Theorem 11.6.1 provides an inversion formula that expresses M in terms of Z.

In Section 11.7 we consider absolute continuity. Absolute continuity implies the existence of a derivative. Let $AC^p[a,b]$ denote the set of all absolute continuous functions ϕ on the interval $[a,b]$ such that the derivative $\dot\phi$ of ϕ satisfies $\int_a^b |\dot\phi(t)|^p dt < \infty$, $p \geq 1$. Theorem 11.7.4 gives necessary and sufficient conditions for an α-stable process, $X(t) = \int_E f(t,x)M(dx) + \eta(t),\ t \in [a,b]$, to have an absolute continuous version in $L^p[a,b],\ p \geq 1$. The conditions involve the functions f and η.

11.1 Existence of a measurable version

Let $\{X(t),\ t \in T\}$ be an α-stable process on a separable metric space (T,ρ). How can we find out whether the process admits a measurable version? The following theorem gives a necessary and sufficient condition. It has a number of important corollaries which will be stated below.

Theorem 11.1.1 *Let $\{X(t),\ t \in T\}$ be an α-stable process, $0 < \alpha < 2$, with an integral representation (10.1.1). The process X has a measurable version if and only if the following conditions hold:*

(a) *There is a (jointly) measurable function g: $T \times E \to \mathbb{R}$ such that for every $t \in T$*

$$m\big(x \in E\colon f(t,x) \neq g(t,x)\big) = 0$$

(i.e., the kernel $f(t,x),\ t \in T,\ x \in E$ has a measurable modification.

(b) *There is a set $E_0 \in \mathcal{E}$ such that the restriction of the control measure m to E_0^c is σ-finite and $m(E_0 \cap \{x \in E\colon f(t,x) \neq 0\}) = 0$ for every $t \in T$ (i.e., the measure m is essentially σ-finite).*

(c) *The function η: $T \to \mathbb{R}$ in (10.1.1) is measurable.*

Moreover, if $\{X(t),\ t \in T\}$ has a measurable version, then the linear subspace of $L^\alpha(E,\mathcal{E},m)$ spanned by the functions $\{f(t,\cdot),\ t \in T\}$ is necessarily separable.

PROOF: Suppose firstly that conditions (a)-(c) of Theorem 11.1.1 hold. Let n be a probability measure on $(E,\mathcal{E})$ equivalent to the measure m restricted to E_0^c and set

$$r(x) = \left(\frac{dm(x)}{dn(x)}\right)^{1/\alpha},\ x \in E_0^c.$$

Proposition 3.5.5 and Theorem 3.10.1 imply

$$\{X(t),\ t \in T\} \stackrel{d}{=} \Big\{C\sum_{j=1}^{\infty}(\gamma_j\Gamma_j^{-1/\alpha}g(t,V_j)r(V_j) - a_j(\alpha,t)) + \eta(t),\ t \in T\Big\}, \tag{11.1.1}$$

where $C > 0$ is a constant, and, as usual, $\{(\substack{V_j\\ \gamma_j}),\ j \geq 1\}$ and $\{\Gamma_j,\ j \geq 1\}$ are independent sequences: the first consists of i.i.d. random elements in $E \times \{-1, 1\}$, and the second of arrival times of a Poisson process of unit rate; V_j has distribution n on E, $P(\gamma_j = 1|V_j) = 1 - P(\gamma_j = -1|V_j) = (1 + \beta(V_j))/2$ and

$$a_j(\alpha, t) = \begin{cases} 0, \\ \left(\int_{1/j}^{1/(j-1)} u^{-2} \sin u \, du - 2^{-1/j} \frac{2}{\pi} \ln \frac{2}{\pi}\right) \int_E g(t, x)\beta(x)m(dx), \\ \frac{\alpha}{\alpha-1}\left(j^{\frac{\alpha-1}{\alpha}} - (j-1)^{\frac{\alpha-1}{\alpha}}\right) \int_E g(t, x)\beta(x)m(dx) \end{cases} \tag{11.1.2}$$

in the cases $0 < \alpha < 1$, $\alpha = 1$ and $1 < \alpha < 2$, respectively. Recall that the series on the right-hand side of (11.1.1) converges with probability 1 for every $t \in T$. For $k = 1, 2, \ldots$, define

$$X_k(t) = C \sum_{j=1}^{k} (\gamma_j \Gamma_j^{-1/\alpha} g(t, V_j) r(V_j) - a_j(\alpha, t)) + \eta(t),\ t \in T,$$

and observe that the functions $a_j(\alpha, \cdot)\colon T \to \mathbb{R}$ are measurable under Condition (a) of the theorem. Therefore the processes $\{X_k(t),\ t \in T\},\ k = 1, 2, \ldots$, are all measurable. Let

$$Y(t, w) = \begin{cases} \lim_{k\to\infty} X_k(t, w) & \text{if the limit exists,} \\ 0 & \text{otherwise.} \end{cases}$$

Then $\{Y(t),\ t \in T\}$ is a measurable process (since it is the pointwise limit of measurable processes) and it is a version of $\{X(t),\ t \in T\}$ by virtue of (11.1.1).

We now turn to the converse and suppose that $\{X(t),\ t \in T\}$ has a measurable version. Let $\{\widetilde{X}(t),\ t \in T\}$ be an independent copy of $\{X(t),\ t \in T\}$ and let $Z(t) = 2^{-1/\alpha}(X(t) - \widetilde{X}(t)),\ t \in T$. Then $\{Z(t),\ t \in T\}$ is a $S\alpha S$ process that has a measurable version and the representation

$$\{Z(t),\ t \in T\} \stackrel{d}{=} \left\{\int_E f(t, x)\widetilde{M}(dx),\ t \in T\right\},$$

where $\widetilde{M}$ is a $S\alpha S$ random measure on $(E, \mathcal{E})$ with the same control measure m as M. Theorem 9.4.2 implies that $\{Z(t),\ t \in T\}$ satisfies Condition S and it follows from Proposition 3.5.1 that the linear subspace of $L^\alpha(E, \mathcal{E}, m)$ spanned by the functions $\{f(t, \cdot),\ t \in T\}$ is separable. Therefore the set $\{f(t, \cdot),\ t \in T\} \subseteq L^\alpha(E, \mathcal{E}, m)$ is separable as well. Let T^* be a countable subset of T such that for every $t \in T$ there is a sequence $\{t_n,\ n = 1, 2, \ldots\} \subseteq T^*$ such that $f(t_n, \cdot) \to f(t, \cdot)$ in $L^\alpha(E, \mathcal{E}, m)$ as $n \to \infty$. It is easy to verify that $E_0 = \{x \in E\colon f(t, x) \neq 0 \text{ for some } t \in T^*\}^c$ satisfies Condition (b) of Theorem 11.1.1 and

$$\{Z(t),\ t \in T\} \stackrel{d}{=} \left\{\int_{E_0^c} f(t, x)\widetilde{M}(dx),\ t \in T\right\}.$$

Let $n(dx)$ and $r(x)$ be defined as before. Theorem 3.10.1 implies that the $S\alpha S$ process

$$W(t) = \sum_{j=1}^{\infty} \epsilon_j \Gamma_j^{-1/\alpha} f(t, V_j) r(V_j),\ t \in T,$$

is (up to a multiplicative factor) a version of $\{Z(t),\ t \in T\}$ and therefore possesses a measurable version. Here, $\{\epsilon_j,\ j \geq 1\}$, $\{\Gamma_j,\ j \geq 1\}$ and $\{V_j,\ j \geq 1\}$ are independent sequences of, respectively, Rademacher random variables, arrival times of a unit rate Poisson process and i.i.d. random elements in E with common law n. Of course, the choice of the space on which these three sequences are defined is arbitrary. Let us choose this space to be $(\Omega_1 \times \Omega_2 \times \Omega_3,\ \mathcal{F}_1 \times \mathcal{F}_2 \times \mathcal{F}_3, P_1 \times P_2 \times P_3)$, such that $\{\epsilon_j,\ j \geq 1\}$ is defined on $(\Omega_1, \mathcal{F}_1, P_1)$, $\{\Gamma_j,\ j \geq 1\}$ is defined on $(\Omega_2, \mathcal{F}_2, P_2)$ and $\{V_j,\ j \geq 1\}$ is defined on $(\Omega_3, \mathcal{F}_3, P_3)$. More specifically, let $(\Omega_3, \mathcal{F}_3, P_3) = (E, \mathcal{E}, n)^{\mathbb{N}}$ so that each V_j is defined on a separate copy of $(E, \mathcal{E}, n)$, and let $V_j(x) = x,\ x \in E$.

Now define $\widetilde{\epsilon}_j = -\epsilon_j,\ j \geq 2,\ \widetilde{\epsilon}_1 = \epsilon_1$, and let

$$\widetilde{W}(t) = \sum_{j=1}^{\infty} \widetilde{\epsilon}_j \Gamma_j^{-1/\alpha} f(t, V_j) r(V_j),\ t \in T.$$

Clearly, $\{\widetilde{W}(t),\ t \in T\}$ is a version of $\{W(t),\ t \in T\}$ and, as such, has a measurable version. But, by Theorem 9.4.2,

$$2\epsilon_1 \Gamma_1^{-1/\alpha} f(t, V_1) r(V_1) = \widetilde{W}(t) + W(t),\ t \in T,$$

has a measurable version as well. It then follows from Theorem 9.4.2 that there is a (jointly) measurable stochastic process $\{U(t, \omega_1, \omega_2, x),\ t \in T\}$ defined on the probability space $(\Omega_1 \times \Omega_2 \times E, \mathcal{F}_1 \times \mathcal{F}_2 \times \mathcal{E}, P_1 \times P_2 \times n)$ such that for any $t \in T$

$$P_1 \times P_2 \times n\{(\omega_1, \omega_2, x) \colon 2\epsilon_1(\omega_1)\Gamma_1(\omega_2)^{-1/\alpha} r(x) f(t, x) \neq U(t, \omega_1, \omega_2, x)\} = 0.$$

In particular, for any $t \in T$ and n-almost every $x \in E$,

$$P_1 \times P_2\{(\omega_1, \omega_2) \colon 2\epsilon_1(\omega_1)\Gamma_1(\omega_2)^{-1/\alpha} r(x) f(t, x) \neq U(t, \omega_1, \omega_2, x)\} = 0. \tag{11.1.3}$$

Define

$$g(t, x) = \int_{\Omega_1} \int_{\Omega_2} \frac{U(t, \omega_1, \omega_2, x)}{2\epsilon_1(\omega_1)\Gamma_1(\omega_2)^{-1/\alpha} r(x)} P_1(d\omega_1) P_2(d\omega_2) \text{ if } x \in E_0^c,$$

and $g(t, x) = 0$ otherwise. Then $g(t, x)$ is a (jointly) measurable function such that for every $t \in T$

$$m\big(x \in E_0^c \colon f(t, x) \neq g(t, x)\big) = n\big(x \in E_0^c \colon f(t, x) \neq g(t, x)\big) = 0$$

and

$$m\big(x \in E_0\colon f(t,x) \neq g(t,x)\big) \leq m\big(x \in E_0\colon f(t,x) \neq 0\big) = 0.$$

Thus $g(t,x)$ satisfies Condition (1) of Theorem 11.1.1.

It follows from the already established sufficiency part of the theorem that the α-stable process $\{\overline{X}(t) = X(t) - \eta(t),\ t \in T\}$ has a measurable version. This means that $\{\eta(t) = X(t) - \overline{X}(t),\ t \in T\}$ has a measurable version and, hence, must be measurable (see Exercise 9.14). The proof of the theorem is now complete. ∎

Remark. As the preceding proof shows, the sufficiency part of the theorem continues to hold when the parameter space $(T, \mathcal{T})$ is any measurable space (not necessarily Borel).

Theorem 11.1.1 has a number of important consequences:

Corollary 11.1.2 *A strictly α-stable, $\alpha \neq 1$ or $S1S$, process $\{X(t),\ t \in T\}$ has a measurable version if and only if it admits an integral representation of the form*

$$\{X(t),\ t \in T\} \stackrel{d}{=} \Big\{\int_0^1 f(t,x) M(dx),\ t \in T\Big\}, \tag{11.1.4}$$

where M is a totally skewed to the right α-stable random measure with Lebesgue control measure and where f is jointly measurable and in L^α for each t.

This corollary is an immediate consequence of Theorem 11.1.1 and the general representation theorem for strictly stable processes in Theorem 13.2.1 given in the chapter on integral representation.

The next corollary follows easily from the proof of Theorem 11.1.1.

Corollary 11.1.3 *Let $\{X(t),\ t \in T\}$ be an α-stable process, $0 < \alpha < 2$, with an integral representation (10.1.1), and suppose that conditions (a)-(c) of Theorem 11.1.1 are satisfied. Then $\{X(t),\ t \in T\}$ has the measurable version $\{Y(t),\ t \in T\}$, where for $t \in T$,*

$$Y(t) = C_\alpha^{1/\alpha} \sum_{j=1}^{\infty} (\gamma_j \Gamma_j^{-1/\alpha} g(t, V_j) r(V_j) - a_j(\alpha, t)) + \eta(t)$$

if the series converges and $Y(t) = 0$ if the series diverges. Here,

$$r(x) = \left(\frac{dm(x)}{dn(x)}\right)^{1/\alpha},\ x \in E_0^c,$$

n is a probability measure on E equivalent to the measure m restricted to E_0^c, $\{\binom{V_j}{\gamma_j},\ j \geq 1\}$ and $\{\Gamma_j,\ j \geq 1\}$ are independent sequences of, respectively, i.i.d. random vectors in $E \times \{-1, 1\}$ and arrival times of a unit rate

Poisson process, V_j has law n,

$$P(\gamma_j = 1|V_j) \;=\; 1 - P(\gamma_j = -1|V_j) \;=\; \frac{1+\beta(V_j)}{2},$$

and C_α and $a_j(\alpha, t)$ are given, respectively, by (1.2.9) and (11.1.2).

Remark. We have discussed measurability of real-valued α-stable processes. The results apply also to the complex-valued case after some obvious modifications.

Example 11.1.4 By applying Theorem 11.1.1, we conclude immediately that the $S\alpha S$ Lévy motion (Example 3.6.1), any $S\alpha S$ moving average (Example 3.6.2) , the well-balanced $S\alpha S$ linear fractional stable motion (Example 3.6.5) and the log-fractional $S\alpha S$ stable motion (Example 3.6.6) all have measurable versions. (This is true as well for their corresponding skewed analogs.) The harmonizable $S\alpha S$ process of (6.5.1) and its real counterpart (Example 6.2.5) also have measurable versions. The sub-Gaussian and sub-stable $S\alpha S$ processes have measurable versions if and only if the underlying Gaussian or stable processes do.

11.2 Integrability of the sample paths of stable processes

Let $\{X(t),\ t \in T\}$ be a *measurable* α-stable process on a measurable space $(T, \mathcal{T})$, let ν be a non-negative σ-finite measure on $(T, \mathcal{T})$ and let p be a positive number. We want to determine whether or not the sample path integral

$$J_p(X) = \int_T |X(t)|^p \nu(dt)$$

converges.

We start with two observations. Firstly, because $\{X(t),\ t \in T\}$ is measurable, Fubini's theorem implies that $J_p(X)$ is a well-defined random variable (taking possibly the value $+\infty$). Secondly, as in the case of sample boundedness and sample continuity, there is a zero-one law for sample integrability.

Theorem 11.2.1 *Let $\{X(t),\ t \in T\}$ be a measurable α-stable process on a measurable space $(T, \mathcal{T})$, let ν be a σ-finite measure on $(T, \mathcal{T})$ and let p be a positive number. Then*

$$P\Big(\omega\colon \int_T |X(t,\omega)|^p \nu(dt) < \infty\Big) = 0 \text{ or } 1,$$

and the left-hand side depends only on the finite-dimensional distributions of $\{X(t),\ t \in T\}$.

PROOF: Let $(\Omega, \mathcal{F}, P)$ be the probability space on which the process $X = \{X(t),\ t \in T\}$ is defined, and let $\mathcal{M}$ be the family of all *measurable* stochastic processes indexed by T defined on the probability space $(\widehat{\Omega}, \widehat{\mathcal{F}}, \widehat{P}) = (\Omega, \mathcal{F}, P)^{\mathbb{N}}$ and depending only on finitely many coordinates of the product space. Say that a measurable function f on T has the property $\mathcal{P}$ if $f \in L^p(T, \mathcal{T}, \nu)$. Now apply Theorem 9.5.4. Of all its conditions, the only one that cannot be trivially verified is Condition (v). To show that it is satisfied, one needs only to prove that for $X \in \mathcal{M}$, the probability $\widehat{P}(\widehat{\omega}\colon \int_T |X(t, \widehat{\omega})|^p \nu(dt) < \infty)$ depends only on the finite-dimensional distributions of X. (Recall that $X \in \mathcal{M}$ implies that X is defined on $(\widehat{\Omega}, \widehat{\mathcal{F}}, \widehat{P})$.)

It is obviously sufficient to consider the case where ν is a probability measure. Let $\tau_1, \tau_2, \ldots$ be a sequence of i.i.d. T-valued random variables with common law ν defined on a different probability space $(\Omega_1, \mathcal{F}_1, P_1)$. Then $\int_T |X(t)|^p \nu(dt) = E_1 |X(\tau_1)|^p$. Exercise 1.24 implies

$$P_1\Big(\omega_1 \in \Omega_1\colon \lim_{n\to\infty} n^{-1/p} X(\tau_n(\omega_1), \widehat{\omega}) = 0\Big) = 1 \tag{11.2.1}$$

for any $\widehat{\omega} \in A$, where $A = \{\widehat{\omega} \in \widehat{\Omega}\colon \int_T |X(t, \widehat{\omega})|^p \nu(dt) < \infty\}$. Therefore, by Fubini's theorem,

$$\widehat{P}(A) = \widehat{P} \times P_1\Big((\widehat{\omega}, \omega_1)\colon \widehat{\omega} \in A,\ \lim_{n\to\infty} n^{1/p} X(\tau_n(\omega_1), \widehat{\omega}) = 0\Big).$$

Exercise 1.24 and Fubini's theorem also imply

$$\widehat{P} \times P_1\Big((\widehat{\omega}, \omega_1)\colon \widehat{\omega} \notin A,\ \lim_{n\to\infty} n^{-1/p} X(\tau_n(\omega_1), \widehat{\omega}) = 0\Big) = 0.$$

Therefore

$$\begin{aligned} \widehat{P}(A) &= \widehat{P} \times P_1\Big((\widehat{\omega}, \omega_1)\colon \lim_{n\to\infty} n^{-1/p} X(\tau_n(\omega_1), \widehat{\omega}) = 0\Big) \\ &= E_1\Big[\widehat{P}(\widehat{\omega} \in \widehat{\Omega}\colon \lim_{n\to\infty} n^{-1/p} X(\tau_n(\omega_1), \widehat{\omega}) = 0)\Big]. \end{aligned}$$

Now, for every fixed $\omega_1 \in \Omega_1$, the probability

$$\widehat{P}\Big(\widehat{\omega} \in \widehat{\Omega}\colon \lim_{n\to\infty} n^{-1/p} X(\tau_n(\omega_1), \widehat{\omega}) = 0\Big)$$

depends only on the finite-dimensional distributions of $\{X(t),\ t \in T\}$ and hence the same is true for $P(A)$. Therefore Theorem 9.5.4 applies and the proof is now complete. ■

We have just seen that the finite-dimensional distributions of a measurable α-stable stochastic process $\{X(t),\ t \in T\}$ determine the probability $P(\int_T |X(t)|^p \nu(dt) < \infty)$. In fact, if that probability equals 1, they determine the distribution of the random variable $\int_T |X(t)|^p \nu(dt)$ (see Exercise 11.3).

11.3 Conditions for integrability

We have seen that the sample path integral is either finite a.s. or infinite a.s. Under what conditions is it a.s. finite?

Let $\{X(t),\ t \in T\}$ be a measurable α-stable process with an integral representation

$$X(t) = \int_E f(t,x)M(dx) + \eta(t),\ t \in T,$$

i.e., it is given in the standard form (10.1.1), and assume that (T, ρ) is a separable metric space. We may and will suppose without loss of generality that the kernel $f\colon T \times E \to \mathbb{R}$ itself is jointly measurable and that conditions (b) and (c) of Theorem 11.1.1 are satisfied. Let ν be a non-negative σ-finite measure on $(T, \mathcal{T})$ where $\mathcal{T}$ is the Borel σ-algebra on T generated by the metric ρ. We shall make these assumptions throughout the remainder of this chapter unless stated otherwise.

The following proposition gives necessary integrability conditions on f for $\{X(t),\ t \in T\}$ to have integrable sample paths.

Proposition 11.3.1 *Let $\{X(t),\ t \in T\}$ be a measurable α-stable process, $0 < \alpha < 2$, with integral representation (10.1.1) and let $p > 0$. If*

$$\int_T |X(t)|^p \nu(dt) < \infty \text{ a.s.,}$$

then

$$\int_T \left(\int_E |f(t,x)|^\alpha m(dx) \right)^{p/\alpha} \nu(dt) < \infty \tag{11.3.1}$$

and

$$\int_E \left(\int_T |f(t,x)|^p \nu(dt) \right)^{\alpha/p} m(dx) < \infty. \tag{11.3.2}$$

PROOF: We may assume without loss of generality that both m and ν are probability measures. Let $(\Omega, \mathcal{F}, P)$ be the probability space on which the process $\{X(t),\ t \in T\}$ is defined, and let $\tau_1, \tau_2, \ldots$ be a sequence of i.i.d. T-valued random variables with common law ν defined on a different probability space $(\Omega_1, \mathcal{F}_1, P_1)$. Then for P-almost every $\omega \in \Omega$, $E_1|X(\tau_1,\omega)|^p < \infty$ and thus Exercise 1.24 implies that $\lim_{n\to\infty} n^{-1/p}X(\tau_n,\omega) = 0$ P_1-a.s., so that, by Fubini's theorem, for P_1-almost every choice of $\tau_1, \tau_2, \ldots,$ we have $P\{\lim_{n\to\infty} n^{-1/p}X(\tau_n) = 0\} = 1$. Appealing to Theorem 10.3.1, we conclude that for P_1-almost every choice of $\tau_1, \tau_2, \ldots,$

$$\int_E \sup_{n\geq 1} \left(n^{-1/p}|f(\tau_n,x)| \right)^\alpha m(dx) < \infty. \tag{11.3.3}$$

Let $Z_1, Z_2, \ldots$ be a sequence of i.i.d. E-valued random variables with common law m living on a still different probability space $(\Omega_2, \mathcal{F}_2, P_2)$. Exercise 1.24 and Fubini's theorem imply that (11.3.3) is equivalent to

$$\sup_{n\geq 1}\sup_{j\geq 1} n^{-1/p} j^{-1/\alpha} |f(\tau_n, Z_j)| < \infty \ P_1 \times P_2\text{-a.s.} \tag{11.3.4}$$

This is the crucial relation. To derive, say (11.3.2), use (11.3.4) and Fubini's theorem to conclude that for P_2-almost every choice of $Z_1, Z_2, \ldots$,

$$\sup_{n\geq 1} n^{-1/p} (\sup_{j\geq 1} j^{-1/\alpha} |f(\tau_n, Z_j)|) < \infty \ P_1\text{-a.s.}.$$

Therefore, by Exercise 1.24, for every such $Z_1, Z_2, \ldots$,

$$\begin{aligned} \infty &> E_1\Big(\sup_{j\geq 1}(j^{-1/\alpha}|f(\tau_n, Z_j)|)\Big)^p \\ &= \int_T \sup_{j\geq 1}\Big(j^{-p/\alpha}|f(t, Z_j)|^p\Big)\nu(dt) \\ &\geq \sup_{j\geq 1} j^{-p/\alpha}\int_T |f(t, Z_j)|^p \nu(dt). \end{aligned}$$

Applying Exercise 1.24 once again, we obtain

$$\int_E \Big(\int_T |f(t,x)|^p \nu(dt)\Big)^{\alpha/p} m(dx) = E_2\Big(\int_T |f(t, Z_1)|^p \nu(dt)\Big)^{\alpha/p} < \infty,$$

proving (11.3.2). The proof of (11.3.1) is identical. ■

Remark. It turns out that both expressions in (11.3.1) and (11.3.2) play an important role in the distribution of the integral $\int_T |X(t)|^p \nu(dt)$ when it is finite. We will return to this point in the sequel.

Let $\log_+ x = \log x$ if $x > 1$ and 0 otherwise.

Theorem 11.3.2 *Let $\{X(t),\ t \in T\}$ be a measurable α-stable process, $0 < \alpha < 2$, with an integral representation (10.1.1). If $\alpha = 1$, we assume that the random measure M is symmetric. Let $p > 0$. Then $\int_T |X(t)|^p \nu(dt) < \infty$ a.s. if and only if*

$$\int_T |\eta(t)|^p \nu(dt) < \infty \tag{11.3.5}$$

and one of the following conditions holds:

$$\int_T \Big(\int_E |f(t,x)|^\alpha m(dx)\Big)^{p/\alpha} \nu(dt) < \infty \text{ when } 0 < p < \alpha; \tag{11.3.6}$$

$$\int_T \int_E |f(t,x)|^\alpha \left[1 + \log_+ A(t,x)\right] m(dx)\nu(dt) < \infty \text{ when } p = \alpha, \quad (11.3.7)$$

where

$$A(t,x) = \frac{|f(t,x)|^\alpha \int_T \int_E |f(u,v)|^\alpha m(dv)\nu(du)}{\int_E |f(t,v)|^\alpha m(dv) \int_T |f(u,x)|^\alpha \nu(du)};$$

$$\int_E \left(\int_T |f(t,x)|^p \nu(dt) \right)^{\alpha/p} m(dx) < \infty \text{ when } p > \alpha. \quad (11.3.8)$$

PROOF: We will not give a proof in the case $p = \alpha$ because the one we know makes use of facts that fall outside the scope of this book. We refer the reader to Rosinski and Woyczyński (1986) and Kwapień and Woyczyński (1987) instead. Suppose $p \neq \alpha$, and assume firstly that $\eta(t) \equiv 0$, and M is a $S\alpha S$ random measure.

Case 1. $0 < p < \alpha$. The necessity of (11.3.6) follows from Proposition 11.3.1. On the other hand, (11.3.6) implies

$$E \int_T |X(t)|^p \nu(dt) = C_{\alpha,p} \int_T \left(\int_E |f(t,x)|^\alpha m(dx) \right)^{p/\alpha} \nu(dt) < \infty,$$

where $C_{\alpha,p}$ is a positive constant depending only on α and p. Thus $\int_T |X(t)|^p \nu(dt) < \infty$ a.s.

Case 2. $p > \alpha$. The necessity of (11.3.8) follows once again from Proposition 11.3.1. On the other hand, suppose that (11.3.8) holds. Then $f(\cdot, x) \in L^p(T, \mathcal{T}, \nu)$ for almost every $x \in E$ and (assuming once again that m is a probability measure),

$$E\|f(\cdot, Z)\|^\alpha_{L^p(T,\mathcal{T},\nu)} < \infty, \quad (11.3.9)$$

where Z is an E-valued random variable with law m. Let $Z_1, Z_2, \ldots$ be i.i.d. copies of Z. Consider

$$S(t) = \sum_{j=1}^\infty \epsilon_j \Gamma_j^{-1/\alpha} f(t, Z_j), \; t \in T,$$

where $\{\epsilon_j, j \geq 1\}$ and $\{\Gamma_j, j \geq 1\}$ are independent sequences of, respectively, Rademacher random variables and arrival times of a unit rate Poisson process, independent of the sequence $\{Z_j, j \geq 1\}$. We know that the series $S(t)$ converges a.s. for every $t \in T$, and that $\{S(t), \; t \in T\}$ is a measurable version of $\{X(t), \; t \in T\}$ up to a multiplicative constant.

It is now sufficient to show that the series

$$\mathbf{S} = \sum_{j=1}^\infty \epsilon_j \Gamma_j^{-1/\alpha} f(\cdot, Z_j)$$

is a well-defined element of $L^p(T,\mathcal{T},\nu)$, i.e., it converges a.s. in $L^p(T,\mathcal{T},\nu)$. Indeed, $\mathbf{S} \in L^p(T,\mathcal{T},\nu)$ a.s. means $\int_T |S(t)|^p \nu(dt) < \infty$ a.s. and, since $\{S(t),\ t \in T\}$ is, up to a constant, a measurable version of $\{X(t),\ t \in T\}$, we have $\int_T |X(t)|^p \nu(dt) < \infty$ a.s. as well by Theorem 11.2.1. Consider two cases:

1. *Case* $\alpha < p \leq 1$. Then

$$\int_T \Big|\sum_{j=N}^{\infty} \epsilon_j \Gamma_j^{-1/\alpha} f(t, Z_j)\Big|^p \nu(dt) \leq \int_T \sum_{j=N}^{\infty} \Gamma_j^{-p/\alpha} |f(t,Z_j)|^p \nu(dt)$$

$$= \sum_{j=N}^{\infty} \Gamma_j^{-p/\alpha} \|f(\cdot, Z_j)\|^p_{L^p(T,\mathcal{T},\nu)} \to 0 \text{ a.s. as } N \to \infty,$$

by (11.3.9) and Theorem 1.4.5.

2. *Case* $p > 1$. We can assume that the sequences $\{\Gamma_j,\ j \geq 1\}$ and $\{Z_j,\ j \geq 1\}$ are defined on a probability space $(\Omega_1, \mathcal{F}_1, P_1)$ and the sequence $\{\epsilon_j,\ j \geq 1\}$ is defined on a different probability space $(\Omega_2, \mathcal{F}_2, P_2)$. By Fubini's theorem it is enough to prove that for P_1-almost any $\omega_1 \in \Omega_1$ the series of *independent* random variables

$$\sum_{j=1}^{\infty} \epsilon_j \Gamma_j^{-1/\alpha}(\omega_1) f(\cdot, Z_j(\omega_1))$$

converges P_2-a.s. in $L^p(T,\mathcal{T},\nu)$. By the Itô-Nisio theorem (Itô and Nisio (1968)), it is enough to prove that we have L^1 convergence in $L^p(T,\mathcal{T},\nu)$, i.e.,

$$\lim_{M,N\to\infty} E_2\left(\Big\|\sum_{j=N}^{M} c_j \Gamma_j^{-1/\alpha}(\omega_1) \int(\cdot, Z_j(\omega_1))\Big\|_{L^p(T,\mathcal{T},\nu)}\right) \quad 0 \quad \Gamma_1 \text{-a.s.} \tag{11.3.10}$$

Suppose, firstly, $1 \leq p \leq 2$. Then

$$E_2\left(\Big\|\sum_{j=N}^{M} \epsilon_j \Gamma_j^{-1/\alpha}(\omega_1) f(\cdot, Z_j(\omega_1))\Big\|_{L^p(T,\mathcal{T},\nu)}\right)$$

$$= E_2\left(\int_T \Big|\sum_{j=N}^{M} \epsilon_j \Gamma_j^{-1/\alpha}(\omega_1) f(t, Z_j(\omega_1))\Big|^p \nu(dt)\right)^{1/p}$$

$$\leq \left(E_2 \int_T \Big|\sum_{j=N}^{M} \epsilon_j \Gamma_j^{-1/\alpha}(\omega_1) f(t, Z_j(\omega_1))\Big|^p \nu(dt)\right)^{1/p}$$

(Jensen's inequality, $p > 1$)

$$\leq \Big(\int_T \Big(E_2\Big|\sum_{j=N}^{M} \epsilon_j \Gamma_j^{-1/\alpha}(\omega_1) f(t, Z_j(\omega_1))\Big|^2\Big)^{p/2} \nu(dt)\Big)^{1/p}$$

(Jensen's inequality, $p \leq 2$)

$$= \Big(\int_T \Big(\sum_{j=N}^{M} \Gamma_j^{-2/\alpha}(\omega_1) f(t, Z_j(\omega_1))^2\Big)^{p/2} \nu(dt)\Big)^{1/p}$$

$$\leq \Big(\int_T \sum_{j=N}^{M} \Gamma_j^{-p/\alpha}(\omega_1) |f(t, Z_j(\omega_1))|^p \nu(dt)\Big)^{1/p} \qquad \text{(since } p \leq 2)$$

$$= \Big(\sum_{j=N}^{M} \Gamma_j^{-p/\alpha}(\omega_1) \|f(\cdot, Z_j(\omega_1))\|^p_{L^p(T,\mathcal{T},\nu)}\Big)^{1/p}$$

which tends to zero P_1-a.s. as $M, N \to \infty$, by (11.3.9) and Theorem 1.4.5.

If $p > 2$,

$$\Big[E_2\Big(\Big\|\sum_{j=N}^{M} \epsilon_j \Gamma_j^{-1/\alpha}(\omega_1) f(\cdot, Z_j(\omega_1))\Big\|_{L^p(T,\mathcal{T},\nu)}\Big)\Big]^2$$

$$\leq \Big(\int_T E_2\Big|\sum_{j=N}^{M} \epsilon_j \Gamma_j^{-1/\alpha}(\omega_1) f(t, Z_j(\omega_1))\Big|^p \nu(dt)\Big)^{2/p}$$

$$\leq \Big(c(p) \int_T \Big(\sum_{j=N}^{M} \Gamma_j^{-2/\alpha}(\omega_1) f(t, Z_j(\omega_1))^2\Big)^{p/2} \nu(dt)\Big)^{2/p}$$

(Khinchine inequality: Exercise 11.1)

$$\leq c(p)^{2/p} \sum_{j=N}^{M} \Big(\int_T \Gamma_j^{-p/\alpha}(\omega_1) |f(t, Z_j(\omega_1))|^p \nu(dt)\Big)^{2/p}$$

(Minkowski's inequality for $p/2$)

$$= c(p)^{2/p} \sum_{j=N}^{M} \Gamma_j^{-2/\alpha}(\omega_1) \|f(\cdot, Z_j(\omega_1))\|^2_{L^p(T,\mathcal{T},\nu)}$$

which tends to zero P_1-a.s. as $M, N \to \infty$, again by (11.3.9) and Theorem 1.4.5. This proves that the series **S** converges a.s. in $L^p(T, \mathcal{T}, \nu)$, and thus the proof of the theorem is complete in the symmetric case.

In the general case, but still assuming $\eta(t) \equiv 0$, let $\{X_1(t),\ t \in T\}$ and $\{X_2(t),\ t \in T\}$ be two independent measurable copies of $\{X(t),\ t \in T\}$. Then $\{Y(t) = 2^{-1/\alpha}(X_1(t) - X_2(t)),\ t \in T\}$ is $S\alpha S$ with an integral representation (10.1.1), but this time the random measure M is symmetric and has the same control measure m as before. Now our claim follows from the easily verifiable fact that $\int_T |X(t)|^p \nu(dt) < \infty$ a.s. if and only if $\int_T |Y(t)|^p \nu(dt) < \infty$ a.s. (see

Exercise 11.2). Finally, the sufficiency part of the theorem in the case $\eta(t) \equiv 0$ implies the necessity of (11.3.5). ■

Remarks

1. The preceding proof shows actually that $\int_T |X(t)|^p \nu(dt) < \infty$ a.s. if and only if (11.3.4) and (11.3.5) hold. The role of (11.3.4) is especially interesting. It is equivalent to $\int_E \sup_{n\geq 1} n^{-1/p} |f(\tau_n, x)|^\alpha m(dx) < \infty$ (Relation (11.3.3)), for P_1-almost every choice of $\tau_1, \tau_2, \ldots$ (using the notation of the proof of Proposition 11.3.1). If $\alpha \geq 1$, this does not imply $P\{\lim_{n\to\infty} n^{-1/p} X(\tau_n) = 0\} = 1$ for any choice of $\tau_1, \tau_2, \ldots$ for which (11.3.3) holds. Of course, $\int_T |X(t)|^p \nu(dt) < \infty$ implies $P(\lim_{n\to\infty} n^{-1/p} X(\tau_n) = 0) = 1$ (Exercise 1.24) only for a P_1-a.s. choice of $\tau_1, \tau_2, \ldots$. Although (11.3.3) is true for a P_1-a.s. choice of $\tau_1, \tau_2, \ldots$ as well, there may be realizations $\tau_1, \tau_2 \ldots$ for which (11.3.3) holds but $P(\lim_{n\to\infty} n^{-1/p} X(\tau_n) = 0) = 1$ does not hold. This seems, nevertheless, to suggest that, even when $\alpha \geq 1$, there are "not too many cases" in which the converse to Theorem 10.3.1 fails.

2. Although proven in the real-valued case, Theorem 11.3.2 remains obviously true for measurable complex-valued $S\alpha S$ processes.

3. In the case $\alpha = p$, one can express (11.3.7) as

$$\int_T \int_E |f(t,x)|^\alpha \left[1 + \log_+ B(t,x)\right] m(dx)\nu(dt) < \infty, \qquad (11.3.11)$$

where

$$B(t,x) = \frac{|f(t,x)|^\alpha}{\int_E |f(t,v)|^\alpha m(dv) \int_T |f(u,x)|^\alpha \nu(du)},$$

because for any constant C, there is a constant a depending only on C such that $|\log_+ Cu - \log_+ u| \leq a$ for any $u \in \mathbb{R}$.

Example 11.3.3 Shepp (1966) proved that if $\{X(t),\ t \geq 0\}$ is Brownian motion, and if $q(t),\ t \geq 1$ is a measurable function such that $\inf_{1\leq t<\tau} q(t) > 0$ for all $\tau > 1$, then, for any $0 < p < \infty$,

$$\int_1^\infty \frac{|X(t)|^p}{q(t)} < \infty \text{ a.s.}$$

if and only if

$$\int_1^\infty \frac{t^{p/2}}{q(t)} dt < \infty.$$

We can now obtain a corresponding result in the α-stable case. Let $\{X(t),\ t \geq 0\}$ be an α-stable Lévy motion, $0 < \alpha < 2$, $X(t) = \int_0^t M(dx)$ where M has Lebesgue control measure. Assume also that M is $S\alpha S$ when $\alpha = 1$. Let $q(t)$ be a non-negative function such that the set $\{t\colon q(t) = 0\}$ has zero Lebesgue measure. The measure ν with density $\nu(dt) = q(t)^{-1}dt$ is thus σ-finite. Then, for any $T \subseteq [0, \infty)$,

$$\int_T \frac{|X(t)|^p}{q(t)} dt < \infty \text{ a.s.}$$

if and only if

$$\int_T \frac{t^{p/\alpha}}{q(t)} dt < \infty \text{ when } 0 < p < \alpha,$$

$$\int_0^\infty \left(\int_{T\cap(x,\infty)} \frac{dt}{q(t)} \right)^{\alpha/p} dx < \infty \text{ when } p > \alpha$$

and,

$$\begin{cases} \int_T \frac{t}{q(t)} dt < \infty, \\ \int_T [\int_0^t \log_+ (t \int_{T\cap(x,\infty)} \frac{du}{q(u)})^{-1} dx] \frac{dt}{q(t)} < \infty, \end{cases}$$

when $p = \alpha$.

Thus, taking $q(t) = t^\eta$, we get for any $0 < \alpha < 2$ and $p > 0$,

$$\int_1^\infty \frac{|X(t)|^p}{t^\eta} dt < \infty \text{ a.s. if and only if } \eta > \frac{p}{\alpha} + 1$$

and

$$\int_0^1 \frac{|X(t)|^p}{t^\eta} dt < \infty \text{ a.s. if and only if } \eta < \frac{p}{\alpha} + 1.$$

Example 11.3.4 Let $\{X(t),\ t \in \mathbb{R}\}$ be a measurable version of the well-balanced $S\alpha S$ linear fractional stable motion of Example 3.6.5. Let

$$Y(t) = \int_0^\infty e^{-u}(X(t) - X(t+u))du,\ t \in \mathbb{R}.$$

It is clear that $\{Y(t),\ t \in \mathbb{R}\}$ is a stationary $S\alpha S$ process *whenever* the integral is defined. Theorem 11.3.2 shows that the latter is true for any $0 < H < 1$ if $\alpha \geq 1$, but only for $\frac{1}{\alpha} - 1 < H < 1$ if $\frac{1}{2} < \alpha < 1$, and for no H if $\alpha \leq \frac{1}{2}$.

Example 11.3.5 Let $\{X(t),\ t \in \mathbb{R}\}$ be a measurable version of the harmonizable $S\alpha S$ process (6.5.1). It follows from Theorem 11.3.2 and the above remark that for any $T > 0$, and $-\infty < a < b$, the integral

$$\int_{-T}^T \frac{e^{-ita} - e^{-itb}}{it} X(t)dt$$

is well defined. This suggests an a.s. inversion formula for the Fourier transform of a $S\alpha S$ noise. We will return to this point in the sequel.

11.4 Changing the order of integration

Let $\{X(t),\ t \in T\}$ be a measurable α-stable process $0 < \alpha < 2$ such that $\int_T |X(t)|\nu(dt) < \infty$ a.s. What is the integral $\int_T X(t)\nu(dt)$ equal to? Let μ be a probability measure equivalent to ν, $h = d\nu/d\mu$, and let $\tau_1, \tau_2, \ldots$ be a sequence of i.i.d. T-valued random variables with common law μ, defined on a probability space $(\Omega_1, \mathcal{F}_1, P_1)$, different from the probability space $(\Omega, \mathcal{F}, P)$ where the process $\{X(t),\ t \in T\}$ is defined. By the strong law of large numbers, for P-almost every $\omega \in \Omega$,

$$P_1\left(\lim_{n\to\infty} \frac{X(\tau_1)h(\tau_1) + \cdots + X(\tau_n)h(\tau_n)}{n} = \int_T X(t)\nu(dt)\right) = 1,$$

and thus, by Fubini's theorem, there are (non-random) points $t_1, t_2, \ldots$ in T such that

$$P\left(\lim_{n\to\infty} \frac{X(t_1)h(t_1) + \cdots + X(t_n)h(t_n)}{n} = \int_T X(t)\nu(dt)\right) = 1.$$

Since $n^{-1}(X(t_1)h(t_1) + \cdots + X(t_n)h(t_n))$ is an α-stable random variable for every $n = 1, 2, \ldots$ (Theorem 2.1.2), the integral $\int_T X(t)\nu(dt)$ is an α-stable random variable as well. What are its parameters? They are easy to find if $\{X(t),\ t \in T\}$ has an integral representation (10.1.1) *and* if we may interchange the deterministic and stochastic integrations. The following theorem, which is a Fubini-type result, justifies such a change of order of integration in the case where $\mathcal{T}$ is a Borel σ-algebra on a separable metric space (T, ρ).

Theorem 11.4.1 *Let $\{X(t),\ t \in T\}$ be a measurable α-stable process on a separable metric space (T, ρ), given by*

$$X(t) = \int_E f(t,x)M(dx) + \eta(t),\ t \in T,$$

where M is an α-stable random measure with control measure m. Assume that M is symmetric when $\alpha = 1$. If $\int_T |X(t)|\nu(dt) < \infty$ a.s., then

$$\int_T X(t)\nu(dt) = \int_E \left(\int_T f(t,x)\nu(dt)\right) M(dx) + \int_T \eta(t)\nu(dt) \text{ a.s.}$$

Thus, in particular, $\int_T f(t,\cdot)\nu(dt) \in L^\alpha(E, \mathcal{E}, m)$.

Remark. We are assuming that $\{X(t),\ t \in T\}$ is *defined* a.s. by its integral representation because the conclusion of Theorem 11.4.1 is an a.s. result. Of course, if (10.1.1) is only a distributional representation of $\{X(t),\ t \in T\}$, then the conclusion of Theorem 11.4.1 should also be understood in the sense of equality of distributions.

PROOF OF THEOREM 11.4.1. We may and will suppose $\eta(t) \equiv 0$. Suppose firstly that $\alpha \geq 1$. Assuming, as usual, and without loss of generality that ν is a probability measure, we choose a sequence of i.i.d. T-valued random variables $\tau_1, \tau_2, \ldots$ with common law ν, defined on a probability space $(\Omega_1, \mathcal{F}_1, P_1)$ different from the probability space $(\Omega, \mathcal{F}, P)$ on which the process $\{X(t),\ t \in T\}$ is defined. By the strong law of large numbers, for almost every $\omega \in \Omega$,

$$\lim_{n\to\infty} \frac{X(\tau_1) + \cdots + X(\tau_n)}{n} = \int_T X(t)\nu(dt)\ P_1\text{-a.s.}$$

Therefore, by Fubini's theorem, for P_1-almost every $\omega_1 \in \Omega_1$,

$$\int_T X(t)\nu(dt) = \lim_{n\to\infty} \frac{1}{n} \sum_{i=1}^{n} X(\tau_i(\omega_1))\ P\text{-a.s.} \tag{11.4.1}$$

Let us now consider the integral $\int_E (\int_T f(t,x)\nu(dt))M(dx)$. We may regard $\{f(\tau_j, x)\ x \in E\},\ j = 1, 2, \ldots$, as a sequence of i.i.d. random vectors taking values in a separable closed subspace S of the Banach space $L^\alpha(E, \mathcal{E}, m)$. (See Theorem 11.1.1.) The norm of these random vectors has a finite mean because

$$E_1 \big\| f(\tau_j, x),\ x \in E \big\|_{L^\alpha(E,\mathcal{E},m)} \equiv E_1 \Big(\int_E |f(\tau_j, x)|^\alpha m(dx) \Big)^{1/\alpha}$$

$$= \int_T \Big(\int_E |f(t,x)|^\alpha m(dx) \Big)^{1/\alpha} \nu(dt) < \infty$$

by Proposition 11.3.1. We can therefore apply the strong law of large numbers for S-valued random vectors to obtain

$$\Big\{ \int_T f(t,x)\nu(dt),\ x \in E \Big\} = E_1 \{ f(\tau_j, x),\ x \in E \}$$

$$= \lim_{n\to\infty} \frac{1}{n} \sum_{j=1}^{\infty} \{ f(\tau_j, x),\ x \in E \}\ P_1\text{-a.s.}$$

in S and thus in $L^\alpha(E, \mathcal{E}, m)$. In other words, for P_1-almost every $\omega_1 \in \Omega_1$,

$$\lim_{n\to\infty} \frac{1}{n} \sum_{j=1}^{n} f(\tau_j(\omega_1), \cdot) = \int_T f(t, \cdot)\nu(dt) \text{ in } L^\alpha(E, \mathcal{E}, m),$$

and thus, by Proposition 3.5.1, we obtain

$$\begin{aligned}\int_E\Big(\int_T f(t,x)\nu(dt)\Big)M(dx) &= \lim_{n\to\infty}\int_E\Big(\frac{1}{n}\sum_{j=1}^{n} f(\tau_j(\omega_1),x)\Big)M(dx)\\ &= \lim_{n\to\infty}\frac{1}{n}\sum_{j=1}^{n} X(\tau_j(\omega_1)). \qquad (11.4.2)\end{aligned}$$

Since both (11.4.1) and (11.4.2) hold on the ω_1-sets of measure 1, there is an $\omega_1 \in \Omega_1$ for which both of them hold. This completes the proof in the case $\alpha \geq 1$.

Consider now the case $0 < \alpha < 1$. Since the control measure m is essentially, σ-finite (Theorem 11.1.1), we may assume that m is actually a probability measure. We will use the following "randomization" lemma due to Kallenberg (1988).

Lemma 11.4.2 *Let S and S' be, respectively, a separable metric space and a Polish (i.e., a complete separable metric) space, and let f: $S' \to S$ be a Borel measurable function. Let $\boldsymbol{\zeta}$ and $\boldsymbol{\theta}$ be, respectively, an S- and a S'-valued random variables defined on a probability space $(\Omega, \mathcal{F}, P)$. If $\boldsymbol{\zeta} \stackrel{d}{=} f(\boldsymbol{\theta})$, then there is an S'-valued random variable $\boldsymbol{\theta}'$ on an extended probability space $(\Omega \times [0,1],\ \mathcal{F} \times \mathcal{B},\ P \times \text{Leb})$ such that $\boldsymbol{\theta}' \stackrel{d}{=} \boldsymbol{\theta}$ and $\boldsymbol{\zeta} = f(\boldsymbol{\theta}')$ $P\times$ Leb a.s.*

This lemma shows that the equality in distribution $\boldsymbol{\zeta} \stackrel{d}{=} f(\boldsymbol{\theta})$ can be realized as an a.s. equality. To complete the proof of Theorem 11.4.1, we apply Lemma 11.4.2 as follows.

Let $S = L^1(T,\mathcal{T},\nu) \times \mathbb{R}$ and $S' = \mathbb{R}^\infty \times \mathbb{R}_+^\infty \times (L^1(T,\mathcal{T},\nu))^\infty \times \mathbb{R}^\infty$. Observe that under our assumption the space $L^1(T,\mathcal{T},\nu)$ is separable (Dunford & Schwartz 1958, Vol 1, Lemma III.8.3).[1] Therefore, both S and S' are Polish spaces when endowed with product topologies. Let $\boldsymbol{\zeta} = (\zeta_1, \zeta_2)$, where $\zeta_1 = \{X(t),\ t \in T\}$ and $\zeta_2 = \int_E(\int_T f(t,x)\nu(dt))M(dx)$. Note that ζ_2 is a well-defined (real-valued) random variable because Theorem 11.3.2 implies $\int_E |\int_T f(t,x)\nu(dt)|^\alpha m(dx) < \infty$. Hence, $\boldsymbol{\zeta}$ is a well-defined S-valued random variable.

Let now $\Gamma_1, \Gamma_2, \ldots$ be a sequence of arrival times of a Poisson process with unit rate independent of the sequence of i.i.d. random vectors $\binom{V_1}{\gamma_1}, \binom{V_2}{\gamma_2}, \ldots,$ where $V_1, V_2, \ldots$ are E-valued with common law m, and

$$P(\gamma_j = 1|V_j) = 1 - P(\gamma_j = -1|V_j) = \frac{1+\beta(V_j)}{2}.$$

[1]Because T is separable, we can choose a countable dense set T_0 and approximate a function $f(\mathbf{x})$ in $L^1(T,\mathcal{T},\nu)$ by simple functions of the form $\sum_{i=1}^n a_i 1(\|\mathbf{x} - \mathbf{c}_i\| < r_i),\ n \geq 1$, where the c_is are in T_0 and the a_is and r_is are rational.

Here, β is the skewness intensity of the random measure M. Define an S'-valued random variable $\boldsymbol{\theta} = (\boldsymbol{\theta}_1, \boldsymbol{\theta}_2, \boldsymbol{\theta}_3, \boldsymbol{\theta}_4)$ by

$$\begin{aligned}
\boldsymbol{\theta}_1 &= (\gamma_1, \gamma_2, \ldots), \\
\boldsymbol{\theta}_2 &= (\Gamma_1, \Gamma_2, \ldots), \\
\boldsymbol{\theta}_3 &= (\{f(t, V_j),\ t \in T\},\ j = 1, 2, \ldots), \\
\boldsymbol{\theta}_4 &= \left(\int_T f(t, V_j)\nu(dt),\ j = 1, 2, \ldots\right).
\end{aligned}$$

Let f: $S' \to S$ be defined by $f(\mathbf{x}_1, \mathbf{x}_2, \mathbf{x}_3, \mathbf{x}_4) = (\mathbf{y}_1, y_2)$, where

$$\mathbf{y}_1 = \begin{cases} C_\alpha^{1/\alpha} \sum_{j=1}^\infty x_1(j) x_2(j)^{-1/\alpha} \mathbf{x}_3(j) & \text{if the sum converges in } L^1(T, \mathcal{T}, \nu), \\ 0 & \text{otherwise,} \end{cases}$$

$$y_2 = \begin{cases} C_\alpha^{1/\alpha} \sum_{j=1}^\infty x_1(j) x_2(j)^{-1/\alpha} x_4(j) & \text{if the sum converges,} \\ 0 & \text{otherwise,} \end{cases}$$

and C_α is given by (1.2.9). Observe that f is clearly a measurable function and, moreover, in the proof of Theorem 11.3.2, we established that the series $\sum_{j=1}^\infty \gamma_j \Gamma_j^{-1/\alpha} f(\cdot, V_j)$ converges a.s. in $L^1(T, \mathcal{T}, \nu)$. Corollary 3.10.2 and Exercise 11.3 therefore imply $\boldsymbol{\zeta} \stackrel{d}{=} f(\boldsymbol{\theta})$.

Applying Lemma 11.4.2, we conclude that there are two independent random sequences, $\Gamma_1', \Gamma_2', \ldots,$ and

$$\begin{pmatrix} \mathbf{r}_1 \\ s_1 \\ \gamma_1' \end{pmatrix}, \begin{pmatrix} \mathbf{r}_2 \\ s_2 \\ \gamma_2' \end{pmatrix}, \ldots$$

such that $(\Gamma_1', \Gamma_2', \ldots) \stackrel{d}{=} (\Gamma_1, \Gamma_2, \ldots)$; $\begin{pmatrix} \mathbf{r}_1 \\ s_1 \\ \gamma_1' \end{pmatrix}, \begin{pmatrix} \mathbf{r}_2 \\ s_2 \\ \gamma_2' \end{pmatrix}, \ldots$ are i.i.d. random vectors, the $\mathbf{r}_j$ are $L^1(T, \mathcal{T}, \nu)$-valued and

$$\begin{pmatrix} \mathbf{r}_j \\ s_j \\ \gamma_j' \end{pmatrix} \stackrel{d}{=} \begin{pmatrix} \{f(t, V_j),\ t \in T\} \\ \int_T f(t, V_j)\nu(dt) \\ \gamma_j \end{pmatrix}. \tag{11.4.3}$$

These sequences are such that

$$\{X(t),\ t \in T\} = \left\{C_\alpha^{1/\alpha} \sum_{j=1}^\infty \gamma_j' (\Gamma_j')^{-1/\alpha} r_j(t)\ t \in T\right\} \text{ a.s. in } L^1(t, \mathcal{T}, \nu) \tag{11.4.4}$$

and

$$\int_E \left(\int_T f(t,x)\nu(dt) \right) M(dx) = C_\alpha^{1/\alpha} \sum_{j=1}^{\infty} \gamma_j' (\Gamma_j')^{-1/\alpha} s_j \text{ a.s.} \tag{11.4.5}$$

Relation (11.4.4) yields

$$\int_T X(t)\nu(dt) = C_\alpha^{1/\alpha} \int_T \sum_{j=1}^{\infty} \gamma_j' (\Gamma_j')^{-1/\alpha} r_j(t)\nu(dt) \text{ a.s.} \tag{11.4.6}$$

Note that

$$\begin{aligned} \int_T \sum_{j=1}^{\infty} (\Gamma_j')^{-1/\alpha} |r_j(t)|\nu(dt) &\stackrel{d}{=} \int_T \sum_{j=1}^{\infty} \Gamma_j^{-1/\alpha} |f(t,V_j)|\nu(dt) \\ &= \sum_{j=1}^{\infty} \Gamma_j^{-1/\alpha} \int_T |f(t,V_j)|\nu(dt) < \infty \text{ a.s.} \end{aligned}$$

by Theorems 11.3.2 and 1.4.5. Applying Fubini's theorem, we obtain by (11.4.3):

$$\begin{aligned} \int_T X(t)\nu(dt) &= C_\alpha^{1/\alpha} \sum_{j=1}^{\infty} \gamma_j' (\Gamma_j')^{-1/\alpha} \int_T r_j(t)\nu(dt) \\ &= C_\alpha^{1/\alpha} \sum_{j=1}^{\infty} \gamma_j' (\Gamma_j')^{-1/\alpha} s_j \text{ a.s.} \end{aligned}$$

This, together with (11.4.5), concludes the proof of the theorem. ■

11.5 Tail behavior of the L^p-norm distribution

We consider now the asymptotic behavior of the distribution of the "L^p-norm"

$$J_p(X)^{1/p} = \left(\int_T |X(t)|^p \nu(dt) \right)^{1/p}.$$

Consider two cases, $p \geq 1$ and $0 < p < 1$. When $p \geq 1$, $(J_p(X))^{1/p}$ is a norm on the sample paths of $\{X(t),\ t \in T\}$. Corollary 6.20 of Araujo and Giné (1980) implies that, if $J_p(X)$ is finite,

$$\lim_{\lambda\to\infty} \lambda^\alpha P(J_p(X)^{1/p} > \lambda) = C_\alpha \int_E \left(\int_T |f(t,x)|^p \nu(dt) \right)^{\alpha/p} m(dx), \tag{11.5.1}$$

where C_α is given in (1.2.9). Compare this limit with condition (11.3.2) in Proposition 11.3.1.

The next theorem states that (11.5.1) remains true even when $0 < p < 1$.

Theorem 11.5.1 *Let $\{X(t),\ t \in T\}$ be a measurable α-stable process, $0 < \alpha < 2$, with an integral representation (10.1.1). If $p > 0$, then $J_p(X) < \infty$ a.s. implies (11.5.1).*

PROOF: This theorem follows from Rosinski and Samorodnitsky (1993). Here, we prove it under some slightly stronger assumptions. We shall assume that the control measure m is finite and there is an $\epsilon > 0$ such that $J_p(X') < \infty$ a.s., where

$$X'(t) = \int_E f(t,x)M'(dx) + \eta(t),\ t \in T,$$

and $f_t \in L^{\alpha+\epsilon}(E, \mathcal{E}, m)$. Here, M' is an $(\alpha+\epsilon)$-stable measure on $(E, \mathcal{E})$ with the same control measure and skewness intensity as M.

Note that $\int_T |X'(t)|^p \nu(dt) < \infty$ a.s. implies $\int_T |X(t)|^p \nu(dt) < \infty$ a.s. and hence $J_p(X) < \infty$ a.s. This is because (11.3.4) holds with α if it holds with $\alpha + \epsilon$. (The conditions (11.3.4) and (11.3.5) are necessary and sufficient for integrability, as noted in the remark following Theorem 11.3.2.) Of course, we may and will assume that m and ν are probability measures. We may also assume $\eta(t) \equiv 0$. By Corollary 3.10.2,

$$J_p(X)^{1/p} \stackrel{d}{=} J := C_\alpha^{1/\alpha}\Big(\int_T \Big|\sum_{j=1}^{\infty}(\gamma_j \Gamma_j^{-1/\alpha} f(t, V_j) - a_j(\alpha, t))\Big|^p \nu(dt)\Big)^{1/p}, \tag{11.5.2}$$

using the notation of Section 11.1. The function $a_j(\alpha, t)$, in particular, is defined in (11.1.2). Consider separately

$$W_1 = C_\alpha^{1/\alpha}\Big(\int_T |\gamma_1 \Gamma_1^{-1/\alpha} f(t, V_1) - a_1(\alpha, t)|^p \nu(dt)\Big)^{1/p} \tag{11.5.3}$$

and

$$W_2 = C_\alpha^{1/\alpha}\Big(\int_T \Big|\sum_{j=2}^{\infty}(\gamma_j \Gamma_j^{-1/\alpha} f(t, V_j) - a_j(\alpha, t))\Big|^p \nu(dt)\Big)^{1/p}. \tag{11.5.4}$$

It follows from Theorem 11.3.2 that $W_1 < \infty$ a.s., and thus $W_2 < \infty$ a.s. as well. We want to show that it is W_1 that determines the asymptotic tail behavior of $J_p(X)$. We have, by Exercise 1.22,

$$\begin{aligned}
&\lim_{\lambda\to\infty} \lambda^\alpha P(W_1 > \lambda) \\
&= \lim_{\lambda\to\infty} \lambda^\alpha P\Big(C_\alpha^{1/\alpha} \int_T \Big|\gamma_1 \Gamma_1^{-1/\alpha} f(t, V_1)\Big|^p \nu(dt)\Big)^{1/p} > \lambda\Big) \\
&= C_\alpha \lim_{\lambda\to\infty} \lambda P\Big(\Gamma_1 \le \lambda^{-1}\Big[\int_T |f(t, V_1)|^p \nu(dt)\Big]^{\alpha/p}\Big)
\end{aligned}$$

$$\begin{aligned}&= C_\alpha E\Big(\int_T |f(t,V_1)|^p \nu(dt)\Big)^{\alpha/p}\\ &= C_\alpha \int_E \Big(\int_T |f(t,x)|^p \nu(dt)\Big)^{\alpha/p} m(dx). \end{aligned} \tag{11.5.5}$$

If we prove

$$EW_2^\alpha < \infty, \tag{11.5.6}$$

then our theorem will follow from (11.5.2), (11.5.5) and (11.5.6).

Let $\{\Gamma_1^{(i)}, \Gamma_2^{(i)}, \ldots, V_1^{(i)}, V_2^{(i)}, \ldots, \gamma_1^{(i)}, \gamma_2^{(i)}, \ldots\}$, $i = 1, 2$, be two independent copies of the random variables determining W_2 and let

$$W_2^{(i)} = C_\alpha^{1/\alpha}\Big(\int_T \Big|\sum_{j=2}^\infty \gamma_j^{(i)} \Gamma_j^{(i)-1/\alpha} f(t, V_j^{(i)}) - a_j(\alpha,t)\Big|^p \nu(dt)\Big)^{1/p}, \ i = 1, 2.$$

It is clearly enough to prove

$$E\big|W_2^{(1)} - W_2^{(2)}\big|^\alpha < \infty \ \text{ if } \ p \geq 1 \tag{11.5.7}$$

and

$$E\big|(W_2^{(1)})^p - (W_2^{(2)})^p\big|^{\alpha/p} < \infty \ \text{ if } 0 < p < 1.$$

We shall treat the case $p \geq 1$. (The case $0 < p < 1$ is identical.)

Let $\epsilon_1, \epsilon_2, \ldots$ be a sequence of i.i.d. random signs independent of all the other random variables. Choose a positive integer m so large that $\alpha/pm \leq 1$ and let

$$A_{j,t} = \gamma_j^{(1)} \Gamma_j^{(1)-1/\alpha} f(t, V_j^{(1)}) - \gamma_j^{(2)} \Gamma_j^{(2)-1/\alpha} f(t, V_j^{(2)}).$$

Then

$$\begin{aligned}
&E\big|W_2^{(1)} - W_2^{(2)}\big|^\alpha \\
&\leq C_\alpha E\left(\int_T \Big|\sum_{j=2}^\infty A_{j,t}\Big|^p \nu(dt)\right)^{\alpha/p} \qquad \text{(by Minkowski, } p \geq 1) \\
&= C_\alpha E\left(\int_T \Big|\sum_{j=2}^\infty \epsilon_j A_{j,t}\Big|^p \nu(dt)\right)^{\alpha/p} \\
&\leq C_\alpha E_{\gamma,\Gamma,V}\left[E_\epsilon \left(\int_T \Big|\sum_{j=2}^\infty \epsilon_j A_{j,t}\Big|^p \nu(dt)\right)^m\right]^{\alpha/pm} \qquad \text{(Hölder, } \alpha/pm \leq 1) \\
&= C_\alpha E_{\gamma,\Gamma,V}\left[\int_{T_1\times\cdots\times T_m} E_\epsilon \prod_{k=1}^m \Big|\sum_{j=2}^\infty \epsilon_j A_{j,t_k}\Big|^p \nu(dt_1)\cdots\nu(dt_m)\right]^{\alpha/pm}
\end{aligned}$$

$$\leq C_\alpha E_{\gamma,\Gamma,V}\left[\int_{T_1\times\cdots\times T_m}\left(\prod_{k=1}^m E_\epsilon\Big|\sum_{j=2}^\infty \epsilon_j A_{j,t_k}\Big|^{pm}\right)^{1/m}\nu(dt_1)\cdots\nu(dt_m)\right]^{\alpha/pm}$$

(by Hölder)

$$\leq CE\left[\int_{T_1\times\cdots\times T_m}\prod_{k=1}^m\Big(\sum_{j=2}^\infty A_{j,t_k}^2\Big)^{p/2}\nu(dt_1)\cdots\nu(dt_m)\right]^{\alpha/pm}$$

(by Exercise 11.1)

$$= CE\left[\int_T\Big(\sum_{j=2}^\infty A_{j,t}^2\Big)^{p/2}\nu(dt)\right]^{\alpha/p}$$

$$\leq C_1E\left[\int_T\Big(\sum_{j=2}^\infty \Gamma_j^{-2/\alpha}f(t,V_j)^2\Big)^{p/2}\nu(dt)\right]^{\alpha/p},$$

where C and C_1 are positive constants.

Now, let $\{X_i'(t),\ t\in T\}$, $i=1,2$, be independent measurable copies of $\{X'(t),\ t\in T\}$. Then $\{Y(t)=2^{-1(\alpha+\epsilon)}(X_1'(t)-X_2'(t)),\ t\in T\}$, is a measurable $S(\alpha+\epsilon)S$ process with an integral representation (10.1.1) where the random measure M has the same control measure m as before, but M is $S(\alpha+\epsilon)S$. Clearly, $\int_T |Y(t)|^p\nu(dt)<\infty$ a.s. By (11.3.3),

$$\int_E \sup_{n\geq 1}(n^{-1/p}|f(\tau_n,x)|)^{\alpha+\epsilon}m(dx)<\infty \tag{11.5.8}$$

for almost every choice of i.i.d. T-valued random variables $\tau_1,\tau_2,\ldots$ with common law ν. Fix $\tau_1,\tau_2,\ldots$ for which (11.5.8) holds. Then $Eg(V_1)^{\alpha+\epsilon}<\infty$, where $g(x)=\sup_{n\geq 1}n^{-1/p}|f(\tau_n,x)|$, $x\in E$, and V_1 is as above. Therefore, letting $\epsilon_1,\epsilon_2,\ldots$ and $\Gamma_1,\Gamma_2,\ldots$ be, as above, independent sequences of i.i.d. random signs and Poisson arrivals, respectively, independent of the i.i.d. sequence $V_1,V_2,\ldots$, we conclude that

$$E\Big|\sum_{j=2}^\infty \epsilon_j\Gamma_j^{-1/\alpha}g(V_j)\Big|^\alpha<\infty.$$

Applying Khinchine's inequality yields

$$\infty > E\Big|\sum_{j=2}^\infty \epsilon_j\Gamma_j^{-1/\alpha}g(V_j)\Big|^\alpha \geq \text{const. } E\Big(\sum_{j=2}^\infty \Gamma_j^{-2/\alpha}g(V_j)^2\Big)^{\alpha/2}$$

$$\geq \text{const. } E\Big(\sup_{n\geq 1} n^{-2/p}\sum_{j=2}^\infty \Gamma_j^{-2/\alpha}f(\tau_n,V_j)^2\Big)^{\alpha/2}.$$

We conclude by Exercise 1.24 that

$$\sup_{i\geq 1} i^{-2/\alpha} \sup_{n\geq 1} n^{-2/p} \sum_{j=2}^{\infty} (\Gamma_j^{(i)})^{-2/\alpha} f(\tau_n, V_j^{(i)})^2 < \infty \text{ a.s.},$$

where $\{\Gamma_j^{(i)}, V_j^{(i)},\ j = 1, 2, \ldots\}$, $i = 1, 2, \ldots$, are i.i.d. copies of $\{\Gamma_j, V_j,\ j = 1, 2, \ldots\}$, independent of the sequence $\tau_1, \tau_2, \ldots$. By Fubini's theorem, for almost every choice of $\{\Gamma_j^{(i)}, V_j^{(i)},\ j = 1, 2, \ldots\}$, $i = 1, 2, \ldots$,

$$\sup_{n\geq 1} n^{-2/p} \left(\sup_{i\geq 1} i^{-2/\alpha} \sum_{j=2}^{\infty} \Gamma_j^{(i)-2/\alpha} f(\tau_n, V_j^{(i)})^2 \right) < \infty \text{ a.s.},$$

and thus, by Exercise 1.24,

$$\begin{aligned} \infty &> E_\tau \left(\sup_{i\geq 1} i^{-2/\alpha} \sum_{j=2}^{\infty} \Gamma_j^{(i)-2/\alpha} f(\tau_n, V_j^{(i)})^2 \right)^{p/2} \\ &\geq \sup_{i\geq 1} i^{-p/\alpha} \int_T \left(\sum_{j=2}^{\infty} \Gamma_j^{(i)-2/\alpha} f(t, V_j^{(i)})^2 \right)^{p/2} \nu(dt). \end{aligned}$$

Applying Exercise 1.24 again, we conclude that

$$E\left(\int_T \left(\sum_{j=2}^{\infty} \Gamma_j^{-2/\alpha} f(t, V_j)^2 \right)^{p/2} \nu(dt) \right)^{\alpha/p} < \infty,$$

proving (11.5.7). This completes the proof of the theorem. ■

11.6 Inversion formula for harmonizable $S\alpha S$ processes

We shall use the results of the previous section to derive an inversion formula for harmonizable $S\alpha S$ processes, $0 < \alpha < 2$.

Let M be a complex-valued $S\alpha S$ random measure on $(\mathbb{R}, \mathcal{B})$ with a finite circular control measure k and control measure m. Let

$$Z(t) = \int_{-\infty}^{\infty} e^{itx} M(dx), \quad -\infty < t < \infty,$$

and

$$X(t) = \text{Re} \int_{-\infty}^{\infty} e^{itx} M(dx), \quad -\infty < t < \infty,$$

be the complex and real $S\alpha S$ harmonizable processes defined, respectively, in (6.5.1) and (6.2.13). It is appealing to regard these two processes as Fourier transforms of the random measure M. If this view is justified, then we would expect these processes to satisfy a kind of inversion formula similar to the one for Fourier transforms of non-random measures. The next result shows that such an inversion formula does exist.

Theorem 11.6.1 *Let $\{Z(t),\ -\infty < t < \infty\}$ and $\{X(t),\ -\infty < t < \infty\}$ be the complex and the real $S\alpha S$ harmonizable processes. Then, for any $-\infty < u < v < \infty$,*

$$\frac{1}{2}M(\{u\}) + \frac{1}{2}M(\{v\}) + M((u,v)) = \text{plim}_{T\to\infty}\frac{1}{2\pi}\int_{-T}^{T}\frac{e^{-iut}-e^{-ivt}}{it}Z(t)dt \tag{11.6.1}$$

and

$$\frac{1}{2}\overline{M}(\{-u\})+\frac{1}{2}\overline{M}(\{-v\})+\overline{M}((-v,-u))+\frac{1}{2}M(\{u\})+\frac{1}{2}M(\{v\})+M((u,v))$$

$$= \text{plim}_{T\to\infty}\frac{1}{\pi}\int_{-T}^{T}\frac{e^{-iut}-e^{-ivt}}{it}X(t)dt. \tag{11.6.2}$$

Remarks

1. It is implicitly assumed that (11.6.1) and (11.6.2) involve measurable versions of the processes $\{Z(t),\ -\infty < t < \infty\}$ and $\{X(t),\ -\infty < t < \infty\}$, respectively.

2. Example 11.3.5 implies that the integrals in the right-hand sides of (11.6.1) and (11.6.2) are well defined for every $T > 0$.

PROOF OF THE THEOREM: Theorem 11.4.1 (in fact, its counterpart in the complex-valued case) implies

$$\frac{1}{2\pi}\int_{-T}^{T}\frac{e^{-iut}-e^{-ivt}}{it}Z(t)dt = \int_{-\infty}^{+\infty} f_T(x)M(dx), \tag{11.6.3}$$

where

$$f_T(x) = \frac{1}{2\pi}\int_{-T}^{T}\frac{e^{i(x-u)t}-e^{i(x-v)t}}{it}dt. \tag{11.6.4}$$

It is well known (and easy to verify) that there is a finite constant K such that $|f_T(x)| \le K$ for every $T > 0$ and $x \in \mathbb{R}$, and, moreover,

$$f(x) = \lim_{T\to\infty} f_T(x) = \frac{1}{2}1_{\{u\}}(x) + \frac{1}{2}1_{\{v\}}(x) + 1_{(u,v)}(x).$$

Proposition 6.2.3 and the bounded convergence theorem imply

$$\text{plim}_{T\to\infty}\frac{1}{2\pi}\int_{-T}^{T}\frac{e^{-iut}-e^{-ivt}}{it}Z(t)dt = \int_{-\infty}^{\infty} f(x)M(dx),$$

proving (11.6.1).

For the second part of the proof, we again apply Theorem 11.4.1 and obtain

$$\frac{1}{\pi}\int_{-T}^{T}\frac{e^{-iut}-e^{-ivt}}{it}X(t)dt$$
$$= \text{Re}\int_{-\infty}^{\infty} g_T^{(1)}(x)M(dx) - i\ \text{Im}\int_{-\infty}^{\infty} g_T^{(2)}(x)M(dx), \qquad (11.6.5)$$

where

$$g_T^{(1)}(x) = \frac{2}{\pi}\int_0^T \frac{\sin vt - \sin ut}{t}\cos xt\, dt,$$

$$g_T^{(2)}(x) = \frac{2}{\pi}\int_0^T \frac{\cos vt - \cos ut}{t}\sin xt\, dt.$$

See Exercise 11.4. It follows from the corresponding properties of the functions f_T in (11.6.4) that $g_T^{(1)}$ and $g_T^{(2)}$ are uniformly bounded (over x and T),

$$\begin{aligned} g^{(1)}(x) &= \lim_{T\to\infty} g_T^{(1)}(x) \\ &= \frac{1}{2}1_{\{-v\}}(x) + 1_{(-v,-u)}(x) + \frac{1}{2}1_{\{-u\}}(x) + \frac{1}{2}1_{\{u\}}(x) \\ &\quad + 1_{(u,v)}(x) + \frac{1}{2}1_{\{v\}}(x) \end{aligned}$$

and

$$\begin{aligned} g^{(2)}(x) &- \lim_{T\to\infty} g_T^{(2)}(x) \\ &= \frac{1}{2}1_{\{-v\}}(x) + 1_{(-v,-u)}(x) + \frac{1}{2}1_{\{-u\}}(x) - \frac{1}{2}1_{\{u\}}(x) \\ &\quad - 1_{(u,v)}(x) - \frac{1}{2}1_{\{v\}}(x). \end{aligned}$$

See Exercise 11.5. Proposition 6.2.3 and the bounded convergence theorem imply

$$\text{plim}_{T\to\infty}\frac{1}{\pi}\int_{-T}^{T}\frac{e^{-iut}-e^{-ivt}}{it}X(t)dt$$
$$= \text{Re}\int_{-\infty}^{\infty} g^{(1)}(x)M(dx) - i\ \text{Im}\int_{-\infty}^{\infty} g^{(2)}(x)M(dx).$$

This proves (11.6.2) and concludes the proof of Theorem 11.6.1 ■

Corollary 11.6.2 *Let M_1 and M_2 be complex-valued $S\alpha S$ random measures on $(\mathcal{R},\mathcal{B})$ with finite circular control measures k_1 and k_2. Let*

$$Z_j(t) = \int_{-\infty}^{\infty} e^{itx}M_j(dx),\ -\infty < t < \infty,$$

and

$$X_j(t) = \text{Re} \int_{-\infty}^{\infty} e^{itx} M_j(dx), \ -\infty < t < \infty,$$

$j = 1, 2$, *be the corresponding complex and real harmonizable processes. Then*

$$\{Z_1(t), \ -\infty < t < \infty\} \stackrel{d}{=} \{Z_2(t), \ -\infty < t < \infty\}$$

if and only if $k_1 = k_2$. *Moreover,*

$$\{X_1(t), \ -\infty < t < \infty\} \stackrel{d}{=} \{X_2(t), \ -\infty < t < \infty\}$$

if and only if $k_1 + \overline{k}_1 = k_2 + \overline{k}_2$ *on* $\mathbb{R}\backslash\{0\}$ *(where* $\overline{k}_j, \ j = 1, 2$, *is defined by (6.5.2)) and* $\int_{S_2} |s_1|^\alpha k_1(\{0\} \times ds) = \int_{S_2} |s_1|^\alpha k_2(\{0\} \times ds)$.

PROOF: Clearly, $k_1 = k_2$ implies $\{Z_1(t), \ -\infty < t < \infty\} \stackrel{d}{=} \{Z_2(t), \ -\infty < t < \infty\}$. For the converse, note that Theorem 11.6.1 implies that for any $-\infty < u < v < \infty$,

$$M_1(\{u\}) + M_1(\{v\}) + 2M_1((u, v)) \stackrel{d}{=} M_2(\{u\}) + M_2(\{v\}) + 2M_2((u, v)).$$

If u and v are continuity points of both control measures $m_j(\cdot) = k_j(\cdot \times S_2), \ j = 1, 2$, then $M_1((u, v)) \stackrel{d}{=} M_2((u, v))$, and hence

$$k_1((u, v) \times A) = k_2((u, v) \times A) \tag{11.6.6}$$

for every Borel set $A \subseteq S_2$. Since such continuity points u and v are dense in $\mathbb{R}$, we conclude that (11.6.6) holds for every $-\infty < u < v < \infty$ and Borel $A \subseteq S_2$, and thus $k_1 = k_2$.

To prove the second part, suppose firstly $k_1 + \overline{k}_1 = k_2 + \overline{k}_2$ on all of $\mathbb{R}$. Let $\overline{M}_j$ be a complex-valued $S\alpha S$ random measure on $(\mathbb{R}, \mathcal{B})$ with circular control measure $\overline{k}_j$, independent of the random measure $M_j, \ j = 1, 2$, and define

$$Y_j(t) = \text{Re} \int_{-\infty}^{\infty} e^{itx} \overline{M}_j(dx), \ -\infty < t < \infty, \ j = 1, 2.$$

It is easy to verify that $\{Y_j(t), \ -\infty < t < \infty\} \stackrel{d}{=} \{X_j(t), \ -\infty < t < \infty\}$, $j = 1, 2$. Therefore, for every $j = 1, 2$,

$$\{X_j(t), \ -\infty < t < \infty\} \stackrel{d}{=} \{2^{-1/\alpha}(X_j(t) + Y_j(t)), \ -\infty < t < \infty\}. \tag{11.6.7}$$

But, for $j = 1, 2$,

$$\{X_j(t) + Y_j(t), \ -\infty < t < \infty\} \stackrel{d}{=} \left\{\text{Re} \int_{-\infty}^{\infty} e^{itx} \widehat{M}_j(dx), \ -\infty < t < \infty\right\}, \tag{11.6.8}$$

where $\widehat{M}_j$ is a complex-valued $S\alpha S$ random measure with circular control measure $k_j + \overline{k}_j$, $j = 1, 2$. Relations (11.6.7) and (11.6.8) imply $\{X_1(t),\ -\infty < t < \infty\} \stackrel{d}{=} \{X_2(t),\ -\infty < t < \infty\}$, because $k_1 + \overline{k}_1 = k_2 + \overline{k}_2$.

If we only have $k_1 + \overline{k}_1 = k_2 + \overline{k}_2$ on $\mathbb{R}\backslash\{0\}$ and $\int_{S_2} |s_1|^\alpha k_1(\{0\} \times ds) = \int_{S_2} |s_1|^\alpha k_2(\{0\} \times ds)$, we obtain instead

$$\left\{\mathrm{Re} \int_{-\infty}^{\infty} e^{itx} 1\{x \neq 0\} M_1(dx),\ -\infty < t < \infty\right\} \stackrel{d}{=} \left\{\mathrm{Re} \int_{-\infty}^{\infty} e^{itx} 1\{x \neq 0\} M_2(dx), -\infty < t < \infty\right\} \tag{11.6.9}$$

and

$$\mathrm{Re}\ M_1(\{0\}) \stackrel{d}{=} \mathrm{Re}\ M_2(\{0\}). \tag{11.6.10}$$

But the left-hand (right-hand) side of (11.6.9) is independent of the left-hand (right-hand) side of (11.6.10) and the two sum to $\{X_1(t),\ -\infty < t < \infty\}$ and $\{X_2(t),\ -\infty < t < \infty\}$, respectively, so (11.6.9) and (11.6.10) yield also $\{X_1(t),\ -\infty < t < \infty\} \stackrel{d}{=} \{X_2(t),\ -\infty < t < \infty\}$.

To establish the converse statement, note that Theorem 11.6.1 implies that for any $-\infty < u < v < \infty$,

$$\begin{aligned}&\overline{M}_1(\{-u\}) + \overline{M}_1(\{-v\}) + 2\overline{M}_1((-v, -u)) + M_1(\{u\}) + M_1(\{v\}) \\ &\quad + 2M_1((u, v)) \\ &\stackrel{d}{=} \overline{M}_2(\{-u\}) + \overline{M}_2(\{-v\}) + 2\overline{M}_2((-v, -u)) + M_2(\{u\}) + M_2(\{v\}) \\ &\quad + 2M_2((u, v)).\end{aligned} \tag{11.6.11}$$

If $0 < u < v$ are such that the four points $u, -u, v, -v$ are continuity points of both m_1 and m_2, then

$$\overline{M}_1((-v, -u)) + M_1((u, v)) \stackrel{d}{=} \overline{M}_2((-v, -u)) + M_2((u, v)),$$

so for u, v as above,

$$(k_1 + \overline{k}_1)((u, v) \times A) = (k_2 + \overline{k}_2)((u, v) \times A) \tag{11.6.12}$$

for every Borel set $A \subseteq S_2$. Since points u, v as above are dense in $(0, \infty)$, (11.6.12) holds for every $0 < u < v$ and every Borel set $A \subseteq S_2$. Since

$$(k_j + \overline{k}_j)((-v, -u) \times A) = (k_j + \overline{k}_j)((u, v) \times A^*),$$

(see (6.5.2)), (11.6.12) extends also to $-\infty < u < v < 0$. Thus, $k_1 + \overline{k}_1 = k_2 + \overline{k}_2$ on $\mathbb{R}\backslash\{0\}$. Finally, choosing $0 = u < v$ in (11.6.11), we obtain $\mathrm{Re}\ M_1(\{0\}) \stackrel{d}{=} \mathrm{Re}\ M_2(\{0\})$, implying $\int_{S_2} |s_1|^\alpha k_1(\{0\} \times ds) = \int_{S_2} |s_1|^\alpha k_2(\{0\} \times ds)$. ∎

11.7 Absolute continuity of stable processes

We want to find conditions for a stable process $\{X(t),\ t \in [a,b]\}$ with integral representation (10.1.1) so that there is a version $\{Y(t),\ t \in [a,b]\}$ whose paths are absolutely continuous. The derivative

$$\dot{Y}(t,\omega) = \frac{dY(t,\omega)}{dt}$$

will then exist at almost every $t \in [a,b]$. We also want to ensure that for a given $p \geq 1$, $\int_a^b |\dot{Y}(t,\omega)|^p dt < \infty$ for all ω.

Let us start with a formal definition of absolute continuity.

Definition 11.7.1 Let $-\infty < a < b < \infty$ be two real numbers. A function ϕ: $[a,b] \to \mathbb{R}$ is called *absolutely continuous* (we write $\phi \in AC([a,b])$) if for every $\epsilon > 0$ there is a $\delta > 0$ such that for any family $(u_k, v_k),\ k = 1, 2, \ldots,$ of disjoint intervals in $[a,b]$,

$$\sum_{k=1}^{m} \left| v_k - u_k \right| < \delta \quad \Rightarrow \quad \sum_{k=1}^{m} \left| \phi(v_k) - \phi(u_k) \right| < \epsilon.$$

It is a well-known fact that ϕ is absolutely continuous if and only if there is a function $\dot{\phi} \in L^1[a,b]$ such that

$$\phi(t) = \phi(a) + \int_a^t \dot{\phi}(s)ds, \quad t \in [a,b]. \tag{11.7.1}$$

An absolutely continuous function is, therefore, continuous on $[a,b]$ and differentiable at almost every point $t \in [a,b]$. More precisely,

$$\frac{d\phi(t)}{dt} = \dot{\phi}(t)$$

for almost every $t \in [a,b]$.

Definition 11.7.2 $AC^p[a,b]$ is the set of all absolutely continuous functions ϕ on $[a,b]$ such that $\dot{\phi} \in L^p[a,b],\ p \geq 1$.

Observe that $AC^p[a,b]$ is a linear space. The following proposition is valid for general stochastic processes, not necessarily α-stable.

Proposition 11.7.3 *Let $\{X(t),\ a \leq t \leq b\}$ be a stochastic process that has a version with all sample paths in $AC^p[a,b],\ p \geq 1$. Let $\{Y(t),\ a \leq t \leq b\}$ be a version of $\{X(t),\ a \leq t \leq b\}$ with* continuous *sample paths, defined on a probability space $(\Omega, \mathcal{F}, P)$. Then there is an event $\Omega_0 \in \mathcal{F}$ with $P(\Omega_0) = 0$*

such that for every $\omega \in \Omega_0^c$*, the function* $Y(\cdot,\omega)\colon [a,b] \to \mathbb{R}$ *is in* $AC^p[a,b]$ *and, moreover,*

$$Y(t,\omega) = Y(a,\omega) + \int_a^t D(s,\omega)ds, \quad a \le t \le b, \tag{11.7.2}$$

for some measurable stochastic process $\{D(t),\ a \le t \le b\}$ *with all sample paths in* $L^p[a,b]$*. Moreover,*

$$D(t,\omega) = \frac{\partial}{\partial t} Y(t,\omega) \ \ (\text{Leb} \times P) - a.s.$$

PROOF: Let $\{Y_1(t),\ a \le t \le b\}$ be a version of $\{X(t),\ a \le t \le b\}$ with all sample paths in $AC^p[a,b]$, $p \ge 1$, defined on a probability space $(\Omega, \mathcal{F}, P)$. Then, every sample path of $\{Y_1(t),\ a \le t \le b\}$ is continuous and, for $j = 1$ and every $\epsilon > 0$,

$$\lim_{\delta \to 0} P\Big(\sup_{m \ge 1} \sup \sum_{i=1}^m |Y_j(v_i) - Y_j(u_i)| \ge \epsilon\Big) = 0 \tag{11.7.3}$$

where the second supremum is over all u_i, v_i rational, $i = 1, \dots, m$, such that

$$a \le u_1 < v_1 \le u_2 < v_2 \le \cdots \le u_m < v_m \le b, \text{ and } \sum_{i=1}^m |v_i - u_i| < \delta.$$

To determine whether the converse holds, let $\{Y_2(t),\ a \le t \le b\}$ be any other version of $\{X(t),\ a \le t \le b\}$ with all sample paths *continuous* (and defined, for simplicity, on the same probability space $(\Omega, \mathcal{F}, P)$). Then (11.7.3) holds with $j = 2$, and the continuity of the sample paths of $\{Y_2(t),\ a \le t \le b\}$ implies that, apart possibly from a null set, the realizations of $\{Y_2(t),\ a \le t \le b\}$ are in $AC[a,b]$. We want to show that, apart from a null set, the realization of $\{Y_2(t),\ a \le t \le b\}$ are, in fact, in $AC^p[a,b]$ as well. Define for $j = 1,2$,

$$A_j = \left\{(t,\omega) \in [a,b] \times \Omega\colon \lim_{h \to 0,\ h \in \mathbb{Q}} \frac{Y_j(t+h,\omega) - Y_j(t,\omega)}{h} \text{ does not exist}\right\}.$$

Since both $\{Y_j(t),\ a \le t \le b\}$, $j = 1,2$, have all sample paths continuous, these two processes are also measurable (see Exercise 9.15) and, therefore, A_1 and A_2 are (product) measurable. Since almost all the realizations of both $\{Y_j(t),\ a \le t \le b\}$, $j = 1,2$, are absolutely continuous, we conclude that for almost every $\omega \in \Omega$, the Lebesgue measure of the section of A_j with ω fixed is 0 for both $j = 1,2$ and, hence, by Fubini's theorem,

$$(\text{Leb} \times P)(A_j) = 0, \ \ j = 1,2.$$

Applying Fubini's theorem once again, we note that there is a Borel set $T_0 \subseteq [a,b]$ of Lebesgue measure 0 such that for every $t \notin T_0$,

$$P\Big(\lim_{h\to 0,\ h\in\mathbb{Q}} \frac{Y_j(t+h)-Y_j(t)}{h} \text{ exists}\Big) = 1, \quad j=1,2. \tag{11.7.4}$$

Now let

$$D_j(t,\omega) = \begin{cases} \lim_{h\to 0,\ h\in\mathbb{Q}} (Y_j(t+h,\omega)-Y_j(t,\omega))/h & \text{if } (t,\omega)\in A_j^c, \\ 0 & \text{otherwise.} \end{cases} \tag{11.7.5}$$

$\{D_1(t),\ a\le t\le b\}$ and $\{D_2(t),\ a\le t\le b\}$ are measurable stochastic processes and Relation (11.7.4) implies that they are versions of each other. It is also clear that, in the notation of (11.7.1), for every $\omega\in\Omega$, $D_1(t,\omega)=\dot{Y}_1(t,\omega)$ for almost every $a\le t\le b$, and so $D_1(\cdot,\omega)\in L^p[a,b]$ for every $\omega\in\Omega$. Since D_1 and D_2 are versions of each other, $D_2(\cdot,\omega)\in L^p[a,b]$ for almost every $\omega\in\Omega$, and, therefore, apart from a null set, all realizations of $\{Y_2(t),\ a\le t\le b\}$ are, indeed, in $L^p[a,b]$. ∎

We now turn to α-stable processes with integral representation (10.1.1). Unless stated otherwise, we assume in the sequel that the control measure m in the integral representation is σ-finite. Note that σ-finiteness is essentially implied by Theorem 11.1.1 since a stochastic process with a version in $AC^p[a,b]$ always has a measurable version (Exercise 9.15).

The following theorem is the main result of this section.

Theorem 11.7.4 *Let $\{X(t),\ a\le t\le b\}$ be an α-stable process $0<\alpha<2$, with an integral representation (10.1.1). If $\alpha=1$, we assume in addition that the random measure M is symmetric. Let $p\ge 1$. Then $\{X(t),\ a\le t\le b\}$ has a version with all sample paths in $AC^p[a,b]$ if and only if*

$$\eta\colon [a,b]\to\mathbb{R} \ \text{ is in } \ AC^p[a,b], \tag{11.7.6}$$

and there is a function $f^0\colon [a,b]\times E\to\mathbb{R}$ such that

(i) *$f^0(t,x)=f(t,x)$ m-almost everywhere for any $a\le t\le b$,*

(ii) *$f^0(\cdot,x)\in AC^p[a,b]$ for every $x\in E$, and, moreover,*

$$f^0(t,x)=f^0(a,x)+\int_a^t \dot{f}^0(s,x)ds, \quad a\le t\le b,$$

where

$$\dot{f}^0\colon [a,b]\times E\to\mathbb{R}$$

is a jointly measurable function satisfying

$$\int_a^b \left(\int_E |\dot f^0(t,x)|^\alpha m(dx)\right)^{p/\alpha} dt < \infty \text{ if } 1 \le p < \alpha, \tag{11.7.7}$$

$$\int_a^b \int_E |\dot f^0(t,x)|^\alpha \left[1 + \log_+ A(t,x)\right] m(dx)du < \infty \text{ if } p = \alpha, \tag{11.7.8}$$

where

$$A(t,x) = \frac{|\dot f^0(t,x)|^\alpha \int_a^b \int_E |\dot f^0(u,v)|^\alpha m(dv)du}{\int_E |\dot f^0(t,v)|^\alpha m(dv) \int_a^b |\dot f^0(u,x)|^\alpha du}$$

and

$$\int_E \left(\int_a^b |\dot f^0(t,x)|^p dt\right)^{\alpha/p} m(dx) < \infty \text{ if } p > \alpha > 0. \tag{11.7.9}$$

PROOF: Suppose firstly that $\{X(t),\ a \le t \le b\}$ has a version with all sample paths in $AC^p[a,b]$. We start with the case when the random measure M is $S\alpha S$ and $\eta \equiv 0$ in (10.1.1). Since $\{X(t),\ a \le t \le b\}$ is sample continuous, it can be written in the form

$$X(t) = \int_{E_1} h(t,x) M(dx) \text{ a.s.}, \quad a \le t \le b, \tag{11.7.10}$$

where E_1 is a measurable subset of E such that the control measure m is σ-finite on E_1, and where $h\colon [a,b] \times E_1 \to \mathbb{R}$ satisfies:

(a) For every $x \in E_1$,

$$\text{the function } h(\cdot, x)\colon [a,b] \to \mathbb{R} \text{ is continuous}, \tag{11.7.11}$$

(b) for every $t \in [a,b]$, $m\big(x \in E_1\colon h(t,x) \ne f(t,x)\big) = 0$,

(c) for every $t \in [a,b]$,

$$m\big(x \in E \backslash E_1\colon f(t,x) \ne 0\big). \tag{11.7.12}$$

(See the discussion after Corollary 10.3.2.)

Let λ be a probability measure on $(E_1, \mathcal{E} \cap E_1)$ equivalent to the control measure m on this space. Then, by Theorem 3.10.1,

$$Z(t) = d_\alpha \sum_{i=1}^\infty \epsilon_i \Gamma_i^{-1/\alpha} h(t, V_i) g(V_i)^{1/\alpha}, \quad a \le t \le b,$$

is a version of $\{X(t),\ a \le t \le b\}$, where $g = dm/d\lambda$ and, as usual, d_α is a constant, and the three independent series of random variables are a Rademacher sequence $\{\epsilon_1, \epsilon_2, \ldots\}$, a sequence of arrival times of a Poisson process with unit arrival rate $\{\Gamma_1, \Gamma_2, \ldots\}$ and a sequence of i.i.d. E_1-valued random variables $\{V_1, V_2, \ldots\}$ with common law λ. Replacing ϵ_i by $-\epsilon_i$ for $i = 2, 3, \ldots$ in the series defining $\{Z(t),\ a \le t \le b\}$, we obtain another version of $\{X(t),\ a \le t \le b\}$, say $\{\widehat{Z}(t),\ a \le t \le b\}$. Both $\{Z(t),\ a \le t \le b\}$ and $\{\widehat{Z}(t),\ a \le t \le b\}$ have versions with all sample paths in $AC^p[a, b]$, and so does their sum,

$$Z(t) + \widehat{Z}(t) = 2d_\alpha \epsilon_1 \Gamma_1^{-1/\alpha} g(V_1)^{1/\alpha} h(t, V_1), \quad a \le t \le b,$$

by Proposition 11.7.3. It follows from (11.7.11) and Proposition 11.7.3 that there is a measurable set $E_2 \subseteq E_1$ with $m(E_1 \backslash E_2) = \lambda(E_1 \backslash E_2) = 0$ such that

$$\text{the function } h(\cdot, x)\colon [a, b] \to \mathbb{R} \text{ is in } AC^p[a, b]\ \forall x \in E_2. \tag{11.7.13}$$

Since (11.7.11)–(11.7.13) hold with E_1 replaced by E_2, we will assume, for notational ease, that $E_2 = E_1$.

Let $\{Y(t),\ a \le t \le b\}$ be a version of $\{X(t),\ a \le t \le b\}$ satisfying (11.7.2), chosen in such a way that $X(t) = Y(t)$ a.s. for every $t \in [a, b]$. It follows from Proposition 11.7.3 that there is a Borel subset T_0 of $[a, b]$ with $\text{Leb}(T_0) = 0$ such that for every $t \notin T_0$, $D(t) = Y'(t)$ a.s. Therefore the stochastic process

$$\widetilde{D}(t) = \begin{cases} D(t) & \text{if } t \notin T_0, \\ 0 & \text{if } t \in T_0, \end{cases}$$

is a measurable $S\alpha S$ process. To determine its integral representation, observe that for every $t_0 \notin T_0$,

$$D(t) = \int_{E_1} \dot{h}(t, x) M(dx) \text{ a.s.}, \tag{11.7.14}$$

where

$$\dot{h}(t, \cdot) = \lim_{\delta \to \infty} \frac{h(t + \delta, \cdot) - h(t, \cdot)}{\delta} \quad \text{in } L^\alpha(E_1, m).$$

Note that according to Theorem 11.1.1, the function $\dot{h}\colon [a, b] \times E_1 \to \mathbb{R}$ (defined $\equiv 0$ for $t \in T_0$) can be chosen to be (jointly) measurable. Also note that for any $t \notin T_0$, there is a sequence $\delta_n \to 0$ such that

$$\dot{h}(t, x) = \lim_{n \to \infty} \frac{h(t + \delta_n, x) - h(t, x)}{\delta_n}$$

for m-almost every $x \in E_1$. In view of (11.7.13), there is a measurable set $E_3 \subseteq E_1$ with $m(E_1 \backslash E_3) = 0$ such that for every $x \in E_3$, $\dot{h}(\cdot, x)\colon [a, b] \to \mathbb{R}$ is in $L^p[a, b]$, and

$$h(t, x) = h(a, x) + \int_a^t \dot{h}(s, x) ds, \quad a \le t \le b.$$

By applying Theorem 11.3.2 to $\{D(t),\ a \le t \le b\}$, which has all its sample paths in $L^p[a, b]$ and setting

$$f^0(t, x) = \begin{cases} h(t, x) & \text{if } x \in E_3, \\ 0 & \text{if } x \notin E_3 \end{cases}$$

and

$$\dot{f}(t, x) = \begin{cases} \dot{h}(t, x) & \text{if } x \in E_3, \\ 0 & \text{if } x \notin E_3, \end{cases}$$

we conclude the proof of the necessity part of the theorem in the $S\alpha S$ case.

The general case follows easily. If $\{X_i(t),\ a \le t \le b\}$, $i = 1, 2$, are independent copies of $\{X(t),\ a \le t \le b\}$, then $\{X_1(t) - X_2(t),\ a \le t \le b\}$ is a $S\alpha S$ stochastic process with a version in $AC^p[a, b]$ and integral representation

$$X_1(t) - X_2(t) = 2^{1/\alpha} \int_E f(t, x) \widetilde{M}(dx), \quad a \le t \le b,$$

where $\widetilde{M}$ is a $S\alpha S$ random measure on $(E, \mathcal{E})$ with control measure m. The necessity part of the theorem (except (11.7.6)) follows from the $S\alpha S$ case considered above. Relation (11.7.6) follows from the sufficiency part of the theorem, which we shall now establish.

To prove sufficiency, suppose that (11.7.6) and conditions (i) and (ii) of the theorem hold. Condition (ii) implies that there is a Borel set $T_0 \subset [a, b]$ of Lebesgue measure 0 such that for every $t \notin T_0$, $\dot{f}^0(t, \cdot) \in L^\alpha(m)$. Define

$$D_1(t) = \begin{cases} \int_E \dot{f}^0(t, x) M(dx) & \text{if } t \notin T_0, \\ 0 & \text{if } t \in T_0. \end{cases}$$

Then $\{D_1(t),\ a \le t \le b\}$ is an α-stable process, and it follows from Theorem 11.3.2 that $\{D_1(t),\ a \le t \le b\}$ has a version $\{D_2(t),\ a \le t \le b\}$ with all sample paths in $L^p[a, b]$. We know from Sections 11.1 and 11.2 that the version $\{D_2(t),\ a \le t \le b\}$ can be chosen such that

$$D_2(t) = \begin{cases} \int_E \dot{f}^0(t, x) M(dx), & \text{if } t \notin T_0, \\ 0 & \text{if } t \in T_0 \end{cases}$$

a.s. for every $t \in [a, b]$. Now let

$$\widetilde{X}(t) = \int_E f^0(a, x)M(dx) + \int_a^t D_2(s)ds, \quad a \le t \le b.$$

Theorem 11.4.1 implies that $\{\widetilde{X}(t), \ a \le t \le b\}$ is a well-defined stochastic process and for any $t \in [a, b]$,

$$\begin{aligned} \widetilde{X}(t) &= \int_E \Big[f^0(a, x) + \int_a^t \dot{f}^0(s, x)ds\Big] M(dx) \\ &= \int_E f^0(t, x)M(dx) \end{aligned}$$

a.s. Obviously, $\{\widetilde{X}(t), \ a \le t \le b\}$ has all sample paths in $AC^p[a, b]$ and therefore $\{\widetilde{X}(t) + \eta(t), \ a \le t \le b\}$ is a version of $\{X(t), \ a \le t \le b\}$ with all sample paths in $AC^p[a, b]$. This completes the proof of the theorem. ∎

The result of Theorem 11.7.4 extends to the complex-valued case, as the following corollary indicates.

Corollary 11.7.5 *Let* $\{X(t), \ a \le t \le b\}$ *be a complex-valued* $S\alpha S$ *process given in the form*

$$X(t) = \int_E f(t, x)M(dx), \quad a \le t \le b,$$

where M *is a complex-valued* $S\alpha S$ *random measure on* $(E, \mathcal{E})$ *with control measure* m *and circular control measure* k, *and* $f(t, \cdot) \in L^\alpha(m)$ *for any* $t \in [a, b]$. *Let* $p \ge 1$. *Then* $\{X(t), \ a \le t \le b\}$ *has a version with sample paths in* $AC^p[a, b]$ *if and only if there is a function* f^0: $[a, b] \times E \to C$ *such that (i) and (ii) of Theorem 11.7.4 hold.*

PROOF: Let $X_1(t) = \text{Re } X(t)$, $X_2(t) = \text{Im } X(t)$, $a \le t \le b$. Then $\{X_1(t), \ a \le t \le b\}$ and $\{X_2(t), \ a \le t \le b\}$ are real-valued $S\alpha S$ processes, and $\{X(t), \ a \le t \le b\}$ has a version with sample paths in $AC^p[a, b]$ if and only if both $\{X_1(t), \ a \le t \le b\}$ and $\{X_2(t), \ a \le t \le b\}$ do. It follows from Proposition 6.2.1 that $\{X_1(t), \ a \le t \le b\}$ and $\{X_2(t), \ a \le t \le b\}$ have the following integral representations (in the sense of joint finite-dimensional distributions):

$$X_1(t) = \int_{E \times S_2} \Big(s_1 f^{(1)}(t, x) - s_2 f^{(2)}(t, x)\Big) \widetilde{M}(ds, dx), \quad a \le t \le b, \quad (11.7.15)$$

$$X_2(t) = \int_{E \times S_2} \Big(s_1 f^{(2)}(t, x) + s_2 f^{(1)}(t, x)\Big) \widetilde{M}(ds, dx), \quad a \le t \le b,$$

where $f^{(1)}(t,x) = \text{Re } f(t,x)$, $f^{(2)}(t,x) = \text{Im } f(t,x)$, and $\widetilde{M}$ is a (real) $S\alpha S$ random measure on $(E \times S_2,\ \mathcal{E} \times \mathcal{B})$ with control measure k. In the sequel, we suppose, without loss of generality, that k (and therefore m) is a probability measure.

Suppose firstly that $\{X(t),\ a \le t \le b\}$ has a version with all sample paths in $AC^p[a,b]$. Then, both $\{X_1(t),\ a \le t \le b\}$ and $\{X_2(t),\ a \le t \le b\}$ so do. Applying Theorem 11.7.4 to the processes $\{X_1(t),\ a \le t \le b\}$ and $\{X_2(t),\ a \le t \le b\}$, we conclude that there are functions $f^{0(i)}\colon [a,b] \times E \to \mathbb{R},\ i = 1,2$, such that $f^{0(i)}(t,x) = f^{(i)}(t,x)$ m-almost everywhere for every $i = 1,2$ and $a \le t \le b$, $f^{0(i)}(\cdot,x) \in AC^p[a,b]$ for every $i = 1,2$ and $x \in E$, and

$$f^{0(i)}(t,x) = f^{0(i)}(a,x) + \int_a^t \dot{f}^{0(i)}(s,x)ds, \quad i = 1,2,\ a \le t \le b,$$

for jointly measurable functions $\dot{f}^{0(i)}\colon [a,b] \times E \to \mathbb{R},\ i = 1,2$. Moreover, the functions $\dot{f}^{0(i)}$ are such that the $S\alpha S$ processes

$$D_1(t) = \int_{E\times S_2} \left(s_1 \dot{f}^{0(1)}(t,x) - s_2 \dot{f}^{0(2)}(t,x)\right) \widetilde{M}(ds, dx), \quad a \le t \le b,$$

and

$$D_2(t) = \int_{E\times S_2} \left(s_1 \dot{f}^{0(2)}(t,x) + s_2 \dot{f}^{0(1)}(t,x)\right) \widetilde{M}(ds, dx), \quad a \le t \le b,$$

have almost all sample paths in $L^p[a,b]$. Let $(\tau_1, \tau_2, \ldots)$ be a sequence of i.i.d. random variables uniformly distributed on (a,b) and $((Z_1, \mathbf{A}_1), (Z_2, \mathbf{A}_2), \ldots)$ be a sequence of i.i.d. $E \times S_2$-valued random vectors with common distribution k, independent of $(\tau_1, \tau_2, \ldots)$. By (11.3.4),

$$\sup_{n\ge1}\sup_{j\ge1} n^{-1/p} j^{-1/\alpha} \left| A_j^{(1)} \dot{f}^{0(1)}(\tau_n, Z_j) - A_j^{(2)} \dot{f}^{0(2)}(\tau_n, Z_j) \right| < \infty \text{ a.s.,} \tag{11.7.16}$$

$$\sup_{n\ge1}\sup_{j\ge1} n^{-1/p} j^{-1/\alpha} \left| A_j^{(1)} \dot{f}^{0(2)}(\tau_n, Z_j) + A_j^{(2)} \dot{f}^{0(1)}(\tau_n, Z_j) \right| < \infty \text{ a.s.,} \tag{11.7.17}$$

where $\mathbf{A}_j = (A_j^{(1)}, A_j^{(2)}),\ j = 1,2,\ldots$. Relations (11.7.16) and (11.7.17) imply

$$\sup_{n\ge1}\sup_{j\ge1} n^{-2/p} j^{-2/\alpha} \left(\dot{f}^{0(1)}(\tau_n, Z_j)^2 + \dot{f}^{0(2)}(\tau_n, Z_j)^2 \right) < \infty \text{ a.s.,}$$

so that

$$\sup_{n\ge1}\sup_{j\ge1} n^{-1/p} j^{-1/\alpha} |\dot{f}^{0(i)}(\tau_n, Z_j)| < \infty \text{ a.s. for } i = 1,2.$$

The $S\alpha S$ processes

$$W_i(t) = \int_E \dot{f}^{0(i)}(t,x)M(dx),\ a \le t \le b,\ i = 1,2,$$

have, therefore, almost all their sample paths in $L^p[a,b]$ (see the remark following Theorem 11.3.2). By Theorem 11.3.2 both $\dot{f}^{0(i)}$, $i = 1,2$, satisfy (11.7.7)–(11.7.9). Then setting $\dot{f}^0 = \dot{f}^{0(1)} + i\dot{f}^{0(2)}$ completes the argument.

The second part of the statement also follows from the corresponding real-valued case: just reverse the preceding steps. ∎

Example 11.7.6 Let $\{Z(t),\ a \le t \le b\}$ be a stationary harmonizable $S\alpha S$ process (see (6.5.1)) and $p \ge 1$. We apply Corollary 11.7.5 to find out when this process has a version with all sample paths in $AC^p[a,b]$. Clearly, the only candidate for f^0 is f itself, i.e., $f^0(t,x) = e^{itx}$, $a \le t \le b$, $x \in \mathbb{R}$. Then $\dot{f}^0(t,x) = ixe^{itx}$, and $|\dot{f}^0(t,x)| = |x|$. A trivial check of (11.7.7)–(11.7.9) shows that these conditions hold if and only if

$$\int_{-\infty}^{\infty} |x|^\alpha m(dx) < \infty. \tag{11.7.18}$$

Therefore (11.7.18) is a necessary and sufficient condition for a stationary harmonizable process to have a version with all sample paths in $AC^p[a,b]$. Note that this condition does not depend on p !

Now let $X(t) = \text{Re}\, Z(t)$, $a \le t \le b$, be a real harmonizable $S\alpha S$ process. Applying Theorem 11.7.4 to the integral representation (11.7.15), we see that for every $p \ge 1$, (11.7.18) is also the necessary and sufficient condition for a real harmonizable $S\alpha S$ process to have a version with all sample paths in $AC^p[a,b]$.

Example 11.7.7 Let $\{X(t),\ -\infty < t < \infty\}$ be a stationary α-stable moving average

$$X(t) = \int_{-\infty}^{\infty} f(t-x)M(dx),\ -\infty < t < \infty,$$

where $f \in L^\alpha(-\infty,+\infty)$ and M is an α-stable random measure on $(\mathbb{R},\mathcal{B})$ with Lebesgue control measure and constant skewness intensity β (we assume $\beta \equiv 0$ if $\alpha = 1$). The discussion in Example 10.4.5 shows that if X is to be sample continuous, let alone have an absolutely continuous version, f must have a continuous modification. We assume, therefore, from the outset that f is continuous.

We now apply Theorem 11.7.4 to find conditions for the moving average process to have a version in $AC^p[a,b]$, $p \ge 1$. The stationarity of the process implies that if such a version exists for some $[a,b]$, then there is a version in $AC^p[a,b]$ for any $a < b$, so we may as well set $a = 0$ and $b = 1$. Since the only candidate for f^0 in Theorem 11.7.4 is f itself, we conclude that the necessary and sufficient conditions for a stationary α-stable moving average to have a version in $AC^p[a,b]$ are:

(i) f is absolutely continuous,

(ii)

$$\int_{-\infty}^{\infty} |f'(x)|^\alpha dx < \infty \text{ if } 1 \le p < \alpha,$$

$$\int_0^1 \int_{-\infty}^{\infty} |f'(x)|^\alpha \left[1 + \log_+ \frac{|f'(x)|^\alpha}{F_\alpha(x+t)}\right] dx dt < \infty \text{ if } \alpha = p,$$

$$\int_{-\infty}^{\infty} F_p(x)^{\alpha/p} dx < \infty \ \text{ if } p > \alpha,$$

where

$$F_p(x) = \int_x^{x+1} |f'(t)|^p dt.$$

In particular, the moving average

$$X(t) = \int_{-\infty}^{\infty} e^{-|t-x|} M(dx), \ -\infty < t < \infty,$$

has a version in $AC^p[a,b]$ for any $0 < \alpha < 2$, $p \ge 1$.

Example 11.7.8 A simple application of Theorem 11.7.4 shows that the well-balanced linear fractional stable motion (3.6.4) with $1 < \alpha < 2$ and $1/\alpha < H < 1$ has no versions in $AC^p[a,b]$. It is, however, sample continuous as we will see in the next chapter.

11.8 Exercises

Exercise 11.1 (i) Prove the Khinchine inequality: for any $p > 0$, any $n = 1, 2, \ldots$,

$$E\Big|\sum_{i=1}^n a_i \epsilon_i\Big|^p \le C(p) \Big(\sum_{i=1}^n a_i^2\Big)^{p/2},$$

where a_i, $i = 1, \ldots, n$ are real numbers, ϵ_i, $i = 1, \ldots, n$ are Rademacher random variables and $C(p)$ is a finite constant depending only on p.

Hint: Prove it for $p = 2m$ first.

(ii) Prove the reverse inequality

$$E\Big|\sum_{i=1}^n a_i \epsilon_i\Big|^p \ge c(p) \Big(\sum_{i=1}^n a_i^2\Big)^{p/2}$$

for $p \geq 2$, where $c(p)$ is a positive constant depending only on p.
(iii) Prove that for any random variable Z with a finite fourth moment and for any $\lambda > 0$,

$$\left(P(|Z| > \lambda)\right)^{1/2} \geq \frac{EZ^2 - \lambda^2}{(EZ^4)^{1/2}}.$$

Hint: Write $EZ^2 = E(Z^2 1(|Z| > \lambda) + Z^2 1(|Z| \leq \lambda))$ and use the Hölder inequality.
(iv) Prove that for any $\rho \in (0,1)$,

$$P\Big(\Big|\sum_{i=1}^n a_i \epsilon_i\Big| > \rho\Big(\sum_{i=1}^n a_i^2\Big)^{1/2}\Big) \geq C(1-\rho^2),$$

where C is a positive constant.
(v) Use (iv) to extend (iii) to all $p > 0$ and to show that the series $\sum_{i=1}^\infty a_i \epsilon_i$ converges a.s. if and only if it converges in L^p for every $p > 0$.
(vi) Suppose that the series $\sum_{i=1}^\infty a_i \epsilon_i$ converges. Then, for every $p > 0$, there are positive constants $c(p)$ and $C(p)$ such that

$$c(p)\Big(\sum_{i=1}^\infty a_i^2\Big)^{1/2} \leq E\Big|\sum_{i=1}^\infty a_i \epsilon_i\Big|^p \leq C(p)\Big(\sum_{i=1}^\infty a_i^2\Big)^{1/2}.$$

Exercise 11.2 Let $\{X_i(t),\ t \in T\}, i = 1, 2$, be independent measurable copies of a strictly α-stable process, $\alpha \neq 1$, defined on a σ-finite measure space $(T, \mathcal{T}, \nu)$. Prove that

$$\int_T |X_1(t)|^p \nu(dt) < \infty \text{ a.s.}$$

if and only if

$$\int_T |Y(t)|^p \nu(dt) < \infty \text{ a.s.},$$

where $Y(t) = 2^{-1/\alpha}(X_1(t) - X_2(t)),\ t \in T$.

Exercise 11.3 Let (T, τ, ν) be a σ-finite measure space and let $X = \{X(t),\ t \in T\}$ and $\{Y = Y(t),\ t \in T\}$ be two measurable α-stable processes with all sample paths in $L^p(T, \tau, \nu)$ and such that

$$\{X(t),\ t \in T\} \stackrel{d}{=} \{Y(t),\ t \in T\}$$

(equality of the finite-dimensional distributions).

Show that

$$\int_T |X(t)|^p \nu(dt) \stackrel{d}{=} \int_T |Y(t)|^p \nu(dt)$$

and, moreover,

$$X \stackrel{d}{=} Y$$

(equality of the distributions of two random elements in $L^p(T, \tau, \nu)$).
Hint: Proceed as in the beginning of Section 11.4 using the law of large numbers to establish convergence to $\int_T |X(t)|^p \nu(dt)$ and $\int_T |Y(t)|^p \nu(dt)$. To prove that $X \stackrel{d}{=} Y$ in L^p, one needs to show that the probabilities that X and Y belong to the same open ball in L^p are equal, i.e., $P(\|X - z\|_p \leq \lambda) = P(\|Y - z\|_p \leq \lambda)$. This, however, follows from the first part of the exercise with X and Y replaced by $X - z$ and $Y - z$, respectively.

Exercise 11.4 Verify (11.6.5).

Exercise 11.5 Verify the properties of the functions $g_T^{(1)}(x)$ and $g_T^{(2)}(x)$ described in the proof of Theorem 11.6.1.

Chapter 12

Boundedness and continuity via metric entropy

In Chapter 10, we obtained necessary conditions for sample boundedness and sample continuity of stable processes of index $0 < \alpha < 2$ in terms of the kernel in the integral representation. We also observed that these conditions are sufficient when $0 < \alpha < 1$ but not when $1 \leq \alpha < 2$ (Example 10.4.1 and Exercise 10.7). No necessary and sufficient conditions are presently known in the case $1 \leq \alpha < 2$. In this chapter, we use "metric entropy" to obtain sufficient conditions and additional necessary conditions for sample boundedness and continuity in the case $1 \leq \alpha < 2$.

Observe that the results of Section 11.7 also provide sufficient conditions for sample path boundedness and continuity when $1 \leq \alpha < 2$. Indeed, Theorem 11.7.4 gives necessary and sufficient conditions for an α-stable process $\{X(t),\ a \leq t \leq b\}$ to have a version with absolutely continuous sample paths; since any such version has continuous sample paths, sufficient conditions for sample continuity follow. There are, however, α-stable processes that are sample continuous but do not have versions with absolutely continuous sample paths, for example, the linear fractional stable motion (Examples 11.7.8 and 12.2.3).

Metric entropy offers an alternative to integral representation techniques. It has been extensively used in the context of Gaussian processes. Because α-stable processes are closely related to Gaussian processes through conditioning, it is natural to check whether the techniques used in the Gaussian case apply. They do, but the "gap" between the necessary and the sufficient conditions for sample regularity is much greater in the α-stable case than in the Gaussian case.

In Section 12.1 we review the Gaussian results and set the scene for the discussion of the α-stable case. Sufficient and necessary conditions in the α-

stable, $1 \leq \alpha \leq 2$, case, are given, respectively, in Sections 12.2 and 12.3.

Finally, in Section 12.4 we collect a number of facts, scattered throughout the book, which are related to boundedness and continuity of α-stable H-self-similar processes.

12.1 Metric entropy

Consider a zero mean Gaussian process $X = \{X(t),\ t \in T\}$. The process X generates the following pseudometric d_X on T:

$$d_X(t,s) = \left(E(X(t) - X(s))^2\right)^{1/2}, \quad t,s \in T. \tag{12.1.1}$$

For any given $\epsilon > 0$, let $N(\epsilon)$ denote the smallest number of open d_X-balls of radius ϵ needed to cover the parameter set T. The number $N(\epsilon)$, which is called *the covering number*, may be infinite. It is, of course, finite for all $\epsilon > 0$ if and only if (T, d_X) has a compact closure. In this case, $N(\epsilon) = 1$ when $\epsilon > \sup_{t,s} d_X(t,s)$. Since $N(\epsilon)$ increases as ϵ decreases, it is the behavior of $N(\epsilon)$ around $\epsilon = 0$ that is of interest.

Definition 12.1.1 The function H: $(0,\infty) \to [0,\infty]$ defined by $H(\epsilon) = \log N(\epsilon)$ is called the *metric entropy (function)* of the process $\{X(t),\ t \in T\}$.

A fundamental theorem of Dudley (1973) states that if

$$\int_0^\infty H(\epsilon)^{1/2} d\epsilon < \infty, \tag{12.1.2}$$

then the Gaussian process $\{X(t),\ t \in T\}$ is both *sample bounded* on T and *sample continuous* in the metric d_X. If T is endowed with a topology that makes d_X a continuous function (i.e., as $n \to \infty$, $t_n \to t \Rightarrow d_X(t_n, t) \to 0$; equivalently $\{X(t),\ t \in T\}$ is continuous in probability), then (12.1.2) also implies that $\{X(t),\ t \in T\}$ is sample continuous in the topology of T. Moreover, if (12.1.2) holds, then a separable version $\{Y(t),\ t \in T\}$ of $\{X(t),\ t \in T\}$ satisfies

$$E \sup_{t \in T} Y(t) \leq K \int_0^\infty H(\epsilon)^{1/2} d\epsilon \tag{12.1.3}$$

where K is a universal constant.

Metric entropy also provides necessary conditions for sample regularity of Gaussian processes. Namely, if $\{X(t), t \in T\}$ is *sample bounded*, then

$$\sup_{\epsilon > 0} \epsilon H(\epsilon)^{1/2} < \infty, \tag{12.1.4}$$

and if (T, d_X) is relatively compact and $\{X(t),\ t \in T\}$ is *sample continuous* in the metric d_X, then

$$\lim_{\epsilon \to 0} \epsilon H(\epsilon)^{1/2} = 0. \tag{12.1.5}$$

Moreover, any separable version $\{Y(t),\ t \in T\}$ of $\{X(t),\ t \in T\}$ satisfies

$$E \sup_{t \in T} Y(t) \geq K^{-1} \sup_{\epsilon > 0} \epsilon H(\epsilon)^{1/2} \tag{12.1.6}$$

where K is a universal constant.

There does not appear to be a big "gap" between the necessary conditions (12.1.4) and (12.1.5) for sample boundedness and continuity and the sufficient condition (12.1.2). In fact, although (12.1.2) is not necessary for sample continuity, it becomes so when (T, d_X) is relatively compact, T is a group and $\{X(t),\ t \in T\}$ is a *stationary* process (so that d_X is a translation invariant metric on T). For a proof of these results see Dudley (1973) and Fernique (1975).

How useful is the metric entropy approach in the case of α-stable processes, $1 \leq \alpha < 2$? It turns out that metric entropy does provide information on sample boundedness and continuity, but the gap between the necessary conditions and the sufficient conditions widens. This can be attributed to the fact that, unlike the Gaussian case, a single number, "the distance" between two points $X(t)$ and $X(s)$, does not tell much about their joint distribution.

Let $\{X(t),\ t \in T\}$ be a $S\alpha S$ process, $1 \leq \alpha < 2$. For $s, t \in T$, define $d_X(s,t)$ as the scale parameter of the $S\alpha S$ random variable $X(t) - X(s)$. Note that

$$d_X(s,t) = \|X(t) - X(s)\|_\alpha$$

is also the covariation norm of $X(t) - X(s)$; see Section 2.8 and Exercises 2.22, 2.23 and 2.24. Let $N(\epsilon)$ and $H(\epsilon)$ be defined as above (but now, of course, in terms of the new metric d_X).

We shall firstly try to obtain sufficient conditions for sample path regularity for $\{X(t).\ t \in T\}$. It is convenient, however, to use a slightly more general framework than one specific to α-stable processes.

Let T be a separable metric space and let $\{X(t),\ t \in T\}$ be a strongly separable stochastic process defined on a probability space $(\Omega, \mathcal{F}, P)$. Consider $L^0(\Omega, \mathcal{F}, P)$, the space of all random variables on $(\Omega, \mathcal{F}, P)$, and let $\|\cdot\|$ be an *extended-valued continuous semi-norm* on $L^0(\Omega, \mathcal{F}, P)$, i.e.,

1. $\|X\| \in [0, \infty]$ for all $X \in L^0(\Omega, \mathcal{F}, P)$,

2. $\|X\| = 0$ iff $X = 0$ a.s.,

3. $\|X + Y\| \leq \|X\| + \|Y\|$ for any $X, Y \in L^0(\Omega, \mathcal{F}, P)$,

4. $\|X_n\| \uparrow \|X\|$ as $n \to \infty$ for any $X, X_n \in L^0(\Omega, \mathcal{F}, P)$, $n = 1, 2, \ldots$, such that $|X_n| \uparrow |X|$ a.s. as $n \to \infty$.

(The linearity condition $\|aX\| = |a| \, \|X\|$ is not imposed.) Define also

$$L^{\|\cdot\|} = \{X \in L^0(\Omega, \mathcal{F}, P) \colon \|X\| < \infty\}.$$

The following theorem is a slight generalization of Theorems 11.2 and 11.6 of Ledoux and Talagrand (1991) whose proof we follow.

Theorem 12.1.2 *Suppose that $X(t) \in L^{\|\cdot\|}$ for any $t \in T$. Let d be a (pseudo)-metric on T and ϕ be a concave non-decreasing function on $(0, \infty)$ such that, for any $A \in \mathcal{F}$ and $s, t \in T$,*

$$\|(X(s) - X(t))1_A\| \le d(s,t) P(A) \phi(1/P(A)). \tag{12.1.7}$$

Then for any $\eta > 0$,

$$\Big\| \sup_{s,t \in T,\ d(s,t) < \eta} |X(s) - X(t)| \, \Big\| \le \eta \phi(N(\delta)^2) + 32 \int_0^{16\delta} \phi(N(\epsilon)) d\epsilon \tag{12.1.8}$$

for any $0 < \delta \le D$ where

$$D = \sup_{s,t \in T} d(s,t)$$

is the diameter of (T, d), and $N(\epsilon)$ is the minimal number of open d-balls of radius ϵ needed to cover T.

We start with a lemma.

Lemma 12.1.3 *Let $X_1, X_2, \ldots, X_N$ be in $L^{\|\cdot\|}$ such that for any $A \in \mathcal{F}$ and $i = 1, \ldots, N$,*

$$\| |X_i| 1_A \| \le P(A) \phi(1/P(A)). \tag{12.1.9}$$

Then, for every $A \in \mathcal{F}$,

$$\Big\| \max_{i=1,\ldots,N} |X_i| 1_A \Big\| \le P(A) \phi(N/P(A)).$$

PROOF: Let $A_i = \{\omega \in A \colon |X_i| \ge \max_{j=1,\ldots,N} |X_j|\}$, $i = 1, \ldots, N$, and let $B_i = A_i \backslash \bigcup_{j=1}^{i-1} A_j$, $i = 1, \ldots, n$. Then, by (12.1.9),

$$\begin{aligned}
\Big\| \max_{i=1,\ldots,N} |X_i| 1_A \Big\| &= \Big\| \sum_{j=1}^N |X_j| 1_{B_j} \Big\| \le \sum_{j=1}^N \| |X_j| 1_{B_j} \| \\
&\le \sum_{j=1}^N P(B_j) \phi(1/P(B_j)) = P(A) \sum_{j=1}^N \frac{P(B_j)}{P(A)} \phi(1/P(B_j)) \\
&\le P(A) \phi \left(\sum_{j=1}^N \frac{P(B_j)}{P(A)} \frac{1}{P(B_j)} \right) = P(A) \phi(N/P(A)),
\end{aligned}$$

using the concavity of ϕ. ■

PROOF OF THEOREM 12.1.2: Let $T^* = \{t_1, t_2, \ldots\}$ be a strong separant for $\{X(t),\ t \in T\}$ and $T_n^* = \{t_1, t_2, \ldots, t_n\}$, $n = 1, 2, \ldots$. We prove that for every $n = 1, 2, \ldots$ and $\eta > 0$,

$$\left\| \max_{\substack{s,t \in T_n^* \\ d(s,t) < \eta}} |X(s) - X(t)| \right\| \leq \eta\phi(N(\delta)^2) + 32 \int_0^{16\delta} \phi(N(\epsilon))d\epsilon \qquad (12.1.10)$$

for every $0 < \delta \leq D$. Relation (12.1.8) then follows because, as $n \to \infty$

$$\max_{\substack{s,t \in T_n^* \\ d(s,t) < \eta}} |X(s) - X(t)| \uparrow \sup_{\substack{s,t \in T^* \\ d(s,t) < \eta}} |X(s) - X(t)| = \sup_{\substack{s,t \in T \\ d(s,t) < \eta}} |X(s) - X(t)|.$$

Fix an $n \geq 1$ and consider T_n^*. For each $m = \ldots, -1, 0, 1, \ldots,$ let S_m be a subset of T_n^* of minimal cardinality such that the open balls of radius 2^{-m} with centers in S_m cover T_n^*. Denoting the cardinality by #, we have $\#(S_m) \leq N(2^{-(m+1)})$ (see Exercise 12.2). For each $t \in T_n^*$, fix an $s_m(t) \in S_m$ such that $d(t, s_m(t)) < 2^{-m}$. Since T_n^* is a finite set, there is an m such that $d(t, s_m(t)) = 0$ for every $t \in T_n^*$. Let m_0 be the smallest such m. For every $m \leq m_0$ and $t \in T_n^*$,

$$X(t) = \sum_{i=m+1}^{m_0} \Big(X(k_i(t)) - X(k_{i-1}(t)) \Big) + X(k_m(t)), \qquad (12.1.11)$$

where for $i \leq m_0$, $k_i(t) = s_i(s_{i+1}(\cdots(s_{m_0}(t))\ldots))$. Then

$$\max_{t \in T_n^*} |X(t) - X(k_m(t))| \leq \sum_{i=m+1}^{m_0} \max_{t \in T_n^*} |X(k_i(t)) - X(k_{i-1}(t))|. \quad (12.1.12)$$

Observe that for any $t \in T_n^*$ and $i \leq m_0$, $d(k_i(t),\ k_{i-1}(t)) \leq 2^{-(i-1)}$. Moreover, the collection $\{X(k_i(t)) - X(k_{i-1}(t)),\ t \in T_n^*\}$ consists of at most $\#(S_i) \leq N(2^{-(i+1)})$ distinct random variables. It follows now from (12.1.7) and Lemma 12.1.3 that for any measurable set $A \in \mathcal{F}$,

$$\left\| 1_A \max_{t \in T_n^*} |X(k_i(t)) - X(k_{i-1}(t))| \right\| \leq 2^{-(i-1)} P(A)\phi(N(2^{-(i+1)})/P(A)),$$

and so, using (12.1.12),

$$\left\| 1_A \max_{t \in T_n^*} |X(t) - X(k_m(t))| \right\| \leq P(A) \sum_{i=m+1}^{m_0} 2^{-(i-1)}\phi(N(2^{-(i+1)})/P(A)).$$

In particular, for $A = \Omega$ and $\delta > 0$ such that $2^{-(m+2)} < \delta \leq 2^{-(m+1)}$,

$$\left\| \max_{t \in T_n^*} |X(t) - X(k_m(t))| \right\| \leq 8 \int_0^{16\delta} \phi(N(\epsilon))d\epsilon. \qquad (12.1.13)$$

Keeping $m \le m_0$ fixed, define for any $\eta > 0$, $U_m(\eta) = \{(x,y) \in S_m \times S_m$: there are $t \in T_n^*$, $s \in T_n^*$ with $d(s,t) < \eta$ such that $k_m(t) = x$, $k_m(s) = y\}$. Obviously, $\#(U_m(\eta)) \le N(2^{-(m+1)})^2 \le N(\delta)^2$. For any $(x,y) \in U_m(\eta)$, fix a $t(x,y) \in T_n^*$ and $s(x,y) \in T_n^*$ such that $d(t(x,y),\ s(x,y)) < \eta$ and $k_m(t(x,y)) = x$, $k_m(s(x,y)) = y$. Applying (12.1.7) and Lemma 12.1.3, we conclude that

$$\Big\| \max_{(x,y)\in U_m(\eta)} |X(t(x,y)) - X(s(x,y))| \Big\| \le \eta\phi(N(\delta)^2). \tag{12.1.14}$$

Take now any $t \in T_n^*$, $s \in T_n^*$ such that $d(t,s) < \eta$. Clearly, $(k_m(t),\ k_m(s)) \in U_m(\eta)$ and thus

$$\begin{aligned}
&|X(s) - X(t)| \\
&\quad \le |X(s) - X(k_m(s))| + |X(k_m(s)) - X(s(k_m(t),\ k_m(s)))| \\
&\qquad + |X(s(k_m(t),\ k_m(s))) - X(t(k_m(t),\ k_m(s)))| \\
&\qquad + |X(t(k_m(t),\ k_m(s))) - X(k_m(t))| + |X(k_m(t)) - X(t)| \\
&\quad \le 4 \max_{t\in T_n^*} |X(t) - X(k_m(t))| + \max_{(x,y)\in U_m(\eta)} |X(t(x,y)) - X(s(x,y))|.
\end{aligned}$$

Using (12.1.13) and (12.1.14), we obtain

$$\Big\| \max_{\substack{s,t\in T_n^* \\ d(s,t)<\eta}} |X(s) - X(t)| \Big\| \le \eta\phi(N(\delta)^2) + 32 \int_0^{16\delta} \phi(N(\epsilon))d\epsilon,$$

which proves (12.1.10) and completes the proof. ∎

12.2 Sufficient conditions in the case $1 \le \alpha < 2$

The following theorem gives sufficient conditions for sample boundedness and continuity of $S\alpha S$ processes, $1 \le \alpha < 2$.

Theorem 12.2.1 *Let* $\{X(t),\ t \in T\}$ *be a* $S\alpha S$ *process,* $1 \le \alpha < 2$. *Let* $d_X(s,t) = \|X(s) - X(t)\|_\alpha$, $s \in T$, $t \in T$, *be the (pseudo) metric generated by the process. Let* $D = \sup_{s,t\in T} d_X(s,t)$ *be the* d_X*-diameter of* T. *If*

$$\int_0^D N(\epsilon)^{1/\alpha} d\epsilon < \infty \ \textit{and}\ 1 < \alpha < 2 \tag{12.2.1}$$

or if

$$\int_0^D |\log\epsilon| N(\epsilon) d\epsilon < \infty \ \textit{and}\ \alpha = 1, \tag{12.2.2}$$

then $\{X(t),\ t \in T\}$ is both sample bounded and sample continuous (in the metric d_X).

Moreover, if $\{Y(t),\ t \in T\}$ is a strong separable version of $\{X(t),\ t \in T\}$ defined on a probability space $(\Omega_1, \mathcal{F}_1, P_1)$ and $1 < \alpha < 2$, then, for any $\eta > 0$,

$$E_1 \sup_{\substack{s,t\in T \\ d_X(s,t)<\eta}} |Y(s) - Y(t)| \le \eta \frac{\alpha}{\alpha-1} r_\alpha N(\delta)^{2/\alpha} + 32\frac{\alpha}{\alpha-1} r_\alpha \int_0^{16\delta} N(\epsilon)^{1/\alpha} d\epsilon \tag{12.2.3}$$

for any $0 \le \delta \le D$. Here, $r_\alpha^\alpha = \sup_{\lambda>0} \lambda^\alpha P(|X_0| > \lambda)$, where $X_0 \sim S_\alpha(1,0,0)$.

PROOF: Let $\{Y(t),\ t \in T\}$ be a strongly separable version of $\{X(t),\ t \in T\}$ defined on a probability space $(\Omega_1, \mathcal{F}_1, P_1)$.

We start with the case $1 < \alpha < 2$. For $X \in L^0(\Omega_1, \mathcal{F}_1, P_1)$, set $\|X\| = E_1|X|$, $d(s,t) = d_X(s,t),\ s,t \in T$, and take $\phi(x) = \alpha' r_\alpha x^{1/\alpha}$, where α' is the conjugate of α (i.e., $1/\alpha + 1/\alpha' = 1$).

Observe that for any $A \in \mathcal{F}_1,\ s,t \in T$, and any $\epsilon > 0$

$$\begin{aligned}
\|1_A(Y(s) - Y(t))\| &= E_1|1_A|Y(s) - Y(t)| \\
&= \int_0^\infty P_1(\{|Y(s) - Y(t)| > \lambda\} \cap A) d\lambda \\
&= \int_0^\epsilon + \int_\epsilon^\infty \\
&\le \epsilon P_1(A) + \int_\epsilon^\infty P_1(|Y(s) - Y(t)| > \lambda) d\lambda \\
&\le \epsilon P_1(A) + \int_\epsilon^\infty d(s,t)^\alpha \lambda^{-\alpha} r_\alpha^\alpha d\lambda \\
&= \epsilon P_1(A) + r_\alpha^\alpha d(s,t)^\alpha (\alpha - 1)^{-1} \epsilon^{-\alpha+1} \\
&= \epsilon[P_1(A) + r_\alpha^\alpha d(s,t)^\alpha (\alpha - 1)^{-1} \epsilon^{-\alpha}].
\end{aligned}$$

Choosing $\epsilon = d(s,t) r_\alpha P_1(A)^{-1/\alpha}$, we see that Condition (12.1.7) of Theorem 12.1.2 holds with $\phi(x) = \frac{\alpha}{\alpha-1} r_\alpha x^{1/\alpha}$. Then (12.2.3) holds as well and we have, under the assumption (12.2.1),

$$\lim_{\eta\to 0} E_1 \sup_{\substack{s,t\in T \\ d_X(s,t)<\eta}} |Y(s) - Y(t)| = 0,$$

so that

$$\lim_{\eta\to 0} \sup_{\substack{s,t\in T \\ d_X(s,t)<\eta}} |Y(s) - Y(t)| = 0 \text{ a.s.}$$

This proves that all sample paths of $\{Y(t),\ t \in T\}$ are continuous in the metric d_X on a set of P_1-measure 1. Since (T, d_X) is relatively compact under the assumption

(12.2.1), all sample paths of $\{Y(t),\ t \in T\}$ on that set are also bounded. Therefore, $\{X(t),\ t \in T\}$ is both sample bounded and sample continuous in the metric d_X.

We now consider the case $\alpha = 1$. We set this time $\|X\| = E_1 \min(1, |X|)$ for $X \in L^0(\Omega_1, \mathcal{F}_1, P_1)$. Assumption (12.2.2) ensures that $D < \infty$. Choose an $a > 0$ large enough so that $f(x) = x \log(a + \frac{1}{x})$ is a positive non-decreasing concave function on $(0, D]$, define $f(0) = 0$ and set $d(s,t) = f(d_X(s,t)),\ s,t \in T$. Observe that $d(s,t)$ is a (pseudo)-metric on T and that $d(s,t) \to 0$ if and only if $d_X(s,t) \to 0$. Choose also a finite positive constant c_1 large enough so that

$$E \min(1, |X|) \le c_1 \sigma \log\Big(a + \frac{1}{\sigma}\Big),$$

for any $S_1(\sigma, 0, 0)$ random variable X with $0 < \sigma < D$. Finally, let $\phi(x) = c_2 x$. Then Condition (12.1.7) of Theorem 12.1.2 holds if $c_2 > 0$ is large enough, and we have, for any $\eta > 0$ and $0 < \delta \le f(D)$,

$$E \min\Big(1, \sup_{\substack{s,t \in T \\ d(s,t) < \eta}} |Y(s) - Y(t)|\Big) \le c_2 \eta \widetilde{N}(\delta)^2 + 32 c_2 \int_0^{16\delta} \widetilde{N}(\epsilon) d\epsilon \quad (12.2.4)$$

where $\widetilde{N}(\epsilon)$ is the minimal number of open d-balls of radius ϵ required to cover T.

Obviously, for any $0 < \epsilon \le f(D)$,

$$\begin{aligned} \widetilde{N}(\epsilon) &= N(f^{-1}(\epsilon)) \\ &\le N\Big(c_3 \epsilon / \log\Big(a + \frac{1}{\epsilon}\Big)\Big) \end{aligned}$$

for some $c_3 \in (0, \infty)$. Substituting this into (12.2.4) and changing the variable of the integration, we conclude

$$E_1 \min\Big(1, \sup_{\substack{s,t \in T \\ d(s,t) < \eta}} |Y(s) - Y(t)|\Big)$$

$$\le c_2 \eta N\Big(c_3 \delta / \log\Big(a + \frac{1}{\delta}\Big)\Big)^2 + c_4 \int_0^{16 c_2 \delta / \log(a + (1/16\delta))} \Big(1 + |\log u|\Big) N(u) du, \quad (12.2.5)$$

where $0 < c_4 < \infty$. It follows now from (12.2.2) that

$$\lim_{\eta \to 0} E_1 \min\Big(1, \sup_{\substack{s,t \in T \\ d(s,t) < \eta}} |Y(s) - Y(t)|\Big) = 0,$$

which implies, as before,

$$\lim_{\eta \to 0} \sup_{\substack{s,t \in T \\ d(s,t) < \eta}} |Y(s) - Y(t)| = 0 \text{ a.s.}$$

Then, on a set of P_1-measure 1, all sample paths of $\{Y(t),\ t \in T\}$ are continuous in the metric d and, therefore, in the metric d_X as well. Hence $\{X(t),\ t \in T\}$ is both sample bounded and sample continuous in the metric d_X. ∎

An analogous result holds in a more general case:

Corollary 12.2.2 *Let $\{X(t),\ t \in T\}$ be a strictly α-stable process, $1 < \alpha < 2$, and let $d_X(s,t)$ be the scale parameter of $\{X(t) - X(s),\ s,t \in T\}$. Suppose that (12.2.1) holds. Then $\{X(t),\ t \in T\}$ is both sample bounded and sample continuous in the metric d_X. Moreover, if $\{Y(t),\ t \in T\}$ is a strongly separable version of $\{X(t),\ t \in T\}$ defined on a probability space $(\Omega_1, \mathcal{F}_1, P_1)$, then for any $\eta > 0$, Relation (12.2.3) holds, provided we multiply its right-hand side by $2^{1/\alpha}\alpha'$, where α' is the conjugate of α (i.e., $1/\alpha + 1/\alpha' = 1$).*

PROOF: Let $Z(t) = 2^{-1/\alpha}(X_1(t) - X_2(t)),\ t \in T$, where $\{X_1(t),\ t \in T\}$, $i = 1, 2$, are independent copies of $\{X(t),\ t \in T\}$. Then $\{Z(t),\ t \in T\}$ is a $S\alpha S$ process generating a metric $d_Z(t,s) = d_X(t,s)$ for every $t, s \in T$. Fix any countable subset C of T. Now (12.2.3) implies

$$E \sup_{t,s \in C} |Z(t) - Z(s)| \ge 2^{-1/\alpha}(1 - 1/\alpha) E \sup_{t,s \in C} |X(t) - X(s)|.$$

Thus, if $\{Y(t),\ t \in T\}$ is a strongly separable version of $\{X(t),\ t \in T\}$ defined on a probability space $(\Omega_1, \mathcal{F}_1, P_1)$, then, by (12.2.3)

$$E_1 \sup_{\substack{t,s \in T \\ d_X(t,s) < \eta}} |Y(t) - Y(s)|$$

$$\le 2^{1/\alpha}\left(1 - \frac{1}{\alpha}\right)^{-1} \frac{\alpha}{\alpha - 1} r_\alpha \left[\eta N(\delta)^{2/\alpha} + 32 \int_0^{16\delta} N(\epsilon)^{1/\alpha} d\epsilon\right]. \tag{12.2.6}$$

Under assumption (12.2.1), Relation (12.2.6) implies, as before, both the sample boundedness and sample continuity in the metric d_X of the α-stable process $\{X(t),\ t \in T\}$. ∎

Remark. In the case $1 < \alpha < 2$, Relation (12.2.3) of Theorem 12.2.1 provides an upper bound on the "size" of the oscillations of the process $\{X(t),\ t \in T\}$. An upper bound on the "size" of the oscillations in the case $\alpha = 1$ is given in the proof of the theorem (see (12.2.5)).

Example 12.2.3 We have seen in Example 10.2.5 that a well-balanced *linear fractional α-stable motion* with $H < 1/\alpha$ is not sample bounded on any interval of positive length. Suppose, now, that $1/\alpha < H < 1$, which is possible only when $1 < \alpha < 2$. We will show that in this case, $\{X(t),\ -\infty < t < \infty\}$ is sample continuous. According to Proposition 9.3.5, it is enough to show that for any $-\infty < a < b < \infty$, $\{X(t),\ a \le t \le b\}$ is sample continuous. Fix $-\infty < a < b < \infty$ and $s, t \in [a, b]$. By the H-self-similarity of the process and the stationarity of its increments (see Exercise 3.9) we have $E|X(t) - X(s)|^p = C_{\alpha,p}|t - s|^{Hp}$ for any $0 < p < \alpha$, and hence the metric generated by the process is of the form

$$d_X(t, s) = c_\alpha |t - s|^H, \quad a \le s, t < b.$$

Apply Theorem 12.2.1. Since

$$N(\epsilon) \le \text{const.}\epsilon^{-1/H}, \quad 0 < \epsilon \le D = c_\alpha (b - a)^H,$$

the assumption $1/\alpha < H$ ensures that Condition (12.2.1) of the theorem holds. Therefore $\{X(t),\ a \le t \le b\}$ is sample continuous.

As we will see in the sequel, there is a big gap between the sufficient conditions for sample path boundedness and continuity given in Theorem 12.2.1 and the corresponding necessary conditions. But in some sense, both type of conditions are the best possible. As far as the sufficient conditions are concerned, this can be seen, for example, by considering the log-fractional stable motion of Example 3.6.6 ($1 < \alpha < 2$). For every $-\infty < a < b < \infty$, this process $\{X(t),\ a \le t \le b\}$ is not a.s. bounded because the necessary conditions of Theorem 10.2.3 fail. But this process is $1/\alpha$-self-similar and has stationary increments so $d_X(s, t) = c_\alpha |s - t|^{1/\alpha}$, $a \le s, t \le b$, and thus $N(\epsilon) \sim \text{const.}\epsilon^{-1/\alpha}$ as $\epsilon \to 0$. The sufficient condition of Theorem 12.2.1 fails (as, of course, it must), but barely (the integral $\int_0^D N(\epsilon)^{1/\alpha} d\epsilon$ has a logarithmic divergence).

In fact, one can show the following:

For any non-increasing function h: $(0, \infty) \to (0, \infty)$ with $h(\theta) \uparrow \infty$ as $\theta \downarrow 0$, there is a $S\alpha S$ process $\{X(t),\ 0 \le t \le 1\}$, $1 \le \alpha < 2$, such that

$$\int_\theta^1 N(\epsilon)^{1/\alpha} d\epsilon \le h(\theta) \ \text{ for every } \ \theta \in (0, 1)$$

and $\{X(t),\ 0 \le t \le 1\}$ is not sample bounded. (See Exercise 12.1.)

12.3 Necessary conditions in the case $1 \le \alpha < 2$

We now turn to necessary conditions for sample path regularity. Here is the main theorem.

Theorem 12.3.1 *Let $\{X(t),\ t \in T\}$ be a $S\alpha S$ process, $1 \le \alpha < 2$. Let $d_X(s, t) = \|X(s) - X(t)\|_\alpha$, $s \in T$, $t \in T$, be the (pseudo)-metric generated* *by the process. Then there is an absolute constant $K_\alpha \in (0, \infty)$ such that for a separable version $\{Y(t),\ t \in T\}$ defined on a probability space $(\Omega_1, \mathcal{F}_1, P_1)$,*

$$E_1 \sup_{t \in T} |Y(t)| \geq K_\alpha \epsilon \left(\log N(\epsilon)\right)^{1/\alpha'} \tag{12.3.1}$$

if $1 < \alpha < 2$, where α' is the conjugate of α, and

$$\left(E_1 \sup_{t \in T} |Y(t)|^{1/2}\right)^2 \geq K_1 \epsilon \log\left(1 + \log N(\epsilon)\right) \tag{12.3.2}$$

if $\alpha = 1$.

In particular, if $\{X(t),\ t \in T\}$ is sample bounded, then

$$\sup_{\epsilon > 0} \epsilon \left(\log N(\epsilon)\right)^{1/\alpha'} < \infty \tag{12.3.3}$$

if $1 < \alpha < 2$ and

$$\sup_{\epsilon > 0} \epsilon \log\left(1 + \log N(\epsilon)\right) < \infty \tag{12.3.4}$$

if $\alpha = 1$. Moreover, if (T, d_X) is relatively compact and $\{X(t),\ t \in T\}$ is sample continuous, then

$$\lim_{\epsilon \to 0} \epsilon \left(\log N(\epsilon)\right)^{1/\alpha'} = 0 \tag{12.3.5}$$

if $1 < \alpha < 2$ and

$$\lim_{\epsilon \to 0} \epsilon \log\left(1 + \log N(\epsilon)\right) = 0 \tag{12.3.6}$$

if $\alpha = 1$.

Remark. Relations (12.2.1), (12.3.3) and (12.3.5) are extensions to the case $1 < \alpha < 2$ of the corresponding Gaussian relations (12.1.2), (12.1.4) and (12.1.5) because $\alpha' = \alpha$ when $\alpha = 2$. One can now clearly notice the "gap" between the necessary conditions for sample path regularity given in Theorem 12.3.1 and the sufficient conditions of Theorem 12.2.1. The gap in the case $1 < \alpha < 2$ is much wider than in the Gaussian case $\alpha = 2$.

PROOF OF THEOREM 12.3.1: We will omit the proof of (12.3.2) and refer the reader to Talagrand (1988) instead. For (12.3.1), fix any $\epsilon > 0$ and use Exercise 12.2 to conclude that there are points $t_1, \ldots, t_{N(\epsilon)}$ in T such that for any $i, j = 1, \ldots, N(\epsilon),\ i \neq j,\ d_X(t_i, t_j) \geq \epsilon$. Obviously, (12.3.1) will follow once we show that for some absolute constant K_α

$$E \max_{i=1,\ldots,N(\epsilon)} |X(t_i)| \geq K_\alpha \epsilon \left(\log N(\epsilon)\right)^{1/\alpha'}. \tag{12.3.7}$$

Applying Proposition 3.11.1 to the $S\alpha S$ random vector $(X(t_1),\ldots,X(t_{N(\epsilon)}))$, we see that it can be represented as a probability mixture of zero mean Gaussian vectors. Namely,

$$(X(t_1),\ldots,X(t_{N(\epsilon)})) \stackrel{d}{=} (Z_1,\ldots,Z_{N(\epsilon)}),$$

where the $S\alpha S$ random vector$\mathbf{Z} = (Z_1,\ldots,Z_{N(\epsilon)})$ is defined on the product of two probability spaces, $(\Omega_1 \times \Omega_2, \mathcal{F}_1 \times \mathcal{F}_2, P_1 \times P_2)$, and for each $\omega_1 \in \Omega$, $\mathbf{Z}(\omega_1,\cdot)$ is a zero mean Gaussian vector. Let

$$d_{\omega_1}(i,j) = (E_2(Z_i(\omega_1,\omega_2) - Z_j(\omega_1,\omega_2))^2)^{1/2}, i,j \in \{1,\ldots,N(\epsilon)\},$$

and note that for any real θ and $i,j \in \{1,\ldots,N(\epsilon)\}$

$$\begin{aligned} e^{-|\theta|^\alpha d_X(t,t_j)^\alpha} &= E\exp\{i\theta(X(t_i) - X(t_j))\} \\ &= E\exp\{i\theta(Z_i - Z_j)\} \\ &= E_1\big(E_2 \exp i\theta(Z_i - Z_j)\big) \\ &= E_1 e^{-\frac{1}{2}\theta^2 d_{\omega_1}(i,j)^2}. \end{aligned} \tag{12.3.8}$$

This identifies $d_{\omega_1}(i,j)^2$ as a positive $\alpha/2$-stable random variable. For a fixed $\omega_1 \times \Omega_1$, let $\Delta(\omega_1) = \min_{i,j\in\{1,\ldots,N(\epsilon)\},\ i\neq j} d_{\omega_1}(i,j)$. We immediately conclude from (12.1.6) and Exercise 12.2 that

$$E_2 \max_{i=1,\ldots,N(\epsilon)} |Z_i| \geq K^{-1}\frac{1}{2}\Delta(\omega_1)(\log N(\epsilon))^{1/2},$$

so that

$$E \max_{i=1,\ldots,N(\epsilon)} |X_i| = E \max_{i=1,\ldots,N(\epsilon)} |Z_i| \geq \frac{1}{2K}(\log N(\epsilon))^{1/2} E_1\Delta(\omega_1). \tag{12.3.9}$$

By Markov's inequality, for any $\delta > 0$ and $\theta > 0$,

$$\begin{aligned} P(d_{\omega_1}(i,j) \leq \delta) &= P\Big(e^{-\frac{1}{2}\theta^2 d_{\omega_1}(i,j)^2} \geq e^{-\frac{1}{2}\theta^2\delta^2}\Big) \\ &\leq \frac{E_1 e^{-\frac{1}{2}\theta^2 d_{\omega_1}(i,j)^2}}{e^{-1/2\theta^2\delta^2}} \\ &= \exp\Big\{\frac{1}{2}\theta^2\delta^2 - |\theta|^\alpha d_X(t_i,t_j)^\alpha\Big\}. \end{aligned}$$

Choosing $\theta = \delta^{-2/(2-\alpha)} d_X(t_i,t_j)^{\alpha/(2-\alpha)}$, we conclude that for any $\delta > 0$

$$P\big(d_{\omega_1}(i,j) \leq \delta\big) \leq \exp\Big\{-\frac{1}{2}(\delta d_X(t_i,t_j))^{-2\alpha/(2-\alpha)}\Big\}.$$

Using Exercise 12.3, we immediately conclude that for a finite positive constant k_α,

$$\begin{aligned} E\Delta(\omega_1) &\ge k_\alpha\Big(1+\log N(\epsilon)^{-(2-\alpha)/(2\alpha)} \min_{i,j\in S1,\dots,N(\epsilon),\ i\neq j} d_X(t_i,t_j)\Big) \\ &\ge k_\alpha\big(1+\log N(\epsilon)\big)^{-(2-\alpha)/(2\alpha)}\epsilon. \end{aligned}$$

Substituting the last estimate into (12.3.9) yields (12.3.7).

Further, if $\{X(t),\ t\in T\}$ is sample bounded, then, by Proposition 10.2.1, $E\sup_{t\in T}|Y(t)|<\infty$, and now (12.3.3) follows from (12.3.1). Similarly, (12.3.4) follows from (12.3.2).

Let (T,d_X) be relatively compact, and assume that $\{X(t),\ t\in T\}$ is sample continuous with $\alpha=1$. It again follows from Proposition 10.2.1 that

$$\lim_{\delta\to 0} E_1 \sup_{d_X(t,s)<\delta}|Y(t)-Y(s)|^{1/2}=0.$$

Fix an $\eta>0$ and choose a $\delta>0$ such that $E\sup_{d_X(t,s)<\delta}|Y(t)-Y(s)|^{1/2}\le\eta$. We can cover T with finitely many open d_X-balls of radius δ, $B_1,\dots,B_m$. Then

$$E_1\sup_{t,s\in B_i}|Y(t)-Y(s)|^{1/2}\le\eta,$$

$i=1,\dots,m$. It follows from (12.3.2) that for each $i=1,\dots,m$ and $\epsilon>0$,

$$\epsilon\log\big(1+\log\widetilde{N}_i(\epsilon)\big)\le K_1^{-1}\eta,$$

where $\widetilde{N}_i(\epsilon)$ is the minimal number of open balls of radius ϵ needed to cover the set $\{(s,t)\colon s\in B_i, t\in B_i\}=B_i\times B_i$ in the metric

$$\tilde{d}((s_1,t_1),(s_2,t_2))=\|(X(t_1)-X(s_1))-(X(t_2)-X(s_2))\|_1.$$

If $N_i(\epsilon)$ denotes the minimal number of open d_X-balls of radius ϵ needed to cover B_i, then, clearly, $N_i(\epsilon)\le\widetilde{N}_i(\epsilon/2)$ (see also Exercise 12.2). Therefore, for any $\epsilon>0$,

$$\epsilon\log\big(1+\log N_i(2\epsilon)\big)\le K_1^{-1}\eta,$$

$i=1,\dots,m$, and so for any $\epsilon>0$,

$$N(\epsilon)\le\sum_{i=1}^m N_i(\epsilon)\le m\left(\exp\left(e^{2K_1^{-1}\eta/\epsilon}-1\right)\right).$$

Therefore

$$\epsilon\log\Big(1+\log N(\epsilon)\Big)\le\epsilon\log\Big(\log m+e^{2K_1^{-1}n/\epsilon}\Big)$$

and

$$\overline{\lim}_{\epsilon \to 0}\, \epsilon \log\left(1 + \log N(\epsilon)\right) \le 2K_1^{-1}\eta.$$

Letting $\eta \to 0$ yields (12.3.6).

If $1 < \alpha < 2$, the same argument (using (12.3.1)) proves (12.3.5). This completes the proof of the theorem. ■

Although the "gap" between the result of Theorems 12.2.1 and 12.3.1 is wide, we saw that the *sufficient* conditions of Theorem 12.2.1 cannot, in general, be improved. Similarly, one cannot hope to improve significantly the *necessary* condition of Theorem 12.3.1. For example, if $\{X(t),\ 0 \le t \le 1\}$ is a real stationary harmonizable $S\alpha S$ process (see Section 6.5), then it is sample bounded and sample continuous if and only if

$$\int_0^1 (\log N(\epsilon))^{1/\alpha'} d\epsilon < \infty, \quad 1 < \alpha < 2, \tag{12.3.10}$$

and

$$\int_0^1 \log\left(1 + \log N(\epsilon)\right) d\epsilon < \infty, \quad \alpha = 1, \tag{12.3.11}$$

(see Marcus and Pisier (1984) and Talagrand (1990)). One has in this case a perfect extension of the necessary and sufficient condition (12.1.2) for sample boundedness and sample continuity of *stationary* Gaussian processes. The relations (12.3.10) and (12.3.11) are in fact *necessary* for the sample boundedness and continuity of any *stationary* $S\alpha S$ process, $1 \le \alpha < 2$ (Nolan 1991).

Finally, we note that even majorizing measures, a more complicated tool than metric entropy, fail to produce necessary and sufficient conditions for sample path continuity and boundedness of $S\alpha S$ processes with $1 \le \alpha < 2$ (Talagrand 1988).

12.4 Boundedness and continuity of self-similar α-stable processes

In this section, we collect a number of facts, scattered throughout the book, which are related to boundedness and continuity of α-stable H-self-similar processes.

By Corollary 7.1.11, the admissible values of (α, H) lie in the domain

$$\mathcal{D} = \{0 < \alpha \le 2,\ 0 < H \le \max(1, \alpha^{-1})\},$$

illustrated in Figure 12.1. Divide this domain into the following regions:

$$A_1 \;=\; \{(H,\alpha) :\ \alpha = 2,\ 0 < H \le 1/2\},$$

$$
\begin{aligned}
A_2 &= \{(H,\alpha):\ 0<\alpha<2,\ 0<H\le 1/2\},\\
B_1 &= \{(H,\alpha):\ \alpha=2,\ 1/2<H\le 1\},\\
B_2 &= \{(H,\alpha):\ 1/H<\alpha<2,\ 1/2<H\le 1\},\\
B_3 &= \{(H,\alpha):\ \alpha=1/H,\ 1/2<H\le 1\},\\
B_4 &= \{(H,\alpha):\ 0<\alpha<1/H,\ 1/2<H\le 1\},\\
C_1 &= \{(H,\alpha):\ \alpha=1/H,\ H>1\},\\
C_2 &= \{(H,\alpha):\ 0<\alpha<1/H,\ H>1\}.
\end{aligned}
$$

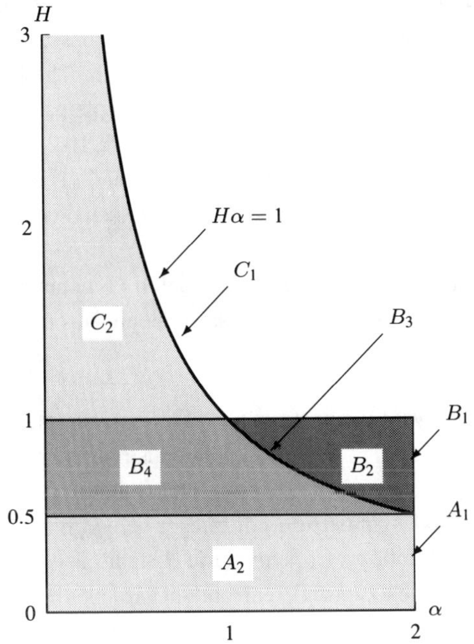

Figure 12.1: The domain $\mathcal{D}$.

The corresponding process can be either:

- sample continuous,
- discontinuous with sample paths which are bounded on any compact interval,
- unbounded on any interval of positive length.

We shall now examine what happens in the various regions of $\mathcal{D}$.

Surprisingly, when $0<\alpha<2$, the region B_2 is the only one where one can make a general statement on sample continuity:

Theorem 12.4.1 *Let $\{X(t),\ t\ge 0\}$ be a strictly α-stable, $1<\alpha<2$, H-sssi process such that*

$$\frac{1}{\alpha}<H\le 1, \tag{12.4.1}$$

that is, suppose $(\alpha, H) \in B_2$. *Then* $\{X(t),\ t \geq 0\}$ *is sample continuous.*

This theorem is a generalization of Example 12.2.3, and its proof is identical to that of the example. In regions A_1 and B_1, the process is fractional Brownian motion, which is, as we know, sample continuous (Exercise 10.1). If A is a $S_{\alpha/2}(1,1,0)$ random variable independent of the H-sssi fractional Brownian motion Z, then the subordinated process $X(t) = A^{1/2}Z(t),\ t \geq 0$, is clearly $S\alpha S$, H-sssi, and sample continuous. This gives examples of sample continuous processes in the regions A_1, B_1, A_2, B_2, B_3 and B_4 of $\mathcal{D}$.

The well-balanced linear fractional stable motions with $H < 1/\alpha$, and the Takenaka process are examples of processes in regions A_2 and B_4 which are not sample bounded on any interval of positive length. The log-fractional stable motion is an example of a process in B_3 whose sample paths are unbounded on any interval of positive length. (See Examples 10.2.5, 10.2.7 and 10.2.6).

Of course, the only $S\alpha S$ H-sssi process in region C_1 is the $S\alpha S$ Lévy motion which is not sample continuous, but is sample bounded on compact intervals (see Exercise 9.5).

The following two examples show that region C_2 contains both sample continuous processes and processes which are unbounded on intervals of positive length.

Example 12.4.2 We want to construct a $S\alpha S$ H-sssi sample continuous process $\{X(t),\ t \geq 0\}$ with (α, H) in region C_2. Our construction is based on a particular family of H-sssi processes due to Kesten and Spitzer (1979). Let $U_1, U_2, \ldots$ be a sequence of (say) symmetric i.i.d. integer-valued random variables in the domain of normal attraction of a symmetric θ-stable law, $1 < \theta \leq 2$, and let $\xi(1), \xi(2), \ldots$ be a sequence of i.i.d. real random variables in the domain of normal attraction of a β-stable law, $0 < \beta \leq 2$. The two sequences are independent. Set $S_k = U_1 + \cdots + U_k$, and let $W_n = \sum_{k=1}^{n} \xi(S_k)$. Then the sequence of stochastic processes $\{D_n(t) = n^{-\delta} W_{nt},\ t \in [0, \infty)\}$ (with W_s defined, for a non-integer s, by linear interpolation) converges, as $n \to \infty$ weakly in $C[0, \infty)$, to a sample continuous δ-self-similar process $\{\Delta(t),\ t \geq 0\}$ with stationary increments where $\delta = 1 - 1/\theta + 1/(\theta\beta)$ (Kesten and Spitzer (1979)).

The law of Δ depends only on θ, β and the tails of U_is and $\xi(j)$s, i.e., on $\lim_{\lambda \to \infty} \lambda^\theta P(U_i > \lambda)$ and $\lim_{\lambda \to \infty} \lambda^\beta P(\pm\xi(j) > \lambda)$. Further, it is easy to show that for every $t > 0$ and every $0 < p < \beta$,

$$E|\Delta(t)|^p < \infty.$$

We now construct a $S\alpha S$ H-sssi process with a given (α, H) in region C_2 as follows. In that region, $0 < \alpha < 1$ and $1 < H < 1/\alpha$. Choose a $\beta \in (\alpha, 1)$ such that

$$\theta = \frac{1-\beta}{\beta}\frac{1}{H-1} \in [1,2).$$

Let $\{\Delta(t),\, t \geq 0\}$ be the corresponding Kesten-Spitzer process with all continuous sample paths, defined, say, on a probability space $(\Omega_1, \mathcal{F}_1, P_1)$. Letting M be a $S\alpha S$ random measure on $(\Omega_1, \mathcal{F}_1)$ with control measure P_1, we now define

$$X(t) = \int_{\Omega_1} \Delta(t, \omega_1) M(d\omega_1),\; t \geq 0. \tag{12.4.2}$$

Then $\{X(t),\, t \geq 0\}$ is a $S\alpha S$ H-sssi process (Theorem 7.9.1). Since $\{\Delta(t),\, t \geq 0\}$ has continuous sample paths and $0 < \alpha < 1$, sample continuity of $\{X(t),\, t \geq 0\}$ will follow from Theorem 10.4.2 once we show that

$$E_1 \sup_{0 \leq t \leq 1} |\Delta(t)|^\alpha < \infty. \tag{12.4.3}$$

In terms of the approximating processes D_n, it is enough, of course, to assume that the $\xi(x)$s are themselves $S\beta S$, and then to show that for every $0 < p < \beta$, there is a constant $C < \infty$ such that, for every $n \geq 1$,

$$E \sup_{0 \leq t \leq 1} |D_n(t)|^p \leq C. \tag{12.4.4}$$

Defining $N_i(x)$ as the number of times the random walk $\{S_k,\, k \geq 1\}$ visits x up to time i, we denote for each $n \geq 1$ and $x \in \mathbb{Z}$, a random vector $\mathbf{N}^{(n,x)} \in C[0,1]$ by $\mathbf{N}^{(n,x)}(i/n) = N_i(x)$ for $i = 0, 1, \ldots, n$, and with linear interpolation between these points. Then

$$\begin{aligned} E \sup_{0 \leq t \leq 1} |D_n(t)|^p &= n^{-p\delta} E \sup_{0 \leq t \leq 1} \Big| \sum_{x=-\infty}^{\infty} \xi(x) N_{[nt]}(x) \Big|^p \\ &\leq n^{-p\delta} E \Big\| \sum_{x=-\infty}^{\infty} \xi(x) \mathbf{N}^{(n,x)} \Big\|^p. \end{aligned} \tag{12.4.5}$$

Write the expectation in the right-hand side of (12.4.5) as $E_{\mathbf{N}} E_\xi$. For fixed $\{\mathbf{N}^{(n,x)},\, x \in \mathbb{Z}\}$, the vector $\mathbf{Y} = \sum_{x=-\infty}^{\infty} \xi(x) \mathbf{N}^{(n,x)}$ is a $S\beta S$ random element of $C[0,1]$ with spectral measure

$$\begin{aligned} \Gamma_{\mathbf{Y}} &= \sum_{x=-\infty}^{\infty} \|\mathbf{N}^{(n,x)}\|^\beta \frac{1}{2} \left(\delta_{\{\mathbf{N}^{(n,x)}/\|\mathbf{N}^{(n,x)}\|\}} + \delta_{\{-\mathbf{N}^{(n,x)}/\|\mathbf{N}^{(n,x)}\|\}} \right) \\ &= \sum_{x=-\infty}^{\infty} N_n(x)^\beta \frac{1}{2} \left(\delta_{\{\mathbf{N}^{(n,x)}/\|\mathbf{N}^{(n,x)}\|\}} + \delta_{\{-\mathbf{N}^{(n,x)}/\|\mathbf{N}^{(n,x)}\|\}} \right). \end{aligned}$$

Thus for any ϕ in the dual space,

$$Ee^{i\phi(\mathbf{Y})} = \exp\Big\{-\int_{B_1} |\phi(\mathbf{x})|^\beta \Gamma_{\mathbf{Y}}(d\mathbf{x})\Big\},$$

where B_1 is the unit ball in $C[0,1]$.

Now $C[0,1]$ is a Banach space and all Banach spaces are of stable type $\beta \in (0,1)$.[1] Applying the inequality (12.4.6) with $E = C[0,1]$, $\|\cdot\| =$ supremum norm and $0 < \beta < 1$, we conclude that given $0 < p < \beta < 1$, there is a constant $C_1 < \infty$ such that

$$\begin{aligned} E_\xi \|\mathbf{Y}\|^p &\equiv E\Big\| \sum_{x=-\infty}^{\infty} \xi(x)\mathbf{N}^{(n,x)}\Big\|^p \\ &\le C_1\Big(\sum_{x=-\infty}^{\infty} \|\mathbf{N}^{(n,x)}\|^\beta\Big)^{p/\beta} \\ &= C_1\Big(\sum_{x=-\infty}^{\infty} N_n(x)^\beta\Big)^{p/\beta}. \end{aligned}$$

Hence, for every $n \ge 1$,

$$E \sup_{0\le t\le 1} |D_n(t)|^p \le C_1 n^{-p\delta} E\Big(\sum_{x=-\infty}^{\infty} N_n(x)^\beta\Big)^{p/\beta}. \tag{12.4.7}$$

Let $K_n = \max_{i\le n} |S_i|$. Then $N_n(x) = 0$ for every $|x| > K_n$. By Hölder's inequality $(0 < \beta < 1)$,

$$\Big(\sum_{x=-\infty}^{\infty} N_n(x)^\beta\Big)^{1/\beta} \le \Big(2K_n+1\Big)^{-1+1/\beta} \sum_{x=-K_n}^{K_n} N_n(x) = \Big(2K_n+1\Big)^{-1+1/\beta} n. \tag{12.4.8}$$

It is now sufficient to show that for every $0 < r < \theta$ there is a constant $C^{(r)} < \infty$ such that for every $n \ge 1$,

$$EK_n^r \le C^{(r)} n^{r/\theta} \tag{12.4.9}$$

---

[1] Let ξ_i be i.i.d. $S\beta S$ random variables with $0 < \beta \le 2$. Then the sum $\sum_{i=0}^\infty \xi_i x_i$ converges a.s. if and only if $\sum_{i=0}^\infty |x_i|^\beta < \infty$. This property does not extend to arbitrary Banach spaces. Let E be a Banach space with norm $\|\cdot\|$. The Banach space E is said to have *stable-type* β if $\sum_{i=1}^\infty \xi_i \mathbf{x}_i$ converges a.s. for all $\mathbf{x}_i$s in E such that $\sum_{i=1}^\infty \|\mathbf{x}_i\|^\beta < \infty$. No $E \ne \{0\}$ has stable type greater than 2. A Hilbert space has stable type 2. All Banach spaces have stable type $\beta \in (0,1)$.

The Banach space E has stable type β if and only if for some (each) $p \in (0,\beta)$,

$$\Big(E\Big\|\sum_{i=1}^n \xi_i \mathbf{x}_i\Big\|^p\Big)^{1/p} \le c\Big(\sum_{i=1}^n \|\mathbf{x}_i\|^\beta\Big)^{1/\beta} \tag{12.4.6}$$

for some $0 < c < \infty$ independent of n and all $\mathbf{x}_1, \ldots, \mathbf{x}_n \in E$. For further details see Linde (1986). For proofs see Schwartz (1981).

because then (12.4.3) follows from (12.4.7), (12.4.8) and (12.4.9) with $r = p(-1 + 1/\beta)$.

To prove (12.4.9), note that by Lévy's inequality,

$$\begin{aligned} EK_n^r &\equiv E \max_{i \le n} |U_1 + \cdots + U_i|^r \\ &\le 2E|U_1 + \cdots + U_n|^r = 2E_\epsilon E_U |\epsilon_i U_1 + \cdots + \epsilon_n U_n|^r, \end{aligned}$$

where the ϵ_is are i.i.d., independent of the U_is, and Rademacher, that is, $P(\epsilon_1 = 1) = P(\epsilon_1 = -1) = 1/2$. By Khinchine's inequality (Exercise 11.1),

$$2E_\epsilon E_U |\epsilon_i U_1 + \cdots + \epsilon_n U_n|^r \le C_2 E(U_1^2 + \cdots + U_n^2)^{r/2}$$

for some $C_2 < \infty$. Since $P(U_1 > \lambda) = O(\lambda^{-\theta})$ as $\lambda \to \infty$, $P(U_1^2 > \lambda) = O(\lambda^{-\theta/2})$ and therefore

$$U_1^2 \overset{\text{st}}{\le} C_3(1 + V_1)$$

(stochastic ordering), where $V_1 \sim S_{\theta/2}(1,1,0)$. Using the stability property of the Vs,

$$\begin{aligned} E(U_1^2 + \cdots + U_n^2)^{r/2} &\le C_3^{r/2} E(n + V_1 + V_2 + \cdots + V_n)^{r/2} \\ &\le C_3^{r/2} E(n + n^{2/\theta} V_1)^{r/2} \\ &\le C_4 n^{r/2} + n^{r/\theta} E\, V_1^{r/2} \\ &\le C_5 n^{r/\theta} \end{aligned}$$

for some $C_5 < \infty$. This concludes the argument.

Example 12.4.3 We shall construct a $S\alpha S$ H-sssi process $\{X(t),\ t \ge 0\}$ with (α, H) in region C_2 which is not sample bounded on intervals of positive length.

Given an (α, H) in region C_2, set

$$\begin{aligned} \alpha_1 &= \left(\frac{\alpha}{H}\right)^{1/2} \in (\alpha, \alpha^{1/2}] \subset (0,1), \\ H_1 &= (\alpha H)^{1/2} \in [\alpha^{1/2}, 1) \subset (0,1). \end{aligned}$$

Let $\{Y(t),\ t \ge 0\}$ be a symmetric linear fractional 1-stable motion with index of self-similarity H_1. Note that this process is not sample bounded on every interval of positive length. Now let $\{Z(t),\ t \in \mathbb{R}\}$ be an increasing strictly α_1-stable Lévy motion, independent of $\{Y(t),\ t \ge 0\}$. The two processes are assumed to be defined on probability spaces $(\Omega_1, \mathcal{F}_1, P_1)$ and $(\Omega_2, \mathcal{F}_2, P_2)$, respectively, and we choose separable and measurable versions of these two processes.

This is always possible by Theorems 9.2.5 and 11.1.1. Define on the product probability space the composition process

$$W(t) = Z(Y(t)),\ t \ge 0.$$

It follows by Exercise 7.1 that $\{W(t),\ t \geq 0\}$ is $H_1 \cdot (1/\alpha_1) = H$-sssi. Further, for every $0 \leq a < b < \infty$ we have $\sup_{a \leq t \leq b} |Y(t)| = \infty$ a.s. Therefore, for P_1-almost every ω_1 there is, say, a countable set $t_1, t_2, \ldots$ in (a, b) such that $|Y(t_n)| \to \infty$ as $n \to \infty$. Thus,

$$P_2\Big(\sup_{a \leq t \leq b} |Z(Y(t))| = \infty\Big) \geq P_2\Big(\sup_{s \in \{Y(t_n),\ n \geq 1\}} |Z(s)| = \infty\Big) = 1.$$

Therefore

$$P_1 \times P_2\Big(\sup_{a \leq t \leq b} |W(t)| = \infty\Big) = 1. \tag{12.4.10}$$

Note also that since $\alpha < \alpha_1$, we have

$$E_1 \times E_2 |W(t)|^\alpha < \infty \tag{12.4.11}$$

for every $t > 0$.

Now again let M be a $S\alpha S$ random measure on $(\Omega_1 \times \Omega_2,\ \mathcal{F}_1 \times \mathcal{F}_2)$ with control measure $P_1 \times P_2$. Relation (12.4.11) ensures that the process

$$X(t) = \int_{\Omega_1 \times \Omega_2} W(t, (\omega_1, \omega_2)) M(d(\omega_1, \omega_2)),\ t \geq 0,$$

is well defined. Moreover, $\{X(t),\ t \geq 0\}$ is a $S\alpha S$ H-sssi process, and since by (12.4.10), $E_1 \times E_2 \sup_{a \leq t \leq b} |W(t)|^\alpha = \infty$ for every $0 \leq a < b < \infty$, it follows from Theorem 10.2.3 that $\{X(t),\ t \geq 0\}$ is not sample bounded on intervals of positive length.

12.5 Exercises

Exercise 12.1 Let h: $(0, \infty) \to (0, \infty)$ be a non-increasing function such that $h(\theta) \uparrow \infty$ as $\theta \downarrow 0$. Show that there is a $S\alpha S$ process $\{X(t),\ 0 \leq t \leq 1\}$, $1 \leq \alpha < 2$, such that $\{X(t),\ 0 \leq t \leq 1\}$ is not sample bounded, and for every $0 < \theta < 1$,

$$\int_\theta^1 N(\epsilon)^{1/\alpha} d\epsilon \leq h(\theta),$$

where $N(\epsilon)$ is the covering number associated with the metric d_X generated by the process.

Hint: Go through the following steps:

(i) Show that it is enough to consider only piecewise constant functions h taking values in the set $\{1, 2, 3, \ldots\}$. Namely, there is a sequence $1 = C_1 \geq C_2 \geq \cdots \to 0$ such that

$$h(\theta) = k \ \text{ if } \ C_{k+1} < \theta \leq C_k,\ k = 1, 2, \ldots.$$

Moreover, the sequence $\{C_k,\ k = 1, 2, \ldots\}$ may be chosen in such a way that $C_k^{-\alpha}$ is a positive integer for any $k = 1, 2, \ldots$.

(ii) Let $X(0) = 0$, and for $\frac{1}{n+1} < t \le \frac{1}{n}$, $X(t) = a_n X_n$, $n = 1, 2, \ldots$, where $X_1, X_2, \ldots$ are i.i.d. $S_\alpha(1, 0, 0)$ random variables, and $\{a_n,\ n = 1, 2, \ldots\}$ is a sequence of positive numbers with $a_1 = 1$, decreasing to 0. Show that $\{X(t),\ 0 \le t \le 1\}$ is sample bounded if and only if $\sum_{n=1}^{\infty} a_n^\alpha < \infty$.

(iii) Show that for any $a_{n+1} < \theta \le a_n$, $N(\theta) = n + 1$, and

$$\int_\theta^1 N(\epsilon)^{1/\alpha} d\epsilon \le \sum_{k=1}^{n} (a_k - a_{k+1})(k+1)^{1/\alpha}, \quad n = 1, 2, \ldots.$$

Let $b_n = h(a_n)$, $n = 1, 2, \ldots$. Conclude that it is enough to find a sequence $\{a_n,\ n = 1, 2, \ldots\}$ such that

$$\sum_{k=1}^{n} (a_k - a_{k+1})(k+1)^{1/\alpha} \le b_n, \quad n = 1, 2, \ldots,$$

and $\sum_{n=1}^{\infty} a_n^\alpha = \infty$.

(iv) For a function h as in (i), let

$$a_n = C_k \text{ if } m_k \le n < m_{k+1}, \quad k = 1, 2, \ldots,$$

where $1 = m_1 \le m_2 \le m_3 \le \ldots$ is a sequence of positive integers that tends to infinity. Then $b_n = k$ if $m_k \le n < k_{m+1}$. Conclude that it is enough to choose the sequence $\{m_k,\ k = 1, 2, \ldots\}$ in such a way that

$$\sum_{i=1}^{k} (C_i - C_{i+1}) m_{i+1}^{1/\alpha} \le k, \quad k = 1, 2, \ldots,$$

and

$$\sum_{k=1}^{\infty} C_k^\alpha (m_{k+1} - m_k) = \infty.$$

(v) Show that the choice $m_1 = 1$,

$$m_k = \sum_{i=1}^{k-1} C_i^{-\alpha}, \quad i = 2, 3, \ldots,$$

works.

Exercise 12.2 Let C be a set, and d a (pseudo) metric on C. For an $\epsilon > 0$ define

$$N(C, \epsilon) \quad = \quad \text{the minimal number of open } d\text{-balls of radius}$$

ϵ centered in C, covering the whole of C,

$M(C, \epsilon)$ = the largest number of points in C such that the distance between any two different points is at least ϵ.

Let $C_1 \subseteq C$. Show that

$$N(C_1, \epsilon) \leq M(C_1, \epsilon) \leq M(C, \epsilon) \leq N(C, \epsilon/2).$$

Exercise 12.3 Let $X_1, X_2, \ldots$ be non-negative random variables such that for some $\beta \geq 1$,

$$P(X_i > x) \leq e^{-x^\beta}, \ x \geq 0, \ i = 1, 2, \ldots.$$

Show that there is a finite number K_β such that for any $n = 1, 2, \ldots$,

$$E \max\{X_1, \ldots, X_n\} \leq K_\beta (1 + \log n)^{1/\beta}.$$

Chapter 13

Integral representation

Let $\{X(t),\ t \in T\}$ be an α-stable process. We have seen in the previous chapters that it is very convenient to study the properties of such a process via an integral representation of the type

$$X(t) = \int_E f(t,x)M(dx) + \eta(t),\ t \in T.$$

Here, our goal is to determine the α-stable processes that can be so represented. We shall also examine whether convenient choices of E and M are possible, for example, $E = (0,1)$ and M having Lebesgue control measure. We may also want M to be symmetric if X is $S\alpha S$ and to have skewness intensity $\beta(\cdot) \equiv 1$ if X is skewed α-stable.

Because the construction of the integral representation involves many steps, we start by describing the main ideas. Suppose firstly that T is countable. The spectral representation of an α-stable random vector $(X_1, X_2, \ldots, X_d)$ in $\mathbb{R}^d$ involves a spectral measure Γ on $\mathbb{R}^d$. Since we want to treat all ds simultaneously, it is best to establish a representation for a vector $\mathbf{X} = (X_1, X_2, \ldots)$ with an infinite number of components. To accomplish this, we must view $\mathbf{X}$ as a vector in a suitable Banach space. We shall suppose $\lim_{i\to\infty} X_i = 0$ a.s., so that $\mathbf{X} \in c_0$, the Banach space of all sequences of real numbers converging to 0, with norm $\|\mathbf{x}\| = \sup_i |x_i|$. Proposition 13.1.1 below states that there is a measure Γ on the unit ball B_1 of c_0 that can be used to represent any finite-dimensional distribution of $\mathbf{X}$.

The resulting representation is defined on $E = B_1$, the unit sphere of c_0, and not on the interval $E = (0,1)$. To pass from B_1 to $(0,1)$, we make a change of variables. This has to be done carefully because the measure Γ may have atoms. Write $\Gamma = \Gamma_a + \Gamma_c$ where Γ_a is the atomic part of the measure. Γ_a is defined on a subset A of B_1 and Γ_c on $B_1 \backslash A$. Firstly, map the set $B_1 \backslash A$ where Γ_c is defined, into the interval $(0, \frac{1}{2}]$. The map is chosen so as to preserve measurability and to transform Γ_c on $B_1 \backslash A$ into Lebesgue measure on $(0, \frac{1}{2}]$. (Theorem 14.3.9 of Royden (1963) is used for this purpose.) Secondly, spread the discrete mass of A uniformly on $(\frac{1}{2}, 1)$. This gives the desired representation, but only for $\mathbf{X}$ satisfying $\lim_{i\to\infty} X_i = 0$ a.s.. To consider a general α-stable $\mathbf{X} = (X_1, X_2, \ldots)$, set $Y_i = a_i X_i$ with $a_i \downarrow 0$ fast enough to ensure that $Y_i \to 0$ a.s. as $i \to \infty$. Apply the previous argument to Y_i to obtain an integral representation involving an integrand, say g_i, then set $f_i = a_i^{-1} g_i$ to obtain the desired integral representation for X_i. The result is formally stated in Theorem 13.1.2 below.

We must finally consider an arbitrary index set T, not merely a countable one. If we insist on $E = (0, 1)$ and Lebesgue control measure, then we must suppose that $\{X(t),\ t \in T\}$ satisfies Condition S. This condition, introduced in Section 3.11, states that $X(t)$ is separable in probability, i.e., there is a countable set $T_0 = \{t_1, t_2, \ldots\} \subseteq T$ so that at each $t \in T,\ X(t)$ can be approximated in probability by $X(t_{i_k})$ where $t_{i_1}, t_{i_2}, \ldots \in T_0$. Since the integral representation holds for $\{X(t_{i_k})\}$, it will hold for $X(t)$. The result is stated in Theorem 13.2.1 below for any α if $\{X(t),\ t \in T\}$ is $S\alpha S$ but only for $\alpha \neq 1$ if $\{X(t),\ t \in T\}$ is skewed. It is not known whether it holds in the skewed $\alpha = 1$ case.

What happens if $\{X(t),\ t \in T\}$ does not satisfy Condition S? Then the choice $E = (0, 1)$ and Lebesgue control measure is impossible, but Theorem 13.2.2 states without proof that in the $S\alpha S$ case, an integral representation exists, for some space E and some control measure m.

Let us now fill in the details.

13.1 Countable parameter space

We suppose in this section that T is countable. The following proposition assumes, in addition, $\lim_{i\to\infty} X_i = 0$ a.s.

Proposition 13.1.1 *Let $\{X_i,\ i = 1, 2, \ldots\}$ be an α-stable process satisfying*

$$\lim_{i\to\infty} X_i = 0 \text{ a.s..}$$

Let c_0 be the Banach space of all sequences of real numbers converging to 0. Then there is a finite measure Γ on the unit sphere B_1 of c_0 and a constant vector $\boldsymbol{\mu} = (\mu_1, \mu_2, \ldots) \in c_0$ such that, for any $n = 1, 2, \ldots$ and real numbers $\theta_1, \ldots, \theta_n$,

$$E \exp\Big\{ i \sum_{j=1}^{n} \theta_j X_j \Big\} =$$

$$\exp\Big\{-\int_{B_1}\Big|\sum_{j=1}^n\theta_jx_j\Big|^\alpha\Big(1-i\,\mathrm{sign}\Big(\sum_{j=1}^n\theta_jx_j\Big)L\Big(\alpha;\sum_{j=1}^n\theta_jx_j\Big)\Big)\Gamma(d\mathbf{x})$$
$$+\,i\sum_{j=1}^n\theta_j\mu_j\Big\}, \tag{13.1.1}$$

where, as usual

$$L(\alpha;a)=\begin{cases}\tan(\pi\alpha/2) & \text{if } \alpha\neq 1,\\ -(2/\pi)\ln|a| & \text{if } \alpha=1.\end{cases} \tag{13.1.2}$$

PROOF: By assumption, $\mathbf{X}=(X_1,X_2,\ldots)$ is a random vector in c_0. We may and shall assume that it is not an a.s. zero vector. Let $P_{\mathbf{X}}$ denote its distribution. For a given $0<r<\alpha$, let $P_{\mathbf{X}}^{(r)}$ be a finite Borel measure on c_0, absolutely continuous with respect to $P_{\mathbf{X}}$, defined by

$$\frac{dP_{\mathbf{X}}^{(r)}}{dP_{\mathbf{X}}}(\mathbf{x})=\|\mathbf{x}\|^r.$$

The measure $P_{\mathbf{X}}^{(r)}$ is finite by Proposition 10.2.1. Now define a finite Borel measure Γ_r on the unit sphere B_1 by

$$\Gamma_r=(\alpha-r)P_{\mathbf{X}}^{(r)}\circ T^{-1}, \tag{13.1.3}$$

where $T\colon c_0\backslash\{0\}\to B_1$, defined by $T\mathbf{x}=\mathbf{x}/\|\mathbf{x}\|$, maps $\mathbf{x}$ into the unit ball B_1. Note that the total mass $|\Gamma_r|$ of Γ_r equals

$$|\Gamma_r|=(\alpha-r)\int_{c_0}\|\mathbf{x}\|^rP_{\mathbf{X}}(d\mathbf{x})=(\alpha-r)E\sup_{i\geq1}|X_i|^r.$$

Applying Proposition 10.2.1 together with the argument leading to Property 1.2.18 gives $0<\inf_{0<r<\alpha}|\Gamma_r|\leq\sup_{0<r<\alpha}|\Gamma_r|<\infty$. Choose now a sequence $\{r_n,\ n\geq1\}$ of positive numbers increasing to α such that $|\Gamma_{r_n}|$ converges to a finite limit as $r_n\uparrow\alpha$, say, $|\Gamma_\alpha|$. Do the measures Γ_{r_n} converge as well? Clearly, $\{\frac{1}{|\Gamma_{r_n}|}\Gamma_{r_n},\ n\geq1\}$ is a family of probability measures on c_0 supported by the unit sphere B_1. We claim that the sequence $m_n=\frac{1}{|\Gamma_{r_n}|}\Gamma_{r_n}$, $n\geq1$, is relatively compact in the topology of weak convergence, i.e., every subsequence of m_n has in turn a convergent subsequence. By Prohorov's theorem, it is enough to show that the sequence m_n is tight. Regard $\{0,1,2,\ldots,\infty\}$ as a compact metric space with a single cluster point. By the Arzelà-Ascoli theorem, it is enough to show that for each positive numbers ϵ and η there are integers I and n_0 such that, for each $n\geq n_0$,

$$m_n\{\mathbf{x}\in c_0\colon\sup_{i\geq I}|x_i|\geq\epsilon\}\leq\eta. \tag{13.1.4}$$

To prove (13.1.4), note that

$$
\begin{aligned}
m_n\Big(\mathbf{x} \in c_0\colon \sup_{i\ge I} |x_i| \ge \epsilon\Big) \\
&= \frac{\alpha - r_n}{|\Gamma_{r_n}|} \int_{\{\mathbf{x}\colon \sup_{i\ge I} |x_i| \ge \epsilon \|\mathbf{x}\|\}} \|\mathbf{x}\|^{r_n} P_{\mathbf{X}}(d\mathbf{x}) \\
&\le \text{const}(\alpha - r_n) E\Big(\|\mathbf{X}\|^{r_n} 1(\sup_{i\ge I} |X_i| \ge \epsilon \|\mathbf{X}\|)\Big) \\
&\le \text{const}\epsilon^{-r_n}(\alpha - r_n) E \sup_{i\ge I} |X_i|^{r_n}. \qquad (13.1.5)
\end{aligned}
$$

The argument used in Property 1.2.18, when applied to Corollary 10.2.2, gives

$$\lim_{I\to\infty} \overline{\lim}_{n\to\infty}(\alpha - r_n) E \sup_{i\ge I} |X_i|^{r_n} = 0.$$

Relation (13.1.4) now follows from (13.1.5), and so the sequence $\{m_n = \frac{1}{|\Gamma_{r_n}|}\Gamma_{r_n}\}$ is, indeed, tight. There is therefore a probability measure Q on c_0 such that for some subsequence $\{r_{n_k},\ k = 1, 2, \ldots\}$, $|\Gamma_{r_{n_k}}|^{-1}\Gamma_{r_{n_k}} \Rightarrow Q$. Since the measures $\Gamma_{r_{n_k}}$ are concentrated on the unit sphere B_1, so is Q. Define

$$\Gamma = (\alpha C_\alpha)^{-1} |\Gamma_\alpha| Q. \qquad (13.1.6)$$

To obtain $\boldsymbol{\mu}$ express the distribution of each X_j as $X_j \sim S_\alpha(\sigma_j, \beta_j, \eta_j)$, and set

$$\mu_j = \begin{cases} \eta_j & \text{if } \alpha \ne 1, \\ \eta_j + \frac{2}{\pi}\int_{B_1} s_j \log|s_j| \Gamma(ds) & \text{if } \alpha = 1, \end{cases} \qquad (13.1.7)$$

for $j = 1, 2, \ldots$. We claim that Γ and $\boldsymbol{\mu}$ constructed in (13.1.6) and (13.1.7) satisfy (13.1.1).

Firstly, we need some notation. Fix $d \in \{1, 2, \ldots\}$. Let $S_d^{\|\cdot\|_\infty} = \{\mathbf{x}\colon \max_{i=1,\ldots,d} |x_i| = 1\}$ be the unit sphere relative to the L^∞ norm on $\mathbb{R}^d$ and let $\Gamma^{(d)}$ and $\boldsymbol{\mu}^{0,(d)}$ be the parameters in the representation of the characteristic function of the α-stable random vector $(X_1, \ldots, X_d)$, given in Proposition 2.3.8. (To be consistent with the notation of Proposition 2.3.8, we should write $\Gamma^{(d)}_{\|\cdot\|_\infty}$ and $\boldsymbol{\mu}^{0,(d)}_{\|\cdot\|_\infty}$ instead of $\Gamma^{(d)}$ and $\boldsymbol{\mu}^{0,(d)}$, but this is too cumbersome.) It is important to distinguish below between the norm $\|\cdot\|$ on c_0 (defined as $\|\mathbf{x}\| = \sup_{i=1,2,\ldots} |x_i|$, $\mathbf{x} \in c_0$) and the norm $\|\cdot\|_\infty$ on $\mathbb{R}^d$ (defined as $\|\mathbf{x}\|_\infty = \max_{i=1,\ldots,d} |x_i|$, $\mathbf{x} \in \mathbb{R}^d$), between the unit spheres $B_1 = \{\mathbf{x} \in c_0\colon \|\mathbf{x}\| = 1\}$ and $S_d^{\|\cdot\|_\infty} = \{\mathbf{x} \in \mathbb{R}^d : \|\mathbf{x}\|_\infty = 1\}$ and between the measures Γ on B_1 and $\Gamma^{(d)}$ on $S_d^{\|\cdot\|_\infty}$. The usual spectral measure Γ_d on the unit sphere S_d (defined with respect to the Euclidean norm) also appears below.

Suppose firstly $\alpha \neq 1$. By virtue of Proposition 2.3.8, it is sufficient to prove

$$\int_{B_1} |\sum_{j=1}^{d} \theta_j x_j|^\alpha \Gamma(d\mathbf{x}) = \int_{S_d^{\|\cdot\|_\infty}} |\sum_{j=1}^{d} \theta_j s_j|^\alpha \Gamma^{(d)}(d\mathbf{s}) \tag{13.1.8}$$

and

$$\int_{B_1} |\sum_{j=1}^{d} \theta_j x_j|^{<\alpha>} \Gamma(d\mathbf{x}) = \int_{S_d^{\|\cdot\|_\infty}} |\sum_{j=1}^{d} \theta_j s_j|^{<\alpha>} \Gamma^{(d)}(d\mathbf{s}). \tag{13.1.9}$$

Since the integrand in the left-hand side of (13.1.8) can be extended to a bounded continuous function on c_0,

$$\begin{aligned}
\int_{B_1} |\sum_{j=1}^{d} \theta_j x_j|^\alpha \Gamma(d\mathbf{x}) &= (\alpha C_\alpha)^{-1} \lim_{k\to\infty} \int_{B_1} |\sum_{j=1}^{d} \theta_j x_j|^\alpha \Gamma_{r_{n_k}}(d\mathbf{x}) \\
&= (\alpha C_\alpha)^{-1} \lim_{k\to\infty} \int_{B_1} |\sum_{j=1}^{d} \theta_j x_j|^{r_{n_k}} \Gamma_{r_{n_k}}(d\mathbf{x}) \\
&= (\alpha C_\alpha)^{-1} \lim_{k\to\infty} (\alpha - r_{n_k}) \int_{c_0} |\sum_{j=1}^{d} \theta_j x_j|^{r_{n_k}} P_{\mathbf{X}}(d\mathbf{x}) \\
&= (\alpha C_\alpha)^{-1} \lim_{k\to\infty} (\alpha - r_{n_k})\, E|\sum_{j=1}^{d} \theta_j X_j|^{r_{n_k}} \\
&= \sigma^\alpha_{\sum_{j=1}^{d} \theta_j X_j} \qquad \text{(Property 1.2.18)} \\
&= \int_{S_d^{\|\cdot\|_\infty}} |\sum_{j=1}^{d} \theta_j s_j|^\alpha \Gamma^{(d)}(d\mathbf{s})
\end{aligned}$$

proving (13.1.8). Since (13.1.9) follows in the same way, this completes the proof of (13.1.1) in the case $\alpha \neq 1$.

Consider, now, the case $\alpha = 1$. With the same notation as before, (13.1.1) will follow from (13.1.8) and

$$\begin{aligned}
&-\frac{2}{\pi} \int_{B_1} \Big(\sum_{j=1}^{d} \theta_j x_j\Big) \log|\sum_{j=1}^{d} \theta_j x_j| \Gamma(d\mathbf{x}) + \sum_{j=1}^{d} \theta_j \mu_j \\
&= -\frac{2}{\pi} \int_{S_d^{\|\cdot\|_\infty}} \Big(\sum_{j=1}^{d} \theta_j s_j\Big) \log|\sum_{j=1}^{d} \theta_j s_j| \Gamma^{(d)}(d\mathbf{s}) + \sum_{j=1}^{d} \theta_j \mu_j^{0,(d)}.
\end{aligned} \tag{13.1.10}$$

Since the preceding proof for (13.1.8) works for $\alpha = 1$ as well, only (13.1.10) must be verified. We use Corollary 4.4.9 and the fact that $\sum_{j=1}^{d} \theta_j s_j \log|\sum_{j=1}^{d} \theta_j s_j|$

is a continuous function on $S_d^{\|\cdot\|_\infty}$. We have

$$
\begin{aligned}
&\int_{B_1}\Big(\sum_{j=1}^d \theta_j x_j\Big)\log|\sum_{j=1}^d \theta_j x_j|\Gamma(d\mathbf{x})\\
&= \frac{\pi}{2}\lim_{k\to\infty}\int_{B_1}\Big(\sum_{j=1}^d \theta_j x_j\Big)\log|\sum_{j=1}^d \theta_j x_j|\Gamma_{r_{n_k}}(d\mathbf{x})\\
&= \frac{\pi}{2}\lim_{k\to\infty}\int_{B_1}\Big(\sum_{j=1}^d \theta_j x_j\Big)^{<r_{n_k}>}\log|\sum_{j=1}^d \theta_j x_j|\Gamma_{r_{n_k}}(d\mathbf{x}) \quad (\text{since } r_{n_k}\uparrow 1)\\
&= \lim_{k\to\infty} a_{r_{n_k}}\int_{C_0}\Big(\sum_{j=1}^d \theta_j x_j\Big)^{<r_{n_k}>}\log\Big|\sum_{j=1}^d \theta_j \frac{x_j}{\|\mathbf{x}\|}\Big|P_{\mathbf{X}}(d\mathbf{x})\\
&\qquad (\text{set } a_r=(\pi/2)(1-r) \text{ and use (13.1.3)})\\
&= \lim_{k\to\infty} a_{r_{n_k}}\int_{C_0}\Big(\sum_{j=1}^d \theta_j x_j\Big)^{<r_{n_k}>}\log\Big|\sum_{j=1}^d \theta_j \frac{x_j}{\max_{i=1,\dots,d}|x_i|}\Big|P_{\mathbf{X}}(d\mathbf{x})\\
&\quad + \lim_{k\to\infty} a_{r_{n_k}}\int_{C_0}\Big(\sum_{j=1}^d \theta_j x_j\Big)^{<r_{n_k}>}\log\Big(\frac{\max_{i=1,\dots,d}|x_i|}{\|\mathbf{x}\|}\Big)P_{\mathbf{X}}(d\mathbf{x})\\
&= \lim_{k\to\infty} a_{r_{n_k}}\int_{\mathbb{R}^d}\Big(\sum_{j=1}^d \theta_j \frac{x_j}{\|\mathbf{x}\|_\infty}\Big)^{<r_{n_k}>}\log\Big|\sum_{j=1}^d \theta_j \frac{x_j}{\|\mathbf{x}\|_\infty}\Big|\\
&\quad \cdot \|\mathbf{x}\|_\infty^{r_{n_k}} P_{(X_1,\dots,X_d)}(d\mathbf{x})\\
&\qquad (\text{here } \|\cdot\|_\infty \text{ is the supremum norm in } \mathbb{R}^d)\\
&\quad + \lim_{k\to\infty} a_{r_{n_k}}\int_{C_0}\Big(\sum_{j=1}^d \theta_j \frac{x_j}{\|\mathbf{x}\|}\Big)^{<r_{n_k}>}\log\Big(\max_{i=1,\dots,d}\frac{|x_i|}{\|\mathbf{x}\|}\Big)\|\mathbf{x}\|^{r_{n_k}}P_{\mathbf{X}}(d\mathbf{x})\\
&= \lim_{k\to\infty} a_{r_{n_k}}\int_{\mathbb{R}^d}\Big(\sum_{j=1}^d \theta_j \frac{x_j}{\|\mathbf{x}\|_\infty}\Big)\log\Big|\sum_{j=1}^d \theta_j \frac{x_j}{\|\mathbf{x}\|_\infty}\Big|\|\mathbf{x}\|_\infty^{r_{n_k}} P_{(X_1,\dots,X_d)}(d\mathbf{x})\\
&\quad + \lim_{k\to\infty} a_{r_{n_k}}\int_{C_o}\Big(\sum_{j=1}^d \theta_j \frac{x_j}{\|\mathbf{x}\|}\Big)\log\Big(\max_{i=1,\dots d}\frac{|x_i|}{\|\mathbf{x}\|}\Big)\|\mathbf{x}\|^{r_{n_k}}P_{\mathbf{X}}(d\mathbf{x})\\
&\qquad (\text{since } r_{n_k}\uparrow 1)\\
&= \int_{S_d^{\|\cdot\|_\infty}}\Big(\sum_{j=1}^d \theta_j s_j\Big)\log|\sum_{j=1}^d \theta_j s_j|\Gamma^{(d)}(d\mathbf{s})\\
&\quad + \int_{B_1}\Big(\sum_{j=1}^d \theta_j x_j\Big)\log(\max_{i=1,\dots d}|x_i|)\Gamma(d\mathbf{x}) \qquad (\text{use (13.1.3)}).
\end{aligned}
$$

From Proposition 2.3.8 and Example 2.3.4, verification of (13.1.10) reduces to showing that for any $j = 1, 2, \ldots, d$,

$$\int_{B_1} x_j \log\left(\frac{\max_{i=1,\ldots,d}|x_i|}{|x_j|}\right)\Gamma(d\mathbf{x}) = \int_{S_d} s_j \log\left(\frac{\max_{i=1,\ldots,d}|s_i|}{|s_j|}\right)\Gamma_d(d\mathbf{s}), \tag{13.1.11}$$

where Γ_d is the spectral measure of the vector $X_1, \ldots, X_d$ (i.e., with respect to the *Euclidean* norm). However, (13.1.11) follows as above by a simple application of Corollary 4.4.9. Thus, we have established (13.1.1) in all cases.

It remains only to show that $\boldsymbol{\mu} \in c_o$. The case $\alpha \neq 1$ being trivial, only the case $\alpha = 1$ must be considered. Applying Theorem 10.3.1, we conclude that

$$\lim_{j\to\infty}\left(\mu_j - \frac{1}{\pi}\int_{B_1} x_j \ln|x_j|\Gamma(d\mathbf{x})\right) = 0,$$

and now the bounded convergence theorem applies and gives $\lim_{j\to\infty}\mu_j = 0$. This completes the proof of the proposition. ∎

We are now in position to prove the representation theorem when T is countable.

Theorem 13.1.2 *Let $T = \{1, 2, \ldots\}$ and $\{X_i,\ i \in T\}$ be an α-stable process.*

(i) There are functions $f_i \in F$, $i = 1, 2, \ldots,$ and real numbers μ_i, $i = 1, 2, \ldots,$ such that

$$\{X_i,\ i = 1, 2, \ldots\} \stackrel{d}{=} \left\{\int_0^1 f_i(x)M(dx) + \mu_i,\ i = 1, 2, \ldots\right\}, \tag{13.1.12}$$

where

$$F = \begin{cases} L^\alpha(0,1) & \text{if } \alpha \neq 1, \\ \{f\colon \int_0^1 |f(x)\ln|f(x)||dx < \infty\} & \text{if } \alpha = 1, \end{cases}$$

and M is an α-stable random measure on $((0,1), \mathcal{B})$ with skewness intensity $\beta \equiv 1$ and Lebesgue control measure.

The functions f_i in (13.1.12) can be chosen uniformly bounded if $\lim_{i\to\infty} X_i = 0$ a.s.

(ii) If $\{X_i,\ i = 1, 2, \ldots\}$ is strictly α-stable and $\alpha \neq 1$ then the representation (13.1.12) holds with $\mu_i \equiv 0$.

(iii) If $\{X_i,\ i = 1, 2, \ldots\}$ is $S\alpha S$, then

$$\{X_i,\ i = 1, 2, \ldots\} \stackrel{d}{=} \left\{\int_0^1 f_i(x)M_S(dx),\ i = 1, 2, \ldots\right\},$$

where $f_i \in L^\alpha(0,1)$, $i = 1, 2, \ldots,$ and M_S is a $S\alpha S$ random measure on $((0,1), \mathcal{B})$ with Lebesgue control measure. The functions f_i can be chosen uniformly bounded if $\lim_{i\to\infty} X_i = 0$ a.s.

PROOF: For $i = 1, 2, \ldots$, choose $a_i > 0$ such that $P(a_i^{-1}|X_i| > i^{-1}) \leq 2^{-i}$. Let $Y_i = a_i^{-1}X_i$, $i = 1, 2, \ldots$. Then $\{Y_i,\ i = 1, 2, \ldots\}$ is an α-stable process satisfying $\lim_{i\to\infty} Y_i = 0$ a.s.. We apply now Proposition 13.1.1 to conclude that there is a finite Borel measure Γ on the unit sphere B_1 of c_0 and a vector $\boldsymbol{\eta} \in c_0$ such that for any $n = 1, 2, \ldots$ and real numbers $\theta_1, \ldots, \theta_n$,

$$-\ln E\exp\Big\{i\sum_{j=1}^{n}\theta_j Y_j\Big\}$$

$$= \int_{B_1} |\sum_{j=1}^{n}\theta_j x_j|^\alpha\Big(1 - i\ \mathrm{sign}\ \Big(\sum_{j=1}^{n}\theta_j x_j\Big)L\Big(\alpha; \sum_{j=1}^{n}\theta_j x_j\Big)\Big)\Gamma(d\mathbf{x}) - i\sum_{j=1}^{n}\theta_j\eta_j$$

with L given by (13.1.2).

We now decompose Γ into "atomic" and "continuous" components. Let $A = \{\mathbf{x}\colon\ \Gamma(\{\mathbf{x}\}) > 0\}$. Since Γ is a finite measure, the set A is at most countable; we will denote its points $\mathbf{x}^{(i)}$, $i = 1, 2, \ldots$. Write $\Gamma = \Gamma_a + \Gamma_c$ where $\Gamma_a(C) = \Gamma(C \cap A)$, $\Gamma_c(C) = \Gamma(C \cap A^c)$ for any Borel set $C \subseteq B_1$. Since Γ_c is an atomless finite Borel measure on B_1, Theorem 15.9 of Royden (1963) applies, and we conclude that there is a Borel set C_0 in B_1 with $\Gamma_c(C_0) = 0$, a Borel set $I_0 \subseteq (0, \frac{1}{2}]$ of Lebesgue measure 0 and a one-to-one map ϕ from $B_1\backslash C_0$ onto $(0, \frac{1}{2}]\backslash I_0$ with both ϕ and ϕ^{-1} Borel measurable, that preserves the measure $(2\Gamma_c(B_1\backslash C_0))^{-1}\,\Gamma_c$. That is, for any Borel set $D \subseteq (0, \frac{1}{2}]\backslash I_0$, we have

$$\lambda(D) = \Gamma_C(\phi^{-1}(D))/2\Gamma_c(B_1),$$

where λ denotes the Lebesgue measure on $(0, \frac{1}{2}]$. For each $j \geq 1$, define a function $g_j^{(1)}\colon (0, \frac{1}{2}] \to \mathbb{R}$ by

$$g_j^{(1)}(y) = \begin{cases} (\phi^{-1}(y))_j(2\Gamma_c(B_1))^{1/\alpha} & \text{if } y \in (0, \frac{1}{2}]\backslash I_0, \\ 0 & \text{if } y \in I_0. \end{cases}$$

Clearly, $g_1^{(1)}, g_2^{(1)}, \ldots$ are measurable uniformly bounded functions.

Construct also the intervals:

$$J_i = \Big(1 - \frac{1}{2\Gamma_a(B_1)}\sum_{k\geq i}\Gamma(\{\mathbf{x}^{(k)}\}),\ 1 - \frac{1}{2\Gamma_a(B_1)}\sum_{k>i}\Gamma(\{\mathbf{x}^{(k)}\})\Big),$$

$i = 1, 2, \ldots$, and use them to define a sequence of piecewise constant functions $g_j^{(2)}\colon (\frac{1}{2}, 1) \to \mathbb{R}$ as follows:

$$g_j^{(2)}(y) = \begin{cases} 0 & \text{if } y \in (\frac{1}{2}, 1)\backslash(\bigcup_{i\geq 1} J_i), \\ (2\Gamma_a(B_1))^{1/\alpha}x_j^{(i)} & \text{if } y \in J_i,\ i = 1, 2, \ldots, \end{cases}$$

where $\mathbf{x}^{(i)} = (x_1^{(i)}, x_2^{(i)}, \ldots)$, $i = 1, 2, \ldots$, are the points of the set A defined above. $g_1^{(2)}, g_2^{(2)}, \ldots$ are measurable uniformly bounded functions because $\mathbf{x}^{(i)} \in B_1$. Now combine the two sequences into a single sequence of measurable bounded functions on $(0, 1)$,

$$g_j(y) = \begin{cases} g_j^{(1)}(y) & \text{if } y \in (0, \frac{1}{2}], \\ \\ g_j^{(2)}(y) & \text{if } y \in (\frac{1}{2}, 1), \end{cases} \tag{13.1.13}$$

$j = 1, 2, \ldots$. Let M be an α-stable random measure on $((0, 1), \mathcal{B})$ with skewness intensity $\beta \equiv 1$ and Lebesgue control measure. We claim that

$$\{Y_j,\ j = 1, 2, \ldots,\} \stackrel{d}{=} \left\{\int_0^1 g_j(y) M(dy) + \widetilde{\mu}_j,\ j = 1, 2, \ldots\right\}, \tag{13.1.14}$$

where

$$\widetilde{\mu}_j = \begin{cases} \eta_j, \\ \\ \eta_j - \frac{2}{\pi}\left[\log(2\Gamma_c(B_1)) \int_{B_1} x_j \Gamma_c(d\mathbf{x}) + \log(2\Gamma_a(B_1)) \int_{B_1} x_j \Gamma_a(d\mathbf{x})\right], \end{cases} \tag{13.1.15}$$

for $\alpha \neq 1$ and $\alpha = 1$, respectively. Indeed, for every $n = 1, 2, \ldots$ and any real $\theta_1, \ldots, \theta_n$,

$$\begin{aligned}
&- \ln E \exp\left\{i \sum_{j=1}^n \theta_j \int_0^1 g_j(y) M(dy)\right\} \\
&= \int_0^1 \Big|\sum_{j=1}^n \theta_j g_j(y)\Big|^\alpha \Big(1 - i \operatorname{sign}\Big(\sum_{j=1}^n \theta_j g_j(y)\Big) L(\alpha; \sum_{j=1}^n \theta_j g_j(y))\Big) dy \\
&= \int_0^{1/2} + \int_{1/2}^1 \\
&= \frac{1}{2\Gamma_c(B_1)} \int_{B_1} \Big|\sum_{j=1}^n \theta_j (2\Gamma_c(B_1))^{1/\alpha} x_j\Big|^\alpha \\
&\quad \times \Big(1 - i \operatorname{sign}\Big(\sum_{j=1}^n \theta_j (2\Gamma_c(B_1))^{1/\alpha} x_j\Big) L\Big(\alpha; \sum_{j=1}^n \theta_j (2\Gamma_c(B_1))^{1/\alpha} x_j\Big)\Big) \Gamma_c(d\mathbf{x}) \\
&\quad + \frac{1}{2\Gamma_a(B_1)} \sum_{k \geq 1} \Gamma(\{\mathbf{x}^{(k)}\}) \Big[\Big|\sum_{j=1}^n \theta_j (2\Gamma_a(B_1))^{1/\alpha} x_j^{(k)}\Big|^\alpha \\
&\quad \times \Big(1 - i \operatorname{sign}\Big(\sum_{j=1}^n \theta_j (2\Gamma_a(B_1))^{1/\alpha} x_j^{(k)}\Big) L\Big(\alpha; \sum_{j=1}^n \theta_j (2\Gamma_a(B_1))^{1/\alpha} x_j^{(k)}\Big)\Big)\Big].
\end{aligned} \tag{13.1.16}$$

If $\alpha \neq 1$, (13.1.16) implies

$$
\begin{aligned}
&-\ln E \exp\Big\{i \sum_{j=1}^{n} \theta_j \Big(\int_0^1 g_j(y) M(dy) + \tilde{\mu}_j \Big)\Big\} \\
&= \int_{B_1} \Big|\sum_{j=1}^{n} \theta_j x_j\Big|^{\alpha} \Big(1 - i \,\text{sign}\, (\sum_{j=1}^{n} \theta_j x_j) \tan \frac{\pi\alpha}{2}\Big) \Gamma_c(d\mathbf{x}) \\
&\quad + \int_{B_1} \Big|\sum_{j=1}^{n} \theta_j x_j\Big|^{\alpha} \Big(1 - i \,\text{sign}\, \Big(\sum_{j=1}^{n} \theta_j x_j\Big) \tan \frac{\pi\alpha}{2}\Big) \Gamma_a(d\mathbf{x}) - i \sum_{j=1}^{n} \theta_j \eta_j \\
&= \int_{B_1} \Big|\sum_{j=1}^{n} \theta_j x_j\Big|^{\alpha} \Big(1 - i \,\text{sign}\, \Big(\sum_{j=1}^{n} \theta_j x_j\Big) \tan \frac{\pi\alpha}{2}\Big) \Gamma(d\mathbf{x}) - i \sum_{j=1}^{n} \theta_j \eta_j \\
&= -\ln E \exp\Big\{i \sum_{j=1}^{n} \theta_j Y_j\Big\},
\end{aligned}
$$

proving (13.1.14) for $\alpha \neq 1$. The case $\alpha = 1$ is similar. Of course, (13.1.14) implies (13.1.12) with $f_j = a_j g_j$, $\mu_j = a_j \eta_j$, $j = 1, 2, \ldots$. (If $\lim_{i\to\infty} X_i = 0$ a.s., then the choice $a_j \equiv 1$ is possible and the functions $f_j = g_j$ are uniformly bounded.) This proves part (i) of the theorem.

The claim of part (ii) follows immediately from part (i) because of Proposition 3.4.3 and Theorem 2.4.1. The proof of part (iii) of is similar and simpler than that of part (i). ∎

Remarks.

1. The representation theorem in $\mathbb{R}^d$ (Theorem 3.5.6) was stated without proof. It is a consequence of Theorem 13.1.2 because a vector $(X_1, \ldots, X_d)$ in $\mathbb{R}^d$ can be viewed as $(X_1, \ldots, X_d, 0, 0, \ldots)$.

2. A $S\alpha S$ vector admits a representation with respect to a totally skewed random measure M as well as a representation with respect to a $S\alpha S$ random measure M_S.

13.2 Arbitrary parameter space

We now consider an arbitrary index set T and determine how widely applicable is an integral representation with $E = (0, 1)$ and M having Lebesgue control measure. We suppose $0 < \alpha < 2$ in the $S\alpha S$ case, but $\alpha \neq 1$ in the skewed case.

Theorem 13.2.1 *Let $\{X(t),\ t \in T\}$ be an α-stable stochastic process satisfying Condition S (see Definition 3.11.2).*

(i) Let $\alpha \neq 1$. Then there are functions $f(t,\cdot) \in L^\alpha(0,1)$, $t \in T$, and real numbers $\mu(t)$, $t \in T$, such that

$$\{X(t),\ t \in T\} \stackrel{d}{=} \left\{\int_0^1 f(t,x)M(dx) + \mu(t),\ t \in T\right\}, \tag{13.2.1}$$

where M is an α-stable random measure on $((0,1),\mathcal{B})$ with skewness intensity $\beta \equiv 1$ and Lebesgue control measure. Moreover, every α-stable process (even with $\alpha = 1$) with an integral representation (13.2.1) must satisfy Condition S.

(ii) Let $\alpha \neq 1$. If $\{X(t),\ t \in T\}$ is strictly α-stable, then the representation (13.2.1) holds with $\mu(t) \equiv 0$.

(iii) If $\{X(t),\ t \in T\}$ is $S\alpha S$, then

$$\{X(t),\ t \in T\} \stackrel{d}{=} \left\{\int_0^1 f(t,x)M_S(dx),\ t \in T\right\}, \tag{13.2.2}$$

where $f(t,\cdot) \in L^\alpha(0,1)$, $t \in T$, and M_S is a $S\alpha S$ random measure on $((0,1),\mathcal{B})$ with Lebesgue control measure. Moreover, every $S\alpha S$ process with an integral representation (13.2.2) must satisfy Condition S.

PROOF: Let $T_0 = \{t_1, t_2, \ldots\}$ be a countable subset of T from the definition of Condition S.

(i) By Theorem 13.1.2 there are functions $f_{t_i} \in L^\alpha(0,1)$, $i = 1,2,\ldots$, and real numbers μ_{t_i}, $i = 1,2,\ldots$, such that

$$\{X(t_i),\ i = 1,2,\ldots,\} \stackrel{d}{=} \left\{\int_0^1 f_{t_i}(x)M(dx) + \mu_{t_i},\ i = 1,2,\ldots\right\}, \tag{13.2.3}$$

where M is an α-stable random measure on $((0,1),\mathcal{B})$ with skewness intensity $\beta \equiv 1$ and Lebesgue control measure.

We extend the integral representation (13.2.3) to the whole of T as follows. Fix $t \in T$. Because $\{X(t),\ t \in T\}$ satisfies Condition S, there is a sequence $\{t_{i_k}\}_{k=1}^\infty$ in T_0 such that $X(t_{i_k}) \to X(t)$ in probability as $k \to \infty$. It follows from Proposition 3.5.1 that the sequence $\{f(t_{i_k},\cdot)\}_{k=1}^\infty$ is Cauchy in $L^\alpha(0,1)$; we let $f(t,\cdot)$ denote its limit in $L^\alpha(0,1)$. The sequence $\{\mu(t_{i_k})\}_{k=1}^\infty$ is a convergent sequence of real numbers; we let $\mu(t)$ be its limit. The family $\{f(t,\cdot),\mu(t)\}$, $t \in T$, obviously satisfies (13.2.1).

For the converse, let $\{X(t),\ t \in T\}$ have the representation (13.2.1). Since the space $L^\alpha(0,1)$ is separable, the set $\{f(t,\cdot) \in T\} \subseteq L^\alpha(0,1)$ is separable as well. There is therefore a countable subset $T_0 \subseteq T$ such that $\{f(t,\cdot),\ t \in T_0\}$ is dense in $\{f(t,\cdot),\ t \in T\}$. Proposition 3.5.1 now implies that the stochastic process

$Y_1(t) = \int_0^1 f(t,x)M(dx)$, $t \in T$, satisfies Condition S. Since the (degenerate) process $Y_2(t) = \mu(t)$, $t \in T$, also satisfies Condition S (Exercise 3.18), it follows (Exercise 3.20) that the process $Y(t) = Y_1(t) + Y_2(t)$, $t \in T$ and hence $\{X(t), t \in T\}$ also satisfy Condition S. Since the case $\alpha = 1$ is similar because the space $L \log L$ (Leb) is also separable, this proves part (i) of the theorem.

Part (ii) now follows from part (i), Proposition 3.4.3 and Theorem 2.4.1.

The proof of part (iii) is identical to that of part (i), but uses part (iii) of Theorem 13.1.2 instead of part (i) of that theorem. ■

The proof of Theorem 13.2.1 implies that an α-stable process $\{X(t), t \in T\}$ that does not satisfy Condition S cannot be represented in the form

$$X(t) = \int_E f(t,x)M(dx) + \eta(t),\ t \in T, \tag{13.2.4}$$

with the function space $L^\alpha(E,\mathcal{E},m)$ (or even its subset $\{f(t,\cdot), t \in T\}$) being separable. Therefore, the only hope for obtaining an integral representation for an α-stable process which does not satisfy Condition S is to use a non-separable space $L^\alpha(E,\mathcal{E},m)$. (The separability of the space depends on E and m.) The following result, due to Bretagnolle, Dacunha-Castelle and Krivine (1966) in the case $1 \le \alpha < 2$ and Schreiber (1972) in the case $0 < \alpha < 1$, shows that this is always possible, at least in the $S\alpha S$ case. It is given here without proof.

Theorem 13.2.2 *Let $\{X(t), t \in T\}$ be a $S\alpha S$ process, $0 < \alpha < 2$. Then there is a measure space $(E,\mathcal{E},m)$ and a family $\{f(t,\cdot), t \in T\} \subseteq L^\alpha(E,\mathcal{E},m)$ such that*

$$\{X(t), t \in T\} \stackrel{d}{=} \left\{\int_E f(t,x)M(dx),\ t \in T\right\},$$

where M is a $S\alpha S$ random measure on $(E,\mathcal{E})$ with control measure m.

Chapter 14

Historical notes and extensions

14.1 Notes to Chapter 1

Section 1.1. Augustin Cauchy discovered in the 1850s that the functions f_α satisfying

$$\int_{-\infty}^{\infty} e^{i\theta x} f_\alpha(x)dx = e^{-\sigma^\alpha|\theta|^\alpha}, \quad \alpha > 0 \tag{14.1.1}$$

have the convolution property

$$(Af_\alpha(A\cdot)) * (Bf_\alpha(B\cdot)) = Cf_\alpha(C\cdot)$$

for some $C = C(A, B)$ and all $A,\ B > 0$ (Cauchy 1853). He was, however, able to show that $f_\alpha(x) \geq 0$ for all x only in the cases $\alpha = 1$ and $\alpha = 2$.

To prove the non-negativity of f_α it is sufficient to show that $\exp\{-|\theta|^\alpha\}$ is a characteristic function. Seventy years after Cauchy's discovery, George Pólya (1923) presented his famous sufficient condition for a function to be the characteristic function of a probability law: $\{\psi(t),\ t \in \mathbb{R}\}$ *is a characteristic function if* $\psi(t)$ *is real, non-negative,* $\psi(0+) = \psi(0) = 1,\ \psi(t) = \psi(-t)$ *and* ψ *is convex on* $(0, \infty)$. The non-negativity of the functions f_α in (14.1.1) with $0 < \alpha < 1$ follows as a simple application.

Paul Lévy (1924) proved that all f_αs in (14.1.1) with $0 < \alpha \leq 2$ are non-negative. Here is a simple argument valid for all $0 < \alpha \leq 2$ found in Durrett (1991), who learned it from Frank Spitzer.

For any β and $|x| < 1,\ (1 - x)^\beta = \sum_{k=0}^{\infty}\binom{\beta}{k}(-x)^k$ where

$$\binom{\beta}{k} = \frac{\beta(\beta - 1)\cdots(\beta - k + 1)}{1 \cdot 2 \cdots k}.$$

Applying this expansion to

$$\psi(\theta) = 1 - (1 - \cos\theta)^{\alpha/2}, \quad 0 < \alpha \leq 2,$$

one obtains

$$\psi(\theta) = \sum_{k=1}^{\infty} c_k (\cos\theta)^k \quad \text{with} \quad c_k = \binom{\alpha/2}{k} (-1)^{k+1} \geq 0$$

since $\alpha \leq 2$. We conclude that $\exp\{-|\theta|^\alpha\}$ is a characteristic function by arguing successively that the functions $\cos\theta$, $(\cos\theta)^k$, ψ, and

$$\lim_{n\to\infty} [\psi(\theta\sqrt{2}n^{-1/\alpha})]^n = e^{-|\theta|^\alpha}$$

are characteristic functions.

That $\exp\{-|\theta|^\alpha\}$ is not a characteristic function when $\alpha > 2$ follows from the following easily verifiable fact:

If $\psi(\theta)$ is a characteristic function of a random variable X and if

$$\lim_{\theta\downarrow 0} \frac{\psi(\theta) - 1}{\theta^2} = -\frac{\sigma^2}{2} > -\infty,$$

then $EX = 0$ and $EX^2 = \sigma^2 < \infty$ (Durrett 1991).

In particular, if $\psi(\theta) = 1 + o(\theta^2)$ as $\theta \to 0$, then $\sigma^2 = 0$ and therefore $\psi(\theta) \equiv 1$. If $\alpha > 2$, $\exp\{-|\theta|^\alpha\} = 1 + o(\theta^2)$ and hence $\exp\{-|\theta|^\alpha\}$ cannot be a characteristic function in this case.

The term *stable* was coined by Paul Lévy. Lévy (1924) gives not only the characteristic functions of all $S\alpha S$ laws but also those of all strictly stable laws. The next step, to pass from strictly to non-strictly stable laws, is of course easy if $\alpha \neq 1$. The case $\alpha = 1$ was settled by Aleksander Yakovlevich Khinchine and Paul Lévy in (1936). It is worth reading in this regard Sections 31, 41 and 42 of Lévy's mathematical biography (Lévy 1970).

Many textbooks contain an introduction to stable laws. See, e.g., Breiman (1968) and Laha and Rohatgi (1979). Lukacs (1970) devotes a number of sections to the analytic properties of stable distributions. There is, in many of these publications, great confusion about the sign of β in (1.1.6) as noted by Hall (1980).

The classic introductions to stable laws have been for many years Gnedenko and Kolmogorov (1954) and Feller (1971). An encylopedic treatment of stable laws on the real line is given by Zolotarev (1983), a work later translated into English (Zolotarev 1986). Rossberg, Jesiak and Siegel (1985) provide a good exposition to the subject. Two more books of interest have recently appeared: Janicki and Weron (1993) and Christoph and Wolf (1993). The second book is about convergence theorems with a stable limit law. It focuses on pseudomoments (moments with respect to $F - S$ where S is a stable distribution and F is a distribution function in the domain of normal attraction of S), Berry-Esseen type inequalities, asymptotic expansions and local central limit theorems.

Definitions 1.1.5 and 1.1.6 view stable laws as a special subclass of infinitely divisible laws. The theory of one-dimensional infinitely divisible laws was developed by Lévy in 1937; see Lévy (1954) and Khinchine (1938).

The $S\alpha S$ distribution with $\alpha = 3/2$ has the name of Holtsmark, a Danish astronomer who described, in 1939, a three-dimensional version of this distribution as a model for the gravitational field of stars (Holtsmark 1939).

Section 1.2. Weron (1984) uses the term "completely asymmetric" for α-stable distributions with $\beta = \pm 1$.

Section 1.3. Hardin (1984) noted the extension of Proposition 1.3.1 to the skewed case and deduces other facts related to skewed stable laws, which were relatively neglected at the time.

Section 1.4. A series representation of infinitely divisible random variables without a Gaussian component (of which the series representation of stable random variables described in the present section is a particular case) was established by Ferguson and Klass (1972) and developed by LePage (1990*b*). It is also considered in Vervaat (1979). An extension of these series representations to more general situations is developed by Rosinski (1990).

Section 1.6. Probability density function tables of the α-stable distribution can be found in Mandelbrot and Zarnfaller (1959) for $\beta = 1, \alpha > 1$, and in Holt and Crow (1973). DuMouchel (1971) and Brothers, DuMouchel and Paulson (1983) provide upper tail probability tables. The tables of Appendix A are taken from the last paper.

Analytic properties of stable densities were a major topic of research in the 1930s, 1940s and 1950s. The following curious relation between stable densities was discovered by Zolotarev (1954): *If* $1/2 < \alpha < 1$ *and*

$$X \sim S_\alpha(\sigma(\alpha,\beta_1),\ \beta_1, 0), \qquad Y \sim S_{1/\alpha}(\sigma(1/\alpha,\beta_2),\beta_2,0),$$

then, for any $x > 0$,

$$f_X(x) = x^{-(\alpha+1)} f_Y(x^{-\alpha}).$$

Here $\sigma(\alpha,\beta) = (1+\beta^2(\tan\frac{\pi\alpha}{2})^2)^{-1/(2\alpha)}$, *and* β_1 *and* β_2 *are related by*

$$\beta_1 = -\left(\tan\frac{\pi\alpha}{2}\tan\frac{\pi\alpha(\gamma+1)}{2}\right)^{-1},$$

$$\beta_2 = \left(\tan\frac{\pi}{2\alpha}\right)^{-1}\left(\tan\frac{\pi\gamma}{2}\right),$$

for some $|\gamma| \le 2 - 1/\alpha$.

The α-stable distributions are all *unimodal*. Establishing this fact was an exciting and difficult problem, whose complete solution escaped probabilists for a long time, leaving a trail of errors and counterexamples. Wintner (1936) proved unimodality of symmetric α-stable laws. Ibragimov and Chernin (1959) claimed that they proved that all stable laws are unimodal, but Kanter (1976) found an error in their proof. Finally, Yamazato (1978) closed the problem by proving that all class L infinitely divisible distributions (including stable distributions as a subclass) are unimodal.

Section 1.7. The *rstab* program in Chambers, Mallows and Stuck (1976) is written in Raftor. It was converted to Fortran by John Nolan.

One must, in general, be careful when calling from S or S-PLUS an external program which uses real variables (i.e., single precision). This is because S does not ordinarily distinguish between the "single" and "double" precision storage modes. When an external function is called from S, the arguments to the function are passed along as well. It seems that the default storage mode for an non-integer numeric field automatically becomes double precision. Thus, if a number, e.g., 0.1, is passed to a Fortran routine which is expecting a real (i.e., single precision) variable, problems will occur. Unfortunately, in many cases these problems are not immediately evident. The program does not crash; it just returns the wrong answers. There are two possible solutions. The simplest is to use only double precision Fortran/C routines, as we do here. Where this is not feasible, the calls from S must explicitly use variables instead of merely numbers, and the variables must be explicitly declared as "single precision ". Chapter 7.2 of Becker, Chambers and Wilks (1988) contains some information on calling external routines from S, and how the variable passing works.

Procedures for estimating the parameter of a stable distribution include: *maximum likelihood* (DuMouchel(1973), Chen(1991)); using *the sample characteristic function* (Paulson, Holcomb and Leitch (1975), Koutrouvelis (1980, 1981) Paulson , Delehanty (1984, 1985)); using *the order statistics* (Hill (1975), Hall and Welsh (1984), Loretan (1991)); using *tabulated quantiles* (Fama and Roll (1968), McCulloch (1986)). For a review, see Mittnik and Rachev (1993).

Section 1.8. Many standard textbooks characterize the *domain of attraction* of a stable law (Theorem 1.8.1). A classic reference is again Gnedenko and Kolmogorov (1954). The situation is more complicated when one considers the convergence of normalized partial sums of stationary *dependent* sequences to stable laws. One class of techniques uses ideas of extreme value limit theory.

A typical assumption is that the normalized sum of an "independent imitation" converges to a stable law. An independent imitation is an i.i.d. sequence with the same marginal distribution as the stationary dependent sequence. For example, Davis (1983) considers $X_j = G(Y_j)$ where Y_j is a stationary Gaussian with mean zero, unit variance and autocovariance $r_n = EY_0Y_n$ and where G is such that the "independent imitation" $\widetilde{X}_j$ of X_j belongs to the domain of attraction of an α-stable law with $0 < \alpha < 1$. He shows that if either $r_n \log n \to 0$ or $\sum_{n=1}^{\infty} r_n^2 < \infty$, then the normalized partial sums of X_j also converge to that stable law. For a recent overview, including various "mixing conditions," see Jakubowski (1991, 1993). Another set of techniques which works well for moving averages is reviewed in Avram and Taqqu (1986*a*). It includes results of Astrauskas (1983), Maejima (1983*a*), and Davis and Resnick (1985). See also Kasahara and Maejima (1986, 1988). Several of these results involve sequences with long-range dependence, a dependence structure considered in Chapter 7.

Sometimes a point process setting is useful for proving functional limit theorems (Resnick 1986). But functional limit theorems can be delicate. Avram and Taqqu (1992), for example, show that if $\{X_j\}$ is a "two-dependent" sequence defined by $X_j = \epsilon_j + \epsilon_{j-1}$, where the ϵ_j are i.i.d., $S\alpha S$, $0 < \alpha < 2$, then as $n \to \infty$, $\{n^{-1/\alpha} \sum_{j=1}^{[nt]} X_j,\ 0 \le t \le 1\}$ does not converge in $D([0,1])$ endowed with the usual Skorokhod topology J_1. Convergence holds if J_1 is replaced by the other Skorokhod topology M_1. Although M_1 is weaker than J_1, it is strong enough to make the widely-used functionals $\max_{0<t<1}$ and $\min_{0<t<1}$ continuous.

14.2 Notes to Chapter 2

Sections 2.1 and 2.2. Dudley and Kanter in (1974) claimed that if all linear combinations of the components of a random vector $\mathbf{X}$ are stable, then $\mathbf{X}$ is itself stable. Their argument is, essentially, that of the second remark following the proof of Theorem 2.1.5. However, this argument works only in the case $1 < \alpha < 2$, as was noted by de Acosta and Kuelbs during a student presentation of the Dudley and Kanter paper in 1981-82 at the University of Wisconsin, Madison.

Dudley remained interested in the problem, and David Marcus (1983), working under his direction, constructed the counterexample of Section 2.2. Marcus's proof was subsequently simplified by Samotij and Zak (1989). The argument of part (i) of Theorem 2.1.5 which is applicable to the range $1 \le \alpha < 2$ is due to Samorodnitsky and Taqqu (1991*c*).

Section 2.3. The characteristic function of a multivariate stable distribution was derived by Feldheim (1937) and presented in Lévy (1954).

Section 2.4. The graphs in Figure 2.1 were obtained by using a Fortran program written by John Nolan. This program calls a Fortran subroutine written by Jarle Berntsen and Terje O. Espelid which evaluates two-dimensional definite integrals. Additional graphs can be found in Byczkowski, Nolan and Rajput (1993).

Section 2.5. Because the term "sub-Gaussian" also refers to a class of stochastic processes with exponential moments, one may sometimes prefer the alternative term "Gaussian scale mixture" to denote the random vector in (2.5.3). Here, we use the term "sub-Gaussian" because it is convenient and has been widely used in the past: cf. Hardin (1982*a*, 1982*b*), Cambanis (1983), Weron (1984), Marques (1987), and Maejima (1989).

The fact that non-degenerate sub-Gaussian $S\alpha S$ random vectors cannot be independent was mentioned (in the case $\alpha \geq 1$) by Bretagnolle, Dacunha-Castelle and Krivine (1966) and proved by Hardin (1982*b*).

Section 2.6. Weron (1984) uses the term "complex $S\alpha S$ random variable" with two different meanings. One coincides with our definition, whereas the other refers to an isotropic $S\alpha S$ random variable.

Section 2.7. The term "covariation" is due to Miller (1978), although the quantity itself has been used even earlier, e.g., by Kanter (1972).

Section 2.8. The covariation norm was introduced and studied by Cambanis and Miller (1980, 1981).

Section 2.9. The relation between covariation and James orthogonality was noted in Cambanis, Hardin and Weron (1988). It first appears in Weron (1984). For a brief review of other notions of orthogonality, see Desbiens (1987).

Section 2.10. The codifference is related to the function (see (4.7.4)) considered by Astrauskas (1983). Lemma 2.10.8 has an interesting history. It is due to Schoenberg (1937) and was reproved by Laurent Schwarz in 1945 (see the footnote in Lévy (1970), p. 134). Our proof follows Cartier (1971).

The following is another measure of dependence considered by Paulauskas (1976). Let (X_1, X_2) be $S\alpha S$, $0 < \alpha < 2$. The distribution of (X_1, X_2) is characterized by the spectral measure Γ on the unit circle S_2. Let (U_1, U_2) be a random vector on S_2 with probability distribution $\widetilde{\Gamma} = \Gamma/\Gamma(S_2)$. Because of the symmetry of Γ, one has $EU_1 = EU_2 = 0$. Paulauskas uses

$$\widetilde{\rho} = \frac{EU_1U_2}{(EU_1^2 EU_2^2)^{1/2}} = \frac{\int_0^\pi \sin 2\phi \Gamma(d\phi)}{2(\int_0^\pi \cos^2 \phi \Gamma(d\phi) \int_0^\pi \sin^2 \phi \Gamma(d\phi))^{1/2}}$$

as a measure of dependence for (X_1, X_2) where, Γ is expressed here in polar coordinates. $\widetilde{\rho}$ has the following properties valid for all $0 < \alpha \leq 2$:

(i) $-1 \leq \widetilde{\rho} \leq 1$.
(ii) If X_1 and X_2 are independent, then $\widetilde{\rho} = 0$.
(iii) If $|\widetilde{\rho}| = 1$ then X_1 and X_2 are linearly dependent.

Note that $\widetilde{\rho}$ is the correlation of a Gaussian distribution whose Γ is the Γ of the original stable distribution.

14.3 Notes to Chapter 3

Section 3.2. It was long known that using stochastic integrals with respect to an independently scattered α-stable random measure is a convenient way to describe α-stable laws; among the first to use these integrals in the symmetric case were Bretagnolle, Dacunha-Castelle and Krivine (1966) and Schilder (1970). For the skewed case, see Hardin (1984).

From a practical viewpoint, a stochastic integral is a linear filter which transforms i.i.d. noise generated by the random measure M. *Multiple stochastic integrals* form non-linear filters. Multiple integrals with respect to a Gaussian measure were defined by Norbert Wiener and Kiyosi Itô in the 1940s. Their work is recorded in Wiener (1958) and Itô (1951) respectively. Whereas the closure of the linear span of these integrals equals the L^2 space generated by the Gaussian measure, no analogous result holds in the α-stable case, $0 < \alpha < 2$. The construction of multiple integrals with respect to an α-stable measure with index $\alpha < 2$ is thus more difficult. Sufficient conditions for integrability are obtained by Surgailis (1981, 1985) for $1 < \alpha < 2$ and by Samorodnitsky and Szulga (1989) for $0 < \alpha < 2$. Kallenberg and Szulga (1989) give conditions for the integrability of iterated Poisson integrals. McConnell and Taqqu (1986) show that the study of double integrals with respect to a stable measure can be reduced to the study of quadratic forms in independent stable variables. Explicit necessary and sufficient conditions were developed by Rosinski and Woyczyński (1986) for double integrals in the case $1 < \alpha < 2$ and extended by Kwapień and Woyczyński (1987) to $0 < \alpha < 2$. McConnell (1986) extends these conditions to triple integrals when $1 < \alpha < 2$. For a summary, see Kwapień and Woyczyński (1992). Extensions to Banach-valued integrands are considered by Krakowiak and Szulga (1988) and Samorodnitsky and Taqqu (1991*b*). For tails of multiple integrals with respect to a $S\alpha S$ random measure $0 < \alpha < 2$, see Samorodnitsky and Szulga (1989) in the case of real-valued integrands and Samorodnitsky and Taqqu (1990*c*) for integrands taking value in a Banach space.

Section 3.5. The result of Theorem 3.5.3 was first noticed (in the symmetric case) by Schilder (1970); his proof is incorrect in the case $1 < \alpha < 2$. The correct proof has been presented by many authors. See, e.g., Hardin (1984) for the proof in the general (skewed) case.

Part (ii) of the representation theorem (Theorem 3.5.6) is due to Schilder (1970), and Part (iii) to Hardin (1984). We present a unified proof in Chapter 13.

Section 3.6. For a discussion of the Ornstein-Uhlenbeck and reversed Ornstein-Uhlenbeck $S\alpha S$ processes and for an analysis of the Markov property in stable processes, see Adler, Cambanis and Samorodnitsky (1990).

Examples 3.6.5 and 3.6.6 are important examples of α-stable self-similar processes and are discussed in detail in Chapter 7. The well-balanced symmetric linear fractional stable motion was introduced by Taqqu and Wolpert (1983) and Maejima (1983*b*). Fractional Brownian motion in the present terminology goes back to Mandelbrot and Van Ness (1968), but a Gaussian process with autocovariance function

$$R(t,s) = C[|t|^a + |s|^a - |t-s|^a]$$

was mentioned in Kolmogorov (1940). (For more details, refer to the notes on Chapter 7.)

Sections 3.7 and 3.8. Scale mixing of stochastic processes means randomizing their scale parameter. Scale-mixed Gaussian and scale-mixed $S\alpha S$ processes are "sub-Gaussian" and "sub-stable" processes, respectively. Their study goes back to Bretagnolle, Dacunha-Castelle and Krivine (1966).

The integral representation of sub-Gaussian and sub-stable $S\alpha S$ processes given in Propositions 3.7.1 and 3.8.2 is due to Hardin (1982*b*).

Section 3.11. The idea of using the conditional normality of $S\alpha S$ processes in order to tap the wealth of information available on the Gaussian processes goes back to Marcus and Pisier (1984). This method has been widely applied ever since. The proof that all $S\alpha S$ processes satisfying Condition S admit the representation (3.11.4) is given in Chapter 13.

Section 3.12. The representation of stable processes as Poisson integrals is sometimes referred to as the Lévy-Itô representation; see Itô (1969).

14.4 Notes to Chapter 4

Section 4.1. Kanter (1972) was the first to notice that the regression $E(X_2|X_1)$ is linear when the vector (X_1, X_2) is $S\alpha S$ with $1 < \alpha < 2$. The simpler proof given here is due to Linde (1986), Lemma 7.1.3. Fix (1949) established linearity in a particular case.

Corollary 4.1.3 on multiple regression is due to Miller (1978). Miller gives also necessary and sufficient conditions for the linearity of the multiple regression in the $S\alpha S$ case. These are stated in Exercise 4.6 but are, in general, not easy to verify.

Corollary 4.1.5 treats the particular case of conditioning on independent random variables. It is due to Kanter (1972). Parts (i) and (ii) of Proposition 4.1.7 which identify the sub-Gaussian $S\alpha S$ random vectors with $1 < \alpha < 2$ as having the multiple regression property are due to Hardin (1982*a*).

Section 4.2. Tortrat (1975) claimed, mistakenly, that for any $S\alpha S$ random vector (X_1, X_2) with $1 < \alpha < 2$ the conditional law of X_2 given X_1 is symmetric around the conditional mean. Linde and Mathé (1983) showed that this is wrong and derived the necessary and sufficient conditions for symmetry presented in this section.

Section 4.3. The results of this section are due to Adler, Cambanis and Samorodnitsky (1990).

Section 4.4. The asymptotic behavior of the distribution tails of order statistics in an α-stable sample given in Theorem 4.4.5 was derived by Samorodnitsky (1988).

Section 4.5. Lemma 4.5.2 in the $S\alpha S$ case is due to Miller (1978). Theorem 4.5.6 which presents necessary and sufficient conditions for the existence of joint moments of α-stable random variables is due to Samorodnitsky and Taqqu (1990*b*).

Section 4.6. See Esary, Proschan and Walkup (1967) for basic properties of associated random variables. Pitt (1982) proved that jointly normal random variables are associated if and only if their correlations are all non-negative (see also Joag-dev, Perlman and Pitt (1983)). The results on association of stable random variables given here are due to Lee, Rachev and Samorodnitsky (1990). For limit theorems, see Dabrowski and Jakubowski (1993).

Slepian inequalities provide ways to compare distributions of random vectors. Slepian inequalities for $S\alpha S$ (and infinitely divisible) random vectors are given in Samorodnitsky and Taqqu (1992*b*, 1994).

Section 4.7. The evaluation of the codifference and, more generally, of the function U defined in (4.7.4) for Ornstein-Uhlenbeck processes (Example 4.7.1) comes from Levy and Taqqu (1991). That paper also contains Theorem 4.7.3 (the moving averages case) and Proposition 4.7.4 (the sub-Gaussian case).

The behavior of the codifference for a stationary $S\alpha S$ process is closely related to the mixing properties of the process where "mixing" is in the sense of ergodic theory. Here is a brief overview.

Let $\Omega = \mathbb{R}^T$ and $\mathcal{F}_X$ be the σ-algebra generated by the stationary process $\{X(t),\ t \in T\}$ defined by $X(t,\omega) = \omega(t)$. For convenience, we assume T is either $\mathbb{R}$ or $\mathbb{R}_+$. Let $S_s : \Omega \to \Omega$ be a shift transformation defined by $(S_s\omega)(t) = \omega(t+s)$. The family $\{S_s\}_{s\in T}$ is a semi-group of measure preserving transformations on Ω. Ergodicity, weak mixing and mixing of the process $\{X(t),\ t \in T\}$ can now be defined in terms of $\{S_s\}_{s\in T}$.

- *Ergodicity or metric transitivity:*

  (a) If $S_s(A) = A$ for all $s \in T$, then $P(A) = 1$ or $P(A) = 0$ (i.e., $\{S_s\}$ has no disjoint components) or, equivalently,

  (b) $\lim_{t\to\infty} \frac{1}{t}\int_0^t P(S_s(A)\cap B)ds = P(A)P(B)$.

- *Weak mixing:*

  (a) $\lim_{t\to\infty} \frac{1}{t}\int_0^t \left|P(S_s(A)\cap B) - P(A)P(B)\right|^2 ds = 0$

  or, equivalently,

  (b) $\lim_{s\to\infty, s\in\Sigma} P(S_s(A)\cap B) = P(A)P(B)$,

  where Σ is a subset of T with density 1, i.e., $\lim_{t\to\infty} t^{-1}|\Sigma\cap[0,t]| = 1$ if $T = \mathbb{R}_+$.

- *Mixing:*

$$\lim_{s\to\infty} P(S_s(A)\cap B) = P(A)P(B).$$

The following implications hold for any stationary process:

$$\text{mixing} \Rightarrow \text{weak mixing} \Rightarrow \text{ergodicity}.$$

For stationary $S\alpha S$ processes, the following holds:

- Ergodicity and weak mixing coincide (Kokoszka & Podgórski 1993)).
- Moving averages are mixing (Cambanis, Hardin, Jr. & Weron 1987).
- Sub-Gaussian and real harmonizable $S\alpha S$ processes are not ergodic (Cambanis et al. 1987). This fact, in the harmonizable case, also follows from Maruyama (1970).
- A necessary and sufficient condition for the process to be mixing is

$$\lim_{t\to\infty} U(\theta_1,\theta_2;t) = 0 \quad \text{for all } \theta_1,\theta_2 \in \mathbb{R} \tag{14.4.1}$$

This result, which is valid for any stationary infinitely divisible process, is due to Maruyama (1970). For stable processes, by (4.7.6), Relation (14.4.1) is equivalent to

$$\lim_{t\to\infty} I(\theta_1,\theta_2;t) = 0 \quad \text{for all } \theta_1,\theta_2 \in \mathbb{R}.$$

This, in fact, provides an alternate proof of Theorem 4.7.3 since, as noted above, moving average $S\alpha S$ processes are mixing and $\tau(t) = -I(-1,1,t)$.

Maruyama's result requires $I(\theta_1,\theta_2;t) \to 0$ as $t \to \infty$ for all $\theta_1,\theta_2 \in \mathbb{R}$ in order to ensure that the process is mixing. Gross (1993) proves that it is not necessary to require $I(\theta_1,\theta_2;t) \to 0$ for all $\theta_1,\theta_2 \in \mathbb{R}$ if the process is $S\alpha S$ and satisfies Condition S (see Section 3.11). It is only necessary to choose for (θ_1,θ_2) the value $(-1,1)$ if $0<\alpha\le 1$ and the values $(-1,1)$ and $(1,1)$ if $1<\alpha<2$. Since $I(\theta_1,\theta_2;t) \to 0$ is equivalent to $U(\theta_1,\theta_2;t) \to 0$, this can also be expressed as $U(-1,1;t) \to 0$ when $0<\alpha\le 1$ and $U(-1,1;t) \to 0$ and $U(1,1;t) \to 0$ when $1<\alpha<2$.

An extension of Gross' result to processes of type G can be found in Kokoszka and Taqqu (1993*b*). Processes of type G extend $S\alpha S$ processes. They are infinitely divisible processes which admit, conditionally, a series representation. In particular, a random variable X is of type G if and only if $X \stackrel{d}{=} ZS$, where Z has the $N(0,1)$ distribution and S is a nonnegative random variable independent of Z such that S^2 is infinitely divisible. (Informally, $X \sim N(0,S^2)$.) For a comprehensive account of distributions and processes of type G, refer to Rosinski (1990, 1991).

A stationary Gaussian process is *ergodic* if and only if its spectral measure (the Fourier transform of the autocovariance function) has no atoms; see, e.g., Rozanov (1967). What about $S\alpha S$ processes?

The shift S_s induces a shift $\widetilde{S}_s$ on random variables as follows: define firstly $\widetilde{S}_s$ on indicator functions by $\widetilde{S}_s 1_A = 1_{S_{-s}A}$, $A \in \mathcal{F}_X$, and then extend it by linearity and continuity in probability to all random variables measurable with respect to $\mathcal{F}_X$. Let $\{X(t),\ t\in T\}$ be a $S\alpha S$ process with index $0<\alpha\le 2$. Consider $\mathcal{L}(\mathcal{F}_X)$, the space of all $S\alpha S$ random variables measurable with respect to $\mathcal{F}_X$. If Y is such a random variable, then $\{\widetilde{S}_s Y\}_{s\in T}$ is a stationary $S\alpha S$ stochastic process. For example, if $Y = 2X(1)+3X(2)$, then $\widetilde{S}_s Y = 2X(1+s)+3X(2+s)$, $s\in T$. Conditions for ergodicity can be formulated in terms of $\mathcal{L}(\mathcal{F}_X)$.

Cambanis, Hardin and Weron (1987) prove that a stationary $S\alpha S$ processes $\{X(t)\}$, $0<\alpha\le 2$, is ergodic if and only if for each random variable $Y \in \mathcal{L}(\mathcal{F}_X)$,

$$\lim_{t\to\infty} \frac{1}{t}\int_0^t \|\widetilde{S}_s Y - Y\|_\alpha^\alpha ds = 2\|Y\|_\alpha^\alpha$$

and

$$\lim_{t\to\infty} \frac{1}{t} \int_0^t \|\widetilde{S}_s Y - Y\|_\alpha^{2\alpha} ds = 4\|Y\|_\alpha^{2\alpha}.$$

Podgórski (1992) gives a different characterization. He proves that the stationary $S\alpha S$ process X is ergodic if and only if for each $Y \in \mathcal{L}(\mathcal{F}_X)$,

$$\lim_{t\to\infty} \frac{1}{t} \int_0^t \exp\Big(2\|Y\|_\alpha^\alpha - \|\widetilde{S}_s Y - Y\|_\alpha^\alpha\Big) ds = 1. \tag{14.4.2}$$

This relation can be rewritten succinctly in terms of the codifference $\tau_Y(s)$ of the stationary process $\{\widetilde{S}_s Y\}_{s\in T}$ as

$$\lim_{t\to\infty} \frac{1}{t} \int_0^t e^{\tau_Y(s)} ds = 1 \tag{14.4.3}$$

or, in terms of the corresponding function U_Y, as

$$\lim_{t\to\infty} \frac{1}{t} \int_0^t U_Y(1,-1;s) ds = 0. \tag{14.4.4}$$

Section 4.8. The first results concerning crossings of stationary Gaussian processes were obtained by Rice (1939, 1944, 1945). More rigorous proofs were given by several authors and can be found in Cramér and Leadbetter (1967) and Leadbetter, Lindgren and Rootzén (1983). See these books for a detailed history. Adler, Samorodnitsky and Gadrich (1993) investigate the asymptotic behavior of the mean number EC_u of crossings of a level u, as $u \to \infty$, in the case where the process X is $S\alpha S$ and stationary. When X is sub-Gaussian, they obtain the result of Theorem 4.8.2. They also consider the case where X is a real harmonizable $S\alpha S$ (as defined in Example 3.6.7). Using the fact that X is a conditionally stationary centered Gaussian process (Proposition 6.6.4), they prove that $EC_0 < \infty$ implies

$$EC_u \sim \frac{\lambda_1 C_\alpha}{\pi} u^{-\alpha}$$

as $u \to \infty$. The constant C_α (do not confuse it with C_u!) is defined in (1.2.9) and λ_1 is the first spectral moment of the process Gaussian G in (6.6.4). Thus, as in the sub-Gaussian case, EC_u is asymptotically proportional to $u^{-\alpha}$. This problem for other $S\alpha S$ processes, e.g., moving averages, is still open.

14.5 Notes to Chapter 5

Section 5.1 Theorem 5.1.2, which relates the finiteness of moments to the behavior of the characteristic function at the origin, is due to Ramachandran (1969); see also Ramachandran and Rao (1968). Our formulation is slightly different from the one stated in Ramachandran (1969).

The relation between the integrability properties of the spectral measure of a $S\alpha S$ random vector and the existence of its conditional moments was discovered by Samorodnitsky and Taqqu (1991*a*). That paper contains the proof of Theorem 5.1.3 in the symmetric case for $0 < p < \min\{\alpha + \nu, 2\}$. The corresponding result in the skewed case and the form of the regression was obtained by Hardin, Samorodnitsky and Taqqu (1991*a*). The extension to $0 < p < \min\{\alpha+\nu,\ 2\alpha+1\}$ is due to Cioczek-Georges and Taqqu (1994*a*, 1994*c*).

LePage (1990*a*) used an alternative approach, taking advantage of the conditional normality of $S\alpha S$ random vectors. While such an approach appears to be more restrictive than the one presented in this chapter, it can be used in some cases to prove the existence of absolute conditional moments of order greater or equal to α.

How sharp is Theorem 5.1.3? If one defines $E(|X_2|^p|X_1 = x)$ as the pth moment of the distribution with characteristic function (5.1.9), then $E(X_2^2|X_1 = 0) < \infty$ is actually equivalent to (5.1.10) with $\nu = 2-\alpha$ (Wu & Cambanis 1991). Cioczek-Georges and Taqqu (1994*b*) extend this result to other values of p. They show under the same conditions that $E(|X_2|^p|X_1 = 0) < \infty$ is equivalent to (5.1.10) with $\nu = p - \alpha$ for

$$\begin{aligned} \alpha < p < 2\alpha+1 \quad &\text{if} \quad 0 < \alpha \le 1/2 \text{ or } 1 < \alpha \le 3/2, \\ \alpha < p \le 2 \quad &\text{if} \quad 1/2 < \alpha \le 1, \\ \alpha < p \le 4 \quad &\text{if} \quad 3/2 < \alpha < 2. \end{aligned}$$

Moreover, in the case $p = \alpha$,

$$E\big(|X_2|^p|X_1 = 0\big) < \infty \quad \text{if and only if} \quad -\int_{S_2} \ln|s_1|\Gamma(ds) < \infty.$$

One can extend Theorem 5.1.3 to the case of conditional moments of the type $E(|X_d|^p|X_1 = x_1, \dots, X_{d-1} = x_{d-1})$ for an α-stable random vector $(X_1, \dots, X_{d-1}, X_d)$ with $d \ge 3$ if one replaces the integrability condition (5.1.10) by the condition

$$P(t_1, \dots, t_{d-1}) = \int_{S_d} \frac{\Gamma(ds)}{|t_1 s_1 + \cdots + t_{d-1}s_{d-1}|^\nu} < \infty \tag{14.5.1}$$

for some $\nu > 0$, for almost all $(t_1, \dots, t_{d-1}) \in S_{d-1}$, and imposes in addition certain integrability conditions on $P(t_1, \dots, t_{d-1})$ over $\mathbb{R}^{d-1}$. Cambanis and Wu (1992), for example, show that in the case $0 < \alpha \le 1$, the multiple regression

$E(|X_d|\,|X_1 = x_1, \ldots, X_{d-1} = x_{d-1})$ exists for almost all $x_1, \ldots, x_d$ if there is a $\nu > 1 - \alpha$ such that

$$\int_{\mathbb{R}^{d-1}} |\phi_{\mathbf{X}}(t_1, \ldots, t_{d-1}, 0)| \int_{S_d} \frac{\Gamma(ds)}{|t_1 s_1 + \cdots + t_{d-1} s_{d-1}|^{\nu}} dt_1 \ldots dt_{d-1} < \infty.$$

Unfortunately, in many cases of interest, either (14.5.1) does not hold for large enough νs or $P(t_1, \ldots, t_{d-1})$ does not possess the required integrability properties. Nevertheless, the integrability condition (5.1.9) for $d = 2$ is still useful when $d > 2$ because

$$E\big(|X_d|^p \big| X_1\big) < \infty \text{ a.s.} \quad \Rightarrow E\big(|X_d|^p \big| X_1, \ldots, X_{d-1}\big) < \infty \text{ a.s.}$$

Section 5.2. In view of the results of Cioczek-Georges and Taqqu (1994*b*, 1994*a*, 1994*c*), Condition (5.2.4) can be replaced by the weaker condition:

$$\int_{S_2} |s_1|^{\alpha-1} \Gamma(ds) < \infty \quad \text{for } \alpha < 1,$$

$$-\int_{S_2} \ln |s_1| \Gamma(ds) < \infty \quad \text{for } \alpha = 1.$$

(See also the remarks following Theorem 5.1.3.) The form of the non-linear regression and the necessary and sufficient condition for linearity of regression presented in this section are due to Hardin, Samorodnitsky and Taqqu (1991*a*).

Wu and Cambanis (1991) provide further insight into the structure of conditional moments of stable laws by computing the conditional variance. Specifically, let (X_1, X_2) be $S\alpha S$, $1 < \alpha < 2$, with spectral measure Γ and suppose that (5.1.10) holds with $\nu = 2 - \alpha$. Then for a.e. $x \in \mathbb{R}$,

$$\text{Var}\,\big(X_2 \big| X_1 = x\big) = C(\alpha, \Gamma) h\Big(\alpha, \frac{x}{\sigma_1}\Big), \tag{14.5.2}$$

where

$$C(\alpha, \Gamma) = \sigma_1^2 \left[\frac{\int_{S_1} |s_1|^{\alpha-2} s_2^2 \Gamma(ds)}{\sigma_1^{\alpha}} - \kappa^2 \right],$$

$$h(\alpha, x) = \frac{\alpha - 1}{f_\alpha(x)} \int_{|x|}^{\infty} u f_\alpha(u) du,$$

and where κ is defined in (5.2.10) and f_α is the density function of the $S_\alpha(1, 0, 0)$ law.

Cioczek-Georges and Taqqu (1993*b*) extend the preceding result to the case $1/2 < \alpha \le 1$. They show that if (5.1.10) holds with $\nu > 2 - \alpha$ for $1/2 < \alpha < 1$ or $\nu = 2 - \alpha$ for $1 \le \alpha < 2$, then (14.5.2) holds with

$$h(\alpha, x) = \frac{\alpha^2}{\pi f_\alpha(x)} \int_0^{\infty} \cos tx \, e^{-t^{\alpha}} t^{2\alpha - 2} dt + x^2.$$

Observe the separation of factors in (14.5.2): the constant $C(\alpha, \Gamma)$ depends on the joint law of (X_1, X_2) but the function $h(\alpha, x)$ involves only the density of X_1.

The paper (Cioczek-Georges & Taqqu 1993*a*) contains necessary and sufficient conditions for asymptotic linearity of $E[Y|X + \mathcal{E} = z]$ where (X, Y) is an α-stable random vector and $\mathcal{E}$ is a random variable, independent of (X, Y), such that $X + \mathcal{E}$ is in the domain of normal attraction of X. Asymptotic linearity does not always hold even when $E[Y|X = x]$ is linear. For some distributions of $\mathcal{E}$, the asymptotic rate of $E[Y|X + \mathcal{E} = z]$ fluctuates.

Cambanis and Fakhre-Zakeri (1993) study the prediction of heavy-tailed AR(1) sequences. The innovations are assumed semistable, i.e. their characteristic function satisfies

$$\phi(r) = \phi^{r}(\lambda r)e^{i\mu r} \text{ for some } 0 < |\lambda| < r,\ \mu \in \mathbb{R}, \text{ and for all } r \in \mathbb{R}.$$

Semistable laws which were introduced by Lévy (1954), p. 203 generalize the stable laws.

Section 5.3. Sub-Gaussian and harmonizable vectors are $S\alpha S$ random vectors whose conditional moments can be studied by the methods of LePage (1990*a*).

Using the comments following the proof of Corollary 5.3.3 (or the necessary conditions in Cioczek-Georges and Taqqu (1994*b*)), part (i) of that corollary can be strengthened as follows:

If $\mathbf{X} = (X_1, X_2)$ *is sub-Gaussian* α*-stable and* X_1 *and* X_2 *are linearly independent, then*

$$E\big(|X_2|^p \big| X_1 = x\big) < \infty \text{ a.e. if and only if } p < \alpha + 1.$$

Section 5.4. The regression graphs of Section 5.4 are from Hardin, Samorodnitsky and Taqqu (1991*a*).

Section 5.5. The paper Hardin, Samorodnitsky and Taqqu (1991*b*) describes the numerical techniques and lists the source code of the software package for computing the non-linear regressions.

14.6 Notes to Chapter 6

Section 6.5. Marcus and Pisier (1984) refer to stationary $S\alpha S$ processes of the form (6.5.1) as strongly stationary.

Section 6.6. The fact that stationary harmonizable $S\alpha S$ processes are conditionally stationary Gaussian was discovered by Marcus and Pisier (1984).

The relation between the classes of stationary harmonizable $S\alpha S$ processes and stationary sub-Gaussian $S\alpha S$ processes was first discussed in Cambanis and Soltani (1984).

14.7 Notes to Chapter 7

Section 7.1. Processes that are self-similar with index H and have stationary increments (H-sssi processes) are sometimes called "statistically self-similar" in order to distinguish them from non-random fractals that are self-similar. Sometimes the term "self-affine" or "statistically self-affine" is used.

Benoit Mandelbrot's pioneering work on self-similarity dates to the early 1960s, e.g., Mandelbrot (1965). John Lamperti, in his seminal paper (1962), showed that self-similar process result from limits of normalized sums. Lamperti used the term "semi-stable" and allowed for shifts.

It is now common to use the letter H to denote the index of self-similarity. H is the first letter of the British Harold Edwin Hurst's last name, who spent 62 years in Egypt, mostly working on projects involving the river Nile. Using the "range" statistic, he discovered empirically that cumulative yearly flows of the Nile obey a power law with $H \cong 0.7$ instead of the expected $H = 0.5$ (Hurst 1951). His work was the basis of Mandelbrot's suggestion that fractional Brownian motion be used to model the yearly levels of the river (see Mandelbrot (1965)).

Proposition 7.1.4 which relates an H-sssi process to a stationary one is due to Lamperti (1962). The proof of Lemma 7.1.9 is adapted from O'Brien and Vervaat (1983) and that of Proposition 7.1.10 comes from Maejima (1986).

Brownian motion, the oldest example of an H-sssi process, is named after the biologist Robert Brown whose research dates to the 1820s. Early in this century, Louis Bachelier, Albert Einstein and Norbert Wiener began developing the mathematical theory of Brownian motion. The construction of Bachelier (1900) was erroneous but it captured many of the essential properties of the process. It is worth reading the short but fascinating account of Bachelier's tribulations in Mandelbrot (1982).

Section 7.2. Andrei Nikolaevich Kolmogorov discovered fractional Brownian motion. The article, written in German, appeared in 1940 in the Soviet "Dokladi." The process is introduced in Kolmogorov (1940) in the context of transformations of curves in Hilbert space. It was then briefly studied by Hunt (1951) and set in a wider framework by Akiva M. Yaglom (1955). It is, however, the seminal paper of Mandelbrot and Van Ness (1968) that investigated the basic properties of fractional Brownian motion and emphasized its relevance to the modeling of natural phenomena. The following papers, for example, Mandelbrot and Wallis

(1969*a*, 1969*c*, 1969*b*), develop applications to hydrology.

The moving average representation (7.2.4) of fractional Brownian motion can be found in Mandelbrot and Van Ness (1968). It involves a fractional-type integral used earlier in a non-stochastic context; see, e.g., Zygmund (1979). Kolmogorov (1940) and Yaglom (1955) used the harmonizable representation (7.2.12). A more general version of Proposition 7.2.7 which relates the two representations can be found in Taqqu (1979).

The terms *fractional Brownian motion* and *fractional Gaussian noise* were coined by Mandelbrot. Fractional Gaussian noise can be defined as the increment of a Gaussian H-sssi process or, directly, as a fixed point of the renormalization transformation. Sinai (1976) takes this second approach and derives its autocovariance (7.2.21) and spectral density (7.2.22). Cox (1984) advocates a similar point of view.

In order to gain modeling flexibility, it is important in applications to be able to replace fractional Gaussian noise by an arbitrary Gaussian moving average whose autocovariance function has the same asymptotic behavior as that of fractional Gaussian noise. Such moving averages are said to belong to the "domain of attraction" of fractional Brownian motion. More generally, a stationary sequence $\{Y_j,\ j = 1, 2, \ldots\}$ belongs to the domain of attraction of a process $X = \{X(t),\ t \geq 0\}$ if $\{\sum_{j=1}^{[Nt]} Y_j,\ t \geq 0\}$ suitably normalized, converges as $N \to \infty$ to the finite-dimensional distribution of X. Theorem 7.2.11 (Taqqu 1975) provides a simple example of a moving average in the domain of attraction of fractional Brownian motion. Other examples can be found in Davydov (1970), Rosenblatt (1961), Taqqu (1979, 1981), Dobrushin and Major (1979) and Major (1981).

There are a number of papers involving statistical estimation. See, e.g., Mandelbrot and Taqqu (1979), Fox and Taqqu (1986), Dahlhaus (1989) and Robinson (1991, 1992*a*, 1992*b*). Beran (1992) provides a good overview. For examples of statistical analysis, see Beran, Sherman, Taqqu and Willinger (1993) and Leland, Taqqu, Willinger and Wilson (1993, 1994). Samarov and Taqqu (1988) and Yajima (1988, 1989) study regression models with Gaussian errors that have long-range dependence.

Section 7.4. The linear fractional stable motion (7.4.1) was introduced by Taqqu and Wolpert (1983) as an integral with respect to a Poisson measure and by Maejima (1983*c*) as an integral with respect to a stable measure. Astrauskas (1982) provides a generalized field version. Theorem 7.4.5 was proved by Cambanis and Maejima (1989) in the case $1 < \alpha < 2$ and $\beta \equiv 0$ and by Samorodnitsky and Taqqu (1989) in the general case.

In the past, a "stable process" meant α-stable Lévy motion. This usage is confusing and should be discontinued because one needs to distinguish between a process with independent increments and the more general stable processes whose increments may be dependent. The importance of α-stable Lévy motion stems from the fact that it is the stable analog of Brownian motion. Theorem 7.5.4, which states that it is the unique $1/\alpha$-sssi process when $0 < \alpha < 1$, is due to Samorodnitsky and Taqqu (1990*a*). The log-fractional stable motion, an $1/\alpha$-sssi process, $1 < \alpha < 2$, different from α-stable Lévy motion, was discovered by Kasahara, Maejima and Vervaat (1988). Other $1/\alpha$-sssi processes with $1 < \alpha < 2$ can be found in Samorodnitsky and Taqqu (1990*a*) (see Exercises 7.10 and 7.12).

Sections 7.7 and 7.8. The real and complex harmonizable fractional stable motions were considered by Cambanis and Maejima (1989). Theorem 7.8.2 is proved in that paper in the case $1 < \alpha < 2$, but the identical proof extends to $0 < \alpha < 2$ once one notices that the process in (7.8.6) is well defined. A good review of α-stable H-sssi processes is given by Kôno and Maejima (1991).

Section 7.10. Linear fractional stable noise is the increment of linear fractional stable motion. Theorem 7.10.1 concerns the asymptotic behavior of the function I and the codifference for linear fractional stable noise. Its proof was sketched by Astrauskas (1983) in the $S\alpha S$ case. The proof in the general α-stable case can be found in Astrauskas, Levy and Taqqu (1991).

Section 7.11. Fractional Gaussian noise has different representations and any one of them can serve as a basis for an "approximate" simulation procedure. Here, we have adapted the "Type 1" approximation of Mandelbrot and Wallis (1969*a*) and applied it to the case of $S\alpha S$ innovations as well. Mandelbrot and Wallis's rougher approximation, called "Type 2", consists of simulating $\sum_{j-M}^{j-1}(j-u)^{d-1}\epsilon_u$, $j \geq 1$, where M is a large number, $d = H - 1/\alpha$, and the ϵ_us are the innovations.

There are also "fast" simulation methods of fractional Gaussian noise. (See for example Mandelbrot (1971) and Granger(1980).) But since that process is Gaussian, it is also possible to use the following "exact" simulation method. Let $C_T = (r(i-j)),\ i,j = 1,\dots T$ be the covariance matrix of $Y_1,\dots,Y_T$ defined in (7.2.21) with $r_0 = 1$. Performing a Cholesky decomposition on C_T, yields $C_T = MM'$, where M is a $T \times T$ lower trianguluar matrix with entries $m_{ju},\ j,u = 1,\dots T$ and where M' denotes the transpose of M. Then, if $\epsilon_u,\ u = 1,\dots,T$ are i.i.d. $N(0,1)$,

$$Y_j = \sigma_0(\sum_{u=1}^{T} m_{ju}\epsilon_u),\quad j = 1,\dots T,$$

is fractional Gaussian noise with variance σ_0^2. This method, which is only practical for moderate values of T (e.g. $T = 200$) is described in McLeod and Hipel (1978).

Section 7.12. There is a large literature on ARMA time series with finite variance. Since it is essentially an L^2- theory, there is often no loss of generality in supposing the innovations Gaussian. Brockwell and Davis (1991) offer a good exposition and also provide a number of results in the case of $S\alpha S$ innovations. Cline (1983), Cline and Brockwell (1985) and Davis and Resnick (1986) consider this latter case as well. Kokoszka and Taqqu (1993*d*) show that the codifference of ARMA time series with stable innovations behaves like the covariance in the Gaussian case: it fluctuates within a band of two exponentially decreasing functions.

A number of statistical results are also available for moving averages X_n of the form (7.12.3). For instance, Klüppelberg and Mikosch (1993*c*) obtain the limit of the normalized periodogram $I_{N,X}(\lambda) = |N^{-1/\alpha} \sum_{n=1}^{N} X_n e^{-in\lambda}|^2$ as $N \to \infty$, when the innovations are in the domain of normal attraction of a $S\alpha S$ distribution, $0 < \alpha < 2$. For fixed $\lambda = 2\pi\omega$, $\omega \in (0, 1/2)$ irrational, the limit is $|\psi(\lambda)|^2 (\alpha^2(\lambda) + \beta^2(\lambda))$ where $\psi(\lambda) = \sum_{j=-\infty}^{\infty} c_j e^{-\lambda j}$ (c_j are the coefficients of the moving average) and $(\alpha(\lambda), \beta(\lambda)) = (A^{1/2}G_1, A^{1/2}G_2)$ is sub-Gaussian. In contrast to the Gaussian case, for $0 < \lambda < \ldots < \lambda_m < \pi$, the periodogram ordinates $I_{N,X}(\lambda_i)$, $i = 1, \ldots, m$, are not asymptotically independent.

Davis and Resnick (1986) derive weak limits of sample autocovariance and autocorrelation functions of X_n under the assumption that the distribution of the innovations has tails $P(|\epsilon_1| > x) = x^{-\alpha}L(x)$, where L is a slowly varying function. Their results have been generalized by Klüppelberg and Mikosch (1993*b*) who also consider the convergence of $\tilde{I}_{N,X}(\lambda) = I_{N,X}(\lambda)/\sum_{n=1}^{N} X_n^2$ as $N \to \infty$. Klüppelberg and Mikosch (1993*a*), study the behavior of the integrated periodogram and apply the results to goodness-of-fit tests.

Estimators for the parameters of the AR(p) model with stable innovations or with innovations in the domain of attraction of a stable distribution have been studied by various authors. Yule-Walker, least square and least deviation estimators (the latter minimizes $\sum_{n=1}^{N} |X_n - \phi_1 X_{n-1} - \cdots - \phi_p X_{n-p}|$) have been considered. See Kanter and Steiger (1974), Hannan and Kanter (1977), Gross and Steiger (1979), An and Chen (1982), Davis and Resnick (1986), Knight (1986, 1987), Liu (1987), and Davis, Knight and Liu (1992).

Mikosch, Gadrich, Klüppelberg, and Adler (1993) also consider the estimation of the parameter vector $(\phi_1, \ldots, \phi_p, \theta_1, \ldots, \theta_q)$ of an ARMA (p, q) sequence whose innovations ϵ_n satisfy the aforementioned conditions and, in the case of $S\alpha S$ innovations, they study a sample periodogram based estimator of the vector $(\phi_1, \ldots, \phi_p, \theta_1, \ldots, \theta_q)$. Bhansali (1993) estimates the coefficients of (7.12.3) when an autoregressive representation exists and the ϵ_js are in the domain of attraction of a $S\alpha S$ law. All these papers assume $\sum_j |c_j| < \infty$ or impose even stricter summability conditions on the sequence $\{c_j\}$ in order to ensure the absolute a.s. convergence of the random series (7.12.3).

Section 7.13. Fractional ARIMA time series extend the realm of ARMA modeling to include long-range dependence. Brockwell and Davis (1991) devote a section of their book to fractional ARIMA with Gaussian innovation. The model was considered by Adenstedt (1974) and developed by Granger and Joyeux (1980) and Hosking (1981). See Samorodnitsky and Taqqu (1992*a*) for a brief description.

Kokoszka and Taqqu (1993*c*) develop the theory of fractional ARIMA when the innovations are $S\alpha S$, $0 < \alpha < 2$, and evaluate the codifference. The proofs of Theorems 7.13.3, 7.13.4, 7.13.5 and Exercise 7.18 can be found in that paper. Kokoszka and Taqqu (1993*e*) extend the results to moving averages with regularly varying coefficients.

Fractional ARIMA is of interest because it presents an alternative way to model long-range dependence. In fact, any stationary sequence of random variables whose normalized partial sums converge weakly to linear fractional stable motion would do as well. See Avram and Taqqu (1986*b*) and Maejima (1989).

14.8 Notes to Chapter 8

The Lévy Brownian motion (Example 8.3.3) is the best known example of an H-sssis Gaussian random field. Paul Lévy (1965) devotes a chapter to its study. Chentsov (1957), in a beautiful short note gives the geometric construction presented in Section 8.3 with V_t equal to the set of hyperplanes separating the origin from the point t. There is also an interesting discussion in Lévy (1966). McKean (1963) and Cartier (1971) show that the field has a kind of Markovian property if the parameter space is $\mathbb{R}^n$ with n odd.

While Lévy Brownian motion is H-sssis with $H = 1/2$, the Lévy fractional Brownian field (Example 8.1.3) is H-sssis for all $H \in (0, 1)$. The Lévy fractional Brownian field was first considered by Yaglom (1957) who characterized its spectral representation. It was then investigated by Gangolli (1967) and Mandelbrot (1975, 1982) among others. Mandelbrot's paper contains the germs of future developments.

Shigeo Takenaka (1987) gives a geometric construction for the Lévy fractional Brownian field with $0 < H \leq 1/2$, using the V_t described in Section 8.4. In Takenaka (1991), he defines the Takenaka and (α, H)-Takenaka fields (he calls them "of Chentsov type"), by considering $S\alpha S$, $0 < \alpha < 2$, random measures instead of merely Gaussian measures. He also introduces the Chentsov random fields which he calls "generalized Chentsov." Mori (1992) gives a general characterization of the Lévy-Chentsov random field in the context of infinitely divisible measures.

The proof of Theorem 8.2.6, which gives sufficient conditions for a Chentsov field to be H-sssis, is due to Cioczek-Georges. An independent proof can be found in Takenaka (1993). The results of Sections 8.5 and 8.6 are due to Yumiko Sato and Shigeo Takenaka (see Sato and Takenaka (1991) and Sato (1992*a*)). Results concerning the codifference and the function $I(\theta_1, \theta_2; u)$ in Section 8.7 were obtained by Kokoszka and Taqqu (1992).[1]

The extension of Theorem 8.8.3 to $\mathbb{R}^1$ and $\mathbb{R}^2$ can be found in Sato (1992*b*). The extension to $\mathbb{R}^n$ (Theorem 8.8.4) is proved in Sato (1991) by relating it to the intersection properties of a family of cones in $\mathbb{R}^n \times \mathbb{R}_+$.

This chapter considered only fields that are H-sssis. If one replaces the H-sssis requirement by H-sssi, then one can include many other random fields. Kokoszka and Taqqu (1993*f*) introduce a large class of such fields, defined by

$$X_p(t) = \int_{\mathbb{R}^n} (p(t-x)^{H-n/\alpha} - p(x)^{H-n/\alpha}) M(dx),\ t \in \mathbb{R}^n, \qquad (14.8.1)$$

where p is a norm on $\mathbb{R}^n$ and M is a $S\alpha S$ random measure on $\mathbb{R}^n$ with Lebesgue control measure. The field X_p is well defined for any $n \geq 1,\ H \in (0,1),\ \alpha \in (0,2]$ and is H-sssi. If $n = 1$, it reduces to the well-balanced linear fractional stable motion (Example 3.6.5). The family (14.8.1) parametrized by p is extremely rich: if two fields X_p and $X_{p'}$ have identical finite-dimensional distribution, then the norms p and p' must be identical. The special choice

$$p = p_{\sigma,r}(x) = \left(\int_{S_n} |(x,s)|^r \sigma(ds) \right)^{1/r},\ x \in \mathbb{R}^n,$$

where $r \geq 1$, S_n is the unit sphere in $\mathbb{R}^n$ and σ is a symmetric finite positive Borel measure on S_n, gives rise to a large class of interesting models. Kokoszka and Taqqu show in (1993*a*) that, for any choice of norm p, the codifference $\tau(u)$ related to X_p is asymptotically proportional to $u^{\alpha H - \alpha}$ as $u \to \infty$ if $n > 1$. Since the codifference related to Chentsov fields is asymptotically proportional to $u^{\alpha H - 2}$ (Theorem 8.7.1), Chentsov fields are different from the fields X_p.

Surgailis, Rosinski, Mandrekar and Cambanis (1993) define mixtures of fields which they call "generalized moving averages." These fields can be represented as

$$X(t) = \int_{\Lambda \times \mathbb{R}^n} f(\lambda, t-x) M(d\lambda, dx),\ \ t \in \mathbb{R}^n,$$

where M is a $S\alpha S$ random measure on $\Lambda \times \mathbb{R}^n$ with control measure $Q \otimes$Leb, i.e., the control measure of M is the product measure of a measure Q on Λ and

[1] The proof of Theorem 3.1 in Kokoszka and Taqqu (1992) is incorrect because increment processes involve three- and not two-dimensional distributions. Whether the theorem itself holds is still an open problem.

Lebesgue measure on $\mathbb{R}^n$. The space (Λ, Q) characterizing the mixture is assumed to be a σ-finite complete measure space and the kernel $f\colon \Lambda\times\mathbb{R}^n \to \mathbb{R}$ is assumed measurable and such that

$$\int_{\Lambda\times\mathbb{R}^n} |f(\lambda, x)|^\alpha Q(d\lambda)dx < \infty.$$

If Q has only one atom or if the space Λ consists of only one point, then the field X becomes the usual $S\alpha S$ moving average. If Q has two atoms, then X is a mixture of two $S\alpha S$ moving averages. Observe that Takenaka fields are mixtures with $\Lambda = \mathbb{R}_+$ and $Q = m_\beta$.

14.9 Notes to Chapter 9

Section 9.2. Stochastic processes defined on uncountable parameter spaces give rise to measurability problems. The idea of using separability to overcome these problems goes back to Doob (1953). We shall use his approach for constructing a separable version of a stochastic process.

Section 9.5. Zero-one laws for α-stable processes with $0 < \alpha < 2$ were first discussed by Dudley and Kanter (1974). The approach we present here is new.

14.10 Notes to Chapter 10

Section 10.2. De Acosta (1975) observed that the tail of the distribution of a measurable semi-norm of an α-stable random vector, $0 < \alpha < 2$, is dominated by the tail of the absolute value of a one-dimensional α-stable random variable (our Proposition 10.2.1 is a particular case of this statement). The following more precise version of the former result was proved in de Acosta (1977):

Let E be a real vector space, and $\mathcal{B}$ be a σ-field generated by a linear vector space of linear functionals on E. A probability measure μ on $(E, \mathcal{B})$ is called α-stable, $0 < \alpha \leq 2$, if for any $A, B > 0$, there is a $\mathbf{D} \in E$ such that

$$A\mathbf{X}_1 + B\mathbf{X}_2 \stackrel{d}{=} (A^\alpha + B^\alpha)^{1/\alpha}\mathbf{X} + \mathbf{D},$$

where $\mathbf{X}_1, \mathbf{X}_2$ are independent copies of $\mathbf{X}$, all with law μ. If q is a measurable semi-norm on E, then the limit

$$l = \lim_{\lambda\to\infty} \lambda^\alpha \mu\big(\mathbf{x} \in E\colon\ q(\mathbf{x}) > \lambda\big)$$

exists, and $l \in [0, \infty)$. (The limit l is positive under certain non-degeneracy conditions.)

Kôno and Maejima (1991) state erroneously that the Takenaka process in Example 10.2.7 has bounded sample paths.

Section 10.3. Rosinski (1986) observed that the sample paths of an α-stable process with $0 < \alpha < 2$ "inherit" the negative properties of a "typical" function $f(t, x)$, $t \in T$, in the integral representation (10.1.1).

Section 10.4. The fact that it is easy to derive necessary and sufficient conditions for sample boundedness and continuity of α-stable processes in the case $0 < \alpha < 1$ but not in the case $1 \leq \alpha < 2$ is closely related to the following fact from the theory of probability on Banach spaces: it is easy to describe all Banach space-valued α-stable random variables as long as the Banach space has the so-called stable type $p > \alpha$; see Linde (1986). Whereas any Banach space has stable type p for any $0 < p < 1$, many important ones do not have a stable type $p \geq 1$, e.g., spaces of bounded or continuous functions.

Sample boundedness and continuity of $S\alpha S$ moving averages with $0 < \alpha < 1$ was considered by Balkema and de Haan (1988).

Section 10.5. Theorem 10.5.1 is adapted from a more general result of Rosinski and Samorodnitsky (1993).

Section 10.6. Oscillation processes were introduced by Itô and Nisio (1968) for processes on the interval $[0, 1]$. Jain and Kallianpur (1972) extended them to stochastic processes indexed by a separable metric space and showed that Gaussian processes that are continuous in probability have non-random oscillations.

Oscillation processes of non-Gaussian infinitely divisible processes were considered by Cambanis, Nolan and Rosinski (1990). Noting that continuity in probability is no longer the appropriate assumption, they introduced instead an analogue of Condition 10.6.4.

14.11 Notes to Chapter 11

Section 11.1. The necessary and sufficient conditions that relate the integral representation of an $S\alpha S$ process, $0 < \alpha < 2$, to existence of a measurable version was discovered by Rosinski and Woyczyński (1986).

Section 11.2. Conditions for integrability of the sample paths of α-stable processes with $0 < \alpha < 2$ covering particular cases are scattered throughout the literature. The (symmetric) case $1 \leq p < \alpha$ was completely described by Cambanis and Miller (1980) and Linde, Mandrekar and Weron (1980), whereas the case $p > \max(\alpha, 1)$ is due to Marcus and Woyczyński (1979) and Linde, Mandrekar and

Weron (1980). The problem in the case $p = \alpha \geq 1$ was solved by Rosinski and Woyczyński (1986).

As most of these results were obtained by working with α-stable measures on L^p spaces, the case $0 < p < 1$ was rarely considered. The case $p = \alpha \in (0, 1)$ has been implicitly solved by Kwapień and Woyczyński (1987), and sufficiency of the integrability conditions in the case $0 < \alpha < p < 1$ can be deduced from Marcus and Woyczyński (1979) and Rosinski and Woyczyński (1985). The unified approach presented here is due to Samorodnitsky (1992).

Section 11.4. The problem of interchanging the order of ordinary and stochastic integration in the $S\alpha S$ case with $1 < \alpha < 2$ was considered by Cambanis and Miller (1981) and McConnell and Taqqu (1984), Lemma 4.4. Rosinski (1986) gave a complete solution in the case $1 \leq \alpha < 2$. Our approach in the case $0 < \alpha < 1$ follows Samorodnitsky (1992).

Section 11.7. Cambanis and Miller (1980) derived necessary and sufficient conditions for a $S\alpha S$ process $\{X(t),\ a \leq t \leq b\}$ to have a version with all sample paths in $AC^1[a, b]$. Rosinski (1986) extended their result to the $S\alpha S$ case with $0 < \alpha < 2$ and $p \geq 1$.

14.12 Notes to Chapter 12

Section 12.1. The importance of metric entropy in the study of the sample path properties of Gaussian processes has been underlined by Dudley (1967, 1973) and Fernique (1975), who established a variety of very precise results.

Pisier (1983) showed that one can derive sufficient conditions for sample boundedness and continuity in a general non-Gaussian framework by using metric entropy. Our approach follows that of Ledoux and Talagrand (1991).

Section 12.2. Another version of the sufficient condition for sample boundedness and sample continuity of α-stable processes is given by Marcus and Pisier (1984). See also Nolan (1989) for a more explicit form.

Section 12.3. The results of Theorem 12.3.1 are due to Marcus and Pisier (1984) in the case $1 < \alpha < 2$ and to Talagrand (1988) in the case $\alpha = 1$.

Section 12.4. The general exposition of this section as well as the breakdown of the set $\mathcal{D}$ follows Kôno and Maejima (1991).

Maejima (1983*a*) was the first to show that the well-balanced linear fractional stable motions are not sample bounded on any integral of positive length if $0 < H < 1/\alpha$. Examples 12.4.2 and 12.4.3 are taken from Samorodnitsky (1993).

A more detailed discussion of sample path properties of certain α-stable self-similar processes can be found in Takashima (1991) and Kôno and Maejima (1991). See also a general discussion in Vervaat (1985).

14.13 Notes to Chapter 13

Bretagnolle, Dacunha-Castelle and Krivine (1966) established the integral representation for $S\alpha S$ processes in the case $\alpha \geq 1$. They showed that for processes continuous in probability, one can choose $E = (0, 1)$ in (13.2.4) and let M have Lebesgue control measure. Schreiber (1972) extended their result to the case $0 < \alpha < 1$.

Schilder (1970), apparently unaware of Bretagnolle, Dacunha-Castelle and Krivine's work, established the integral representation in the case $0 < \alpha < 2$ and T finite, noting that such a representation "for even countably many [time points] xs has eluded the most vigorous efforts of the author." Kuelbs (1973) reproved the result for countable T and extended it to the case of $S\alpha S$ processes satisfying Condition S. Hardin (1984) then extended the integral representation to the skewed α-stable processes with $\alpha \neq 1$, satisfying Condition S.

The approach we follow here is close to the one in Ledoux and Talagrand (1991) in the $S\alpha S$ case; it does not use the theory of infinitely divisible random vectors in Banach space.

Appendix A

Tables of symmetric α-stable fractiles

Let $X \sim S(\alpha, 0, 0)$ and let $f = P\{X \leq x_f\}$ be the upper fractile corresponding to x_f. The following tables display x_f as a function of

$$f = 0.500\ (0.025)\ 0.900\ (0.010)\ 0.970\ (0.005)\ 0.995,\ 0.998.$$

for

$$\alpha = 0.1\ (0.1)\ 1.9,$$

and also as a function of the extreme upper fractiles[1]

$$f = 0.9990\ (0.0001)\ 0.9999 \quad \text{for} \quad \alpha = 0.5\ (0.1)\ 1.9.$$

Since X is symmetric, entries for $f < 0.5$ are obtained by using the relation

$$f = P(X \leq x_f) = 1 - P(X \leq x_{1-f}).$$

These tables are reproduced from Brothers, DuMouchel and Paulson (1983), with the kind permission of Albert Paulson. They are based, for the most part, on Zolotarev's integral representation of the cumulative distribution function (see Section 1.6). We refer the reader to the aforementioned paper for details on the computational techniques. The listing for the symmetric Cauchy ($\alpha = 1$) is obtained by using the relation

$$x_f = \tan \pi(f - 1/2),$$

which is a consequence of (1.1.14). The listing for $\alpha = 2$ is obtained by using the $N(0, \nu^2)$ distribution, with $\nu = \sqrt{2}$.

[1] There are a few discrepancies with Paulson and Delehanty (1993) in the third and fourth significant digits for $f = 0.9998$ and $f = 0.9999$.

Table of fractiles of the $S\alpha S$ law

f	0.500	0.525	0.550	0.575	0.600
$\alpha = 2.0$	0.	.08868	.17771	.26745	.35829
$\alpha = 1.9$	0.	.88570 − 01	.17751	.26720	.35804
$\alpha = 1.8$	0.	.88383 − 01	.17716	.26672	.35752
$\alpha = 1.7$	0.	.88094 − 01	.17661	.26597	.35664
$\alpha = 1.6$	0.	.87674 − 01	.17580	.26484	.35529
$\alpha = 1.5$	0.	.87082 − 01	.17465	.26323	.35334
$\alpha = 1.4$	0.	.86262 − 01	.17307	.26097	.35059
$\alpha = 1.3$	0.	.85139 − 01	.17088	.25787	.34677
$\alpha = 1.2$	0.	.83609 − 01	.16791	.25363	.34155
$\alpha = 1.1$	0.	.81530 − 01	.16387	.24787	.33447
$\alpha = 1.0$	0.	.07870	.15838	.07870	0.3249
$\alpha = 0.9$	0.	.74848 − 01	.15093	.22955	.31212
$\alpha = 0.8$	0.	.69591 − 01	.14082	.21539	.29515
$\alpha = 0.7$	0.	.62436 − 01	.12720	.19659	.27298
$\alpha = 0.6$	0.	.52820 − 01	.10917	.17211	.24459
$\alpha = 0.5$	0.	.40371 − 01	.86244 − 01	.14126	.20889
$\alpha = 0.4$	0.	.25641 − 01	.59055 − 01	.10396	.16464
$\alpha = 0.3$	0.	.11236 − 01	.30558 − 01	.61802 − 01	.11086
$\alpha = 0.2$	0.	.19652 − 02	.79023 − 02	.21616 − 01	.50394 − 01
$\alpha = 0.1$	0.	.88650 − 05	.12744 − 03	.90029 − 03	.47190 − 02

f	0.625	0.650	0.675	0.700	0.725
$\alpha = 2.0$	45062	.54493	.64172	.74161	.84536
$\alpha = 1.9$	.45047	.54497	.64212	.74258	.84714
$\alpha = 1.8$	.45000	.54469	.64221	.74327	.84875
$\alpha = 1.7$	.44912	.54398	.64189	.74362	.85016
$\alpha = 1.6$	.44771	.54271	.64102	.74352	.85131
$\alpha = 1.5$	.44560	.54070	.63945	.74285	.85214
$\alpha = 1.4$	.44258	.53774	.63698	.74146	.85262
$\alpha = 1.3$	.43836	.53354	.63337	.73921	.85278
$\alpha = 1.2$	.43257	.52774	.62834	.73598	.85276
$\alpha = 1.1$	.42473	.51991	.62157	.73171	.85294
$\alpha = 1.0$	.41421	.50953	.61280	.72654	.85408
$\alpha = 0.9$	.40030	.49606	.60187	.72093	.85751
$\alpha = 0.8$	.38219	.47903	.58882	.71570	.86529
$\alpha = 0.7$	.35904	.45796	.57383	.71214	.88054
$\alpha = 0.6$	.32990	.43221	.55703	.71205	.90828
$\alpha = 0.5$	.29339	.40049	.53821	.71818	.95778
$\alpha = 0.4$	.24709	.36000	.51630	.73577	1.0496
$\alpha = 0.3$	.18695	.30477	.48823	.77739	1.2416
$\alpha = 0.2$	.10814	.22178	.44466	.88558	1.7750
$\alpha = 0.1$	.21181 − 01	.87379 − 01	.34590	1.3549	5.3862

f	0.750	0.775	0.800	0.825	0.850
$\alpha = 2.0$	.95387	1.0683	1.1902	1.3217	1.4657
$\alpha = 1.9$	.95680	1.0729	1.1970	1.3316	1.4801
$\alpha = 1.8$	.95976	1.0777	1.2045	1.3429	.14968
$\alpha = 1.7$	.96274	1.0830	1.2130	1.3561	1.5167
$\alpha = 1.6$	.96577	1.0888	1.2229	1.3718	1.5411
$\alpha = 1.5$	.96893	1.0954	1.2346	1.3910	1.5715
$\alpha = 1.4$	.97237	1.1033	1.2491	1.4154	1.6107
$\alpha = 1.3$	.97638	1.1132	1.2678	1.4472	1.6625
$\alpha = 1.2$	.98154	1.1263	1.2928	1.4900	1.7325
$\alpha = 1.1$	.98885	1.1445	1.3274	1.5491	.18286
$\alpha = 1.0$	1.000	1.1709	1.3764	1.6319	1.9626
$\alpha = 0.9$	1.0176	1.2098	1.4470	1.7499	2.1532
$\alpha = 0.8$	1.0455	1.2681	1.5508	1.9226	2.4332
$\alpha = 0.7$	1.0901	1.3573	1.7082	2.1858	2.8661
$\alpha = 0.6$	1.1621	1.4989	1.9595	2.6140	3.5905
$\alpha = 0.5$	1.2838	1.7395	2.3975	3.3879	4.9611
$\alpha = 0.4$	1.5090	2.2008	3.2817	5.0517	8.1347
$\alpha = 0.3$	2.0060	3.3072	5.6199	9.9695	18.789
$\alpha = 0.2$	3.6241	7.6349	16.845	39.665	102.33
$\alpha = 0.1$	22.255	98.014	473.89	2612.0	17294.

f	0.875	0.900	0.910	0.920	0.930
$\alpha = 2.0$	1.6268	1.8124	1.8961	1.9871	2.0871
$\alpha = 1.9$	1.6476	1.8430	1.9322	2.0300	2.1386
$\alpha = 1.8$	1.6724	1.8803	1.9765	2.0831	2.2031
$\alpha - 1.7$	1.7025	1.9265	2.0320	2.1503	2.2856
$\alpha = 1.6$	1.7400	1.9853	2.1031	2.2372	2.3933
$\alpha = 1.5$	1.7878	2.0615	2.1959	2.3515	2.5360
$\alpha = 1.4$	1.8502	2.1622	2.3192	2.5039	2.7269
$\alpha = 1.3$	1.9333	2.2971	2.4846	2.7084	2.9828
$\alpha = 1.2$	2.0458	2.4796	2.7081	2.9845	3.3284
$\alpha = 1.1$	2.1998	2.7293	3.0139	3.3628	3.8026
$\alpha = 1.0$	2.4142	3.0777	1.4715	3.8947	1.9626
$\alpha = 0.9$	2.7203	3.5805	4.0641	4.6743	5.4677
$\alpha = 0.8$	3.1754	4.3439	5.0191	5.8868	7.0381
$\alpha = 0.7$	3.8957	5.5918	6.6056	7.9390	9.7538
$\alpha = 0.6$	5.1460	7.8640	9.5619	11.863	15.103
$\alpha = 0.5$	7.6487	12.741	16.121	20.898	27.942
$\alpha = 0.4$	13.976	26.460	35.515	49.139	70.675
$\alpha = 0.3$	38.622	90.392	133.80	206.26	334.82
$\alpha = 0.2$	300.84	1074.9	1934.4	3699.6	7646.3
$\alpha = 0.1$	.14876 + 06	.18909 + 07	.61138 + 07	.22329 + 08	.95242 + 08

f	0.940	0.950	0.960	0.970	0.975
$\alpha = 2.0$	2.1988	2.3262	2.4758	2.6598	2.7718
$\alpha = 1.9$	2.2616	2.4043	2.5761	2.7951	2.9338
$\alpha = 1.8$	2.3411	2.5049	2.7079	2.9785	3.1584
$\alpha = 1.7$	2.4443	2.6373	2.8850	3.2314	3.4731
$\alpha = 1.6$	2.5804	2.8143	3.1247	3.5780	3.9056
$\alpha = 1.5$	2.7622	3.0519	3.4476	4.0427	4.4814
$\alpha = 1.4$	3.0055	3.3699	3.8778	4.6571	5.2391
$\alpha = 1.3$	3.3314	3.7947	4.4512	5.4757	6.2507
$\alpha = 1.2$	3.7712	4.3687	5.2289	6.5953	7.6445
$\alpha = 1.1$	4.3773	5.1646	6.3181	8.1887	9.6509
$\alpha = 1.0$	3.0777	6.3138	7.9158	10.579	12.706
$\alpha = 0.9$	6.5404	8.0679	10.408	14.418	17.708
$\alpha = 0.8$	8.6301	10.956	14.631	21.170	26.716
$\alpha = 0.7$	12.336	16.235	22.640	34.607	45.198
$\alpha = 0.6$	19.889	27.440	40.504	66.550	90.942
$\alpha = 0.5$	38.913	57.304	91.522	166.24	241.92
$\alpha = 0.4$	106.96	173.58	311.80	657.79	1051.7
$\alpha = 0.3$	581.72	1109.4	2422.2	6553.8	12253.
$\alpha = 0.2$	17500.	46060.	.14852 + 06	.66067 + 06	.16884 + 07
$\alpha = 0.1$	.49816 + 09	.34462 + 10	.35786 + 11	.70717 + 12	.46157 + 13

f	0.980	0.985	0.990	0.995	0.998
$\alpha = 2.0$	2.9044	3.0690	3.2900	3.6428	4.0703
$\alpha = 1.9$	3.1049	3.3309	3.6691	4.3676	5.9412
$\alpha = 1.8$	3.3915	3.7221	4.2768	5.6428	8.7634
$\alpha = 1.7$	3.8003	4.2893	5.1519	7.2896	12.030
$\alpha = 1.6$	4.3601	5.0524	6.2841	9.3323	16.158
$\alpha = 1.5$	5.0967	6.0421	7.7364	11.983	21.735
$\alpha = 1.4$	6.0628	7.3404	9.6588	15.595	29.707
$\alpha = 1.3$	7.3590	9.0997	12.313	20.774	41.773
$\alpha = 1.2$	9.1641	11.589	16.160	28.630	61.216
$\alpha = 1.1$	11.801	15.300	22.071	41.348	94.961
$\alpha = 1.0$	15.895	21.205	31.821	63.657	159.15
$\alpha = 0.9$	22.754	31.409	49.411	106.99	296.57
$\alpha = 0.8$	35.476	51.060	85.139	203.38	640.98
$\alpha = 0.7$	62.569	94.972	170.56	461.94	1716.5
$\alpha = 0.6$	133.01	216.61	429.22	1373.6	6355.9
$\alpha = 0.5$	381.98	686.14	1559.7	6302.5	39630.
$\alpha = 0.4$	1861.9	3872.9	10813.	61965.	.61710 + 06
$\alpha = 0.3$	26241.	69680.	.27395 + 06	.28095 + 07	.60197 + 08
$\alpha = 0.2$	52907 + 07	.22887 + 08	.17838 + 09	.58571 + 10	.58083 + 12
$\alpha = 0.1$	.45292 + 14	.84711 + 15	.51424 + 17	.55411 + 20	.54473 + 24

Table of extreme upper fractiles of the $S\alpha S$ law

f	.9990	.9991	.9992	.9993	.9994
$\alpha = 2.0$	4.3702	4.4143	4.4631	4.5179	4.5805
$\alpha = 1.9$	8.084	8.498	8.993	9.596	10.35
$\alpha = 1.8$	12.59	13.32	14.18	15.24	16.57
$\alpha = 1.7$	17.85	18.97	20.30	21.93	23.99
$\alpha = 1.6$	24.72	26.38	28.37	30.81	33.91
$\alpha = 1.5$	34.32	36.80	39.78	43.46	48.14
$\alpha = 1.4$	48.57	52.35	56.92	62.60	69.87
$\alpha = 1.3$	71.04	77.02	84.30	93.41	105.1
$\alpha = 1.2$	108.9	118.9	131.2	146.6	166.7
$\alpha = 1.1$	178.2	196.1	218.3	246.4	283.5
$\alpha = 1$	318.3	353.6	397.9	454.7	530.5
$\alpha = 0.9$	640.9	720.5	821.3	952.7	1131.
$\alpha = 0.8$	1526.	1741.	2017.	2384.	2890.
$\alpha = 0.7$	4626.	5378.	6365.	7703.	9602.
$\alpha = 0.6$	.2021 + 05	.2409 + 05	.2933 + 05	.3664 + 05	.4738 + 05
$\alpha = 0.5$	.1588 + 06	.1961 + 06	.2483 + 06	.3244 + 06	.4416 + 06

f	.9995	.9996	.9997	.9998	.9999
$\alpha = 2.0$	4.6535	4.7416	4.8530	5.0064	5.2595
$\alpha = 1.9$	11.33	12.68	14.67	18.06	25.86
$\alpha = 1.8$	18.29	20.66	24.18	30.22	44.28
$\alpha = 1.7$	26.67	30.37	35.92	45.53	68.31
$\alpha = 1.6$	37.97	43.61	52.16	67.15	103.4
$\alpha = 1.5$	54.34	63.02	76.30	99.93	158.4
$\alpha = 1.4$	79.56	93.27	114.5	152.9	250.4
$\alpha = 1.3$	121.0	143.6	179.1	244.6	416.0
$\alpha = 1.2$	194.0	233.6	296.9	416.1	739.7
$\alpha = 1.1$	334.6	409.8	532.3	769.4	1441.
$\alpha = 1$	636.6	795.8	1061.	1591.	3171.
$\alpha = 0.9$	1385.	1774.	2443.	3833.	8236.
$\alpha = 0.8$	3631.	4799.	6876.	11412.	26943.
$\alpha = 0.7$	.1246 + 05	.1714 + 05	.2585 + 05	.4612 + 05	.1228 + 06
$\alpha = 0.6$	.6422 + 05	.9316 + 05	.1505 + 06	.2955 + 06	.9392 + 06
$\alpha = 0.5$	.6360 + 06	.9939 + 06	.1767 + 07	.3964 + 07	.1590 + 08

Bibliography

Adenstedt, R. (1974), 'On large-sample estimation for the mean of a stationary random sequence', *The Annals of Statistics* **2**(6), 1095–1107.

Adler, R. J. (1990), An introduction to continuity extrema and related topics for general Gaussian processes, *in* 'IMS Lectures Notes', Institute of Mathematical Statistics, Hayward. Vol. 12.

Adler, R. J., Cambanis, S. & Samorodnitsky, G. (1990), 'On stable Markov processes', *Stochastic Processes and their Applications* **34**, 1–17.

Adler, R. J., Samorodnitsky, G. & Gadrich, T. (1993), 'The expected number of level crossings for stationary, harmonizable, symmetric, stable processes', *The Annals of Applied Probability* **3**, 553–575.

Ahlfors, L. V. (1979), *Complex Analysis*, 3rd edn, McGraw-Hill Book Co.

Alam, K. & Saxena, K. M. L. (1982), 'Positive dependence in multivariate distributions', *Communications in Statistics* **A10**, 1183–1186.

An, H. Z. & Chen, Z. G. (1982), 'On convergence of LAD estimates in autoregression with infinite variance', *Journal of Multivariate Analysis* **12**, 335–345.

Araujo, A. & Giné, E. (1980), *The Central Limit Theorem for Real and Banach Valued Random Variables*, Wiley, New York.

Astrauskas, A. (1982), 'Stable self-similar fields', *Lithuanian Mathematical Journal* **22**(3), 215–221.

Astrauskas, A. (1983), 'Limit theorems for sums of linearly generated random variables', *Lithuanian Mathematical Journal* **23**(2), 127–134.

Astrauskas, A., Levy, J. B. & Taqqu, M. S. (1991), 'The asymptotic dependence structure of the linear fractional Lévy motion', *Lietuvos Matematikos Rinkinys (Lithuanian Mathematical Journal)* **31**(1), 1–28.

Avram, F. & Taqqu, M. S. (1986*a*), 'Symmetric polynomials of random variables attracted to an infinitely divisible law', *Probability Theory and Related Fields* **71**, 491–500.

Avram, F. & Taqqu, M. S. (1986*b*), 'Weak convergence of moving averages with infinite variance', *Dependence in Probability and Statistics* pp. 399–415. E. Eberlein and M. S. Taqqu, eds., Boston:Birkhauser.

Avram, F. & Taqqu, M. S. (1992), 'Weak convergence of sums of moving averages in the α-stable domain of attraction', *The Annals of Probability* **20**, 483–503.

Bachelier, L. (1900), 'Théorie de la spéculation', *Annales Scientifiques de l'Ecole Normale Supérieure* **III-17**, 21–86. Translated in *The random character of stock market prices*, Ed. P.H. Cootner, pp. 17-78, Cambridge, MIT Press, 1964.

Balkema, A. A. & de Haan, L. (1988), 'Almost sure continuity of stable moving averages processes with index <1', *The Annals of Probability* **16**, 333–343.

Becker, R. A., Chambers, J. & Wilks, A. (1988), *The New S Language*, Wadsworth, Pacific Grove, Calif.

Belyaev, K. Y. (1961), Continuity and Hölder's conditions for sample functions of stationary gaussian processes, *in* 'Proceedings of the 4th Berkeley Symposium Mathematical Statistics and Probability', University of California Press, pp. 23–33. Volume 2.

Beran, J. (1992), 'Statistical methods for data with long-range dependence', *Statistical Science* **7**(4), 404–416. With discussions and rejoinder, pages 404-427.

Beran, J., Sherman, R., Taqqu, M. S. & Willinger, W. (1993), 'Variable-bit-rate video traffic and long-range dependence', *IEEE Transactions on Communications*. Accepted subject to revisions.

Bhansali, R. J. (1993), 'Estimation of the impulse response coefficients of a linear process with infinite variance', *Journal of Multivariate Analysis* **45**, 274–290.

Billingsley, P. (1986), *Probability and Measure*, 2nd edn, Wiley, New York.

Box, G. E. P. & Muller, M. E. (1958), 'A note on the generation of random normal deviates', *The Annals of Mathematical Statistics* **29**, 610–611.

Breiman, L. (1968), *Probability*, Addison-Wesley, Reading, MA.

Bretagnolle, J., Dacunha-Castelle, D. & Krivine, J. L. (1966), 'Lois stables et espaces L^p', *Annales de l'Institut Henri Poincaré* **2**, 231–259. Section B.

Brindley, J. E. C. & Thompson, J. W. A. (1972), 'Dependence and aging aspects of multivariate survival', *Journal of the American Statistical Association* **67**, 821–829.

Brockwell, P. J. & Davis, R. A. (1991), *Time Series: Theory and Methods*, 2nd edn, Springer-Verlag, New York.

Brothers, K. M., DuMouchel, W. H. & Paulson, A. S. (1983), Fractiles of the stable laws, Technical report, Rensselaer Polytechnic Institute, Troy.

Brown, G. W. & Tukey, J. W. (1946), 'Some distribution of sample means', *The Annals of Mathematical Statistics* **17**, 1–12.

Byczkowski, T., Nolan, J. & Rajput, B. (1993), 'Approximation of multidimensional stable densities', *Journal of Multivariate Analysis* **46**, 13–31.

Cambanis, S. (1983), Complex symmetric stable variables and processes, *in* P. Sen, ed., 'Contribution to Statistics, Essays in Honor of Norman L. Johnson', North-Holland, pp. 63–79.

Cambanis, S. & Fakhre-Zakeri, I. (1993), On prediction of heavy-tailed autoregressive sequences: forward versus reversed time, Technical Report 383, Center for Stochastic Processes at the University of North Carolina, Chapel Hill.

Cambanis, S. & Maejima, M. (1989), 'Two classes of self-similar stable processes with stationary increments', *Stochastic Processes and their Applications* **32**, 305–329.

Cambanis, S. & Miller, G. (1980), 'Some path properties of *p*th order and stable processes', *The Annals of Probability* **6**, 1148–1156.

Cambanis, S. & Miller, G. (1981), 'Linear problems in *p*th order and symmetric stable processes', *SIAM Journal on Applied Mathematics* **41**, 43–69.

Cambanis, S. & Soltani, A. R. (1984), 'Prediction of stable processes: spectral and moving average representation', *Zeitschrift für Wahrscheinlichkeitstheorie und verwandte Gebiete* **66**, 593–612.

Cambanis, S. & Wu, W. (1992), 'Multiple regression on stable vectors', *Journal of Multivariate Analysis* **41**, 243–272.

Cambanis, S., Hardin, Jr., C. D. & Weron, A. (1987), 'Ergodic properties of stationary stable processes', *Stochastic Processes and their Applications* **24**, 1–18.

Cambanis, S., Hardin, Jr., C. D. & Weron, A. (1988), 'Innovations and Wold decompositions of stable sequences', *Probability Theory and Related Fields* **79**, 1–27.

Cambanis, S., Nolan, J. P. & Rosinski, J. (1990), 'On the oscillation of infinitely divisible processes', *Stochastic Processes and their Applications* **35**, 87–97.

Cartier, P. (1971), Introduction à l'étude des mouvements browniens à plusieurs paramètres, *in* 'Séminaire de Probabilités V', Springer-Verlag, New York, pp. 58–75. Lecture Notes in Mathematics, Vol. 191.

Cauchy, A. (1853), 'Sur les résultats moyens d'observations de même nature, et sur les résultats les plus probables', *Comptes Rendus de l'Académie des Sciences de Paris* **37**, 198–206.

Chambers, J. M., Mallows, C. & Stuck, B. W. (1976), 'A method for simulating stable random variables', *Journal of the American Statistical Association* **71**(354), 340–344. Theory and Methods Section.

Chen, R. & Shepp, L. (1983), 'On the sum of symmetric random variables', *American Statistician* **37**, 237.

Chen, Y. (1991), Distributions for Asset Returns, PhD thesis, SUNY-Stony Brook, Dept. of Economics.

Chentsov, N. N. (1957), 'Lévy's Brownian motion of several parameters and generalized white noise', *Theory of Probability and its Applications* **2**, 265–266.

Choquet-Bruhat, Y. (1967), *Problems and solutions in mathematical physics*, Holden-Day, San Francisco.

Christoph, G. & Wolf, W. (1993), *Convergence Theorems with a Stable Limit Law*, Akademie Verlag, Berlin.

Cioczek-Georges, R. & Taqqu, M. S. (1993*a*), 'Does asymptotic linearity of the regression extend to stable domains of attraction?', *Journal of Multivariate Analysis*. To appear.

Cioczek-Georges, R. & Taqqu, M. S. (1993*b*), Form of the conditional variance for symmetric stable random variables, Preprint.

Cioczek-Georges, R. & Taqqu, M. S. (1994*a*), 'How do conditional moments of stable vectors depend on the spectral measure?', *Stochastic Processes and their Applications*. To appear.

Cioczek-Georges, R. & Taqqu, M. S. (1994*b*), Necessary conditions for the existence of conditional moments of stable random variables, Preprint.

Cioczek-Georges, R. & Taqqu, M. S. (1994*c*), Sufficient conditions for the existence of conditional moments of stable random variables, Preprint.

Cline, D. B. H. (1983), Estimation and linear prediction for regression, autoregression and ARMA with infinite variance data, PhD thesis, Colorado State University.

Cline, D. B. H. & Brockwell, P. J. (1985), 'Linear prediction of ARMA processes with infinite variance', *Stochastic Processes and their Applications* **19**, 281–296.

Cohn, D. L. (1972), 'Measurable choice of limit points and the existence of separable and measurable processes', *Zeitschrift für Wahrscheinlichkeitstheorie und verwandte Gebiete* **22**, 161–165.

Cox, D. R. (1984), Long-range dependence: a review, *in* H. David & H. David, eds, 'Statistics: An Appraisal', Iowa State University Press, pp. 55–74.

Cramér, H. & Leadbetter, M. R. (1967), *Stationary and Related Stochastic Processes*, Wiley, New York.

Dabrowski, A. R. & Jakubowski, A. (1993), 'Stable limits for associated random variables', *The Annals of Probability*. To appear.

Dahlhaus, R. (1989), 'Efficient parameter estimation for self similar processes', *The Annals of Statistics* **17**(4), 1749–1766.

Davis, R. A. (1983), 'Stable limits for partial sums of dependent random variables', *The Annals of Probability* **11**, 262–269.

Davis, R. A. & Resnick, S. I. (1985), 'Limit theory for moving averages of random variables with regularly varying tail probabilities', *The Annals of Probability* **13**(1), 179–195.

Davis, R. A. & Resnick, S. I. (1986), 'Limit theory for the sample covariance and correlation functions of moving averages', *The Annals of Statistics* **14**(2), 533–558.

Davis, R. A., Knight, K. & Liu, J. (1992), 'M-estimation for autoregressions with infinite variance', *Stochastic Processes and Their Applications* **40**(1), 145–180.

Davydov, Y. A. (1970), 'The invariance principle for stationary processes', *Theory of Probability and its Applications* **15**, 487–498.

de Acosta, A. (1975), 'Stable measures and seminorms', *The Annals of Probability* **3**, 865–875.

de Acosta, A. (1977), 'Asymptotic behavior of stable measures', *The Annals of Probability* **5**, 494–499.

de Acosta, A. (1980), 'Exponential moments of vector-valued random series and triangular arrays', *The Annals of Probability* **8**, 381–389.

Desbiens, J. (1987), 'Caractérisation d'un espace préhilbertien au moyen de la relation d'orthogonalite de Birkhoff-James', *Les Annales des Sciences Mathématiques du Québec* **21**, 295–303.

Dobrushin, R. L. & Major, P. (1979), 'Non-central limit theorems for non-linear functions of Gaussian fields', *Zeitschrift für Wahrscheinlichkeitstheorie und verwandte Gebiete* **50**, 27–52.

Doob, J. L. (1953), *Stochastic Processes*, Wiley, New York.

Dudley, R. M. (1967), 'The sizes of compact subsets of Hilbert space and continuity of Gaussian processes', *Journal of Functional Analysis* **1**, 290–330.

Dudley, R. M. (1973), 'Sample functions of the Gaussian process', *The Annals of Probability* **1**, 66–103.

Dudley, R. M. & Kanter, M. (1974), 'Zero-one laws for stable measures', *Proceedings of the American Mathematical Society* **45**, 245–252.

DuMouchel, W. H. (1971), Stable distributions in statistical inference, PhD thesis, Univ. Ann Arbor, Univ. Microfilms, Ann Arbor, Mich.

DuMouchel, W. H. (1973), 'On the asymptotic normality of the maximum-likelihood estimate when sampling from a stable distribution', *The Annals of Statistics* **1**, 948–957.

Dunford, N. & Schwartz, J. T. (1958), *Linear Operators, Part 1: General Theory*, Wiley, New York.

Durrett, R. (1991), *Probability: Theory and Examples*, Wadsworth & Brooks/Cole, Pacific Grove, California.

Esary, J. D., Proschan, F. & Walkup, D. W. (1967), 'Association of random variables, with applications', *The Annals of Mathematical Statistics* **38**, 1466–1474.

Evertsz, J. G. & Mandelbrot, B. B. (1991), Left-sided multifractal measures, *in* A. Bunde & S. Havlin, eds, 'Fractals and Disordered Systems', Springer-Verlag, Heidelberg, pp. 322–344.

Evertsz, J. G. & Mandelbrot, B. B. (1992), Multifractal measures, *in* H. O. Peitgen, H. Jürgens & D. Saupe, eds, 'Chaos and Fractals', Springer-Verlag, New York, pp. 921–953. Appendix B.

Fama, E. & Roll, R. (1968), 'Some properties of symmetric stable distributions', *Journal of the American Statistical Association* **63**, 817–836.

Feldheim, E. (1937), Etude de la stabilité des lois de probabilité, PhD thesis, Thése de la Faculté des Sciences de Paris.

Feller, W. (1957), *An Introduction to Probability Theory and its Applications*, Vol. 1, 3rd edn, Wiley, New York.

Feller, W. (1966), *An Introduction to Probability Theory and its Applications*, Vol. 2, 1st edn, Wiley, New York.

Feller, W. (1971), *An Introduction to Probability Theory and its Applications*, Vol. 2, 2nd edn, Wiley, New York.

Ferguson, T. & Klass, M. (1972), 'A representation theorem of independent increment processes without Gaussian component.', *The Annals of Mathematical Statistics* **43**, 1634–1643.

Fernique, X. (1975), Regularité des trajectoires des fonctions aléatoires gaussiennes, *in* 'Lecture Notes in Mathematics, Vol. 480', Springer-Verlag, New York, pp. 1–96.

Fix, E. (1949), Distributions which lead to linear regression, *in* 'Proceedings of the Berkeley Symposium on Mathematical Statistics and Probability 1945-46', University of California Press, Berkeley.

Fox, R. & Taqqu, M. S. (1986), 'Large-sample properties of parameter estiamtes for strongly dependent stationary Gaussian time series', *The Annals of Statistics* **14**, 517–532.

Gangolli, R. (1967), 'Lévy's brownian motion of several parameters', *Annales de l'Institut Henri Poincaré* **3B**, 121–226.

Giné, E. & Hahn, M. G. (1983), 'On stability of probability laws with univariate stable marginals', *Zeitschrift für Wahrscheinlichkeitstheorie und verwandte Gebiete* **64**, 157–165.

Gnedenko, B. V. & Kolmogorov, A. N. (1954), *Limit distributions for sums of independent random variables*, Addison-Wesley, Reading, MA.

Gradshteyn, I. S. & Ryzhik, I. M. (1980), *Table of Integrals, Series, and Products*, Academic Press, New York.

Granger, C. W. J. (1980), 'Long memory relationships and aggregationof dynamic models', *Journal of Econometrics* **14**, 227–238.

Granger, C. W. J. & Joyeux, R. (1980), 'An introduction to long-memory time series and fractional differencing', *Journal of Time Series Analysis* **1**, 15–30.

Gross, A. (1993), 'Some mixing conditions for stationary symmetric stable stochastic processes', *Stochastic Processes and their Applications.* To appear.

Gross, S. & Steiger, W. L. (1979), 'Least absolute deviation estimates in autoregression with infinite variance', *Journal of Applied Probability* **16**, 104–116.

Gupta, V. K. & Waymire, E. (1990), 'Multiscaling properties of spatial rainfall and river flow distributions', *Journal of Geophysical Research* **95**(D3), 1999–2009.

Hall, P. (1980), 'A comedy of errors: the canonical form for a stable characteristic function', *Bulletin of the London Mathematical Society* **13**, 23–27.

Hall, P. & Welsh, A. H. (1984), 'Best attainable rates of convergence for estimates of parameters of regular variation', *The Annals of Statistics* **12**(3), 1079–1084.

Hannan, E. J. & Kanter, M. (1977), 'Autoregressive processes with infinite variance', *Journal of Applied Probability* **14**, 411–415.

Hardin Jr., C. D. (1982*a*), 'On the linearity of regression', *Zeitschrift für Wahrscheinlichkeitstheorie und verwandte Gebiete* **61**, 293–302.

Hardin Jr., C. D. (1982*b*), 'On the spectral representation of symmetric stable processes', *Journal of Multivariate Analysis* **12**, 385–401.

Hardin Jr., C. D. (1984), Skewed stable variables and processes, Technical Report 79, Center for Stochastic Processes at the University of North Carolina, Chapel Hill.

Hardin Jr., C. D., Samorodnitsky, G. & Taqqu, M. S. (1991*a*), 'Non-linear regression of stable random variables', *The Annals of Applied Probability* **1**, 582–612.

Hardin Jr., C. D., Samorodnitsky, G. & Taqqu, M. S. (1991*b*), Numerical computations of non-linear stable regression functions, *in* S. Cambanis, G. Samorodnitsky & M. S. Taqqu, eds, 'Stable Processes and Related Topics', Vol. 25 of *Progress in Probability*, Birkhäuser, Boston, pp. 143–180.

Hill, B. M. (1975), 'A simple general approach to inference about the tail of a distribution', *The Annals of Statistics* **3**, 1163–1174.

Hoffmann-Jørgensen, J. (1973), 'Existence of a measurable modification of stochastic processes', *Zeitschrift für Wahrscheinlichkeitstheorie und verwandte Gebiete* **25**, 205–207.

Holt, D. R. & Crow, E. L. (1973), 'Tables and graphs of the stable probability density functions', *Journal of Research of the National Bureau of Standards* **77B**, 143–198.

Holtsmark, J. (1939), 'Über die Verbreiterung von Spektrallinien', *Annalen der Physik* **58**, 577–630.

Hosking, J. R. M. (1981), 'Fractional differencing', *Biometrika* **68**(1), 165–176.

Hunt, G. A. (1951), 'Random Fourier transforms', *Transactions of the American Society of Civil Engineers* **116**, 770–808.

Hurst, H. E. (1951), 'Long-term storage capacity of reservoirs', *Transactions of the American Society of Civil Engineers* **116**, 770–808.

Ibragimov, I. A. & Chernin, K. E. (1959), 'On the unimodality of stable laws', *Theory of Probability and its Applications* **4**, 417–419.

Ibragimov, I. A. & Linnik, Y. V. (1971), *Independent and stationary sequences of random variables*, Wolters-Nordhoff, The Netherlands.

Itô, K. (1951), 'Multiple Wiener integral', *Journal of the Mathematical Society of Japan* **3**, 157–169.

Itô, K. (1969), *Stochastic Processes*, Vol. 16, Aarhus University Lecture Notes Series.

Itô, K. & Nisio, N. (1968), 'On the oscillation functions of Gaussian processes', *Mathematica Scandinavica* **22**, 209–223.

Jain, N. C. & Kallianpur, G. (1972), 'Oscillation functions of a multiparameter Gaussian process', *Nagoya Mathematical Journal* **47**, 15–28.

Jakubowski, A. (1991), *Asymptotic Independent Representations for Sums and Order Statistics of Stationary Sequences*, Uniwersytet Mikołaja Kopernika, Toruń, Poland.

Jakubowski, A. (1993), 'Minimal conditions in p-stable limit theorems', *Stochastic Processes and their Applications* **44**, 291–327.

James, R. C. (1947), 'Inner products in normed linear spaces', *Bulletin of the American Mathematical Society* pp. 559–566.

Janicki, A. & Weron, A. (1993), *Simulation and Chaotic Behavior of α-stable Stochastic Processes*, Marcel Dekker. To appear.

Joag-dev, K., Perlman, M. D. & Pitt, L. D. (1983), 'Association of normal random variables and Slepian's inequality', *The Annals of Probability* **11**, 451–455.

Kallenberg, O. (1988), 'Spreading and predictable sampling in exchangeable sequences and processes', *The Annals of Probability* **16**, 508–534.

Kallenberg, O. & Szulga, J. (1989), 'Multiple integration with respect to Poisson and lévy processes', *Probability Theory and Related Fields* **83**, 101–134.

Kanter, M. (1972), 'Linear sample spaces and stable processes', *Journal of Functional Analysis* **9**, 441–459.

Kanter, M. (1973), 'The L^p norm of sums of translates of a function.', *Transactions of the American Mathematical Society* **179**, 35–47.

Kanter, M. (1976), 'On the unimodality of stable densities', *The Annals of Probability* **4**, 1006–1008.

Kanter, M. & Steiger, W. L. (1974), 'Regression and autoregression with infinite variance', *Advances in Applied Probability* **6**, 768–783.

Kasahara, Y. & Maejima, M. (1986), 'Functional limit theorems for weighted sums of i.i.d. random variables', *Probability Theory and Related Fields* **72**, 161–183.

Kasahara, Y. & Maejima, M. (1988), 'Weighted sums of i.i.d. random variables attracted to integrals of stable processes', *Probability Theory and Related Fields* **78**, 75–96.

Kasahara, Y., Maejima, M. & Vervaat, W. (1988), 'Log-fractional stable processes', *Stochastic Processes and their Applications* **30**, 329–339.

Kawata, T. (1972), *Fourier Analysis in Probability Theory*, Academic Press, New York.

Kelley, J. L. (1955), *General Topology*, Vol. 27 of *Graduate Texts in Mathematics*, Springer-Verlag, New York.

Kesten, H. & Spitzer, F. (1979), 'A limit theorem related to a new class of self-similar processes', *Zeitschrift für Wahrscheinlichkeitstheorie und verwandte Gebiete* **50**, 5–25.

Khinchine, A. (1938), *Predelnye zakony dlya summ nezavisimykh sluchainykh velichin*, GONTI, Moscow-Leningrad.

Khinchine, A. & Lévy, P. (1936), 'Sur les lois stables', *Comptes Rendus de l'Académie des Sciences de Paris* **202**, 374–376.

Klüppelberg, C. & Mikosch, T. (1993*a*), The integrated periodogram for stable processes, Preprint.

Klüppelberg, C. & Mikosch, T. (1993*b*), 'Some limit theory for the self-normalized periodogram of stable processes', *Scandinavian Journal of Statistics*. To appear.

Klüppelberg, C. & Mikosch, T. (1993*c*), 'Spectral estimates and stable processes', *Stochastic Processes and their Applications* **47**, 323–344.

Knight, K. (1986), Estimation for infinite variance autoregressive processes, PhD thesis, University of Washington, Seattle, Washington.

Knight, K. (1987), 'Rate of convergence of centered estimates of autoregressive parameters for infinite variance autoregressions', *Journal of Time Series Analysis* **8**, 51–60.

Kokoszka, P. S. & Podgórski, K. (1993), 'Ergodicity and weak mixing of semistable processes', *Probability and Mathematical Statistics*. To appear.

Kokoszka, P. S. & Taqqu, M. S. (1992), Asymptotic dependence of stable self-similar processes of Chentsov type, *in* R. M. Dudley, M. G. Hahn & J. Kuelbs, eds, 'Probability in Banach Spaces, 8: Proceedings of the Eighth International Conference', Birkhäuser, Boston, pp. 152–165.

Kokoszka, P. S. & Taqqu, M. S. (1993*a*), 'Asymptotic dependence of moving average type self-similar stable random fields', *Nagoya Mathematical Journal* **130**, 85–100.

Kokoszka, P. S. & Taqqu, M. S. (1993*b*), A characterization of mixing processes of type G, Preprint.

Kokoszka, P. S. & Taqqu, M. S. (1993*c*), Fractional ARIMA with stable innovations, Preprint.

Kokoszka, P. S. & Taqqu, M. S. (1993*d*), 'Infinite variance stable ARMA processes', *Journal of Time Series Analysis*. To appear.

Kokoszka, P. S. & Taqqu, M. S. (1993*e*), Infinite variance stable moving averages with long memory, Preprint.

Kokoszka, P. S. & Taqqu, M. S. (1993*f*), 'New classes of self-similar symmetric stable random fields', *Journal of Theoretical Probability*. To appear.

Kolmogorov, A. N. (1940), 'Wienersche Spiralen und einige andere interessante Kurven im Hilbertschen raum', *Comptes Rendus (Doklady) de l'Académie des Sciences de l' URSS (N.S.)* **26**, 115–118.

Kôno, N. & Maejima, M. (1991), Self-similar stable processes with stationary increments, *in* S. Cambanis, G. Samorodnitsky & M. S. Taqqu, eds, 'Stable Processes and Related Topics', Vol. 25 of *Progress in Probability*, Birkhäuser, Boston, pp. 275–295.

Koutrouvelis, I. A. (1980), 'Regression-type estimation of the parameters of stable laws', *Journal of the American Statistical Association* **75**, 918–928.

Koutrouvelis, I. A. (1981), 'An iterative procedure for the estimation of the parameters of the stable law', *Communication in Statistics-Simulation and Computation* **10**, 17–28.

Krakowiak, W. & Szulga, J. (1988), 'A multiple stochastic integral with respect to a strictly p-stable measure', *The Annals of Probability* **16**, 764–777.

Kuelbs, J. (1973), 'A representation theorem for symmetric stable processes and stable measures on H', *Zeitschrift für Wahrscheinlichkeitstheorie und verwandte Gebiete* **26**, 259–271.

Kwapień, S. & Woyczyński, N. A. (1987), 'Double stochastic integrals, random quadratic forms and random series in Orlicz spaces.', *The Annals of Probability* **15**, 1072–1096.

Kwapień, S. & Woyczyński, N. A. (1992), *Random Series and Stochastic Integrals: Single and Multiple*, Birkhäuser, Boston.

Laha, R. G. & Rohatgi, V. K. (1979), *Probability Theory*, Wiley & Sons, New York.

Lamperti, J. W. (1962), 'Semi-stable stochastic processes', *Transaction of the American Mathematical Society* **1962**, 62–78.

Leadbetter, M. R., Lindgren, G. & Rootzén, H. (1983), *Extremes and Related Properties of Random Sequences and Processes*, Springer-Verlag, New York.

Ledoux, M. & Talagrand, M. (1991), *Probability in Banach Spaces: Isoperimetry and Processes*, Springer-Verlag, New York.

Lee, M.-L. T., Rachev, S. T. & Samorodnitsky, G. (1990), 'Association of stable random variables', *The Annals of Probability* **18**, 1759–1764.

Leland, W. E., Taqqu, M. S., Willinger, W. & Wilson, D. V. (1993), 'On the self-similar nature of Ethernet traffic', *ACM/SIGCOMM Computer Communications Review* **23**, 183–193. Proceedings of the ACM/SIGCOMM93, San Francisco, September 1993.

Leland, W. E., Taqqu, M. S., Willinger, W. & Wilson, D. V. (1994), 'Self-similarity in high-speed packet traffic: analysis and modeling of Ethernet traffic measurements', *Statistical Science*. To appear.

LePage, R. (1990*a*), Conditional moments for coordinates of stable vectors, *in* S. Cambanis & A. Weron, eds, 'Probability Theory on Vector Spaces IV. Proceedings 1987', Springer-Verlag, New York, pp. 148–152.

LePage, R. (1990*b*), Multidimensional infinitely divisible variables and processes. Part I: Stable Case, *in* S. Cambanis & A. Weron, eds, 'Probability Theory on Vector Spaces IV. Proceedings 1987', Springer-Verlag, New York, pp. 153–163.

Levy, J. B. & Taqqu, M. S. (1991), A characterization of the asymptotic behavior of stationary stable processes, *in* S. Cambanis, G. Samorodnitsky & M. S. Taqqu, eds, 'Stable Processes and Related Topics', Vol. 25 of *Progress in Probability*, Birkhäuser, Boston, pp. 181–198.

Lévy, P. (1924), 'Théorie des erreurs la loi de Gauss et les lois exceptionelles', *Bulletin de la Société Mathématique de France* **52**, 49–85.

Lévy, P. (1954), *Théorie de l'addition des variables aléatoires*, 2nd edn, Gauthier-Villars, Paris. First edition appeared in 1937.

Lévy, P. (1965), *Processus stochastiques et mouvement brownien*, 2nd edn, Gauthier-Villars, Paris. First edition appeared in 1948.

Lévy, P. (1966), Fonctions browniennes dans l'espace Euclidien et dans l'espace de Hilbert, *in* F. N. David, ed., 'Research Papers in Statistics', Wiley, pp. 189–223.

Lévy, P. (1970), *Quelques aspects de la pensée d'un mathématicien*, Albert Blanchard, 9 rue de Médicis, Paris 6e.

Linde, W. (1986), *Probability in Banach Spaces-Stable and Infinitely Divisible Distributions*, Wiley, Chichester.

Linde, W. & Mathé, P. (1983), 'Conditional symmetries of stable measures on $\mathbf{R}^n$', *Annales de l'Institut Henri Poincaré* **19**, 57–69.

Linde, W., Mandrekar, V. & Weron, A. (1980), p-stable measures and absolutely summing operators, *in* A. Weron, ed., 'Probability Theory on Vector Spaces II. Proceedings', Springer-Verlag, pp. 167–178. Lecture Notes in Mathematics. Vol. 828.

Liu, J. (1987), Infinite variance and nonlinear time series models, PhD thesis, Colorado State University, Department of Statistics, Fort Collins, Colorado.

Loretan, M. (1991), Testing Covariance Stationarity of Heavy-Tailed Economic Time Series, PhD thesis, Yale University, Department of Economics.

Lukacs, E. (1970), *Characteristic functions*, Charles Griffin & Co., London.

Maejima, M. (1983*a*), 'On a class of self-similar processes', *Zeitschrift für Wahrscheinlichkeitstheorie und verwandte Gebiete* **62**, 235–245.

Maejima, M. (1983*b*), 'A self-similar process with nowhere bounded sample paths', *Zeitschrift für Wahrscheinlichkeitstheorie und verwandte Gebiete* **65**, 115–119.

Maejima, M. (1983*c*), 'Some limit theorems for sojourn times of strongly dependent Gaussian processes', *Zeitschrift für Wahrscheinlichkeitstheorie und verwandte Gebiete* **62**, 235–245.

Maejima, M. (1986), 'A remark on self-similar processes with stationary increments', *Canadian Journal of Statistics* **14**, 81–82.

Maejima, M. (1989), 'Self-similar processes and limit theorems', *Sugaku Expositions* **2**, 103–123.

Major, P. (1981), *Multiple Wiener-Itô Integrals*, Vol. 849 of *Springer Lecture Notes in Mathematics*, Springer-Verlag, New York.

Mandelbrot, B. B. (1965), 'Une classe de processus stochastiques homothetiques a soi; application a loi climatologique de H.E. Hurst', *Comptes Rendus de l'Académie des Sciences de Paris* **240**, 3274–3277.

Mandelbrot, B. B. (1971), 'A fast fractional Gaussian noise generator', *Water Resources Research* **7**, 543–553.

Mandelbrot, B. B. (1975), 'Fonctions aléatoires pluri-temporelles: approximation poissonienne du cas brownien et généralisations', *Comptes Rendus de l'Académie des Sciences de Paris* **280A**, 1075–1078.

Mandelbrot, B. B. (1982), *The Fractal Geometry of Nature*, W.H. Freeman and Co., San Francisco.

Mandelbrot, B. B. & Taqqu, M. S. (1979), Robust R/S analysis of long-run serial correlation, *in* 'Proceedings of the 42nd Session of the International Statistical Institute', Bulletin of the International Statistical Institute, Manila. Vol.48, Book 2, pp. 69-104.

Mandelbrot, B. B. & Van Ness, J. W. (1968), 'Fractional Brownian motions, fractional noises and applications', *SIAM Review* **10**, 422–437.

Mandelbrot, B. B. & Wallis, J. R. (1968), 'Noah, Joseph and operational hydrology', *Water Resources Research* **4**, 909–918.

Mandelbrot, B. B. & Wallis, J. R. (1969*a*), 'Computer experiments with fractional Gaussian noises, Parts 1,2,3', *Water Resources Research* **5**, 228–267.

Mandelbrot, B. B. & Wallis, J. R. (1969*b*), 'Robustness of the rescaled range r/s in the measurement of noncyclic long-run statistical dependence', *Water Resources Research* **5**, 967–988.

Mandelbrot, B. B. & Wallis, J. R. (1969*c*), 'Some long-run properties of geophysical records', *Water Resources Research* **5**, 321–340.

Mandelbrot, B. B. & Zarnfaller, F. (1959), Five place tables of certain stable distributions, Technical Report RC-421, IBM Thomas J. Watson Research Center, Yorktown Heights, New York.

Marcus, D. J. (1983), 'Non-stable laws with all projections stable', *Zeitschrift für Wahrscheinlichkeitstheorie und verwandte Gebiete* **64**, 139–156.

Marcus, M. B. & Pisier, G. (1984), 'Characterization of almost surely continuous p-stable random Fourier series and strongly stationary processes', *Acta Mathematica* **152**, 245–301.

Marcus, M. B. & Woyczynski, W. (1979), 'Stable measures and central limit theorems in spaces of stable type', *Transactions of the American Mathematical Society* **251**, 71–102.

Marques, M. (1987), A study on Lebesgue decomposition of measures induced by stable processes, PhD thesis, University of North Carolina, Chapel Hill.

Maruyama, G. (1970), 'Infinitely divisible processes', *Theory of Probability and its Applications* **15**(1), 1–22.

McConnell, T. R. (1986), Triple integration with respect to stable processes., Preprint.

McConnell, T. R. & Taqqu, M. S. (1984), Double integration with respect to symmetric stable processes., Technical Report 618, Cornell University.

McConnell, T. R. & Taqqu, M. S. (1986), 'Dyadic approximations of double integrals with respect to symmetric stable processes', *Stochastic Processes and their Applications* **22**, 323–331.

McCulloch, J. H. (1986), 'Simple consistent estimators of stable distribution parameters', *Communications in Statistics-Computation and Simulation* **15**, 1109–1136.

McKean, Jr., H. P. (1963), 'Brownian motion with a several dimensional time', *Theory of Probability and its Applications* **8**, 357–378.

McLeod, A. I. & Hipel, K. (1978), 'Preservation of the rescaled adjusted range, Parts 1,2,3', *Water Resources Research* **14**, 491–518.

Mijnheer, J. L. (1975), Sample path properties of Stable Processes, *in* 'Mathematical Centre Tracts', Mathematical Centrum, Amsterdam. Vol. 59.

Mikosch, T., Gadrich, T., Klüppelberg, C. & Adler, R. J. (1993), 'Parameter estimation for ARMA models with infinite variance innovations', *The Annals of Statistics*. To appear.

Miller, G. (1978), 'Properties of certain symmetric stable distributions', *Journal of Multivariate Analysis* **8**, 346–360.

Mittnik, S. & Rachev, S. T. (1993), 'Modeling asset returns with alternative stable distributions', *Econometric Review*. To appear.

Mori, T. (1992), 'Representational of linearly additive random fields', *Probability Theory and Related Fields* **92**, 91–115.

Nolan, J. P. (1989), 'Continuity of symmetric stable processes', *Journal of Multivariate Analysis* **29**, 84–93.

Nolan, J. P. (1991), Bounded stationary stable processes and entropy, *in* S. Cambanis, G. Samorodnitsky & M. S. Taqqu, eds, 'Stable Processes and Related Topics', Vol. 25 of *Progress in Probability*, Birkhäuser, Boston, pp. 101–105.

O'Brien, G. L. & Vervaat, W. (1983), 'Marginal distributions of self-similar processes with stationary increments', *Zeitschrift für Wahrscheinlichkeitstheorie und verwanddte Gebiete* **64**, 129–138.

Paulauskas, V. J. (1976), 'Some remarks on multivariate stable distributions', *Journal of Multivariate Analysis* **6**, 356–368.

Paulson, A. S. & Delehanty, T. A. (1984), 'Some properties of modified integrated squared error estimators for the stable laws', *Communications in Statistics* **13**(3), 337–365.

Paulson, A. S. & Delehanty, T. A. (1985), 'Modified weighted squared error estimation procedures with special emphasis on the stable laws', *Communications in Statististics* **14**(4), 927–972.

Paulson, A. S. & Delehanty, T. A. (1993), Tables of the fractiles of the stable law, Preprint.

Paulson, A. S., Holcomb, E. W. & Leitch, R. A. (1975), 'The estimation of the parameters of the stable law', *Biometrika* **62**, 163–170.

Pisier, G. (1983), Some applications of the metric entropy condition to harmonic analysis, *in* 'Banach Spaces, Harmonic Analysis and Probability', University of Connecticut, Springer-Verlag, pp. 123–154. Lecture Notes in Mathematics Vol. 995.

Pitt, L. D. (1982), 'Positively correlated normal variables are associated', *The Annals of Probability* **10**, 496–499.

Podgórski, K. (1992), 'A note on ergodic symmetric symmetric stable processes', *Stochastic Processes and their Applications* **43**, 355–362.

Pólya, G. (1923), 'Herleitung des Gausschen Fehlergesetzes aus einer Funktionalgleichung', *Mathematische Zeitschrift* **18**, 96–108.

Press, S. J. (1972), *Applied Multivariate Analysis*, Holt, Rinehart and Winston.

Ramachandran, B. (1969), 'On characteristic functions and moments', *Sankhya* **31**(Series A), 1–12.

Ramachandran, B. & Rao, C. R. (1968), 'Some results on characteristic functions and characterizations of the normal and generalized stable laws.', *Sankhya* **30**(Series A), 125–140.

Reed, M. & Simon, B. (1972), *Methods of Modern Mathematical Physics. I: Functional Analysis*, Academic Press, New York.

Resnick, S. I. (1986), 'Point processes, regular variation and weak convergence', *Advances in Applied Probability* **18**, 66–138.

Resnick, S. I. (1987), *Extreme Values, Regular Variation and Point Processes*, Springer-Verlag, New York.

Rice, S. O. (1939), 'The distribution of the maxima of a random curve', *American Journal of Mathematics* **61**, 409–416.

Rice, S. O. (1944), 'Mathematical analysis of random noise', *Bell System Technical Journal* **23**, 282–332.

Rice, S. O. (1945), 'Mathematical analysis of random noise', *Bell System Technical Journal* **24**, 46–156.

Robinson, P. M. (1991), Nonparametric function estimation for long memory time series, *in* W. Barnett, J. Powell & G. Tauchen, eds, 'Nonparametric and Semiparametric Methods in Econometrics and Statistics: Proceedings of the Fifth International Symposium in Economic Theory and Econometrics', Cambridge University Press, pp. 437–457.

Robinson, P. M. (1992*a*), Log-periodogram regression of time series with long range dependence, Preprint.

Robinson, P. M. (1992*b*), Rates of convergence and optimal bandwidth in spectral analysis of processes with long range dependence, Preprint.

Rosenblatt, M. (1961), Independence and dependence, *in* 'Proceedings of the 4th Berkeley Symposium Mathematical Statistics and Probability', University of California Press, pp. 431–443.

Rosinki, J. & Woyczynski, W. A. (1985), Moment inequalities for real and vector-valued p-stable stochastic integrals, *in* A. B. et al., ed., 'Probability Theory in Banach Spaces V', Springer-Verlag, pp. 369–387. Lecture Notes in Mathematics. Vol. 1153.

Rosinski, J. (1986), 'On stochastic integral representation of stable processes with sample paths in Banach spaces', *Journal of Multivariate Analysis* **20**, 277–302.

Rosinski, J. (1990), 'On series representation of infinitely divisible random vectors', *The Annals of Probability* **18**, 405–430.

Rosinski, J. (1991), On a class of infinitely divisible processes represented as mixtures of Gaussian processes, *in* S. Cambanis, G. Samorodnitsky & M. S. Taqqu, eds, 'Stable Processes and Related Topics', Birkhäuser, Boston, pp. 27–41.

Rosinski, J. & Samorodnitsky, G. (1993), 'Distributions of subadditive functionals of sample paths of infinitely divisible processes', *The Annals of Probability* **21**, 996–1014.

Rosinski, J. & Woyczyński, W. A. (1986), 'On Ito stochastic integration with respect to p-stable motion: inner clock, integrability of sample paths, double and multiple integrals.', *The Annals of Probability* **14**, 271–286.

Ross, S. M. (1985), *Introduction to Probability Models*, 3rd edn, Academic Press, Orlando.

Rossberg, H. J., Jesiak, B. & Siegel, G. (1985), *Analytic Methods of Probability Theory*, Akademie-Verlag, Berlin.

Royden, H. L. (1963), *Real Analysis*, Macmillan.

Rozanov, Y. A. (1967), *Stationary Random Processes*, Holden-Day, San Francisco.

Samarov, A. & Taqqu, M. S. (1988), 'On the efficiency of the sample mean in long memory noise', *Journal of Time Series Analysis* **9**, 191–200.

Samorodnitsky, G. (1988), 'Extrema of skewed stable processes.', *Stochastic Processes and their Applications* **30**, 17–39.

Samorodnitsky, G. (1992), 'Integrability of stable processes', *Probability and Mathematical Statistics* **13**, 191–204.

Samorodnitsky, G. (1993), 'Possible sample paths of α-stable self-similar processes', *Statistics and Probability Letters*. To appear.

Samorodnitsky, G. & Szulga, J. (1989), 'An asymptotic evaluation of the tail of a multiple symmetric α-stable integral.', *The Annals of Probability* **17**, 1503–1520.

Samorodnitsky, G. & Taqqu, M. S. (1989), The various linear fractional Lévy motions, *in* T. W. Anderson, K. B. Athreya & D. L. Iglehart, eds, 'Probability, Statistics and Mathematics: Papers in Honor of Samuel Karlin', Academic Press, Boston, pp. 261–270.

Samorodnitsky, G. & Taqqu, M. S. (1990*a*), '$(1/\alpha)$-self-similar processes with stationary increments', *Journal of Multivariate Analysis* **35**, 308–313.

Samorodnitsky, G. & Taqqu, M. S. (1990*b*), 'Existence of joint moments of stable random variables', *Statistics and Probability Letters* **10**, 167–172.

Samorodnitsky, G. & Taqqu, M. S. (1990*c*), 'Multiple stable integrals of Banach-valued functions', *Journal of Theoretical Probability* **3**, 267–287.

Samorodnitsky, G. & Taqqu, M. S. (1991*a*), 'Conditional moments and linear regression for stable random variables', *Stochastic Processes and their Applications* **39**, 183–199.

Samorodnitsky, G. & Taqqu, M. S. (1991*b*), Construction of multiple stable measures and integrals using LePage representation, *in* S. Cambanis, G. Samorodnitsky & M. S. Taqqu, eds, 'Stable Processes and Related Topics', Vol. 25 of *Progress in Probability*, Birkhäuser, Boston, pp. 121–141.

Samorodnitsky, G. & Taqqu, M. S. (1991*c*), 'Probability laws with 1-stable marginals are 1-stable', *The Annals of Probability* **19**, 1777–1780.

Samorodnitsky, G. & Taqqu, M. S. (1992*a*), Linear models with long-range dependence and finite or infinite variance, *in* D. Brillinger, P. Caines, J. Geweke, E. Parzen, M. Rosenblatt & M. S. Taqqu, eds, 'New Directions in Time Series Analysis, Part II', IMA Volumes in Mathematics and its Applications, Volume 46, Springer-Verlag, New York, pp. 325–340.

Samorodnitsky, G. & Taqqu, M. S. (1992*b*), 'Stochastic monotonicity and Slepian-type inequalities for infinitely divisible and stable random vectors', *The Annals of Probability* **21**, 143–160.

Samorodnitsky, G. & Taqqu, M. S. (1994), 'Lévy measures of infinitely divisible random vectors and Slepian inequalities', *The Annals of Probability*. To appear.

Samotij, K. & Zak, T. (1989), 'A note on Marcus' example of a nonstable law with all projections stable', *Bulletin de l'Académie Polonaise des Sciences, Série Mathématique* **37**, 199–202.

Sato, Y. (1991), 'Distributions of stable random fields of Chentsov type', *Nagoya Mathematical Journal* **123**, 119–139.

Sato, Y. (1992*a*), Structure of bivariate distributions of linear fractional and Chentsov type stable processes, *in* A. N. Shiryaev, V. S. Korolyuk, S. Watanabe & M. Fukushima, eds, 'Probability Theory and Mathematical Statistics', Proceedings of the Sixth USSR-Japan Symposium, World Scientific, Singapore, pp. 306–316.

Sato, Y. (1992*b*), 'Structure of Lévy measures of stable random fields of Chentsov type', *Probability and Mathematical Statistics* **13**, 165–176.

Sato, Y. & Takenaka, S. (1991), On determinism of symmetric α-stable processes of generalized Chentsov type, *in* K. Itô & T. Hida, eds, 'Gaussian Random Fields', World Scientific, Singapore, pp. 332–345. Series on Probability and Statistics, Vol.1.

Schertzer, D. & Lovejoy, S. (1993), Scaling and multifractal processes, *in* 'Nonlinear Variability in Geophysics 3', Institut d'Etudes Scientifiques de Cargèse. Lecture notes.

Schilder, M. (1970), 'Some structure theorems for the symmetric stable laws.', *The Annals of Mathematical Statistics* **41**, 412–421.

Schoenberg, I. J. (1937), 'On certain metric spaces arising from Euclidean spaces by a change of metric and their imbedding in Hilbert space', *Annals of Mathematics* **38**(4), 787–793.

Schreiber, M. (1972), 'Quelques remarques sur les caractéristiques des espaces L^p, $0 \leq p < 1$', *Annales de l'Institut Henri Poincaré* **8**, 83–92.

Schwartz, L. (1981), Geometry and probability in Banach spaces, *in* 'Lecture Notes in Mathematics', Springer-Verlag, New York.

Shepp, L. A. (1966), 'Radon-Nikodym derivatives of Gaussian measures', *The Annals of Mathematical Statistics* **37**, 321–354.

Sinai, Y. G. (1976), 'Self-similar probability distributions', *Theory of Probability and its Applications* **21**, 64–80.

Surgailis, D. (1981), On L^2 and non-L^2 multiple stochastic integration, *in* 'Stochastic Differential Systems', Lecture Notes in Control and Information Sciences, 36, Springer-Verlag, New York, pp. 212–226.

Surgailis, D. (1985), 'On the multiple stable integral', *Zeitschrift für Wahrscheinlichkeitstheorie und verwandte Gebiete* **70**, 621–632.

Surgailis, D., Rosinski, J., Mandrekar, V. & Cambanis, S. (1993), 'Stable generalized moving averages', *Probability Theory and Related Fields.* To appear.

Takashima, K. (1991), 'Sample path properties of ergodic self-similar processes', *Osaka Journal of Mathematics* **26**, 159–189.

Takenaka, S. (1987), 'Representations of Euclidean random field', *Nagoya Mathematical Journal* **105**, 19–31.

Takenaka, S. (1991), 'Integral-geometric construction of self-similar stable processes', *Nagoya Mathematical Journal* **123**, 1–12.

Takenaka, S. (1993), Examples of self-similar stable processes, *in* S. Cambanis, J. K. Ghosh, R. L. Karandikar & P. K. Sen, eds, 'Stochastic Processes. A Festschrift in Honour of Gopinath Kallianpur', Springer-Verlag, New York, pp. 303–311.

Talagrand, M. (1988), 'Necessary conditions for sample boundedness of p-stable processes', *The Annals of Probability* **16**, 1584–1595.

Talagrand, M. (1990), 'Characterization of almost surely continuous 1-stable random Fourier series an strongly stationary processes', *The Annals of Probability* **18**, 85–91.

Taqqu, M. S. (1975), 'Weak convergence to fractional Brownian motion and to the Rosenblatt process', *Zeitschrift für Wahrscheinlichkeitstheorie und verwandte Gebiete* **31**, 287–302.

Taqqu, M. S. (1979), 'Convergence of integrated processes of arbitrary Hermite rank', *Zeitschrift für Wahrscheinlichkeitstheorie und verwandte Gebiete* **50**, 53–83.

Taqqu, M. S. (1981), Self-similar processes and related ultraviolet and infrared catastrophes, *in* 'Random Fields: Rigorous Results in Statistical Mechanics and Quantum Field Theory', Colloquia Mathematica Societatis Janos Bolyai, North-Holland, Amsterdam, pp. 1027–1096. Vol.27 Book 2.

Taqqu, M. S. (1986), A bibliographical guide to self-similar processes and long-range dependence, *in* E. Eberlein & M. S. Taqqu, eds, 'Dependence in Probability and Statistics', Birkhäuser, Boston, pp. 137–162.

Taqqu, M. S. (1988), Self-similar processes, *in* S. Kotz & N. Johnson, eds, 'Encyclopedia of Statistical Sciences', Wiley, New York, pp. 352–357. Volume 8.

Taqqu, M. S. & Wolpert, R. (1983), 'Infinite variance self-similar processes subordinate to a Poisson measure', *Zeitschrift für Wahrscheinlichkeitstheorie und verwandte Gebiete* **62**, 53–72.

Titchmarsh, E. C. (1986), *Introduction to the Theory of Fourier Integrals*, 3rd edn, Chelsea, New York.

Tortrat, A. (1975), 'Pseudo-martingales et lois stables', *Comptes Rendus de l'Académie des Sciences de Paris* **281**(Ser. A), 463–465.

Varadhan, S. R. S. (1962), 'Limit theorems for sums of independent random variables with values in a Hilbert space', *Sankhya* **24**(Ser.A), 213–238.

Vervaat, W. (1979), 'On a stochastic difference equation and a representation of non-negative infinitely divisible random variables', *Advances in Applied Probability* **11**, 750–783.

Vervaat, W. (1985), 'Sample paths of self-similar processes with stationary increments', *The Annals of Probability* **13**, 1–27.

Weron, A. (1984), Stable processes and measures: A survey. Probability Theory on Vector Spaces III, *in* D. Szynal & A. Weron, eds, 'Lecture Notes in Mathematics, Vol. 1080', Springer-Verlag, Berlin, pp. 306–364.

Wiener, N. (1958), *Nonlinear Problems in Random Theory*, MIT Press, Cambridge.

Wintner, A. (1936), 'On a class of Fourier transforms', *American Journal of Mathematics* **58**, 45–90.

Wu, W. & Cambanis, S. (1991), Conditional variance of symmetric stable variables, *in* S. Cambanis, G. Samorodnitsky & M. S. Taqqu, eds, 'Stable Processes and Related Topics', Vol. 25 of *Progress in Probability*, Birkhäuser, Boston, pp. 85–99.

Yaglom, A. M. (1955), 'Correlation theory of processes with stationary random increments of order n', *Matematicheski Sbornik* **37**, 141–196. English translation in *Am. Math. Soc. Translations Ser. 2* **8**(1958), 87-141.

Yaglom, A. M. (1957), 'Some classes of random fields in n-dimensional space, related to stationary random processes', *Theory of Probability and its Applications* **2**, 273–320.

Yajima, Y. (1988), 'On estimation of a regression model with long-memory stationary errors', *The Annals of Statistics* **16**, 791–807.

Yajima, Y. (1989), 'A central limit theorem of Fourier transforms of strongly dependent stationary processes', *Journal of Time Series Analysis* **10**, 375–383.

Yamazato, M. (1978), 'Unimodality of infinitely divisible distribution functions of class l', *The Annals of Probability* **6**, 523–531.

Zolotarev, V. M. (1954), 'Vyrazhenie plotnosti ustoichivogo raspredeleniya s pokazatelem α bolshim edinitsy cherez plotnost s pokazatelem $1/\alpha$', *Doklady Akademii Nauk USSR (N.S.)* **98**, 735–738.

Zolotarev, V. M. (1983), *Odnomernye ustoichivye raspredeleniya*, Nauka, Moscow. Subsequently translated as "One-dimensional Stable Distributions", American Mathematical Society, 1986.

Zolotarev, V. M. (1986), *One-dimensional Stable Distributions*, Vol. 65 of "Translations of mathematical monographs", American Mathematical Society. Translation from the original 1983 Russian edition.

Zygmund, A. (1979), *Trigonometric Series*, Cambridge University Press, Cambridge. Volumes I,II.

[illegible], A. (1975). Pseudo-martingales et lois stables [illegible] *Comptes Rendus [illegible] de Paris* **281**, Ser. A, 463–465.

[illegible] (19[illegible]). Limit theorems for sums of independent random variables with values in a Hilbert space [illegible]

[illegible] (1979). On a stochastic differential equation and a representation of non-negative [illegible] random variables. [illegible]

[illegible], W. [illegible] of self-similar processes [illegible] *The Annals of Probability* [illegible]

[illegible] A survey. Probability Theory [illegible]

[illegible] MIT Press [illegible]

[illegible]

[illegible]

Subject index

Author index